MATHEMATICS
FOR ECONOMICS AND BUSINESS

Visit the *Mathematics for Economics and Business, fifth edition*, Companion Website at **www.pearsoned.co.uk/jacques** to find valuable **student** learning material including:

- Multiple choice questions to test your understanding

PEARSON
Education

fifth edition

MATHEMATICS FOR ECONOMICS AND BUSINESS

IAN JACQUES

FT Prentice Hall
FINANCIAL TIMES

An imprint of Pearson Education
Harlow, England • London • New York • Boston • San Francisco • Toronto • Sydney • Singapore • Hong Kong
Tokyo • Seoul • Taipei • New Delhi • Cape Town • Madrid • Mexico City • Amsterdam • Munich • Paris • Milan

Pearson Education Limited
Edinburgh Gate
Harlow
Essex CM20 2JE
England

and Associated Companies throughout the world

Visit us on the World Wide Web at:
www.pearsoned.co.uk

First published 1991
Second edition 1994
Third edition 1999
Fourth edition 2003
Fifth edition published 2006

ISBN-10 0-273-70195-9
ISBN-13 978-0-273-70195-8

British Library Cataloguing-in-Publication Data
A catalogue record for this book is available from the British Library

Library of Congress Cataloging-in-Publication Data
A catalog record for this book is available from the Library of Congress

10 9 8 7 6 5 4 3 2 1
10 09 08 07 06

Typeset in 10/12.5pt Minion Reg by 35
Printed and bound by Mateu-Cromo Artes Graficas, Spain

The publisher's policy is to use paper manufactured from sustainable forests.

To my mother, and in memory of my father

Supporting resources

Visit **www.pearsoned.co.uk/jacques** to find valuable online resources

Companion Website for students

- Multiple choice questions to test your understanding

For instructors

- Complete, downloadable Instructor's Manual containing teaching hints plus over a hundred additional problems with solutions and marking schemes
- Downloadable PowerPoint slides of figures from the book

Also: The Companion Website provides the following features:

- Search tool to help locate specific items of content
- E-mail results and profile tools to send results of quizzes to instructors
- Online help and support to assist with website usage and troubleshooting

For more information please contact your local Pearson Education sales representative or visit **www.pearsoned.co.uk/jacques**

Contents

Preface

This book is intended primarily for students on economics, business studies and management courses. It assumes very little prerequisite knowledge, so it can be read by students who have not undertaken a mathematics course for some time. The style is informal and the book contains a large number of worked examples. Students are encouraged to tackle problems for themselves as they read through each section. Detailed solutions are provided so that all answers can be checked. Consequently, it should be possible to work through this book on a self-study basis. The material is wide ranging, and varies from elementary topics such as percentages and linear equations, to more sophisticated topics such as constrained optimization of multivariate functions. The book should therefore be suitable for use on both low- and high-level quantitative methods courses. Examples and exercises are included which make use of the computer software packages Excel and Maple.

This book was first published in 1991. The prime motivation for writing it then was to try and produce a textbook that students could actually read and understand for themselves. This remains the guiding principle and the most significant change for this, the fifth edition, is in the design, rather than content. I was brought up with the fixed idea that mathematics textbooks were written in a small font with many equations crammed on to a page. However, I fully accept that these days books need to look attractive and be easy to negotiate. I hope that the new style will encourage more students to read it and will reduce the 'fear factor' of mathematics. In response to anonymous reviewers' comments, I have included additional problems for several exercises together with two new appendices on implicit differentiation and Hessian matrices. Finally, I have also included the highlighted key terms at the end of each section and in a glossary at the end of the book.

The book now has an accompanying website that is intended to be rather more than just a gimmick. I hope that the commentary in the Instructor's Manual will help tutors using the book for the first time. It also contains about a hundred new questions. Although a few of these problems are similar to those in the main book, the majority of questions are genuinely different. There are roughly two test exercises per chapter, which are graded to accommodate different levels of student abilities. These are provided on the website so that they can easily be cut, pasted and edited to suit. Fully worked solutions and marking schemes are included. Tutors can also control access. The website has a a section containing multiple-choice tests. These can be given to students for further practice or used for assessment. The multiple choice questions can be marked online with the results automatically transferred to the tutor's markbook if desired.

Ian Jacques

Introduction

Getting Started

Notes for students: how to use this book

I am always amazed by the mix of students on first-year economics courses. Some have not acquired any mathematical knowledge beyond elementary algebra (and even that can be of a rather dubious nature), some have never studied economics before in their lives, while others have passed preliminary courses in both. Whatever category you are in, I hope that you will find this book of value. The chapters covering algebraic manipulation, simple calculus, finance and matrices should also benefit students on business studies and accountancy courses.

The first few chapters are aimed at complete beginners and students who have not taken mathematics courses for some time. I would like to think that these students once enjoyed mathematics and had every intention of continuing their studies in this area, but somehow never found the time to fit it into an already overcrowded academic timetable. However, I suspect that the reality is rather different. Possibly they hated the subject, could not understand it and dropped it at the earliest opportunity. If you find yourself in this position, you are probably horrified to discover that you must embark on a quantitative methods course with an examination looming on the horizon. However, there is no need to worry. My experience is that every student, no matter how innumerate, is capable of passing a mathematics examination. All that is required is a commitment to study and a willingness to suspend any prejudices about the subject gained at school. The fact that you have bothered to buy this book at all suggests that you are prepared to do both.

To help you get the most out of this book, let me compare the working practices of economics and engineering students. The former rarely read individual books in any great depth. They tend to visit college libraries (usually several days after an essay was due to be handed in) and to skim through a large number of books picking out the relevant information. Indeed, the ability to read selectively and to compare various sources of information is an important skill that all arts and social science students must acquire. Engineering students, on the other hand, are more likely to read just a few books in any one year. They read each of these from cover to cover and attempt virtually every problem *en route*. Even though you are most definitely not an engineer, it is the engineering approach that you need to adopt while studying mathematics. There are several reasons for this. Firstly, a mathematics book can never be described, even by its most ardent admirers, as a good bedtime read. It can take an hour or two of concentrated effort to understand just a few pages of a mathematics text. You are therefore recommended to work through this book systematically in short bursts rather than to attempt to read whole chapters. Each section is designed to take between one and two hours to complete and this is quite sufficient for a single session. Secondly, mathematics is a hierarchical subject in which one topic follows on from the next. A construction firm building an office block is hardly likely to erect the fiftieth storey without making sure that the intermediate floors and foundations are securely in place. Likewise, you cannot 'dip' into the middle of a mathematics book and expect to follow it unless you have satisfied the prerequisites for that topic. Finally, you actually need to do mathematics yourself before you can understand it. No matter how wonderful your lecturer is, and no matter how many problems are discussed in class, it is only by solving problems yourself that you are ever going to become confident in using and applying mathematical techniques. For this reason, several problems are interspersed within the text and you are encouraged to tackle these as you go along. You will require writing paper, graph paper, pens and a calculator for this. There is no need to buy an expensive calculator unless you are feeling particularly wealthy at the moment. A bottom-of-the-range **scientific** calculator should be good enough. Detailed solutions are provided at the end of this book so that you can check your answers. However, please avoid the temptation to look at them until you have made an honest attempt at each one. Remember that in the future you may well have to sit down in an uncomfortable chair, in front of a blank sheet of paper, and be expected to produce solutions to examination questions of a similar type.

At the end of each section there are some further practice problems to try. You may prefer not to bother with these and to work through them later as part of your revision. Ironically, it is those students who really ought to try more problems who are most likely to miss them out. Human psychology is such that, if students do not at first succeed in solving problems, they are then deterred from trying additional problems. However, it is precisely these people who need more practice.

The chapter dependence is shown in Figure I.1. If you have studied some advanced mathematics before then you will discover that parts of Chapters 1, 2 and 4 are familiar. However, you may find that the sections on economics applications contain new material. You are best advised to test yourself by attempting a selection of problems in each section to see if you need to read through it as part of a refresher course. Economics students in a desperate hurry to experience the delights of calculus can miss out Chapter 3 without any loss of continuity and move straight on to Chapter 4. The mathematics of finance is probably more relevant to business and accountancy students, although you can always read it later if it is part of your economics syllabus.

Figure I.1

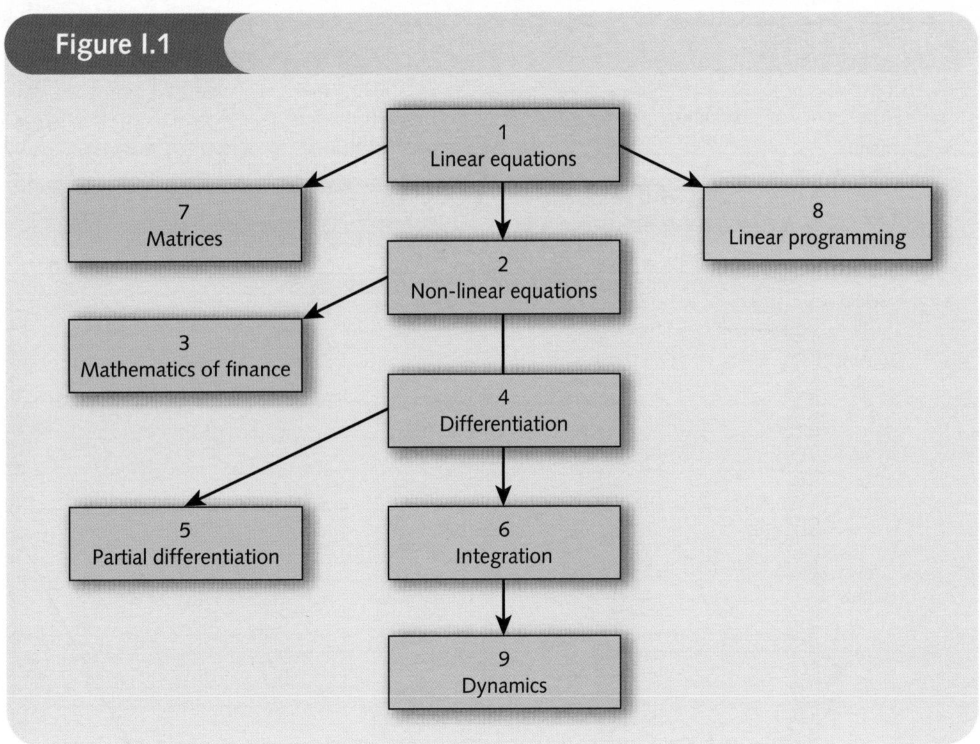

I hope that this book helps you to succeed in your mathematics course. You never know, you might even enjoy it. Remember to wear your engineer's hat while reading the book. I have done my best to make the material as accessible as possible. The rest is up to you!

Getting started with Excel

Excel is the Microsoft® spreadsheet package that we shall be using in some of our worked examples. If you are already familiar with this product, you may be able to skip some, or all, of this introductory section.

A spreadsheet is simply an array of boxes, or cells, into which tables of data can be inserted. This can consist of normal text, numerical data or a formula, which instructs the spreadsheet package to perform a calculation. The joy about getting the spreadsheet to perform the calculation is that it not only saves us some effort, but also detects any subsequent changes we make to the table, and recalculates its values automatically without waiting to be asked.

To get the most out of this section, it is advisable to work through it on your own computer, as there is no substitute for having a go. When you enter the Excel package, either by double-clicking the icon on your desktop, or by selecting it from the list of programs, a blank worksheet will be displayed, as shown in Figure I.2 (overleaf).

Each cell is identified uniquely by its column and row label. The current cell is where the cursor is positioned. In Figure I.2, the cursor is in the top left-hand corner: the cell is highlighted, and it can be identified as cell A1.

Figure I.2

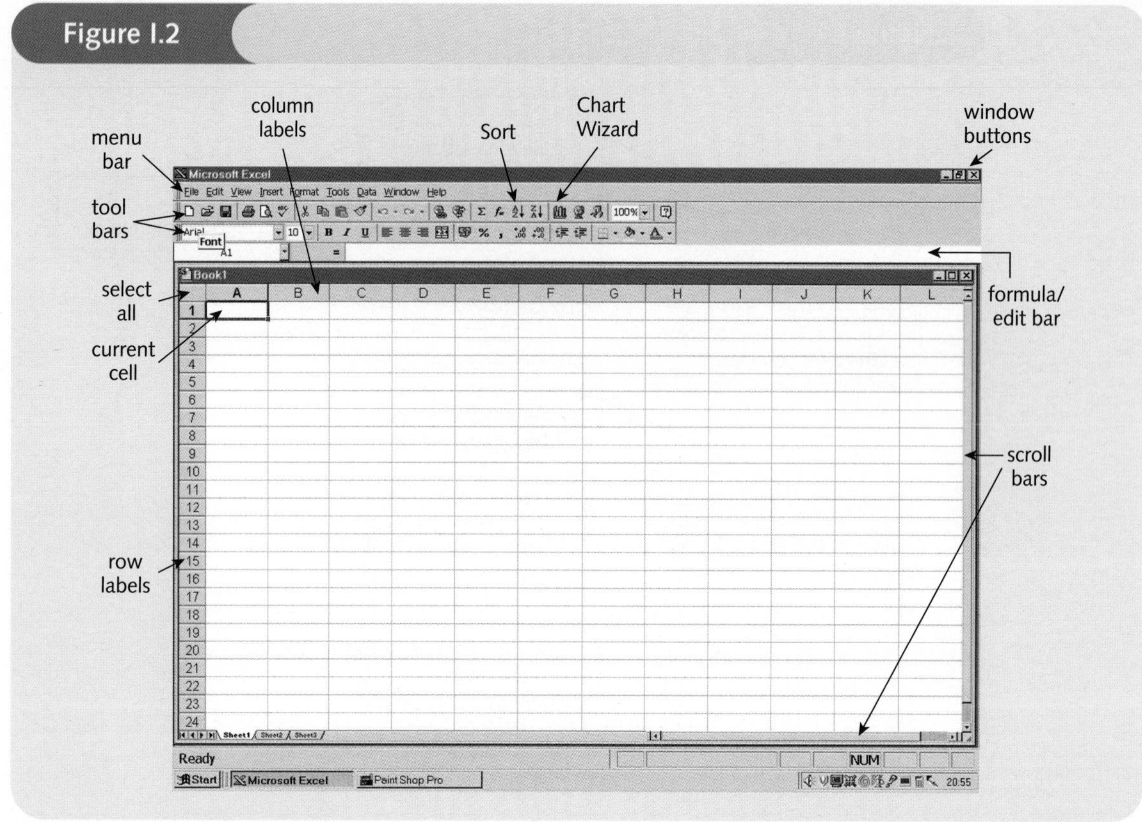

Example EXCEL

A shop audits its toy department to see how much profit it makes from sales of its five best-selling lines. Table I.1 shows the wholesale price (which is the cost to the shop of buying the toy from the manufacturer), the retail price (which is the price that customers pay for each toy), and sales (which is the total number of toys of each type that are sold during the year).

(a) Enter the information in this table into a blank spreadsheet, with the title, Annual Profit, in the first row.

(b) In a fifth column, calculate the annual profit generated by each toy and hence find the total profit made from all five toys.

(c) Format and print the completed spreadsheet.

Table I.1

Item	Wholesale price ($)	Retail price ($)	Sales
Badminton racket	28	58	236
Doll	36	85	785
Silly Putty	1	2	472
Paddling pool	56	220	208
Building bricks	8	26	582

Solution

(a) Entering the data

You can move between the different cells on the spreadsheet using the tab keys or arrow keys, or by positioning the cursor in the required cell and clicking the left mouse button. Have a go at this on your blank sheet to get the feel of it before we begin to enter the data.

To give the spreadsheet a title, we position the cursor in cell A1, and type Annual Profit. Don't worry that the text has run into the next cell. This does not matter, as we are not going to put anything more in this row.

Leaving the next row blank, we type in the column headings for the spreadsheet in row 3. To do this, we position the cursor in cell A3 and type Item; we then move the cursor to cell B3, and type Wholesale price ($). At this stage, the spreadsheet looks like:

Book1

	A	B	C
1	Annual Profit		
2			
3	Item	Wholesale price ($)	
4			
5			

This text has also run into the next cell. Although it looks as if we are positioned in C3 now, we are actually still in B3, as shown by the highlighting. The cursor can be positioned in cell C3 by using the tab, or right arrow key to give:

Book1

	A	B	C
1	Annual Profit		
2			
3	Item	Wholesale	price ($)
4			
5			

Notice that the next cell is highlighted, even though it still contains our previous typing. We can ignore this, and enter Retail price ($). As soon as you start entering this, the previous typing disappears. It is actually still there, but hidden from view as its own cell is not large enough to show all of its contents:

Book1

	A	B	C	D
1	Annual Profit			
2				
3	Item	Wholesale	Retail price ($)	
4				
5				

There is no need to worry about the hidden typing. We will sort this out when we format our spreadsheet in part (c). Finally, we position the cursor in cell D3 and type in the heading Sales.

We can now enter the names of the five items in cells A4 to A8, together with the prices and sales in columns B, C and D to create the spreadsheet:

Book1

	A	B	C	D	E
1	Annual Profit				
2					
3	Item	Wholesale	Retail pric	Sales	
4	Badminton	28	58	236	
5	Doll	36	85	785	
6	Silly Putty	1	2	472	
7	Paddling p	56	220	208	
8	Building br	8	26	582	
9					
10					

If you subsequently return to modify the contents of any particular cell, you will find that when you start typing, the original contents of the cell are deleted, and replaced. If you simply want to amend, rather than replace the text, highlight the relevant cell, and then position the cursor at the required position in the original text, *which is displayed on the edit bar*. You can then edit the text as normal.

(b) Calculating profit

In order to create a fifth column containing the profits, we first type the heading Profit in cell E3. Excel is capable of performing calculations and entering the results in particular cells. This is achieved by typing mathematical formulae into these cells. In this case, we need to enter an appropriate formula for profit in cells E4 to E8.

The profit made on each item is the difference between the wholesale price and retail price. For example, the shop buys a badminton racket from the manufacturer for $28 and sells it to the customers at $58. The profit made on the sale of a single racket is therefore

$58 - 28 = 30$

During the year the shop sells 236 badminton rackets, so the annual profit is

$30 \times 236 = 7080$

In other words, the profit on the sale of badminton rackets is worked out from

$(58 - 28) \times 236$

Looking carefully at the spreadsheet, notice that the numbers 58, 28 and 236 are contained in cells C4, B4 and D4, respectively. Hence annual profit made from the sale of badminton rackets is given by the formula

(C4-B4)*D4

in Excel the multiplication sign is *

We would like the result of this calculation to appear underneath the heading Profit, in column 5, so in cell E4 we type

=(C4-B4)*D4

in Excel always start a formula with =

If you move the cursor down to cell E5, you will notice that the formula has disappeared, and the answer, 7080, has appeared in its place. To get back to the formula, click on cell E4, and the formula is displayed in the formula bar, where it can be edited if necessary.

We would like a similar formula to be entered into every cell in column E, to work out the profit generated by each type of toy. To avoid having to re-enter a similar formula for every cell, it is possible to replicate the one we just put into E4 down the whole column. The spreadsheet will automatically change the cell identities as we go.

To do this, position the cursor in E4, and move the mouse very carefully towards the bottom right-hand corner of the *cell* until the cursor changes from a ✣ to a ✚. Hold down the left mouse button and drag the cell down the column to E8. When the mouse button is released, the values of the profit will appear in the relevant cells.

To put the total profit into cell E9, we need to sum up cells E4 to E8. This can be done by typing

=SUM(E4:E8)

into E9. Pressing the Enter key will then display the answer, 90 605, in this position.

The spreadsheet is displayed in Figure I.3.

Figure I.3

Book1

	A	B	C	D	E	F
1	Annual Profit					
2						
3	Item	Wholesale	Retail pric	Sales	Profit	
4	Badminton	28	58	236	7080	
5	Doll	36	85	785	38465	
6	Silly Putty	1	2	472	472	
7	Paddling p	56	220	208	34112	
8	Building br	8	26	582	10476	
9					90605	
10						
11						
12						

(c) Formatting and printing the spreadsheet

Before we can print the spreadsheet we need to format it, to make it look more attractive to read. In particular, we must alter the column widths to reveal the partially hidden headings. If necessary, we can also insert or delete rows and columns. Perhaps the most useful function is the Undo, which reverses the previous action. If you do something wrong and want to go back a stage, simply click on the ↶ button, which is located towards the middle of the toolbar.

Here is a list of four useful activities that we can easily perform to tidy up the spreadsheet.

Adjusting the column widths to fit the data

Excel can automatically adjust the width of each column to reveal the hidden typing. You can either select an individual column by clicking on its label, or select all the columns at once by clicking the Select All button in the top left-hand corner (see Figure I.2 earlier). From the menu bar we then select **Format: Column: Autofit Selection.** The text that was obscured, because it was too long to fit into the cells, will now be displayed.

Shading and borders

Although the spreadsheet appears to have gridlines around each of the cells, these will not appear on the final printout unless we explicitly instruct Excel to do so. This can be done by highlighting the cells A3 to E8 by first clicking on cell A3, and then with the left mouse button held down, dragging the cursor across the table until all the cells are highlighted. We then release the mouse button, and select **Format: Cells** via the menu bar. Click on the **Border** tab, choose a style, and click on the boxes so that each cell is surrounded on all four sides by gridlines.

Sorting data into alphabetical order

It is sometimes desirable to list items in alphabetical order. To do this, highlight cells A4 to E8, by clicking and dragging, and then click the A → Z button on the toolbar.

Printing the spreadsheet

Before printing a spreadsheet, it is a good idea to select **File: Print Preview** from the menu bar to give you some idea of what it will look like. To change the orientation of the paper, select **File: Page Setup**. Additional

➔

Figure I.4

Annual Profit

Item	Wholesale price ($)	Retail price ($)	Sales	Profit
Badminton racket	28	58	236	7080
Building bricks	8	26	582	10476
Doll	36	85	785	38465
Paddling pool	56	220	208	34112
Silly Putty	1	2	472	472
			Total:	90605

features can be introduced such as headers, footers, column headings repeated at the top of every page, and so on. You might like to experiment with some of these to discover their effect. When you are happy, either click on the Print button, or select **File: Print** from the menu bar.

The final printout is shown in Figure I.4. As you can see, we have chosen to type in the text Total: in cell D9 and have also put gridlines around cells D9 and E9, for clarity.

Practice Problem

EXCEL

1 An economics examination paper is in two sections. Section A is multiple choice and marked out of 40, whereas Section B consists of essay questions and is marked out of 60. Table I.2 shows the marks awarded in each section to six candidates.

Table I.2

Candidate	Section A mark	Section B mark
Fofaria	20	17
Bull	38	12
Eoin	34	38
Arefin	40	52
Cantor	29	34
Devaux	30	49

(a) Enter the information in this table into a blank spreadsheet, with the title, Economics Examination Marks, in the first row.

(b) In a fourth column, calculate the total mark awarded to each candidate.

(c) Use Excel to calculate the average examination mark of these six candidates and give it an appropriate heading.

(d) Format and print the spreadsheet, putting the names of the candidates in alphabetical order.

(e) The second candidate, Bull, asks for a re-mark. Although the Section A mark is correct, the Section B mark is raised to 42. Produce a new spreadsheet based on the correct results.

Getting started with Maple

The second computer package that will be used in this book is Maple. This is a symbolic algebra system. It not only performs numerical calculations but also manipulates mathematical symbols. In effect, it obligingly does the mathematics for you. There are other similar packages available, such as Matlab, Derive and Mathcad, and most of the Maple examples and exercises given in this book can be tackled just as easily using these packages instead. This is not the place to show you the full power of Maple, but hopefully the examples given in this book will give you a flavour of what can be achieved, and why it is such a valuable tool in mathematical modelling.

It is not possible in this introductory section to use Maple to solve realistic problems because you need to learn some mathematics first. However, we will show you how to use it as a calculator, and how to type in mathematical formulae correctly. Figure I.5 shows a typical worksheet which appears on the screen when you double-click on the Maple icon. If you ignore the toolbar at the top of the screen, you can think of it as a blank sheet of paper on which to do some mathematics. You type this after the '>' prompt and end each instruction with a semi-colon ';'. Pressing the Enter key will then make Maple perform your instruction and give you an answer. For example, if you want Maple to work out $3 + 4 \times 2$ you type:

Figure I.5

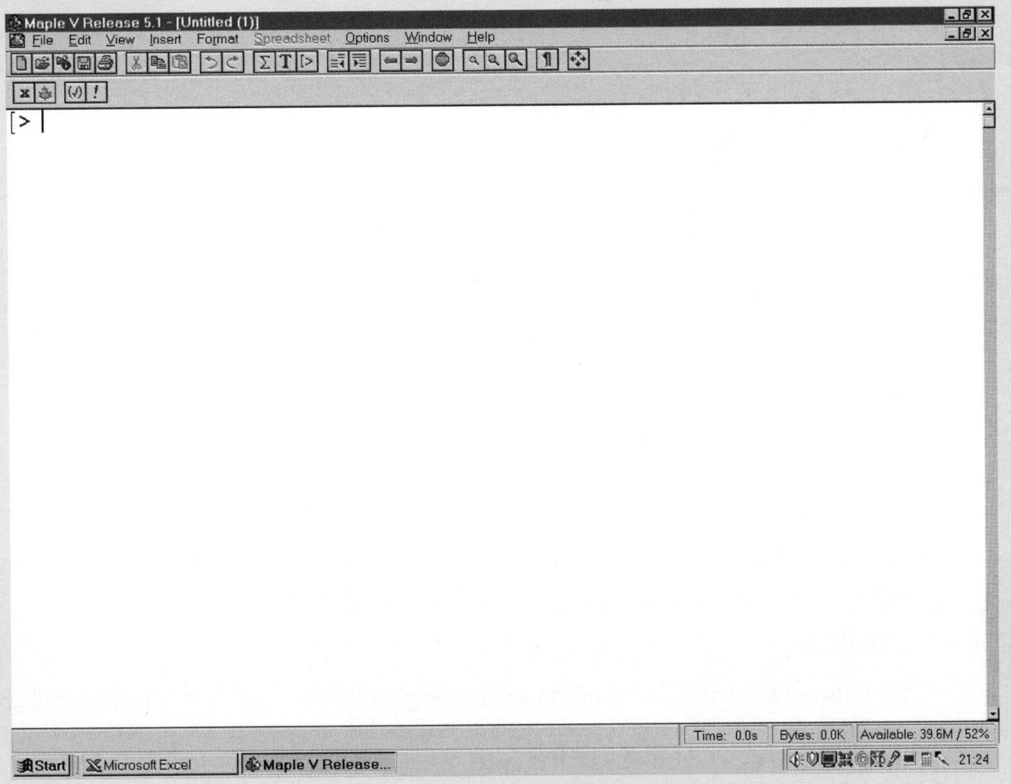

After pressing the Enter key, the package will respond with the answer of 11. Try it now.

Notice that to get this answer, Maple must have performed the multiplication first (to get 8) before adding on the 3. This is because, like the rest of the mathematical world, Maple follows the BIDMAS convention:

	B	(Brackets first)	
	I	(Indices second)	
then	D	(Division)	(a tie for third place)
and	M	(Multiplication)	
finally	A	(Addition)	(a tie for fourth place)
and	S	(Subtraction)	

Since multiplication has a higher priority than addition, Maple works out 4×2 first. If you really want to work out $3 + 4$ before multiplying by 2, you put in brackets:

```
>(3+4)*2;
```

don't forget to put the semi-colon at the end

which gives 14. Notice that in Maple (like Excel), it is necessary to use * rather than × as the multiplication symbol. Similarly, Maple and Excel both use/(instead of ÷) for division, and ^ as the instruction to 'raise to the power of'. So, to work out $\frac{70}{7} + 5^2$ you type:

```
>70/7+5^2;
```

which gives 35. Again, using BIDMAS, since indices have a higher priority than addition, 5^2 has been evaluated first as 25, before adding on $70/7 = 10$, to give a final answer of 35.

Practice Problem — MAPLE

2 Use Maple to work out each of the following:

(a) $12 + 18 \div 9$ **(b)** $3^3 - 4^2$ **(c)** $(7 + 3) \div 2$

Suppose now that you wish to work out all of the following sums:

$3 \times 5^2 - 2 \times 5$

$3 \times 6^2 - 2 \times 6$

$3 \times 7^2 - 2 \times 7$

$3 \times 6.4^2 - 2 \times 6.4$

$3 \times 92.5^2 - 2 \times 92.5$

You could, of course, just type all five calculations, one after the other to get the answers. However, there is a common pattern. They each take the form

$3x^2 - 2x$

for various values of x, and it makes sense to exploit this fact. As a first step, we shall give this expression a name. We could call it Fred or Wilma, but in practice, we prefer to give it a name that relates to the context in which it arises. A mathematical expression that contains a square term like this is called a quadratic so let us name this particular one quad1. To do this, type:

```
>quad1:=3*x^2-2*x;
```

The symbol ':=' tells Maple that you wish to *define* quad1 to be $3x^2 - 2x$.

To work out $3 \times 5^2 - 2 \times 5$ we substitute $x = 5$ into this expression: that is, we replace the symbol x by the number 5. In Maple, this is achieved by typing

```
>subs(x=5,quad1);
```

Maple responds with the answer 65. To perform the other four calculations, all we need do is to edit the Maple instruction, change the 5 to a new value, move the cursor to the right of the semi-colon, and press the Enter key. You might like to try this for yourself.

Hopefully, this brief introduction has given you some idea of how to use Maple to perform simple calculations. However, you may be wondering what all the fuss is about. Surely we could have performed these calculations just as easily on an ordinary calculator? Well, the honest answer is probably yes. The real advantage of using Maple is as a tool for solving complex mathematical problems. We shall meet some examples in the second half of this book. The following problem gives a brief glimpse at the sort of things that it can do.

Practice Problem — MAPLE

3 Type in the following after the '>' prompt. Before doing so, you might like to guess what each instruction is likely to do.

(a) `solve(2*x-8=0,x);`

(b) `plot(2*x-8,x=0..10);`

(c) `expand((x+2)^2);`

(d) `simplify(2*x+6+5*x-2);`

(e) `plot3d(x^3-3*x+x*y,x=-2..2,y=-2..2);`

Key Terms

The key terms in each section are shown in *colour* in the text when they are first used. Their definitions are shown at the end of each section in a coloured box. In addition, all key terms are repeated in the glossary at the end of the book.

chapter 1
Linear Equations

The main aim of this chapter is to introduce the mathematics of linear equations. This is an obvious first choice in an introductory text, since it is an easy topic which has many applications. There are six sections, which are intended to be read in the order that they appear.

Sections 1.1, 1.2, 1.4 and 1.5 are devoted to mathematical methods. They serve to revise the rules of arithmetic and algebra, which you probably met at school but may have forgotten. In particular, the properties of negative numbers and fractions are considered. A reminder is given on how to multiply out brackets and how to manipulate mathematical expressions. You are also shown how to solve simultaneous linear equations. Systems of two equations in two unknowns can be solved using graphs, which are described in Section 1.1. However, the preferred method uses elimination, which is considered in Section 1.2. This algebraic approach has the advantage that it always gives an exact solution and it extends readily to larger systems of equations.

The remaining two sections are reserved for applications in microeconomics and macroeconomics. You may be pleasantly surprised by how much economic theory you can analyse using just the basic mathematical tools considered here. Section 1.3 introduces the fundamental concept of an economic function and describes how to calculate equilibrium prices and quantities in supply and demand theory. Section 1.6 deals with national income determination in simple macroeconomic models.

The first five sections underpin the rest of the book and are essential reading. The final section is not quite as important and can be omitted if desired.

section 1.1

Graphs of linear equations

Objectives

At the end of this section you should be able to:

- Plot points on graph paper given their coordinates.
- Add, subtract, multiply and divide negative numbers.
- Sketch a line by finding the coordinates of two points on the line.
- Solve simultaneous linear equations graphically.
- Sketch a line by using its slope and intercept.

Consider the two straight lines shown in Figure 1.1. The horizontal line is referred to as the x ***axis*** and the vertical line is referred to as the y ***axis***. The point where these lines intersect is known as the ***origin*** and is denoted by the letter O. These lines enable us to identify uniquely any point, P, in terms of its ***coordinates*** (x, y). The first number, x, denotes the horizontal distance along the x axis and the second number, y, denotes the vertical distance along the y axis. The arrows on the axes indicate the positive direction in each case.

Figure 1.1

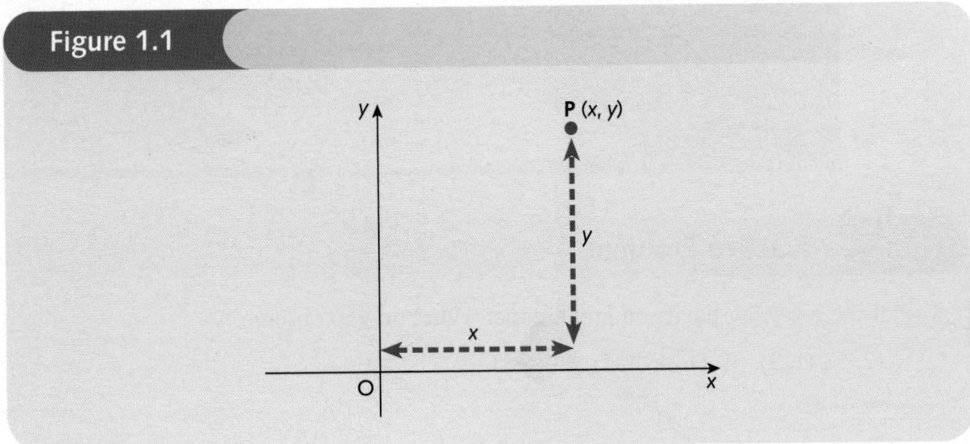

Example

Plot the points A(2, 3), B(−1, 4), C(−3, −1), D(3, −2) and E(5, 0).

Solution

The point A with coordinates (2, 3) is obtained by starting at the origin, moving 2 units to the right and then moving 3 units vertically upwards. Similarly, the point B with coordinates (−1, 4) is located 1 unit to the left of O (because the x coordinate is negative) and 4 units up. These points, together with C(−3, −1), D(3, −2) and E(5, 0) are plotted in Figure 1.2.

Note that the point C lies in the bottom left-hand quadrant since its x and y coordinates are both negative. It is also worth noticing that E actually lies on the x axis since its y coordinate is zero. Likewise, a point with coordinates of the form (0, y) for some number y would lie somewhere on the y axis. Of course, the point with coordinates (0, 0) is the origin, O.

Advice

The best way for you to understand mathematics is to practise the techniques yourself. For this reason, problems are included within the text as well as at the end of every section. Please stop reading the book, pick up a pencil and a ruler, and attempt these problems as you go along. You should then check your answers honestly with those given at the back of the book.

Practice Problem

1 Plot the following points on graph paper. What do you observe?

(2, 5), (1, 3), (0, 1), (−2, −3), (−3, −5)

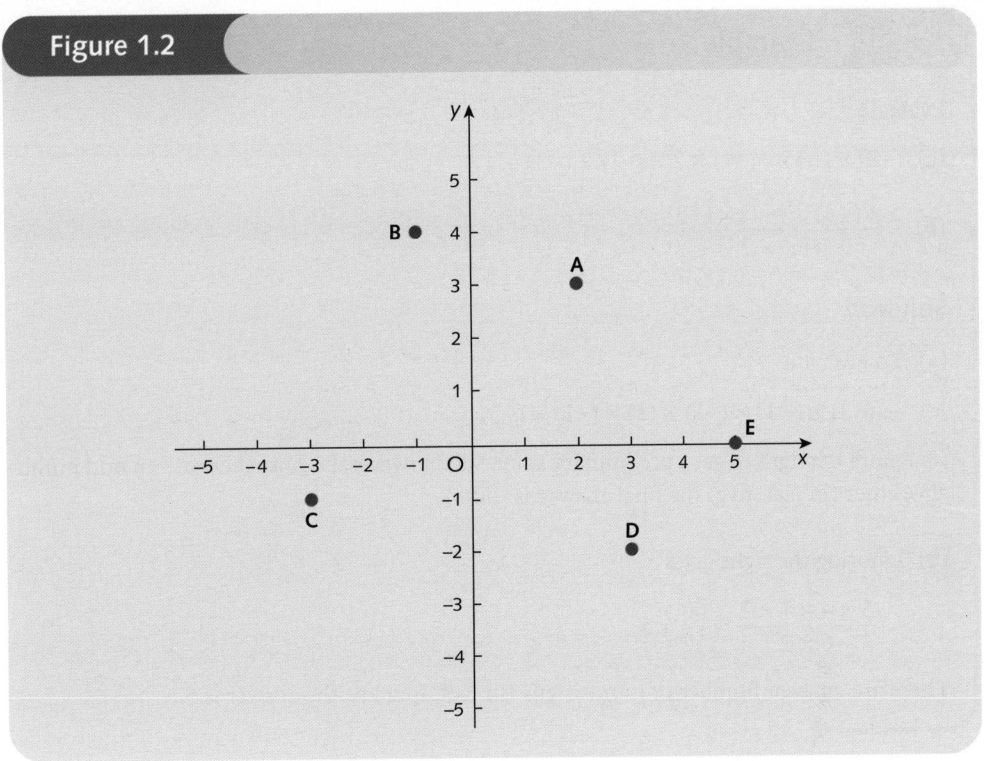

Before we can continue the discussion of graphs it is worthwhile revising the properties of negative numbers. The rules for the multiplication of negative numbers are

negative	×	negative	=	positive
negative	×	positive	=	negative

It does not matter in which order two numbers are multiplied, so

positive	×	negative	=	negative

These rules produce

$$(-2) \times (-3) = 6$$
$$(-4) \times 5 = -20$$
$$7 \times (-5) = -35$$

respectively. Also, because division is the same sort of operation as multiplication (it just undoes the result of multiplication and takes you back to where you started), exactly the same rules apply when one number is divided by another. For example,

$$(-15) \div (-3) = 5$$
$$(-16) \div 2 = -8$$
$$2 \div (-4) = -1/2$$

In general, to multiply or divide lots of numbers it is probably simplest to ignore the signs to begin with and just to work the answer out. The final result is negative if the total number of minus signs is odd and positive if the total number is even.

Example

Evaluate

(a) $(-2) \times (-4) \times (-1) \times 2 \times (-1) \times (-3)$

(b) $\dfrac{5 \times (-4) \times (-1) \times (-3)}{(-6) \times 2}$

Solution

(a) To evaluate

$$(-2) \times (-4) \times (-1) \times (2) \times (-1) \times (-3)$$

we ignore the signs to get a preliminary value, 48. However, because there are an odd number of minus signs altogether (in fact, five) the final answer is -48.

(b) Ignoring the signs gives

$$\frac{5 \times 4 \times 1 \times 3}{6 \times 2} = \frac{60}{12} = 5$$

There are an even number of minus signs (in fact, four) so the answer is 5.

Advice

Attempt the following problem yourself both with and without a calculator. On most machines a negative number such as −6 is entered by pressing the button labelled (−) followed by 6.

Practice Problem

2 (1) Without using a calculator evaluate

(a) $5 \times (-6)$ **(b)** $(-1) \times (-2)$ **(c)** $(-50) \div 10$

(d) $(-5) \div (-1)$ **(e)** $2 \times (-1) \times (-3) \times 6$ **(f)** $\dfrac{2 \times (-1) \times (-3) \times 6}{(-2) \times 3 \times 6}$

(2) Confirm your answer to part (1) using a calculator.

To add or subtract negative numbers it helps to think in terms of a picture of the x axis:

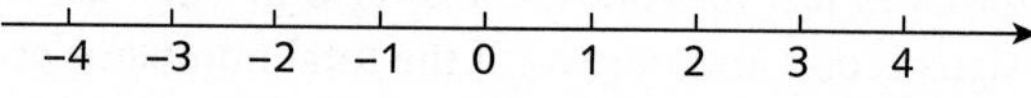

If b is a positive number then

$$a - b$$

can be thought of as an instruction to start at a and to move b units to the left. For example,

$$1 - 3 = -2$$

because if you start at 1 and move 3 units to the left, you end up at −2:

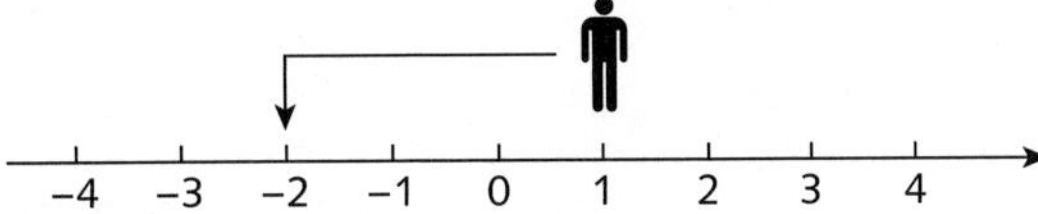

Similarly,

$$-2 - 1 = -3$$

because 1 unit to the left of −2 is −3.

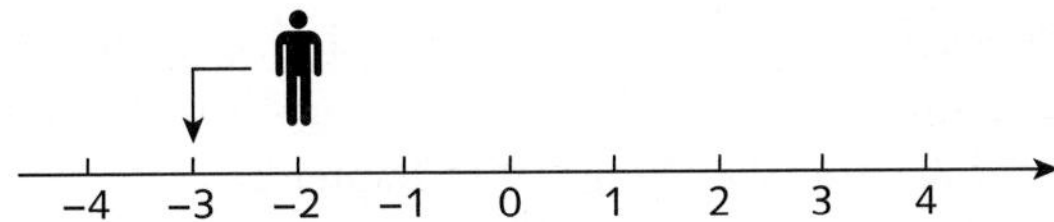

On the other hand,

$$a - (-b)$$

is taken to be $a + b$. This follows from the rule for multiplying two negative numbers, since

$$-(-b) = (-1) \times (-b) = b$$

Consequently, to evaluate

$$a - (-b)$$

you start at a and move b units to the right (that is, in the positive direction). For example,

$$-2 - (-5) = -2 + 5 = 3$$

because if you start at −2 and move 5 units to the right you end up at 3.

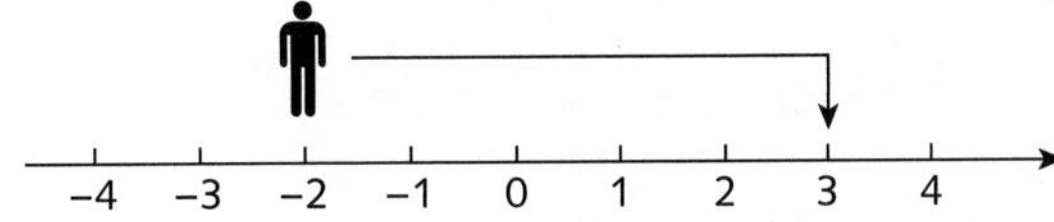

Example

Evaluate

(a) $-32 - 4$

(b) $-68 - (-62)$

Solution

(a) $-32 - 4 = -36$ because 4 units to the left of −32 is −36.

(b) $-68 - (-62) = -68 + 62 = -6$ because 62 units to the right of −68 is −6.

Practice Problem

3 (1) Without using a calculator evaluate

(a) $1 - 2$ **(b)** $-3 - 4$ **(c)** $1 - (-4)$

(d) $-1 - (-1)$ **(e)** $-72 - 19$ **(f)** $-53 - (-48)$

(2) Confirm your answer to part (1) using a calculator.

We now return to the problem of graphs. In economics we need to do rather more than just plot individual points on graph paper. We would like to be able to sketch curves represented by equations and to deduce information from such a picture. Incidentally, it is sometimes more appropriate to label axes using letters other than x and y. For example, in the analysis of supply and demand, the variables involved are the quantity and price of a good. It is then convenient to use Q and P instead of x and y. This helps us to remember which variable we have used on which axis. However, in this section, only the letters x and y are used. Also, we restrict our attention to those equations whose graphs are straight lines, deferring consideration of more general curve sketching until Chapter 2.

In Practice Problem 1 you will have noticed that the five points (2, 5), (1, 3), (0, 1), (−2, −3) and (−3, −5) all lie on a straight line. In fact, the equation of this line is

$$-2x + y = 1$$

Any point lies on this line if its x and y coordinates satisfy this equation. For example, (2, 5) lies on the line because when the values $x = 2$ and $y = 5$ are substituted into the left-hand side of the equation we obtain

$$-2(2) + 5 = -4 + 5 = 1$$

which is the right-hand side of the equation. The other points can be checked similarly (Table 1.1).

Table 1.1

Point	Check	
(1, 3)	$-2(1) + 3 = -2 + 3 = 1$	✓
(0, 1)	$-2(0) + 1 = 0 + 1 = 1$	✓
(−2, −3)	$-2(-2) - 3 = 4 - 3 = 1$	✓
(−3, −5)	$-2(-3) - 5 = 6 - 5 = 1$	✓

Notice how the rules for manipulating negative numbers have been used in the calculations.

The general equation of a straight line takes the form

$$\boxed{\text{a multiple of } x} + \boxed{\text{a multiple of } y} = \boxed{\text{a number}}$$

that is,

$$dx + ey = f$$

for some given numbers d, e and f. Consequently, such an equation is called a ***linear equation***. The numbers d and e are referred to as the ***coefficients***. The coefficients of the linear equation,

$$-2x + y = 1$$

are -2 and 1 (the coefficient of y is 1 because y can be thought of as $1 \times y$).

Advice

If you have forgotten about the order in which operations are performed, you should read the section on BIDMAS on page 10.

Example

Decide which of the following points lie on the line $5x - 2y = 6$:

A(0, −3), B(2, 2), C(−10, −28) and D(4, 8)

Solution

$$5(0) - 2(-3) = 0 - (-6) = 0 + 6 = 6$$

$$5(2) - 2(2) = 10 - 4 = 6$$

$$5(-10) - 2(-28) = -50 - (-56) = -50 + 56 = 6$$

$$5(4) - 2(8) = 20 - 16 = 4 \neq 6$$

Hence points A, B and C lie on the line, but D does not lie on the line.

Practice Problem

4 Check that the points

$$(-1, 2), (-4, 4), (5, -2), (2, 0)$$

all lie on the line

$$2x + 3y = 4$$

and hence sketch this line on graph paper. Does the point (3, −1) lie on this line?

In general, to sketch a line from its mathematical equation, it is sufficient to calculate the coordinates of any two distinct points lying on it. These two points can be plotted on graph paper and a ruler used to draw the line passing through them. One way of finding the coordinates of a point on a line is simply to choose a numerical value for x and to substitute it into the equation. The equation can then be used to deduce the corresponding value of y. The whole process can be repeated to find the coordinates of the second point by choosing another value for x.

Example

Sketch the line

$$4x + 3y = 11$$

Solution

For the first point, let us choose $x = 5$. Substitution of this number into the equation gives

$$4(5) + 3y = 11$$
$$20 + 3y = 11$$

The problem now is to find the value of y which satisfies this equation. A naïve approach might be to use trial and error: that is, we could just keep guessing values of y until we find the one that works. Can you guess what y is in this case? However, a more reliable and systematic approach is actually to solve this equation using the rules of mathematics. In fact, the only rule that we need is this:

you can apply whatever mathematical operation you like to an equation,
provided that you do the same thing to both sides

There is only one exception to this rule: you must never divide both sides by zero. This should be obvious because a number such as 11/0 does not exist. (If you do not believe this, try dividing 11 by 0 on your calculator.)

The first obstacle that prevents us from writing down the value of y immediately is the number 20, which is added on to the left-hand side. This can be removed by subtracting 20 from the left-hand side. In order for this to be legal, we must also subtract 20 from the right-hand side to get

$$3y = 11 - 20$$
$$3y = -9$$

The second obstacle is the number 3, which is multiplying the y. This can be removed by dividing the left-hand side by 3. Of course, we must also divide the right-hand side by 3 to get

$$y = -9/3 = -3$$

Consequently, the coordinates of one point on the line are $(5, -3)$.

For the second point, let us choose $x = -1$. Substitution of this number into the equation gives

$$4(-1) + 3y = 11$$
$$-4 + 3y = 11$$

This can be solved for y as follows:

$$3y = 11 + 4 = 15 \qquad \text{(add 4 to both sides)}$$
$$y = 15/3 = 5 \qquad \text{(divide both sides by 3)}$$

Hence $(-1, 5)$ lies on the line, which can now be sketched on graph paper as shown in Figure 1.3.

Practice Problem

5 Find the coordinates of two points on the line

$$3x - 2y = 4$$

by taking $x = 2$ for the first point and $x = -2$ for the second point. Hence sketch its graph.

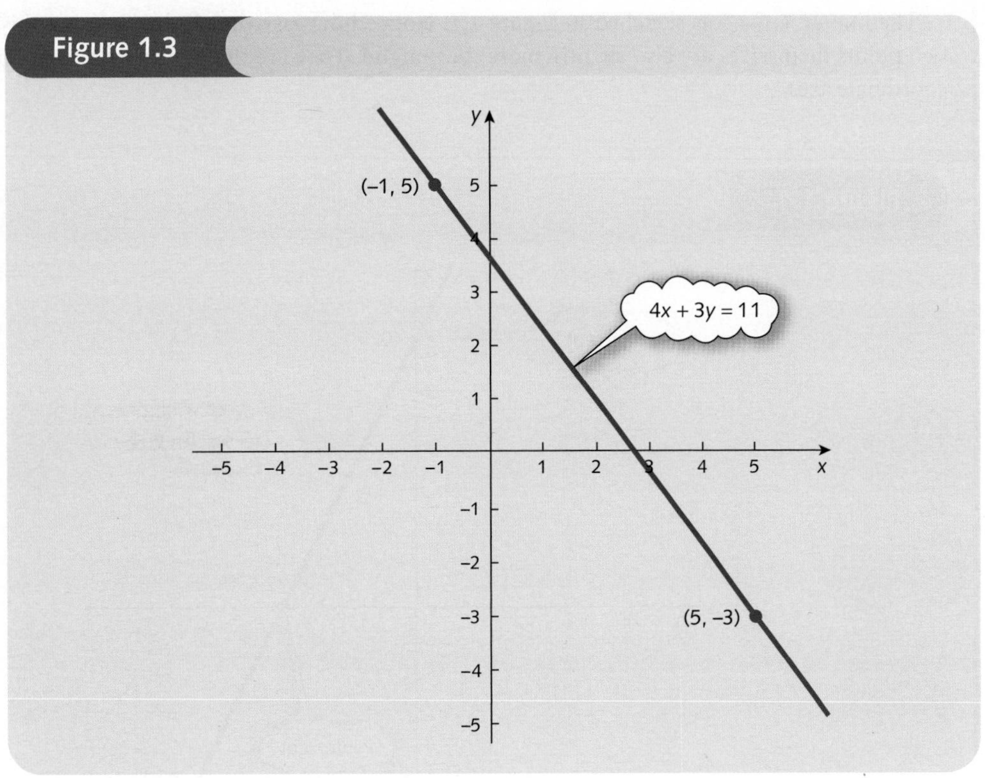

In this example we arbitrarily picked two values of x and used the linear equation to work out the corresponding values of y. There is nothing particularly special about the variable x. We could equally well have chosen values for y and solved the resulting equations for x. In fact, the easiest thing to do (in terms of the amount of arithmetic involved) is to put $x = 0$ and find y and then to put $y = 0$ and find x.

Example

Sketch the line

$$2x + y = 5$$

Solution

Setting $x = 0$ gives

$$2(0) + y = 5$$

$$0 + y = 5$$

$$y = 5$$

Hence (0, 5) lies on the line.

Setting $y = 0$ gives

$$2x + 0 = 5$$

$$2x = 5$$

$$x = 5/2 \quad \text{(divide both sides by 2)}$$

Hence (5/2, 0) lies on the line.

The line $2x + y = 5$ is sketched in Figure 1.4. Notice how easy the algebra is using this approach. The two points themselves are also slightly more meaningful. They are the points where the line intersects the coordinate axes.

Figure 1.4

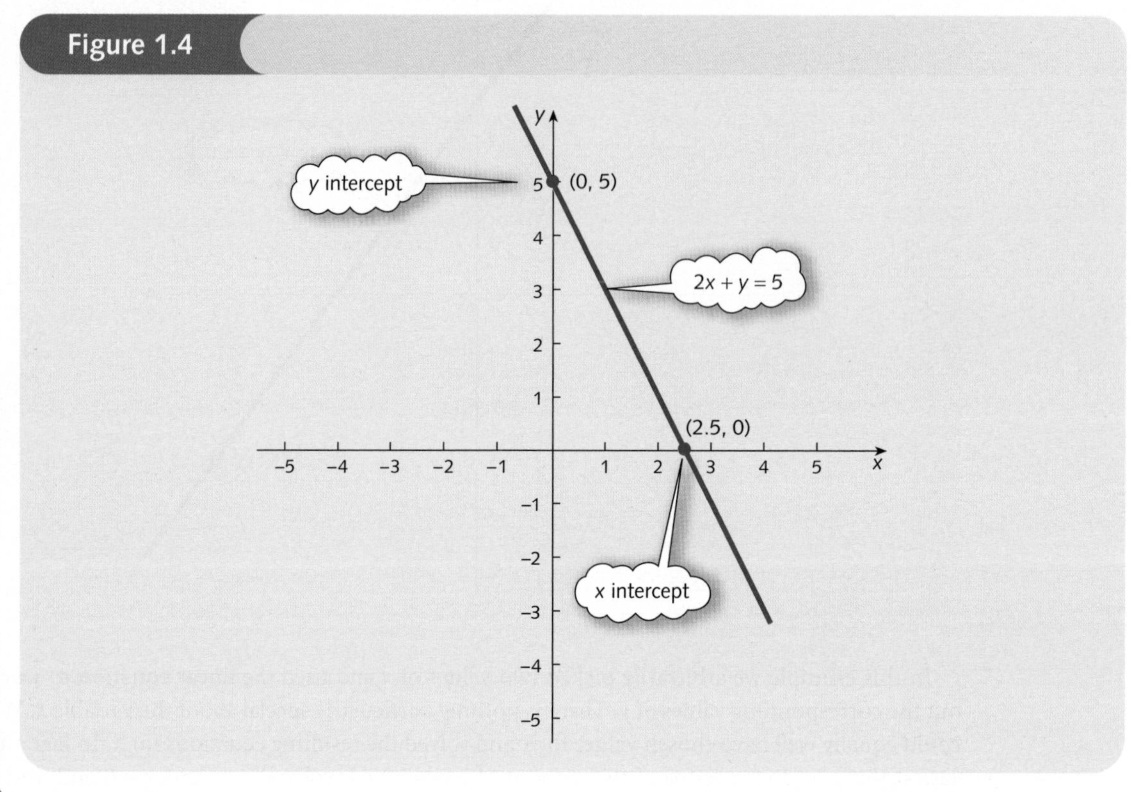

Practice Problem

6 Find the coordinates of the points where the line

$$x - 2y = 2$$

intersects the axes. Hence sketch its graph.

In economics it is sometimes necessary to handle more than one equation at the same time. For example, in supply and demand analysis we are interested in two equations, the supply equation and the demand equation. Both involve the same variables Q and P, so it makes sense to sketch them on the same diagram. This enables the market equilibrium quantity and price to be determined by finding the point of intersection of the two lines. We shall return to the analysis of supply and demand in Section 1.3. There are many other occasions in economics and business studies when it is necessary to determine the coordinates of points of intersection. The following is a straightforward example which illustrates the general principle.

Example

Find the point of intersection of the two lines

$$4x + 3y = 11$$

$$2x + y = 5$$

Solution

We have already seen how to sketch these lines in the previous two examples. We discovered that

$$4x + 3y = 11$$

passes through (5, –3) and (–1, 5), and that

$$2x + y = 5$$

passes through (0, 5) and (5/2, 0).

These two lines are sketched on the same diagram in Figure 1.5, from which the point of intersection is seen to be (2, 1).

It is easy to verify that we have not made any mistakes by checking that (2, 1) lies on both lines. It lies on

$4x + 3y = 11$ because $4(2) + 3(1) = 8 + 3 = 11$ ✓

and lies on $2x + y = 5$ because $2(2) + 1 = 4 + 1 = 5$ ✓

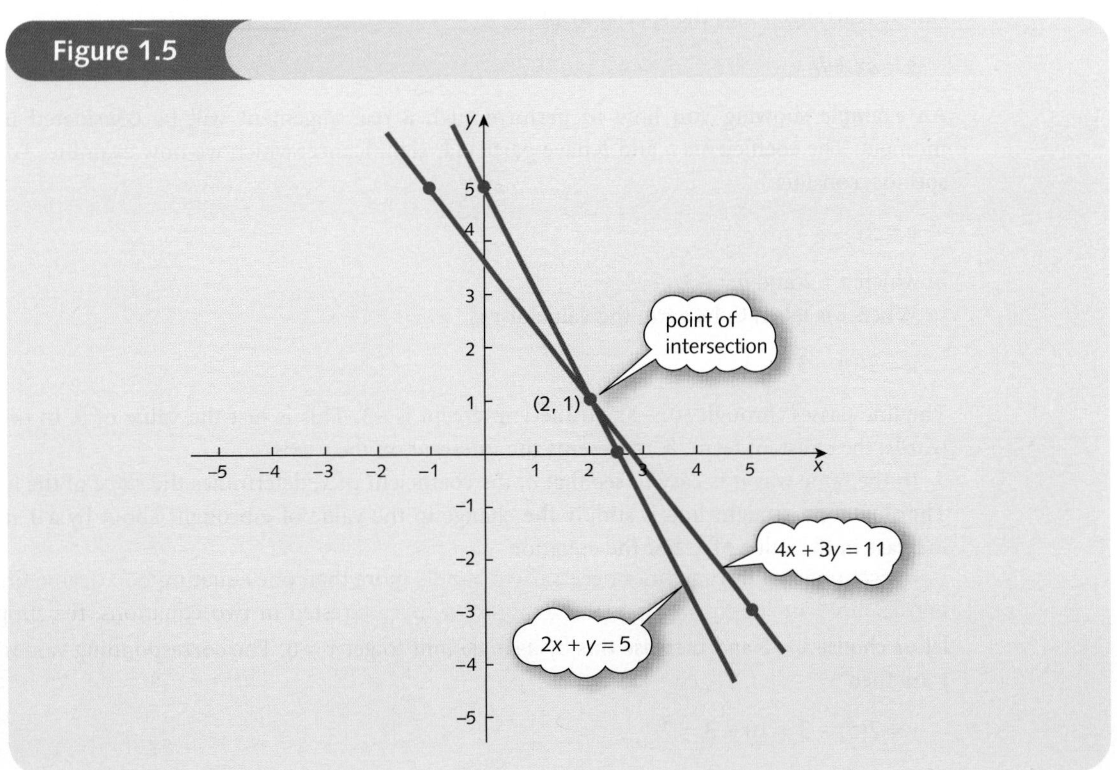

Figure 1.5

→

For this reason, we say that $x = 2$, $y = 1$, is the solution of the *simultaneous linear equations*

$$4x + 3y = 11$$

$$2x + y = 5$$

Practice Problem

7 Find the point of intersection of

$$3x - 2y = 4$$

$$x - 2y = 2$$

[Hint: you might find your answers to Problems 5 and 6 useful.]

Quite often it is not necessary to produce an accurate plot of an equation. All that may be required is an indication of the general shape together with a few key points or features. It can be shown that, provided e is non-zero, any equation given by

$$dx + ey = f$$

can be rearranged into the special form

$$y = ax + b$$

An example showing you how to perform such a rearrangement will be considered in a moment. The coefficients a and b have particular significance, which we now examine. To be specific, consider

$$y = 2x - 3$$

in which $a = 2$ and $b = -3$.

When x is taken to be zero, the value of y is

$$y = 2(0) - 3 = -3$$

The line passes through (0, −3), so the y intercept is −3. This is just the value of b. In other words, the constant term, b, represents the ***intercept*** on the y axis.

In the same way it is easy to see that a, the coefficient of x, determines the ***slope*** of the line. The slope of a straight line is simply the change in the value of y brought about by a 1 unit increase in the value of x. For the equation

$$y = 2x - 3$$

let us choose $x = 5$ and increase this by a single unit to get $x = 6$. The corresponding values of y are then

$$y = 2(5) - 3 = 10 - 3 = 7$$

$$y = 2(6) - 3 = 12 - 3 = 9$$

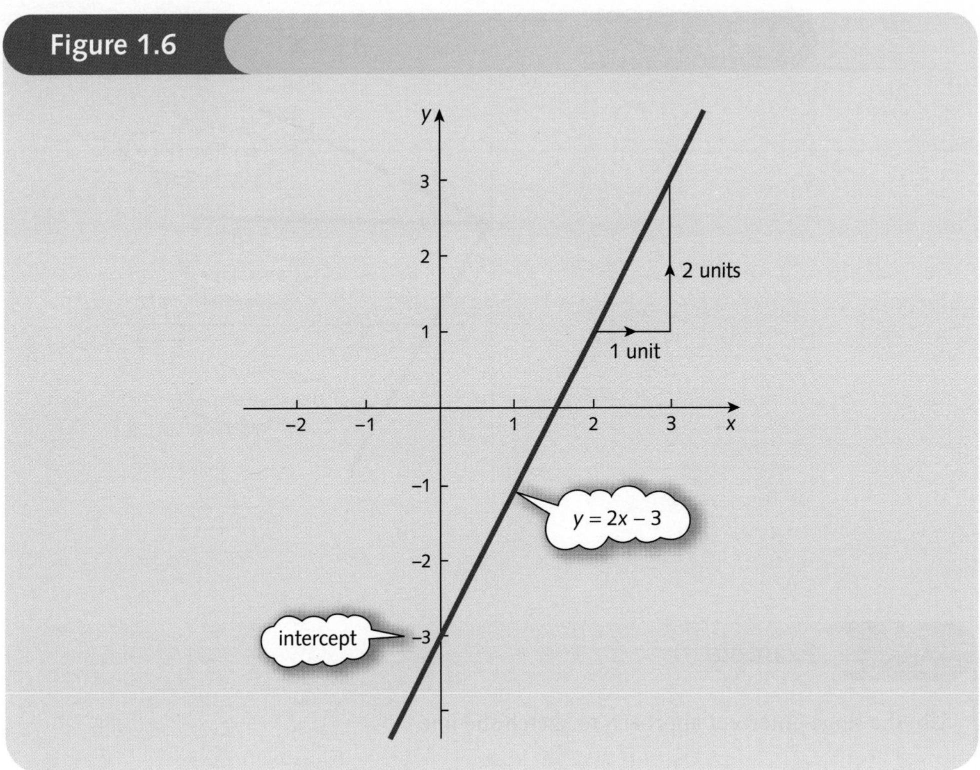

respectively. The value of y increases by 2 units when x rises by 1 unit. The slope of the line is therefore 2, which is the value of a. The slope of a line is fixed throughout its length, so it is immaterial which two points are taken. The particular choice of $x = 5$ and $x = 6$ was entirely arbitrary. You might like to convince yourself of this by choosing two other points, such as $x = 20$ and $x = 21$, and repeating the previous calculations.

A graph of the line

$$y = 2x - 3$$

is sketched in Figure 1.6. This is sketched using the information that the intercept is −3 and that for every 1 unit along we go 2 units up. In this example the coefficient of x is positive. This does not have to be the case. If a is negative then for every increase in x there is a corresponding decrease in y, indicating that the line is downhill. If a is zero then the equation is just

$$y = b$$

indicating that y is fixed at b and the line is horizontal. The three cases are illustrated in Figure 1.7 (overleaf).

It is important to appreciate that in order to use the slope–intercept approach it is necessary for the equation to be written as

$$y = ax + b$$

If a linear equation does not have this form, it is usually possible to perform a preliminary rearrangement to isolate the variable y on the left-hand side, as the following example demonstrates.

Figure 1.7

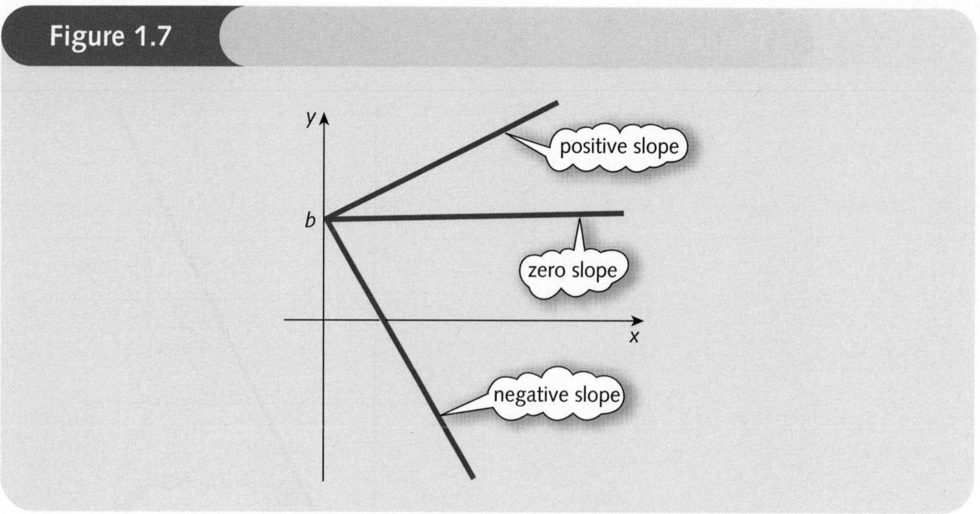

Example

Use the slope–intercept approach to sketch the line

$$2x + 3y = 12$$

Solution

We can remove the x term on the left-hand side of

$$2x + 3y = 12$$

by subtracting $2x$. As usual, to balance the equation we must also subtract $2x$ from the right-hand side to get

$$3y = 12 - 2x$$

We now just divide through by 3 to get

$$y = 4 - {}^{2}/_{3}x$$

This is now in the required form with $a = -2/3$ and $b = 4$. The line is sketched in Figure 1.8. A slope of $-2/3$ means that, for every 1 unit along, we go 2/3 units down (or, equivalently, for every 3 units along, we go 2 units down). An intercept of 4 means that it passes through (0, 4).

Practice Problem

8 Use the slope–intercept approach to sketch the lines

(a) $y = x + 2$ **(b)** $4x + 2y = 1$

Figure 1.8

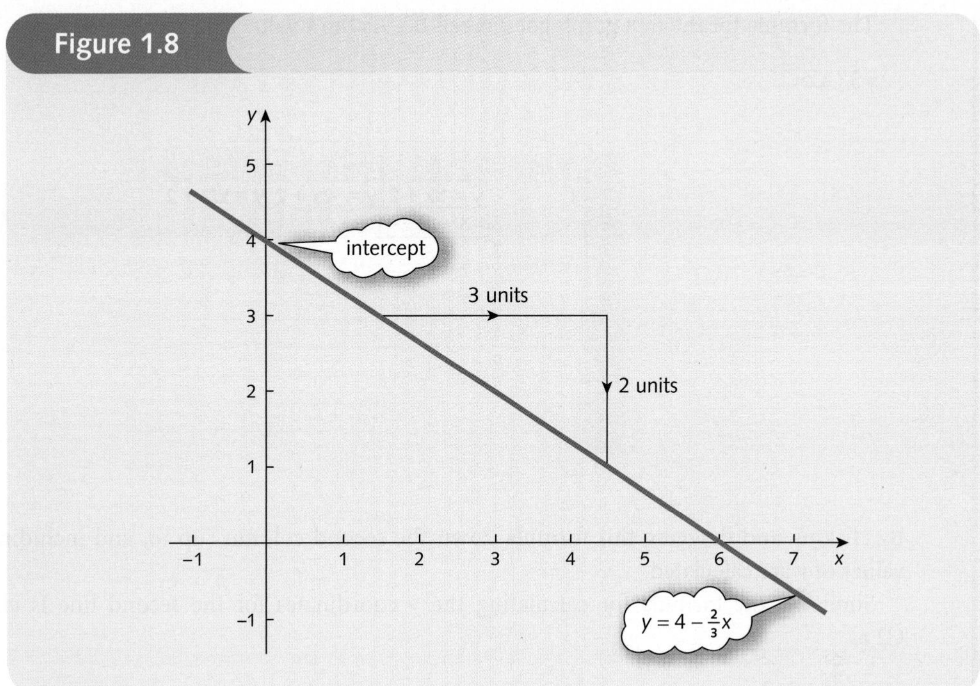

Example

EXCEL

(a) Use Excel to draw the graphs of

$$y = 3x + 2$$

$$y = -2x + 2$$

$$y = {}^1\!/_2x + 2$$

on the same set of axes, taking values of x between −3 and +3.

(b) On another set of axes, use Excel to draw the graphs of

$$y = 2x$$

$$y = 2x - 3$$

$$y = 2x + 1$$

for $-3 \leq x \leq 3$.

(c) What do you notice about the two sets of graphs?

Solution

(a) To draw graphs with Excel, we first have to set up a table of values. By giving a title to each column, we will be able to label the graphs at a later stage, so we type the headings x, $y = 3x + 2$, $y = -2x + 2$ and $y = x/2 + 2$ in cells A1, B1, C1 and D1 respectively.

The x values are now typed into the first column, as shown in the diagram overleaf. In the next three columns, we generate the corresponding values for y by entering formulae for each of the three lines.

➔

The formula for the first graph goes in cell B2. As the x value is in cell A2, we type

=3*A2+2

	A	B	C	D
1	x	y = 3x + 2	y = -2x + 2	y = x/2 + 2
2	-3	=3*A2+2		
3	-2			
4	-1			
5	0			
6	1			
7	2			
8	3			
9				
10				

By clicking and dragging this formula down the second column (up to, and including, cell B8), the values of y are calculated.

Similarly, the formula for calculating the y coordinates for the second line is entered into cell C2 as

=−2*A2+2

and the formula for the third line is entered into cell D2 as

=A2/2+2

To plot these points on a graph, we highlight all the cells in the table, including the column titles, and click on the Chart Wizard button on the toolbar. The Chart Wizard box will appear:

	A	B	C	D
1	x	y = 3x + 2	y = -2x + 2	y = x/2 + 2
2	-3	-7	8	0.5
3	-2	-4	6	1
4	-1	-1	4	1.5
5	0	2	2	2
6	1	5	0	2.5
7	2	8	-2	3
8	3	11	-4	3.5

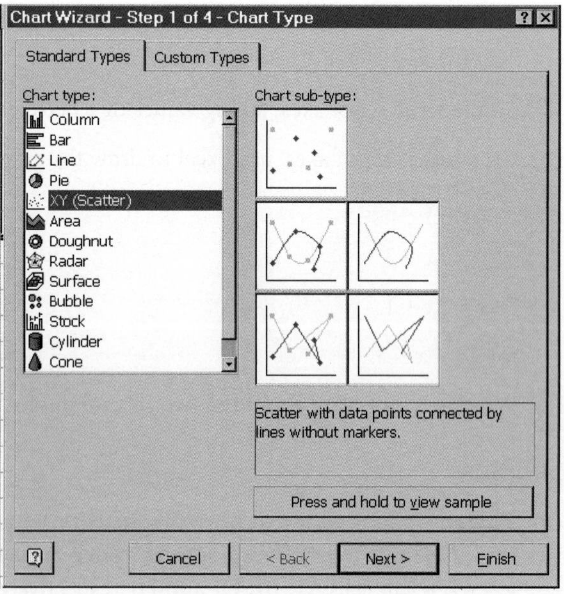

Chart Wizard - Step 1 of 4 - Chart Type

Standard Types | Custom Types

Chart type: Column, Bar, Line, Pie, XY (Scatter), Area, Doughnut, Radar, Surface, Bubble, Stock, Cylinder, Cone

Chart sub-type:

Scatter with data points connected by lines without markers.

Press and hold to view sample

Cancel | < Back | Next > | Finish

From the list of chart types, we choose **XY (Scatter)**, and then choose an appropriate sub-type. As we are plotting straight lines, we have selected **Scatter with data points connected by lines without markers.**

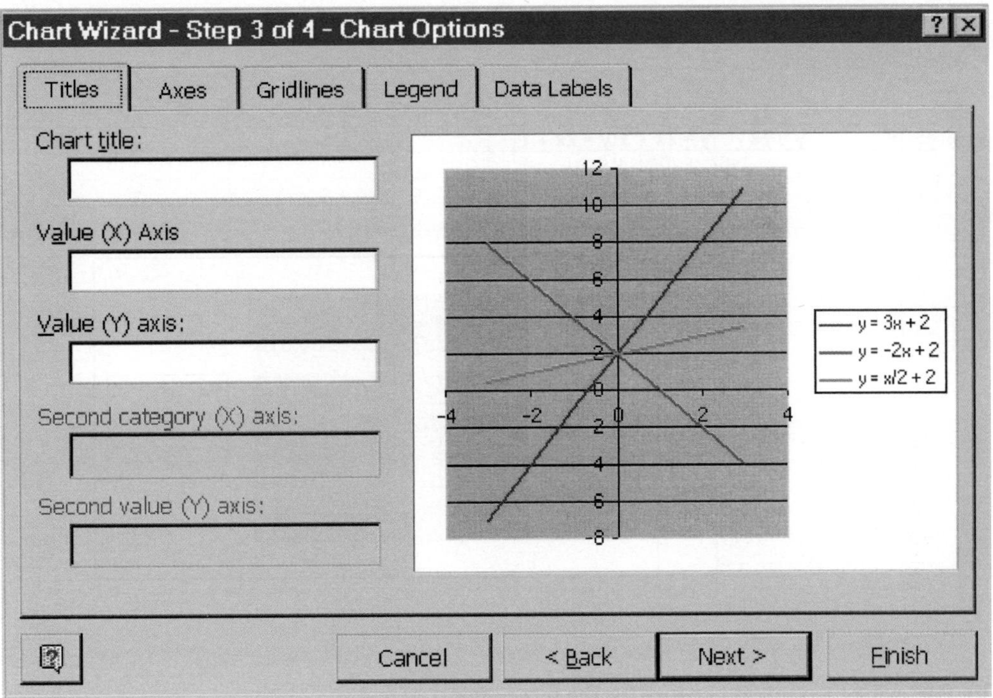

The Next screen shows you a preview of the graph, with the option to change the range of the cells that have been plotted. If the graph looks wrong, it is usually because the wrong cells have been highlighted before going into Chart Wizard, so go back and check this, rather than altering the range.

The third screen allows you to label your graph, and alter its gridlines. You should always label your axes, but you could, for example, delete the Legend if you feel it is inappropriate. Adding gridlines can make it easier to read values off the graph.

Finally, we click Next and Finish, to transfer the graph on to the spreadsheet, as shown in Figure 1.9. Notice that Excel provides a key showing which line is which.

Figure 1.9

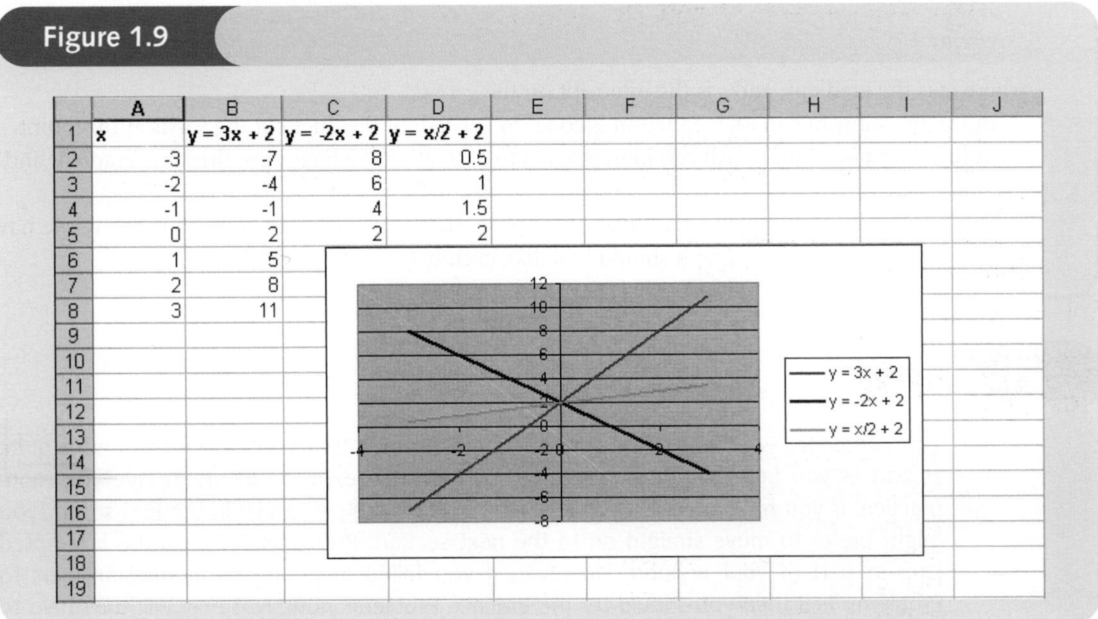

	A	B	C	D
1	x	y = 3x + 2	y = -2x + 2	y = x/2 + 2
2	-3	-7	8	0.5
3	-2	-4	6	1
4	-1	-1	4	1.5
5	0	2	2	2
6	1	5		
7	2	8		
8	3	11		

Figure 1.10

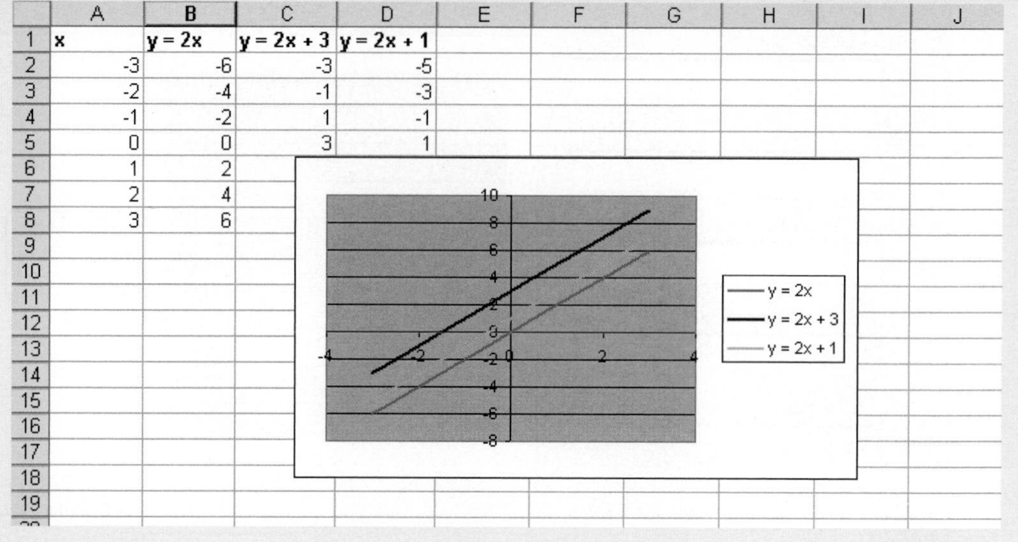

	A	B	C	D
1	x	y = 2x	y = 2x + 3	y = 2x + 1
2	-3	-6	-3	-5
3	-2	-4	-1	-3
4	-1	-2	1	-1
5	0	0	3	1
6	1	2		
7	2	4		
8	3	6		

(b) Following the same procedure for the three lines

$y = 2x$

$y = 2x - 3$

$y = 2x + 1$

produces a graph as shown in Figure 1.10.

(c) Notice that in part (a), all of the graphs cut the y axis at the point (0, 2). In part (b), the graphs are parallel, which means that they have the same gradient.

This illustrates the fact that every straight line has an equation of the form

$y = mx + c$

where m is the gradient, and c is the intercept on the y axis.

In (a), the value of c in each equation is equal to 2, so all of the lines cut the y axis at this point.

In (b), the value of m in each equation is equal to 2, so all of the lines have the same gradient and are parallel.

It is very useful to be able to recognize these properties from the equations, as it means we have a fairly good idea of what our graph should look like even before we draw it.

Advice

This completes your first piece of mathematics. I hope that you have not found it quite as bad as you first thought. There now follow a few extra problems to give you more practice. If you have successfully tackled all of the problems given in the text so far, you might prefer to move straight on to the next section. Problems 9–16 can be attempted later as part of your revision. However, if you failed to obtain the correct answers to Problems 1–8 then you should try the Practice Problems now. Not only will they help to strengthen your mathematical skills, but also they should improve your overall confidence.

Key Terms

Coefficient A numerical multiplier of the variables in an algebraic term, such as the numbers 4 and 7 in the expression, $4x + 7yz^2$.

Coordinates A set of numbers that determine the position of a point relative to a set of axes.

Intercept The point(s) where a graph crosses one of the coordinate axes.

Linear equation An equation of the form $y = ax + b$.

Origin The point where the coordinate axes intersect.

Simultaneous linear equations A set of linear equations in which there are (usually) the same number of equations and unknowns. The solution consists of values of the unknowns which satisfy all of the equations at the same time.

Slope of a line Also known as the gradient, it is the change in the value of y when x increases by 1 unit.

x *axis* The horizontal coordinate axis pointing from left to right.

y *axis* The vertical coordinate axis pointing upwards.

Practice Problems

9 Plot the following points on graph paper:

P (4, 0), Q (−2, 9), R (5, 8), S (−1, −2)

Hence find the coordinates of the point of intersection of the line passing through P and Q, and the line passing through R and S.

10 Without using a calculator evaluate

(a) $10 \times (-2)$ **(b)** $(-1) \times (-3)$ **(c)** $(-8) \div 2$

(d) $(-5) \div (-5)$ **(e)** $5 - 6$ **(f)** $-1 - 2$

(g) $7 - (-4)$ **(h)** $-9 - (-9)$ **(i)** $\frac{(-3) \times (-6) \times (-1)}{2 - 3}$

11 If $x = 2$ and $y = -3$ evaluate

(a) $2x + y$ **(b)** $x - y$ **(c)** $3x + 4y$

(d) xy **(e)** $5xy$ **(f)** $4x - 6xy$

12 Solve the following equations:

(a) $2x = 1$ **(b)** $2x + 5 = 13$ **(c)** $3 - x = 7$

(d) $3 - 4x = 5$ **(e)** $7x = 0$ **(f)** $\frac{3}{x} + 7 = 10$

13 If $4x + 3y = 24$, complete the following table and hence sketch this line

x	y
0	
	0
3	

→

14 Solve the following pairs of simultaneous linear equations graphically:

(a) $-2x + y = 2$ $\quad 2x + y = -6$

(b) $3x + 4y = 12$ $\quad x + 4y = 8$

(c) $2x + y = 4$ $\quad 4x - 3y = 3$

(d) $x + y = 1$ $\quad 6x + 5y = 15$

15 Use the slope–intercept approach to sketch the lines

(a) $y = -x$ **(b)** $x - 2y = 6$

16 **(a)** Without using a calculator, work out the value of $(-4)^2$.

(b) Press the following key sequence on your calculator:

 4 

Explain carefully why this does not give the same result as part (a) and give an alternative key sequence that *does* give the correct answer.

section 1.2

Algebraic solution of simultaneous linear equations

Objectives

At the end of this section you should be able to:

- Solve a system of two simultaneous linear equations in two unknowns using elimination.
- Detect when a system of equations does not have a solution.
- Detect when a system of equations has infinitely many solutions.
- Solve a system of three simultaneous linear equations in three unknowns using elimination.

In Section 1.1 a graphical method was described for the solution of simultaneous linear equations. Both lines are sketched on the same piece of graph paper and the coordinates of the point of intersection are then simply read off from the diagram. Unfortunately this approach has several drawbacks. It is not always easy to decide on a suitable scale for the axes. Even if the scale allows all four points (two from each line) to fit on the diagram, there is no guarantee that the point of intersection itself also lies on it. You may have encountered this difficulty when solving Practice Problem 14(d) in the previous section. When this happens you have no alternative but to throw away your graph paper and to start again, choosing a smaller scale in the hope that the solution will now fit. The second drawback concerns the accuracy of the graphical solution. All of the problems in Section 1.1 were deliberately chosen so that the answers had nice numbers in them; whole numbers such as −1, 2 and 5 or at worst simple fractions such as $\frac{1}{2}$, $2\frac{1}{2}$ and $-\frac{1}{4}$. In practice, the coefficients of the equations may well involve decimals and we might expect a decimal solution. Indeed, even if the coefficients are whole numbers the solution itself could involve nasty fractions such as 7/8 or perhaps something like 231/571. A moment's thought should convince you that in these circumstances it is virtually impossible to obtain the solution graphically, even if we use a really large scale and our sharpest HB pencil in the process. The final drawback concerns the nature of the problem itself. Quite frequently in economics we need to solve three equations in three unknowns or maybe four equations in four unknowns. Unfortunately, the graphical method of solution does not extend to these cases.

In this section an alternative method of solution is described which relies on algebra. It is called the ***elimination method***, since each stage of the process eliminates one (or more) of the unknowns. This method always produces the exact solution and can be applied to systems of equations larger than just two equations in two unknowns. In order to illustrate the method, we return to the simple example considered in the previous section:

$$4x + 3y = 11 \tag{1}$$

$$2x + y = 5 \tag{2}$$

The coefficient of x in equation (1) is 4 and the coefficient of x in equation (2) is 2. If these numbers had turned out to be exactly the same then we could have eliminated the variable x by subtracting one equation from the other. However, we can arrange for this to be the case by multiplying the left-hand side of the second equation by 2. Of course, we must also remember to multiply the right-hand side of the second equation by 2 in order for this operation to be valid. The second equation then becomes

$$4x + 2y = 10 \tag{3}$$

We may now subtract equation (3) from (1) to get

$$y = 1$$

You may like to think of this in terms of the usual layout for the subtraction of two ordinary numbers: that is,

$$\begin{array}{r} 4x + 3y = 11 \\ 4x + 2y = 10\ - \\ \hline y = \ \ 1 \end{array}$$

the x's cancel
when you subtract

This number can now be substituted into one of the original equations to deduce x. From equation (1)

$$\begin{aligned} 4x + 3(1) &= 11 && \text{(substitute } y = 1) \\ 4x + 3 &= 11 \\ 4x &= 8 && \text{(subtract 3 from both sides)} \\ x &= 2 && \text{(divide both sides by 4)} \end{aligned}$$

Hence the solution is $x = 2$, $y = 1$. As a check, substitution of these values into the other original equation (2) gives

$$2(2) + 1 = 5 \quad \checkmark$$

The method of elimination can be summarized as follows.

Step 1

Add/subtract a multiple of one equation to/from a multiple of the other to eliminate x.

Step 2

Solve the resulting equation for y.

Step 3

Substitute the value of y into one of the original equations to deduce x.

Step 4

Check that no mistakes have been made by substituting both x and y into the other original equation.

Advice

The following example involves fractions. If you have forgotten how to perform calculations with fractions, you might like to read through Section 1.4.3 (pp. 76–81) now, since this section contains a review of their basic properties.

Example

Solve the system of equations

$$3x + 2y = 1 \tag{1}$$

$$-2x + y = 2 \tag{2}$$

Solution

Step 1

The coefficients of x in equations (1) and (2) are 3 and -2 respectively. We can arrange for these to be the same size (but of opposite sign) by multiplying equation (1) by 2 and multiplying (2) by 3. The new equations will then have x coefficients of 6 and -6, so we can eliminate x this time by adding the equations together. The details are as follows.

Doubling the first equation produces

$$6x + 4y = 2 \tag{3}$$

Tripling the second equation produces

$$-6x + 3y = 6 \tag{4}$$

If equation (4) is added to equation (3) then

$$\begin{array}{r} 6x + 4y = 2 \\ \underline{-6x + 3y = 6} + \\ 7y = 8 \end{array} \tag{5}$$

the x's cancel when you add

Step 2

Equation (5) can be solved by dividing both sides by 7 to get

$$y = 8/7$$

Step 3

If 8/7 is substituted for y in equation (1) then

$$3x + 2\left(\frac{8}{7}\right) = 1$$

$$3x + \frac{16}{7} = 1$$

$$3x = 1 - \frac{16}{7} \quad \text{(subtract 16/7 from both sides)}$$

$$3x = \frac{7-16}{7} \quad \text{(put over a common denominator)}$$

$$3x = -\frac{9}{7}$$

$$x = \frac{1}{3} \times \left(-\frac{9}{7}\right) \quad \text{(divide both sides by 3)}$$

$$x = -\frac{3}{7}$$

The solution is therefore $x = -3/7$, $y = 8/7$.

Step 4

As a check, equation (2) gives

$$-2\left(-\frac{3}{7}\right) + \frac{8}{7} = \frac{6}{7} + \frac{8}{7} = \frac{6+8}{7} = \frac{14}{7} = 2 \quad \checkmark$$

Advice

In the general description of the method, we suggested that the variable x is eliminated in step 1. There is nothing special about x. We could equally well eliminate y at this stage and then solve the resulting equation in step 2 for x.

You might like to solve the above example using this alternative strategy. You need to double equation (2) and then subtract from (1).

Practice Problem

1 (a) Solve the equations

$$3x - 2y = 4$$
$$x - 2y = 2$$

by eliminating one of the variables.

(b) Solve the equations

$$3x + 5y = 19$$
$$-5x + 2y = -11$$

by eliminating one of the variables.

The following examples provide further practice in using the method and illustrate some special cases which may occur.

Example

Solve the system of equations

$$x - 2y = 1$$
$$2x - 4y = -3$$

Solution

Step 1

The variable x can be eliminated by doubling the first equation and subtracting the second:

both the x's and the y's cancel!

$$\begin{aligned} 2x - 4y &= 2 \\ 2x - 4y &= -3 \; - \\ \hline 0 &= 5 \end{aligned}$$

The statement '0 = 5' is clearly nonsense and something has gone seriously wrong. To understand what is going on here, let us try and solve this problem graphically.

The line $x - 2y = 1$ passes through the points (0, −1/2) and (1, 0) (check this). The line $2x - 4y = -3$ passes through the points (0, 3/4) and (−3/2, 0) (check this). Figure 1.11 shows that these lines are parallel and so they do not intersect. It is therefore not surprising that we were unable to find a solution using algebra, because this system of equations does not have one. We could have deduced this before when subtracting the equations. The equation that only involves y in step 2 can be written as

$$0y = 5$$

and the problem is to find a value of y for which this equation is true. No such value exists, since

$$\boxed{\text{zero}} \times \boxed{\text{any number}} = \boxed{\text{zero}}$$

and so the original system of equations does not have a solution.

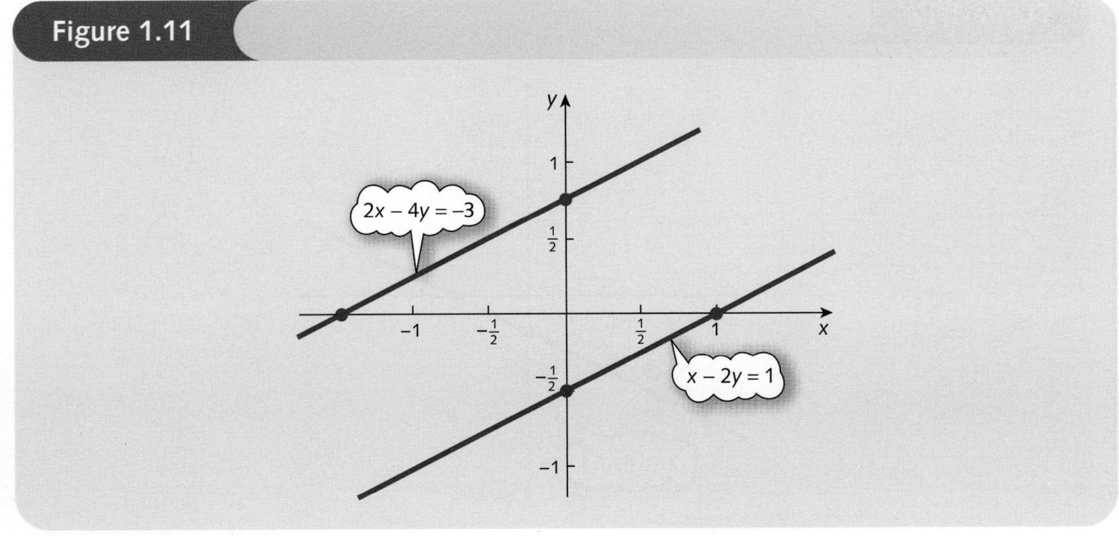

Example

Solve the equations

$$2x - 4y = 1$$
$$5x - 10y = 5/2$$

Solution

Step 1

The variable x can be eliminated by multiplying the first equation by 5, multiplying the second equation by 2 and subtracting

$$\begin{array}{r} 10x - 20y = 5 \\ 10x - 20y = 5 \; - \\ \hline 0 = 0 \end{array}$$

everything cancels including the right-hand side!

Again, it is easy to explain this using graphs. The line $2x - 4y = 1$ passes through $(0, -1/4)$ and $(1/2, 0)$. The line $5x - 10y = 5/2$ passes through $(0, -1/4)$ and $(1/2, 0)$. Consequently, both equations represent the same line. From Figure 1.12 the lines intersect along the whole of their length and any point on this line is a solution. This particular system of equations has infinitely many solutions. This can also be deduced algebraically. The equation involving y in step 2 is

$$0y = 0$$

which is true for any value of y.

Figure 1.12

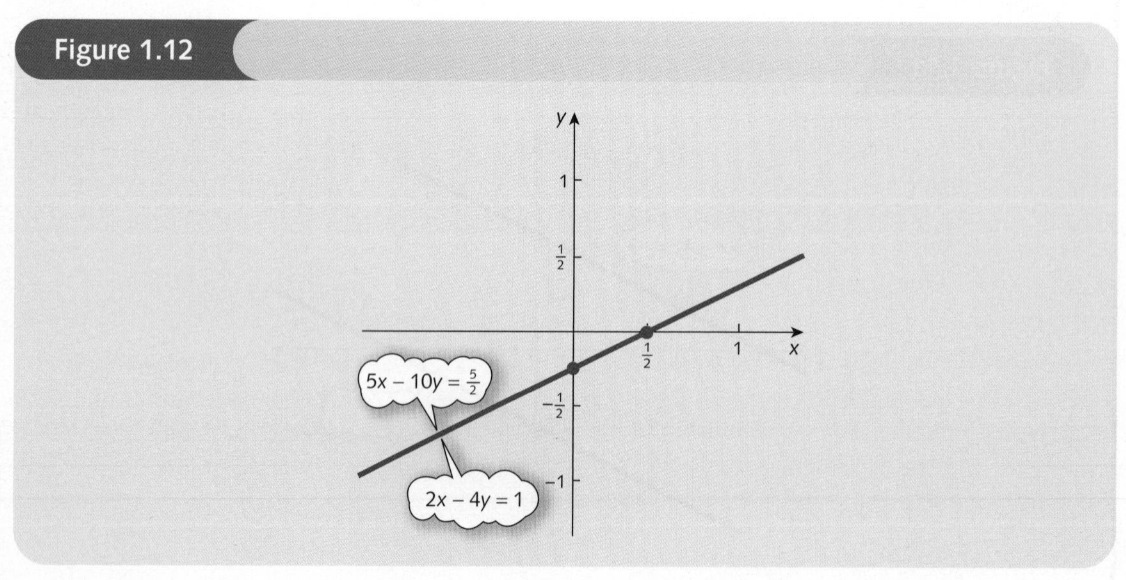

These examples show that a system of equations can possess a unique solution, no solution or infinitely many solutions. Algebraically, this can be detected in step 2. If the equation resulting from the elimination of x looks like

$$\boxed{\text{any non-zero number}} \times \boxed{y} = \boxed{\text{any number}}$$

then the equations have a unique solution, or if it looks like

$$\boxed{\text{zero}} \times \boxed{y} = \boxed{\text{any non-zero number}}$$

then the equations have no solution, or if it looks like

$$\boxed{\text{zero}} \times \boxed{y} = \boxed{\text{zero}}$$

then the equations have infinitely many solutions.

It is interesting to notice how the graphical approach 'saved the day' in the previous two examples. They show how useful pictures are as an aid to understanding in mathematics.

Practice Problem

2 Attempt to solve the following systems of equations

(a) $3x - 6y = -2$
$-4x + 8y = -1$

(b) $-5x + y = 4$
$10x - 2y = -8$

Comment on the nature of the solution in each case.

We now show how the algebraic method can be used to solve three equations in three unknowns. As you might expect, the details are more complicated than for just two equations, but the principle is the same. We begin with a simple example to illustrate the general method. Consider the system

$$x + 3y - z = 4 \quad (1)$$

$$2x + y + 2z = 10 \quad (2)$$

$$3x - y + z = 4 \quad (3)$$

The objective is to find three numbers x, y and z which satisfy these equations simultaneously. Our previous work suggests that we should begin by eliminating x from all but one of the equations.

The variable x can be eliminated from the second equation by multiplying equation (1) by 2 and subtracting equation (2):

$$\begin{array}{r} 2x + 6y - 2z = 8 \\ \underline{2x + y + 2z = 10} \; - \\ 5y - 4z = -2 \end{array} \quad (4)$$

Similarly, we can eliminate x from the third equation by multiplying equation (1) by 3 and subtracting equation (3):

$$\begin{array}{r} 3x + 9y - 3z = 12 \\ \underline{3x - y + z = 4} \; - \\ 10y - 4z = 8 \end{array} \quad (5)$$

At this stage the first equation is unaltered but the second and third equations of the system have changed to equations (4) and (5) respectively, so the current equations are

$$x + 3y - z = 4 \quad (1)$$

$$5y - 4z = -2 \quad (4)$$

$$10y - 4z = 8 \quad (5)$$

Notice that the last two equations constitute a system of just two equations in two unknowns, y and z. This, of course, is precisely the type of problem that we already know how to solve. Once y and z have been calculated, the values can be substituted into equation (1) to deduce x.

We can eliminate y in the last equation by multiplying equation (4) by 2 and subtracting equation (5):

$$\begin{array}{r} 10y - 8z = -4 \\ 10y - 4z = 8 - \\ \hline -4z = -12 \end{array} \quad (6)$$

Collecting together the current equations gives

$$x + 3y - z = 4 \quad (1)$$

$$5y - 4z = -2 \quad (4)$$

$$-4z = -12 \quad (6)$$

From the last equation,

$$z = \frac{-12}{-4} = 3 \qquad \text{(divide both sides by } -4\text{)}$$

If this is substituted into equation (4) then

$$5y - 4(3) = -2$$

$$5y - 12 = -2$$

$$5y = 10 \qquad \text{(add 12 to both sides)}$$

$$y = 2 \qquad \text{(divide both sides by 5)}$$

Finally, substituting $y = 2$ and $z = 3$ into equation (1) produces

$$x + 3(2) - 3 = 4$$

$$x + 3 = 4$$

$$x = 1 \qquad \text{(subtract 3 from both sides)}$$

Hence the solution is $x = 1$, $y = 2$, $z = 3$.

As usual, it is possible to check the answer by putting these numbers back into the original equations (1), (2) and (3)

$$1 + 3(2) - 3 = 4 \quad ✓$$

$$2(1) + 2 + 2(3) = 10 \quad ✓$$

$$3(1) - 2 + 3 = 4 \quad ✓$$

The general strategy may be summarized as follows. Consider the system

$$?x + ?y + ?z = ?$$

$$?x + ?y + ?z = ?$$

$$?x + ?y + ?z = ?$$

where ? denotes some numerical coefficient.

Step 1

Add/subtract multiples of the first equation to/from multiples of the second and third equations to eliminate x. This produces a new system of the form

$$\begin{aligned} ?x + ?y + ?z &= ? \\ ?y + ?z &= ? \\ ?y + ?z &= ? \end{aligned}$$

Step 2

Add/subtract a multiple of the second equation to/from a multiple of the third to eliminate y. This produces a new system of the form

$$\begin{aligned} ?x + ?y + ?z &= ? \\ ?y + ?z &= ? \\ ?z &= ? \end{aligned}$$

Step 3

Solve the last equation for z. Substitute the value of z into the second equation to deduce y. Finally, substitute the values of both y and z into the first equation to deduce x.

Step 4

Check that no mistakes have been made by substituting the values of x, y and z into the original equations.

It is possible to adopt different strategies from that suggested above. For example, it may be more convenient to eliminate z from the last equation in step 2 rather than y. However, it is important to notice that we use the second equation to do this, not the first. Any attempt to use the first equation in step 2 would reintroduce the variable x into the equations, which is the last thing we want to do at this stage.

Example

Solve the equations

$$4x + y + 3z = 8 \quad (1)$$

$$-2x + 5y + z = 4 \quad (2)$$

$$3x + 2y + 4z = 9 \quad (3)$$

Solution

Step 1

To eliminate x from the second equation we multiply it by 2 and add to equation (1):

$$\begin{aligned} 4x + y + 3z &= 8 \\ -4x + 10y + 2z &= 8 \; + \\ \hline 11y + 5z &= 16 \end{aligned} \quad (4)$$

To eliminate x from the third equation we multiply equation (1) by 3, multiply equation (3) by 4 and subtract:

$$\begin{array}{r} 12x + 3y + 9z = 24 \\ 12x + 8y + 16z = 36 \; - \\ \hline -5y - 7z = -12 \end{array} \qquad (5)$$

This produces a new system:

$$4x + y + 3z = 8 \qquad (1)$$

$$11y + 5z = 16 \qquad (4)$$

$$-5y - 7z = -12 \qquad (5)$$

Step 2

To eliminate y from the new third equation (that is, equation (5)) we multiply equation (4) by 5, multiply equation (5) by 11 and add:

$$\begin{array}{r} 55y + 25z = 80 \\ -55y - 77z = -132 \; + \\ \hline -52z = -52 \end{array} \qquad (6)$$

This produces a new system

$$4x + y + 3z = 8 \qquad (1)$$

$$11y + 5z = 16 \qquad (4)$$

$$-52z = -52 \qquad (6)$$

Step 3

The last equation gives

$$z = \frac{-52}{-52} = 1 \quad \text{(divide both sides by } -52\text{)}$$

If this is substituted into equation (4) then

$$11y + 5(1) = 16$$

$$11y + 5 = 16$$

$$11y = 11 \quad \text{(subtract 5 from both sides)}$$

$$y = 1 \quad \text{(divide both sides by 11)}$$

Finally, substituting $y = 1$ and $z = 1$ into equation (1) produces

$$4x + 1 + 3(1) = 8$$

$$4x + 4 = 8$$

$$4x = 4 \quad \text{(subtract 4 from both sides)}$$

$$x = 1 \quad \text{(divide both sides by 4)}$$

Hence the solution is $x = 1$, $y = 1$, $z = 1$.

Step 4

As a check the original equations (1), (2) and (3) give

$$4(1) + 1 + 3(1) = 8 \quad ✓$$

$$-2(1) + 5(1) + 1 = 4 \quad ✓$$

$$3(1) + 2(1) + 4(1) = 9 \quad ✓$$

respectively.

Practice Problem

3 Solve the following system of equations:

$$2x + 2y - 5z = -5 \qquad (1)$$

$$x - y + z = 3 \qquad (2)$$

$$-3x + y + 2z = -2 \qquad (3)$$

As you might expect, it is possible for three simultaneous linear equations to have either no solution or infinitely many solutions. An illustration of this is given in Practice Problem 8. The method described in this section has an obvious extension to larger systems of equations. However, the calculations are extremely tedious to perform by hand. Fortunately there are many computer packages available which are capable of solving large systems accurately and efficiently (a matter of a few seconds to solve 10 000 equations in 10 000 unknowns).

Advice

We shall return to the solution of simultaneous linear equations in Chapter 7 when we describe how matrix theory can be used to solve them. This does not depend on any subsequent chapters in this book, so you might like to read through this material now. Two techniques are suggested. A method based on inverse matrices is covered in Section 7.2 and an alternative using Cramer's rule can be found in Section 7.3.

Key Terms

Elimination method The method in which variables are removed from a system of simultaneous equations by adding (or subtracting) a multiple of one equation to (or from) a multiple of another.

Practice Problems

4 Use the method of elimination to solve the systems of equations given in Section 1.1, Problem 14.

5 Sketch the following lines on the same diagram:

$$2x - 3y = 6, \quad 4x - 6y = 18, \quad x - \frac{3}{2}y = 3$$

Hence comment on the nature of the solutions of the following systems of equations:

(a) $2x - 3y = 6$

$x - \frac{3}{2}y = 3$

(b) $4x - 6y = 18$

$x - \frac{3}{2}y = 3$

6 Use the elimination method to attempt to solve the following systems of equations. Comment on the nature of the solution in each case.

(a) $-3x + 5y = 4$

$9x - 15y = -12$

(b) $6x - 2y = 3$

$15x - 5y = 4$

7 Solve the following systems of equations:

(a)
$$x - 3y + 4z = 5 \quad (1)$$
$$2x + y + z = 3 \quad (2)$$
$$4x + 3y + 5z = 1 \quad (3)$$

(b)
$$3x + 2y - 2z = -5 \quad (1)$$
$$4x + 3y + 3z = 17 \quad (2)$$
$$2x - y + z = -1 \quad (3)$$

8 Attempt to solve the following systems of equations. Comment on the nature of the solution in each case.

(a)
$$x - 2y + z = -2 \quad (1)$$
$$x + y - 2z = 4 \quad (2)$$
$$-2x + y + z = 12 \quad (3)$$

(b)
$$2x + 3y - z = 13 \quad (1)$$
$$x - 2y + 2z = -3 \quad (2)$$
$$3x + y + z = 10 \quad (3)$$

section 1.3

Supply and demand analysis

Objectives

At the end of this section you should be able to:

- Use the function notation, $y = f(x)$.
- Identify the endogenous and exogenous variables in an economic model.
- Identify and sketch a linear demand function.
- Identify and sketch a linear supply function.
- Determine the equilibrium price and quantity for a single-commodity market both graphically and algebraically.
- Determine the equilibrium price and quantity for a multicommodity market by solving simultaneous linear equations.

Microeconomics is concerned with the analysis of the economic theory and policy of individual firms and markets. In this section we focus on one particular aspect known as market equilibrium, in which the supply and demand balance. We describe how the mathematics introduced in the previous two sections can be used to calculate the equilibrium price and quantity. However, before we do this it is useful to explain the concept of a function. This idea is central to nearly all applications of mathematics in economics.

A *function*, f, is a rule which assigns to each incoming number, x, a uniquely defined outgoing number, y. A function may be thought of as a 'black box' that performs a dedicated arithmetic calculation. As an example, consider the rule 'double and add 3'. The effect of this rule on two specific incoming numbers, 5 and -17, is illustrated in Figure 1.13 (overleaf).

Unfortunately, such a representation is rather cumbersome. There are, however, two alternative ways of expressing this rule which are more concise. We can write either

$$y = 2x + 3 \quad \text{or} \quad f(x) = 2x + 3$$

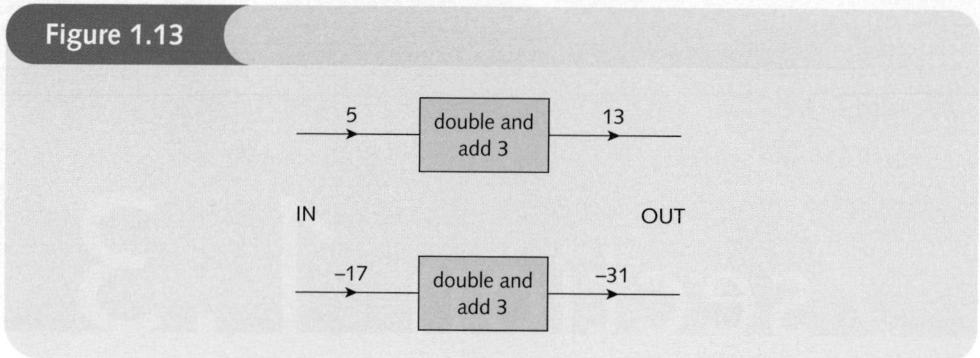

Figure 1.13

The first of these is familiar to you from our previous work; corresponding to any incoming number, x, the right-hand side tells you what to do with x to generate the outgoing number, y. The second notation is also useful. It has the advantage that it involves the label f, which is used to name the rule. If, in a piece of economic theory, there are two or more functions, we can use different labels to refer to each one. For example, a second function might be

$$g(x) = -3x + 10$$

and we subsequently identify the respective functions simply by referring to them by name: that is, as either f or g.

The new notation also enables the information conveyed in Figure 1.13 to be written

$f(5) = 13$ (read 'f of 5 equals 13')

$f(-17) = -31$ (read 'f of −17 equals −31')

The number inside the brackets is the incoming value, x, and the right-hand side is the corresponding outgoing value, y.

Example

(a) If $f(x) = 2x^2 - 3x$ find the value of $f(5)$.

(b) If $g(Q) = \dfrac{3}{5 + 2Q}$ find the value of $g(2)$.

Solution

(a) Substituting $x = 5$ into $2x^2 - 3x$ gives

$$\begin{aligned} f(5) &= 2 \times 5^2 - 3 \times 5 \\ &= 2 \times 25 - 3 \times 5 \quad \text{(BIDMAS)} \\ &= 50 - 15 = 35 \end{aligned}$$

(b) Although the letter Q is used instead of x, the procedure is the same.

$$g(2) = \frac{3}{5 + 2 \times 2} = \frac{3}{9} = \frac{1}{3}$$

Practice Problem

1 Evaluate

(a) $f(25)$ **(b)** $f(1)$ **(c)** $f(17)$ **(d)** $g(0)$ **(e)** $g(48)$ **(f)** $g(16)$

for the two functions

$$f(x) = -2x + 50$$

$$g(x) = -\tfrac{1}{2}x + 25$$

Do you notice any connection between f and g?

The incoming and outgoing variables are referred to as the ***independent*** and ***dependent*** variables respectively. The value of y clearly 'depends' on the actual value of x that is fed into the function. For example, in microeconomics the quantity demanded, Q, of a good depends on the market price, P. We might express this as

$$Q = f(P)$$

Such a function is called a ***demand*** function. Given any particular formula for $f(P)$ it is then a simple matter to produce a picture of the corresponding demand curve on graph paper. There is, however, a difference of opinion between mathematicians and economists on how this should be done. If your quantitative methods lecturer is a mathematician then he or she is likely to plot Q on the vertical axis and P on the horizontal axis. Economists, on the other hand, normally plot them the other way round with Q on the horizontal axis. In doing so, we are merely noting that since Q is related to P then, conversely, P must be related to Q, and so there is a function of the form

$$P = g(Q)$$

The two functions, f and g, are said to be ***inverse*** functions: that is, f is the inverse of g and, equivalently, g is the inverse of f. We adopt the economists' approach in this book. In subsequent chapters we shall investigate other microeconomic functions such as total revenue, average cost, and profit. It is conventional to plot each of these against Q (that is, with Q on the horizontal axis), so it makes sense to be consistent and to do the same here.

Written in the form $P = g(Q)$, the demand function tells us that P is a function of Q but it gives us no information about the precise relationship between these two variables. To find this we need to know the form of the function which can be obtained either from economic theory or from empirical evidence. For the moment we hypothesize that the function is linear so that

$$P = aQ + b$$

for some appropriate constants (called ***parameters***), a and b. Of course, in reality, the relationship between price and quantity is likely to be much more complicated than this. However, the use of linear functions makes the mathematics nice and easy, and the result of any analysis at least provides a first approximation to the truth. The process of identifying the key features of the real world and making appropriate simplifications and assumptions is known as ***modelling***. Models are based on economic laws and help to explain and predict the behaviour of real-world situations. Inevitably there is a conflict between mathematical ease and the model's accuracy. The closer the model comes to reality, the more complicated the mathematics is likely to be.

A graph of a typical linear demand function is shown in Figure 1.14 (overleaf). Elementary theory shows that demand usually falls as the price of a good rises and so the slope of the line is negative. Mathematically, P is then said to be a ***decreasing*** function of Q.

Figure 1.14

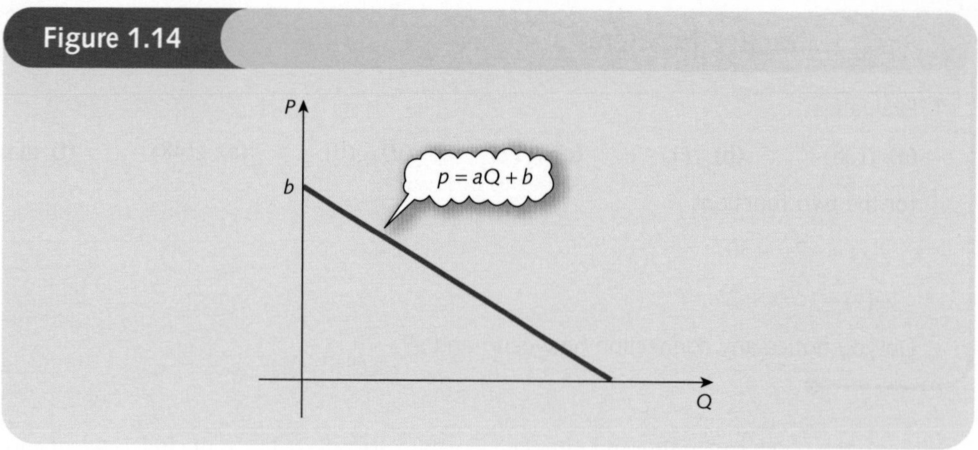

In symbols we write

$a < 0$ read '*a* is less than zero'

It is also apparent from the graph that the intercept, *b*, is positive: that is,

$b > 0$ read '*b* is greater than zero'

In fact, it is possible in theory for the demand curve to be horizontal with $a = 0$. This corresponds to perfect competition and we shall return to this special case in Chapter 4.

Example

Sketch a graph of the demand function

$$P = -2Q + 50$$

Hence, or otherwise, determine the value of

(a) P when $Q = 9$

(b) Q when $P = 10$

Solution

For the demand function

$$P = -2Q + 50$$

$a = -2$, $b = 50$, so the line has a slope of -2 and an intercept of 50. For every 1 unit along, the line goes down by 2 units, so it must cross the horizontal axis when $Q = 25$. (Alternatively, note that when $P = 0$ the equation reads $0 = -2Q + 50$, with solution $Q = 25$.) The graph is sketched in Figure 1.15.

(a) Given any quantity, Q, it is straightforward to use the graph to find the corresponding price, P. A line is drawn vertically upwards until it intersects the demand curve and the value of P is read off from the vertical axis. From Figure 1.15, when $Q = 9$ we see that $P = 32$. This can also be found by substituting $Q = 9$ directly into the demand function to get

$$P = -2(9) + 50 = 32$$

(b) Reversing this process enables us to calculate Q from a given value of P. A line is drawn horizontally until it intersects the demand curve and the value of Q is read off from the horizontal axis. Figure 1.15 indicates that $Q = 20$ when $P = 10$. Again this can be found by calculation. If $P = 10$ then the equation reads

$$10 = -2Q + 50$$

$$-40 = -2Q \qquad \text{(subtract 50 from both sides)}$$

$$20 = Q \qquad \text{(divide both sides by } -2\text{)}$$

Figure 1.15

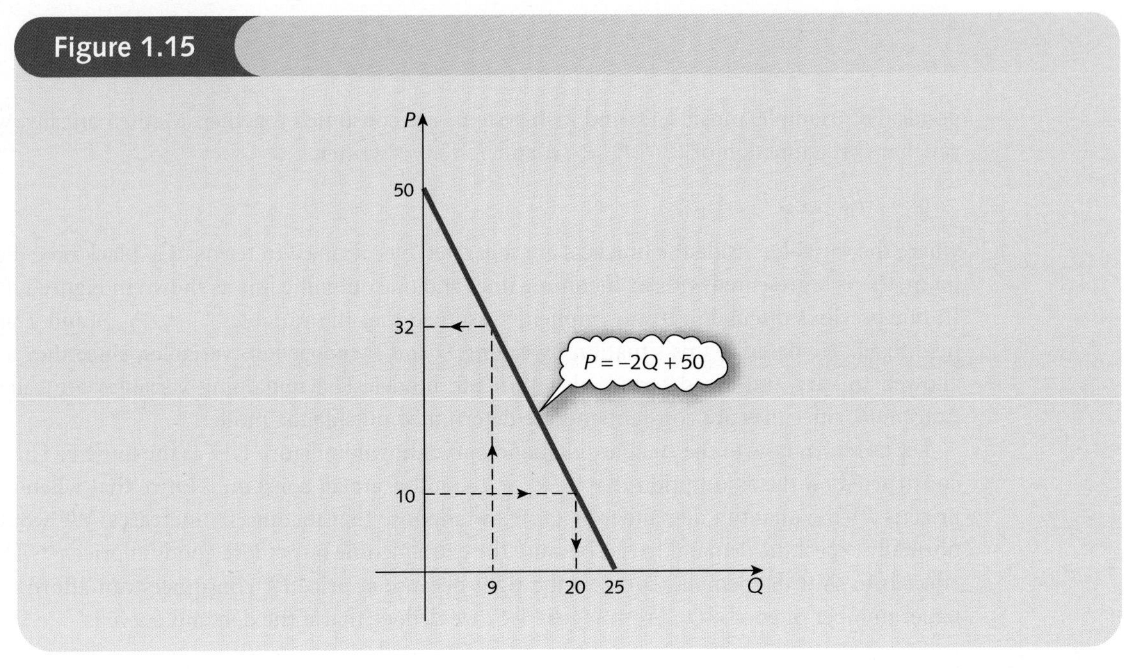

Practice Problem

2 Sketch a graph of the demand function

$$P = -3Q + 75$$

Hence, or otherwise, determine the value of

(a) P when $Q = 23$

(b) Q when $P = 18$

The model of consumer demand given so far is fairly crude in that it assumes that quantity depends solely on the price, P, of the good being considered. In practice, Q depends on other factors as well. These include the incomes of consumers, Y, the price of substitutable goods, P_S, the price of complementary goods, P_C, advertising expenditure, A, and consumers' tastes, T. A ***substitutable*** good is one that could be consumed instead of the good under consideration. For example, in the transport industry, buses and taxis could obviously be substituted for each other in urban areas. A ***complementary*** good is one that is used in conjunction with other

Figure 1.16

goods. For example, music CDs and hi-fi systems are consumed together. Mathematically, we say that Q is a function of P, Y, P_S, P_C, A and T. This is written

$$Q = f(P, Y, P_S, P_C, A, T)$$

where the variables inside the brackets are separated by commas. In terms of a 'black box' diagram, this is represented with six incoming lines and one outgoing line as shown in Figure 1.16. In our previous discussion it was implicitly assumed that the variables Y, P_S, P_C, A and T are held fixed. We describe this situation by calling Q and P ***endogenous*** variables, since they are allowed to vary and are determined within the model. The remaining variables are called ***exogenous***, since they are constant and are determined outside the model.

Let us return now to the standard demand curve shown in Figure 1.17 as the line EF. This is constructed on the assumption that Y, P_S, P_C, A and T are all constant. Notice that when the price is P^* the quantity demanded is Q_1. Now suppose that income, Y, increases. We would normally expect the demand to rise because the extra income buys more goods at price P^*. The effect is to shift the demand curve to the right because at price P^* consumers can afford the larger number of goods, Q_2. From Figure 1.17 we deduce that if the demand curve is

Figure 1.17

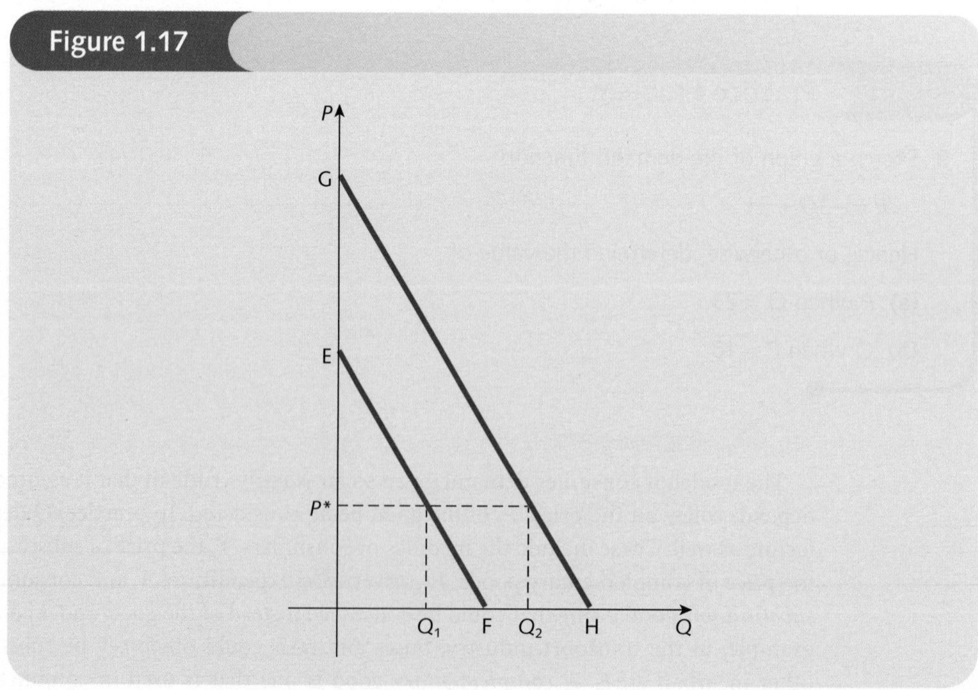

$$P = aQ + b$$

then a rise in income causes the intercept, b, to increase.

We conclude that if one of the exogenous variables changes then the whole demand curve moves, whereas if one of the endogenous variables changes, we simply move along the fixed curve.

Incidentally, it is possible that, for some goods, an increase in income actually causes the demand curve to shift to the left. In the 1960s and 1970s, most western economies saw a decline in the domestic consumption of coal as a result of an increase in income. In this case, higher wealth meant that more people were able to install central heating systems which use alternative forms of energy. Under these circumstances the good is referred to as an ***inferior*** good. On the other hand, a ***superior*** good is one whose demand rises as income rises. Cars and electrical goods are obvious examples of superior goods. Currently, concern about global warming is also reducing demand for coal. This factor can be incorporated as part of taste, although it is difficult to handle mathematically since it is virtually impossible to quantify taste and so to define T numerically.

The ***supply*** function is the relation between the quantity, Q, of a good that producers plan to bring to the market and the price, P, of the good. A typical linear supply curve is indicated in Figure 1.18. Economic theory indicates that, as the price rises, so does the supply. Mathematically, P is then said to be an ***increasing*** function of Q. A price increase encourages existing producers to raise output and entices new firms to enter the market. The line shown in Figure 1.18 has equation

$$P = aQ + b$$

with slope $a > 0$ and intercept $b > 0$. Note that when the market price is equal to b the supply is zero. It is only when the price exceeds this threshold level that producers decide that it is worth supplying any good whatsoever.

Again this is a simplification of what happens in the real world. The supply function does not have to be linear and the quantity supplied, Q, is influenced by things other than price. These exogenous variables include the prices of factors of production (that is, land, capital, labour and enterprise), the profits obtainable on alternative goods, and technology.

Figure 1.18

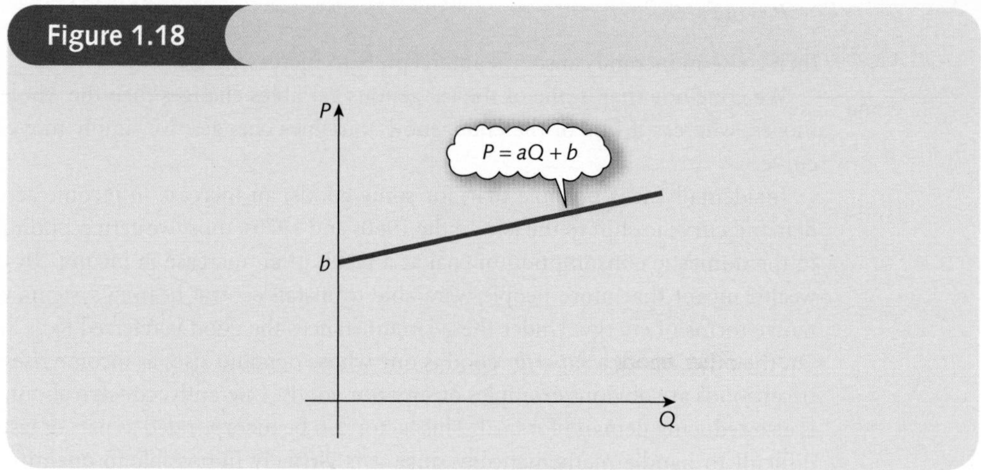

In microeconomics we are concerned with the interaction of supply and demand. Figure 1.19 shows typical supply and demand curves sketched on the same diagram. Of particular significance is the point of intersection. At this point the market is in ***equilibrium*** because the quantity supplied exactly matches the quantity demanded. The corresponding price, P_0, and quantity, Q_0, are called the equilibrium price and quantity.

In practice, it is often the deviation of the market price away from the equilibrium price that is of most interest. Suppose that the market price, P^*, exceeds the equilibrium price, P_0. From Figure 1.19 the quantity supplied, Q_S, is greater than the quantity demanded, Q_D, so there is excess supply. There are stocks of unsold goods, which tend to depress prices and cause firms to cut back production. The effect is for 'market forces' to shift the market back down towards equilibrium. Likewise, if the market price falls below equilibrium price then demand exceeds supply. This shortage pushes prices up and encourages firms to produce more goods, so the market drifts back up towards equilibrium.

Figure 1.19

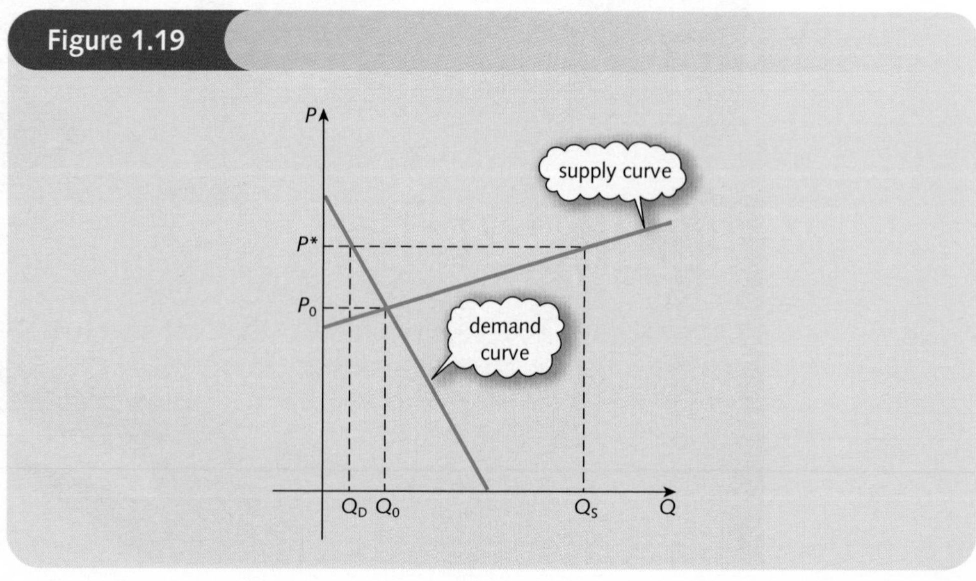

Example

The demand and supply functions of a good are given by

$$P = -2Q_D + 50$$

$$P = \tfrac{1}{2}Q_S + 25$$

where P, Q_D and Q_S denote the price, quantity demanded and quantity supplied respectively.

(a) Determine the equilibrium price and quantity.

(b) Determine the effect on the market equilibrium if the government decides to impose a fixed tax of \$5 on each good.

Solution

(a) The demand curve has already been sketched in Figure 1.15. For the supply function

$$P = \tfrac{1}{2}Q_S + 25$$

we have $a = \tfrac{1}{2}$, $b = 25$, so the line has a slope of $\tfrac{1}{2}$ and an intercept of 25. It therefore passes through (0, 25). For a second point, let us choose $Q_S = 20$, say. The corresponding value of P is

$$P = \tfrac{1}{2}(20) + 25 = 35$$

so the line also passes through (20, 35). The points (0, 25) and (20, 35) can now be plotted and the supply curve sketched. Figure 1.20 shows both the demand and supply curves sketched on the same diagram. The point of intersection has coordinates (10, 30), so the equilibrium quantity is 10 and the equilibrium price is 30.

It is possible to calculate these values using algebra. In equilibrium, $Q_D = Q_S$. If this common value is denoted by Q then the demand and supply equations become

$$P = -2Q + 50 \quad \text{and} \quad P = \tfrac{1}{2}Q + 25$$

Figure 1.20

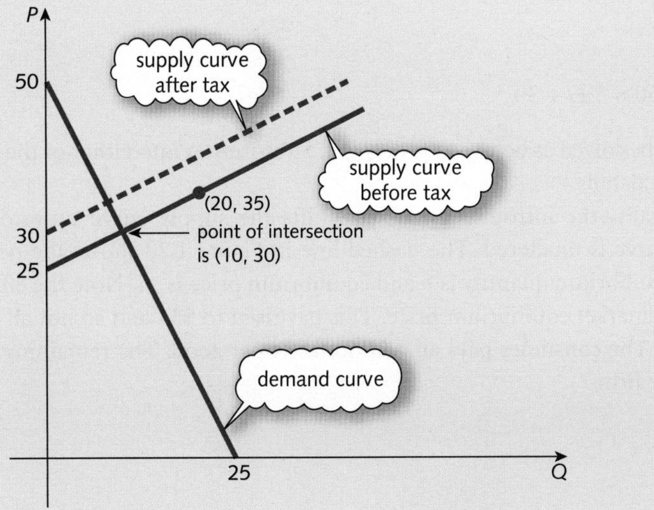

This represents a pair of simultaneous equations for the two unknowns P and Q, and so could be solved using the elimination method described in the previous section. However, this is not strictly necessary because it follows immediately from the above equations that

$$-2Q + 50 = \tfrac{1}{2}Q + 25$$

since both sides are equal to P. This can be rearranged to calculate Q:

$$-2\tfrac{1}{2}Q + 50 = 25 \quad \text{(subtract } \tfrac{1}{2}Q \text{ from both sides)}$$

$$-2\tfrac{1}{2}Q = -25 \quad \text{(subtract 50 from both sides)}$$

$$Q = 10 \quad \text{(divide both sides by } -2\tfrac{1}{2}\text{)}$$

Finally, P can be found by substituting this value into either of the original equations.

The demand equation gives

$$P = -2(10) + 50 = 30$$

As a check, the supply equation gives

$$P = \tfrac{1}{2}(10) + 25 = 30 \quad ✓$$

(b) If the government imposes a fixed tax of \$5 per good then the money that the firm actually receives from the sale of each good is the amount, P, that the consumer pays, less the tax, 5: that is, $P - 5$. Mathematically, this problem can be solved by replacing P by $P - 5$ in the supply equation to get the new supply equation

$$P - 5 = \tfrac{1}{2}Q_S + 25$$

that is,

$$P = \tfrac{1}{2}Q_S + 30$$

The remaining calculations proceed as before. In equilibrium, $Q_D = Q_S$. Again setting this common value to be Q gives

$$P = -2Q + 50$$

$$P = \tfrac{1}{2}Q + 30$$

Hence

$$-2Q + 50 = \tfrac{1}{2}Q + 30$$

which can be solved as before to give $Q = 8$. Substitution into either of the above equations gives $P = 34$. (Check the details.)

Graphically, the introduction of tax shifts the supply curve upwards by 5 units. Obviously the demand curve is unaltered. The dashed line in Figure 1.20 shows the new supply curve, from which the new equilibrium quantity is 8 and equilibrium price is 34. Note the effect that government taxation has on the market equilibrium price. This has risen to \$34 and so not all of the tax is passed on to the consumer. The consumer pays an additional \$4 per good. The remaining \$1 of tax must, therefore, be paid by the firm.

Practice Problem

3 The demand and supply functions of a good are given by

$$P = -4Q_D + 120$$

$$P = \tfrac{1}{3}Q_S + 29$$

where P, Q_D and Q_S denote the price, quantity demanded and quantity supplied respectively.

(a) Calculate the equilibrium price and quantity.

(b) Calculate the new equilibrium price and quantity after the imposition of a fixed tax of \$13 per good. Who pays the tax?

Example

EXCEL

The demand and supply functions of a good are given by

$$P = -\tfrac{1}{2}Q_D + 20$$

$$P = \tfrac{1}{3}(Q_S + 10)$$

The government imposes a fixed tax, $\$\alpha$, on each good. Determine the equilibrium price and quantity in the case when

(a) $\alpha = 0$

(b) $\alpha = 5$

(c) $\alpha = 10$

(d) $\alpha = 2.50$

In each case, calculate the tax paid by the consumer and comment on these values.

Solution

(a) In the case when $\alpha = 0$, there is no tax and the demand and supply functions are as given above. In equilibrium, $Q_D = Q_S$, so by writing this value as Q, we can find the equilibrium position by solving the simultaneous equations

$$P = -\tfrac{1}{2}Q + 20$$

$$P = \tfrac{1}{3}(Q + 10)$$

In Excel, we first set up a table of values for Q. In Figure 1.21 (overleaf), the label Q has been put in cell A1, and values from 0 to 40 (going up in steps of 10) occupy cells A2 to A6. At this stage, we need to enter a formula for calculating the corresponding values of P using each of the equations in turn. As the first value of Q is in cell A2, we type

=–A2/2 + 20

in cell B2 for the demand function, and

=1/3*(A2 + 10)

in cell C2 for the supply function. By clicking and dragging down the columns, Excel will generate corresponding values for demand and supply.

➔

Figure 1.21

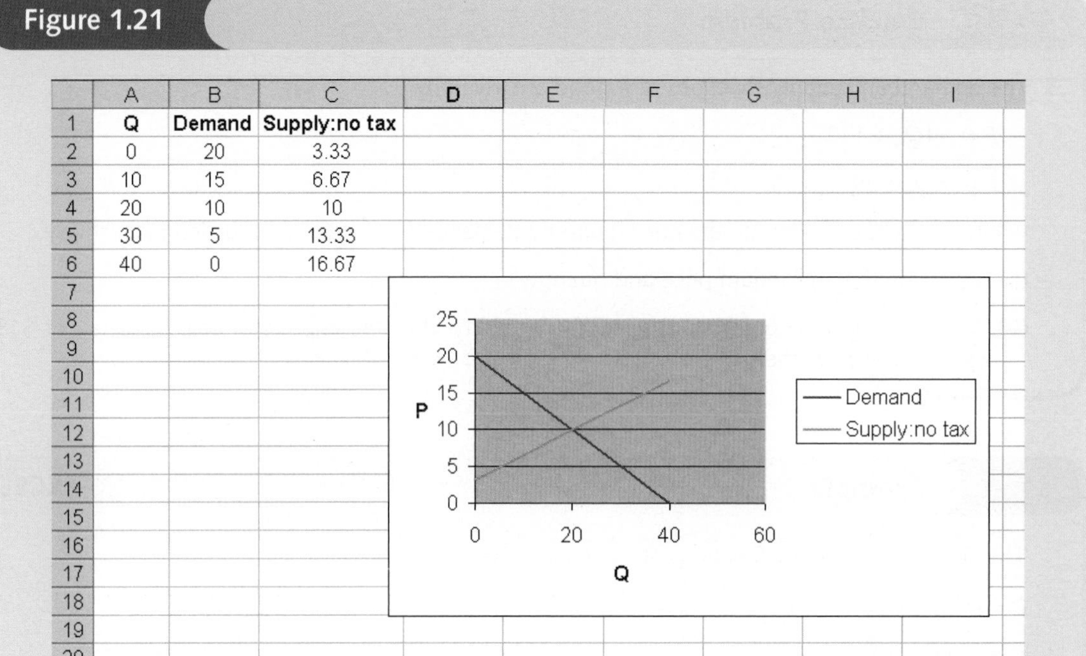

	A	B	C	D	E	F	G	H	I
1	Q	Demand	Supply:no tax						
2	0	20	3.33						
3	10	15	6.67						
4	20	10	10						
5	30	5	13.33						
6	40	0	16.67						

You will find that the values in the third column look very unfriendly to start with, as they have lots of figures after the decimal point. However, these can be removed by highlighting these numbers, and clicking on the Decrease Decimal icon, which can be found on the toolbar:

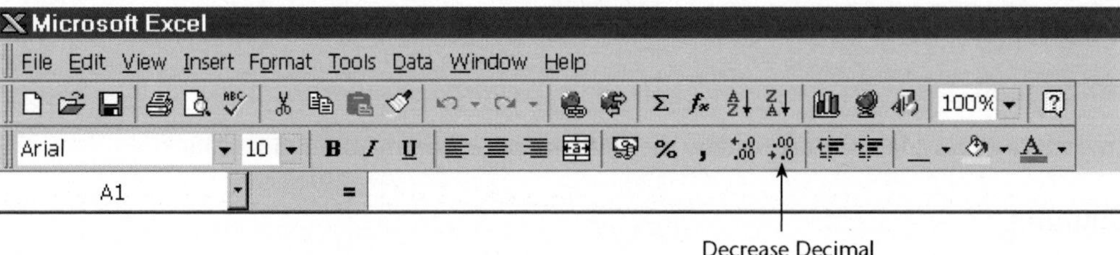

This has the effect of reducing the number of decimal places by rounding. Each time you click the icon, another decimal place is removed. As we are dealing with money, we round to 2 decimal places.

Finally, we highlight the contents of the first three columns and use Chart Wizard to create a diagram showing the demand and supply functions, as shown in Figure 1.21.

From the graph, it can be seen that the lines cross when $Q = 20$ and $P = 10$, which gives us the equilibrium position. This can also be seen by looking at the fourth row of the table.

(b) If the government imposes a tax of \$5 per item, the company producing the goods now receives \$5 less per item sold. The supply equation now becomes

$$P - 5 = \tfrac{1}{3}(Q_S + 10)$$

so that

$$P = \tfrac{1}{3}(Q_S + 10) + 5 \text{ (add 5 to both sides)}$$

The demand function remains unchanged. We can extend our spreadsheet from part (a) to include an extra column for this amended supply function, and we can then plot this extra line on the same graph. This can be done by typing

=1/3*(A2+10)+5

in cell D2, and dragging down to D6.

Figure 1.22

	A	B	C	D	E	F	G	H	I
1	Q	Demand	Supply:no tax	Supply:tax = $5					
2	0	20	3.33	8.33					
3	10	15	6.67	11.67					
4	20	10	10	15					
5	30	5	13.33	18.33					
6	40	0	16.67	21.67					
7									
8									
9									
10									
11									
12									
13									
14									
15									
16									
17									
18									
19									
20									

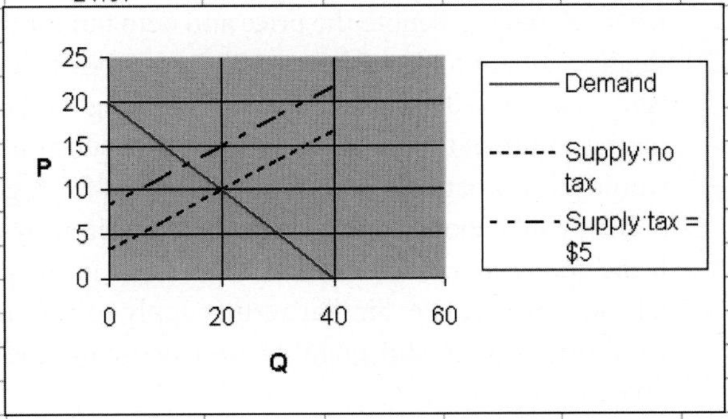

It is possible to alter the type of line drawn by the Chart Wizard by clicking on the line you wish to change. This should highlight the points that were plotted. Select **Format** from the menu bar, and then click on **Data Series**, **Patterns** and finally scroll through the styles of lines available and select the one required. Figure 1.22 shows the new spreadsheet.

Notice that the effect of the tax is to move the supply line up by 5, and the position of equilibrium has moved to (14, 13). This means that the price has increased from $10 to $13, with the consumer paying an additional $3 in tax. The remaining $2 is therefore paid by the company.

(c) The calculations in (b) can obviously be repeated by editing the formula for the supply equation in cell D2 to

=1/3*(A2+10)+10

The equilibrium quantity and price are now

$Q = 8$ and $P = 16$

so the consumer pays $6 of this tax.

(d) Changing D2 to

=1/3*(A2+10)+2.5

gives equilibrium values

$Q = 17$ and $P = 11.5$

so the consumer pays $1.50 of the tax.

Notice that, as expected, the consumer pays increasing amounts of tax as the value of α increases. More significantly, notice that the *fraction* of the tax paid by the consumer is the same in each case. For example, in part (c), the consumer pays $6 of the $10 tax, which is $^3/_5$ of the tax. You might like to check that in cases (b) and (d) the tax is also split in the ratio of 3:2. We shall investigate this further in Practice Problem 16 at the end of this section.

We conclude this section by considering a more realistic model of supply and demand, taking into account substitutable and complementary goods. Let us suppose that there are two goods in related markets, which we call good 1 and good 2. The demand for either good depends on the prices of both good 1 and good 2. If the corresponding demand functions are linear then

$$Q_{D_1} = a_1 + b_1P_1 + c_1P_2$$

$$Q_{D_2} = a_2 + b_2P_1 + c_2P_2$$

where P_i and Q_{D_i} denote the price and demand for the ith good and a_i, b_i, c_i are parameters. For the first equation, $a_1 > 0$ because there is a positive demand when the prices of both goods are zero. Also, $b_1 < 0$ because the demand of a good falls as its price rises. The sign of c_1 depends on the nature of the goods. If the goods are substitutable then an increase in the price of good 2 would mean that consumers would switch from good 2 to good 1, causing Q_{D_1} to increase. Substitutable goods are therefore characterized by a positive value of c_1. On the other hand, if the goods are complementary then a rise in the price of either good would see the demand fall, so c_1 is negative. Similar results apply to the signs of a_2, b_2 and c_2. The calculation of the equilibrium price and quantity in a two-commodity market model is demonstrated in the following example.

Example

The demand and supply functions for two interdependent commodities are given by

$$Q_{D_1} = 10 - 2P_1 + P_2$$

$$Q_{D_2} = 5 + 2P_1 - 2P_2$$

$$Q_{S_1} = -3 + 2P_1$$

$$Q_{S_2} = -2 + 3P_2$$

where Q_{D_i}, Q_{S_i} and P_i denote the quantity demanded, quantity supplied and price of good i respectively. Determine the equilibrium price and quantity for this two-commodity model.

Solution

In equilibrium, we know that the quantity supplied is equal to the quantity demanded for each good, so that

$$Q_{D_1} = Q_{S_1} \quad \text{and} \quad Q_{D_2} = Q_{S_2}$$

Let us write these respective common values as Q_1 and Q_2. The demand and supply equations for good 1 then become

$$Q_1 = 10 - 2P_1 + P_2$$

$$Q_1 = -3 + 2P_1$$

Hence

$$10 - 2P_1 + P_2 = -3 + 2P_1$$

since both sides are equal to Q_1. It makes sense to tidy this equation up a bit by collecting all of the unknowns on the left-hand side and putting the constant terms on to the right-hand side:

$$10 - 4P_1 + P_2 = -3 \quad \text{(subtract } 2P_1 \text{ from both sides)}$$

$$-4P_1 + P_2 = -13 \quad \text{(subtract 10 from both sides)}$$

We can perform a similar process for good 2. The demand and supply equations become

$$Q_2 = 5 + 2P_1 - 2P_2$$
$$Q_2 = -2 + 3P_2$$

because $Q_{D_2} = Q_{S_2} = Q_2$ in equilibrium. Hence

$$5 + 2P_1 - 2P_2 = -2 + 3P_2$$
$$5 + 2P_1 - 5P_2 = -2 \quad \text{(subtract } 3P_2 \text{ from both sides)}$$
$$2P_1 - 5P_2 = -7 \quad \text{(subtract 5 from both sides)}$$

We have therefore shown that the equilibrium prices, P_1 and P_2, satisfy the simultaneous linear equations

$$-4P_1 + P_2 = -13 \quad (1)$$
$$2P_1 - 5P_2 = -7 \quad (2)$$

which can be solved by elimination. Following the steps described in Section 1.2 we proceed as follows.

Step 1

Double equation (2) and add to equation (1) to get

$$\begin{array}{r} -4P_1 + P_2 = -13 \\ 4P_1 - 10P_2 = -14 + \\ \hline -9P_2 = -27 \end{array} \quad (3)$$

Step 2

Divide both sides of equation (3) by −9 to get $P_2 = 3$.

Step 3

If this is substituted into equation (1) then

$$-4P_1 + 3 = -13$$
$$-4P_1 = -16 \quad \text{(subtract 3 from both sides)}$$
$$P_1 = 4 \quad \text{(divide both sides by } -4\text{)}$$

Step 4

As a check, equation (2) gives

$$2(4) - 5(3) = -7 \quad ✓$$

Hence $P_1 = 4$ and $P_2 = 3$.

Finally, the equilibrium quantities can be deduced by substituting these values back into the original supply equations. For good 1,

$$Q_1 = -3 + 2P_1 = -3 + 2(4) = 5$$

For good 2,

$$Q_2 = -2 + 3P_2 = -2 + 3(3) = 7$$

As a check, the demand equations also give

$$Q_1 = 10 - 2P_1 + P_2 = 10 - 2(4) + 3 = 5 \quad ✓$$
$$Q_2 = 5 + 2P_1 - 2P_2 = 5 + 2(4) - 2(3) = 7 \quad ✓$$

Practice Problem

4 The demand and supply functions for two interdependent commodities are given by

$$Q_{D_1} = 40 - 5P_1 - P_2$$

$$Q_{D_2} = 50 - 2P_1 - 4P_2$$

$$Q_{S_1} = -3 + 4P_1$$

$$Q_{S_2} = -7 + 3P_2$$

where Q_{D_i}, Q_{S_i} and P_i denote the quantity demanded, quantity supplied and price of good i respectively. Determine the equilibrium price and quantity for this two-commodity model. Are these goods substitutable or complementary?

For a two-commodity market the equilibrium prices and quantities can be found by solving a system of two simultaneous equations. Exactly the same procedure can be applied to a three-commodity market, which requires the solution of a system of three simultaneous equations.

Advice

An example of a three-commodity model can be found in Practice Problem 13. Alternative methods and further examples are described in Chapter 7. In general, with n goods it is necessary to solve n equations in n unknowns and, as pointed out in Section 1.2, this is best done using a computer package whenever n is large.

Key Terms

Complementary goods A pair of goods consumed together. As the price of either goes up, the demand for both goods goes down.

Decreasing function A function, $y = f(x)$, in which y decreases as x increases.

Demand function A relationship between the quantity demanded and various factors that affect demand, including price.

Dependent variable A variable whose value is determined by that taken by the independent variables; in $y = f(x)$, the dependent variable is y.

Endogenous variable A variable whose value is determined within a model.

Equilibrium This state occurs when quantity supplied and quantity demanded are equal.

Exogenous variable A variable whose value is determined outside a model.

Function A rule that assigns to each incoming number, x, a uniquely defined outgoing number, y.

Increasing function A function, $y = f(x)$, in which y increases as x increases.

Independent variable A variable whose value determines that of the dependent variable; in $y = f(x)$, the independent variable is x.

Inferior good A good whose demand decreases as income increases.

Inverse function A function, written f^{-1}, which reverses the effect of a given function, f, so that $x = f^{-1}(y)$ when $y = f(x)$.

Modelling The creation of piece of mathematical theory which represents (a simplification of) some aspect of practical economics.

Parameter A constant whose value affects the specific values but not the general form of a mathematical expression such as the constants a, b and c in $ax^2 + bx + c$.

Substitutable goods A pair of goods that are alternatives to each other. As the price of one of them goes up, the demand for the other rises.

Superior good A good whose demand increases as income increases.

Supply function A relationship between the quantity supplied and various factors that affect supply, including price.

Practice Problems

5 If $f(x) = 3x + 15$ and $g(x) = \frac{1}{3}x - 5$, evaluate

(a) $f(2)$ **(b)** $f(10)$ **(c)** $f(0)$

(d) $g(21)$ **(e)** $g(45)$ **(f)** $g(15)$

What word describes the relationship between f and g?

6 Sketch a graph of the supply function

$$P = \frac{1}{3}Q + 7$$

Hence, or otherwise, determine the value of

(a) P when $Q = 12$

(b) Q when $P = 10$

(c) Q when $P = 4$

7 Describe the effect on the demand curve due to an increase in

(a) the price of substitutable goods, P_S

(b) the price of complementary goods, P_C

(c) advertising expenditure, A

8 The demand function of a good is

$$Q = 100 - P + 2Y + \frac{1}{2}A$$

where Q, P, Y and A denote quantity demanded, price, income and advertising expenditure respectively.

(a) Calculate the demand when $P = 10$, $Y = 40$ and $A = 6$. Assuming that price and income are fixed, calculate the additional advertising expenditure needed to raise demand to 179 units.

(b) Is this good inferior or superior?

9 The demand, Q, for a certain good depends on its own price, P, and the price of an alternative good, P_A, according to

$$Q = 30 - 3P + P_A$$

(a) Find Q if $P = 4$ and $P_A = 5$.

(b) Is the alternative good substitutable or complementary? Give a reason for your answer.

(c) Determine the value of P if $Q = 23$ and $P_A = 11$.

10 The demand and supply functions of a good are given by

$$P = -5Q_D + 80$$

$$P = 2Q_S + 10$$

where P, Q_D and Q_S denote price, quantity demanded and quantity supplied respectively.

(1) Find the equilibrium price and quantity

(a) graphically **(b)** algebraically

(2) If the government deducts, as tax, 15% of the market price of each good, determine the new equilibrium price and quantity.

11 The supply and demand equations of a good are given by

$$P = Q_S + 8$$

$$P = -3Q_D + 80$$

where P, Q_S and Q_D denote price, quantity supplied and quantity demanded respectively.

(a) Find the equilibrium price and quantity if the government imposes a fixed tax of £36 on each good.

(b) Find the corresponding value of the government's tax revenue.

12 The demand and supply functions for two interdependent commodities are given by

$$Q_{D_1} = 100 - 2P_1 + P_2$$

$$Q_{D_2} = 5 + 2P_1 - 3P_2$$

$$Q_{S_1} = -10 + P_1$$

$$Q_{S_2} = -5 + 6P_2$$

where Q_{D_i}, Q_{S_i} and P_i denote the quantity demanded, quantity supplied and price of good i respectively. Determine the equilibrium price and quantity for this two-commodity model.

13 The demand and supply functions for three interdependent commodities are

$$Q_{D_1} = 15 - P_1 + 2P_2 + P_3$$

$$Q_{D_2} = 9 + P_1 - P_2 - P_3$$

$$Q_{D_3} = 8 + 2P_1 - P_2 - 4P_3$$

$$Q_{S_1} = -7 + P_1$$

$$Q_{S_2} = -4 + 4P_2$$

$$Q_{S_3} = -5 + 2P_3$$

where Q_{D_i}, Q_{S_i} and P_i denote the quantity demanded, quantity supplied and price of good i respectively. Determine the equilibrium price and quantity for this three-commodity model.

14 If the demand function of a good is

$$2P + 3Q_D = 60$$

where P and Q_D denote price and quantity demanded respectively, find the largest and smallest values of P for which this function is economically meaningful.

15 **(Excel)** The demand function of a good is given by

$$Q = 100 - 2P + Y - 3P_A$$

where Q, P, Y and P_A denote quantity, price, income and price of an alternative good, respectively. For each of the following cases, tabulate values of Q, when P is 0, 20, 40, 60. Hence sketch all three demand curves on the same diagram.

(a) $Y = 20$, $P_A = 10$

(b) $Y = 50$, $P_A = 10$

(c) $Y = 20$, $P_A = 16$

Is the good inferior or superior?
Is the alternative good substitutable or complementary?
Give reasons for your answers.

16 **(Excel)** The supply and demand functions of a good are given by

$$P = -Q_D + 240$$

$$P = 60 + 2Q_S$$

where P, Q_D and Q_S denote price, quantity demanded and quantity supplied, respectively. Sketch graphs of both functions on the same diagram, on the range $0 \le Q \le 80$ and hence find the equilibrium price. The government now imposes a fixed tax, $60, on each good. Draw the new supply equation on the same diagram and hence find the new equilibrium price. What fraction of the $60 tax is paid by the consumer?

Consider replacing the demand function by the more general equation

$$P = -kQ_D + 240$$

By repeating the calculations above, find the fraction of the tax paid by the consumer for the case when k is

(a) 2 **(b)** 3 **(c)** 4

State the connection between this fraction and the value of k. Use this connection to predict how much tax is paid by the consumer when $k = 6$.

section 1.4

Algebra

Objectives

At the end of this section you should be able to:

- Recognize the symbols $<$, $>$, $\leq$ and $\geq$.
- Manipulate inequalities.
- Multiply out brackets.
- Add, subtract, multiply and divide numerical fractions.
- Add, subtract, multiply and divide algebraic fractions.

ALGEBRA IS BORING

There is no getting away from the fact that algebra *is* boring. Doubtless there are a few enthusiasts who get a kick out of algebraic manipulation, but economics students are rarely to be found in this category! Indeed, the mere mention of the word 'algebra' is enough to strike fear into the heart of many a first-year student. Unfortunately, you cannot get very far with mathematics unless you have completely mastered this topic. An apposite analogy is the game of chess. Before you can begin to play a game of chess it is necessary to go through the tedium of learning the moves of individual pieces. In the same way it is essential that you learn the rules of algebra before you can enjoy the 'game' of mathematics. Of course, just because you know the rules does not mean that you are going to excel at the game and no one is expecting you to become a grandmaster of mathematics. However, you should at least be able to follow the mathematics presented in economics books and journals, as well as being able to solve simple problems for yourself.

Advice

In the introduction to this book you were advised to attempt all questions in the text. This applies particularly to the material in this section. This section is likely to be heavy going if you have not studied mathematics for some time. To make life easy for you it is split up into four subsections:

- Inequalities
- Brackets
- Fractions
- Equations

You might like to work through these subsections on separate occasions to enable the ideas to sink in. To rush this topic now is likely to give you only a half-baked understanding, which will result in hours of frustration when you study the later chapters of this book.

1.4.1 Inequalities

So far we have repeatedly made use of a picture of the x axis:

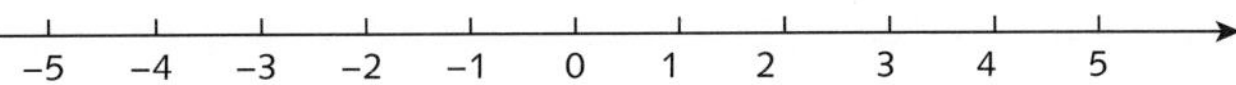

Now, although only whole numbers are marked on this diagram, it is implicitly assumed that it can also be used to indicate fractions and decimal numbers as well. To each point on the line there corresponds a particular number. Conversely, every number can be represented by a particular point on the line. For this reason, the line is sometimes referred to as a ***number line***. For example, $-2\frac{1}{2}$ lies exactly halfway between –3 and –2. Similarly, $4\frac{7}{8}$ lies $\frac{7}{8}$ths of the way between 4 and 5. In theory, we can even find a point on the line corresponding to a number such as $\sqrt{2}$, although it may be difficult to sketch such a point accurately in practice. My calculator gives the value of $\sqrt{2}$ to be 1.414 213 56 to eight decimal places. This number therefore lies just less than halfway between 1 and 2.

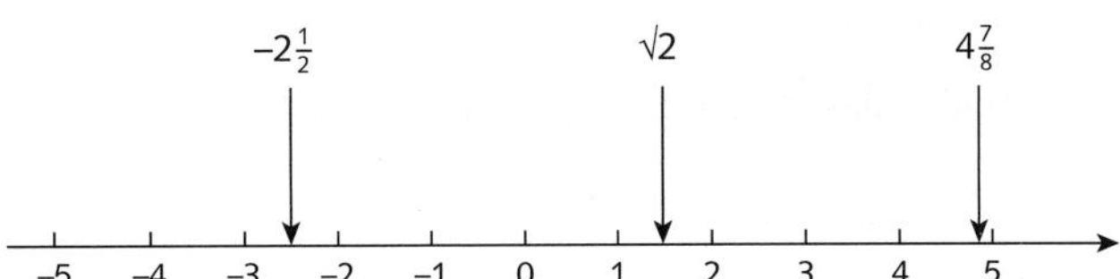

A number line can be used to decide whether or not one number is greater or less than another number. We say that a number a is greater than a number b if a lies to the right of b on the line and write this as

$$a > b$$

(In Section 1.3 the notation $a > 0$ was used to indicate that a is positive: that is, a is greater than zero.) Likewise, we say that a is less than b if a lies to the left of b and write this as

$$a < b$$

From the diagram we see that

$$-2 > -4$$

because −2 lies to the right of −4. This is equivalent to the statement

$-4 < -2$

Similarly,

$0 > -1$ (or equivalently $-1 < 0$)

$2 > -2\frac{1}{2}$ (or equivalently $-2\frac{1}{2} < 2$)

$4\frac{7}{8} > \sqrt{2}$ (or equivalently $\sqrt{2} < 4\frac{7}{8}$)

There are occasions when we would like the letters a and b to stand for mathematical expressions rather than actual numbers. In this situation we sometimes use the symbols ≥ and ≤ to mean 'greater than or equal to' and 'less than or equal to' respectively.

We have already seen that we can manipulate equations in any way we like, provided that we do the same thing to both sides. An obvious question to ask is whether this rule extends to inequalities. To investigate this, consider the following example.

Example

Starting with the true statement

$1 < 3$

decide which of the following are valid operations when performed on both sides:

(a) add 4 **(b)** add −5 **(c)** multiply by 2 **(d)** multiply by −6

Solution

(a) If we add 4 to both sides of the inequality

$$1 < 3 \quad (1)$$

then we obtain

$5 < 7$

which is a true statement.

(b) If we add −5 to both sides of inequality (1) then we obtain

$-4 < -2$

which is also true.

(c) If we multiply both sides of inequality (1) by 2 then we obtain

$2 < 6$

which is again true.

(d) So far so good, but if we now multiply both sides of inequality (1) by −6 then we obtain

$-6 < -18$

which is false. In fact, quite the reverse is true, since −6 lies to the right of −18 on the number line and so −6 is actually greater than −18. This indicates that the rule needs modifying before we can extend it to inequalities and that we need to be careful when manipulating such things.

Practice Problem

1 Starting with the true statement

$$6 > 3$$

decide which of the following are valid operations when performed on both sides:

(a) add 6 **(b)** multiply by 2 **(c)** subtract 3

(d) add –3 **(e)** divide by 3 **(f)** multiply by –4

(g) multiply by –1 **(h)** divide by –3 **(i)** add –10

These examples show that the usual rule does apply to inequalities with the important proviso that

if both sides are multiplied or divided by a negative number then the sense of the inequality is reversed

By this we mean that '>' changes to '<', '≤' changes to '≥' and so on.

Example

Simplify the inequality

$$2x + 3 < 4x + 7$$

Solution

The first problem is to decide what is meant by the word 'simplify'. At the moment there are x's on both sides of the inequality sign and it would obviously look neater if these were collected together. We do this by subtracting $4x$ from both sides to get

$$-2x + 3 < 7$$

We can also put all of the constant terms on to the right-hand side by subtracting 3 from both sides to get

$$-2x < 4$$

This is certainly an improvement, but we can go further to make the inequality even more meaningful. We may divide both sides by –2 to get

$$x > -2$$

Notice that the sense has been reversed at this stage because we have divided by a negative number. We have therefore shown that any number x satisfies the original inequality provided that it lies to the right of the number –2 on the number line.

Advice

You should check your answer using a couple of test values. Substituting $x = 1$ (which lies to the right of -2, so should work) into both sides of the original inequality $2x + 3 < 4x + 7$ gives $5 < 11$, which is true. On the other hand, substituting $x = -3$ (which lies to the left of -2, so should fail) gives $-3 < -5$, which is false.

Of course, just checking a couple of numbers like this does not prove that the final inequality is correct, but it should protect you against gross blunders.

Practice Problem

2 Simplify the inequalities

(a) $2x < 3x + 7$ **(b)** $21x - 19 \geq 4x + 15$

1.4.2 Brackets

Brackets are used to avoid any misunderstanding about the way an expression is to be evaluated. Suppose that a group of students is asked to find the value of

$$1 - 3 + 5$$

One suspects that the majority of students would say that the answer is 3, which is found by first subtracting 3 from 1 and then adding 5. However, there is a fair chance that some might produce -7, thinking that they should first add 3 and 5 and then subtract the result from 1. (There may be other students who obtain different values entirely, but we had better forget about them!) In a sense both answers are correct since, as it stands, the expression is ambiguous. To overcome this, brackets are introduced, using the convention that things inside brackets are evaluated first. Hence we would either write

$$(1 - 3) + 5$$

to indicate that subtraction is performed first, or write

$$1 - (3 + 5)$$

to indicate that addition is performed first. In fact, brackets have already been used in Section 1.1 in the context of multiplying negative numbers. For example, on page 18 we wrote

$$(-2) \times (-4) \times (-1) \times 2 \times (-1) \times (-3)$$

which is much easier to interpret than its bracketless counterpart

$$-2 \times -4 \times -1 \times 2 \times -1 \times -3$$

It is also conventional to suppress the multiplication sign when multiplying brackets together, so the above product could be written as

$$(-2)(-4)(-1)(2)(-1)(-3)$$

Similarly, the multiplication sign is implied in

$$(5 - 2)(7 + 1)$$

which is the product of 3 and 8.

Incidentally, if you really are confronted with the calculation

$1 - 3 + 5$

you should perform the subtraction first. Of course, in BIDMAS (see page 10) the operations of addition and subtraction have equal precedence. However, it is generally accepted that in these circumstances you work from left to right, which in this case means working out $1 - 3$ first, before adding on the 5, to get the answer of 3.

Example

Evaluate

(a) $(12 - 8) - (6 - 5)$ **(b)** $12 - (8 - 6) - 5$ **(c)** $12 - 8 - 6 - 5$

Solution

(a) $(12 - 8) - (6 - 5) = 4 - 1 = 3$

(b) $12 - (8 - 6) - 5 = 12 - 2 - 5 = 10 - 5 = 5$

(c) $12 - 8 - 6 - 5 = 4 - 6 - 5 = -2 - 5 = -7$

The following problem gives you an opportunity to try out these conventions for yourself and to use the brackets facility on your calculator.

Practice Problem

3 (1) Without using your calculator evaluate

(a) $(1 - 3) + 10$ **(b)** $1 - (3 + 10)$ **(c)** $2(3 + 4)$

(d) $8 - 7 + 3$ **(e)** $(15 - 8)(2 + 6)$ **(f)** $((2 - 3) + 7) \div 6$

(2) Confirm your answer to part (1) using a calculator.

In mathematics it is necessary to handle expressions in which some of the terms involve letters as well as numbers. It is useful to be able to take an expression containing brackets and to rewrite it as an equivalent expression without brackets and vice versa. The process of removing brackets is called 'expanding the brackets' or 'multiplying out the brackets'. This is based on the *distributive law*, which states that for any three numbers a, b and c

$$\boxed{a(b + c) = ab + ac}$$

It is easy to verify this law in simple cases. For example, if $a = 2$, $b = 3$ and $c = 4$ then the left-hand side is

$$2(3 + 4) = 2 \times 7 = 14$$

However,

$$ab = 2 \times 3 = 6 \quad \text{and} \quad ac = 2 \times 4 = 8$$

and so the right-hand side is 6 + 8, which is also 14.

This law can be used when there are any number of terms inside the brackets. We have

$$a(b + c + d) = ab + ac + ad$$

$$a(b + c + d + e) = ab + ac + ad + ae$$

and so on.

It does not matter in which order two numbers are multiplied, so we also have

$$(b + c)a = ab + ac + ad$$

$$(b + c + d)a = ba + ca + da$$

$$(b + c + d + e)a = ba + ca + da + ea$$

Example

Multiply out the brackets in

(a) $x(x - 2)$

(b) $2(x + y - z) + 3(z + y)$

(c) $x + 3y - (2y + x)$

Solution

(a) The use of the distributive law to multiply out $x(x - 2)$ is straightforward. This gives

$$x(x - 2) = xx - x2$$

It is usual in mathematics to abbreviate xx to x^2. It is also standard practice to put the numerical coefficient in front of the variable, so $x2$ is usually written $2x$. Hence

$$x(x - 2) = x^2 - 2x$$

(b) To expand

$$2(x + y - z) + 3(z + y)$$

we need to apply the distributive law twice. We have

$$2(x + y - z) = 2x + 2y - 2z$$

$$3(z + y) \quad = 3z + 3y$$

Adding together gives

$$2(x + y - z) + 3(z + y) = 2x + 2y - 2z + 3z + 3y$$

We could stop at this point. Note, however, that some of the terms are similar. Towards the beginning of the expression there is a term $2y$, whereas at the end there is a like term $3y$. Obviously these can be collected together to make a total of $5y$. A similar process can be applied to the terms involving z. The expression simplifies to

$$2x + 5y + z$$

(c) It may not be immediately apparent how to expand

$$x + 3y - (2y + x)$$

However, note that

$$-(2y + x)$$

is the same as

$$(-1)(2y + x)$$

which expands to give

$$(-1)(2y) + (-1)x = -2y - x$$

Hence

$$x + 3y - (2y + x) = x + 3y - 2y - x = y$$

after collecting like terms.

Advice

In this example the solutions are written out in painstaking detail. This is done to show you precisely how the distributive law is applied. The solutions to all three parts could have been written down in only one or two steps of working. You are, of course, at liberty to compress the working in your own solutions, but please do not be tempted to overdo this. You might want to check your answers at a later date and may find it difficult if you have tried to be too clever.

Practice Problem

4 Multiply out the brackets, simplifying your answer as far as possible.

(a) $(5 - 2z)z$ **(b)** $6(x - y) + 3(y - 2x)$ **(c)** $x - y + z - (x^2 + x - y)$

We conclude our discussion of brackets by describing how to multiply two brackets together. This is based on the result

$$(a + b)(c + d) = ac + ad + bc + bd$$

At first sight this formula might appear to be totally unmemorable. In fact, all you have to do is to multiply each term in the first pair of brackets by each term in the second in all possible combinations: that is,

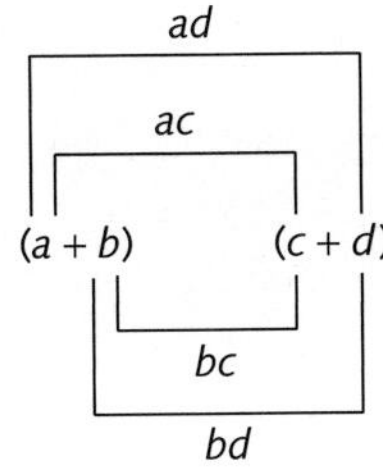

This rule then extends to brackets with more than two terms. For example, to multiply out

$$(a + b)(c + d + e)$$

notice that the first pair of brackets has two terms and the second has three terms. So, to form each individual product, we can pick from one of two terms in the first pair of brackets and from one of three terms in the second. There are then six possibilities in total, giving

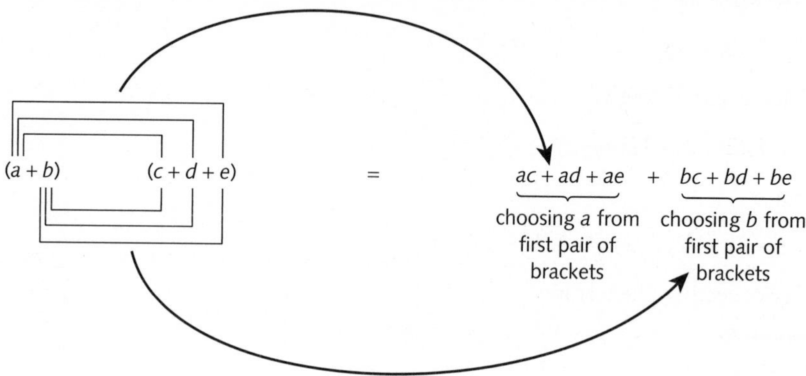

Example

Multiply out the brackets

(a) $(x + 1)(x + 2)$ **(b)** $(x + 5)(x - 5)$ **(c)** $(2x - y)(x + y - 6)$

simplifying your answer as far as possible.

Solution

(a) $(x + 1)(x + 2) = xx + x2 + 1x + (1)(2)$

If we use the abbreviation x^2 for xx and the convention that numerical coefficients are placed in front of the variable then this can be written as

$$x^2 + 2x + x + 2$$

Finally, collecting like terms gives

$$x^2 + 3x + 2$$

If you find it difficult to remember how to multiply out brackets, you might like to look at the 'smiley face'

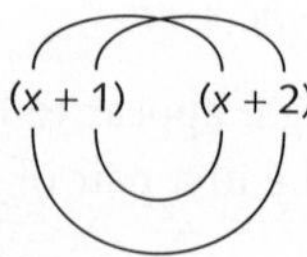

You get

left eyebrow $= x \times x = x^2$

right eyebrow $= 1 \times 2 = 2$

nose $= 1 \times x = x$

mouth $= x \times 2 = 2x$

giving a total of $x^2 + 3x + 2$.

(b)

x^2 -25

$(x + 5)\quad(x - 5) = x^2 - 25$

$5x$

$-5x$

More formally,

$$(x + 5)(x - 5) = x^2 - 5x + 5x - 25$$
$$= x^2 - 25$$

the x's cancel

(c) $(2x - y)(x + y - 6) = 2xx + 2xy + 2x(-6) - yx - yy - y(-6)$

$$= 2x^2 + 2xy - 12x - yx - y^2 + 6y$$

It might seem that there are no like terms, but since it does not matter in which order two numbers are multiplied, yx is the same as xy. The terms $2xy$ and $-yx$ can therefore be combined to give

$$2xy - xy = xy$$

Hence the simplified expression is

$$2x^2 + xy - 12x - y^2 + 6y$$

Practice Problem

5 Multiply out the brackets.

(a) $(x + 3)(x - 2)$

(b) $(x + y)(x - y)$

(c) $(x + y)(x + y)$

(d) $(5x + 2y)(x - y + 1)$

Looking back at part (b) of the previous worked example, notice that

$$(x + 5)(x - 5)$$

can be written as

$$x^2 - 5^2$$

Quite generally

$$(a + b)(a - b) = a^2 - ab + ba - b^2$$
$$= a^2 - ab + ab - b^2$$
$$= a^2 - b^2$$

The result

$$\boxed{a^2 - b^2 = (a + b)(a - b)}$$

is called the **difference of two squares** formula. It provides a quick way of factorizing an expression: that is, of producing an equivalent expression with brackets.

Example

Factorize the following expressions:

(a) $x^2 - 16$ (b) $9x^2 - 100$

Solution

(a) Noting that

$$x^2 - 16 = x^2 - 4^2$$

we can use the difference of two squares formula to deduce that

$$x^2 - 16 = (x + 4)(x - 4)$$

(b) Noting that

$$9x^2 - 100 = (3x)^2 - (10)^2$$

$(3x)^2 = 3x \times 3x$
$= 9x^2$

we can use the difference of two squares formula to deduce that

$$9x^2 - 100 = (3x + 10)(3x - 10)$$

Practice Problem

6 Factorize the following expressions:

(a) $x^2 - 64$ **(b)** $4x^2 - 81$

1.4.3 Fractions

For a numerical fraction such as

$$\frac{7}{8}$$

the number 7, on the top, is called the *numerator* and the number 8, on the bottom, is called the *denominator*. In this book we are also interested in the case when the numerator and denominator involve letters as well as numbers. These are referred to as *algebraic fractions*. For example,

$$\frac{1}{x^2 - 2} \quad \text{and} \quad \frac{2x^2 - 1}{y + z}$$

are both algebraic fractions. The letters x, y and z are used to represent numbers, so the rules for the manipulation of algebraic fractions are the same as those for ordinary numerical fractions. The rules for multiplication and division are as follows.

to multiply fractions you multiply their corresponding numerators and denominators

In symbols,

$$\frac{a}{b} \times \frac{c}{d} = \frac{a \times c}{b \times d} = \frac{ac}{bd}$$

to divide by a fraction you turn it upside down and multiply

In symbols,

$$\frac{a}{b} \div \frac{c}{d} = \frac{a}{b} \times \frac{d}{c}$$

turn the divisor upside down

$$= \frac{ad}{bc}$$

rule for multiplying fractions

Example

Calculate

(a) $\frac{2}{3} \times \frac{5}{4}$

(b) $2 \times \frac{6}{13}$

(c) $\frac{6}{7} \div \frac{4}{21}$

(d) $\frac{1}{2} \div 3$

Solution

(a) The multiplication rule gives

$$\frac{2}{3} \times \frac{5}{4} = \frac{2 \times 5}{3 \times 4} = \frac{10}{12}$$

We could leave the answer like this, although it can be simplified by dividing top and bottom by 2 to get $^5/_6$.

The two answers are equivalent. If a cake is cut into 6 pieces and you eat 5 of them then you eat just as much as someone who cuts the cake into 12 pieces and eats 10 (although it might appear that you are not such a glutton). It is also valid to 'cancel' by 2 at the very beginning: that is,

$$\frac{{}^{1}\not{2}}{3} \times \frac{5}{\not{4}_{2}} = \frac{1 \times 5}{3 \times 2} = \frac{5}{6}$$

(b) The whole number 2 is equivalent to the fraction $^2/_1$, so

$$2 \times \frac{6}{13} = \frac{2}{1} \times \frac{6}{13} = \frac{2 \times 6}{1 \times 13} = \frac{12}{13}$$

➔

(c) To calculate

$$\frac{6}{7} \div \frac{4}{21}$$

the divisor is turned upside down to get $^{21}/_{4}$ and then multiplied to get

$$\frac{6}{7} \div \frac{4}{21} = \frac{{}^{3}\cancel{6}}{{}_{1}\cancel{7}} \times \frac{\cancel{21}^{3}}{\cancel{4}_{2}} = \frac{3 \times 3}{1 \times 2} = \frac{9}{2}$$

(d) We write 3 as $^{3}/_{1}$, so

$$\frac{1}{2} \div 3 = \frac{1}{2} \div \frac{3}{1} = \frac{1}{2} \times \frac{1}{3} = \frac{1}{6}$$

Practice Problem

7 (1) Without using a calculator evaluate

(a) $\frac{1}{2} \times \frac{3}{4}$ **(b)** $7 \times \frac{1}{14}$ **(c)** $\frac{2}{3} \div \frac{8}{9}$ **(d)** $\frac{8}{9} \div 16$

(2) Confirm your answer to part (1) using a calculator.

The rules for addition and subtraction are as follows:

to add (or subtract) two fractions you put them over a common denominator and add (or subtract) their numerators

Example

Calculate

(a) $\frac{1}{5} + \frac{2}{5}$ **(b)** $\frac{1}{4} + \frac{2}{3}$ **(c)** $\frac{7}{12} - \frac{5}{8}$

Solution

(a) The fractions $^{1}/_{5}$ and $^{2}/_{5}$ already have the same denominator, so to add them we just add their numerators to get

$$\frac{1}{5} + \frac{2}{5} = \frac{1+2}{5} = \frac{3}{5}$$

(b) The fractions $^{1}/_{4}$ and $^{2}/_{5}$ have denominators 4 and 3. One number that is divisible by both 3 and 4 is 12, so we choose this as the common denominator. Now 4 goes into 12 exactly 3 times, so

$$\frac{1}{4} = \frac{1 \times 3}{4 \times 3} = \frac{3}{12}$$

multiply top and bottom by 3

and 3 goes into 12 exactly 4 times, so

$$\frac{2}{3} = \frac{2 \times 4}{3 \times 4} = \frac{8}{12}$$

multiply top and bottom by 4

Hence

$$\frac{1}{4} + \frac{2}{3} = \frac{3}{12} + \frac{8}{12} = \frac{3+8}{12} = \frac{11}{12}$$

(c) The fractions $^7/_{12}$ and $^5/_8$ have denominators 12 and 8. One number that is divisible by both 12 and 8 is 24, so we choose this as the common denominator. Now 12 goes into 24 exactly twice, so

$$\frac{7}{12} = \frac{7 \times 2}{24} = \frac{14}{24}$$

and 8 goes into 24 exactly 3 times, so

$$\frac{5}{8} = \frac{5 \times 3}{24} = \frac{15}{24}$$

Hence

$$\frac{7}{12} - \frac{5}{8} = \frac{14}{24} - \frac{15}{24} = -\frac{1}{24}$$

It is not essential that the lowest common denominator is used. Any number will do provided that it is divisible by the two original denominators. If you are stuck then you could always multiply the original two denominators together. In part (c) the denominators multiply to give 96, so this can be used instead. Now

$$\frac{7}{12} = \frac{7 \times 8}{96} = \frac{56}{96}$$

and

$$\frac{5}{8} = \frac{5 \times 12}{96} = \frac{60}{96}$$

so

$$\frac{7}{12} - \frac{5}{8} = \frac{56}{96} - \frac{60}{96} = \frac{56-60}{96} = \frac{-\not{4}^{1}}{\not{96}_{24}} = -\frac{1}{24}$$

as before.

Practice Problem

8 (1) Without using a calculator evaluate

(a) $\frac{3}{7} - \frac{1}{7}$ **(b)** $\frac{1}{3} + \frac{2}{5}$ **(c)** $\frac{7}{18} - \frac{1}{4}$

(2) Confirm your answer to part (1) using a calculator.

Provided that you can manipulate ordinary fractions, there is no reason why you should not be able to manipulate algebraic fractions just as easily, since the rules are the same.

Example

Find expressions for each of the following:

(a) $\frac{x}{x-1} \times \frac{2}{x(x+4)}$

(b) $\frac{2}{x-1} \div \frac{x}{x-1}$

(c) $\frac{x+1}{x^2+2} + \frac{x-6}{x^2+2}$

(d) $\frac{x}{x+2} - \frac{1}{x+1}$

Solution

(a) To multiply two fractions we multiply their corresponding numerators and denominators, so

$$\frac{x}{x-1} \times \frac{2}{x(x+4)} = \frac{2x}{(x-1)x(x+4)} = \frac{2}{(x-1)(x+4)}$$

the *x*'s cancel top and bottom

(b) To divide by

$$\frac{x}{x-1}$$

we turn it upside down and multiply, so

$$\frac{2}{x-1} \div \frac{x}{x-1} = \frac{2}{x-1} \times \frac{x-1}{x} = \frac{2}{x}$$

the (*x* – 1)'s cancel top and bottom

(c) The fractions

$$\frac{x+1}{x^2+2} \quad \text{and} \quad \frac{x-6}{x^2+2}$$

already have the same denominator, so to add them we just add their numerators to get

$$\frac{x+1}{x^2+2} + \frac{x-6}{x^2+2} = \frac{x+1+x-6}{x^2+2} = \frac{2x-5}{x^2+2}$$

(d) The fractions

$$\frac{x}{x+2} \quad \text{and} \quad \frac{1}{x+1}$$

have denominators $x + 2$ and $x + 1$. An obvious common denominator is given by their product, $(x+2)(x+1)$. Now $x+2$ goes into $(x+2)(x+1)$ exactly $x+1$ times, so

$$\frac{x}{x+2} = \frac{x(x+1)}{(x+2)(x+1)}$$

multiply top and bottom by (*x* + 1)

Also $x + 1$ goes into $(x + 2)(x + 1)$ exactly $x + 2$ times, so

$$\frac{1}{x+1} = \frac{x(x+1)}{(x+2)(x+1)}$$

multiply top and bottom by $(x + 2)$

Hence

$$\frac{x}{x+2} - \frac{1}{x+1} = \frac{x(x+1)}{(x+2)(x+1)} - \frac{(x+2)}{(x+2)(x+1)} = \frac{x(x+1)-(x+2)}{(x+2)(x+1)}$$

It is worth multiplying out the brackets on the top to simplify: that is,

$$\frac{x^2+x-x-2}{(x+2)(x+1)} = \frac{x^2-2}{(x+2)(x+1)}$$

Practice Problem

9 Find expressions for the following algebraic fractions, simplifying your answers as far as possible.

(a) $\frac{5}{x-1} \times \frac{x-1}{x+2}$ **(b)** $\frac{x^2}{x+10} \div \frac{x}{x+1}$ **(c)** $\frac{4}{x+1} + \frac{1}{x+1}$ **(d)** $\frac{2}{x+1} - \frac{1}{x+2}$

1.4.4 Equations

In the course of working through the first section of this chapter, you will have learnt how to solve simple equations such as

$$3x + 1 = 13$$

Such problems are solved by 'doing the same thing to both sides'. For this particular equation, the steps would be

$3x = 12$ (subtract 1 from both sides)

$x = 4$ (divide both sides by 3)

We now show how to solve more complicated equations. The good news is that the basic method is the same.

Example

Solve

(a) $6x + 1 = 10x - 9$ **(b)** $3(x - 1) + 2(2x + 1) = 4$

(c) $\frac{20}{3x-1} = 7$ **(d)** $\frac{9}{x+2} = \frac{7}{2x+1}$ **(e)** $\sqrt{\frac{2x}{x-6}} = 2$

Solution

(a) To solve

$$6x + 1 = 10x - 9$$

the strategy is to collect terms involving x on one side of the equation, and to collect all of the number terms on to the other side. It does not matter which way round this is done. In this particular case, there are more x's on the right-hand side than there are on the left-hand side. Consequently, to avoid negative numbers, you may prefer to stack the x terms on the right-hand side. The details are as follows:

$$1 = 4x - 9 \quad \text{(subtract } 6x \text{ from both sides)}$$

$$10 = 4x \quad \text{(add 9 to both sides)}$$

$$\frac{10}{4} = x \quad \text{(divide both sides by 4)}$$

Hence $x = 5/2 = 2\frac{1}{2}$.

(b) The novel feature of the equation

$$3(x - 1) + 2(2x + 1) = 4$$

is the presence of brackets. To solve it, we first remove the brackets by multiplying out, and then collect like terms:

$$3x - 3 + 4x + 2 = 4 \quad \text{(multiply out the brackets)}$$

$$7x - 1 = 4 \quad \text{(collect like terms)}$$

Note that this equation is now of the form that we know how to solve:

$$7x = 5 \quad \text{(add 1 to both sides)}$$

$$x = \frac{5}{7} \quad \text{(divide both sides by 7)}$$

(c) The novel feature of the equation

$$\frac{20}{3x - 1} = 7$$

is the fact that it involves an algebraic fraction. This can easily be removed by multiplying both sides by the bottom of the fraction:

$$\frac{20}{3x - 1} \times (3x - 1) = 7(3x - 1)$$

which cancels down to give

$$20 = 7(3x - 1)$$

The remaining steps are similar to those in part (b):

$$20 = 21x - 7 \quad \text{(multiply out the brackets)}$$

$$27 = 21x \quad \text{(add 7 to both sides)}$$

$$\frac{27}{21} = x \quad \text{(divide both sides by 21)}$$

Hence $x = 9/7 = 1\frac{2}{7}$.

(d) The next equation,

$$\frac{9}{x+2}=\frac{7}{2x+1}$$

looks particularly daunting since there are fractions on both sides. However, these are easily removed by multiplying both sides by the denominators, in turn:

$$9=\frac{7(x+2)}{2x+1} \qquad \text{(multiply both sides by } x+2\text{)}$$

$$9(2x+1)=7(x+2) \qquad \text{(multiply both sides by } 2x+1\text{)}$$

With practice you can do these two steps simultaneously and write this as the first line of working. The procedure of going straight from

$$\frac{9}{x+2}=\frac{7}{2x+1}$$

to

$$9(2x+1)=7(x+2)$$

is called 'cross-multiplication'. In general, if

$$\frac{a}{b}=\frac{c}{d}$$

then

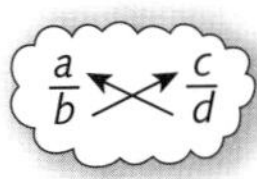

$$ad=bc$$

The remaining steps are similar to those used in the earlier parts of this example:

$$18x+9=7x+14 \qquad \text{(multiply out the brackets)}$$

$$11x+9=14 \qquad \text{(subtract } 7x \text{ from both sides)}$$

$$11x=5 \qquad \text{(subtract 9 from both sides)}$$

$$x=\frac{5}{11} \qquad \text{(divide both sides by 11)}$$

(e) The left-hand side of the final equation

$$\sqrt{\frac{2x}{x-6}}=2$$

is surrounded by a square root, which can easily be removed by squaring both sides to get

$$\frac{2x}{x-6}=4$$

The remaining steps are 'standard':

$$2x=4(x-6) \qquad \text{(multiply both sides by } x-6\text{)}$$

$$2x=4x-24 \qquad \text{(multiply out the brackets)}$$

$$-2x=-24 \qquad \text{(subtract } 4x \text{ from both sides)}$$

$$x=12 \qquad \text{(divide both sides by } -2\text{)}$$

Looking back over each part of the previous example, notice that there is a common strategy. In each case, the aim is to convert the given equation into one of the form

$$ax + b = c$$

which is the sort of equation that we can easily solve. If the original equation contains brackets then remove them by multiplying out. If the equation involves fractions then remove them by cross-multiplying.

Advice

If you have the time, it is always worth checking your answer by substituting your solution back into the original equation. For the last part of the above example, putting $x = 12$ into $\sqrt{\frac{2x}{x-6}}$ gives

$$\sqrt{\frac{2 \times 12}{12 - 6}} = \sqrt{\frac{24}{6}} = \sqrt{4} = 2 \quad ✓$$

Practice Problem

10 Solve each of the following equations. Leave your answer as a fraction, if necessary.

(a) $4x + 5 = 5x - 7$ **(b)** $3(3 - 2x) + 2(x - 1) = 10$

(c) $\frac{4}{x-1} = 5$ **(d)** $\frac{3}{x} = \frac{5}{x-1}$

Key Terms

Algebraic fraction Ratio of two expressions; $p(x)/q(x)$ where $p(x)$ and $q(x)$ are algebraic expressions such as $ax^2 + bx + c$ or $dx + e$.

Denominator The number (or expression) on the bottom of a fraction.

Distributive law (for multiplication over addition) The rule which states that $a(b + c) = ab + ac$, for any numbers, a, b and c.

Number line An infinite line on which the points represent real numbers by their (signed) distance from the origin.

Numerator The number (or expression) on the top of a fraction.

Practice Problems

11 Which of the following inequalities are true?

(a) $-2 < 1$ **(b)** $-6 > -4$ **(c)** $3 < 3$

(d) $3 \le 3$ **(e)** $-21 \ge -22$ **(f)** $4 < \sqrt{25}$

12 Simplify the following inequalities:

(a) $2x > x + 1$

(b) $7x + 3 \leq 9 + 5x$

(c) $x - 5 > 4x + 4$

(d) $x - 1 < 2x - 3$

13 Without using a calculator work out

(a) $(5 - 2)^2$ **(b)** $5^2 - 2^2$

Is it true in general that $(a - b)^2 = a^2 - b^2$?

14 Use your calculator to work out the following. Round your answer, if necessary, to 2 decimal places.

(a) $5.31 \times 8.47 - 1.01^2$ **(b)** $(8.34 + 2.27)/9.41$

(c) $9.53 - 3.21 + 4.02$ **(d)** $2.41 \times 0.09 - 1.67 \times 0.03$

(e) $45.76 - (2.55 + 15.83)$ **(f)** $(3.45 - 5.38)^2$

(g) $4.56(9.02 + 4.73)$ **(h)** $6.85/(2.59 + 0.28)$

15 Simplify the following expressions by collecting together like terms:

(a) $2x + 3y + 4x - y$ **(b)** $2x^2 - 5x + 9x^2 + 2x - 3$

(c) $5xy + 2x + 9yx$ **(d)** $7xyz + 3yx - 2zyx + yzx - xy$

(e) $2(5a + b) - 4b$ **(f)** $5(x - 4y) + 6(2x + 7y)$

(g) $5 - 3(p - 2)$ **(h)** $\frac{2x + 3}{4} + \frac{x - 1}{2}$

16 Multiply out the brackets.

(a) $7(x - y)$ **(b)** $(5x - 2y)z$ **(c)** $y + 2z - 2(x + 3y - z)$

(d) $(x - 5)(x - 2)$ **(e)** $x(x - y + 7)$ **(f)** $x(x + 1)(x + 2)$ **(g)** $(x - 1)(x + 1 - y)$

17 **(1)** Use the formula for the difference of two squares to factorize

(a) $x^2 - 4$ **(b)** $x^2 - y^2$ **(c)** $9x^2 - 100y^2$ **(d)** $a^2b^2 - 25$

(2) Use the formula for the difference of two squares to evaluate the following without using a calculator:

(a) $50\,563^2 - 49\,437^2$ **(b)** $90^2 - 89.99^2$

(c) $759^2 - 541^2$ **(d)** $123\,456\,789^2 - 123\,456\,788^2$

18 **(1)** Without using your calculator evaluate

(a) $\frac{4}{5} \times \frac{25}{28}$ **(b)** $\frac{2}{7} \times \frac{14}{25} \times \frac{30}{48}$ **(c)** $\frac{9}{16} \div \frac{3}{8}$ **(d)** $\frac{2}{5} \times \frac{1}{12} \div \frac{8}{25}$

(e) $\frac{10}{13} - \frac{2}{13}$ **(f)** $\frac{5}{9} + \frac{2}{3}$ **(g)** $2\frac{3}{5} + 1\frac{3}{7}$ **(h)** $5\frac{9}{10} - \frac{1}{2} + 1\frac{2}{5}$

(i) $3\frac{3}{4} \times 1\frac{3}{5}$ **(j)** $\frac{3}{5} \times \left(\frac{2}{3} + \frac{1}{2}\right)$ **(k)** $\frac{5}{6} \times \left(2\frac{1}{3} - 1\frac{2}{5}\right)$ **(l)** $\left(3\frac{1}{3} \div 2\frac{1}{6}\right) \div \frac{5}{13}$

(2) Confirm your answer to part (1) using a calculator.

➔

19 Find expressions for the following fractions:

(a) $\dfrac{x^2+6x}{x-2} \times \dfrac{x-2}{x}$ **(b)** $\dfrac{1}{x} \div \dfrac{1}{x+1}$ **(c)** $\dfrac{2}{xy} + \dfrac{3}{xy}$

(d) $\dfrac{x}{2} + \dfrac{x+1}{3}$ **(e)** $\dfrac{3}{x} + \dfrac{4}{x+1}$ **(f)** $\dfrac{3}{x} + \dfrac{5}{x^2}$

(g) $x - \dfrac{2}{x+1}$ **(h)** $\dfrac{5}{x(x+1)} - \dfrac{2}{x} + \dfrac{3}{x+1}$

20 Solve the following equations:

(a) $5(2x+1) = 3(x-2)$ **(b)** $5(x+2) + 4(2x-3) = 11$

(c) $5(1-x) = 4(10+x)$ **(d)** $3(3-2x) - 7(1-x) = 10$

(e) $9 - 5(2x-1) = 6$ **(f)** $\dfrac{3}{2x+1} = 2$

(g) $\dfrac{2}{x-1} = \dfrac{3}{5x+4}$ **(h)** $\dfrac{x}{2} + 3 = 7$

(i) $5 - \dfrac{x}{3} = 2$ **(j)** $\dfrac{5(x-3)}{2} = \dfrac{2(x-1)}{5}$

(k) $\sqrt{(2x-5)} = 3$ **(l)** $(x+3)(x-1) = (x+4)(x-3)$

(m) $(x+2)^2 + (2x-1)^2 = 5x(x+1)$ **(n)** $\dfrac{2x+7}{3} = \dfrac{x-4}{6} + \dfrac{1}{2}$

(o) $\sqrt{\dfrac{45}{2x-1}} = 3$ **(p)** $\dfrac{4}{x} - \dfrac{3}{4} = \dfrac{1}{4x}$

section 1.5

Transposition of formulae

Objectives

At the end of this section you should be able to:

- Manipulate formulae.
- Draw a flow chart representing a formula.
- Use a reverse flow chart to transpose a formula.
- Change the subject of a formula involving several letters.

Mathematical modelling involves the use of formulae to represent the relationship between economic variables. In microeconomics we have already seen how useful supply and demand formulae are. These provide a precise relationship between price and quantity. For example, the connection between price, P, and quantity, Q, might be modelled by

$$P = -4Q + 100$$

Given any value of Q it is trivial to deduce the corresponding value of P by merely replacing the symbol Q by a number. A value of $Q = 2$, say, gives

$$\begin{aligned} P &= -4 \times 2 + 100 \\ &= -8 + 100 \\ &= 92 \end{aligned}$$

On the other hand, given P, it is necessary to solve an equation to deduce Q. For example, when $P = 40$, the equation is

$$-4Q + 100 = 40$$

which can be solved as follows:

$$\begin{aligned} -4Q &= -60 \quad \text{(subtract 100 from both sides)} \\ Q &= 15 \quad \text{(divide both sides by } -4\text{)} \end{aligned}$$

This approach is reasonable when only one or two values of P are given. However, if we are given many values of P, it is clearly tedious and inefficient for us to solve the equation each time to find Q. The preferred approach is to ***transpose*** the formula for P. In other words, we rearrange the formula

$$P = \text{an expression involving } Q$$

into

$$Q = \text{an expression involving } P$$

Written this way round, the formula enables us to find Q by replacing P by a number. For the specific formula

$$-4Q + 100 = P$$

the steps are

$$-4Q = P - 100 \quad \text{(subtract 100 from both sides)}$$

$$Q = \frac{P - 100}{-4} \quad \text{(divide both sides by } -4\text{)}$$

Notice that

$$\frac{P - 100}{-4} = \frac{P}{-4} - \frac{100}{-4}$$
$$= -{}^1\!/\!_4P + 25$$

so the rearranged formula simplifies to

$$Q = -{}^1\!/\!_4P + 25$$

If we now wish to find Q when $P = 40$, we immediately get

$$Q = -{}^1\!/\!_4 \times 40 + 25$$
$$= -10 + 25$$
$$= 15$$

The important thing to notice about the algebra is that the individual steps are identical to those used previously for solving the equation

$$-4Q + 100 = 40$$

i.e. the operations are again

'subtract 100 from both sides'

followed by

'divide both sides by −4'

Example

Make x the subject of the formula

$$\frac{1}{7}x - 2 = y$$

Solution

If you needed to solve an equation such as $\frac{1}{7}x - 2 = 4$, say, you would first add 2 to both sides and then multiply both sides by 7. Performing the same operations to the general equation

$$\frac{1}{7}x - 2 = y$$

gives

$$\frac{1}{7}x = y + 2 \qquad \text{(add 2 to both sides)}$$

$$x = 7(y + 2) \qquad \text{(multiply both sides by 7)}$$

If desired, you can multiply out the brackets to give the alternative version:

$$x = 7y + 14$$

Practice Problem

1 (a) Solve the equation

$$\tfrac{1}{2}Q + 13 = 17$$

State clearly exactly what operation you have performed to both sides at each stage of your solution.

(b) By performing the same operations as part (a), rearrange the formula

$$\tfrac{1}{2}Q + 13 = P$$

into the form

$$Q = \text{an expression involving } P$$

(c) By substituting $P = 17$ into the formula derived in part (b), check that this agrees with your answer to part (a).

In general, there are two issues concerning formula transposition. Firstly, we need to decide what to do to both sides of the given formula and the order in which they should be performed. Secondly, we need to carry out these steps accurately. The first of these is often the more difficult. However, there is a logical strategy that can be used to help. To illustrate this, consider the task of making Q the subject of

$$P = \tfrac{1}{3}Q + 5$$

that is, of rearranging this formula into the form

$$Q = \text{an expression involving } P$$

Imagine starting with a value of Q and using a calculator to work out P from

$$P = \tfrac{1}{3}Q + 5$$

The diagram overleaf shows that two operations are required and indicates the order in which they must be done. This diagram is called a *flow chart*.

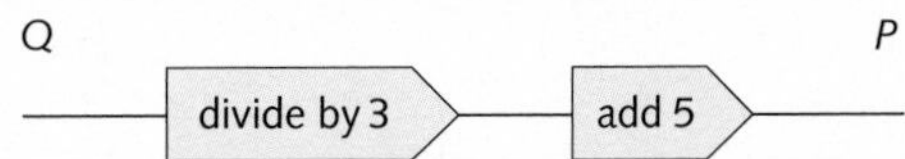

To go backwards from P to Q we need to undo these operations. Now the reverse of 'divide by 3' is 'multiply by 3' and the reverse of 'add 5' is 'subtract 5', so the operations needed to transpose the formula are as follows:

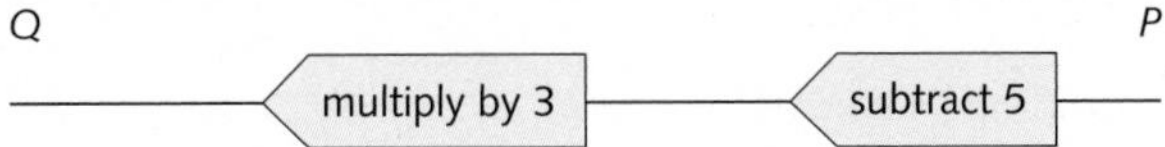

This diagram is called a *reverse flow chart*. The process is similar to that of unwrapping a parcel (or peeling an onion); you start by unwrapping the outer layer first and work inwards. If we now actually perform these steps in the order specified by the reverse flow chart, we get

$$\tfrac{1}{3}Q + 5 = P$$

$$\tfrac{1}{3}Q = P - 5 \quad \text{(subtract 5 from both sides)}$$

$$Q = 3(P - 5) \quad \text{(multiply both sides by 3)}$$

The rearranged formula can be simplified by multiplying out the brackets to give

$$Q = 3P - 15$$

Incidentally, if you prefer, you can actually use the reverse flow chart itself to perform the algebra for you. All you have to do is to pass the letter P through the reverse flow chart. Working from right to left gives

$Q = 3(P - 5)$ $P - 5$ P

multiply by 3 —— subtract 5

Notice that by taking P as the input to the box 'subtract 5' gives the output $P - 5$, and if the whole of this is taken as the input to the box 'multiply by 3', the final output is the answer, $3(P - 5)$. Hence

$$Q = 3(P - 5)$$

Example

Make x the subject of

(a) $y = \sqrt{\dfrac{x}{5}}$ **(b)** $y = \dfrac{4}{2x + 1}$

Solution

(a) To go from x to y the operations are

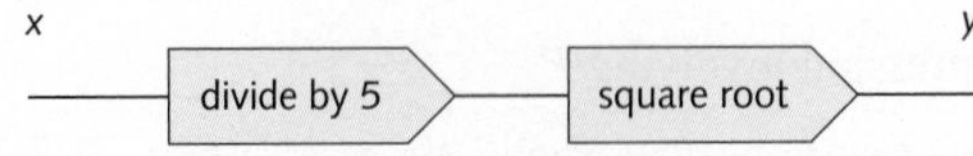

so the steps needed to transpose the formula are

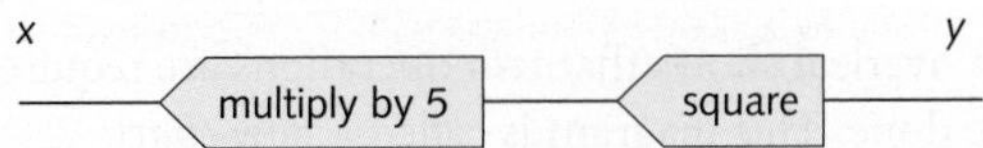

The algebraic details are as follows:

$$\sqrt{\frac{x}{5}} = y$$

$$\frac{x}{5} = y^2 \quad \text{(square both sides)}$$

$$x = 5y^2 \quad \text{(multiply both sides by 5)}$$

Hence the transposed formula is

$$x = 5y^2$$

Alternatively, if you prefer, the reverse flow chart can be used directly to obtain

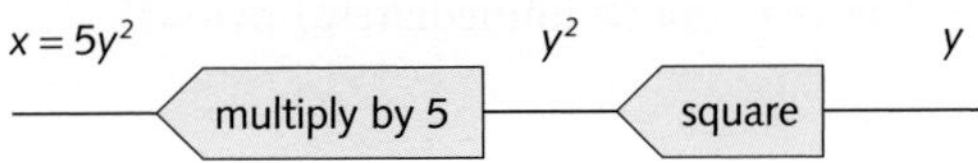

Hence

$$x = 5y^2$$

(b) The forwards flow chart is

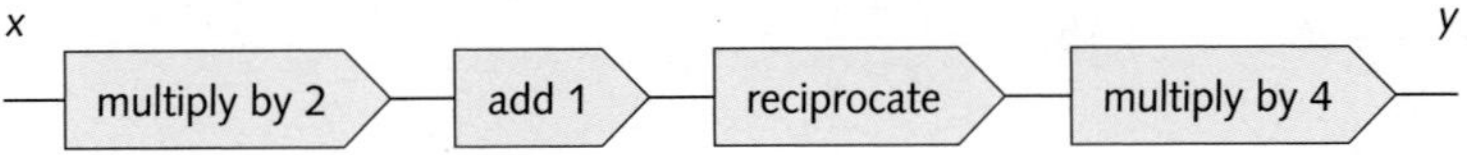

so the reverse flow chart is

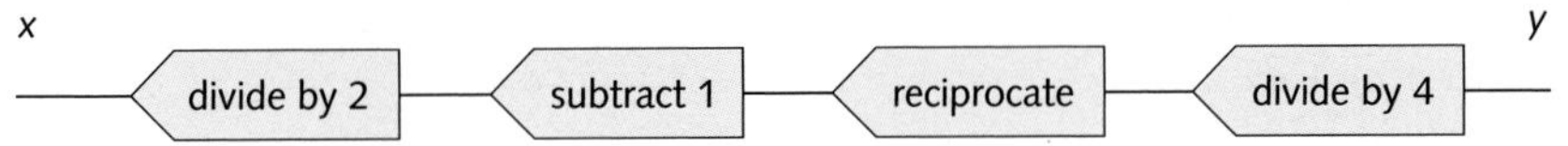

The algebraic details are as follows:

$$\frac{4}{2x+1} = y$$

$$\frac{1}{2x+1} = \frac{y}{4} \quad \text{(divide both sides by 4)}$$

$$2x + 1 = \frac{4}{y} \quad \text{(reciprocate both sides)}$$

$$2x = \frac{4}{y} - 1 \quad \text{(subtract 1 from both sides)}$$

$$x = \frac{1}{2}\left(\frac{4}{y} - 1\right) \quad \text{(divide both sides by 2)}$$

which can be simplified, by multiplying out the brackets, to give

$$x = \frac{2}{y} - \frac{1}{2}$$

Again, the reverse flow chart can be used directly to obtain

$x = \frac{1}{2}(\frac{4}{y} - 1)$ ← divide by 2 ← $\frac{4}{y} - 1$ ← subtract 1 ← $\frac{4}{y}$ ← reciprocate ← $\frac{y}{4}$ ← divide by 4 ← y

Practice Problem

2 Use flow charts to make x the subject of the following formulae:

(a) $y = 6x^2$ **(b)** $y = \frac{1}{7x - 1}$

The following example contains two difficult instances of transposition. In both cases the letter x appears more than once on the right-hand side. If this happens, the technique based on flow charts cannot be used. However, it may still be possible to perform the manipulation even if some of the steps may not be immediately obvious.

Example

Transpose the following equations to express x in terms of y:

(a) $ax = bx + cy + d$ **(b)** $y = \frac{x + 1}{x - 2}$

Solution

(a) In the equation

$$ax = bx + cy + d$$

there are terms involving x on both sides and since we are hoping to rearrange this into the form

$$x = \text{an expression involving } y$$

it makes sense to collect the x's on the left-hand side. To do this we subtract bx from both sides to get

$$ax - bx = cy + d$$

Notice that x is a common factor of the left-hand side, so the distributive law can be applied 'in reverse' to take the x outside the brackets: that is,

$$(a - b)x = cy + d$$

Finally, both sides are divided by $a - b$ to get

$$x = \frac{cy + d}{a - b}$$

which is of the desired form.

(b) It is difficult to see where to begin with the equation

$$y = \frac{x + 1}{x - 2}$$

because there is an x in both the numerator and the denominator. Indeed, the thing that is preventing us getting started is precisely the fact that the expression is a fraction. We can, however, remove the fraction simply by multiplying both sides by the denominator to get

$$(x - 2)y = x + 1$$

and if we multiply out the brackets then

$$xy - 2y = x + 1$$

We want to rearrange this into the form

x = an expression involving y

so we collect the x's on the left-hand side and put everything else on to the right-hand side. To do this we first add $2y$ to both sides to get

$$xy = x + 1 + 2y$$

and then subtract x from both sides to get

$$xy - x = 1 + 2y$$

The distributive law can now be applied 'in reverse' to take out the common factor of x: that is,

$$(y - 1)x = 1 + 2y$$

Finally, dividing through by $y - 1$ gives

$$x = \frac{1 + 2y}{y - 1}$$

Advice

This example contains some of the hardest algebraic manipulation seen so far in this book. I hope that you managed to follow the individual steps. However, it all might appear as if we have 'pulled rabbits out of hats'. You may feel that, if left on your own, you are never going to be able to decide what to do at each stage. Unfortunately there is no watertight strategy that always works, although the following five-point plan is worth considering if you get stuck.

To transpose a given equation of the form

y = an expression involving x

into an equation of the form

x = an expression involving y

you proceed as follows:

Step 1 Remove fractions.
Step 2 Multiply out the brackets.
Step 3 Collect all of the x's on to the left-hand side.
Step 4 Take out a factor of x.
Step 5 Divide by the coefficient of x.

You might find it helpful to look back at the previous example in the light of this strategy. In part (b) it is easy to identify each of the five steps. Part (a) also used this strategy, starting with the third step.

Example

Make x the subject of

$$y = \sqrt{\frac{ax + b}{cx + d}}$$

Solution

In this formula there is a square root symbol surrounding the right-hand side. This can be removed by squaring both sides to get

$$y^2 = \frac{ax + b}{cx + d}$$

We now apply the five-step strategy:

Step 1 $(cx + d)y^2 = ax + b$

Step 2 $cxy^2 + dy^2 = ax + b$

Step 3 $cxy^2 - ax = b - dy^2$

Step 4 $(cy^2 - a)x = b - dy^2$

Step 5 $x = \dfrac{b - dy^2}{cy^2 - a}$

Practice Problem

3 Transpose the following formulae to express x in terms of y:

(a) $x - ay = cx + y$

(b) $y = \dfrac{x - 2}{x + 4}$

Key Terms

Flow chart A diagram consisting of boxes of instructions indicating a sequence of operations and their order.

Reverse flow chart A flow chart indicating the inverse of the original sequence of operations in reverse order.

Transpose a formula The rearrangement of a formula to make one of the other letters the subject.

Practice Problems

4 Make Q the subject of

$$P = 2Q + 8$$

Hence find the value of Q when $P = 52$.

5 Write down the formula representing each of the following flow charts

(a) x → y

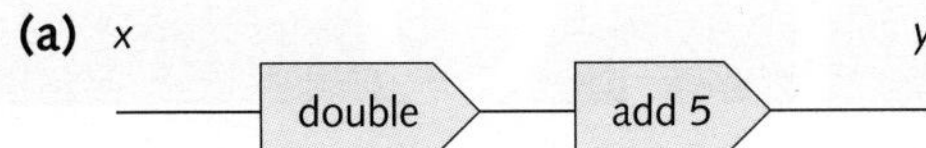

(b) x → add 5 → double → y

(c) x → square → reciprocate → multiply by 5 → y

(d) x → add 4 → square → multiply by 2 → subtract 3 → y

6 Draw flow charts for each of the following formulae:

(a) $y = 5x + 3$ **(b)** $y = 5(x + 3)$ **(c)** $y = 4x^2 - 6$ **(d)** $y = \dfrac{4}{x^2 + 8}$

7 Make x the subject of each of the following formulae:

(a) $y = 9x - 6$ **(b)** $y = (x + 4)/3$ **(c)** $y = \dfrac{x}{2}$

(d) $y = \dfrac{x}{5} + 8$ **(e)** $y = \dfrac{1}{x + 2}$ **(f)** $y = \dfrac{4}{3x - 7}$

8 Transpose the formulae:

(a) $Q = aP + b$ to express P in terms of Q

(b) $Y = aY + b + I$ to express Y in terms of I

(c) $Q = \dfrac{1}{aP + b}$ to express P in terms of Q

(d) $V = \dfrac{5t + 1}{t - 1}$ to express t in terms of V

9 Make x the subject of the following formulae:

(a) $\dfrac{a}{x} + b = \dfrac{c}{x}$ **(b)** $a - x = \dfrac{b + x}{a}$ **(c)** $e + \sqrt{x + f} = g$

(d) $a\sqrt{\left(\dfrac{x - n}{m}\right)} = \dfrac{a^2}{b}$ **(e)** $\dfrac{\sqrt{x - m}}{n} = \dfrac{1}{m}$ **(f)** $\dfrac{\sqrt{x} + a}{\sqrt{x} - b} = \dfrac{b}{a}$

section 1.6

National income determination

Objectives

At the end of this section you should be able to:

- Identify and sketch linear consumption functions.
- Identify and sketch linear savings functions.
- Set up simple macroeconomic models.
- Calculate equilibrium national income.
- Analyse IS and LM schedules.

Macroeconomics is concerned with the analysis of economic theory and policy at a national level. In this section we focus on one particular aspect known as national income determination. We describe how to set up simple models of the national economy which enable equilibrium levels of income to be calculated. Initially we assume that the economy is divided into two sectors, households and firms. Firms use resources such as land, capital, labour and raw materials to produce goods and services. These resources are known as ***factors of production*** and are taken to belong to households. ***National income*** represents the flow of income from firms to households given as payment for these factors. Households can then spend this money in one of two ways. Income can be used for the consumption of goods produced by firms or it can be put into savings. Consumption, C, and savings, S, are therefore functions of income, Y: that is,

$$C = f(Y)$$

$$S = g(Y)$$

Figure 1.23

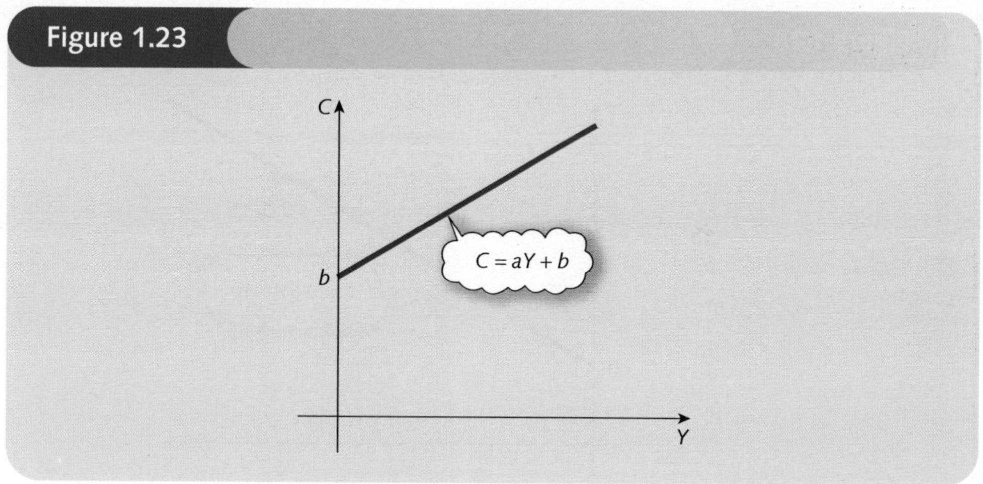

for some appropriate consumption function, f, and savings function, g. Moreover, C and S are normally expected to increase as income rises, so f and g are both increasing functions.

We begin by analysing the ***consumption function***. As usual we need to quantify the precise relationship between C and Y. If this relationship is linear then a graph of a typical consumption function is shown in Figure 1.23. It is clear from this graph that if

$$C = aY + b$$

then $a > 0$ and $b > 0$. The intercept b is the level of consumption when there is no income (that is, when $Y = 0$) and is known as ***autonomous consumption***. The slope, a, is the change in C brought about by a 1 unit increase in Y and is known as the ***marginal propensity to consume*** (MPC). As previously noted, income is used up in consumption and savings so that

$$Y = C + S$$

It follows that only a proportion of the 1 unit increase in income is consumed; the rest goes into savings. Hence the slope, a, is generally smaller than 1: that is, $a < 1$. It is standard practice in mathematics to collapse the two separate inequalities $a > 0$ and $a < 1$ into the single inequality

$$0 < a < 1$$

The relation

$$Y = C + S$$

enables the precise form of the savings function to be determined from any given consumption function. This is illustrated in the following example.

Example

Sketch a graph of the consumption function

$$C = 0.6Y + 10$$

Determine the corresponding savings function and sketch its graph.

Solution

The graph of the consumption function

$$C = 0.6Y + 10$$

Figure 1.24

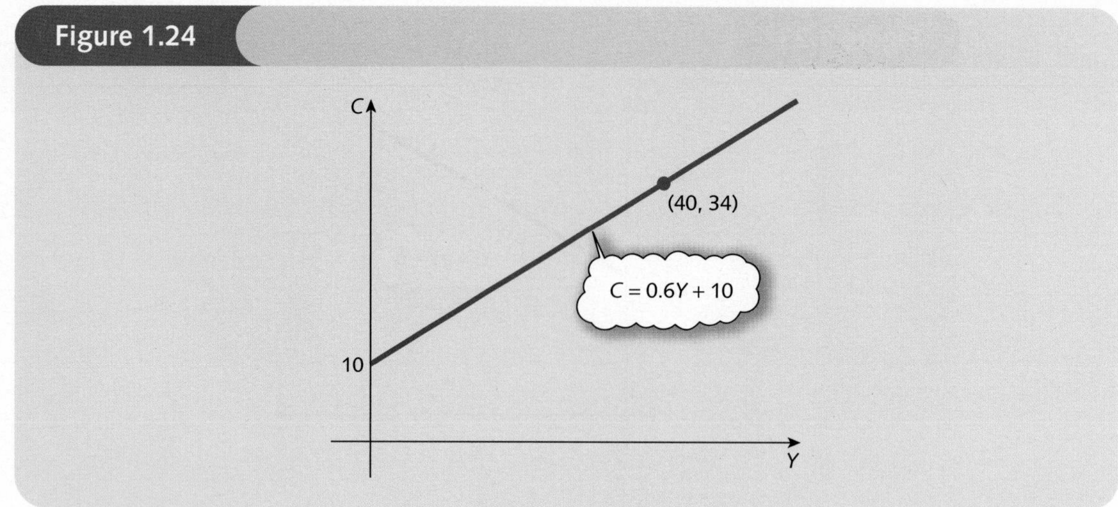

has intercept 10 and slope 0.6. It passes through (0, 10). For a second point, let us choose $Y = 40$, which gives $C = 34$. Hence the line also passes through (40, 34). The consumption function is sketched in Figure 1.24.

To find the savings function we use the relation

$$Y = C + S$$

which gives

$$\begin{aligned} S &= Y - C && \text{(subtract } C \text{ from both sides)} \\ &= Y - (0.6Y + 10) && \text{(substitute } C\text{)} \\ &= Y - 0.6Y - 10 && \text{(multiply out the brackets)} \\ &= 0.4Y - 10 && \text{(collect terms)} \end{aligned}$$

The savings function is also linear. Its graph has intercept −10 and slope 0.4. This is sketched in Figure 1.25 using the fact that it passes through (0, −10) and (25, 0).

Figure 1.25

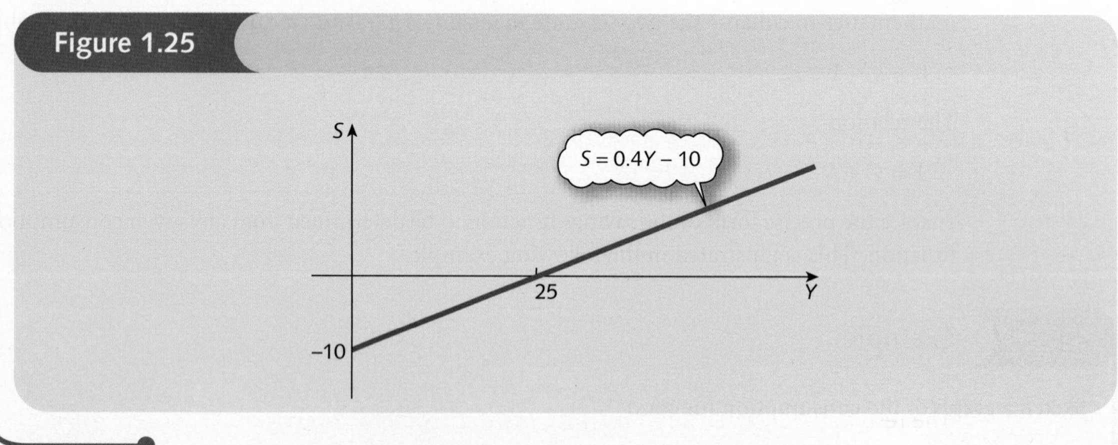

Practice Problem

1 Determine the savings function that corresponds to the consumption function

$$C = 0.8Y + 25$$

Figure 1.26

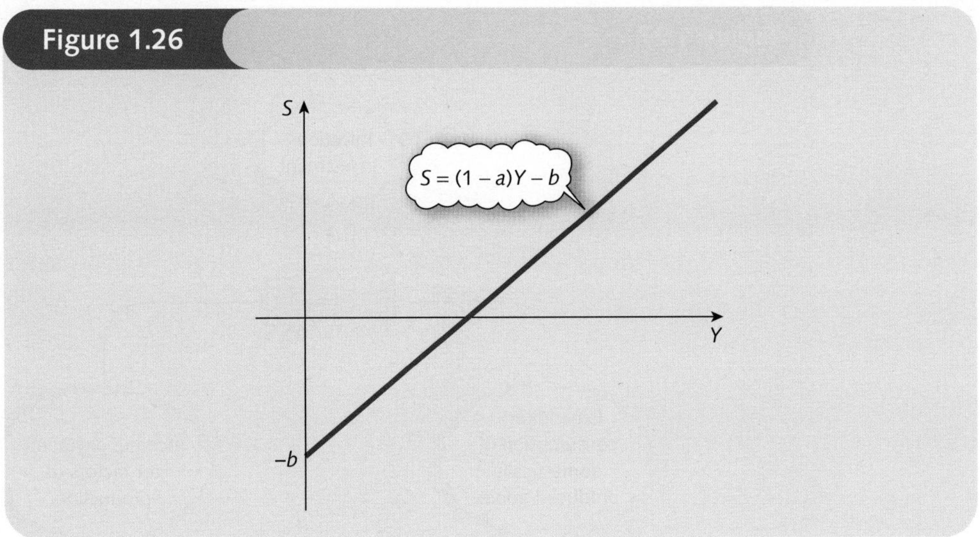

For the general consumption function

$$C = aY + b$$

we have

$$\begin{aligned} S &= Y - C \\ &= Y - (aY + b) \quad \text{(substitute } C\text{)} \\ &= Y - aY - b \quad \text{(multiply out the brackets)} \\ &= (1 - a)Y - b \quad \text{(take out a common factor of } Y\text{)} \end{aligned}$$

The slope of the savings function is called the ***marginal propensity to save*** (MPS) and is given by $1 - a$: that is,

$$\text{MPS} = 1 - a = 1 - \text{MPC}$$

Moreover, since $a < 1$ we see that the slope, $1 - a$, is positive. Figure 1.26 shows the graph of this savings function. One interesting feature, which contrasts with other economic functions considered so far, is that it is allowed to take negative values. In particular, note that ***autonomous savings*** (that is, the value of S when $Y = 0$) are equal to $-b$, which is negative because $b > 0$. This is to be expected because whenever consumption exceeds income, households must finance the excess expenditure by withdrawing savings.

Advice

The result, MPC + MPS = 1, is always true, even if the consumption function is non-linear. A proof of this generalization can be found on page 274.

The simplest model of the national economy is illustrated in Figure 1.27, which shows the circular flow of income and expenditure. This is fairly crude, since it fails to take into account government activity or foreign trade. In this diagram ***investment***, I, is an injection into the circular flow in the form of spending on capital goods.

Figure 1.27

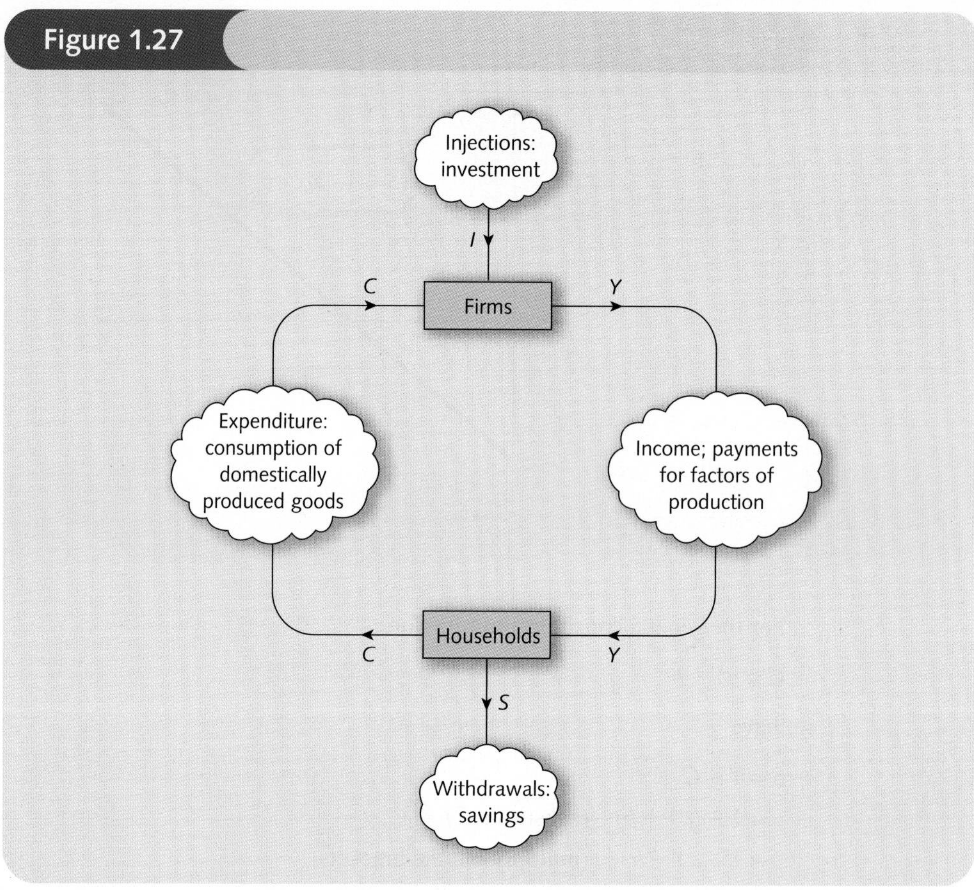

Let us examine this more closely and represent the diagrammatic information in symbols. Consider first the box labelled 'Households'. The flow of money entering this box is Y and the flow leaving it is $C + S$. Hence we have the familiar relation

$$Y = C + S$$

For the box labelled 'Firms' the flow entering it is $C + I$ and the flow leaving it is Y, so

$$Y = C + I$$

Suppose that the level of investment that firms plan to inject into the economy is known to be some fixed value, I^*. If the economy is in equilibrium, the flow of income and expenditure balance so that

$$Y = C + I^*$$

From the assumption that the consumption function is

$$C = aY + b$$

for given values of a and b these two equations represent a pair of simultaneous equations for the two unknowns Y and C. In these circumstances C and Y can be regarded as endogenous variables, since their precise values are determined within the model, whereas I^* is fixed outside the model and is exogenous.

Example

Find the equilibrium level of income and consumption if the consumption function is

$$C = 0.6Y + 10$$

and planned investment $I = 12$.

Solution

We know that

$$Y = C + I \quad \text{(from theory)}$$
$$C = 0.6Y + 10 \quad \text{(given in problem)}$$
$$I = 12 \quad \text{(given in problem)}$$

If the value of I is substituted into the first equation then

$$Y = C + 12$$

The expression for C can also be substituted to give

$$Y = 0.6Y + 10 + 12$$
$$Y = 0.6Y + 22$$
$$0.4Y = 22 \quad \text{(subtract } 0.6Y \text{ from both sides)}$$
$$Y = 55 \quad \text{(divide both sides by 0.4)}$$

The corresponding value of C can be deduced by putting this level of income into the consumption function to get

$$C = 0.6(55) + 10 = 43$$

The equilibrium income can also be found graphically by plotting expenditure against income. In this example the aggregate expenditure, $C + I$, is given by $0.6Y + 22$. This is sketched in Figure 1.28 using the fact that it passes through (0, 22) and (80, 70). Also sketched is the '45° line', so called because it makes an angle of 45° with the horizontal. This line passes through the points (0, 0), (1, 1), . . . , (50, 50) and so on. In other words, at any point on this line expenditure and income are in balance. The equilibrium income can therefore be found by inspecting the point of intersection of this line and the aggregate expenditure line, $C + I$. From Figure 1.28 this occurs when $Y = 55$, which is in agreement with the calculated value.

Practice Problem

2 Find the equilibrium level of income if the consumption function is

$$C = 0.8Y + 25$$

and planned investment $I = 17$. Calculate the new equilibrium income if planned investment rises by 1 unit.

To make the model more realistic let us now include ***government expenditure***, G, and ***taxation***, T, in the model. The injections box in Figure 1.27 now includes government expenditure in addition to investment, so

Figure 1.28

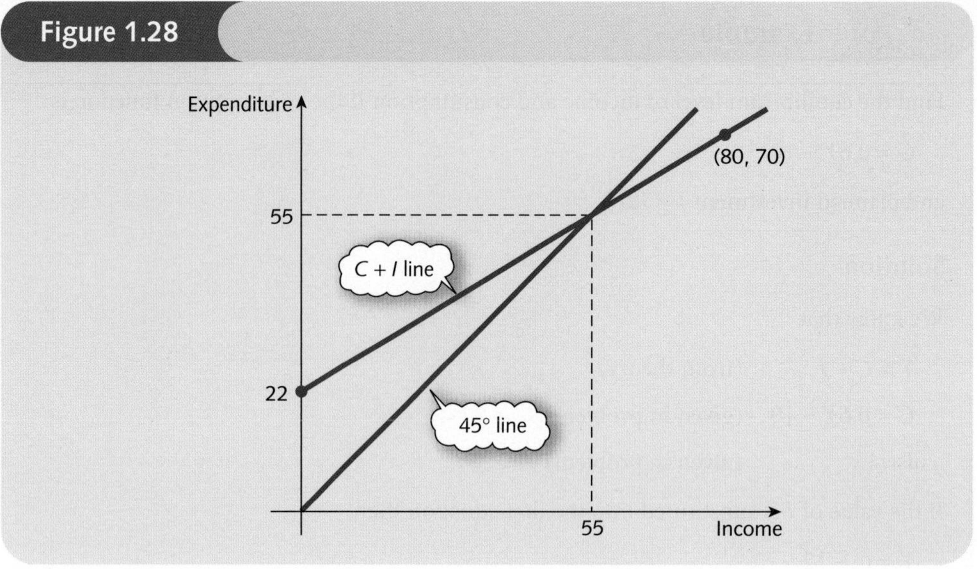

$$Y = C + I + G$$

We assume that planned government expenditure and planned investment are autonomous with fixed values G^* and I^* respectively, so that in equilibrium

$$Y = C + I^* + G^*$$

The withdrawals box in Figure 1.27 now includes taxation. This means that the income that households have to spend on consumer goods is no longer Y but rather $Y - T$ (income less tax), which is called ***disposable income***, Y_d. Hence

$$C = aY_d + b$$

with

$$Y_d = Y - T$$

In practice, the tax will either be autonomous ($T = T^*$ for some lump sum T^*) or be a proportion of national income ($T = tY$ for some proportion t), or a combination of both ($T = tY + T^*$).

Example

Given that

$$G = 20$$

$$I = 35$$

$$C = 0.9Y_d + 70$$

$$T = 0.2Y + 25$$

calculate the equilibrium level of national income.

Solution

At first sight this problem looks rather forbidding, particularly since there are so many variables. However, all we have to do is to write down the relevant equations and to substitute systematically one equation into another until only Y is left.

We know that

$Y = C + I + G$ (from theory) (1)

$G = 20$ (given in problem) (2)

$I = 35$ (given in problem) (3)

$C = 0.9Y_d + 70$ (given in problem) (4)

$T = 0.2Y + 25$ (given in problem) (5)

$Y_d = Y - T$ (from theory) (6)

This represents a system of six equations in six unknowns. The obvious thing to do is to put the fixed values of G and I into equation (1) to get

$$Y = C + 35 + 20 = C + 55 \tag{7}$$

This has at least removed G and I, so there are only three more variables (C, Y_d and T) left to eliminate. We can remove T by substituting equation (5) into (6) to get

$$\begin{aligned} Y_d &= Y - (0.2Y + 25) \\ &= Y - 0.2Y - 25 \\ &= 0.8Y - 25 \end{aligned} \tag{8}$$

and then remove Y_d by substituting equation (8) into (4) to get

$$\begin{aligned} C &= 0.9(0.8Y - 25) + 70 \\ &= 0.72Y - 22.5 + 70 \\ &= 0.72Y + 47.5 \end{aligned} \tag{9}$$

We can eliminate C by substituting equation (9) into (7) to get

$$\begin{aligned} Y &= C + 55 \\ &= 0.72Y + 47.5 + 55 \\ &= 0.72Y + 102.5 \end{aligned}$$

Finally, solving for Y gives

$0.28Y = 102.5$ (subtract $0.72Y$ from both sides)

$Y = 366$ (divide both sides by 0.28)

Practice Problem

3 Given that

$$\begin{aligned} G &= 40 \\ I &= 55 \\ C &= 0.8Y_d + 25 \\ T &= 0.1Y + 10 \end{aligned}$$

calculate the equilibrium level of national income.

To conclude this section we return to the simple two-sector model:

$$Y = C + I$$

$$C = aY + b$$

Previously, the investment, I, was taken to be constant. It is more realistic to assume that planned investment depends on the rate of interest, r. As the interest rate rises, so investment falls and we have a relationship

$$I = cr + d$$

where $c < 0$ and $d > 0$. Unfortunately, this model consists of three equations in the four unknowns Y, C, I and r, so we cannot expect it to determine national income uniquely. The best we can do is to eliminate C and I, say, and to set up an equation relating Y and r. This is most easily understood by an example. Suppose that

$$C = 0.8Y + 100$$

$$I = -20r + 1000$$

We know that the commodity market is in equilibrium when

$$Y = C + I$$

Substitution of the given expressions for C and I into this equation gives

$$\begin{aligned} Y &= (0.8Y + 100) + (-20r + 1000) \\ &= 0.8Y - 20r + 1100 \end{aligned}$$

which rearranges as

$$0.2Y + 20r = 1100$$

This equation, relating national income, Y, and interest rate, r, is called the ***IS schedule***.

We obviously need some additional information before we can pin down the values of Y and r. This can be done by investigating the equilibrium of the money market. The money market is said to be in equilibrium when the supply of money, M_S, matches the demand for money, M_D: that is, when

$$M_S = M_D$$

There are many ways of measuring the ***money supply***. In simple terms it can be thought of as consisting of the notes and coins in circulation, together with money held in bank deposits. The level of M_S is assumed to be controlled by the central bank and is taken to be autonomous, so that

$$M_S = M_S^*$$

for some fixed value M_S^*.

The demand for money comes from three sources: transactions, precautions and speculations. The ***transactions demand*** is used for the daily exchange of goods and services, whereas the ***precautionary demand*** is used to fund any emergencies requiring unforeseen expenditure. Both are assumed to be proportional to national income. Consequently, we lump these together and write

$$L_1 = k_1 Y$$

where L_1 denotes the aggregate transaction–precautionary demand and k_1 is a positive constant. The ***speculative demand*** for money is used as a reserve fund in case individuals or firms decide to invest in alternative assets such as government bonds. In Chapter 3 we show that, as interest rates rise, speculative demand falls. We model this by writing

$$L_2 = k_2 r + k_3$$

where L_2 denotes speculative demand, k_2 is a negative constant and k_3 is a positive constant. The total demand, M_D, is the sum of the transaction–precautionary demand and speculative demand: that is,

$$\begin{aligned} M_D &= L_1 + L_2 \\ &= k_1 Y + k_2 r + k_3 \end{aligned}$$

If the money market is in equilibrium then

$$M_S = M_D$$

that is,

$$M_S^* = k_1 Y + k_2 r + k_3$$

This equation, relating national income, Y, and interest rate, r, is called the ***LM schedule***. If we assume that equilibrium exists in both the commodity and money markets then the IS and LM schedules provide a system of two equations in two unknowns, Y and r. These can easily be solved either by elimination or by graphical methods.

Example

Determine the equilibrium income and interest rate given the following information about the commodity market

$$C = 0.8Y + 100$$
$$I = -20r + 1000$$

and the money market

$$M_S = 2375$$
$$L_1 = 0.1Y$$
$$L_2 = -25r + 2000$$

What effect would a decrease in the money supply have on the equilibrium levels of Y and r?

Solution

The IS schedule for these particular consumption and investment functions has already been obtained in the preceding text. It was shown that the commodity market is in equilibrium when

$$0.2Y + 20r = 1100 \quad (1)$$

➔

For the money market we see that the money supply is

$$M_S = 2375$$

and that the total demand for money (that is, the sum of the transaction–precautionary demand, L_1, and the speculative demand, L_2) is

$$M_D = L_1 + L_2 = 0.1Y - 25r + 2000$$

The money market is in equilibrium when

$$M_S = M_D$$

that is,

$$2375 = 0.1Y - 25r + 2000$$

The LM schedule is therefore given by

$$0.1Y - 25r = 375 \qquad (2)$$

Equations (1) and (2) constitute a system of two equations for the two unknowns Y and r. The steps described in Section 1.2 can be used to solve this system:

Step 1

Double equation (2) and subtract from equation (1) to get

$$\begin{array}{r} 0.2Y + 20r = 1100 \\ 0.2Y - 50r = 750 \quad - \\ \hline 70r = 350 \end{array} \qquad (3)$$

Step 2

Divide both sides of equation (3) by 70 to get

$$r = 5$$

Step 3

Substitute $r = 5$ into equation (1) to get

$$0.2Y + 100 = 1100$$

$$0.2Y = 1000 \quad \text{(subtract 100 from both sides)}$$

$$Y = 5000 \quad \text{(divide both sides by 0.2)}$$

Step 4

As a check, equation (2) gives

$$0.1(5000) - 25(5) = 375 \quad ✓$$

The equilibrium levels of Y and r are therefore 5000 and 5 respectively.

To investigate what happens to Y and r as the money supply falls, we could just take a smaller value of M_S such as 2300 and repeat the calculations. However, it is more instructive to perform the investigation graphically. Figure 1.29 shows the IS and LM curves plotted on the same diagram with r on the horizontal axis and Y on the vertical axis. These lines intersect at (5, 5000), confirming the equilibrium levels of interest rate and income obtained by calculation. Any change in the money supply will obviously have no effect on the IS curve. On the other hand, a change in the money supply does affect the LM curve. To see this, let us return to the general LM schedule

$$k_1Y + k_2r + k_3 = M_S^*$$

and transpose it to express Y in terms of r:

$$k_1Y = -k_2r - k_3 + M_S^* \qquad \text{(subtract } k_2r + k_3 \text{ from both sides)}$$

$$Y = \left(\frac{-k_2}{k_1}\right)r + \frac{-k_3 + M_S^*}{k_1} \qquad \text{(divide both sides by } k_1\text{)}$$

Expressed in this form, we see that the LM schedule has slope $-k_2/k_1$ and intercept $(-k_3 + M_S^*)/k_1$.

Any decrease in M_S^* therefore decreases the intercept (but not the slope) and the LM curve shifts downwards. This is indicated by the dashed line in Figure 1.29. The point of intersection shifts both downwards and to the right. We deduce that, as the money supply falls, interest rates rise and national income decreases (assuming that both the commodity and money markets remain in equilibrium).

Figure 1.29

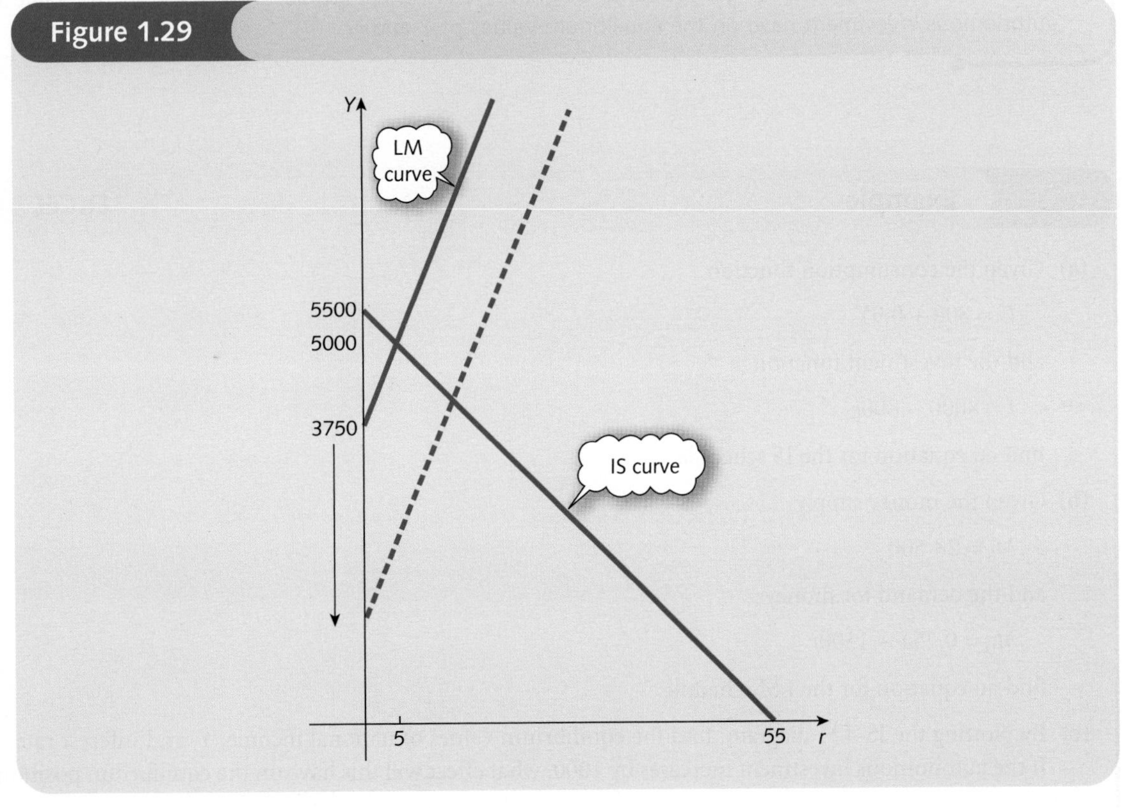

Advice

It is possible to produce general formulae for the equilibrium level of income in terms of various parameters used to specify the model. As you might expect, the algebra is a little harder but it does allow for a more general investigation into the effects of varying these parameters. We will return to this in Section 5.3.

Practice Problem

4 Determine the equilibrium income, *Y*, and interest rate, *r*, given the following information about the commodity market

$$C = 0.7Y + 85$$

$$I = -50r + 1200$$

and the money market

$$M_S = 500$$

$$L_1 = 0.2Y$$

$$L_2 = -40r + 230$$

Sketch the IS and LM curves on the same diagram. What effect would an increase in the value of autonomous investment have on the equilibrium values of *Y* and *r*?

Example

EXCEL

(a) Given the consumption function

$$C = 800 + 0.9Y$$

and the investment function

$$I = 8000 - 800r$$

find an equation for the IS schedule.

(b) Given the money supply

$$M_S = 28\ 500$$

and the demand for money

$$M_D = 0.75Y - 1500r$$

find an equation for the LM schedule.

(c) By plotting the IS–LM diagram, find the equilibrium values of national income, *Y*, and interest rate, *r*. If the autonomous investment increases by 1000, what effect will this have on the equilibrium position?

Solution

(a) The IS schedule is given by an equation relating national income, *Y*, and interest rate, *r*.

In equilibrium, $Y = C + I$. By substituting the equations given in (a) into this equilibrium equation, we eliminate *C* and *I*, giving

$$Y = 800 + 0.9Y + 8000 - 800r$$

$$0.1Y = 8800 - 800r \quad \text{(subtract } 0.9Y \text{ from both sides)}$$

$$Y = 88\ 000 - 8000r \quad \text{(divide both sides by 0.1)}$$

(b) The LM schedule is also given by an equation relating *Y* and *r*, but this time, it is derived from the equilibrium of the money markets: that is, when $M_S = M_D$. Substituting the equations given in (b) into this equilibrium equation gives

$$0.75Y - 1500r = 28\ 500$$

$$0.75Y = 28\ 500 + 1500r \quad \text{(add } 1500r \text{ to both sides)}$$

$$Y = 38\ 000 + 2000r \quad \text{(divide both sides by 0.75)}$$

(c) To find the equilibrium position, we plot these two lines on a graph using Excel in the usual way. We need to choose values for r and then work out corresponding values for Y. It is most likely that r will lie somewhere between 0 and 10, so values of r are tabulated between 0 and 10, going up in steps of 2. We type the formula

=88000–8000*A2

in cell B2 and type

=38000+2000*A2

in cell C2. The values of Y are then generated by clicking and dragging down the columns. Figure 1.30 shows the completed Excel screen.

Placing the cursor at the point of intersection tells us that the lines cross when

$$r = 5\% \quad \text{and} \quad Y = 48\ 000$$

If the autonomous investment increases by 1000, the equation for the IS schedule will change, as the equation for investment now becomes

$$I = 9000 - 800r$$

giving

$$Y = 98\ 000 - 8000r$$

The new IS schedule can be plotted on the same graph by adding a column of figures into the spreadsheet, as shown in Figure 1.31.

Notice that the point of intersection has shifted both upwards and to the right. The equilibrium position has now changed, resulting in a rise in interest rates to 6% and an increase in income to 50 000.

Figure 1.30

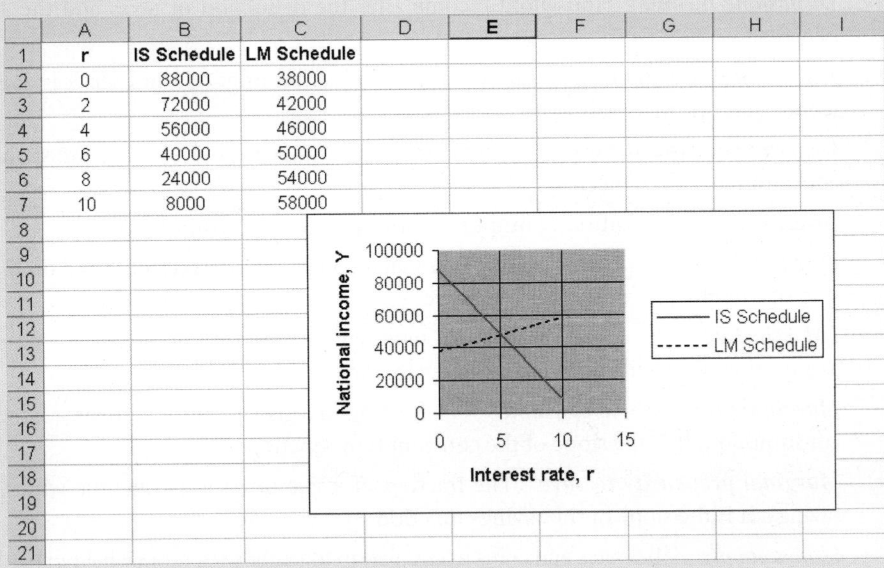

	A	B	C
1	r	IS Schedule	LM Schedule
2	0	88000	38000
3	2	72000	42000
4	4	56000	46000
5	6	40000	50000
6	8	24000	54000
7	10	8000	58000

Figure 1.31

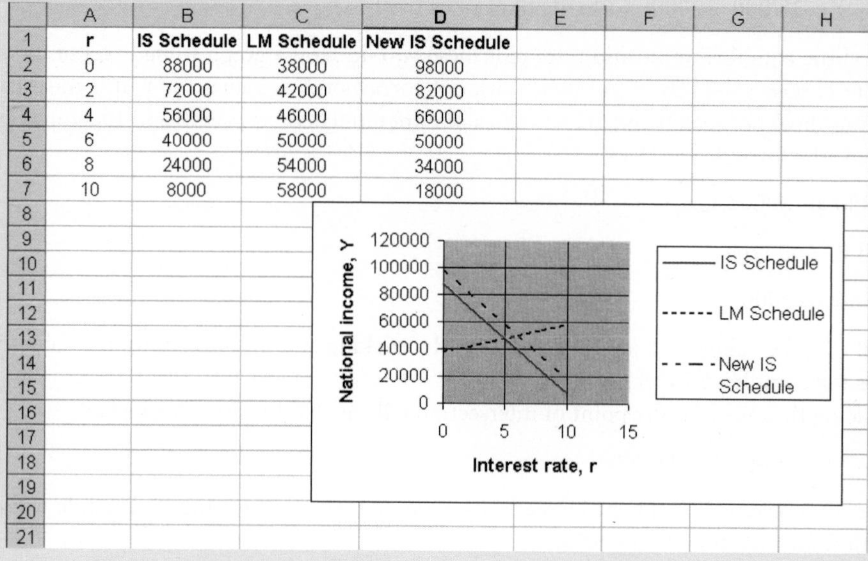

	A	B	C	D	E	F	G	H
1	r	IS Schedule	LM Schedule	New IS Schedule				
2	0	88000	38000	98000				
3	2	72000	42000	82000				
4	4	56000	46000	66000				
5	6	40000	50000	50000				
6	8	24000	54000	34000				
7	10	8000	58000	18000				

Key Terms

Autonomous consumption The level of consumption when there is no income.

Autonomous savings The withdrawals from savings when there is no income.

Consumption The flow of money from households to firms as payment for goods and services.

Consumption function The relationship between national income and consumption.

Disposable income Household income after the deduction of taxes and the addition of benefits.

Factors of production The inputs to the production of goods and services: land, capital, labour and raw materials.

Government expenditure The total amount of money spent by government on defence, education, health, police, etc.

Investment The creation of output not for immediate consumption.

IS schedule The equation relating national income and interest rate based on the assumption of equilibrium in the goods market.

LM schedule The equation relating national income and interest rate based on the assumption of equilibrium in the money market.

Marginal propensity to consume The fraction of a rise in national income which goes on consumption. It is the slope of the consumption function.

Marginal propensity to save The fraction of a rise in national income which goes into savings. It is the slope of the savings function.

Money supply The notes and coins in circulation together with money held in bank deposits.

National income The flow of money from firms to households.

Precautionary demand for money Money held in reserve by individuals or firms to fund unforeseen future expenditure.

Speculative demand for money Money held back by firms or individuals for the purpose of investing in alternative assets, such as government bonds, at some future date.

Taxation Money paid to government based on an individual's income and wealth (direct taxation) together with money paid by suppliers of goods or services based on expenditure (indirect taxation).

Transactions demand for money Money used for everyday transactions of goods and services.

Practice Problems

5 If the consumption function is given by

$$C = 0.7Y + 40$$

state the values of

(a) autonomous consumption

(b) marginal propensity to consume

Transpose this formula to express Y in terms of C and hence find the value of Y when $C = 110$.

6 Write down expressions for the savings function given that the consumption function is

(a) $C = 0.9Y + 72$ **(b)** $C = \dfrac{Y^2 + 500}{Y + 10}$

7 For a closed economy with no government intervention the consumption function is

$$C = 0.6Y + 30$$

and planned investment is

$$I = 100$$

Calculate the equilibrium level of

(a) national income

(b) consumption

(c) savings

8 If

$$C = aY + b$$
$$Y = C + I$$
$$I = I^*$$

show that

$$Y = \frac{b + I^*}{1 - a}$$

and obtain a similar expression for C in terms of a, b and I^*.

9 An open economy is in equilibrium when

$$Y = C + I + G + X - M$$

➔

where

Y = national income
C = consumption
I = investment
G = government expenditure
X = exports
M = imports

Determine the equilibrium level of income given that

$C = 0.8Y + 80$
$I = 70$
$G = 130$
$X = 100$
$M = 0.2Y + 50$

10 Given that

consumption, $C = 0.8Y + 60$
investment, $I = -30r + 740$
money supply, $M_S = 4000$
transaction–precautionary demand for money, $M_S = 0.15Y$
speculative demand for money, $L_2 = -20r + 3825$

determine the values of national income, Y, and interest rate, r, on the assumption that both the commodity and the money markets are in equilibrium.

11 **(Excel)** Consider the consumption function

$C = 120 + 0.8Y_d$

where Y_d is disposable income.

Write down expressions for C, in terms of national income, Y, when there is

(a) no tax
(b) a lump sum tax of \$100
(c) a proportional tax in which the proportion is 0.25

Sketch all three functions on the same diagram, over the range $0 \le Y \le 800$, and briefly describe any differences or similarities between them.

Sketch the 45 degree line, $C = Y$, on the same diagram, and hence estimate equilibrium levels of national income in each case.

12 **(Excel)** If the consumption function is

$C = 0.9Y + 20$

and planned investment $I = 10$, write down an expression for the aggregate expenditure, $C + I$, in terms of Y.

Draw graphs of aggregate expenditure, and the 45 degree line, on the same diagram, over the range $0 \le Y \le 500$. Deduce the equilibrium level of national income.

Describe what happens to the aggregate expenditure line in the case when

(a) the marginal propensity to consume falls to 0.8
(b) planned investment rises to 15

and find the new equilibrium income in each case.

chapter 2

Non-linear Equations

The main aim of this chapter is to describe the mathematics of non-linear equations. The approach is similar to that of Chapter 1. There are four sections. Section 2.1 should be read before Section 2.2, and Section 2.3 should be read before Section 2.4.

The first section investigates the simplest non-linear equation, known as a quadratic. A quadratic equation can easily be solved either by factorizing it as the product of two linear factors or by using a special formula. You are also shown how to sketch the graphs of quadratic functions. The techniques are illustrated by finding the equilibrium price and quantity for quadratic supply and demand functions.

Section 2.2 introduces additional functions in microeconomics, including revenue and profit. There is very little new material in this section. It mainly consists of applying the ideas of Section 2.1 to sketch graphs of quadratic revenue and profit functions and to find their maximum values.

Finally, the topic of algebra, which we started in Chapter 1, is completed by investigating the rules of indices and logarithms. The basic concepts are covered in Section 2.3. The notation and rules of indices are extremely important and are used frequently in subsequent chapters. Section 2.4 focuses on two specific functions, namely the exponential and natural logarithm functions. If you run into difficulty, or are short of time, then this section could be omitted, particularly if you do not intend to study the next chapter on the mathematics of finance.

section 2.1

Quadratic functions

Objectives

At the end of this section you should be able to:

- Solve a quadratic equation using 'the formula'.
- Solve a quadratic equation given its factorization.
- Sketch the graph of a quadratic function using a table of function values.
- Sketch the graph of a quadratic function by finding the coordinates of the intercepts.
- Determine equilibrium price and quantity given a pair of quadratic demand and supply functions.

The first chapter considered the topic of linear mathematics. In particular, we described how to sketch the graph of a linear function and how to solve a linear equation (or system of simultaneous linear equations). It was also pointed out that not all economic functions are of this simple form. In assuming that the demand and supply graphs are straight lines, we are certainly making the mathematical analysis easy, but we may well be sacrificing realism. It may be that the demand and supply graphs are curved and, in these circumstances, it is essential to model them using more complicated functions. The simplest non-linear function is known as a *quadratic* and takes the form

$$f(x) = ax^2 + bx + c$$

for some parameters a, b and c. (In fact, even if the demand function is linear, functions derived from it, such as total revenue and profit, turn out to be quadratic. We investigate these functions in the next section.) For the moment we concentrate on the mathematics of quadratics and show how to sketch graphs of quadratic functions and how to solve quadratic equations.

Consider the elementary equation

$$x^2 - 9 = 0$$

It is easy to see that the expression on the left-hand side is a special case of the above with $a = 1$, $b = 0$ and $c = -9$. To solve this equation we add 9 to both sides to get

$$x^2 = 9$$

x^2 is an abbreviation for $x \times x$

so we need to find a number, x, which when multiplied by itself produces the value 9. A moment's thought should convince you that there are exactly two numbers that work, namely 3 and −3 because

$$3 \times 3 = 9 \quad \text{and} \quad (-3) \times (-3) = 9$$

These two solutions are called the ***square roots*** of 9. The symbol $\sqrt{}$ is reserved for the positive square root, so in this notation the solutions are $\sqrt{9}$ and $-\sqrt{9}$. These are usually combined and written $\pm\sqrt{9}$. The equation

$$x^2 - 9 = 0$$

is trivial to solve because the number 9 has obvious square roots. In general, it is necessary to use a calculator to evaluate square roots. For example, the equation

$$x^2 - 2 = 0$$

can be written as

$$x^2 = 2$$

and so has solutions $x = \pm\sqrt{2}$. My calculator gives 1.414 213 56 (correct to 8 decimal places) for the square root of 2, so the above equation has solutions

$$1.414\,213\,56 \quad \text{and} \quad -1.414\,213\,56$$

Example

Solve the following quadratic equations:

(a) $5x^2 - 80 = 0$ **(b)** $x^2 + 64 = 0$ **(c)** $(x + 4)^2 = 81$

Solution

(a) $5x^2 - 80 = 0$

$5x^2 = 80$ (add 80 to both sides)

$x^2 = 16$ (divide both sides by 5)

$x = \pm 4$ (square root both sides)

(b) $x^2 + 64 = 0$

$x^2 = -64$ (subtract 64 from both sides)

This equation does not have a solution because you cannot square a real number and get a negative answer.

(c) $(x + 4)^2 = 81$

$x + 4 = \pm 9$ (square root both sides)

The two solutions are obtained by taking the + and – signs separately. Taking the + sign,

$$x + 4 = 9 \quad \text{so} \quad x = 9 - 4 = 5$$

Taking the – sign,

$$x + 4 = -9 \quad \text{so} \quad x = -9 - 4 = -13$$

The two solutions are 5 and –13.

Problem

1 Solve the following quadratic equations. (Round your solutions to 2 decimal places if necessary.)

(a) $x^2 - 100 = 0$ **(b)** $2x^2 - 8 = 0$ **(c)** $x^2 - 3 = 0$ **(d)** $x^2 - 5.72 = 0$

(e) $x^2 + 1 = 0$ **(f)** $3x^2 + 6.21 = 0$ **(g)** $x^2 = 0$

All of the equations considered in Problem 1 are of the special form

$$ax^2 + c = 0$$

in which the coefficient of x is zero. To solve more general quadratic equations we use a formula that enables the solutions to be calculated in a few lines of working. It can be shown that

$$ax^2 + bx + c = 0$$

has solutions

$$x = \frac{-b \pm \sqrt{(b^2 - 4ac)}}{2a}$$

The following example describes how to use this formula. It also illustrates the fact (which you have already discovered in Practice Problem 1) that a quadratic equation can have two solutions, one solution or no solutions.

Example

Solve the quadratic equations

(a) $2x^2 + 9x + 5 = 0$

(b) $x^2 - 4x + 4 = 0$

(c) $3x^2 - 5x + 6 = 0$

Solution

(a) For the equation

$$2x^2 + 9x + 5 = 0$$

we have $a = 2$, $b = 9$ and $c = 5$. Substituting these values into the formula

$$x = \frac{-b \pm \sqrt{(b^2 - 4ac)}}{2a}$$

→

gives

$$x = \frac{-9 \pm \sqrt{(9^2 - 4(2)(5))}}{2(2)}$$

$$= \frac{-9 \pm \sqrt{(81 - 40)}}{4}$$

$$= \frac{-9 \pm \sqrt{41}}{4}$$

The two solutions are obtained by taking the + and – signs separately: that is,

$$\frac{-9 + \sqrt{41}}{4} = -0.649 \quad \text{(correct to 3 decimal places)}$$

$$\frac{-9 - \sqrt{41}}{4} = -3.851 \quad \text{(correct to 3 decimal places)}$$

It is easy to check that these are solutions by substituting them into the original equation. For example, putting $x = -0.649$ into

$$2x^2 + 9x + 5$$

gives

$$2(-0.649)^2 + 9(-0.649) + 5 = 0.001\,402$$

which is close to zero, as required. We cannot expect to produce an exact value of zero because we rounded $\sqrt{41}$ to 3 decimal places. You might like to check for yourself that –3.851 is also a solution.

(b) For the equation

$$x^2 - 4x + 4 = 0$$

we have $a = 1$, $b = -4$ and $c = 4$. Substituting these values into the formula

$$x = \frac{-b \pm \sqrt{(b^2 - 4ac)}}{2a}$$

gives

$$x = \frac{-(-4) \pm \sqrt{((-4^2) - 4(1)(4))}}{2(1)}$$

$$= \frac{4 \pm \sqrt{(16 - 16)}}{2}$$

$$= \frac{4 \pm \sqrt{0}}{2}$$

$$= \frac{4 \pm 0}{2}$$

Clearly we get the same answer irrespective of whether we take the + or the – sign here. In other words, this equation has only one solution, $x = 2$. As a check, substitution of $x = 2$ into the original equation gives

$$(2)^2 - 4(2) + 4 = 0$$

(c) For the equation

$$3x^2 - 5x + 6 = 0$$

we have $a = 3$, $b = -5$ and $c = 6$. Substituting these values into the formula

$$x = \frac{-b \pm \sqrt{(b^2 - 4ac)}}{2a}$$

gives

$$x = \frac{-(-5) \pm \sqrt{((-5^2) - 4(3)(6))}}{2(3)}$$

$$= \frac{5 \pm \sqrt{(25 - 72)}}{6}$$

$$= \frac{5 \pm \sqrt{(-47)}}{6}$$

The number under the square root sign is negative and, as you discovered in Practice Problem 1, it is impossible to find the square root of a negative number. We conclude that the quadratic equation

$$3x^2 - 5x + 6 = 0$$

has no solutions.

This example demonstrates the three cases that can occur when solving quadratic equations. The precise number of solutions that an equation can have depends on whether the number under the square root sign is positive, zero or negative. The number $b^2 - 4ac$ is called the ***discriminant*** because the sign of this number discriminates between the three cases that can occur.

- If $b^2 - 4ac > 0$ then there are two solutions

  $$x = \frac{-b + \sqrt{(b^2 - 4ac)}}{2a} \quad \text{and} \quad x = \frac{-b - \sqrt{(b^2 - 4ac)}}{2a}$$

- If $b^2 - 4ac = 0$ then there is one solution

  $$x = \frac{-b \pm \sqrt{0}}{2a} = \frac{-b}{2a}$$

- If $b^2 - 4ac < 0$ then there are no solutions because $\sqrt{(b^2 - 4ac)}$ does not exist.

Practice Problem

2 Solve the following quadratic equations (where possible):

(a) $2x^2 - 19x - 10 = 0$ **(b)** $4x^2 + 12x + 9 = 0$

(c) $x^2 + x + 1 = 0$ **(d)** $x^2 - 3x + 10 = 2x + 4$

You may be familiar with another method for solving quadratic equations. This is based on the factorization of a quadratic into the product of two linear factors. Section 1.4 described how to multiply out two brackets. One of the examples in that section showed that

$$(x + 1)(x + 2) = x^2 + 3x + 2$$

Consequently, the solutions of the equation

$$x^2 + 3x + 2 = 0$$

are the same as those of

$$(x + 1)(x + 2) = 0$$

Now the only way that two numbers can be multiplied together to produce a value of zero is when (at least) one of the numbers is zero.

if $ab = 0$ then either $a = 0$ or $b = 0$ (or both)

It follows that either

$x + 1 = 0$ with solution $x = -1$

or

$x + 2 = 0$ with solution $x = -2$

The quadratic equation

$$x^2 + 3x + 2 = 0$$

therefore has two solutions, $x = -1$ and $x = -2$.

The difficulty with this approach is that it is impossible, except in very simple cases, to work out the factorization from any given quadratic, so the preferred method is to use the formula. However, if you are lucky enough to be given the factorization, or perhaps clever enough to spot the factorization for yourself, then it does provide a viable alternative.

Example

Write down the solutions to the following quadratic equations:

(a) $x(3x - 4) = 0$ **(b)** $(x - 7)^2 = 0$

Solution

(a) If $x(3x - 4) = 0$ then either $x = 0$ or $3x - 4 = 0$

The first gives the solution $x = 0$ and the second gives $x = 4/3$.

(b) If $(x - 7)(x - 7) = 0$ then either $x - 7 = 0$ or $x - 7 = 0$

Both options lead to the same solution, $x = 7$.

Practice Problem

3 Write down the solutions to the following quadratic equations. (There is no need to multiply out the brackets.)

(a) $(x - 4)(x + 3) = 0$

(b) $x(10 - 2x) = 0$

(c) $(2x - 6)^2 = 0$

One important feature of linear functions is that their graphs are always straight lines. Obviously the intercept and slope vary from function to function, but the shape is always the same. It turns out that a similar property holds for quadratic functions. Now, whenever you are asked to produce a graph of an unfamiliar function, it is often a good idea to tabulate the function, to plot these points on graph paper and to join them up with a smooth curve. The precise number of points to be taken depends on the function but, as a general rule, between 5 and 10 points usually produce a good picture.

Example

Sketch a graph of the square function, $f(x) = x^2$.

Solution

A table of values for the simple square function

$$f(x) = x^2$$

is given by

x	−3	−2	−1	0	1	2	3
$f(x)$	9	4	1	0	1	4	9

The first row of the table gives a selection of 'incoming' numbers, x, while the second row shows the corresponding 'outgoing' numbers, y. Points with coordinates (x, y) are then plotted on graph paper to produce the curve shown in Figure 2.1. For convenience, different scales are used on the x and y axes.

Mathematicians call this curve a ***parabola***, whereas economists refer to it as ***U-shaped***. Notice that the graph is symmetric about the y axis with a minimum point at the origin; if a mirror is placed along the y axis then the left-hand part is the image of the right-hand part.

Figure 2.1

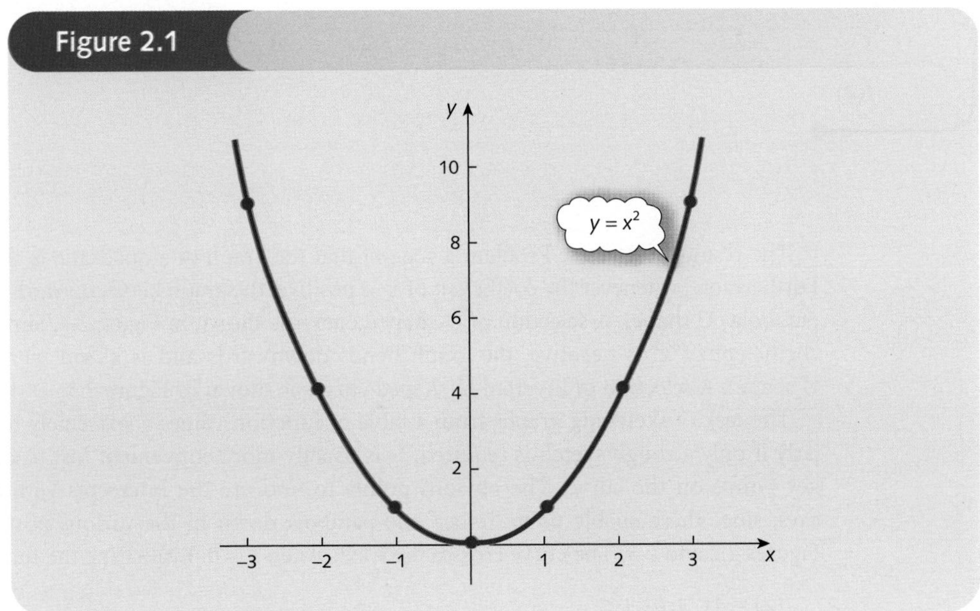

Advice

The following problem is designed to give you an opportunity to tabulate and sketch graphs of more general quadratic functions. Please remember that when you substitute numbers into a formula you must use BIDMAS to decide the order of the operations. For example, in part (a) you need to substitute $x = -1$ into $4x^2 - 12x + 5$. You get

$$4(-1)^2 - 12(-1) + 5$$
$$= 4 + 12 + 5$$
$$= 21$$

Note also that when using a calculator you must use brackets when squaring negative numbers. In this case a possible sequence of key presses might be

[4] [(] [(−)] [1] [)] [x^2] [−] [12] [×] [(−)] [1] [+] [5] [=]

Practice Problem

4 Complete the following tables of function values and hence sketch a graph of each quadratic function.

(a) $f(x) = 4x^2 - 12x + 5$

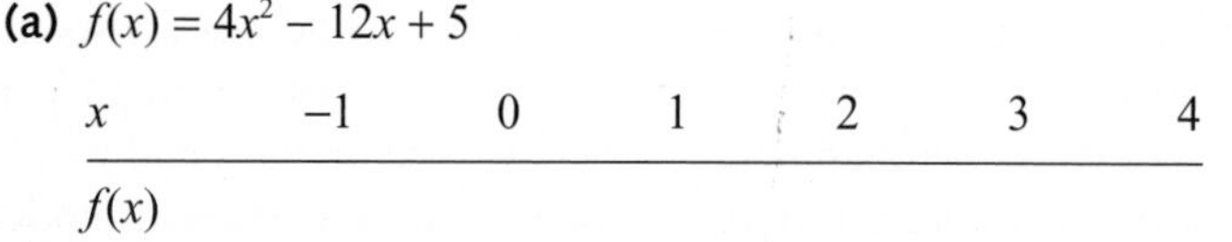

x	−1	0	1	2	3	4
$f(x)$						

(b) $f(x) = -x^2 + 6x - 9$

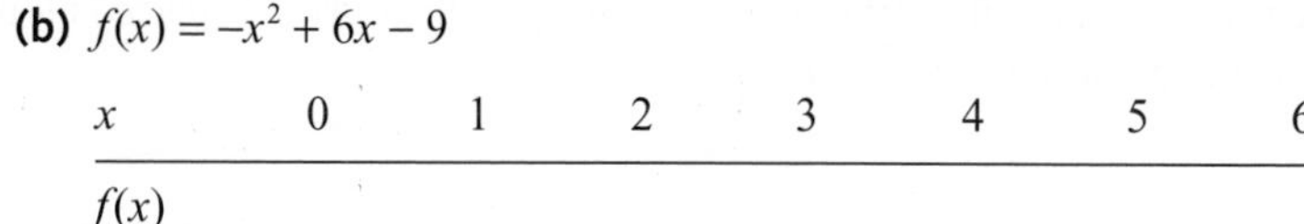

x	0	1	2	3	4	5	6
$f(x)$							

(c) $f(x) = -2x^2 + 4x - 6$

x	−2	−1	0	1	2	3	4
$f(x)$							

The results of Practice Problem 4 suggest that the graph of a quadratic is always parabolic. Furthermore, whenever the coefficient of x^2 is positive, the graph bends upwards and is a 'happy' parabola (U shape). A selection of U-shaped curves is shown in Figure 2.2. Similarly, when the coefficient of x^2 is negative, the graph bends downwards and is a 'sad' parabola (inverted U shape). A selection of inverted U-shaped curves is shown in Figure 2.3.

The task of sketching graphs from a table of function values is extremely tedious, particularly if only a rough sketch is required. It is usually more convenient just to determine a few key points on the curve. The obvious points to find are the intercepts with the coordinate axes, since these enable us to 'tether' the parabola down in the various positions shown in Figures 2.2 and 2.3. The curve crosses the y axis when $x = 0$. Evaluating the function

$$f(x) = ax^2 + bx + c$$

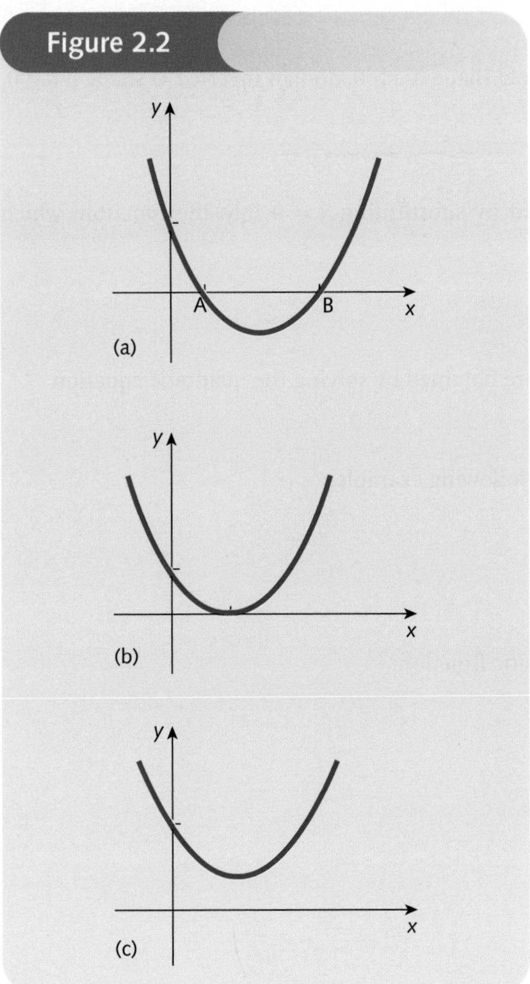

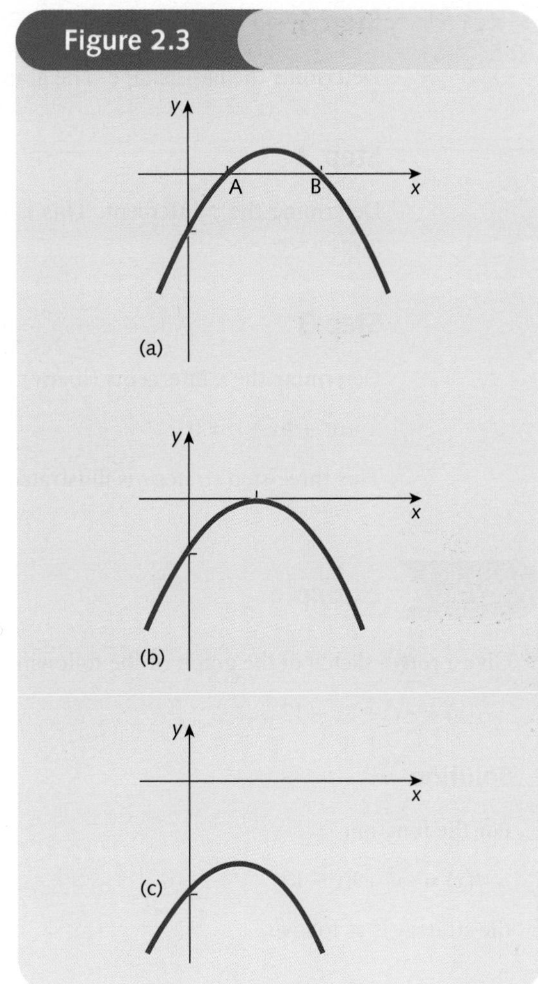

at $x = 0$ gives

$$f(0) = a(0)^2 + b(0) + c = c$$

so the constant term determines where the curve cuts the vertical axis (as it did for linear functions). The curve crosses the x axis when $y = 0$ or, equivalently, when $f(x) = 0$, so we need to solve the quadratic equation

$$ax^2 + bx + c = 0$$

This can be done using 'the formula' and the solutions are the points where the graph cuts the horizontal axis. In general, a quadratic equation can have two, one or no solutions and these possibilities are illustrated in cases (a), (b) and (c) in Figures 2.2 and 2.3. In case (a) the curve crosses the x axis at A, turns round and crosses it again at B, so there are two solutions. In case (b) the curve turns round just as it touches the x axis, so there is only one solution. Finally, in case (c) the curve turns round before it has a chance to cross the x axis, so there are no solutions.

The strategy for sketching the graph of a quadratic function

$$f(x) = ax^2 + bx + c$$

may now be stated.

Step 1

Determine the basic shape. The graph has a U shape if $a > 0$, and an inverted U shape if $a < 0$.

Step 2

Determine the y intercept. This is obtained by substituting $x = 0$ into the function, which gives $y = c$.

Step 3

Determine the x intercepts (if any). These are obtained by solving the quadratic equation

$$ax^2 + bx + c = 0$$

This three-step strategy is illustrated in the following example.

Example

Give a rough sketch of the graph of the following quadratic function:

$$f(x) = -x^2 + 8x - 12$$

Solution

For the function

$$f(x) = -x^2 + 8x - 12$$

the strategy is as follows.

Step 1

The coefficient of x^2 is -1, which is negative, so the graph is a 'sad' parabola with an inverted U shape.

Step 2

The constant term is -12, so the graph crosses the vertical axis at $y = -12$.

Step 3

For the quadratic equation

$$-x^2 + 8x - 12 = 0$$

the formula gives

$$x = \frac{-8 \pm \sqrt{(8^2 - 4(-1)(-12))}}{2(-1)} = \frac{-8 \pm \sqrt{(64 - 48)}}{-2}$$

$$= \frac{-8 \pm \sqrt{16}}{-2} = \frac{-8 \pm 4}{-2}$$

so the graph crosses the horizontal axis at

$$x = \frac{-8 + 4}{-2} = 2$$

and

$$x = \frac{-8 - 4}{-2} = 6$$

The information obtained in steps 1–3 is sufficient to produce the sketch shown in Figure 2.4.

In fact, we can go even further in this case and locate the coordinates of the turning point – that is, the maximum point – on the curve. By symmetry, the x coordinate of this point occurs exactly halfway between $x = 2$ and $x = 6$: that is, at

$$x = \tfrac{1}{2}(2 + 6) = 4$$

The corresponding y coordinate is found by substituting $x = 4$ into the function to get

$$f(4) = -(4)^2 + 8(4) - 12 = 4$$

The maximum point on the curve therefore has coordinates (4, 4).

Figure 5.2

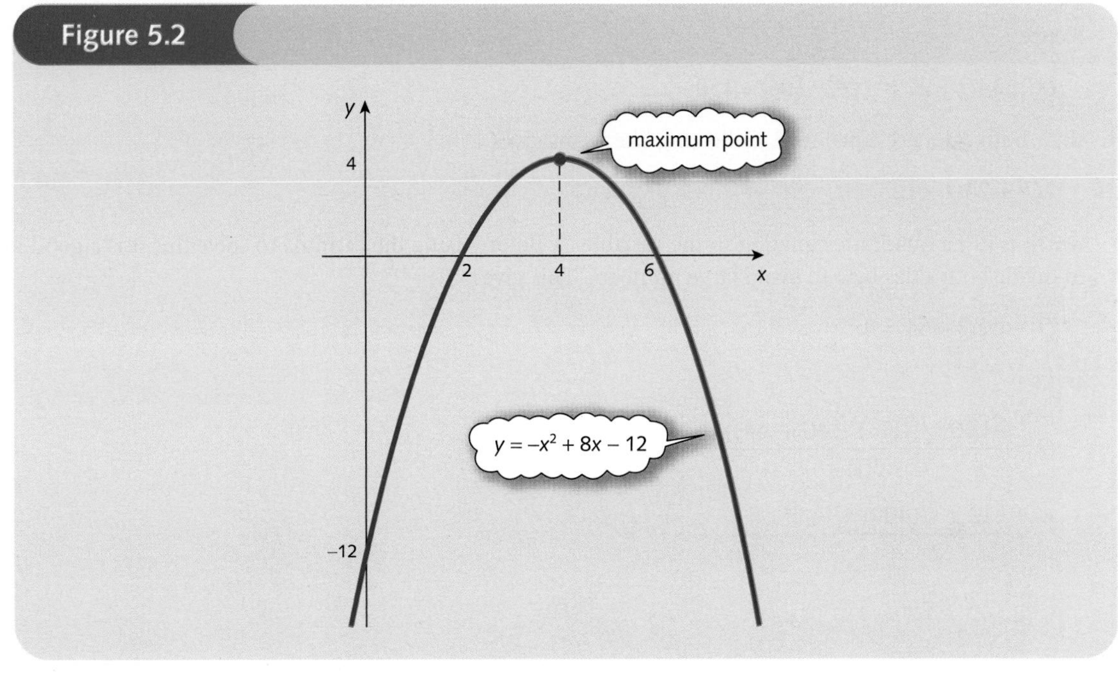

Practice Problem

5 Use the three-step strategy to produce rough graphs of the following quadratic functions:

(a) $f(x) = 2x^2 - 11x - 6$ **(b)** $f(x) = x^2 - 6x + 9$

We conclude this section by seeing how to solve a particular problem in microeconomics. In Section 1.3 the concept of market equilibrium was introduced and in each of the problems the supply and demand functions were always given to be linear. The following example shows this to be an unnecessary restriction and indicates that it is almost as easy to manipulate quadratic supply and demand functions.

Example

Given the supply and demand functions

$$P = Q_S^2 + 14Q_S + 22$$
$$P = -Q_D^2 - 10Q_D + 150$$

calculate the equilibrium price and quantity.

Solution

In equilibrium, $Q_S = Q_D$, so if we denote this equilibrium quantity by Q, the supply and demand functions become

$$P = Q^2 + 14Q + 22$$
$$P = -Q^2 - 10Q + 150$$

Hence

$$Q^2 + 14Q + 22 = -Q^2 - 10Q + 150$$

since both sides are equal to P. Collecting like terms gives

$$2Q^2 + 24Q - 128 = 0$$

which is just a quadratic equation in the variable Q. Before using the formula to solve this it is a good idea to divide both sides by 2 to avoid large numbers. This gives

$$Q^2 + 12Q - 64 = 0$$

and so

$$Q = \frac{-12 \pm \sqrt{((12^2) - 4(1)(-64))}}{2(1)}$$
$$= \frac{-12 \pm \sqrt{(400)}}{2}$$
$$= \frac{-12 \pm 20}{2}$$

The quadratic equation has solutions $Q = -16$ and $Q = 4$. Now the solution $Q = -16$ can obviously be ignored because a negative quantity does not make sense. The equilibrium quantity is therefore 4. The equilibrium price can be calculated by substituting this value into either the original supply or demand equation.

From the supply equation,

$$P = 4^2 + 14(4) + 22 = 94$$

As a check, the demand equation gives

$$P = -(4)^2 - 10(4) + 150 = 94 \quad ✓$$

You might be puzzled by the fact that we actually obtain two possible solutions, one of which does not make economic sense. The supply and demand curves are sketched in Figure 2.5. This shows that there are indeed two points of intersection confirming the mathematical solution. However, in economics the quantity and price are both positive, so the functions are only defined in the top right-hand (that is, positive) quadrant. In this region there is just one point of intersection, at (4, 94).

Figure 2.5

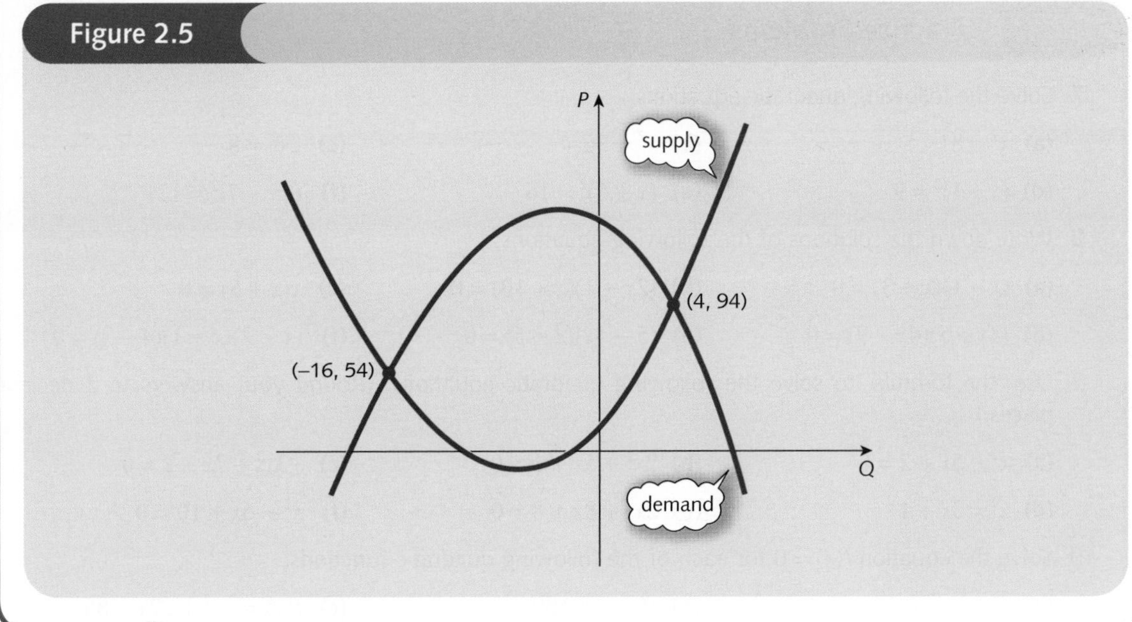

Practice Problem

6 Given the supply and demand functions

$$P = 2Q_S^2 + 10Q_S + 10$$

$$P = -Q_D^2 - 5Q_D + 52$$

calculate the equilibrium price and quantity.

Key Terms

Discriminant The number $b^2 - 4ac$ which is used to indicate the number of solutions of the quadratic equation $ax^2 + bx + c = 0$.

Parabola The shape of the graph of a quadratic function.

Quadratic function A function of the form $f(x) = ax^2 + bx + c$ where $a \neq 0$.

Square root A number that when multiplied by itself equals a given number; the solutions of the equation $x^2 = c$ which are written $\pm\sqrt{c}$.

U-shaped curve A term used by economists to describe a curve, such as a parabola, which bends upwards, like the letter U.

Practice Problems

7 Solve the following quadratic equations:

(a) $x^2 = 81$ **(b)** $x^2 = 36$ **(c)** $2x^2 = 8$

(d) $(x - 1)^2 = 9$ **(e)** $(x + 5)^2 = 16$ **(f)** $(2x - 7)^2 = 121$

8 Write down the solutions of the following equations:

(a) $(x - 1)(x + 3) = 0$ **(b)** $(2x - 1)(x + 10) = 0$ **(c)** $x(x + 5) = 0$

(d) $(3x + 5)(4x - 9) = 0$ **(e)** $(5 - 4x)(x - 5) = 0$ **(f)** $(x - 2)(x + 1)(4 - x) = 0$

9 Use 'the formula' to solve the following quadratic equations. (Round your answers to 2 decimal places.)

(a) $x^2 - 5x + 2 = 0$ **(b)** $2x^2 + 5x + 1 = 0$ **(c)** $-3x^2 + 7x + 2 = 0$

(d) $x^2 = 3x + 1$ **(e)** $2x^2 + 8x + 8 = 0$ **(f)** $x^2 - 6x + 10 = 0$

10 Solve the equation $f(x) = 0$ for each of the following quadratic functions:

(a) $f(x) = x^2 - 16$ **(b)** $f(x) = x(100 - x)$ **(c)** $f(x) = -x^2 + 22x - 85$

(d) $f(x) = x^2 - 18x + 81$ **(e)** $f(x) = 2x^2 + 4x + 3$

11 Sketch the graphs of the quadratic functions given in Practice Problem 10.

12 One solution of the quadratic equation

$$x^2 - 8x + c = 0$$

is known to be $x = 2$. By substituting this into the equation, find the value of c and hence obtain the second solution.

13 Given the quadratic supply and demand functions

$$P = Q_S^2 + 2Q_S + 12$$

$$P = -Q_D^2 - 4Q_D + 68$$

determine the equilibrium price and quantity.

14 Given the supply and demand functions

$$P = Q_S^2 + 2Q_S + 7$$

$$P = -Q_D + 25$$

determine the equilibrium price and quantity.

section 2.2

Revenue, cost and profit

Objectives

At the end of this section you should be able to:

- Sketch the graphs of the total revenue, total cost, average cost and profit functions.
- Find the level of output that maximizes total revenue.
- Find the level of output that maximizes profit.
- Find the break-even levels of output.

The main aim of this section is to investigate one particular function in economics, namely profit. By making reasonable simplifying assumptions, the profit function is shown to be quadratic and so the methods developed in Section 2.1 can be used to analyse its properties. We describe how to find the levels of output required for a firm to break even and to maximize profit. The ***profit*** function is denoted by the Greek letter π (pi, pronounced 'pie') and is defined to be the difference between total revenue, TR, and total cost, TC: that is,

$$\pi = \text{TR} - \text{TC}$$

This definition is entirely sensible because TR is the amount of money received by the firm from the sale of its goods and TC is the amount of money that the firm has to spend to produce these goods. We begin by considering the total revenue and total cost functions in turn.

The ***total revenue*** received from the sale of Q goods at price P is given by

$$\text{TR} = PQ$$

For example, if the price of each good is \$70 and the firm sells 300 then the revenue is

$$\$70 \times 300 = \$21\,000$$

Given any particular demand function, expressing P in terms of Q, it is a simple matter to obtain a formula for TR solely in terms of Q. A graph of TR against Q can then be sketched.

Example

Given the demand function

$$P = 100 - 2Q$$

express TR as a function of Q and hence sketch its graph.

(**a**) For what values of Q is TR zero?

(**b**) What is the maximum value of TR?

Solution

Total revenue is defined by

$$\text{TR} = PQ$$

and, since $P = 100 - 2Q$, we have

$$\text{TR} = (100 - 2Q)Q = 100Q - 2Q^2$$

This function is quadratic and so its graph can be sketched using the strategy described in Section 2.1.

Step 1

The coefficient of Q^2 is negative, so the graph has an inverted U shape.

Step 2

The constant term is zero, so the graph crosses the TR axis at the origin.

Step 3

To find where the curve crosses the horizontal axis, we could use 'the formula'. However, this is not necessary, since it follows immediately from the factorization

$$\text{TR} = (100 - 2Q)Q$$

that $\text{TR} = 0$ when either $100 - 2Q = 0$ or $Q = 0$. In other words, the quadratic equation has two solutions, $Q = 0$ and $Q = 50$.

The total revenue curve is shown in Figure 2.6.

- From Figure 2.6 the total revenue is zero when $Q = 0$ and $Q = 50$.
- By symmetry, the parabola reaches its maximum halfway between 0 and 50, that is at $Q = 25$. The corresponding total revenue is given by

$$\text{TR} = 100(25) - 2(25)^2 = 1250$$

Practice Problem

1 Given the demand function

$$P = 1000 - Q$$

express TR as a function of Q and hence sketch a graph of TR against Q. What value of Q maximizes total revenue and what is the corresponding price?

Figure 2.6

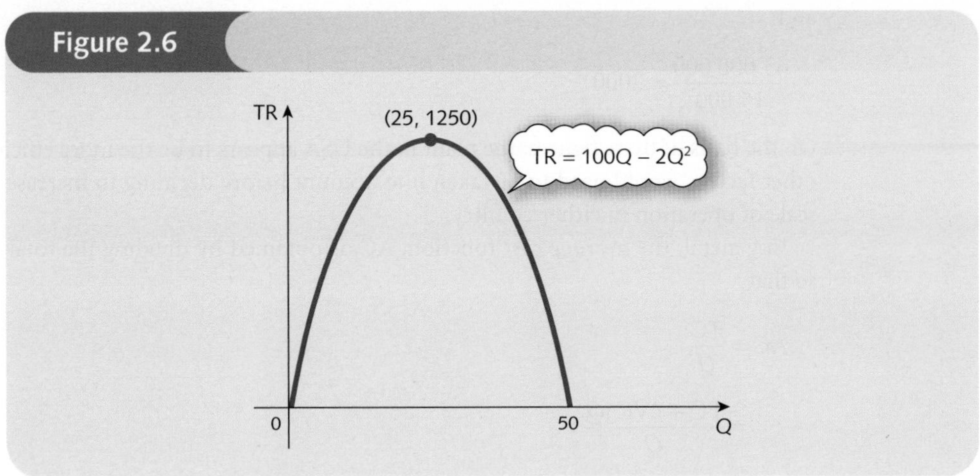

In general, given the linear demand function

$$P = aQ + b \ (a < 0, b > 0)$$

the total revenue function is

$$\begin{aligned} \text{TR} &= PQ \\ &= (aQ + b)Q \\ &= aQ^2 + bQ \end{aligned}$$

This function is quadratic in Q and, since $a < 0$, the TR curve has an inverted U shape. Moreover, since the constant term is zero, the curve always intersects the vertical axis at the origin. This fact should come as no surprise to you; if no goods are sold, the revenue must be zero.

We now turn our attention to the ***total cost*** function, TC, which relates the production costs to the level of output, Q. As the quantity produced rises, the corresponding cost also rises, so the TC function is increasing. However, in the short run, some of these costs are fixed. ***Fixed costs***, FC, include the cost of land, equipment, rent and possibly skilled labour. Obviously, in the long run all costs are variable, but these particular costs take time to vary, so can be thought of as fixed in the short run. ***Variable costs***, on the other hand, vary with output and include the cost of raw materials, components, energy and unskilled labour. If VC denotes the variable cost per unit of output then the total variable cost, TVC, in producing Q goods is given by

$$\text{TVC} = (\text{VC})Q$$

The total cost is the sum of the contributions from the fixed and variable costs, so is given by

$$\text{TC} = \text{FC} + (\text{VC})Q$$

Now although this is an important economic function, it does not always convey the information necessary to compare individual firms. For example, suppose that an international car company operates two plants, one in the USA and one in Europe, and suppose that the total annual costs are known to be \$200 million and \$45 million respectively. Which of these two plants is regarded as the more efficient? Unfortunately, unless we also know the total number of cars produced it is impossible to make any judgement. The significant function here is not the total cost, but rather the average cost per car. If the plants in the USA and Europe manufacture 80 000 and 15 000 cars, respectively, their corresponding average costs are

$$\frac{200\,000\,000}{80\,000} = 2500$$

and

$$\frac{45\,000\,000}{15\,000} = 3000$$

On the basis of these figures, the plant in the USA appears to be the more efficient. In practice, other factors would need to be taken into account before deciding to increase or decrease the scale of operation in either country.

In general, the ***average cost*** function, AC, is obtained by dividing the total cost by output, so that

$$\begin{aligned} \text{AC} &= \frac{\text{TC}}{Q} \\ &= \frac{\text{FC} + (\text{VC})Q}{Q} \\ &= \frac{\text{FC}}{Q} + \frac{(\text{VC})Q}{Q} \\ &= \frac{\text{FC}}{Q} + \text{VC} \end{aligned}$$

Example

Given that fixed costs are 1000 and that variable costs are 4 per unit, express TC and AC as functions of Q. Hence sketch their graphs.

Solution

We are given that FC = 1000 and VC = 4, so

$$\text{TC} = 1000 + 4Q$$

and

$$\text{AC} = \frac{\text{TC}}{Q} = \frac{1000 + 4Q}{Q} = \frac{1000}{Q} + 4$$

The graph of the total cost function is easily sketched. It is a straight line with intercept 1000 and slope 4. It is sketched in Figure 2.7. The average cost function is of a form that we have not met before, so we have no prior knowledge about its basic shape. Under these circumstances it is useful to tabulate the function. The tabulated values are then plotted on graph paper and a smooth curve obtained by joining the points together. One particular table of function values is

Q	100	250	500	1000	2000
AC	14	8	6	5	4.5

These values are readily checked. For instance, when $Q = 100$

$$\text{AC} = \frac{1000}{100} + 4 = 10 + 4 = 14$$

A graph of the average cost function, based on this table, is sketched in Figure 2.8. This curve is known as a ***rectangular hyperbola*** and is sometimes referred to by economists as being ***L-shaped***.

Figure 2.7

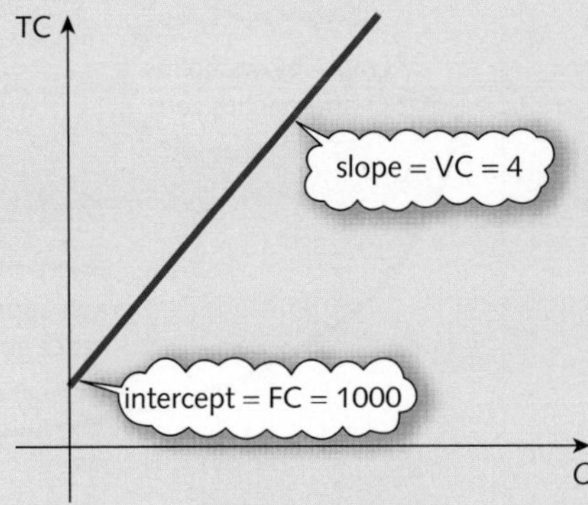

Figure 2.8

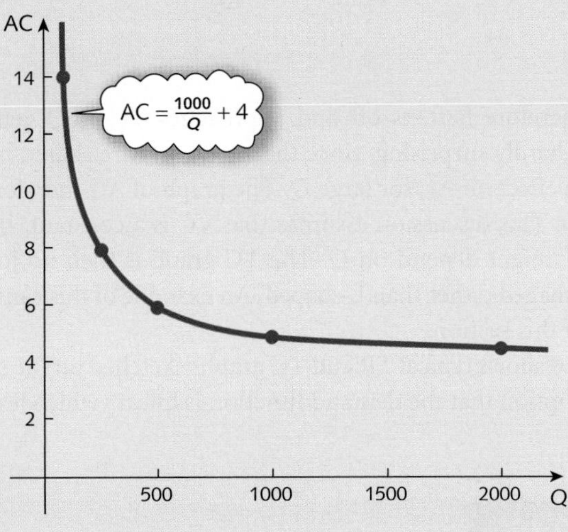

Practice Problem

2 Given that fixed costs are 100 and that variable costs are 2 per unit, express TC and AC as functions of Q. Hence sketch their graphs.

In general, whenever the variable cost, VC, is a constant the total cost function,

$$\text{TC} = \text{FC} + (\text{VC})Q$$

is linear. The intercept is FC and the slope is VC. For the average cost function

$$\text{AC} = \frac{\text{FC}}{Q} + \text{VC}$$

note that if Q is small, then FC/Q is large, so the graph bends sharply upwards as Q approaches zero. As Q increases, FC/Q decreases and eventually tails off to zero for large values of Q. The

Figure 2.9

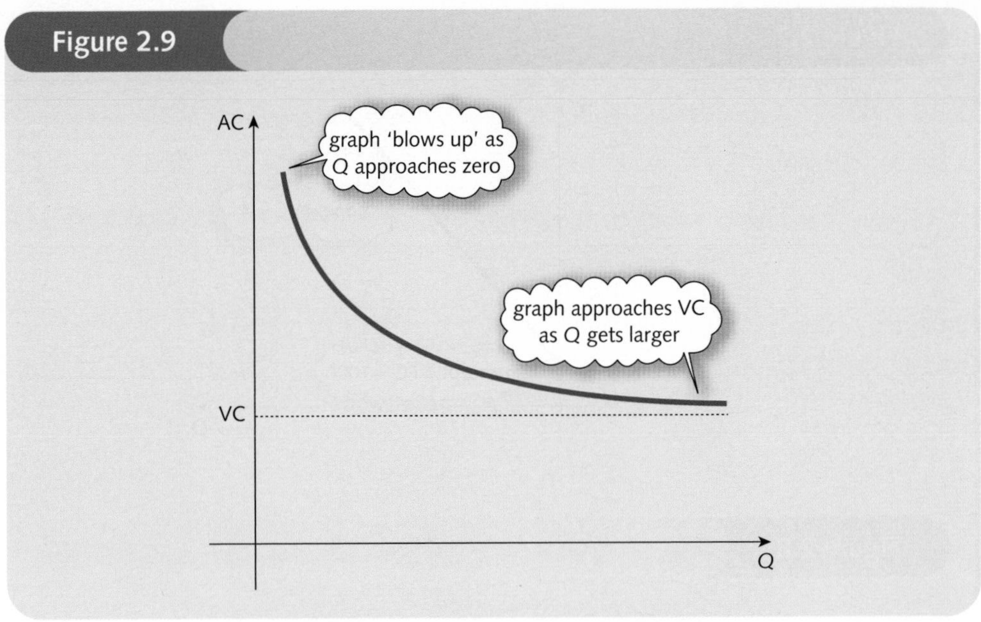

AC curve therefore flattens off and approaches VC as Q gets larger and larger. This phenomenon is hardly surprising, since the fixed costs are shared between more and more goods, so have little effect on AC for large Q. The graph of AC therefore has the basic L shape shown in Figure 2.9. This discussion assumes that VC is a constant. In practice, this may not be the case and VC might depend on Q. The TC graph is then no longer linear and the AC graph becomes U-shaped rather than L-shaped. An example of this can be found in Practice Problem 7 at the end of this section.

Figure 2.10 shows typical TR and TC graphs sketched on the same diagram. These are drawn on the assumption that the demand function is linear (which leads to a quadratic total revenue

Figure 2.10

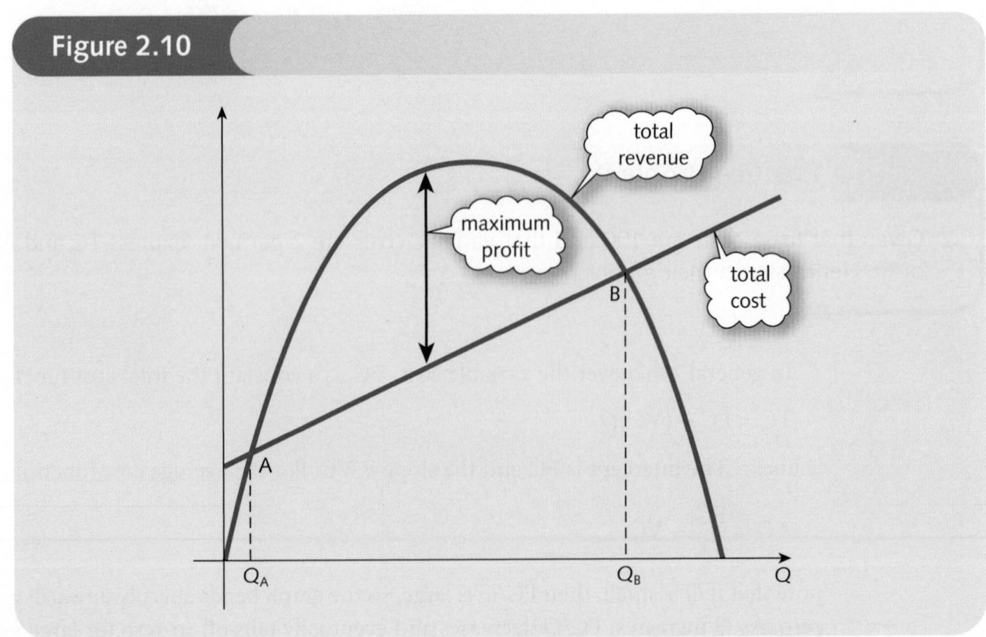

function) and that the variable costs are constant (which leads to a linear total cost function). The horizontal axis represents quantity, Q. Strictly speaking the label Q means different things for the two functions. For the revenue function, Q denotes the quantity of goods actually sold, whereas for the cost function it denotes the quantity produced. In sketching both graphs on the same diagram we are implicitly assuming that these two values are the same and that the firm sells all of the goods that it produces.

The two curves intersect at precisely two points, A and B, corresponding to output levels Q_A and Q_B. At these points the cost and revenue are equal and the firm breaks even. If $Q < Q_A$ or $Q > Q_B$ then the TC curve lies above that of TR, so cost exceeds revenue. For these levels of output the firm makes a loss. If $Q_A < Q < Q_B$ then revenue exceeds cost and the firm makes a profit that is equal to the vertical distance between the revenue and cost curves. The maximum profit occurs where the gap between them is largest. The easiest way of calculating maximum profit is to obtain a formula for profit directly in terms of Q using the defining equation

$$\pi = \text{TR} - \text{TC}$$

Example

If fixed costs are 4, variable costs per unit are 1 and the demand function is

$$P = 10 - 2Q$$

obtain an expression for π in terms of Q and hence sketch a graph of π against Q.

(a) For what values of Q does the firm break even?

(b) What is the maximum profit?

Solution

We begin by obtaining expressions for the total cost and total revenue. For this problem, FC = 4 and VC = 1, so

$$\text{TC} = \text{FC} + (\text{VC})Q = 4 + Q$$

The given demand function is

$$P = 10 - 2Q$$

so

$$\begin{aligned} \text{TR} &= PQ \\ &= (10 - 2Q)Q \\ &= 10Q - 2Q^2 \end{aligned}$$

Hence the profit is given by

$$\begin{aligned} \pi &= \text{TR} - \text{TC} \\ &= (10 - 2Q^2) - (4 + Q) \\ &= 10Q - 2Q^2 - 4 - Q \\ &= -2Q^2 + 9Q^2 - 4 \end{aligned}$$

To sketch a graph of the profit function we follow the strategy described in Section 2.1.

Step 1

The coefficient of Q^2 is negative, so the graph has an inverted U shape.

Step 2

The constant term is -4, so the graph crosses the vertical axis when $\pi = -4$.

Step 3

The graph crosses the horizontal axis when $\pi = 0$, so we need to solve the quadratic equation

$$-2Q^2 + 9Q - 4 = 0$$

This can be done using 'the formula' to get

$$Q = \frac{-9 \pm \sqrt{(81 - 32)}}{2(-2)}$$

$$= \frac{-9 \pm 7}{-4}$$

so $Q = 0.5$ and $Q = 4$.

The profit curve is sketched in Figure 2.11.

(a) From Figure 2.11 we see that profit is zero when $Q = 0.5$ and $Q = 4$.

(b) By symmetry, the parabola reaches its maximum halfway between 0.5 and 4: that is, at

$$Q = \tfrac{1}{2}(0.5 + 4) = 2.25$$

The corresponding profit is given by

$$\pi = -2(2.25)^2 + 9(2.25) - 4 = 6.125$$

Figure 2.11

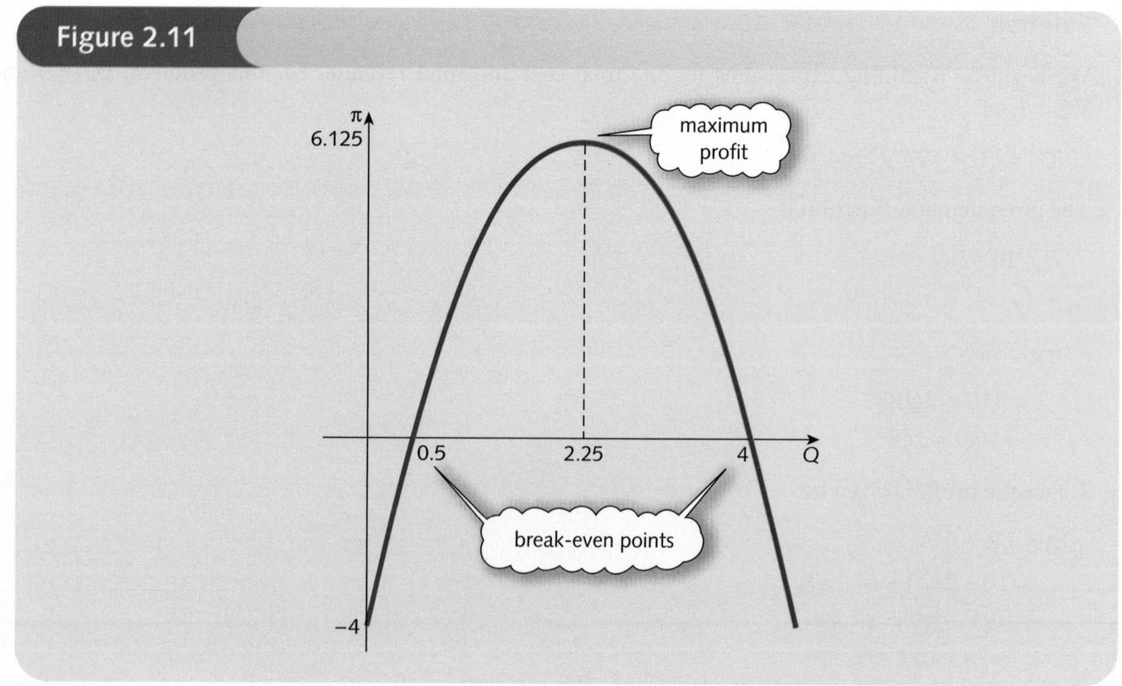

Advice

It is important to notice the use of brackets in the previous derivation of π. A common student mistake is to forget to include the brackets and just to write down

$$\begin{aligned}\pi &= \text{TR} - \text{TC}\\ &= 10Q - 2Q^2 - 4 + Q\\ &= -2Q^2 + 11Q - 4\end{aligned}$$

This cannot be right, since the whole of the total cost needs to be subtracted from the total revenue, not just the fixed costs. You might be surprised to learn that many economics students make this sort of blunder, particularly under examination conditions. I hope that if you have carefully worked through Section 1.4 on algebraic manipulation then you will not fall into this category!

Practice Problem

3 If fixed costs are 25, variable costs per unit are 2 and the demand function is

$$P = 20 - Q$$

obtain an expression for π in terms of Q and hence sketch its graph.

(a) Find the levels of output which give a profit of 31.

(b) Find the maximum profit and the value of Q at which it is achieved.

Example

EXCEL

A firm's profit function is given by

$$\pi = -Q^3 + 21Q - 18$$

Draw a graph of π against Q, over the range $0 \le Q \le 5$, and hence estimate

(a) the interval in which $\pi \ge 0$

(b) the maximum profit

Solution

Figure 2.12 (overleaf) shows the tabulated values of Q which have been entered in cells A4 to A8. In the second column, cell B4 contains the formula to work out the corresponding values of π:

=–(A4)^3+21*A4–18

This has been replicated down the profit column by clicking and dragging in the usual way.

It can be seen that the maximum profit occurs somewhere between 2 and 4, so it makes sense to add a few extra entries in here so that the graph can be plotted more accurately in this region.

→

Figure 2.12

	A	B
1	Profit function	
2		
3	Q	Profit
4	0	-18
5	1	2
6	2	16
7	3	18
8	4	2
9	5	-38
10		

Initially, inserting extra rows for $Q = 2.5$ and 3.5 shows that the maximum occurs between 2.5 and 3. Inserting a few more rows enables us to pinpoint the maximum profit value more accurately, as shown in Figure 2.13. At this stage, we can be confident that the maximum profit occurs between $Q = 2.6$ and $Q = 2.7$. The graph of the firm's profit function based on this table of values can now be drawn using Chart Wizard, as shown in Figure 2.13.

This diagram shows that

(a) the firm makes a profit for values of Q between 0.9 and 4.1

(b) the maximum profit is about 19.

Figure 2.13

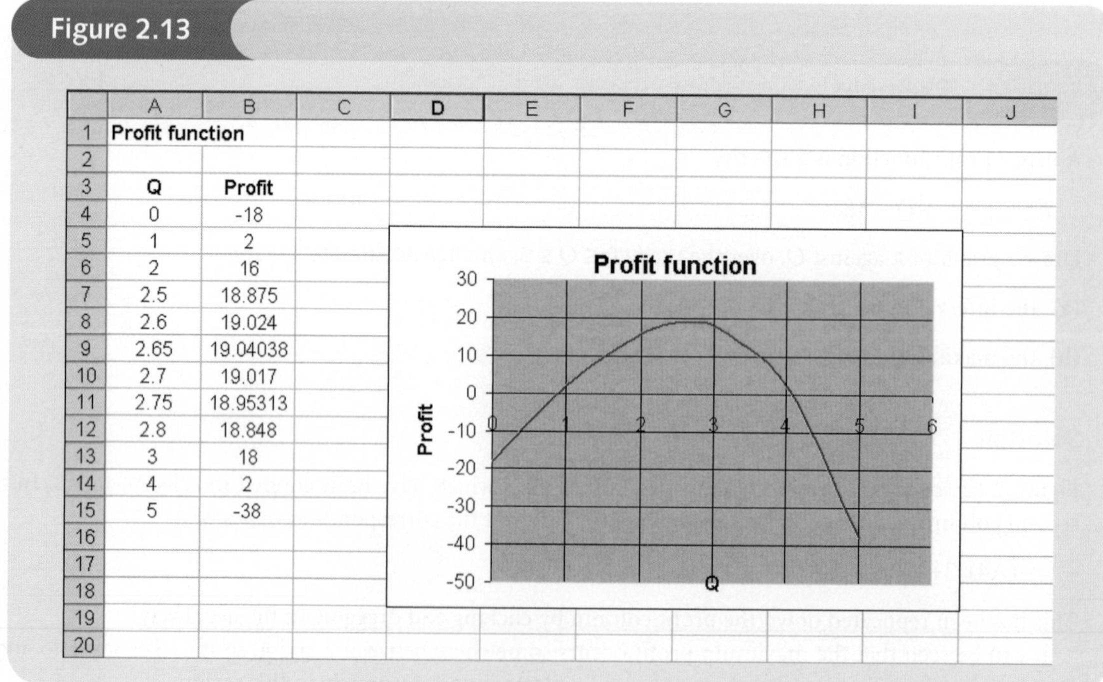

	A	B
1	Profit function	
2		
3	Q	Profit
4	0	-18
5	1	2
6	2	16
7	2.5	18.875
8	2.6	19.024
9	2.65	19.04038
10	2.7	19.017
11	2.75	18.95313
12	2.8	18.848
13	3	18
14	4	2
15	5	-38

Key Terms

Average cost Total cost per unit of output: AC = TC/Q.

Fixed costs Total costs that are independent of output.

L-shaped curve A term used by economists to describe the graph of a function, such as $f(x) = a + \frac{b}{x}$, which bends roughly like the letter L.

Profit Total revenue minus total cost: π = TR − TC.

Rectangular hyperbola A term used by mathematicians to describe the graph of a function, such as $f(x) = a + \frac{b}{x}$, which is a hyperbola with horizontal and vertical asymptotes.

Total cost The sum of the total variable and fixed costs: TC = TVC + FC.

Total revenue A firm's total earnings from the sales of a good: TR = *PQ*.

Variable costs Total costs that change according to the amount of output produced.

Practice Problems

4 Given the following demand functions, express TR as a function of Q and hence sketch the graphs of TR against Q.

(a) $P = 4$ **(b)** $P = 7/Q$ **(c)** $P = 10 - 4Q$

5 Given the following total revenue functions, find the corresponding demand functions

(a) $\text{TR} = 50Q - 4Q^2$ **(b)** $\text{TR} = 10$

6 Given that fixed costs are 500 and that variable costs are 10 per unit, express TC and AC as functions of Q. Hence sketch their graphs.

7 Given that fixed costs are 1 and that variable costs are Q + 1 per unit, express TC and AC as functions of Q. Hence sketch their graphs.

8 Find an expression for the profit function given the demand function

$$2Q + P = 25$$

and the average cost function

$$\text{AC} = \frac{32}{Q} + 5$$

Find the values of Q for which the firm

(a) breaks even

(b) makes a loss of 432 units

(c) maximizes profit

9 Sketch, on the same diagram, graphs of the total revenue and total cost functions,

$$\text{TR} = -2Q^2 + 14Q$$

$$\text{TC} = 2Q + 10$$

➔

(1) Use your graphs to estimate the values of Q for which the firm

(a) breaks even **(b)** maximizes profit

(2) Confirm your answer to part (1) using algebra.

10 The profit function of a firm is of the form

$$\pi = aQ^2 + bQ + c$$

If it is known that $\pi = 9$, 34 and 19 when $Q = 1$, 2 and 3 respectively, write down a set of three simultaneous equations for the three unknowns, a, b and c. Solve this system to find a, b and c. Hence find the profit when $Q = 4$.

11 **(Excel)** Tabulate values of the total cost function

$$\text{TC} = 0.01Q^3 + 0.5Q^2 + Q + 1000$$

when Q is 0, 2, 4, . . . , 30 and hence plot a graph of this function on the range $0 \leq Q \leq 30$. Use this graph to estimate the value of Q for which TC = 1400.

12 **(Excel)** A firm's total revenue and total cost functions are given by

$$\text{TR} = -0.5Q^2 + 24Q$$

$$\text{TC} = Q\sqrt{Q} + 100$$

Sketch these graphs on the same diagram on the range $0 \leq Q \leq 48$. Hence estimate the values of Q for which the firm

(a) breaks even **(b)** maximizes profit

section 2.3

Indices and logarithms

Objectives

At the end of this section you should be able to:

- Evaluate b^n in the case when n is positive, negative, a whole number or fraction.
- Simplify algebraic expressions using the rules of indices.
- Investigate the returns to scale of a production function.
- Evaluate logarithms in simple cases.
- Use the rules of logarithms to solve equations in which the unknown occurs as a power.

Advice

This section is quite long with some important ideas. If you are comfortable using the rules of indices and already know what a logarithm is, you should be able to read through the material in one sitting, concentrating on the applications. However, if your current understanding is hazy (or non-existent), you should consider studying this topic on separate occasions. To help with this, the material in this section has been split into the following convenient sub-sections:

- index notation
- rules of indices
- logarithms
- summary

2.3.1 Index notation

We have already used b^2 as an abbreviation for $b \times b$. In this section we extend the notation to b^n for any value of n, positive, negative, whole number or fraction. In general, if

$$M = b^n$$

we say that b^n is the ***exponential form of* M *to base* b**. The number n is then referred to as the ***index, power*** or ***exponent***. An obvious way of extending

$$b^2 = b \times b$$

to other positive whole-number powers, n, is to define

$$b^3 = b \times b \times b$$

$$b^4 = b \times b \times b \times b$$

and, in general,

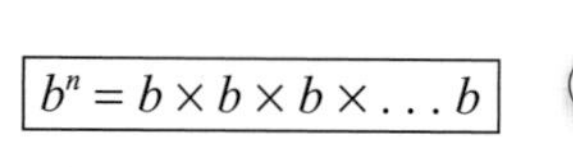

$$\boxed{b^n = b \times b \times b \times \ldots b}$$

a total of n b's multiplied together

To include the case of negative powers, consider the following table of values of 2^n.

2^{-3}	2^{-2}	2^{-1}	2^0	2^1	2^2	2^3	2^4
?	?	?	?	2	4	8	16

To work from left to right along the completed part of the table, all you have to do is to multiply each number by 2. Equivalently, if you work from right to left, you simply divide by 2. It makes sense to continue this pattern beyond $2^1 = 2$. Dividing this by 2 gives

$$2^0 = 2 \div 2 = 1$$

and dividing again by 2 gives

$$2^{-1} = 1 \div 2 = \tfrac{1}{2}$$

and so on. The completed table is then

2^{-3}	2^{-2}	2^{-1}	2^0	2^1	2^2	2^3	2^4
$\tfrac{1}{8}$	$\tfrac{1}{4}$	$\tfrac{1}{2}$	1	2	4	8	16

Notice that

$$2^{-1} = \frac{1}{2} = \frac{1}{2^1}$$

$$2^{-2} = \frac{1}{4} = \frac{1}{2^2}$$

$$2^{-3} = \frac{1}{8} = \frac{1}{2^3}$$

In other words, negative powers are evaluated by taking the reciprocal of the corresponding positive power. Motivated by this particular example, we define

$$\boxed{b^0 = 1}$$

and

$$\boxed{b^{-n} = \frac{1}{b^n}}$$

where n is any positive whole number.

Example

Evaluate

(a) 3^2 **(b)** 4^3 **(c)** 7^0 **(d)** 5^1 **(e)** 5^{-1}

(f) $(-2)^6$ **(g)** 3^{-4} **(h)** $(-2)^{-3}$ **(i)** $(1.723)^0$

Solution

Using the definitions

$$b^n = b \times b \times b \times \ldots \times b$$

$$b^0 = 1$$

$$b^{-n} = \frac{1}{b^n}$$

we obtain

(a) $3^2 = 3 \times 3 = 9$

(b) $4^3 = 4 \times 4 \times 4 = 64$

(c) $7^0 = 1$

(d) $5^1 = 5$

(e) $5^{-1} = \frac{1}{5^1} = \frac{1}{5}$

(f) $(-2)^6 = (-2) \times (-2) \times (-2) \times (-2) \times (-2) \times (-2) = 64$

where the answer is positive because there are an even number of minus signs.

(g) $3^{-4} = \frac{1}{3^4} = \frac{1}{3 \times 3 \times 3 \times 3} = \frac{1}{81}$

(h) $(-2)^{-3} = \frac{1}{(-2)^3} = \frac{1}{(-2) \times (-2) \times (-2)} = -\frac{1}{8}$

where the answer is negative because there are an odd number of minus signs.

(i) $(1.723)^0 = 1$

because any number raised to the power of zero equals 1.

Practice Problem

1 **(1)** Without using a calculator evaluate

(a) 10^2 **(b)** 10^1 **(c)** 10^0 **(d)** 10^{-1} **(e)** 10^{-2} **(f)** $(-1)^{100}$

(g) $(-1)^{99}$ **(h)** 7^{-3} **(i)** $(-9)^2$ **(j)** $(72\ 101)^1$ **(k)** $(2.718)^0$

(2) Confirm your answer to part (1) using a calculator.

We handle fractional powers in two stages. We begin by defining b^m where m is a reciprocal such as $\frac{1}{2}$ or $\frac{1}{8}$ and then consider more general fractions such as $\frac{3}{4}$ or $\frac{3}{8}$ later. Assuming that n is a positive whole number, we define

$$\boxed{b^{1/n} = n\text{th root of } b}$$

By this we mean that $b^{1/n}$ is a number which, when raised to the power n, produces b. In symbols, if $c = b^{1/n}$ then $c^n = b$. Using this definition,

$9^{1/2}$ = square root of 9 = 3 (because $3^2 = 9$)

$8^{1/3}$ = cube root of 8 = 2 (because $2^3 = 8$)

$625^{1/4}$ = fourth root of 625 = 5 (because $5^4 = 625$)

Of course, the nth root of a number may not exist. There is no number c satisfying $c^2 = -4$, for example, and so $(-4)^{1/2}$ is not defined. It is also possible for some numbers to have more than one nth root. For example, there are two values of c which satisfy $c^4 = 16$, namely $c = 2$ and $c = -2$. In these circumstances it is standard practice to take the positive root, so $16^{1/4} = 2$.

We now turn our attention to the case of b^m, where m is a general fraction of the form p/q for some whole numbers p and q. What interpretation are we going to put on a number such as $16^{3/4}$? To be consistent with our previous definitions, the numerator, 3, can be thought of as an instruction for us to raise 16 to the power of 3, and the denominator tells us to take the fourth root. In fact, it is immaterial in which order these two operations are carried out. If we begin by cubing 16 we get

$$16^3 = 16 \times 16 \times 16 = 4096$$

and taking the fourth root of this gives

$$16^{3/4} = (4096)^{1/4} = 8 \quad \text{(because } 8^4 = 4096\text{)}$$

On the other hand, taking the fourth root first gives

$$16^{1/4} = 2 \quad \text{(because } 2^4 = 16\text{)}$$

and cubing this gives

$$16^{3/4} = 2^3 = 8$$

which is the same answer as before. We therefore see that

$$(16^3)^{1/4} = (16^{1/4})^3$$

This result holds for any base b and fraction p/q (provided that q is positive), so we define

$$\boxed{b^{p/q} = (b^p)^{1/q} = (b^{1/q})^p}$$

Example

Evaluate

(a) $8^{4/3}$ (b) $25^{-3/2}$

Solution

(a) To evaluate $8^{4/3}$ we need both to raise the number to the power of 4 and to find a cube root. Choosing to find the cube root first,

$$8^{4/3} = (8^{1/3})^4 = 2^4 = 16$$

(b) Again it is easy to find the square root of 25 first before raising the number to the power of −3, so

$$25^{-3/2} = (25^{1/2})^{-3} = 5^{-3} = \frac{1}{5^3} = \frac{1}{125}$$

For this particular exponential form we have actually carried out three distinct operations. The minus sign tells us to reciprocate, the fraction $^1/_2$ tells us to take the square root and the 3 tells us to cube. You might like to check for yourself that you get the same answer irrespective of the order in which these three operations are performed.

Advice

Given that we are allowed to perform these operations in any order, it is usually easier to find the qth root first to avoid having to spot roots of large numbers.

Practice Problem

2 (1) Without using your calculator, evaluate

(a) $16^{1/2}$ **(b)** $27^{1/3}$ **(c)** $4^{5/2}$ **(d)** $8^{-2/3}$ **(e)** $1^{-17/25}$

(2) Confirm your answer to part (1) using a calculator.

2.3.2 Rules of indices

There are two reasons why the exponential form is useful. Firstly, it is a convenient shorthand for what otherwise might be a very lengthy number. The exponential form

$$9^8$$

is much easier to write down than either of the equivalent forms

$$9 \times 9 \times 9 \times 9 \times 9 \times 9 \times 9 \times 9$$

or

$$43\ 046\ 721$$

Secondly, there are four basic rules of indices which facilitate the manipulation of such numbers. The four rules may be stated as follows:

Rule 1 $b^m \times b^n = b^{m+n}$
Rule 2 $b^m \div b^n = b^{m-n}$
Rule 3 $(b^m)^n = b^{mn}$
Rule 4 $(ab)^n = a^n b^n$

It is certainly not our intention to provide mathematical proofs in this book. However, it might help you to remember these rules if we give you a justification based on some simple examples. We consider each rule in turn.

Rule 1

Suppose we want to multiply together 2^2 and 2^5. Now $2^2 = 2 \times 2$ and $2^5 = 2 \times 2 \times 2 \times 2 \times 2$, so

$$2^2 \times 2^5 = (2 \times 2) \times (2 \times 2 \times 2 \times 2 \times 2)$$

Notice that we are multiplying together a total of seven 2s and so by definition this is just 2^7: that is,

$$2^2 \times 2^5 = 2^7 = 2^{2+5}$$

This confirms rule 1, which tells you that if you multiply two numbers, all you have to do is to add the indices.

Rule 2

Suppose we want to divide 2^2 by 2^5. This gives

$$\frac{\not{2} \times \not{2}}{2 \times 2 \times 2 \times \not{2} \times \not{2}} = \frac{1}{2 \times 2 \times 2} = \frac{1}{2^3}$$

Now, by definition, reciprocals are denoted by negative indices, so this is just 2^{-3}: that is,

$$2^2 \div 2^5 = 2^{-3} = 2^{2-5}$$

This confirms rule 2, which tells you that if you divide two numbers, all you have to do is to subtract the indices.

Rule 3

Suppose we want to raise 10^2 to the power 3. By definition, for any number b,

$$b^3 = b \times b \times b$$

so replacing b by 10^2 we have

$$(10^2)^3 = 10^2 \times 10^2 \times 10^2 = (10 \times 10) \times (10 \times 10) \times (10 \times 10) = 10^6$$

because there are six 10s multiplied together: that is,

$$(10^2)^3 = 10^6 = 10^{2\times3}$$

This confirms rule 3, which tells you that if you take a 'power of a power', all you have to do is to multiply the indices.

Rule 4

Suppose we want to raise 2×3 to the power 4. By definition,

$$b^4 = b \times b \times b \times b$$

so replacing b by 2×3 gives

$$(2 \times 3)^4 = (2 \times 3) \times (2 \times 3) \times (2 \times 3) \times (2 \times 3)$$

and, because it does not matter in which order numbers are multiplied, this can be written as

$$(2 \times 2 \times 2 \times 2) \times (3 \times 3 \times 3 \times 3)$$

that is,

$$(2 \times 3)^4 = 2^4 \times 3^4$$

This confirms rule 4, which tells you that if you take the power of a product of two numbers, all you have to do is to take the power of each number separately and multiply.

A word of warning is in order regarding these laws. Notice that in rules 1 and 2 the bases of the numbers involved are the same. These rules do not apply if the bases are different. For example, rule 1 gives no information about

$$2^4 \times 3^5$$

Similarly, please notice that in rule 4 the numbers a and b are multiplied together. For some strange reason, some business and economics students seem to think that rule 4 also applies to addition, so that

$$(a + b)^n = a^n + b^n \qquad \text{This statement is } \textbf{NOT TRUE}.$$

It would make algebraic manipulation a whole lot easier if it were true, but I am afraid to say that it is definitely false! If you need convincing of this, note, for example, that

$$(1 + 2)^3 = 3^3 = 27$$

which is not the same as

$$1^3 + 2^3 = 1 + 8 = 9$$

One variation of rule 4 which is true is

$$\left(\frac{a}{b}\right)^n = \frac{a^n}{b^n} \quad (b \neq 0)$$

This is all right because division (unlike addition or subtraction) is the same sort of operation as multiplication. In fact,

$$\left(\frac{a}{b}\right)^n$$

can be thought of as

$$\left(a \times \frac{1}{b}\right)^n$$

so applying rule 4 to this product gives

$$a^n \times \left(\frac{1}{b}\right)^n = \frac{a^n}{b^n}$$

as required.

Advice

There might be occasions (such as in examinations!) when you only half remember a rule or perhaps think that you have discovered a brand new rule for yourself. If you are ever worried about whether some rule is legal or not, you should always check it out by trying numbers, just as we did for $(a + b)^n$. Obviously, one numerical example which actually works does not prove that the rule will always work. However, one example which fails is good enough to tell you that your supposed rule is rubbish.

The following example demonstrates how rules 1–4 are used to simplify algebraic expressions.

Example

Simplify

(a) $x^{1/4} \times x^{3/4}$ **(b)** $\dfrac{x^2y^3}{x^4y}$ **(c)** $(x^2y^{-1/3})^3$

Solution

(a) The expression

$$x^{1/4} \times x^{3/4}$$

represents the product of two numbers in exponential form with the same base. From rule 1 we may add the indices to get

$$x^{1/4} \times x^{3/4} = x^{1/4+3/4} = x^1$$

which is just x.

(b) The expression

$$\frac{x^2y^3}{x^4y}$$

is more complicated than that in part (a) since it involves numbers in exponential form with two different bases, x and y. From rule 2,

$$\frac{x^2}{x^4}$$

may be simplified by subtracting indices to get

$$x^2 \div x^4 = x^{2-4} = x^{-2}$$

Similarly,

$$\frac{y^3}{y} = y^3 \div y^1 = y^{3-1} = y^2$$

Hence

$$\frac{x^2y^3}{x^4y} = x^{-2}y^2$$

It is not possible to simplify this any further, because x^{-2} and y^2 have different bases. However, if you prefer, this can be written as

$$\frac{y^2}{x^2}$$

because negative powers denote reciprocals.

(c) An obvious first step in the simplification of

$$(x^2y^{-1/3})^3$$

is to apply rule 4, treating x^2 as the value of a and $y^{-1/3}$ as b to get

$$(x^2y^{-1/3})^3 = (x^2)^3(y^{-1/3})^3$$

Rule 3 then allows us to write

$$(x^2)^3 = x^{2\times 3} = x^6$$

$$(y^{-1/3})^3 = y^{(-1/3)\times 3} = y^{-1}$$

Hence

$$(x^2y^{-1/3})^3 = x^6y^{-1}$$

As in part (b), if you think it looks neater, you can write this as

$$\frac{x^6}{y}$$

because negative powers denote reciprocals.

Practice Problem

3 Simplify

(a) $(x^{3/4})^8$ (b) $\frac{x^2}{x^{3/2}}$ (c) $(x^2y^4)^3$ (d) $\sqrt{x}(x^{5/2} + y^3)$

[Hint: in part (d) note that $\sqrt{x} = x^{1/2}$ and multiply out the brackets.]

There are occasions throughout this book when we use the rules of indices and definitions of b^n. For the moment, we concentrate on one specific application where we see these ideas in action. The output, Q, of any production process depends on a variety of inputs, known as ***factors of production***. These comprise land, capital, labour and enterprise. For simplicity we restrict our attention to capital and labour. ***Capital***, K, denotes all man-made aids to production such as buildings, tools and plant machinery. ***Labour***, L, denotes all paid work in the production process. The dependence of Q on K and L may be written

$$Q = f(K, L)$$

which is called a ***production function***. Once this relationship is made explicit, in the form of a formula, it is straightforward to calculate the level of production from any given combination of inputs. For example, if

$$Q = 100K^{1/3}L^{1/2}$$

then the inputs $K = 27$ and $L = 100$ lead to an output

$$\begin{aligned} Q &= 100(27)^{1/3}(100)^{1/2} \\ &= 100(3)(10) \\ &= 3000 \end{aligned}$$

Of particular interest is the effect on output when inputs are scaled in some way. If capital and labour both double, does the production level also double, does it go up by more than double or does it go up by less than double? For the particular production function,

$$Q = 100K^{1/3}L^{1/2}$$

we see that, when K and L are replaced by $2K$ and $2L$, respectively,

$$Q = 100(2K)^{1/3}(2L)^{1/2}$$

Now, by rule 4,

$$(2K)^{1/3} = 2^{1/3}K^{1/3} \quad \text{and} \quad (2L)^{1/2} = 2^{1/2}L^{1/2}$$

so

$$\begin{aligned} Q &= 100(2^{1/3}K^{1/3})(2^{1/2}L^{1/2}) \\ &= (2^{1/3}2^{1/2})(100K^{1/3}L^{1/2}) \end{aligned}$$

The second term, $100K^{1/3}L^{1/2}$, is just the original value of Q, so we see that the output is multiplied by

$$2^{1/3}2^{1/2}$$

Using rule 1, this number may be simplified by adding the indices to get

$$2^{1/3}2^{1/2} = 2^{5/6}$$

Moreover, because 5/6 is less than 1, the scale factor is smaller than 2. In fact, my calculator gives

$$2^{5/6} = 1.78 \quad \text{(to 2 decimal places)}$$

so output goes up by just less than double.

It is important to notice that the above argument does not depend on the particular value, 2, that is taken as the scale factor. Exactly the same procedure can be applied if the inputs, K and L, are scaled by a general number λ (where λ is a Greek letter pronounced 'lambda'). Replacing K and L by λK and λL respectively in the formula

$$Q = 100K^{1/3}L^{1/2}$$

gives

$$\begin{aligned} Q &= 100(\lambda K)^{1/3}(\lambda L)^{1/2} \\ &= 100\lambda^{1/3}K^{1/3}\lambda^{1/2}L^{1/2} && \text{(rule 4)} \\ &= (\lambda^{1/3}\lambda^{1/2})(100K^{1/3}L^{1/2}) \\ &= \lambda^{5/6}(100K^{1/3}L^{1/2}) && \text{(rule 1)} \end{aligned}$$

We see that the output gets scaled by $\lambda^{5/6}$, which is smaller than λ since the power, 5/6, is less than 1. We describe this by saying that the production function exhibits decreasing returns to scale.

In general, a function

$$Q = f(K, L)$$

is said to be *homogeneous* if

$$f(\lambda K, \lambda L) = \lambda^n f(K, L)$$

for some number, n. This means that when both variables K and L are multiplied by λ we can pull out all of the λs as a common factor, λ^n. The power, n, is called the **degree of homogeneity**. In the previous example we showed that

$$f(\lambda K, \lambda L) = \lambda^{5/6} f(K, L)$$

and so it is homogeneous of degree 5/6. In general, if the degree of homogeneity, n, satisfies:

- $n < 1$, the function is said to display *decreasing returns to scale*
- $n = 1$, the function is said to display *constant returns to scale*
- $n > 1$, the function is said to display *increasing returns to scale*.

Example

Show that the following production function is homogeneous and find its degree of homogeneity:

$$Q = 2K^{1/2}L^{3/2}$$

Does this function exhibit decreasing returns to scale, constant returns to scale or increasing returns to scale?

Solution

We are given that

$$f(K, L) = 2K^{1/2}L^{3/2}$$

so replacing K by λK and L by λL gives

$$f(\lambda K, \lambda L) = 2(\lambda K)^{1/2}(\lambda L)^{3/2}$$

➔

We can pull out all of the λs by using rule 4 to get

$$2\lambda^{1/2}K^{1/2}\lambda^{3/2}L^{3/2}$$

and then using rule 1 to get

$$\lambda^2(2K^{1/2}L^{3/2})$$

$\lambda^{1/2}\lambda^{3/2} = \lambda^{1/2+3/2} = \lambda^2$

We have therefore shown that

$$f(\lambda K, \lambda L) = \lambda^2 f(K, L)$$

and so the function is homogeneous of degree 2. Moreover, since $2 > 1$ we deduce that it has increasing returns to scale.

Practice Problem

4 Show that the following production functions are homogeneous and comment on their returns to scale:

(a) $Q = 7KL^2$ **(b)** $Q = 50K^{1/4}L^{3/4}$

You may well have noticed that all of the production functions considered so far are of the form

$$Q = AK^{\alpha}L^{\beta}$$

for some positive constants, A, α and β. (The Greek letters α and β are pronounced 'alpha' and 'beta' respectively.) Such functions are called ***Cobb–Douglas*** production functions. It is easy to see that they are homogeneous of degree $\alpha + \beta$ because if

$$f(K, L) = AK^{\alpha}L^{\beta}$$

then

$$\begin{aligned} f(\lambda K, \lambda L) &= A(\lambda K)^{\alpha}(\lambda L)^{\beta} \\ &= A\lambda^{\alpha}K^{\alpha}\lambda^{\beta}L^{\beta} \quad \text{(rule 4)} \\ &= \lambda^{\alpha+\beta}(AK^{\alpha}L^{\beta}) \quad \text{(rule 1)} \\ &= \lambda^{\alpha+\beta}f(K, L) \end{aligned}$$

Consequently, Cobb–Douglas production functions exhibit

- decreasing returns to scale, if $\alpha + \beta < 1$
- constant returns to scale, if $\alpha + \beta = 1$
- increasing returns to scale, if $\alpha + \beta > 1$.

By the way, not all production functions are of this type. Indeed, it is not even necessary for a production function to be homogeneous. Some examples illustrating these cases are given in Practice Problems 13 and 14 at the end of this section. We shall return to the topic of production functions in Chapter 5.

2.3.3 Logarithms

At the beginning of this section we stated that if a number, M, is expressed as

$$M = b^n$$

then b^n is called the exponential form of M to base b. The approach taken so far has simply been to evaluate M from any given values of b and n. In practice, it may be necessary to reverse this process and to find n from known values of M and b. To solve the equation

$$32 = 2^n$$

we need to express 32 as a power of 2. In this case it is easy to work out n by inspection. Simple trial and error easily gives $n = 5$ because

$$2^5 = 32$$

We describe this expression by saying that the logarithm of 32 to base 2 is 5. In symbols we write

$$\log_2 32 = 5$$

Quite generally,

if $M = b^n$ then $\log_b M = n$

when n is called the *logarithm of* M *to base* b.

Advice

Students have been known to regard logarithms as something rather abstract and difficult to understand. There is, however, no need to worry about logarithms, since they simply provide an alternative way of thinking about numbers such as b^n. Read through the following example and try Practice Problem 5 for yourself. You might discover that they are easier than you expect!

Example

Evaluate

(a) $\log_3 9$ **(b)** $\log_4 2$ **(c)** $\log_7 {}^1\!/\!_7$

Solution

(a) To find the value of $\log_3 9$ we convert the problem into one involving powers. From the definition of a logarithm to base 3 we see that the statement

➔

$\log_3 9 = n$

is equivalent to

$9 = 3^n$

The problem of finding the logarithm of 9 to base 3 is exactly the same as that of writing 9 as a power of 3. The solution of this equation is clearly $n = 2$ since

$9 = 3^2$

Hence $\log_3 9 = 2$.

(b) Again to evaluate $\log_4 2$ we merely rewrite

$\log_4 2 = n$

in exponential form as

$2 = 4^n$

The problem of finding the logarithm of 2 to base 4 is exactly the same as that of writing 2 as a power of 4. The value of 2 is obtained from 4 by taking the square root, which involves raising 4 to the power of $1/2$, so

$2 = 4^{1/2}$

Hence $\log_2 4 = 1/2$.

(c) If

$\log_7 1/7 = n$

then

$1/7 = 7^n$

The value of $1/7$ is found by taking the reciprocal of 7, which involves raising 7 to the power of -1: that is,

$1/7 = 7^{-1}$

Hence $\log_7 1/7 = -1$.

Practice Problem

5 **(1)** Write down the values of *n* which satisfy

(a) $1000 = 10^n$ **(b)** $100 = 10^n$ **(c)** $10 = 10^n$

(d) $1 = 10^n$ **(e)** $\frac{1}{10} = 10^n$ **(f)** $\frac{1}{100} = 10^n$

(2) Use your answer to part (1) to write down the values of

(a) $\log_{10} 1000$ **(b)** $\log_{10} 100$ **(c)** $\log_{10} 10$

(d) $\log_{10} 1$ **(e)** $\log_{10} 1/10$ **(f)** $\log_{10} 1/100$

(3) Confirm your answer to part (2) using a calculator.

Given the intimate relationship between exponentials and logarithms, you should not be too surprised to learn that logarithms satisfy three rules that are comparable with those for indices. The rules of logarithms are as follows.

Rule 1 $\log_b(x \times y) = \log_b x + \log_b y$
Rule 2 $\log_b(x \div y) = \log_b x - \log_b y$
Rule 3 $\log_b x^m = m\log_b x$

A long time ago, before the pocket calculator was invented, people used tables of logarithms to perform complicated arithmetic calculations. It was generally assumed that everyone could add or subtract numbers using pen and paper, but that people found it hard to multiply and divide. The first two rules gave a means of converting calculations involving multiplication and division into easier calculations involving addition and subtraction. For example, to work out

$$1.765\ 12 \times 25.329\ 71$$

we would first look up the logarithms of 1.765 12 and 25.329 71 using tables and then add these logarithms together on paper. According to rule 1, the value obtained is just the logarithm of the answer. Finally, using tables of antilogarithms (which in effect raised the base to an appropriate power), the result of the calculation was obtained. Fortunately for us, this is all history and we can now perform arithmetic calculations in a fraction of the time it took our predecessors to multiply or divide two numbers. This might suggest that logarithms are redundant. However, the idea of a logarithm remains an important one. The logarithm function itself – that is,

$$f(x) = \log_b(x)$$

is of value and we shall investigate its properties later in the book. For the time being we first show how to use the laws of logarithms in algebra and then demonstrate how logarithms can be used to solve algebraic equations in which the unknown appears as a power. This technique will be of particular use in the next chapter when we solve compound interest problems.

Example

Use the rules of logarithms to express each of the following as a single logarithm:

(a) $\log_b x + \log_b y - \log_b z$ **(b)** $2\log_b x - 3\log_b y$

Solution

(a) The first rule of logs shows that the *sum* of two logs can be written as the log of a *product*, so

$$\log_b x + \log_b y - \log_b z = \log_b(xy) - \log_b z$$

Also, according to rule 2, the *difference* of two logs is the log of a *quotient*, so we can simplify further to get

$$\log_b\left(\frac{xy}{z}\right)$$

(b) Given any combination of logs such as

$$2\log_b x - 3\log_b y$$

➔

the trick is to use the third rule to 'get rid' of the coefficients. Since

$$2\log_b x = \log_b x^2 \text{ and } 3\log_b y = \log_b y^3$$

we see that

$$2\log_b x - 3\log_b y = \log_b x^2 - \log_b y^3$$

Only now can we use the second rule of logs, which allows us to write the expression as the single logarithm

$$\log_b\left(\frac{x^2}{y^3}\right)$$

Practice Problem

6 Use the rules of logs to express each of the following as a single logarithm:

(a) $\log_b x - \log_b y + \log_b z$ **(b)** $4\log_b x + 2\log_b y$

Example

Find the value of x which satisfies

(a) $200(1.1)^x = 20\,000$ (b) $5^x = 2(3)^x$

Solution

(a) An obvious first step in the solution of

$$200(1.1)^x = 20\,000$$

is to divide both sides by 200 to get

$$(1.1)^x = 100$$

In Chapter 1 it was pointed out that we can do whatever we like to an equation, provided that we do the same thing to both sides. In particular, we may take logarithms of both sides to get

$$\log(1.1)^x = \log(100)$$

Now by rule 3 we have

$$\log(1.1)^x = x\log(1.1)$$

so the equation becomes

$$x\log(1.1) = \log(100)$$

Notice the effect that rule 3 has on the equation. It brings the unknown down to the same level as the rest of the expression. This is the whole point of taking logarithms, since it converts an equation in which the unknown appears as a power into one which can be solved using familiar algebraic methods. Dividing both sides of the equation

$$x\log(1.1) = \log(100)$$

by log(1.1) gives

$$x = \frac{\log(100)}{\log(1.1)}$$

So far no mention has been made of the base of the logarithm. The above equation for x is true no matter what base is used. It makes sense to use logarithms to base 10 because all scientific calculators have this facility as one of their function keys. Using base 10, my calculator gives

$$x = \frac{\log(100)}{\log(1.1)} = \frac{2}{0.041\ 392\ 685} = 48.32$$

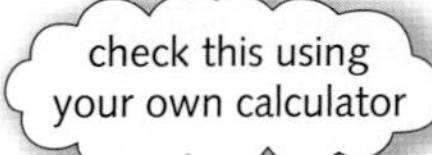

to 2 decimal places.

As a check, if this number is substituted back into the original equation, then

$$200(1.1)^x = 200(1.1)^{48.32} = 20\ 004 \quad ✓$$

We cannot expect to obtain the exact answer, because we rounded x to only two decimal places.

(b) To solve

$$5^x = 2(3)^x$$

we take logarithms of both sides to get

$$\log(5^x) = \log(2 \times 3^x)$$

The right-hand side is the logarithm of a product and, according to rule 1, can be written as the sum of the logarithms, so the equation becomes

$$\log(5^x) = \log(2) + \log(3^x)$$

As in part (a) the key step is to use rule 3 to 'bring down the powers'. If rule 3 is applied to both $\log(5^x)$ and $\log(3^x)$ then the equation becomes

$$x\log(5) = \log(2) + x\log(3)$$

This is now the type of equation that we know how to solve. We collect x's on the left-hand side to get

$$x\log(5) - x\log(3) = \log(2)$$

and then pull out a common factor of x to get

$$x[\log(5) - \log(3)] = \log(2)$$

Now, by rule 2, the difference of two logarithms is the same as the logarithm of their quotient, so

$$\log(5) - \log(3) = \log(5 \div 3)$$

Hence the equation becomes

$$x\log\left(\frac{5}{3}\right) = \log(2)$$

so

$$x = \frac{\log(2)}{\log(5/3)}$$

Finally, taking logarithms to base 10 using a calculator gives

$$x = \frac{0.301\ 029\ 996}{0.221\ 848\ 750} = 1.36$$

to 2 decimal places.

➔

As a check, the original equation

$$5^x = 2(3)^x$$

becomes

$$5^{1.36} = 2(3)^{1.36}$$

that is,

$$8.92 = 8.91 \quad ✓$$

Again the slight discrepancy is due to rounding errors in the value of x.

Practice Problem

7 Solve the following equations for x:

(a) $3^x = 7$ **(b)** $5(2)^x = 10^x$

Advice

In this section we have met a large number of definitions and rules concerning indices and logarithms. For convenience, we have collected these together in the form of a summary. The facts relating to indices are particularly important and you should make every effort to memorize these before proceeding with the rest of this book.

2.3.4 Summary

Indices

If n is a positive whole number then

$$b^n = b \times b \times \ldots \times b$$

$$b^0 = 1$$

$$b^{-n} = 1/b^n$$

$$b^{1/n} = n\text{th root of } b$$

Also, if p and q are whole numbers with $q > 0$ then

$$b^{p/q} = (b^p)^{1/q} = (b^{1/q})^p$$

The four rules of indices are:

Rule 1 $b^m \times b^n = b^{m+n}$

Rule 2 $b^m \div b^n = b^{m-n}$

Rule 3 $(b^m)^n = b^{mn}$

Rule 4 $(ab)^n = a^n b^n$

Logarithms

If $M = b^n$ then $n = \log_b M$. The three rules of logarithms are:

Rule 1 $\log_b(x \times y) = \log_b x + \log_b y$

Rule 2 $\log_b(x \div y) = \log_b x - \log_b y$

Rule 3 $\log_b x^m = m\log_b x$

Key Terms

Capital Man-made assets used in the production of goods and services.

Cobb–Douglas production function A production function of the form: $Q = AK^{\alpha}L^{\beta}$.

Constant returns to scale Exhibited by a production function when a given percentage increase in input leads to the same percentage increase in output: $f(\lambda K, \lambda L) = \lambda f(K, L)$.

Decreasing returns to scale Exhibited by a production function when a given percentage increase in input leads to a smaller percentage increase in output: $f(\lambda K, \lambda L) = \lambda^n f(K, L)$ where $0 < n < 1$.

Exponent A superscript attached to a variable; the number 5 is the exponent in the expression, $2x^5$.

Exponential form A representation of a number which is written using powers. For example, 2^5 is the exponential form of the number 32.

Factors of production The inputs into the production of goods and services: labour, land, capital and raw materials.

Homogeneous functions A function with the property that when all of the inputs are multiplied by a constant, λ, the output is multiplied by λ^n where n is the degree of homogeneity.

Increasing returns to scale Exhibited by a production function when a given percentage increase in input leads to a larger percentage increase in output: $f(\lambda K, \lambda L) = \lambda^n f(K, L)$ where $n > 1$.

Index Another word for exponent.

Labour All forms of human input to the production process.

Logarithm The power to which a base must be raised to yield a particular number.

Power Another word for exponent. If this is a positive integer then it gives the number of times a number is multiplied by itself.

Production function The relationship between the output of a good and the inputs used to produce it.

Practice Problems

8 **(1)** Without using your calculator evaluate

(a) 8^2 **(b)** 2^1 **(c)** 3^{-1} **(d)** 17^0 **(e)** $1^{1/5}$ **(f)** $36^{1/2}$ **(g)** $8^{2/3}$ **(h)** $49^{-3/2}$

(2) Confirm your answer to part (1) using a calculator.

9 Evaluate the following without using a calculator

(a) $32^{3/5}$ **(b)** $64^{-5/6}$ **(c)** $\left(\frac{1}{125}\right)^{-4/3}$ **(d)** $\left(3\frac{3}{8}\right)^{2/3}$ **(e)** $\left(2\frac{1}{4}\right)^{-1/2}$

➔

10 Use the rules of indices to simplify

(a) $y^{3/2} \times y^{1/2}$ **(b)** $\frac{x^2y}{xy^{-1}}$ **(c)** $(xy^{1/2})^4$

(d) $(p^2)^{1/3} \div (p^{1/3})^2$ **(e)** $(24q)^{1/3} \div (3q)^{1/3}$ **(f)** $(25p^2q^4)^{1/2}$

11 Write the following expressions using index notation

(a) $\frac{1}{x^4}$ **(b)** $5\sqrt{x}$ **(c)** $\frac{1}{\sqrt{x}}$ **(d)** $2x\sqrt{x}$ **(e)** $\frac{8}{x(\sqrt[3]{x})}$

12 For the production function, $Q = 200K^{1/4}L^{2/3}$ find the output when

(a) $K = 16, L = 27$ **(b)** $K = 10\,000, L = 1000$

13 Which of the following production functions are homogeneous? For those functions which are homogeneous write down their degrees of homogeneity and comment on their returns to scale.

(a) $Q = 500K^{1/3}L^{1/4}$

(b) $Q = 3LK + L^2$

(c) $Q = L + 5L^2K^3$

14 Show that the production function

$$Q = A[bK^{\alpha} + (1 - b)L^{\alpha}]^{1/\alpha}$$

is homogeneous and displays constant returns to scale.

15 Write down the values of x which satisfy each of the following equations:

(a) $5^x = 25$ **(b)** $3^x = \frac{1}{3}$ **(c)** $2^x = \frac{1}{8}$

(d) $2^x = 64$ **(e)** $100^x = 10$ **(f)** $8^x = 16$

16 Solve the following equations:

(a) $2^{3x} = 4$ **(b)** $4 \times 2^x = 32$ **(c)** $8^x = 2 \times \left(\frac{1}{2}\right)^x$

17 Write down the value of

(a) $\log_b b^2$ **(b)** $\log_b b$ **(c)** $\log_b 1$ **(d)** $\log_b \sqrt{b}$ **(e)** $\log_b(1/b)$

18 Use the rules of logs to express each of the following as a single log:

(a) $\log_b(xy) - \log_b x - \log_b y$

(b) $3\log_b x - 2\log_b y$

(c) $\log_b y + 5\log_b x - 2\log_b z$

19 Express the following in terms of $\log_b x$, $\log_b y$ and $\log_b z$:

(a) $\log_b(x^2y^3z^4)$

(b) $\log_b\left(\frac{x^4}{y^2z^5}\right)$

(c) $\log_b\left(\frac{x}{\sqrt{yz}}\right)$

20 If $\log_b 2 = p$, $\log_b 3 = q$ and $\log_b 10 = r$, express the following in terms of p, q and r:

(a) $\log_b\left(\frac{1}{3}\right)$ **(b)** $\log_b 12$ **(c)** $\log_b 0.0003$ **(d)** $\log_b 600$

21 Solve the following equations:

(a) $10(1.07)^x = 2000$ **(b)** $10^{x-1} = 3$ **(c)** $5^{x-2} = 5$ **(d)** $2(7)^{-x} = 3^x$

22 Solve the inequalities

(a) $3^{2x+1} \le 7$ **(b)** $0.8^x < 0.04$

23 Solve the equation

$$\log_{10}(x+2) + \log_{10}x - 1 = \log_{10}\left(\frac{3}{2}\right)$$

section 2.4

The exponential and natural logarithm functions

Objectives

At the end of this section you should be able to:

- Sketch graphs of general exponential functions.
- Understand how the number e is defined.
- Use the exponential function to model growth and decay.
- Use log graphs to find unknown parameters in simple models.
- Use the natural logarithm function to solve equations.

In the previous section we described how to define numbers of the form b^x, and discussed the idea of a logarithm, $\log_b x$. It turns out that there is one base (the number e = 2.718 281 . . .) that is particularly important in mathematics. The purpose of this present section is to introduce you to this strange number and to consider a few simple applications.

Example

Sketch the graphs of the functions

(a) $f(x) = 2^x$ **(b)** $g(x) = 2^{-x}$

Comment on the relationship between these graphs.

Solution

(a) As we pointed out in Section 2.3, a number such as 2^x is said to be in exponential form. The number 2 is called the base and x is called the exponent. Values of this function are easily found either by pressing the power key $\boxed{x^y}$ on a calculator or by using the definition of b^n given in Section 2.3. A selection of these are given in the following table:

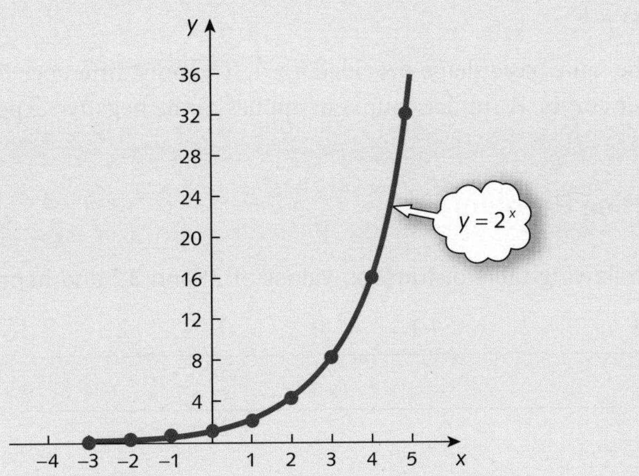

x	−3	−2	−1	0	1	2	3	4	5
2^x	0.125	0.25	0.5	1	2	4	8	16	32

A graph of $f(x)$ based on this table is sketched in Figure 2.14. Notice that the graph approaches the x axis for large negative values of x and it rises rapidly as x increases.

(b) The negative exponential

$$g(x) = 2^{-x}$$

has values

x	−5	−4	−3	−2	−1	0	1	2	3
2^{-x}	32	16	8	4	2	1	0.5	0.25	0.125

This function is sketched in Figure 2.15. It is worth noticing that the numbers appearing in the table of 2^{-x} are the same as those of 2^x but arranged in reverse order. Hence the graph of 2^{-x} is obtained by reflecting the graph of 2^x in the y axis.

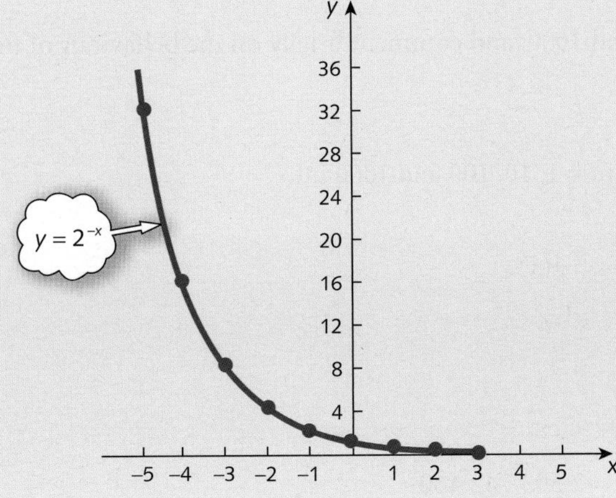

Figure 2.14 displays the graph of a particular exponential function, 2^x. Quite generally, the graph of any exponential function

$$f(x) = b^x$$

has the same basic shape provided $b > 1$. The only difference is that larger values of b produce steeper curves. A similar comment applies to the negative exponential, b^{-x}.

Practice Problem

1 Complete the following table of function values of 3^x and 3^{-x} and hence sketch their graphs.

x	-3	-2	-1	0	1	2	3
3^x							
3^{-x}							

Obviously there is a whole class of functions, each corresponding to a different base, b. Of particular interest is the case when b takes the value

2.718 281 828 459 . . .

This number is written as e and the function

$$f(x) = e^x$$

is referred to as *the exponential function*. In fact, it is not necessary for you to understand where this number comes from. All scientific calculators have an e^x button and you may simply wish to accept the results of using it. However, it might help your confidence if you have some appreciation of how it is defined. To this end consider the following example and subsequent problem.

Example

Evaluate the expression

$$\left(1 + \frac{1}{m}\right)^m$$

where $m = 1$, 10, 100 and 1000, and comment briefly on the behaviour of this sequence.

Solution

Substituting the values $m = 1$, 10, 100 and 1000 into

$$\left(1 + \frac{1}{m}\right)^m$$

gives

$$\left(1 + \frac{1}{1}\right)^1 = 2^1 = 2$$

$$\left(1 + \frac{1}{10}\right)^{10} = (1.1)^{10} = 2.593\ 742\ 460$$

$$\left(1+\frac{1}{100}\right)^{100} = (1.01)^{100} = 2.704\,813\,829$$

$$\left(1+\frac{1}{1000}\right)^{1000} = (1.001)^{1000} = 2.716\,923\,932$$

The numbers are clearly getting bigger as m increases. However, the rate of increase appears to be slowing down, suggesting that numbers are converging to some fixed value.

The following problem gives you an opportunity to continue the sequence and to discover for yourself the limiting value.

Practice Problem

2 (a) Use the power key x^y on your calculator to evaluate

$$\left(1+\frac{1}{m}\right)^m$$

where m = 10 000, 100 000 and 1 000 000.

(b) Use your calculator to evaluate e^1 and compare with your answer to part (a).

Hopefully, the results of Practice Problem 2 should convince you that as m gets larger, the value of

$$\left(1+\frac{1}{m}\right)^m$$

approaches a limiting value of 2.718 281 828 . . . , which we choose to denote by the letter e. In symbols we write

$$e = \lim_{m\to\infty}\left(1+\frac{1}{m}\right)^m$$

The significance of this number can only be fully appreciated in the context of calculus, which we study in Chapter 4. However, it is useful at this stage to consider some preliminary examples. These will give you practice in using the e^x button on your calculator and will give you some idea how this function can be used in modelling.

Advice

The number e has a similar status in mathematics as the number π and is just as useful. It arises in the mathematics of finance, which we discuss in the next chapter. You might like to glance through Section 3.2 now if you need convincing of the usefulness of e.

Example

The percentage, y, of households possessing refrigerators, t years after they have been introduced in a developed country, is modelled by

$$y = 100 - 95e^{-0.15t}$$

→

(1) Find the percentage of households that have refrigerators

  (a) at their launch
  (b) after 1 year
  (c) after 10 years
  (d) after 20 years

(2) What is the market saturation level?

(3) Sketch a graph of y against t and hence give a qualitative description of the growth of refrigerator ownership over time.

Solution

(1) To calculate the percentage of households possessing refrigerators now and in 1, 10 and 20 years' time, we substitute $t = 0$, 1, 10 and 20 into the formula

$$y = 100 - 95e^{-0.15t}$$

to get

  (a) $y(0) = 100 - 95e^{0} = 5\%$
  (b) $y(1) = 100 - 95e^{-0.15} = 18\%$

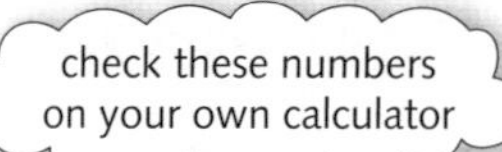

  (c) $y(10) = 100 - 95e^{-1.5} = 79\%$
  (d) $y(20) = 100 - 95e^{-3.0} = 95\%$

(2) To find the saturation level we need to investigate what happens to y as t gets ever larger. We know that the graph of a negative exponential function has the basic shape shown in Figure 2.15. Consequently, the value of $e^{-0.15t}$ will eventually approach zero as t increases. The market saturation level is therefore given by

$$y = 100 - 95(0) = 100\%$$

(3) A graph of y against t, based on the information obtained in parts (1) and (2), is sketched in Figure 2.16.

This shows that y grows rapidly to begin with, but slows down as the market approaches saturation level. A saturation level of 100% indicates that eventually all households are expected to possess refrigerators, which is not surprising given the nature of the product.

Figure 2.16

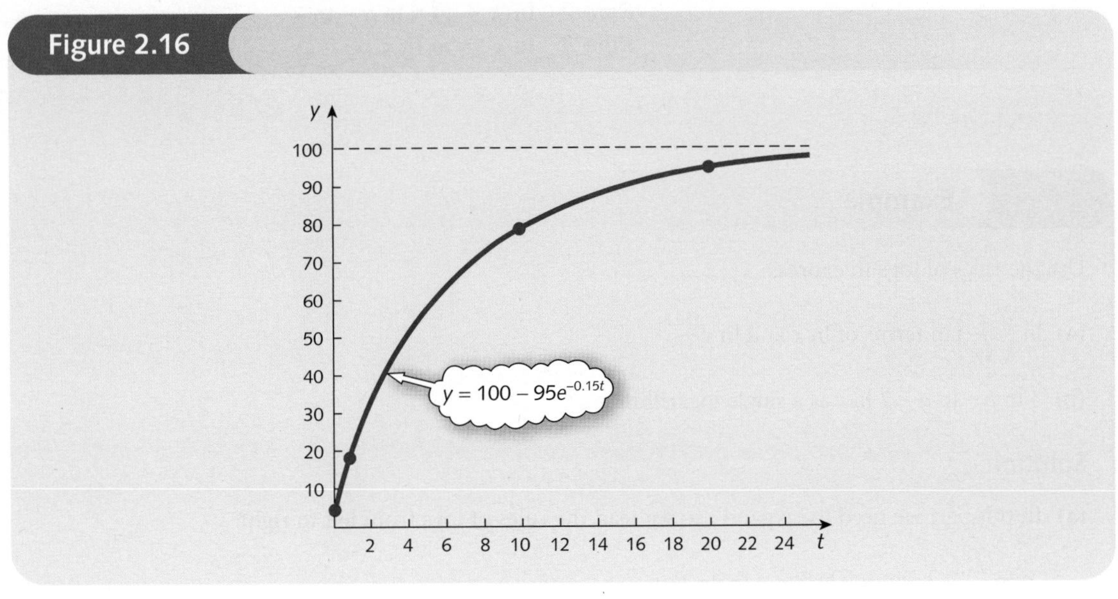

Practice Problem

3 The percentage, y, of households possessing camcorders t years after they have been launched is modelled by

$$y = \frac{55}{1 + 800e^{-0.3t}}$$

(1) Find the percentage of households that have camcorders

(a) at their launch

(b) after 10 years

(c) after 20 years

(d) after 30 years

(2) What is the market saturation level?

(3) Sketch a graph of y against t and hence give a qualitative description of the growth of camcorder ownership over time.

In Section 2.3 we noted that if a number M can be expressed as b^n then n is called the logarithm of M to base b. In particular, for base e,

$$\text{if} \quad M = e^n \quad \text{then} \quad n = \log_e M$$

We call logarithms to base e *natural logarithms*. These occur sufficiently frequently to warrant their own notation. Rather than writing $\log_e M$ we simply put $\ln M$ instead. The three rules of logs can then be stated as

Rule 1 $\ln(x \times y) = \ln x + \ln y$
Rule 2 $\ln(x \div y) = \ln x - \ln y$
Rule 3 $\ln x^m = m \ln x$

Example

Use the rules of logs to express

(a) $\ln\left(\frac{x}{\sqrt{y}}\right)$ in terms of $\ln x$ and $\ln y$

(b) $3 \ln p + \ln q - 2 \ln r$ as a single logarithm

Solution

(a) In this part we need to 'expand', so we read the rules of logs from left to right:

$$\begin{aligned}\ln\left(\frac{x}{\sqrt{y}}\right) &= \ln x - \ln \sqrt{y} && \text{(rule 2)}\\ &= \ln x - \ln y^{1/2} && \text{(fractional powers denote roots)}\\ &= \ln x - \frac{1}{2}\ln y && \text{(rule 3)}\end{aligned}$$

(b) In this part we need to reverse this process and so read the rules from right to left:

$$\begin{aligned}3 \ln p + \ln q - 2 \ln r &= \ln p^3 + \ln q - \ln r^2 && \text{(rule 3)}\\ &= \ln(p^3q) - \ln r^2 && \text{(rule 1)}\\ &= \ln\left(\frac{p^3q}{r^2}\right) && \text{(rule 2)}\end{aligned}$$

Practice Problem

4 Use the rules of logs to express

(a) $\ln(a^2b^3)$ in terms of $\ln a$ and $\ln b$

(b) $\frac{1}{2} \ln x - 3 \ln y$ as a single logarithm

As we pointed out in Section 2.3, logs are particularly useful for solving equations in which the unknown occurs as a power. If the base is the number e then the equation can be solved by using natural logarithms.

Example

An economy is forecast to grow continuously so that the gross national product (GNP), measured in billions of dollars, after t years is given by

$$\text{GNP} = 80e^{0.02t}$$

After how many years is GNP forecast to be $88 billion?

Solution

We need to solve

$$88 = 80e^{0.02t}$$

for t. Dividing through by 80 gives

$$1.1 = e^{0.02t}$$

Using the definition of natural logarithms we know that

$$\text{if} \quad M = e^n \quad \text{then} \quad n = \ln M$$

If we apply this definition to the equation

$$1.1 = e^{0.02t}$$

we deduce that

$$0.02t = \ln 1.1 = 0.095\ 31 \ldots \quad \text{(check this using your own calculator)}$$

so

$$t = \frac{0.095\ 31}{0.02} = 4.77$$

We therefore deduce that GNP reaches a level of $88 billion after 4.77 years.

Practice Problem

5 During a recession a firm's revenue declines continuously so that the revenue, TR (measured in millions of dollars), in t years' time is modelled by

$$\text{TR} = 5e^{-0.15t}$$

(a) Calculate the current revenue and also the revenue in 2 years' time.

(b) After how many years will the revenue decline to $2.7 million?

One important (but rather difficult) problem in modelling is to extract a mathematical formula from a table of numbers. If this relationship is of the form of an exponential then it is possible to estimate values for some of the parameters involved.

Advice

The following example shows how to find such a formula from data points. This is an important skill. However, it is not crucial to your understanding of subsequent material in this book. You may wish to miss this out on first reading and move straight on to the Practice Problems at the end of this chapter.

Example

The values of GNP, g, measured in billions of dollars, over a period of t years was observed to be

t (years)	2	5	10	20
g (billions of dollars)	12	16	27	74

Model the growth of GNP using a formula of the form

$$g = Be^{At}$$

for appropriate values of A and B. Hence estimate the value of GNP after 15 years.

Solution

Figure 2.17 shows the four points plotted with g on the vertical axis and t on the horizontal axis. The basic shape of the curve joining these points certainly suggests that an exponential function is likely to provide a reasonable model, but it gives no information about what values to use for the parameters A and B. However, since one of the unknown parameters, A, occurs as a power in the relation

$$g = Be^{At}$$

it is a good idea to take natural logs of both sides to get

$$\ln g = \ln(Be^{At})$$

Figure 2.17

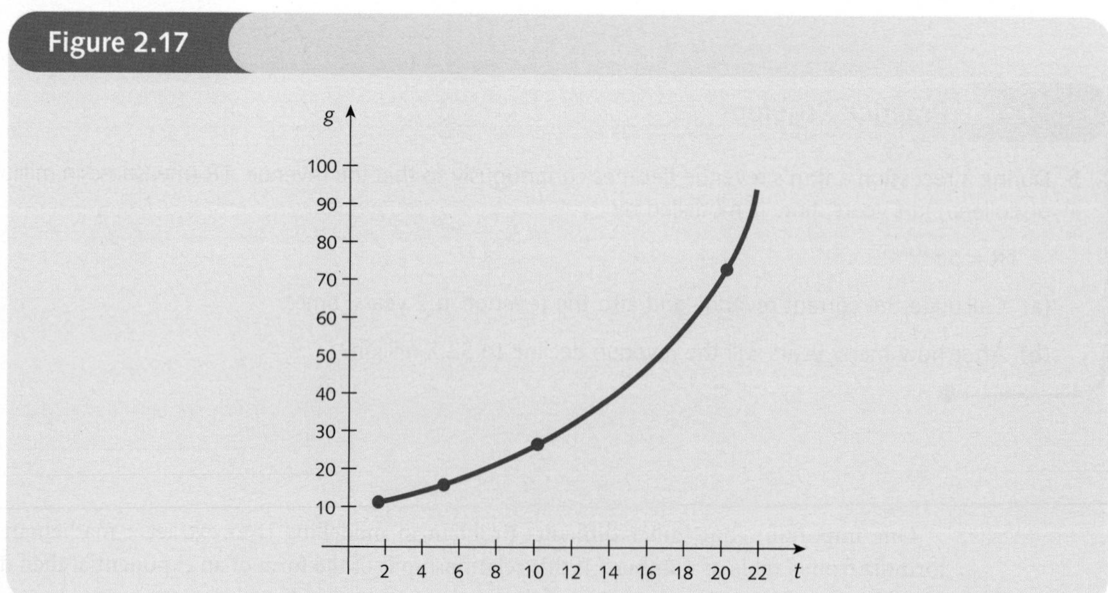

The rules of logs enable us to expand the right-hand side to get

$$\ln(Be^{At}) = \ln B + \ln(e^{At}) \quad \text{(rule 1)}$$
$$= \ln B + At \quad \text{(definition of a log to base e)}$$

Hence

$$\ln g = At + \ln B$$

Although this does not look like it at first sight, this relation is actually the equation of a straight line! To see this recall that the usual equation of a line is $y = ax + b$. The log equation is indeed of this form if we put

$y = \ln g$ and $x = t$

The equation then becomes

$$y = Ax + \ln B$$

so a graph of $\ln g$ plotted on the vertical axis with t plotted on the horizontal axis should produce a straight line with slope A and with an intercept on the vertical axis of $\ln B$.

Figure 2.18 shows this graph based on the table of values

$x = t$	2	5	10	20
$y = \ln g$	2.48	2.77	3.30	4.30

As one might expect, the points do not exactly lie on a straight line, since the formula is only a model. However, the line sketched in Figure 2.18 is a remarkably good fit. The slope can be calculated as

$$A = \frac{4 - 3}{18.6 - 7.6} = 0.09$$

and the vertical intercept can be read off the graph as 2.25. This is $\ln B$ and so

$$B = e^{2.25} = 9.49$$

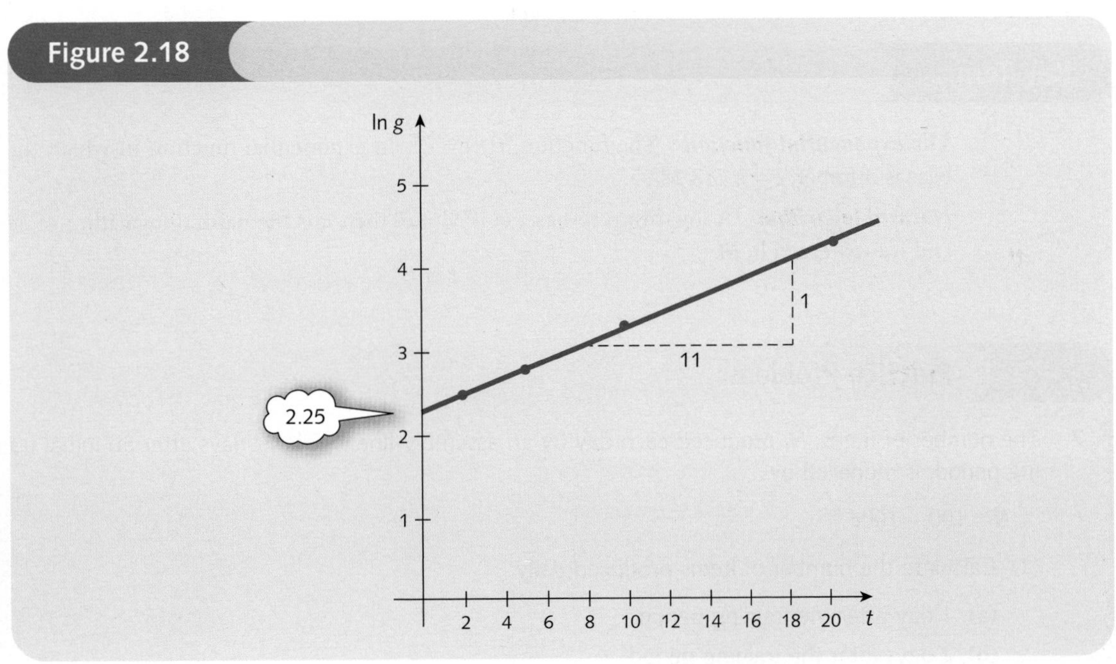

Hence the formula for the approximate relation between g and t is

$$g = 9.49e^{0.09t}$$

Finally, an estimate of the GNP after 15 years can be obtained by substituting $t = 15$ into this formula to get

$$g = 36.6 \quad \text{(billion dollars)}$$

Practice Problem

6 Immediately after the launch of a new product, the monthly sales figures (in thousands) are as follows:

t (months)	1	3	6	12
s (sales)	1.8	2.7	5.0	16.5

(1) Complete the following table of values of ln s

t	1	3	6	12
ln s	0.59		1.61	

(2) Plot these points on graph paper with the values of ln s on the vertical axis and t on the horizontal axis. Draw a straight line passing close to these points. Write down the value of the vertical intercept and calculate the slope.

(3) Use your answers to part (2) to estimate the values of A and B in the relation $s = Be^{At}$.

(4) Use the exponential model derived in part (3) to estimate the sales when

(a) $t = 9$ **(b)** $t = 60$

Which of these estimates would you expect to be the more reliable? Give a reason for your answer.

Key Terms

The exponential function The function, $f(x) = e^x$; an exponential function in which the base is number, e = 2.718 28 . . .

Natural logarithm A logarithm to base, e; if $M = e^n$ then n is the natural logarithm of M and we write, $n = \ln M$

Practice Problems

7 The number of items, N, produced each day by an assembly-line worker, t days after an initial training period, is modelled by

$$N = 100 - 100e^{-0.4t}$$

(1) Calculate the number of items produced daily

(a) 1 day after the training period

(b) 2 days after the training period

(c) 10 days after the training period

(2) What is the worker's daily production in the long run?

(3) Sketch a graph of N against t and explain why the general shape might have been expected.

8 Use the rules of logs to expand each of the following:

(a) $\ln xy$ **(b)** $\ln xy^4$ **(c)** $\ln(xy)^2$

(d) $\ln \dfrac{x^5}{y^7}$ **(e)** $\ln \sqrt{\dfrac{x}{y}}$ **(f)** $\ln \sqrt{\dfrac{xy^3}{z}}$

9 Use the rules of logs to express each of the following as a single logarithm:

(a) $\ln x + 2 \ln x$ **(b)** $4 \ln x - 3 \ln y + 5 \ln z$

10 Solve each of the following equations. (Round your answer to 2 decimal places.)

(a) $e^x = 5.9$ **(b)** $e^x = 0.45$ **(c)** $e^x = -2$

(d) $e^{3x} = 13.68$ **(e)** $e^{-5x} = 0.34$ **(f)** $4e^{2x} = 7.98$

11 The value of a second-hand car reduces exponentially with age, so that its value $\$y$ after t years can be modelled by the formula

$$y = Ae^{-ax}$$

If the car was \$50 000 when new and was worth \$38 000 after 2 years, find the values of A and a, correct to 3 decimal places.

Use this model to predict the value of the car

(a) when the car is 5 years old

(b) in the long run

12 A team of financial advisers guiding the launch of a national newspaper has modelled the future circulation of the newspaper by the equation

$$N = c(1 - e^{-kt})$$

where N is the daily circulation after t days of publication, and c and k are positive constants. Transpose this formula to show that

$$t = \frac{1}{k} \ln\left(\frac{c}{c - N}\right)$$

When the paper is launched, audits show that

$$c = 700\ 000 \quad \text{and} \quad k = \frac{1}{30} \ln 2$$

(a) Calculate the daily circulation after 30 days of publication.

(b) After how many days will the daily circulation first reach 525 000?

(c) What advice can you give the newspaper proprietor if it is known that the paper will break even only if the daily circulation exceeds 750 000?

13 A Cobb–Douglas production function is given by

$$Q = 3L^{1/2}K^{1/3}$$

By taking logs of both sides of this equation, show that

$$\ln Q = \ln 3 + \frac{1}{2} \ln L + \frac{1}{3} \ln K$$

If a graph were to be sketched of $\ln Q$ against $\ln K$ (for varying values of Q and K but with L fixed), explain briefly why the graph will be a straight line and state its slope and vertical intercept.

➔

14 The following table gives data relating a firm's output, Q and labour, L.

L	1	2	3	4	5
Q	0.50	0.63	0.72	0.80	0.85

The firm's short-run production function is believed to be of the form

$Q = AL^n$

(a) Show that

$\ln Q = n \ln L + \ln A$

(b) Using the data supplied, copy and complete the following table:

$\ln L$		0.69		1.39	
$\ln Q$	−0.69		−0.33		−0.16

Plot these points with ln *L* on the horizontal axis and ln Q on the vertical axis. Draw a straight line passing as close as possible to all five points.

(c) By finding the slope and vertical intercept of the line sketched in part (b), estimate the values of the parameters *n* and *A*.

15 **(Excel)** Tabulate values of the following functions for $x = 0, 0.2, 0.4, \ldots, 2$. Hence sketch graphs of these functions, on the same diagram, over the range $0 \le x \le 2$. Discuss, in qualitative terms, any differences or similarities between these functions.

(a) $y = x$ **(b)** $y = x^2$ **(c)** $y = x^3$ **(d)** $y = \sqrt{x}$ **(e)** $y = e^x$

[In Excel, e^x is typed EXP(x)]

16 **(Excel)** Tabulate values of the functions ln *x*, $\log_{10} x$ and $\log_6 x$ for $x = 0.2, 0.4, 0.6, 0.8, 1.0, 2, 3, 4, \ldots, 8$. Hence sketch graphs of these functions on the same diagram, over the range $0.2 \le x \le 8$. Briefly comment on any similarities and differences between them.

[In Excel, natural logs and logs to base 10 are typed as LN(x) and LOG(x) respectively. In general, to find the logarithm of a number *x* to base *n*, type LOG(x, n)]

17 **(Excel)** The demand function of a good can be modelled approximately by

$$P = 100 - \frac{2}{3}Q^n$$

(a) Show that if this relation is exact then a graph of ln(150 − 1.5*P*) against ln Q will be a straight line passing through the origin with slope *n*.

(b) For the data given below, tabulate the values of ln(150 − 1.5*P*) and ln Q. Find the line of best fit and hence estimate the value of *n* correct to 1 decimal place.

Q	10	50	60	100	200	400
P	95	85	80	70	50	20

chapter 3
Mathematics of Finance

This chapter provides an understanding of the way in which financial calculations are worked out. There are four sections, which should be read in the order that they appear.

Section 3.1 revises work on percentages. In particular, a quick method of dealing with percentage increase and decrease calculations is described. This enables an overall percentage change to be deduced easily from a sequence of individual changes. Percentages are used to calculate and interpret index numbers, and to adjust value data for inflation.

Section 3.2 shows how to calculate the future value of a lump sum which is invested to earn interest. This interest can be added to the investment annually, semi-annually, quarterly or even more frequently. The exponential function is used to solve problems in which interest is compounded continuously.

A wide variety of applications are considered in Sections 3.3 and 3.4. In Section 3.3 a mathematical device known as a geometric progression, which is used to calculate the future value of a savings plan and the monthly repayments of a loan, is introduced. Section 3.4 describes the opposite problem of calculating the present value given a future value. The process of working backwards is called discounting. It can be used to decide how much money to invest today in order to achieve a specific target sum in a few years' time. Discounting can be used to appraise different investment projects. On the macroeconomic level, the relationship between interest rates and speculative demand for money is investigated.

The material in this chapter will be of greatest benefit to students on business studies and accountancy courses. This chapter could be omitted without affecting your understanding of the rest of this book.

section 3.1

Percentages

Objectives

At the end of this section you should be able to:

- Understand what a percentage is.
- Solve problems involving a percentage increase or decrease.
- Write down scale factors associated with percentage changes.
- Work out overall percentage changes.
- Calculate and interpret index numbers.
- Adjust value data for inflation.

Advice

The first part of this section provides a leisurely revision of the idea of a percentage as well as reminding you about how to use scale factors to cope with percentage changes. These ideas are crucial to any understanding of financial mathematics. However, if you are already confident in using percentages, you may wish to miss this out and move straight on to the applications covered in sub-sections 3.1.1 and 3.1.2.

In order to be able to handle financial calculations, it is necessary to use percentages proficiently. The word 'percentage' literally means 'per cent', i.e. per hundredth, so that whenever we speak of r% of something, we simply mean the fraction $(r/100)$ths of it.

For example,

25% is the same as $\frac{25}{100} = \frac{1}{4}$

30% is the same as $\frac{30}{100} = \frac{3}{10}$

50% is the same as $\frac{50}{100} = \frac{1}{2}$

Example

Calculate

(a) 15% of 12 **(b)** 98% of 17 **(c)** 150% of 290

Solution

(a) 15% of 12 is the same as

$$\frac{15}{100} \times 12 = 0.15 \times 12 = 1.8$$

(b) 98% of 17 is the same as

$$\frac{98}{100} \times 17 = 0.98 \times 17 = 16.66$$

(c) 150% of 290 is the same as

$$\frac{150}{100} \times 290 = 1.5 \times 290 = 435$$

Practice Problem

1 Calculate

(a) 10% of \$2.90 **(b)** 75% of \$1250 **(c)** 24% of \$580

Whenever any numerical quantity increases or decreases, it is customary to refer to this change in percentage terms. The following example serves to remind you how to perform calculations involving percentage changes.

Example

(a) An investment rises from \$2500 to \$3375. Express the increase as a percentage of the original.

(b) At the beginning of a year, the population of a small village is 8400. If the annual rise in population is 12%, find the population at the end of the year.

(c) In a sale, all prices are reduced by 20%. Find the sale price of a good originally costing \$580.

Solution

(a) The rise in the value of the investment is

$$3375 - 2500 = 875$$

As a fraction of the original this is

$$\frac{875}{2500} = 0.35$$

This is the same as 35 hundredths, so the percentage rise is 35%.

(b) As a fraction

$$12\% \text{ is the same as } \frac{12}{100} = 0.12$$

so the rise in population is

$$0.12 \times 8400 = 1008$$

Hence the final population is

$$8400 + 1008 = 9408$$

(c) As a fraction

$$20\% \text{ is the same as } \frac{20}{100} = 0.2$$

so the fall in price is

$$0.2 \times 580 = 116$$

Hence the final price is

$$580 - 116 = \$464$$

Practice Problem

2 **(a)** A firm's annual sales rise from 50 000 to 55 000 from one year to the next. Express the rise as a percentage of the original.

(b) The government imposes a 15% tax on the price of a good. How much does the consumer pay for a good priced by a firm at $1360?

(c) Investments fall during the course of a year by 7%. Find the value of an investment at the end of the year if it was worth $9500 at the beginning of the year.

In the previous example and in Practice Problem 2, the calculations were performed in two separate stages. The actual rise or fall was first worked out, and these changes were then applied to the original value to obtain the final answer. It is possible to obtain this answer in a single calculation, and we now describe how this can be done. Not only is this new approach quicker, but it also enables us to tackle more difficult problems. To be specific, let us suppose that the price of good is set to rise by 9%, and that its current price is $78. The new price consists of the original (which can be thought of as 100% of the \$78) plus the increase (which is 9% of \$78). The final price is therefore

$$100\% + 9\% = 109\% \text{ (of the \$78)}$$

which is the same as

$$\frac{109}{100} = 1.09$$

In other words, in order to calculate the final price all we have to do is to multiply by the *scale factor*, 1.09. Hence the new price is

$$1.09 \times 78 = \$85.02$$

One advantage of this approach is that it is then just as easy to go backwards and work out the original price from the new price. To go backwards in time we simply *divide* by the scale factor. For example, if the final price of a good is \$1068.20 then before a 9% increase the price would have been

$$1068.20 \div 1.09 = \$980$$

In general, if the percentage rise is r% then the final value consists of the original (100%) together with the increase (r%), giving a total of

$$\frac{100}{100} + \frac{r}{100} = 1 + \frac{r}{100}$$

To go forwards in time we multiply by this scale factor, whereas to go backwards we divide.

Example

(a) If the annual rate of inflation is 4%, find the price of a good at the end of a year if its price at the beginning of the year is \$25.

(b) The cost of a good is \$799 including 17.5% VAT (value added tax). What is the cost excluding VAT?

(c) Express the rise from 950 to 1007 as a percentage.

Solution

(a) The scale factor is

$$1 + \frac{4}{100} = 1.04$$

We are trying to find the price *after* the increase, so we *multiply* to get

$$25 \times 1.04 = \$26$$

(b) The scale factor is

$$1 + \frac{17.5}{100} = 1.175$$

This time we are trying to find the price *before* the increase, so we *divide* by the scale factor to get

$$799 \div 1.175 = \$680$$

(c) The scale factor is

$$\frac{\text{new value}}{\text{old value}} = \frac{1007}{950} = 1.06$$

which can be thought of as

$$1 + \frac{6}{100}$$

so the rise is 6%.

Practice Problem

3 (a) The value of a good rises by 13% in a year. If it was worth \$6.5 million at the beginning of the year, find its value at the end of the year.

(b) The GNP of a country has increased by 63% over the past 5 years and is now \$124 billion. What was the GNP 5 years ago?

(c) Sales rise from 115 000 to 123 050 in a year. Find the annual percentage rise.

It is possible to use scale factors to solve problems involving percentage decreases. To be specific, suppose that an investment of \$76 falls by 20%. The new value is the original (100%) less the decrease (20%), so is 80% of the original. The scale factor is therefore 0.8, giving a new value of

$$0.8 \times 76 = \$60.80$$

In general, the scale factor for an r% decrease is

$$\frac{100}{100} - \frac{r}{100} = 1 - \frac{r}{100}$$

Once again, you multiply by this scale factor when going forwards in time and divide when going backwards.

Example

(a) The value of a car depreciates by 25% in a year. What will a car, currently priced at \$43 000, be worth in a year's time?

(b) After a 15% reduction in a sale, the price of a good is \$39.95. What was the price before the sale began?

(c) The number of passengers using a rail link fell from 190 205 to 174 989. Find the percentage decrease.

Solution

(a) The scale factor is

$$1 - \frac{25}{100} = 0.75$$

→

so the new price is

$$43\ 000 \times 0.75 = \$32\ 250$$

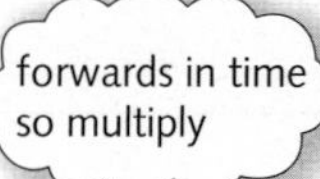

(b) The scale factor is

$$1 - \frac{15}{100} = 0.85$$

so the original price was

$$39.95 \div 0.85 = \$47$$

backwards in time
so divide

(c) The scale factor is

$$\frac{\text{new value}}{\text{old value}} = \frac{174\ 989}{190\ 205} = 0.92$$

which can be thought of as

$$1 - \frac{8}{100}$$

so the fall is 8%. *not* 92%!

Practice Problem

4 (a) Current monthly output from a factory is 25 000. In a recession, this is expected to fall by 65%. Estimate the new level of output.

(b) As a result of a modernization programme, a firm is able to reduce the size of its workforce by 24%. If it now employs 570 workers, how many people did it employ before restructuring?

(c) Shares originally worth \$10.50 fall in a stock market crash to \$2.10. Find the percentage decrease.

The final application of scale factors that we consider is to the calculation of overall percentage changes. It is often the case that over various periods of time the price of a good is subject to several individual percentage changes. It is useful to be able to replace these by an equivalent single percentage change spanning the entire period. This can be done by simply multiplying together successive scale factors.

Example

(a) Share prices rise by 32% during the first half of the year and rise by a further 10% during the second half. What is the overall percentage change?

(b) Find the overall percentage change in the price of a good if it rises by 5% in a year but is then reduced by 30% in a sale.

Solution

(a) To find the value of shares at the end of the first 6 months we would multiply by

$$1 + \frac{32}{100} = 1.32$$

and at the end of the year we would multiply again by the scale factor

$$1 + \frac{10}{100} = 1.1$$

The net effect is to multiply by their product

$$1.32 \times 1.1 = 1.452$$

which can be thought of as

$$1 + \frac{45.2}{100}$$

so the overall change is 45.2%.

Notice that this is not the same as

$$32\% + 10\% = 42\%$$

This is because during the second half of the year we not only get a 10% rise in the original value, but we also get a 10% rise on the gain accrued during the first 6 months.

(b) The individual scale factors are 1.05 and 0.7, so the overall scale factor is

$$1.05 \times 0.7 = 0.735$$

The fact that this is less than 1 indicates that the overall change is a decrease. Writing

$$0.735 = 1 - 0.265 = 1 - \frac{26.5}{100}$$

we see that this scale factor represents a 26.5% decrease.

Practice Problem

5 Find the single percentage increase or decrease equivalent to

(a) an increase of 30% followed by an increase of 40%

(b) a decrease of 30% followed by a decrease of 40%

(c) an increase of 10% followed by a decrease of 50%.

We conclude this section by describing two applications of percentages in macroeconomics:

- index numbers
- inflation.

We consider each of these in turn.

3.1.1 Index numbers

Economic data often take the form of a *time series*; values of economic indicators are available on an annual, quarterly or monthly basis, and we are interested in analysing the rise and fall of these numbers over time. *Index numbers* enable us to identify trends and relationships in the data. The following example shows you how to calculate index numbers and how to interpret them.

Example

Table 3.1 shows the values of household spending (in billions of dollars) during a 5-year period. Calculate the index numbers when 2000 is taken as the base year and give a brief interpretation.

Table 3.1

	Year				
	1999	2000	2001	2002	2003
Household spending	686.9	697.2	723.7	716.6	734.5

Solution

When finding index numbers, a base year is chosen and the value of 100 is allocated to that year. In this example, we are told to take 2000 as the base year, so the index number of 2000 is 100. To find the index number of the year 2001 we work out the scale factor associated with the change in household spending from the base year, 2000 to 2001, and then multiply the answer by 100.

$$\boxed{\text{index number} = \text{scale factor from base year} \times 100}$$

In this case, we get

$$\frac{723.7}{697.2} \times 100 = 103.8$$

This shows that the value of household spending in 2001 was 103.8% of its value in 2000. In other words, household spending increased by 3.8% during 2001.

For the year 2002, the value of household spending was 716.6, giving an index number

$$\frac{716.6}{697.2} \times 100 = 102.8$$

This shows that the value of household spending in 2002 was 102.8% of its value in 2000. In other words, household spending increased by 2.8% between 2000 and 2002. Notice that this is less than that calculated for 2001, reflecting the fact that spending actually fell slightly during 2002. The remaining two index numbers are calculated in a similar way and are shown in Table 3.2.

Table 3.2

	Year				
	1999	2000	2001	2002	2003
Household spending	686.9	697.2	723.7	716.6	734.5
Index number	98.5	100	103.8	102.8	105.3

Practice Problem

6 Calculate the index numbers for the data shown in Table 3.1, this time taking 1999 as the base year.

Index numbers themselves have no units. They merely express the value of some quantity as a percentage of a base number. This is particularly useful, since it enables us to compare how values of quantities, of varying magnitudes, change in relation to each other. The following example shows the rise and fall of two share prices during an 8-month period. The prices (in dollars) listed for each share are taken on the last day of each month. Share A is exceptionally cheap. Investors often include this type of share in their portfolio, since they can occasionally make spectacular gains. This was the case with many dot.com shares at the end of the 1990s. The second share is more expensive and corresponds to a larger, more established firm.

Example

Find the index numbers of each share price shown in Table 3.3, taking April as the base month. Hence compare the performances of these two share prices during this period.

Table 3.3

Month	Jan	Feb	Mar	Apr	May	Jun	Jul	Aug
Share A	0.31	0.28	0.31	0.34	0.40	0.39	0.45	0.52
Share B	6.34	6.40	6.45	6.52	6.57	6.43	6.65	7.00

Solution

The index numbers have been listed in Table 3.4. Notice that both shares are given the same index number of 100 in April, which is the base month. This is despite the fact that the values of the two shares are very different. This creates 'a level playing-field', enabling us to monitor the relative performance of the two shares. The index numbers show quite clearly that share A has outperformed share B during this period. Indeed, if an investor had spent $1000 on shares of type A in January, they could have bought 3225 of them, which would be worth $1677 in August, making a profit of $677. The corresponding profit for share B is only $99.

Table 3.4

Month	Jan	Feb	Mar	Apr	May	Jun	Jul	Aug
Index of share price A (April = 100)	91.2	82.3	91.2	100	117.6	114.7	132.4	152.9
Index of share price B (April = 100)	97.2	98.2	98.9	100	100.8	98.6	102.0	107.4

Incidentally, if the only information you have about the time series is the set of index numbers, then it is possible to work out the percentage changes between any pair of values. Table 3.5 shows the index numbers of the output of a particular firm for the years 2004 and 2005.

Table 3.5

	Output							
	04Q1	04Q2	04Q3	04Q4	05Q1	05Q2	05Q3	05Q4
Index	89.3	98.1	105.0	99.3	100	106.3	110.2	105.7

The table shows that the base quarter is the first quarter of 2005 because the index number is 100 in 05Q1. It is, of course, easy to find the percentage change from this quarter to any subsequent quarter. For example, the index number associated with the third quarter of 2005 is 110.2, so we know immediately that the percentage change in output from 05Q1 to 05Q3 is 10.2%. However, it is not immediately obvious what the percentage change is from, say, 04Q2 to 05Q2. To work this out, notice that the index number has increased by

$$106.3 - 98.1 = 8.2$$

so the percentage increase is

$$\frac{8.2}{98.1} \times 100 = 8.4\%$$

Alternatively, note that the scale factor of this change is

$$\frac{106.3}{98.1} = 1.084$$

which corresponds to an 8.4% increase.

Similarly, the scale factor of the change from 04Q3 to 05Q1 is

$$\frac{100}{105} = 0.952$$

This is less than 1, reflecting the fact that output has fallen. To find the percentage change we write the scale factor as

$$1 - 0.048$$

which shows that the percentage decrease is 4.8%.

Practice Problem

7 Use the index numbers listed in Table 3.5 to find the percentage change in output from

(a) 05Q1 to 05Q4

(b) 04Q1 to 05Q4

(c) 04Q1 to 05Q1

Example

EXCEL

Table 3.6 shows the unit costs of labour, energy, communications and raw materials during a 3-year period. In year 0, a firm used 70 units of labour, 25 units of energy, 10 units of communication and 140 units of raw materials. Taking year 0 as the base year, calculate an appropriate index number for years 1 and 2.

Table 3.6

	Year 0	Year 1	Year 2
Labour	16	23	28
Energy	7	10	9
Communications	12	14	10
Raw materials	5	9	12

Solution

We are told to take year 0 as the base year, so the index number for year 0 is 100. One way of calculating the index number for subsequent years would be to work out the totals of each column in Table 3.6 and find the associated scale factors of these. On this basis, the index number for year 1 would be calculated as

$$\frac{23 + 10 + 14 + 9}{16 + 7 + 12 + 5} \times 100 = 140$$

However, this fails to take into account the fact, for example, that we use twice as many units of raw materals than labour. It is important that each item is weighted according to how many units of each type are used. To do this, all we have to do is to multiply each of the unit costs by the associated quantities. The appropriate index number is then worked out as

$$\frac{23 \times 70 + 10 \times 25 + 14 \times 10 + 9 \times 140}{16 \times 70 + 7 \times 25 + 12 \times 10 + 5 \times 140} \times 100 = 154.1$$

The fact that this number is greater than before is to be expected because the unit price of raw materials has nearly doubled, and the firm uses a greater proportion of these in its total costs. Index numbers that are weighted according to the quantity consumed in the base year are called *Laspeyre indices*. Spreadsheets provide an easy way of presenting the calculations. For each year, we simply include an extra column in the table, for the products P_nQ_0 where Q_0 denotes the quantities used in the base year, and P_n denotes the unit prices in year n. The Laspeyre index for year n is then worked out as

$$\frac{\text{total of column } P_nQ_0}{\text{total of column } P_0Q_0} \times 100$$

Figure 3.1 (overleaf) shows the completed spreadsheet. The Laspeyre indices for years 1 and 2 are seen to be 154.1 and 187.5, respectively. Notice that this index has increased rapidly over this period, in spite of the fact that communication and energy costs have hardly changed. This is because expenditure is dominated by labour and raw material costs, which have both increased substantially during this time.

Figure 3.1

Calculation of the Laspeyre Index

	Year 0		Year 1	Year 2	P_0Q_0	P_1Q_0	P_2Q_0
Input	**Quantity**	**Price**	**Price**	**Price**			
Labour	70	16	23	28	1120	1610	1960
Energy	25	7	10	9	175	250	225
Communications	10	12	14	10	120	140	100
Raw materials	140	5	9	12	700	1260	1680
				Totals:	2115	3260	3965
				Laspeyre Index =		154.1	187.5

Advice

There are other methods for assessing the variation of a 'basket' of values over time. Index numbers that are weighted according to the quantity consumed in the current year (instead of the base year) are called ***Paasche indices***. An example of this is given in Practice Problem 22 at the end of this section.

3.1.2 Inflation

Over a period of time, the prices of many goods and services usually increase. The *annual rate of inflation* is the average percentage change in a given selection of these goods and services, over the previous year. Seasonal variations are taken into account, and the particular basket of goods and services is changed periodically to reflect changing patterns of household

expenditure. The presence of inflation is particularly irritating when trying to interpret a time series that involves a monetary value. It is inevitable that this will be influenced by inflation during any year, and what is of interest is the fluctuation of a time series 'over and above' inflation. Economists deal with this by distinguishing between nominal and real data. ***Nominal data*** are the original, raw data such as those listed in tables in the previous sub-section. These are based on the prices that prevailed at the time. ***Real data*** are the values that have been adjusted to take inflation into account. The standard way of doing this is to pick a year and then convert the values for all other years to the level that they would have had in this base year. This may sound rather complicated, but the idea and calculations involved are really quite simple as the following example demonstrates.

Example

Table 3.7 shows the price (in thousands of dollars) of an average house in a certain town during a 5-year period. The price quoted is the value of the house at the end of each year. Use the annual rates of inflation given in Table 3.8 to adjust the prices to those prevailing at the end of 1991. Compare the rise in both the nominal and real values of house prices during this period.

Table 3.7

	Year				
	1990	1991	1992	1993	1994
Average house price	72	89	93	100	106

Table 3.8

	Year			
	1991	1992	1993	1994
Annual rate of inflation	10.7%	7.1%	3.5%	2.3%

Solution

The raw figures shown in Table 3.7 give the impression that houses increased steadily in value throughout this period, with a quite substantial gain during the first year. However, if inflation had been very high then the gain in real terms would have been quite small. Indeed, if the rate of inflation were to exceed the percentage rise of this nominal data, then the price of a house would actually fall in real terms. To analyse this situation we will use Table 3.8, which shows the rates of inflation during this period. Notice that since the house prices listed in Table 3.8 are quoted at the end of each year, we are not interested in the rate of inflation during 1990.

We are told in the question to choose 1991 as the base year and calculate the value of the house at '1991 prices'. The value of the house at the end of 1991 is obviously \$89 000, since no adjustment need be made. At the end of 1992, the house is worth \$93 000. However, during that year inflation was 7.1%. To adjust this price to '1991 prices' we simply divide by the scale factor 1.071, since we are going backwards in time. We get

$$\frac{93\ 000}{1.071} = 86\ 835$$

In real terms the house has fallen in value by over \$2000.

To adjust the price of the house in 1993 we first need to divide by 1.035 to backtrack to the year 1992, and then divide again by 1.071 to reach 1991. We get

$$\frac{100\,000}{1.035 \times 1.071} = 90\,213$$

In real terms there has at least been some gain during 1993. However, this is less than impressive, and from a purely financial point of view, there would have been more lucrative ways of investing this capital.

For the 1994 price, the adjusted value is

$$\frac{106\,000}{1.023 \times 1.035 \times 1.071} = 93\,476$$

and, for 1990, the adjusted value is

$$72\,000 \times 1.107 = 79\,704$$

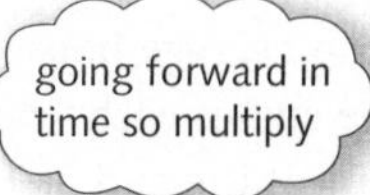

Table 3.9 lists both the nominal and the 'constant 1991' values of the house (rounded to the nearest thousand) for comparison. It shows quite clearly that, apart from the gain during 1991, the increase in value has, in fact, been quite modest.

Table 3.9

	Year				
	1990	1991	1992	1993	1994
Nominal house price	72	89	94	100	106
1991 house price	80	89	87	90	93

Practice Problem

8 Table 3.10 shows the average annual salary (in thousands of dollars) of employees in a small firm, together with the annual rate of inflation for that year. Adjust these salaries to the prices prevailing at the end of 1991 and so give the real values of the employees' salaries at constant '1991 prices'. Comment on the rise in earnings during this period.

Table 3.10

	Year				
	1990	1991	1992	1993	1994
Salary	17.3	18.1	19.8	23.5	26.0
Inflation		4.9	4.3	4.0	3.5

Key Terms

Annual rate of inflation The percentage increase in the level of prices over a 12-month period.

Index numbers The scale factor of a variable measured from the base year multiplied by 100.

Laspeyre index An index number for groups of data which are weighted by the quantities used in the base year.

Nominal data Monetary values prevailing at the time that they were measured.

Paasche index An index number for groups of data which are weighted by the quantities used in the current year.

Real data Monetary values adjusted to take inflation into account.

Scale factor The multiplier that gives the final value in percentage problems.

Time series A sequence of numbers indicating the variation of data over time.

Practice Problems

9 Express the following percentages as fractions in their simplest form:

(a) 35% **(b)** 88% **(c)** 250% **(d)** $17\frac{1}{2}$% **(e)** 0.2%

10 Calculate each of the following:

(a) 5% of 24 **(b)** 8% of 88 **(c)** 48% of 4563 **(d)** 112% of 56

11 Write down the scale factors corresponding to

(a) an increase of 19%

(b) an increase of 250%

(c) a decrease of 2%

(d) a decrease of 43%

12 Find the new quantities when

(a) £16.25 is increased by 12%

(b) the population of a town, currently at 113 566, rises by 5%

(c) a good priced by a firm at £87.90 is subject to value added tax at 15%

(d) a good priced at £2300 is reduced by 30% in a sale

(e) a car, valued at £23 000, depreciates by 32%

13 A shop sells books at '20% below the recommended retail price (r.r.p.)'. If it sells a book for £12.40 find

(a) the r.r.p.

(b) the cost of the book after a further reduction of 15% in a sale

(c) the overall percentage discount obtained by buying the book from the shop in the sale compared with the manufacturer's r.r.p.

➔

14 A TV costs £940 including 17.5% VAT. Find the new price if VAT is reduced to 8%.

15 An antiques dealer tries to sell a vase at 45% above the £18 000 which the dealer paid at auction.

(a) What is the new sale price?

(b) By what percentage can the dealer now reduce the price before making a loss?

16 Find the single percentage increase or decrease equivalent to

(a) a 10% increase followed by a 25% increase

(b) a 34% decrease followed by a 65% increase

(c) a 25% increase followed by a 25% decrease

Explain in words why the overall change in part (c) is not 0%.

17 Table 3.11 shows the index numbers associated with transport costs during a 20-year period. The public transport costs reflect changes to bus and train fares, whereas private transport costs include purchase, service, petrol, tax and insurance costs of cars.

(1) Which year is chosen as the base year?

(2) Find the percentage increases in the cost of public transport from

(a) 1985 to 1990 **(b)** 1990 to 1995 **(c)** 1995 to 2000 **(d)** 2000 to 2005

(3) Repeat part (2) for private transport.

(4) Comment briefly on the relative rise in public and private transport costs during this 20-year period.

Table 3.11

	Year				
	1985	**1990**	**1995**	**2000**	**2005**
Public transport	100	130	198	224	245
Private transport	100	125	180	199	221

18 Table 3.12 shows the number of items (in thousands) produced from a factory production line during the course of a year. Taking the second quarter as the base quarter, calculate the associated index numbers. Suggest a possible reason for the fluctuations in output.

Table 3.12

	Quarter			
	Q1	**Q2**	**Q3**	**Q4**
Output	13.5	1.4	2.5	10.5

19 Table 3.13 shows government expenditure (in billions of dollars) on education for four consecutive years, together with the rate of inflation for each year.

(a) Taking 1994 as the base year, work out the index numbers of the nominal data given in the second row of the table.

(b) Find the values of expenditure at constant 1994 prices and hence recalculate the index numbers of real government expenditure.

(c) Give an interpretation of the index numbers calculated in part (b).

Table 3.13

	Year			
	1994	**1995**	**1996**	**1997**
Spending	236	240	267	276
Inflation		4.7	4.2	3.4

20 Index numbers associated with the growth of unemployment during an 8-year period are shown in Table 3.14.

(a) What are the base years for the two indices?

(b) If the government had not switched to index 2, what would be the values of index 1 in years 7 and 8?

(c) What values would index 2 have been in years 1, 2, 3, 4 and 5?

(d) If unemployment was 1.2 million in year 4, how many people were unemployed in years 1 and 8?

Table 3.14

	Year							
	1	**2**	**3**	**4**	**5**	**6**	**7**	**8**
Index 1	100	95	105	110	119	127		
Index 2						100	112	118

21 **(Excel)** Table 3.15 shows the annual salaries (in thousands of dollars) of four categories of employee during a 3-year period. In year 0 the firm employed 24, 250, 109 and 7 people of types A, B, C and D, respectively. Calculate the Laspeyre index of the total wage bill in years 1 and 2 taking year 0 as the base year. Comment briefly on these values.

Table 3.15

	Year 0	**Year 1**	**Year 2**
Type A	12	13	13
Type B	26	28	29
Type C	56	56	64
Type D	240	340	560

22 **(Excel)** In the Laspeyre index, the quantities used for the weights are those of the base year. If these are replaced by quantities for the current year, then the index is called the ***Paasche index***. In Problem 21, suppose that the numbers of employees of types A, B, C and D in year 1 are 30, 240, 115 and 8 respectively. For year 2 the corresponding figures are 28, 220, 125 and 20. By adding two extra columns for the products P_1Q_1 and P_2Q_2 to the spreadsheet of Problem 21, calculate the Paasche index for years 1 and 2. Compare with the Laspeyre index calculated in Problem 21.

State one advantage and one disadvantage of using the Laspeyre and Paasche methods for the calculation of combined index numbers.

section 3.2

Compound interest

Objectives

At the end of this section you should be able to:

- Understand the difference between simple and compound interest.
- Calculate the future value of a principal under annual compounding.
- Calculate the future value of a principal under continuous compounding.
- Determine the annual percentage rate of interest given a nominal rate of interest.

Today, businesses and individuals are faced with a bewildering array of loan facilities and investment opportunities. In this section we explain how these financial calculations are carried out to enable an informed choice to be made between the various possibilities available. We begin by considering what happens when a single lump sum is invested and show how to calculate the amount accumulated over a period of time.

Suppose that someone gives you the option of receiving \$500 now or \$500 in 3 years' time. Which of these alternatives would you accept? Most people would take the money now, partly because they may have an immediate need for it, but also because they recognize that \$500 is worth more today than in 3 years' time. Even if we ignore the effects of inflation, it is still better to take the money now, since it can be invested and will increase in value over the 3-year period. In order to work out this value we need to know the rate of interest and the basis on which it is calculated. Let us begin by assuming that the \$500 is invested for 3 years at 10% interest compounded annually. What exactly do we mean by '10% interest compounded annually'? Well, at the end of each year, the interest is calculated and is added on to the amount currently invested. If the original amount is \$500 then after 1 year the interest is 10% of \$500, which is

$$\frac{10}{100} \times \$500 = \frac{1}{10} \times \$500 = \$50$$

so the amount rises by \$50 to \$550.

What happens to this amount at the end of the second year? Is the interest also \$50? This would actually be the case with ***simple interest***, when the amount of interest received is the same for all years. However, with ***compound interest***, we get 'interest on the interest'. Nearly all financial investments use compound rather than simple interest, because investors need to be rewarded for not taking the interest payment out of the fund each year. Under annual compounding the interest obtained at the end of the second year is 10% of the amount invested at the start of that year. This not only consists of the original \$500, but also the \$50 already received as interest on the first year's investment. Consequently, we get an additional

$$\frac{1}{10} \times \$550 = \$55$$

raising the sum to \$605. Finally, at the end of the third year, the interest is

$$\frac{1}{10} \times \$605 = \$60.50$$

so the investment is \$665.50. You are therefore \$165.50 better off by taking the \$500 now and investing it for 3 years. The calculations are summarized in Table 3.16.

Table 3.16

End of year	Interest (\$)	Investment (\$)
1	50	550
2	55	605
3	60.50	665.50

Example

Find the value, in 4 years' time, of \$10 000 invested at 5% interest compounded annually.

Solution

- At the end of year 1 the interest is $0.05 \times 10\,000 = 500$, so the investment is 10 500.
- At the end of year 2 the interest is $0.05 \times 10\,500 = 525$, so the investment is 11 025.
- At the end of year 3 the interest is $0.05 \times 11\,025 = 551.25$, so the investment is 11 576.25.
- At the end of year 4 the interest is $0.05 \times 11\,576.25 = 578.81$ rounded to 2 decimal places, so the final investment is \$12 155.06 to the nearest cent.

Practice Problem

1 Find the value, in 10 years' time, of \$1000 invested at 8% interest compounded annually.

The previous calculations were performed by finding the interest earned each year and adding it on to the amount accumulated at the beginning of the year. As you may have discovered in Practice Problem 1, this can be rather laborious, particularly if the money is invested over a long period of time. What is really needed is a method of calculating the investment after, say, 10 years without having to determine the amount for the 9 intermediate years. This can be done using the scale factor approach discussed in the previous section. To illustrate this, let us return to the problem of investing \$500 at 10% interest compounded annually. The original sum of money is called the ***principal*** and is denoted by P, and the final sum is called the ***future value*** and is denoted by S. The scale factor associated with an increase of 10% is

$$1 + \frac{10}{100} = 1.1$$

so at the end of 1 year the total amount invested is $P(1.1)$.

After 2 years we get

$$P(1.1) \times (1.1) = P(1.1)^2$$

and after 3 years the future value is

$$S = P(1.1)^2 \times (1.1) = P(1.1)^3$$

Setting $P = 500$, we see that

$$S = 500(1.1)^3 = \$665.50$$

which is, of course, the same as the amount calculated previously.

In general, if the interest rate is r% compounded annually then the scale factor is

$$1 + \frac{r}{100}$$

so, after n years,

$$\boxed{S = P\left(1 + \frac{r}{100}\right)^n}$$

Given the values of r, P and n it is trivial to evaluate S using the power key $\boxed{x^y}$ on a calculator. To see this let us rework the previous example using this formula.

Example

Find the value, in 4 years' time, of \$10 000 invested at 5% interest compounded annually.

Solution

In this problem, $P = 10\,000$, $r = 5$ and $n = 4$, so the formula $S = P\left(1 + \frac{r}{100}\right)^n$ gives

$$S = 10\,000\left(1 + \frac{5}{100}\right)^4 = 10\,000(1.05)^4 = 12\,155.06$$

which is, of course, the same answer as before.

Practice Problem

2 Use the formula

$$S = P\left(1 + \frac{r}{100}\right)^n$$

to find the value, in 10 years' time, of \$1000 invested at 8% interest compounded annually. [You might like to compare your answer with that obtained in Practice Problem 1.]

The compound interest formula derived above involves four variables, r, n, P and S. Provided that we know any three of these, we can use the formula to determine the remaining variable. This is illustrated in the following example.

Example

A principal of \$25 000 is invested at 12% interest compounded annually. After how many years will the investment first exceed \$250 000?

Solution

We want to save a total of \$250 000 starting with an initial investment of \$25 000. The problem is to determine the number of years required for this on the assumption that the interest is fixed at 12% throughout this time. The formula for compound interest is

$$S = P\left(1 + \frac{r}{100}\right)^n$$

We are given that

$$P = 25\ 000,\ S = 250\ 000,\ r = 12$$

so we need to solve the equation

$$250\ 000 = 25\ 000\left(1 + \frac{12}{100}\right)^n$$

for n.

One way of doing this would just be to keep on guessing values of n until we find the one that works. However, a more mathematical approach is to use logarithms, because we are being asked to solve an equation in which the unknown occurs as a power. Following the method described in Section 2.3, we first divide both sides by 25 000 to get

$$10 = (1.12)^n$$

Taking logarithms of both sides gives

$$\log(10) = \log(1.12)^n$$

and if you apply rule 3 of logarithms you get

$$\log(10) = n \log(1.12)$$

$\log_b x^m = m \log_b x$

Hence

$$n = \frac{\log(10)}{\log(1.12)}$$

$$= \frac{1}{0.049\ 218\ 023} \quad \text{(taking logarithms to base 10)}$$

$$= 20.3 \quad \text{(to 1 decimal place)}$$

Now we know that n must be a whole number because interest is only added on at the end of each year. We assume that the first interest payment occurs exactly 12 months after the initial investment and every

12 months thereafter. The answer, 20.3, tells us that after only 20 years the amount is less than \$250 000, so we need to wait until 21 years have elapsed before it exceeds this amount. In fact, after 20 years

$$S = \$25\,000(1.12)^{20} = \$241\,157.33$$

and after 21 years

$$S = \$25\,000(1.12)^{21} = \$270\,096.21$$

In this example we calculated the time taken for \$25 000 to increase by a factor of 10. It can be shown that this time depends only on the interest rate and not on the actual value of the principal. To see this, note that if a general principal, P, increases tenfold then its future value is $10P$. If the interest rate is 12%, then we need to solve

$$10P = P\left(1 + \frac{12}{100}\right)^n$$

for n. The Ps cancel (indicating that the answer is independent of P) to produce the equation

$$10 = (1.12)^n$$

This is identical to the equation obtained in the previous example and, as we have just seen, has the solution $n = 20.3$.

Practice Problem

3 A firm estimates that its sales will rise by 3% each year and that it needs to sell at least 10 000 goods each year in order to make a profit. Given that its current annual sales are only 9000, how many years will it take before the firm breaks even?

You may have noticed that in all of the previous problems it is assumed that the interest is compounded annually. It is possible for interest to be added to the investment more frequently than this. For example, suppose that a principal of \$500 is invested for 3 years at 10% interest compounded quarterly. What do we mean by '10% interest compounded quarterly'? Well, it does *not* mean that we get 10% interest every 3 months. Instead, the 10% is split into four equal portions, one for each quarter. Every 3 months the interest accrued is

$$\frac{10\%}{4} = 2.5\%$$

so after the first quarter the investment gets multiplied by 1.025 to give

$$500(1.025)$$

and after the second quarter it gets multiplied by another 1.025 to give

$$500(1.025)^2$$

and so on. Moreover, since there are exactly 12 three-month periods in 3 years we deduce that the future value is

$$500(1.025)^{12} = \$672.44$$

Notice that this is greater than the sum obtained at the start of this section under annual compounding. (Why is this?)

This example highlights the fact that the compound interest formula

$$S = P\left(1 + \frac{r}{100}\right)^n$$

derived earlier for annual compounding can also be used for other types of compounding. All that is needed is to reinterpret the symbols r and n. The variable r now represents the rate of interest per time period and n represents the total number of periods.

Example

A principal of $10 is invested at 12% interest for 1 year. Determine the future value if the interest is compounded

(a) annually **(b)** semi-annually **(c)** quarterly **(d)** monthly **(e)** weekly

Solution

The formula for compound interest gives

$$S = 10\left(1 + \frac{r}{100}\right)^n$$

(a) If the interest is compounded annually then $r = 12$, $n = 1$, so

$$S = \$10(1.12)^1 = \$11.20$$

(b) If the interest is compounded semi-annually then an interest of 12/2 = 6% is added on every 6 months and, since there are two 6-month periods in a year,

$$S = \$10(1.06)^2 = \$11.24$$

(c) If the interest is compounded quarterly then an interest of 12/4 = 3% is added on every 3 months and, since there are four 3-month periods in a year,

$$S = \$10(1.03)^4 = \$11.26$$

(d) If the interest is compounded monthly then an interest of 12/12 = 1% is added on every month and, since there are 12 months in a year,

$$S = \$10(1.01)^{12} = \$11.27$$

(e) If the interest is compounded weekly then an interest of 12/52 = 0.23% is added on every week and, since there are 52 weeks in a year,

$$S = \$10(1.0023)^{52} = \$11.27$$

In the above example we see that the future value rises as the frequency of compounding rises. This is to be expected because the basic feature of compound interest is that we get 'interest on the interest'. However, one important observation that you might not have expected is that, although the future values increase, they appear to be approaching a fixed value. It can be shown that this always occurs. The type of compounding in which the interest is added on with increasing frequency is called *continuous compounding*. In theory, we can find the future value of a principal under continuous compounding using the approach taken in the previous example. We work with smaller and smaller time periods until the numbers settle down to a fixed value. However, it turns out that there is a special formula that can be used to compute this directly. The future value, S, of a principal, P, compounded continuously for t years at an annual rate of r% is

$$\boxed{S = Pe^{rt/100}}$$

where e is the number

2.718 281 828 459 045 235 36 (to 20 decimal places)

If $r = 12$, $t = 1$ and $P = 10$ then this formula gives

$$S = \$10e^{12\times1/100} = \$10e^{0.12} = \$11.27$$

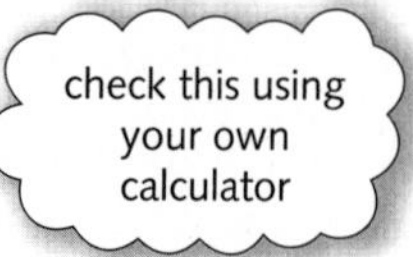

which is in agreement with the limiting value obtained in the previous example.

Advice

The number e and the related natural logarithm function were first introduced in Section 2.4. If you missed this section out, you should go back and read through this work now before proceeding. The link between the number e and the above formula for continuous compounding is given in Practice Problem 23 at the end of this section. However, you may prefer to accept it without justification and concentrate on the applications.

Example

A principal of $2000 is invested at 10% interest compounded continuously. After how many days will the investment first exceed $2100?

Solution

We want to save a total of $2100 starting with an initial investment of $2000. The problem is to determine the number of days required for this on the assumption that the interest rate is 10% compounded continuously. The formula for continuous compounding is

$$S = Pe^{rt/100}$$

We are given that

$$S = 2100,\ P = 2000,\ r = 10$$

so we need to solve the equation

$$2100 = 2000e^{10t/100}$$

for t. Dividing through by 2000 gives

$$1.05 = e^{0.1t}$$

As explained in Section 2.4, equations such as this can be solved using natural logarithms. Recall that

if $M = e^n$ then $n = \ln M$

If we apply this definition to the equation

$$1.05 = e^{0.1t}$$

with $M = 1.05$ and $n = 0.1t$ then

$$0.1t = \ln(1.05)$$
$$= 0.048\,790\,2$$

and so $t = 0.488$ to three decimal places.

The variable t which appears in the formula for continuous compounding is measured in years, so to convert it to days we multiply by 365 (assuming that there are 365 days in a year). Hence

$$t = 365 \times 0.488 = 178.1 \text{ days}$$

We deduce that the amount invested first exceeds $2100 some time during the 179th day.

Practice Problems

4 **(1)** A principal, $30, is invested at 6% interest for 2 years. Determine the future value if the interest is compounded

(a) annually **(b)** semi-annually **(c)** quarterly

(d) monthly **(e)** weekly **(f)** daily

(2) Use the formula

$$S = Pe^{rt/100}$$

to determine the future value of $30 invested at 6% interest compounded continuously for 2 years. Confirm that it is in agreement with the results of part (1).

5 Determine the rate of interest required for a principal of $1000 to produce a future value of $4000 after 10 years compounded continuously.

Example

EXCEL

A principal of $10 000 is invested at one of the following banks:

(a) Bank A offers 4.75% interest, compounded annually.

(b) Bank B offers 4.70% interest, compounded semi-annually.

(c) Bank C offers 4.65% interest, compounded quarterly.

(d) Bank D offers 4.6% interest, compounded continuously.

For each bank, tabulate the value of the investment at the end of every year, for the next 10 years. Which of these banks would you recommend?

→

Figure 3.2

Book1

	A	B	C	D	E	F
1	Compound Interest					
2						
3	Year	Bank A	Bank B	Bank C	Bank D	
4	0	10000	10000	10000	10000	
5	1					
6	2					
7	3					
8	4					
9	5					
10	6					
11	7					
12	8					
13	9					
14	10					
15						
16						

Solution

In Figure 3.2 the numbers 0 to 10 have been entered in the first column, together with appropriate headings. The initial investment is the same for each bank, so the value 10 000 is typed into cells B4 to E4.

For banks A, B and C, the future values can be worked out using the formula

$$10\,000\left(1+\frac{r}{100}\right)^n$$

for appropriate values of r and n.

(a) In Bank A, the interest rate is 4.75% compounded annually, so at the end of year 1, the investment is

$$10\,000(1 + 0.0475)^1$$

The reason for writing it to the power of 1 is so that when we enter it into Excel, and copy down the first column, the power will automatically increase in accordance with the changing years. We type

=10000*(1+0.0475)^A5

in cell B5, and then click and drag down to cell B14.

(b) Bank B offers a return of 4.7% compounded semi-annually, so that at the end of year 1, the investment is

$$10\,000(1 + 0.047/2)^{2\times 1}$$

In general, after t years, the investment is

$$10\,000(1 + 0.0475/2)^{2\times t}$$

so in Excel, we type

=10000*(1+0.0475/2)^(2*A5)

in cell C5, and copy down this column in the usual way.

Figure 3.3

	A	B	C	D	E	F
1	Compound Interest					
2						
3	Year	Bank A	Bank B	Bank C	Bank D	
4	0	10000	10000	10000	10000	
5	1	10475.00	10480.64	10473.17	10470.74	
6	2	10972.56	10984.38	10968.73	10963.65	
7	3	11493.76	11512.34	11487.74	11479.76	
8	4	12039.71	12065.67	12031.31	12020.16	
9	5	12611.60	12645.59	12600.60	12586.00	
10	6	13210.65	13253.39	13196.82	13178.48	
11	7	13838.16	13890.40	13821.26	13798.85	
12	8	14495.47	14558.03	14475.24	14448.42	
13	9	15184.00	15257.75	15160.16	15128.57	
14	10	15905.24	15991.10	15877.50	15840.74	
15						
16						

(c) Bank C offers a return of 4.65% compounded quarterly, so we type

=10000*(1+0.0465/4)^(4*A5)

in cell D5, and copy down the column.

(d) Bank D offers a return of 4.6% compounded continuously, so after t years the future value is given by

$$10\,000e^{4.6t/100} = 10\,000e^{0.046t}$$

The corresponding values are calculated in column E by typing

=10000*EXP(0.046*A5) in cell E5 and copying down the column.

The completed spreadsheet is shown in Figure 3.3. The amounts have been rounded to 2 decimal places by highlighting cells B5 through to E14, and using the Decrease Decimal icon on the toolbar.

Figure 3.3 shows that there is very little to choose between these banks and that, in practice, other issues (such as any conditions or penalties attached to future withdrawals from the account) may well influence our recommendation. However, from a purely monetary point of view, we should advise the investor to put the money into Bank B, as this offers the greatest return. Notice that Bank B is not the one with the highest rate of interest. This example highlights the importance of taking into account the frequency of compounding, as well as the actual rate of interest.

Given that there are so many ways of calculating compound interest, people often find it difficult to appraise different investment opportunities. What is needed is a standard 'benchmark' that enables an individual to compare different forms of savings or credit schemes on an equal basis. The one that is commonly used is annual compounding. All firms offering investment or loan facilities are required to provide the effective annual rate. This is often referred to as the ***annual percentage rate***, which is abbreviated to APR. The APR is the rate of interest which, when compounded annually, produces the same yield as the nominal (that is, the stated) rate of interest. The phrase 'annual equivalent rate' (AER) is frequently used when applied to savings. However, in this book we shall use APR for both savings and loans.

Example

Determine the annual percentage rate of interest of a deposit account that has a nominal rate of 8% compounded monthly.

Solution

The APR is the overall rate of interest, which can be calculated using scale factors. If the account offers a return of 8% compounded monthly then each month the interest is

$$\frac{8}{12} = \frac{2}{3} = 0.67\%$$

of the amount invested at the beginning of that month. The monthly scale factor is

$$1 + \frac{0.67}{100} = 1.0067$$

so in a whole year the principal gets multiplied by

$$(1.0067)^{12} = 1.0834$$

which can be written as

$$1 + \frac{8.34}{100}$$

so the APR is 8.34%.

Practice Problem

6 Determine the annual percentage rate of interest if the nominal rate is 12% compounded quarterly.

Although the aim of this chapter is to investigate the mathematics of finance, the mathematical techniques themselves are more widely applicable. We conclude this section with two examples to illustrate this.

Example

A country's annual GNP (gross national product), currently at $25 000 million, is predicted to grow by 3.5% each year. The population is expected to increase by 2% a year from its current level of 40 million. After how many years will GNP per capita (that is, GNP per head of population) reach $700?

Solution

The per capita value of GNP is worked out by dividing GNP by the size of the population. Initially, this is

$$\frac{25\,000\,000\,000}{40\,000\,000} = \$625$$

During the next few years, GNP is forecast to grow at a faster rate than the population so this value will increase.

The scale factor associated with a 3.5% increase is 1.035, so after n years GNP (in millions of dollars) will be

$$\text{GNP} = 25\,000 \times (1.035)^n$$

Similarly, the population (also in millions) will be

$$\text{population} = 40 \times (1.02)^n$$

Hence GNP per capita is

$$\frac{25\,000 \times (1.035)^n}{40 \times (1.02)^n} = \frac{25\,000}{40} \times \frac{(1.035)^n}{(1.02)^n} = 625 \times \left(\frac{1.035}{1.02}\right)^n$$

We want to find the number of years required for this to reach 700, so we need to solve the equation

$$625 \times \left(\frac{1.035}{1.02}\right)^n = 700$$

for n. Dividing both sides by 625 gives

$$\left(\frac{1.035}{1.02}\right)^n = 1.12$$

and after taking logs of both sides we get

$$\log\left(\frac{1.035}{1.02}\right)^n = \log(1.12)$$

$$n \log\left(\frac{1.035}{1.02}\right) = \log(1.12) \quad \text{(rule 3 of logs)}$$

so that

$$n = \frac{\log(1.12)}{\log(1.035/1.02)} = 7.76$$

We deduce that the target figure of \$700 per capita will be achieved after 8 years.

Example

A firm decides to increase output at a constant rate from its current level of 50 000 to 60 000 during the next 5 years. Calculate the annual rate of increase required to achieve this growth.

Solution

If the rate of increase is r% then the scale factor is $1 + \frac{r}{100}$ so after 5 years, output will be

$$50\,000\left(1 + \frac{r}{100}\right)^5$$

To achieve a final output of 60 000, the value of r is chosen to satisfy the equation

$$50\,000\left(1 + \frac{r}{100}\right)^5 = 60\,000$$

Dividing both sides by 50 000 gives

$$\left(1 + \frac{r}{100}\right)^5 = 1.2$$

→

The difficulty in solving this equation is that the unknown, r, is trapped inside the brackets, which are raised to the power of 5. This is analogous to the problem of solving an equation such as

$$x^2 = 5.23$$

which we would solve by taking square roots of both sides to find x. This suggests that we can find r by taking fifth roots of both sides of

$$\left(1 + \frac{r}{100}\right)^5 = 1.2$$

to get

$$1 + \frac{r}{100} = (1.2)^{1/5} = 1.037$$

Hence $r = 3.7\%$.

Practice Problem

7 The turnover of a leading supermarket chain, A, is currently \$560 million and is expected to increase at a constant rate of 1.5% a year. Its nearest rival, supermarket B, has a current turnover of \$480 million and plans to increase this at a constant rate of 3.4% a year. After how many years will supermarket B overtake supermarket A?

Key Terms

Annual percentage rate The equivalent annual interest paid for a loan, taking into account the compounding over a variety of time periods.

Compound interest The interest which is added on to the initial investment, so that this will itself gain interest in subsequent time periods.

Continuous compounding The limiting value when interest is compounded with ever-increasing frequency.

Future value The final value of an investment after one or more time periods.

Principal The value of the original sum invested.

Simple interest The interest which is paid direct to the investor instead of being added to the original amount.

Practice Problems

8 A bank offers a return of 7% interest compounded annually. Find the future value of a principal of \$4500 after 6 years. What is the overall percentage rise over this period?

9 Find the future value of \$20 000 in 2 years' time if compounded quarterly at 8% interest.

10 Midwest Bank offers a return of 5% compounded annually for each and every year. The rival BFB offers a return of 3% for the first year and 7% in the second and subsequent years (both compounded annually). Which bank would you choose to invest in if you decided to invest a principal for **(a)** 2 years; **(b)** 3 years?

11 The value of an asset, currently priced at \$100 000, is expected to increase by 20% a year.

(a) Find its value in 10 years' time.

(b) After how many years will it be worth \$1 million?

12 How long will it take for a sum of money to double if it is invested at 5% interest compounded annually?

13 A piece of machinery depreciates in value by 5% a year. Determine its value in 3 years' time if its current value is \$50 000.

14 A principal, \$7000, is invested at 9% interest for 8 years. Determine its future value if the interest is compounded

(a) annually **(b)** semi-annually **(c)** monthly **(d)** continuously

15 A car depreciates by 40% in the first year, 30% in the second year and 20% thereafter. I buy a car for \$14 700 when it is 2 years old.

(a) How much did it cost when new?

(b) After how many years will it be worth less than 25% of the amount that I paid for it?

16 Find the future value of \$100 compounded continuously at an annual rate of 6% for 12 years.

17 How long will it take for a sum of money to triple in value if invested at an annual rate of 3% compounded continuously?

18 If a piece of machinery depreciates continuously at an annual rate of 4%, how many years will it take for the value of the machinery to halve?

19 A department store has its own credit card facilities, for which it charges interest at a rate of 2% each month. Explain briefly why this is not the same as an annual rate of 24%. What is the annual percentage rate?

20 Determine the APR if the nominal rate is 7% compounded continuously.

21 Current annual consumption of energy is 78 billion units and this is expected to rise at a fixed rate of 5.8% each year. The capacity of the industry to supply energy is currently 104 billion units.

(a) Assuming that the supply remains steady, after how many years will demand exceed supply?

(b) What constant rate of growth of energy production would be needed to satisfy demand for the next 50 years?

22 The population of a country is currently at 56 million and is forecast to rise by 3.7% each year. It is capable of producing 2500 million units of food each year, and it is estimated that each member of the population requires a minimum of 65 units of food each year. At the moment, the extra food needed to satisfy this requirement is imported, but the government decides to increase food production at a constant rate each year, with the aim of making the country self-sufficient after 10 years. Find the annual rate of growth required to achieve this.

23 If a principal, P, is invested at r% interest compounded annually then its future value, S, after n years is given by

$$S = P\left(1 + \frac{r}{100}\right)^n$$

(a) Use this formula to show that if an interest rate of r% is compounded k times a year then after t years

$$S = P\left(1 + \frac{r}{100k}\right)^{kt}$$

(b) Show that if $m = 100k/r$ then the formula in part (1) can be written as

$$S = P\left(\left(1 + \frac{1}{m}\right)^{m}\right)^{rt/100}$$

(c) Use the definition

$$e = \lim_{m \to \infty}\left(1 + \frac{1}{m}\right)^{m}$$

to deduce that if the interest is compounded with ever-increasing frequency (that is, continuously) then

$$S = Pe^{rt/100}$$

24 **(Excel)** The sum of \$100 is invested at 12% interest for 20 years. Tabulate the value of the investment at the end of each year, if the interest is compounded

(a) annually **(b)** quarterly **(c)** monthly **(d)** continuously

Draw graphs of these values on the same diagram. Comment briefly on any similarities and differences between these graphs.

25 **(Excel)** A department store charges interest on any outstanding debt at the end of each month. It decides to produce a simple table of APRs for its customers, based on a variety of monthly rates. Use a spreadsheet to produce such a table for monthly interest rates of 0.5%, 0.6%, 0.7%, . . . , 3%. Plot a graph of APR against monthly rate and comment briefly on its basic shape.

section 3.3

Geometric series

Objectives

At the end of this section you should be able to:

- Recognize a geometric progression.
- Evaluate a geometric series.
- Calculate the total investment obtained from a regular savings plan.
- Calculate the instalments needed to repay a loan.

Consider the following sequence of numbers:

2, 6, 18, 54, . . .

One obvious question, often asked in intelligence tests, is what is the next term in the sequence? All that is required is for you to spot the pattern so that it can be used to generate the next term. In this case, successive numbers are obtained by multiplying by 3, so the fifth term is

$$54 \times 3 = 162$$

the sixth term is

$$162 \times 3 = 486$$

and so on. Any sequence in which terms are calculated by multiplying their predecessor by a fixed number is called a ***geometric progression*** and the multiplicative factor itself is called a ***geometric ratio***. The sequence above is a geometric progression with geometric ratio 3. The reason for introducing these sequences is not to help you to answer intelligence tests, but rather to analyse compound interest problems. You may well have noticed that all of the problems given in the previous section produced such a sequence. For example, if a principal, \$500, is invested at 10% interest compounded annually, then the future values in successive years are

$$500(1.1),\ 500(1.1)^2,\ 500(1.1)^3,\ \ldots$$

which we recognize as a geometric progression with geometric ratio 1.1.

Example

Which of the following sequences are geometric progressions? For those sequences that are of this type, write down their geometric ratios.

(a) 1000, −100, 10, −1, . . . (b) 2, 4, 6, 8, . . . (c) $a, ar, ar^2, ar^3, \ldots$

Solution

(a) 1000, −100, 10, −1, . . . is a geometric progression with geometric ratio, $-\frac{1}{10}$.

(b) 2, 4, 6, 8, . . . is not a geometric progression because to go from one term to the next you *add* 2. Such a sequence is called an ***arithmetic progression*** and is of little interest in business and economics.

(c) $a, ar, ar^2, ar^3, \ldots$ is a geometric progression with geometric ratio, r.

Practice Problem

1 Decide which of the following sequences are geometric progressions. For those sequences that are of this type, write down their geometric ratios.

(a) 3, 6, 12, 24, . . . **(b)** 5, 10, 15, 20, . . . **(c)** 1, −3, 9, −27, . . .

(d) 8, 4, 2, 1, ½, . . . **(e)** 500, 500(1.07), $500(1.07)^2$, . . .

All of the problems considered in Section 3.2 involved a single lump-sum payment into an investment account. The task was simply to determine its future value after a period of time when it is subject to a certain type of compounding. In this section, we extend this to include multiple payments. This situation occurs whenever individuals save regularly or when businesses take out a loan that is paid back using fixed monthly or annual instalments. To tackle these problems we need to be able to sum (that is, to add together) consecutive terms of a geometric progression. Such an expression is called a ***geometric series***. Suppose that we want to sum the first six terms of the geometric progression given by the sequence

$$2, 6, 18, 54, \ldots \qquad (*)$$

The easiest way of doing this is to write down these six numbers and add them together to get

$$2 + 6 + 18 + 54 + 162 + 486 = 728$$

There is, however, a special formula to sum a geometric series which is particularly useful when there are lots of terms or when the individual terms are more complicated to evaluate. It can be shown that the sum of the first n terms of a geometric progression in which the first term is a, and the geometric ratio is r, is equal to

$$a\left(\frac{r^n - 1}{r - 1}\right) \quad (r \neq 1)$$

Use of the symbol r to denote both the interest rate and the geometric ratio is unfortunate but fairly standard. In practice, it is usually clear from the context what this symbol represents, so no confusion should arise.

A proof of this formula is given in Practice Problem 11 at the end of this section. As a check, let us use it to determine the sum of the first six terms of sequence (*) above. In this case the first term $a = 2$, the geometric ratio $r = 3$ and the number of terms $n = 6$, so the geometric series is equal to

$$2\left(\frac{3^6 - 1}{3 - 1}\right) = 3^6 - 1 = 728$$

which agrees with the previous value found by summing the terms longhand. In this case there is no real benefit in using the formula. However, in the following example, its use is almost essential. You might like to convince yourself of the utility of the formula by evaluating this series longhand and comparing the computational effort involved!

Example

Evaluate the geometric series

$$500(1.1) + 500(1.1)^2 + 500(1.1)^3 + \ldots + 500(1.1)^{25}$$

Solution

We have $a = 500(1.1)$, $r = 1.1$, $n = 25$, so the geometric series is equal to

$$500(1.1)\left(\frac{(1.1)^{25} - 1}{1.1 - 1}\right) = 54\,090.88$$

Practice Problem

2 (a) Write down the next term in the sequence

1, 2, 4, 8, . . .

and hence find the sum of the first five terms. Check that this agrees with the value obtained using

$$a\left(\frac{r^n - 1}{r - 1}\right)$$

(b) Evaluate the geometric series

$$100(1.07) + 100(1.07)^2 + \ldots + 100(1.07)^{20}$$

There are two particular applications of geometric series that we now consider, involving savings and loans. We begin by analysing savings plans. In the simplest case, an individual decides to invest a regular sum of money into a bank account. This is sometimes referred to as a ***sinking fund*** and is used to meet some future financial commitment. It is assumed that he or she saves an equal amount and that the money is put into the account at the same time each year (or month). We further assume that the interest rate does not change. The latter may not be an entirely realistic assumption, since it can fluctuate wildly in volatile market conditions. Indeed, banks offer a variety of rates of interest depending on the notice required for withdrawal and on the actual amount of money saved. Practice Problem 9 at the end of this section considers what happens when the interest rate rises as the investment goes above certain threshold levels.

Example

A person saves \$100 in a bank account at the beginning of each month. The bank offers a return of 12% compounded monthly.

(a) Determine the total amount saved after 12 months.

(b) After how many months does the amount saved first exceed \$2000?

Solution

(a) During the year a total of 12 regular savings of \$100 are made. Each \$100 is put into an account that gives a return of 12% compounded monthly, or equivalently, a return of 1% each month. However, each payment is invested for a different period of time. For example, the first payment is invested for the full 12 months, whereas the final payment is invested for 1 month only. We need to work out the future value of each payment separately and add them together.

The first payment is invested for 12 months, gaining a monthly interest of 1%, so its future value is

$$100(1.01)^{12}$$

The second payment is invested for 11 months, so its future value is

$$100(1.01)^{11}$$

Likewise, the third payment yields

$$100(1.01)^{10}$$

and so on. The last payment is invested for 1 month, so its future value is

$$100(1.01)^{1}$$

The total value of the savings at the end of 12 months is then

$$100(1.01)^{12} + 100(1.01)^{11} + \ldots + 100(1.01)^{1}$$

If we rewrite this series in the order of ascending powers, we then have the more familiar form

$$100(1.01)^{1} + \ldots + 100(1.01)^{11} + 100(1.01)^{12}$$

This is equal to the sum of the first 12 terms of a geometric progression in which the first term is 100(1.01) and the geometric ratio is 1.01. Its value can therefore be found by using

$$a\left(\frac{r^n - 1}{r - 1}\right)$$

with $a = 100(1.01)$, $r = 1.01$ and $n = 12$, which gives

$$\$100(1.01)\left(\frac{(1.01)^{12} - 1}{1.01 - 1}\right) = \$1280.93$$

(b) In part (a) we showed that after 12 months the total amount saved is

$$100(1.01) + 100(1.01)^2 + \ldots + 100(1.01)^{12}$$

Using exactly the same argument, it is easy to see that after n months the account contains

$$100(1.01) + 100(1.01)^2 + \ldots + 100(1.01)^n$$

The formula for the sum of the first n terms of a geometric progression shows that this is the same as

$$100(1.01)\left(\frac{1.01^n - 1}{1.01 - 1}\right) = 10\,100(1.01^n - 1)$$

The problem here is to find the number of months needed for total savings to rise to \$2000. Mathematically, this is equivalent to solving the equation

$$10\,100(1.01^n - 1) = 2000$$

for n. Following the strategy described in Section 2.3 gives

$$1.01^n - 1 = 0.198$$

$$1.01^n = 1.198$$

$$\log(1.01)^n = \log(1.198)$$

$$n \log(1.01) = \log(1.198)$$

$$n = \frac{\log(1.198)}{\log(1.01)}$$

$$= 18.2$$

It follows that after 18 months savings are less than \$2000, whereas after 19 months savings exceed this amount. The target figure of \$2000 is therefore reached at the end of the 19th month.

Practice Problem

3 An individual saves \$1000 in a bank account at the beginning of each year. The bank offers a return of 8% compounded annually.

(a) Determine the amount saved after 10 years.

(b) After how many years does the amount saved first exceed \$20 000?

We now turn our attention to loans. Many businesses finance their expansion by obtaining loans from a bank or other financial institution. Banks are keen to do this provided that they receive interest as a reward for lending money. Businesses pay back loans by monthly or annual repayments. The way in which this repayment is calculated is as follows. Let us suppose that interest is calculated on a monthly basis and that the firm repays the debt by fixed monthly instalments at the end of each month. The bank calculates the interest charged during the first month based on the original loan. At the end of the month, this interest is added on to the original loan and the repayment is simultaneously deducted to determine the amount owed. The bank then charges interest in the second month based on this new amount and the process is repeated. Provided that the monthly repayment is greater than the interest charged each month, the amount owed decreases and eventually the debt is cleared. In practice, the period during which the loan is repaid is fixed in advance and the monthly repayments are calculated to achieve this end.

Example

Determine the monthly repayments needed to repay a \$100 000 loan which is paid back over 25 years when the interest rate is 8% compounded annually.

Solution

In this example the time interval between consecutive repayments is 1 month, whereas the period during which interest is charged is 1 year. This type of financial calculation typifies the way in which certain types of housing loan are worked out. The interest is compounded annually at 8%, so the amount of interest charged during the first year is 8% of the original loan: that is,

$$\frac{8}{100} \times 100\,000 = 8000$$

This amount is added on to the outstanding debt at the end of the first year. During this time, 12 monthly repayments are made, so if each instalment is \$$x$, the outstanding debt must decrease by $12x$. Hence, at the end of the first year, the amount owed is

$$100\,000 + 8000 - 12x = 108\,000 - 12x$$

In order to be able to spot a pattern in the annual debt, let us write this as

$$100\,000(1.08) - 12x$$

where the first part simply reflects the fact that 8% interest is added on to the original sum of \$100 000. At the end of the second year, a similar calculation is performed. The amount owed rises by 8% to become

$$[100\,000(1.08) - 12x](1.08) = 100\,000(1.08)^2 - 12x(1.08)$$

and we deduct $12x$ for the repayments to get

$$100\,000(1.08)^2 - 12x(1.08) - 12x$$

This is the amount owed at the end of the second year. Each year we multiply by 1.08 and subtract $12x$, so at the end of the third year we owe

$$[100\,000(1.08)^2 - 12x(1.08) - 12x](1.08) - 12x = 100\,000(1.08)^3 - 12x(1.08)^2 - 12x(1.08) - 12x$$

and so on. These results are summarized in Table 3.17. If we continue the pattern, we see that after 25 years the amount owed is

$$\begin{aligned} &100\,000(1.08)^{25} - 12x(1.08)^{24} - 12x(1.08)^{23} - \ldots - 12x \\ &= 100\,000(1.08)^{25} - 12x[1 + 1.08 + (1.08)^2 + \ldots + (1.08)^{24}] \end{aligned}$$

where we have taken out a common factor of $12x$ and rewritten the powers of 1.08 in ascending order. The first term is easily evaluated using a calculator to get

$$100\,000(1.08)^{25} = 684\,847.520$$

Table 3.17

End of year	Outstanding debt
1	$100\,000(1.08)^1 - 12x$
2	$100\,000(1.08)^2 - 12x(1.08)^1 - 12x$
3	$100\,000(1.08)^3 - 12x(1.08)^2 - 12x(1.08)^1 - 12x$

The geometric series inside the square brackets can be worked out from the formula

$$a\left(\frac{r^n - 1}{r - 1}\right)$$

The first term $a = 1$, the geometric ratio $r = 1.08$, and we are summing the first 25 terms, so $n = 25$. (Can you see why there are actually 25 terms in this series rather than 24?) Hence

$$[1 + 1.08 + (1.08)^2 + \ldots + (1.08)^{24}] = \frac{1.08^{25} - 1}{1.08 - 1} = 73.106$$

The amount owed at the end of 25 years is therefore

$$684\,847.520 - 12x(73.106) = 684\,847.520 - 877.272x$$

In this expression, x denotes the monthly repayment, which is chosen so that the debt is completely cleared after 25 years. This will be so if x is the solution of

$$684\,847.520 - 877.272x = 0$$

Hence

$$x = \frac{684\,847.520}{877.272} = \$780.66$$

The monthly repayment on a 25-year loan of $100 000 is $780.66, assuming that the interest rate remains fixed at 8% throughout this period.

It is interesting to substitute this value of x into the expressions for the outstanding debt given in Table 3.17. The results are listed in Table 3.18. What is so depressing about these figures is that the debt only falls by about $1500 to begin with, in spite of the fact that over $9000 is being repaid each year!

Table 3.18

End of year	Outstanding debt
1	$98 632.08
2	$97 154.73
3	$95 559.18

Practice Problem

4 A person requests an immediate bank overdraft of $2000. The bank generously agrees to this, but insists that it should be repaid by 12 monthly instalments and charges 1% interest every month on the outstanding debt. Determine the monthly repayment.

The mathematics used in this section for problems on savings and loans can be used for other time series. Reserves of non-renewable commodities such as minerals, oil and gas continue to decline, and geometric series can be used to estimate the year in which these stocks are likely to run out.

Example

Total reserves of a non-renewable resource are 250 million tonnes. Annual consumption, currently at 20 million tonnes per year, is expected to rise by 2% a year. After how many years will stocks be exhausted?

Solution

In the first year, consumption will be 20 million tonnes. In the second year, this will rise by 2%, so consumption will be 20(1.02) million tonnes. In the third year, this will again rise by 2% to become $20(1.02)^2$ million tonnes. The total consumption (in millions of tonnes) during the next n years will be

$$20 + 20(1.02) + 20(1.02)^2 + \ldots + 20(1.02)^{n-1}$$

This represents the sum of n terms of a geometric series with first term $a = 20$ and geometric ratio $r = 1.02$, so is equal to

$$20\left(\frac{1.02^n - 1}{1.02 - 1}\right) = 1000(1.02^n - 1)$$

Reserves will run out when this exceeds 250 million, so we need to solve the equation

$$1000(1.02^n - 1) = 250$$

for n. This is easily solved using logarithms:

$$1.02^n - 1 = 0.25 \qquad \text{(divide both sides by 1000)}$$

$$1.02^n = 1.25 \qquad \text{(add 1 to both sides)}$$

$$\log(1.02^n) = \log(1.25) \qquad \text{(take logs of both sides)}$$

$$n \log(1.02) = \log(1.25) \qquad \text{(rule 3 of logs)}$$

$$n = \frac{\log(1.25)}{\log(1.02)} \qquad \text{(divide both sides by log(1.02))}$$

$$= 11.27$$

so the reserves will be completely exhausted after 12 years.

Practice Problem

5 It is estimated that world reserves of oil currently stand at 2625 billion units. Oil is currently extracted at an annual rate of 45.5 billion units and this is set to increase by 2.6% a year. After how many years will oil reserves run out?

Example

Current annual extraction of a non-renewable resource is 40 billion units and this is expected to fall at a rate of 5% each year. Estimate the current minimum level of reserves if this resource is to last in perpetuity (that is, for ever).

Solution

In the first year 40 billion units are extracted. In the second year this falls by 5% to 40(0.95) billion units. In the third year this goes down by a further 5% to $40(0.95)^2$. After n years the total amount extracted will be

$$40 + 40(0.95) + 40(0.95)^2 + \ldots + 40(0.95)^{n-1}$$

Using the formula for the sum of a geometric progression gives

$$40\left(\frac{0.95^n - 1}{0.95 - 1}\right)$$

$$= 40\left(\frac{0.95^n - 1}{-0.05}\right)$$

$$= 800(1 - 0.95^n)$$

To see what happens in perpetuity we need to investigate the behaviour 0.95^n as n tends to infinity. Now since the magnitude of 0.95 is less than unity, it is easy to see that 0.95^n converges to zero and so the total amount will be 800 billion units.

Key Terms

Arithmetic progression A sequence of numbers with a constant difference between consecutive terms; the nth term takes the form, $a + bn$.

Geometric progression A sequence of numbers with a constant ratio between consecutive terms; the nth term takes the form, ar^{n-1}.

Geometric ratio The constant multiplier in a geometric series.

Geometric series A sum of the consecutive terms of a geometric progression.

Sinking fund A fixed sum of money saved at regular intervals which is used to fund some future financial commitment.

Practice Problems

6 Find the value of the geometric series

$$1000 + 1000(1.03) + 1000(1.03)^2 + \ldots + 1000(1.03)^9$$

7 A regular saving of \$500 is made into a sinking fund at the start of each year for 10 years. Determine the value of the fund at the end of the tenth year on the assumption that the rate of interest is

(a) 11% compounded annually

(b) 10% compounded continuously

8 Monthly sales figures for January are 5600. This is expected to fall for the following 9 months at a rate of 2% each month. Thereafter sales are predicted to rise at a constant rate of 4% each month. Estimate total sales for the next 2 years (including the first January).

9 A bank has three different types of account in which the interest rate depends on the amount invested. The 'ordinary' account offers a return of 6% and is available to every customer. The 'extra' account offers 7% and is available only to customers with \$5000 or more to invest. The 'superextra' account

➔

offers 8% and is available only to customers with \$20 000 or more to invest. In each case, interest is compounded annually and is added to the investment at the end of the year.

A person saves \$4000 at the beginning of each year for 25 years. Calculate the total amount saved on the assumption that the money is transferred to a higher-interest account at the earliest opportunity.

10 Determine the monthly repayments needed to repay a \$50 000 loan that is paid back over 25 years when the interest rate is 9% compounded annually. Calculate the increased monthly repayments needed in the case when

(a) the interest rate rises to 10%

(b) the period of repayment is reduced to 20 years

11 If

$$S_n = a + ar + ar^2 + \ldots + ar^{n-1}$$

write down an expression for rS_n and deduce that

$$rS_n - S_n = ar^n - a$$

Hence show that the sum of the first n terms of a geometric progression with first term a and geometric ratio r is given by

$$a\left(\frac{r^n - 1}{r - 1}\right)$$

provided that $r \neq 1$.

12 A prize fund is set up with a single investment of \$5000 to provide an annual prize of \$500. The fund is invested to earn interest at a rate of 7% compounded annually. If the first prize is awarded 1 year after the initial investment, find the number of years for which the prize can be awarded before the fund falls below \$500.

13 The current extraction of a certain mineral is 12 million tonnes a year and this is expected to fall at a constant rate of 6% each year. Estimate the current minimum level of world reserves if the extraction is to last in perpetuity.

14 At the beginning of a month, a customer owes a credit card company \$8480. In the middle of the month, the customer repays \$*A*, where $A <$ \$8480, and at the end of the month the company adds interest at a rate of 6% of the outstanding debt. This process is repeated with the customer continuing to pay off the same amount, \$*A*, each month.

(a) Find the value of *A* for which the customer still owes \$8480 at the start of each month.

(b) If $A = 1000$, calculate the amount owing at the end of the eighth month.

(c) Show that the value of *A* for which the whole amount owing is exactly paid off after the *n*th payment is given by

$$A = \frac{8480R^{n-1}(R - 1)}{R^n - 1} \quad \text{where} \quad R = 1.06$$

(d) Find the value of *A* if the debt is to be paid off exactly after 2 years.

15 **(Excel)** A bank decides to produce a simple table for its customers, indicating the monthly repayments of a \$5000 loan that is paid back over different periods of time. Produce such a table, with 13 rows corresponding to monthly interest rates of 0.5%, 0.525%, 0.55%, 0.575%, . . . , 0.8%, and 9 columns corresponding to a repayment period of 12, 18, 24, . . . , 60 months.

16 **(Excel)** Determine the monthly repayments needed to repay a \$500 000 loan that is paid back over 25 years when the interest rate is 7.25% compounded annually. Use a spreadsheet to tabulate the outstanding debt at the end of 1, 2, . . . , 25 years.

17 **(Excel)** Estimates of reserves of an oil field are 60 billion barrels. Current annual production of 4 billion barrels is set to rise at a constant rate of r% a year. Show that the value of r required to exhaust this oil over 10 years (including the current year) satisfies the equation

$$\left(1 + \frac{r}{100}\right)^{10} - 0.15r - 1 = 0$$

By tabulating values of the left-hand side for $r = 8.00, 8.05, 8.10, 8.15, \ldots$ calculate the value of r correct to 1 decimal place.

section 3.4

Investment appraisal

Objectives

At the end of this section you should be able to:

- Calculate present values under discrete and continuous compounding.
- Use net present values to appraise investment projects.
- Calculate the internal rate of return.
- Calculate the present value of an annuity.
- Use discounting to compare investment projects.
- Calculate the present value of government securities.

In Section 3.2 the following two formulas were used to solve compound interest problems

$$S = P\left(1 + \frac{r}{100}\right)^t \qquad (1)$$

$$S = Pe^{rt/100} \qquad (2)$$

The first of these can be applied to any type of compounding in which the interest is added on to the investment at the end of discrete time intervals. The second formula is used when the interest is added on continuously. Both formulas involve the variables

P = principal

S = future value

r = interest rate

t = time

In the case of discrete compounding, the letter t represents the number of time periods. (In Section 3.2 this was denoted by n.) For continuous compounding, t is measured in years. Given any three of these variables it is possible to work out the value of the remaining variable. Various examples were considered in Section 3.2. Of particular interest is the case where S, r and t are given, and P is the unknown to be determined. In this situation we know the future value, and we want to work backwards to calculate the original principal. This process is called ***discounting*** and the principal, P, is called the ***present value***. The rate of interest is sometimes referred to as the ***discount rate***. Equations (1) and (2) are easily rearranged to produce explicit formulas for the present value under discrete and continuous compounding:

$$P = \frac{S}{(1 + r/100)^t} = S\left(1 + \frac{r}{100}\right)^{-t}$$

$$P = \frac{S}{e^{rt/100}} = Se^{-rt/100}$$

reciprocals are denoted by negative powers

Example

Find the present value of \$1000 in 4 years' time if the discount rate is 10% compounded

(a) semi-annually

(b) continuously

Solution

(a) The discount formula for discrete compounding is

$$P = S\left(1 + \frac{r}{100}\right)^{-t}$$

If compounding occurs semi-annually then $r = 5$ since the interest rate per 6 months is $10/2 = 5$, and $t = 8$ since there are eight 6-month periods in 4 years. We are given that the future value is \$1000, so

$$P = \$1000(1.05)^{-8} = \$676.84$$

(b) The discount formula for continuous compounding is

$$P = Se^{-rt/100}$$

In this formula, r is the annual discount rate, which is 10, and t is measured in years, so is 4. Hence the present value is

$$P = \$1000e^{-0.4} = \$670.32$$

Notice that the present value in part (b) is smaller than that in part (a). This is to be expected because continuous compounding always produces a higher yield. Consequently, we need to invest a smaller amount under continuous compounding to produce the future value of \$1000 after 4 years.

Practice Problem

1 Find the present value of \$100 000 in 10 years' time if the discount rate is 6% compounded

(a) annually **(b)** continuously

Present values are a useful way of appraising investment projects. Suppose that you are invited to invest \$600 today in a business venture that is certain to produce a return of \$1000 in 5 years' time. If the discount rate is 10% compounded semi-annually then part (a) of the previous example shows that the present value of this return is \$676.84. This exceeds the initial outlay of \$600, so the venture is regarded as profitable. We quantify this profit by calculating the difference between the present value of the revenue and the present value of the costs, which is known as the *net present value* (NPV). In this example, the net present value is

$$\$676.84 - \$600 = \$76.84$$

Quite generally, a project is considered worthwhile when the NPV is positive. Moreover, if a decision is to be made between two different projects then the one with the higher NPV is the preferred choice.

An alternative way of assessing individual projects is based on the *internal rate of return* (IRR). This is the annual rate which, when applied to the initial outlay, yields the same return as the project after the same number of years. The investment is considered worthwhile provided the IRR exceeds the market rate. Obviously, in practice, other factors such as risk need to be considered before a decision is made.

The following example illustrates both NPV and IRR methods and shows how a value of the IRR itself can be calculated.

Example

A project requiring an initial outlay of \$15 000 is guaranteed to produce a return of \$20 000 in 3 years' time. Use the

(a) net present value

(b) internal rate of return

methods to decide whether this investment is worthwhile if the prevailing market rate is 5% compounded annually. Would your decision be affected if the interest rate were 12%?

Solution

(a) The present value of \$20 000 in 3 years' time, based on a discount rate of 5%, is found by setting $S = 20\ 000$, $t = 3$ and $r = 5$ in the formula

$$P = S\left(1 + \frac{r}{100}\right)^{-t}$$

This gives

$$P = \$20\ 000(1.05)^{-3} = \$17\ 276.75$$

The NPV is therefore

$$\$17\ 276.75 - \$15\ 000 = \$2276.75$$

The project is to be recommended because this value is positive.

(b) To calculate the IRR we use the formula

$$S = P\left(1 + \frac{r}{100}\right)^{t}$$

We are given $S = 20\ 000$, $P = 15\ 000$ and $t = 3$, so we need to solve

$$20\ 000 = 15\ 000\left(1 + \frac{r}{100}\right)^{3}$$

for r. An obvious first step is to divide both sides of this equation by 15 000 to get

$$\frac{4}{3} = \left(1 + \frac{r}{100}\right)^3$$

We can extract r by taking cube roots of both sides of

$$\left(1 + \frac{r}{100}\right)^3 = \frac{4}{3}$$

to get

$$1 + \frac{r}{100} = \left(\frac{4}{3}\right)^{1/3} = 1.1$$

Hence

$$\frac{r}{100} = 1.1 - 1 = 0.1$$

and so the IRR is 10%. The project is therefore to be recommended because this value exceeds the market rate of 5%.

For the last part of the problem we are invited to consider whether our advice would be different if the market rate were 12%. Using the NPV method, we need to repeat the calculations, replacing 5 by 12. The corresponding net present value is then

$$\$20\ 000(1.12)^{-3} - \$15\ 000 = -\$764.40$$

This time the NPV is negative, so the project leads to an effective loss and is not to be recommended. The same conclusion can be reached more easily using the IRR method. We have already seen that the internal rate of return is 10% and can deduce immediately that you would be better off investing the $15 000 at the market rate of 12%, since this gives the higher yield.

Practice Problem

2 An investment project requires an initial outlay of $8000 and will produce a return of $17 000 at the end of 5 years. Use the

(a) net present value

(b) internal rate of return

methods to decide whether this is worthwhile if the capital could be invested elsewhere at 15% compounded annually.

Advice

This problem illustrates the use of two different methods for investment appraisal. It may appear at first sight that the method based on the IRR is the preferred approach, particularly if you wish to consider more than one interest rate. However, this is not usually the case. The IRR method can give wholly misleading advice when *comparing* two or more projects, and you must be careful when interpreting the results of this method. The following example highlights the difficulty.

Example

Suppose that it is possible to invest in only one of two different projects. Project A requires an initial outlay of \$1000 and yields \$1200 in 4 years' time. Project B requires an outlay of \$30 000 and yields \$35 000 after 4 years. Which of these projects would you choose to invest in when the market rate is 3% compounded annually?

Solution

Let us first solve this problem using net present values.

For Project A

$$\text{NPV} = \$1200(1.03)^{-4} - \$1000 = \$66.18$$

For Project B

$$\text{NPV} = \$35\ 000(1.03)^{-4} - \$30\ 000 = \$1097.05$$

Both projects are viable as they produce positive net present values. Moreover, the second project is preferred, since it has the higher value. You can see that this recommendation is correct by considering how you might invest \$30 000. If you opt for Project A then the best you can do is to invest \$1000 of this amount to give a return of \$1200 in 4 years' time. The remaining \$29 000 could be invested at the market rate of 3% to yield

$$\$29\ 000(1.03)^4 = \$32\ 639.76$$

The total return is then

$$\$1200 + \$32\ 639.76 = \$33\ 839.76$$

On the other hand, if you opt for Project B then the whole of the \$30 000 can be invested to yield \$35 000. In other words, in 4 years' time you would be

$$\$35\ 000 - \$33\ 639.76 = \$1160.24$$

better off by choosing Project B, which confirms the advice given by the NPV method.

However, this is contrary to the advice given by the IRR method. For Project A, the internal rate of return, r_A, satisfies

$$1200 = 1000\left(1 + \frac{r_A}{100}\right)^4$$

Dividing by 1000 gives

$$\left(1 + \frac{r_A}{100}\right)^4 = 1.2$$

and if we take fourth roots we get

$$1 + \frac{r_A}{100} = (1.2)^{1/4} = 1.047$$

so $r_A = 4.7\%$.

For Project B the internal rate of return, r_B, satisfies

$$35\ 000 = 30\ 000\left(1 + \frac{r_B}{100}\right)^4$$

This can be solved as before to get $r_B = 3.9\%$.

Project A gives the higher internal rate of return even though, as we have seen, Project B is the preferred choice.

The results of this example show that the IRR method is an unreliable way of comparing investment opportunities when there are significant differences between the amounts involved. This is because the IRR method compares percentages, and obviously a large percentage of a small sum could give a smaller profit than a small percentage of a larger sum.

Practice Problem

3 A firm needs to choose between two projects, A and B. Project A involves an initial outlay of \$13 500 and yields \$18 000 in 2 years' time. Project B requires an outlay of \$9000 and yields \$13 000 after 2 years. Which of these projects would you advise the firm to invest in if the annual market rate of interest is 7%?

So far in this section we have calculated the present value of a single future value. We now consider the case of a sequence of payments over time. The simplest cash flow of this type is an *annuity*, which is a sequence of regular equal payments. It can be thought of as the opposite of a sinking fund. This time a lump sum is invested and, subsequently, equal amounts of money are withdrawn at fixed time intervals. Provided that the payments themselves exceed the amount of interest gained during the time interval between payments, the fund will decrease and eventually become zero. At this point the payments cease. In practice, we are interested in the value of the original lump sum needed to secure a regular income over a known period of time. This can be done by summing the present values of the individual payments.

Example

Find the present value of an annuity that yields an income of \$10 000 at the end of each year for 10 years, assuming that the interest rate is 7% compounded annually.

What would the present value be if the annuity yields this income in perpetuity?

Solution

The first payment of \$10 000 is made at the end of the first year. Its present value is calculated using the formula

$$P = S\left(1 + \frac{r}{100}\right)^{-t}$$

with $S = 10\,000$, $r = 7$ and $t = 1$, so

$$P = \$10\,000(1.07)^{-1} = \$9345.79$$

This means that if we want to take out \$10 000 from the fund in 1 year's time then we need to invest \$9345.79 today. The second payment of \$10 000 is made at the end of the second year, so its present value is

$$\$10\,000(1.07)^{-2} = \$8734.39$$

This is the amount of money that needs to be invested now to cover the second payment from the fund. In general, the present value of \$10 000 in t years' time is

$$10\,000(1.07)^{-t}$$

→

so the total present value is

$$10\,000(1.07)^{-1} + 10\,000(1.07)^{-2} + \ldots + 10\,000(1.07)^{-10}$$

This is a geometric series, so we may use the formula

$$a\left(\frac{r^n - 1}{r - 1}\right)$$

In this case, $a = 10\,000(1.07)^{-1}$, $r = 1.07^{-1}$ and $n = 10$, so the present value of the annuity is

$$\$10\,000(1.07)^{-1}\left(\frac{1.07^{-10} - 1}{1.07^{-1} - 1}\right) = \$70\,235.82$$

This represents the amount of money that needs to be invested now so that a regular annual income of $10 000 can be withdrawn from the fund for the next 10 years.

If the income stream is to continue for ever then we need to investigate what happens to the formula

$$a\left(\frac{r^n - 1}{r - 1}\right)$$

as n gets bigger and bigger. In this case $r = 1.07^{-1} < 1$, so as n increases, r^n decreases and tends towards zero. This behaviour can be seen clearly from the table:

n	1	10	100
1.07^{-n}	0.9346	0.5083	0.0012

Setting $r^n = 0$ in the formula for the sum of geometric series shows that if the series goes on for ever then eventually the sum approaches

$$\frac{a}{1 - r}$$

so that the present value of the annuity in perpetuity is

$$\frac{10\,000(1.07)^{-1}}{1 - 1.07^{-1}} = \$142\,857.14$$

Practice Problem

4 Find the present value of an annuity that yields an income of $2000 at the end of each month for 10 years, assuming that the interest rate is 6% compounded monthly.

The argument used in the previous example can be used to calculate the net present value. For instance, suppose that a business requires an initial investment of $60 000, which is guaranteed to return a regular payment of $10 000 at the end of each year for the next 10 years. If the discount rate is 7% compounded annually then the previous example shows that the present value is $70 235.82. The net present value of the investment is therefore

$$\$70\,235.82 - \$60\,000 = \$10\,235.82$$

A similar procedure can be used when the payments are irregular, although it is no longer possible to use the formula for the sum of a geometric progression. Instead the present value of each individual payment is calculated and the values are then summed longhand.

Example

A small business has a choice of investing $20 000 in one of two projects. The revenue flows from the two projects during the next 4 years are listed in Table 3.19. If the interest rate is 11% compounded annually, which of these two projects would you advise the company to invest in?

Table 3.19

	Revenue ($)	
End of year	**Project A**	**Project B**
1	6 000	10 000
2	3 000	6 000
3	10 000	9 000
4	8 000	1 000
Total	27 000	26 000

Solution

If we simply add together all of the individual receipts, it appears that Project A is to be preferred, since the total revenue generated from Project A is $1000 greater than that from Project B. However, this naïve approach fails to take into account the time distribution.

From Table 3.19 we see that both projects yield a single receipt of $10 000. For Project A this occurs at the end of year 3, whereas for Project B this occurs at the end of year 1. This $10 000 is worth more in Project B because it occurs earlier in the revenue stream and, once received, could be invested for longer at the prevailing rate of interest. To compare these projects we need to discount the revenue stream to the present value. The present values obtained depend on the discount rate. Table 3.20 shows the present values based on the given rate of 11% compounded annually. These values are calculated using the formula

$$P = S(1.11)^{-t}$$

For example, the present value of the $10 000 revenue in Project A is given by

$$\$10\,000(1.11)^{-3} = \$7311.91$$

The net present values for Project A and Project B are given by

$$\$20\,422.67 - \$20\,000 = \$422.67$$

Table 3.20

	Discounted revenue ($)	
End of year	**Project A**	**Project B**
1	5 405.41	9 000.01
2	2 434.87	4 869.73
3	7 311.91	6 580.72
4	5 269.85	658.73
Total	20 422.04	21 109.19

and

$21 109.19 – $20 000 = $1109.19

respectively. Consequently, if it is possible to invest in only one of these projects, the preferred choice is Project B.

Practice Problem

5 A firm has a choice of spending $10 000 today on one of two projects. The revenue obtained from these projects is listed in Table 3.21. Assuming that the discount rate is 15% compounded annually, which of these two projects would you advise the company to invest in?

Table 3.21

	Revenue ($)	
End of year	**Project 1**	**Project 2**
1	2000	1000
2	2000	1000
3	3000	2000
4	3000	6000
5	3000	4000

It is sometimes useful to find the internal rate of return of a project yielding a sequence of payments over time. However, as the following example demonstrates, this can be difficult to calculate, particularly when there are more than two payments.

Example

(a) Calculate the IRR of a project which requires an initial outlay of $20 000 and produces a return of $8000 at the end of year 1 and $15 000 at the end of year 2.

(b) Calculate the IRR of a project which requires an initial outlay of $5000 and produces returns of $1000, $2000 and $3000 at the end of years 1, 2 and 3, respectively.

Solution

(a) In the case of a single payment, the IRR is the annual rate of interest, r, which, when applied to the initial outlay, P, yields a known future payment, S. If this payment is made after t years then

$$S = P\left(1 + \frac{r}{100}\right)^t$$

or, equivalently

$$P = S\left(1 + \frac{r}{100}\right)^{-t}$$

Note that the right-hand side of this last equation is just the present value of S. Consequently, the IRR can be thought of as the rate of interest at which the present value of S equals the initial outlay P.

The present value of \$8000 in 1 year's time is

$$8000\left(1+\frac{r}{100}\right)^{-1}$$

where r is the annual rate of interest. Similarly, the present value of \$15 000 in 2 years' time is

$$15\,000\left(1+\frac{r}{100}\right)^{-2}$$

If r is to be the IRR then the sum of these present values must equal the initial investment of \$20 000. In other words, the IRR is the value of r that satisfies the equation

$$20\,000 = 8000\left(1+\frac{r}{100}\right)^{-1} + 15\,000\left(1+\frac{r}{100}\right)^{-2}$$

The simplest way of solving this equation is to multiply both sides by $(1 + r/100)^2$ to remove all negative indices. This gives

$$20\,000\left(1+\frac{r}{100}\right)^{2} = 8000\left(1+\frac{r}{100}\right) + 15\,000$$

$b^m \times b^n = b^{m+n}$
$b^0 = 1$

Now

$$\left(1+\frac{r}{100}\right)^{2} = \left(1+\frac{r}{100}\right)\left(1+\frac{r}{100}\right) = 1+\frac{r}{50}+\frac{r^2}{10\,000}$$

so if we multiply out the brackets, we obtain

$$20\,000 + 400r + 2r^2 = 8000 + 80r + 15\,000$$

Collecting like terms gives

$$2r^2 + 320r - 3000 = 0$$

This is a quadratic in r, so can be solved using the formula described in Section 2.1 to get

$$r = \frac{-320 \pm \sqrt{((320)^2 - 4(2)(-3000))}}{2(2)}$$

$$= \frac{-320 \pm 355.5}{4}$$

$$= 8.9\% \text{ or } -168.9\%$$

We can obviously ignore the negative solution, so can conclude that the IRR is 8.9%.

(b) If an initial outlay of \$5000 yields \$1000, \$2000 and \$3000 at the end of years 1, 2 and 3, respectively, then the internal rate of return, r, satisfies the equation

$$5000 = 1000\left(1+\frac{r}{100}\right)^{-1} + 2000\left(1+\frac{r}{100}\right)^{-2} + 3000\left(1+\frac{r}{100}\right)^{-3}$$

A sensible thing to do here might be to multiply through by $(1 + (r/100))^{-3}$. However, this produces an equation involving r^3 (and lower powers of r), which is no easier to solve than the original. Indeed, a moment's thought should convince you that in general, when dealing with a sequence of payments over n years, the IRR will satisfy an equation involving r^n (and lower powers of r). Under these circumstances

it is virtually impossible to obtain the exact solution. The best way of proceeding would be to use a non-linear equation-solver routine on a computer, particularly if it is important that an accurate value of r is obtained. However, if all that is needed is a rough approximation then this can be done by systematic trial and error. We merely substitute likely solutions into the right-hand side of the equation until we find the one that works. For example, putting $r = 5$ gives

$$\frac{1000}{1.05} + \frac{2000}{(1.05)^2} + \frac{3000}{(1.05)^3} = 5358$$

Other values of the expression

$$1000\left(1 + \frac{r}{100}\right)^{-1} + 2000\left(1 + \frac{r}{100}\right)^{-2} + 3000\left(1 + \frac{r}{100}\right)^{-3}$$

corresponding to $r = 6, 7, \ldots, 10$ are listed in the following table:

r	6	7	8	9	10
value	5242	5130	5022	4917	4816

Given that we are trying to find r so that this value is 5000, this table indicates that r is somewhere between 8% (which produces a value greater than 5000) and 9% (which produces a value less than 5000).

If a more accurate estimate of IRR is required then we simply try further values between 8% and 9%. For example, it is easy to check that putting $r = 8\frac{1}{2}$ gives 4969, indicating that the exact value of r is between 8% and $8\frac{1}{2}$%. We conclude that the IRR is 8% to the nearest percentage.

Practice Problem

6 A project requires an initial investment of \$12 000. It has a guaranteed return of \$8000 at the end of year 1 and a return of \$2000 each year at the end of years 2, 3 and 4.

Estimate the IRR to the nearest percentage. Would you recommend that someone invests in this project if the prevailing market rate is 8% compounded annually?

Problem 6 should have convinced you how tedious it is to calculate the internal rate of return 'by hand' when there are more than two payments in a revenue flow. A computer spreadsheet provides the ideal tool for dealing with this. Chart Wizard can be used to sketch a graph from which a rough estimate of IRR can be found. A more accurate value can be found using a 'finer' tabulation in the vicinity of this estimate.

Example

EXCEL

A proposed investment project costs \$11 600 today. The expected revenue flow (in thousands of dollars) for the next 4 years is

Year	1	2	3	4
Revenue flow	2	3.7	3.8	4.5

Use a graphical method to determine the IRR to the nearest whole number. By tabulating further values, estimate the IRR correct to 1 decimal place.

Figure 3.4

	A	B	C	D	E	F	G	H
1	Calculation of the Internal Rate of Return							
2								
3				Interest Rate				
4								
5	Year	Revenue	2%	4%	6%	8%	10%	
6	0	-11600	-11600	-11600	-11600	-11600	-11600	
7	1	2000						
8	2	3700						
9	3	3800						
10	4	4500						
11								

Solution

Before we tackle this particular example, it will be useful to review the definition of the internal rate of return. So far, we have taken it to be the rate of interest at which the total present values of the revenue stream equal the initial outlay. Of course, this is the same as saying that the difference between present values and the initial outlay is zero. In other words, the internal rate of return is the interest rate which gives a net present value (NPV) of zero. We shall exploit this fact by plotting a graph of net present values against interest rate (r). The IRR is the value of r at which the graph crosses the horizontal axis.

We begin by typing suitable headings together with values of the years and revenue flows for this project into a spreadsheet, as shown in Figure 3.4.

Notice that the initial investment in the project has been input as a negative number, since this represents an outflow of funds. The present value of this is also −11 600, since this occurs in year 0. The columns represent interest rates of 2%, 4%, . . . , 10%. The values in the body of the table will be the present values of the revenue flows, calculated at each of these rates of interest. For example, the entry in cell C7 will be the present value of the $2000 received at the end of year 1 when the interest rate is 2%. From the formula

$$P = S\left(1 + \frac{r}{100}\right)^{-t}$$

this is

$$2000\left(1 + \frac{2}{100}\right)^{-1}$$

Notice that the numbers 2000 and 1 appear in cells B7 and A7, respectively, so the formula that we need for cell C7 is

=B7*(1.02)^(-A7)

By clicking and dragging this formula down to C10, we complete the present values for the 2% interest rate.

For the next column, we simply change the scale factor 1.02 to 1.04, so we type

=B7*(1.04)^(-A7)

in cell D7 and repeat the process. We can obviously continue in this way along the rest of the table. Finally, we calculate the net present values by summing the entries for the present values in each column. For example, to find the NPV for the 2% interest rate we type

➔

=SUM(C6:C10)

in cell C11. (A quicker way of doing this is just to highlight cells C6 to C11 and click on the Σ icon on the toolbar. This is the Greek letter sigma, which mathematicians use as an abbreviation for SUM. Excel will then sum these five cells and put the answer in C11.) Figure 3.5 shows the completed spreadsheet. The values have been rounded to 2 decimal places using the Decrease Decimal button on the toolbar.

Figure 3.5

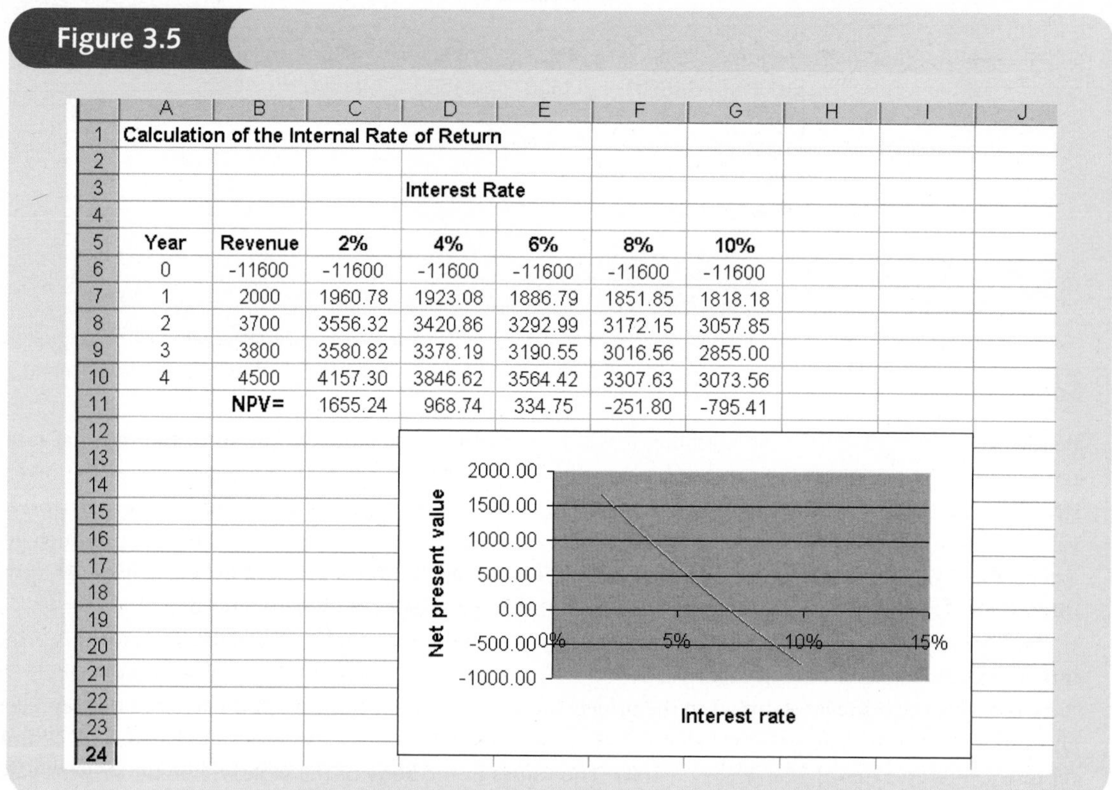

	A	B	C	D	E	F	G	H	I	J
1	Calculation of the Internal Rate of Return									
2										
3				Interest Rate						
4										
5	Year	Revenue	2%	4%	6%	8%	10%			
6	0	-11600	-11600	-11600	-11600	-11600	-11600			
7	1	2000	1960.78	1923.08	1886.79	1851.85	1818.18			
8	2	3700	3556.32	3420.86	3292.99	3172.15	3057.85			
9	3	3800	3580.82	3378.19	3190.55	3016.56	2855.00			
10	4	4500	4157.30	3846.62	3564.42	3307.63	3073.56			
11		NPV=	1655.24	968.74	334.75	-251.80	-795.41			
12										
13										
14										
15										
16										
17										
18										
19										
20										
21										
22										
23										
24										

A graph of NPV against r is shown in Figure 3.5. This is plotted by highlighting the cells in rows 5 and 11, and using Chart Wizard. (Click and drag from C5 to G5 then, holding down the Ctrl key, click and drag from C11 to G11.) The graph shows that to the nearest whole number, the internal rate of return is 7%.

To obtain a more accurate estimate of IRR, we return to the spreadsheet and add more columns for interest rates near 7%. By looking at the bottom row of the table in Figure 3.6 we see that, to 1 decimal place, the IRR is 7.1%.

Figure 3.6

7%	7.1%	7.15%
-11600	-11600	-11600
1869.16	1867.41	1866.54
3231.72	3225.69	3222.68
3101.93	3093.25	3088.92
3433.03	3420.22	3413.85
35.84	6.58	-8.01

We conclude this section by using the theory of discounting to explain the relationship between interest rates and the speculative demand for money. This was first introduced in Section 1.6 in the analysis of LM schedules. Speculative demand consists of money held in reserve to take advantage of changes in the value of alternative financial assets, such as government bonds. As their name suggests, these issues can be bought from the government at a certain price. In return, the government pays out interest on an annual basis for a prescribed number of years. At the end of this period the bond is redeemed and the purchaser is repaid the original sum. Now these bonds can be bought and sold at any point in their lifetime. The person who chooses to buy one of these bonds part-way through this period is entitled to all of the future interest payments, together with the final redemption payment. The value of existing securities clearly depends on the number of years remaining before redemption, together with the prevailing rate of interest.

Example

A 10-year bond is originally offered by the government at \$5000 with an annual return of 9%. Assuming that the bond has 4 years left before redemption, calculate its present value assuming that the prevailing interest rate is

(a) 5% **(b)** 7% **(c)** 9% **(d)** 11% **(e)** 13%

Solution

The government pays annual interest of 9% on the \$5000 bond, so agrees to pay the holder \$450 every year for 10 years. At the end of the 10 years, the bond is redeemed by the government and \$5000 is paid back to the purchaser. If there are just 4 years left between now and the date of redemption, the future cash flow that is paid on the bond is summarized in the second column of Table 3.22. This is similar to that of an annuity except that in the final year an extra payment of \$5000 is received when the government pays back the original investment. The present value of this income stream is calculated in Table 3.22 using the given discount rates of 5%, 7%, 9%, 11% and 13% compounded annually. The total present value in each case is given in the last row of this table and varies from \$5710 when the interest rate is 5% to \$4405 when it is 13%.

Notice from the table in the previous example that the value of a bond falls as interest rates rise. This is entirely to be expected, since the formula we use to calculate individual present values is

$$P = \frac{S}{(1 + r/100)^t}$$

and larger values of r produce smaller values of P.

Table 3.22

		Present values				
End of year	**Cash flow**	**5%**	**7%**	**9%**	**11%**	**13%**
1	450	429	421	413	405	398
2	450	408	393	379	365	352
3	450	389	367	347	329	312
4	5450	4484	4158	3861	3590	3343
Total present value		5710	5339	5000	4689	4405

The effect of this relationship on financial markets can now be analysed. Let us suppose that the interest rate is high at, say, 13%. As you can see from Table 3.22, the price of the bond is relatively low. Moreover, one might reasonably expect that, in the future, interest rates are likely to fall, thereby increasing the present value of the bond. In this situation an investor would be encouraged to buy this bond in the expectation of not only receiving the cash flow from holding the bond but also receiving a capital gain on its present value. Speculative balances therefore decrease as a result of high interest rates because money is converted into securities. Exactly the opposite happens when interest rates are low. The corresponding present value is relatively high, and, with an expectation of a rise in interest rates and a possible capital loss, investors are reluctant to invest in securities, so speculative balances are high.

Practice Problem

7 A 10-year bond is originally offered by the government at $1000 with an annual return of 7%. Assuming that the bond currently has 3 years left before redemption and that the prevailing interest rate is 8% compounded annually, calculate its present value.

Key Terms

Annuity A lump sum investment designed to produce a sequence of equal regular payments over time.

Discount rate The interest rate that is used when going backwards in time to calculate the present value from a future value.

Discounting The process of working backwards in time to find the present values from a future value.

Internal rate of return The interest rate for which the net present value is zero.

Net present value The present value of a revenue flow minus the original cost.

Present value The amount that is invested initially to produce a specified future value after a given period of time.

Practice Problems

8 Determine the present value of $7000 in 2 years' time if the discount rate is 8% compounded

(a) quarterly **(b)** continuously

9 A small business promises a profit of $8000 on an initial investment of $20 000 after 5 years.

(a) Calculate the internal rate of return.

(b) Would you advise someone to invest in this business if the market rate is 6% compounded annually?

10 You are given the opportunity of investing in one of three projects. Projects A, B and C require initial outlays of $20 000, $30 000 and $100 000 and are guaranteed to return $25 000, $37 000 and $117 000, respectively, in 3 years' time. Which of these projects would you invest in if the market rate is 5% compounded annually?

11 Determine the present value of an annuity that pays out \$100 at the end of each year

(a) for 5 years **(b)** in perpetuity

if the interest rate is 10% compounded annually.

12 A firm decides to invest in a new piece of machinery which is expected to produce an additional revenue of \$8000 at the end of every year for 10 years. At the end of this period the firm plans to sell the machinery for scrap, for which it expects to receive \$5000. What is the maximum amount that the firm should pay for the machine if it is not to suffer a net loss as a result of this investment? You may assume that the discount rate is 6% compounded annually.

13 During the next 3 years a business decides to invest \$10 000 at the *beginning* of each year. The corresponding revenue that it can expect to receive at the *end* of each year is given in Table 3.23. Calculate the net present value if the discount rate is 4% compounded annually.

Table 3.23

End of year	Revenue (\$)
1	5 000
2	20 000
3	50 000

14 A project requires an initial investment of \$50 000. It produces a return of \$40 000 at the end of year 1 and \$30 000 at the end of year 2. Find the exact value of the internal rate of return.

15 A government bond that originally cost \$500 with a yield of 6% has 5 years left before redemption. Determine its present value if the prevailing rate of interest is 15%.

16 An annuity pays out \$20 000 per year in perpetuity. If the interest rate is 5% compounded annually, find

(a) the present value of the whole annuity

(b) the present value of the annuity for payments received, starting from the end of the thirtieth year

(c) the present value of the annuity of the first 30 years

17 An engineering company needs to decide whether or not to build a new factory. The costs of building the factory are \$150 million initially, together with a further \$100 million at the end of the next 2 years. Annual operating costs are \$5 million commencing at the end of the third year. Annual revenue is predicted to be \$50 million commencing at the end of the third year. If the interest rate is 6% compounded annually, find

(a) the present value of the building costs

(b) the present value of the operating costs at the end of n years ($n > 2$)

(c) the present value of the revenue after n years ($n > 2$)

(d) the minimum value of n for which the net present value is positive

18 An annuity pays out \$$a$ per year in perpetuity. If the interest rate is r% compounded annually, show that the present value of the whole annuity is $100a/r$.

19 **(Excel)** A proposed investment project costs \$970 000 today, and is expected to generate revenues (in thousands of dollars) at the end of each of the following four years of 280, 450, 300, 220

respectively. Sketch a graph of net present values against interest rates, r, over the range $0 \leq r \leq 14$. Use this graph to estimate the internal rate of return, to the nearest whole number. Use a spreadsheet to perform more calculations in order to calculate the value of the IRR, correct to 1 decimal place.

20 **(Excel)** A civil engineering company needs to buy a new excavator. Model A is expected to make a loss of $60 000 at the end of the first year, but is expected to produce revenues of $24 000 and $72 000 for the second and third years of operation. The corresponding figures for model B are $96 000, $12 000 and $120 000, respectively. Use a spreadsheet to tabulate the revenue flows (using negative numbers for the losses in the first year), together with the corresponding present values based on a discount rate of 8% compounded annually. Find the net present value for each model. Which excavator, if any, would you recommend buying?

What difference does it make if the discount rate is 8% compounded continuously?

chapter 4
Differentiation

This chapter provides a simple introduction to the general topic of calculus. In fact, 'calculus' is a Latin word and a literal translation of it is 'stone'. Unfortunately, all too many students interpret this as meaning a heavy millstone that they have to carry around with them! However, as we shall see, the techniques of calculus actually provide us with a quick way of performing calculations. (The process of counting was originally performed using stones a long time ago.)

There are eight sections, which should be read in the order that they appear. It should be possible to omit Sections 4.5 and 4.7 at a first reading and Section 4.6 can be read any time after Section 4.3.

Section 4.1 provides a leisurely introduction to the basic idea of differentiation. The material is explained using pictures, which will help you to understand the connection between the underlying mathematics and the economic applications in later sections.

There are six rules of differentiation, which are evenly split between Sections 4.2 and 4.4. Section 4.2 considers the easy rules that all students will need to know. However, if you are on a business studies or accountancy course, or are on a low-level economics route, then the more advanced rules in Section 4.4 may not be of relevance and could be ignored. As far as possible, examples given in later sections and chapters are based on the easy rules only so that such students are not disadvantaged. However, the more advanced rules are essential to any proper study of mathematical economics and their use in deriving general results is unavoidable.

Sections 4.3 and 4.5 describe standard economic applications. Marginal functions associated with revenue, cost, production, consumption and savings functions are all discussed in Section 4.3. The important topic of elasticity is described in Section 4.5. The distinction is made between price elasticity along an arc and price elasticity at a point. Familiar results involving general linear demand functions and the relationship between price elasticity of demand and revenue are derived.

Sections 4.6 and 4.7 are devoted to the topic of optimization, which is used to find the maximum and minimum values of economic functions. In the first half of Section 4.6 we concentrate on the mathematical technique. The second half contains four examination-type problems, all taken from economics, which are solved in detail. In Section 4.7, mathematics is used to derive general results relating to the optimization of profit and production functions.

The final section revises two important mathematical functions, namely the exponential and natural logarithm functions. We describe how to differentiate these functions and illustrate their use in economics.

Differentiation is probably the most important topic in the whole book, and one that we shall continue in Chapters 5 and 6, since it provides the necessary background theory for much of mathematical economics. You are therefore advised to make every effort to attempt the problems given in each section. The prerequisites include an understanding of the concept of a function together with the ability to manipulate algebraic expressions. These are covered in Chapters 1 and 2, and if you have worked successfully through this material, you should find that you are in good shape to begin calculus.

section 4.1

The derivative of a function

Objectives

At the end of this section you should be able to:

- Find the slope of a straight line given any two points on the line.
- Detect whether a line is uphill, downhill or horizontal using the sign of the slope.
- Recognize the notation $f'(x)$ and $\mathrm{d}y/\mathrm{d}x$ for the derivative of a function.
- Estimate the derivative of a function by measuring the slope of a tangent.
- Differentiate power functions.

This introductory section is designed to get you started with differential calculus in a fairly painless way. There are really only three things that we are going to do. We discuss the basic idea of something called a derived function, give you two equivalent pieces of notation to describe it and finally show you how to write down a formula for the derived function in simple cases.

In Chapter 1 the slope of a straight line was defined to be the change in the value of y brought about by a 1 unit increase in x. In fact, it is not necessary to restrict the change in x to a 1 unit increase. More generally, the *slope*, or *gradient*, of a line is taken to be the change in y divided by the corresponding change in x as you move between any two points on the line. It is customary to denote the change in y by Δy, where Δ is the Greek letter 'delta'.

Likewise, the change in x is written Δx. In this notation we have

$$\boxed{\text{slope} = \frac{\Delta y}{\Delta x}}$$

Example

Find the slope of the straight line passing through

(a) A (1, 2) and B (3, 4) **(b)** A (1, 2) and C (4, 1) **(c)** A (1, 2) and D (5, 2)

Solution

(a) Points A and B are sketched in Figure 4.1. As we move from A to B, the y coordinate changes from 2 to 4, which is an increase of 2 units, and the x coordinate changes from 1 to 3, which is also an increase of 2 units. Hence

$$\text{slope} = \frac{\Delta y}{\Delta x} = \frac{4-2}{3-1} = \frac{2}{2} = 1$$

Figure 4.1

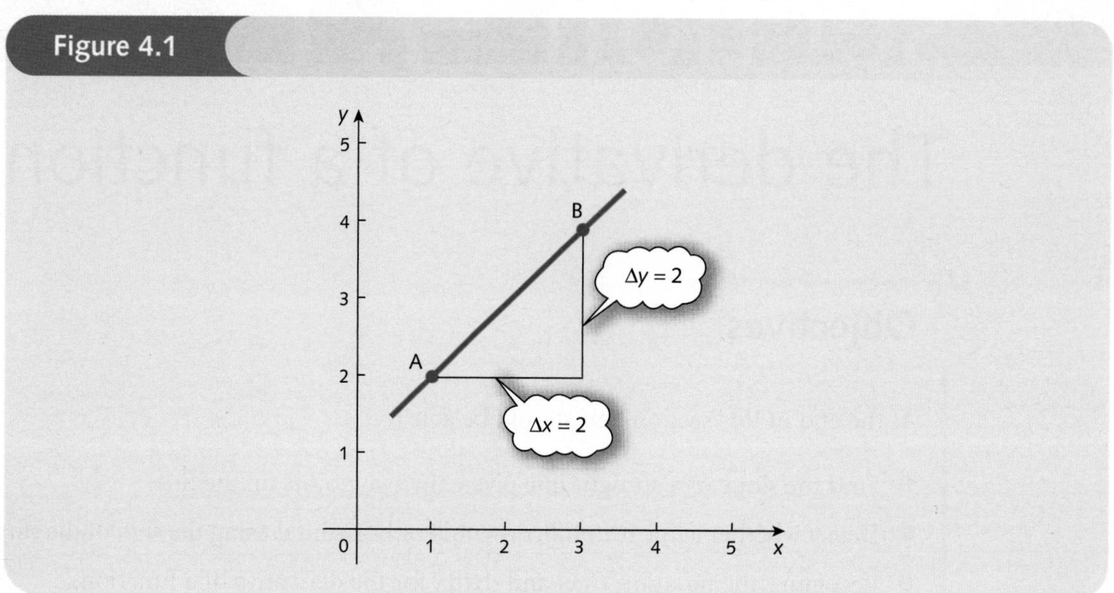

(b) Points A and C are sketched in Figure 4.2. As we move from A to C, the y coordinate changes from 2 to 1, which is a decrease of 1 unit, and the x coordinate changes from 1 to 4, which is an increase of 3 units. Hence

$$\text{slope} = \frac{\Delta y}{\Delta x} = \frac{1-2}{4-1} = \frac{-1}{3}$$

Figure 4.2

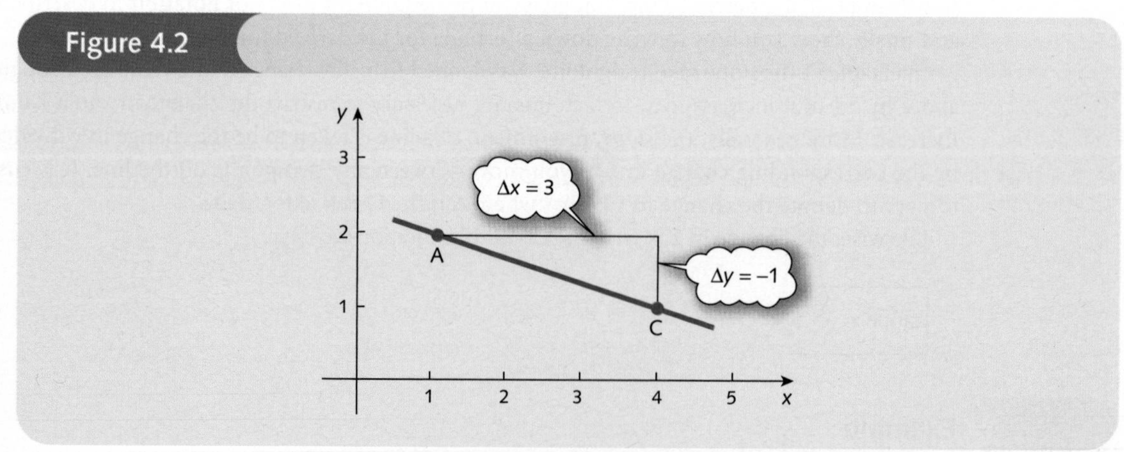

(c) Points A and D are sketched in Figure 4.3. As we move from A to D, the y coordinate remains fixed at 2, and the x coordinate changes from 1 to 5, which is an increase of 4 units. Hence

Figure 4.3

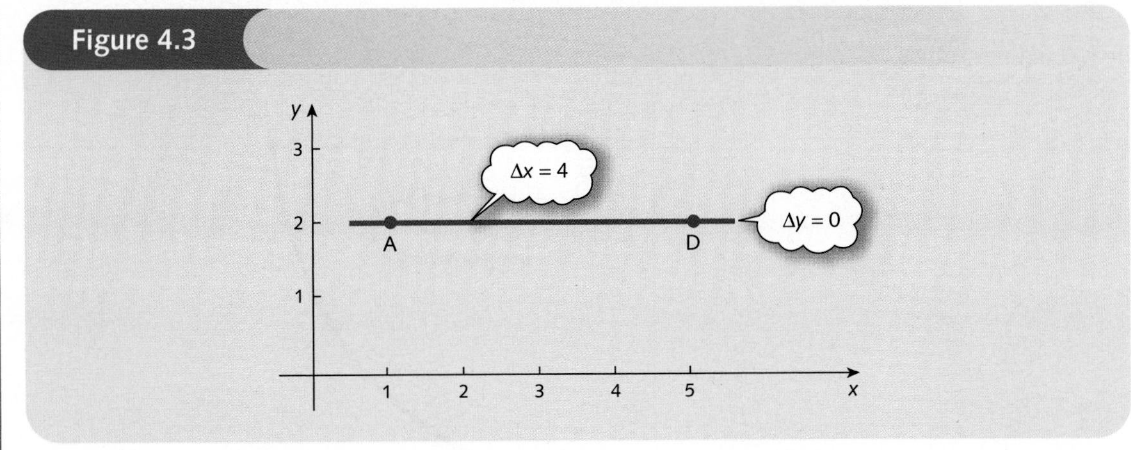

$$\text{slope} = \frac{\Delta y}{\Delta x} = \frac{2 - 2}{5 - 1} = \frac{0}{4} = 0$$

Practice Problem

1 Find the slope of the straight line passing through

(a) E (−1, 3) and F (3, 11) **(b)** E (−1, 3) and G (4, −2) **(c)** E (−1, 3) and H (49, 3)

From these examples we see that the gradient is positive if the line is uphill, negative if the line is downhill and zero if the line is horizontal.

Unfortunately, not all functions in economics are linear, so it is necessary to extend the definition of slope to include more general curves. To do this we need the idea of a tangent, which is illustrated in Figure 4.4.

A straight line which passes through a point on a curve and which just touches the curve at this point is called a *tangent*. The slope, or gradient, of a curve at $x = a$ is then defined to be that of the tangent at $x = a$. Since we have already seen how to find the slope of a straight line, this

Figure 4.4

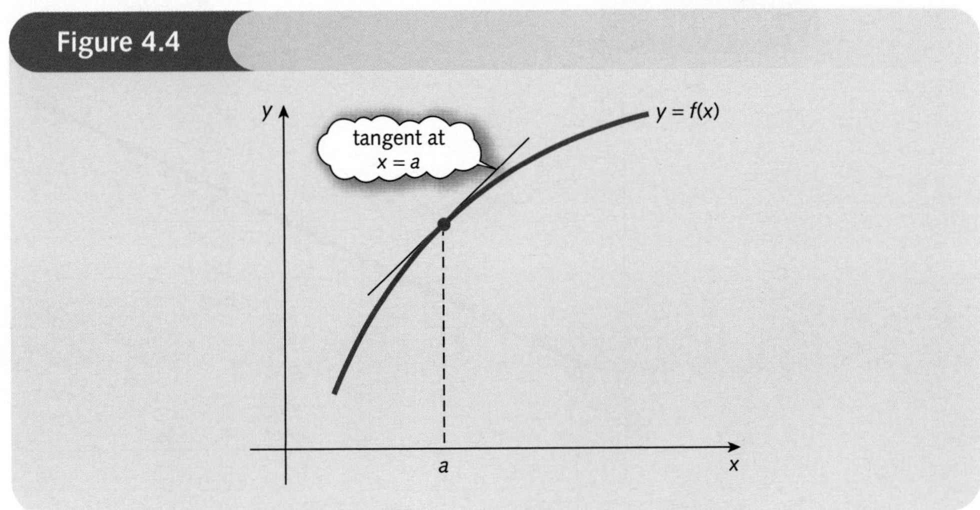

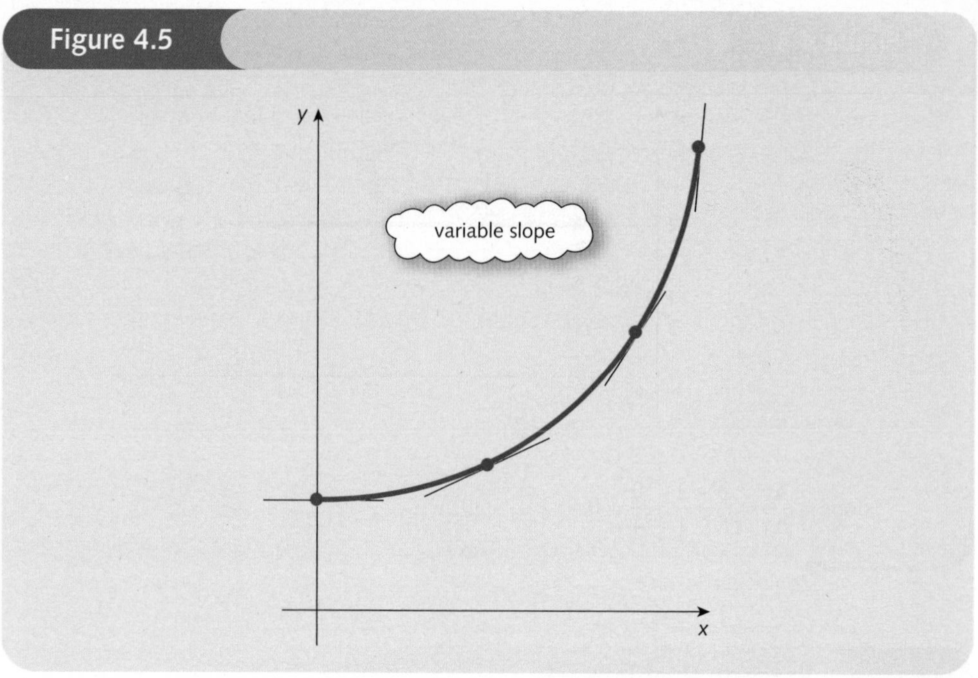

Figure 4.5

gives us a precise way of measuring the slope of a curve. A simple curve together with a selection of tangents at various points is shown in Figure 4.5. Notice how each tangent passes through exactly one point on the curve and strikes a glancing blow. In this case, the slopes of the tangents increase as we move from left to right along the curve. This reflects the fact that the curve is flat at $x = 0$ but becomes progressively steeper further away.

This highlights an important difference between the slope of a straight line and the slope of a curve. In the case of a straight line, the gradient is fixed throughout its length and it is immaterial which two points on a line are used to find it. For example, in Figure 4.6 all of the ratios $\Delta y/\Delta x$ have the value $\frac{1}{2}$. However, as we have just seen, the slope of a curve varies as we move along it. In mathematics we use the symbol

$f'(a)$ 

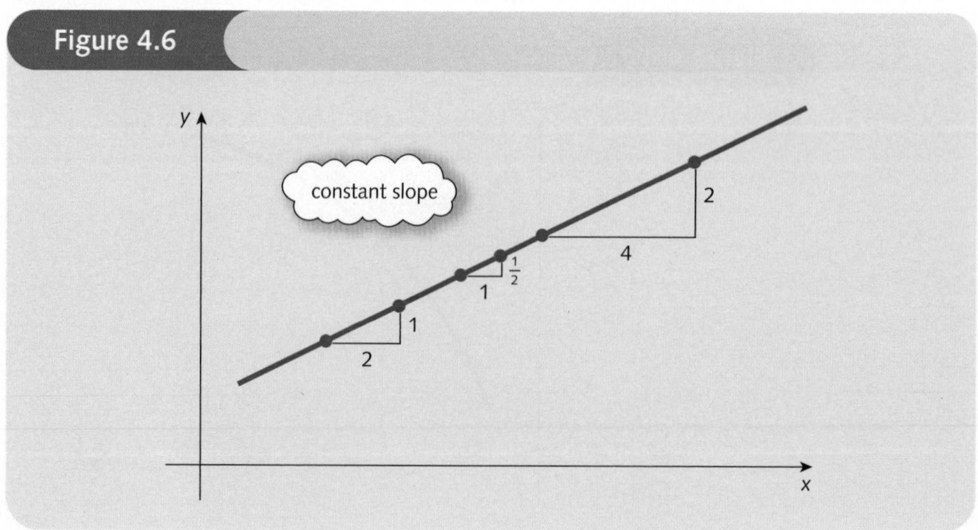

Figure 4.6

to represent the slope of the graph of a function f at $x = a$. This notation conveys the maximum amount of information with the minimum of fuss. As usual, we need the label f to denote which function we are considering. We certainly need the a to tell us at which point on the curve the gradient is being measured. Finally, the 'prime' symbol ′ is used to distinguish the gradient from the function value. The notation $f(a)$ gives the height of the curve above the x axis at $x = a$, whereas $f'(a)$ gives the gradient of the curve at this point.

The slope of the graph of a function is called the ***derivative*** of the function. It is interesting to notice that corresponding to each value of x there is a uniquely defined derivative $f'(x)$. In other words, the rule 'find the slope of the graph of f at x' defines a function. This slope function is usually referred to as the ***derived function***. An alternative notation for the derived function is

$$\frac{dy}{dx}$$

read 'dee y by dee x'

Historically, this symbol arose from the corresponding notation $\Delta y/\Delta x$ for the gradient of a straight line; the letter 'd' is the English equivalent of the Greek letter Δ. However, it is important to realize that

$$\frac{dy}{dx}$$

does not mean 'dy divided by dx'. It should be thought of as a single symbol representing the derivative of y with respect to x. It is immaterial which notation is used, although the context may well suggest which is more appropriate. For example, if we use

$$y = x^2$$

to identify the square function then it is natural to use

$$\frac{dy}{dx}$$

for the derived function. On the other hand, if we use

$$f(x) = x^2$$

then $f'(x)$ seems more appropriate.

Example

Complete the following table of function values and hence sketch an accurate graph of $f(x) = x^2$.

x	−2.0	−1.5	−1.0	−0.5	0.0	0.5	1.0	1.5	2.0
$f(x)$									

Draw the tangents to the graph at $x = -1.5, -0.5, 0, 0.5$ and 1.5. Hence estimate the values of $f'(-1.5)$, $f'(-0.5)$, $f'(0)$, $f'(0.5)$ and $f'(1.5)$.

Solution

Using a calculator we obtain

x	−2.0	−1.5	−1.0	−0.5	0.0	0.5	1.0	1.5	2.0
$f(x)$	4	2.25	1	0.25	0	0.25	1	2.25	4

➔

Figure 4.7

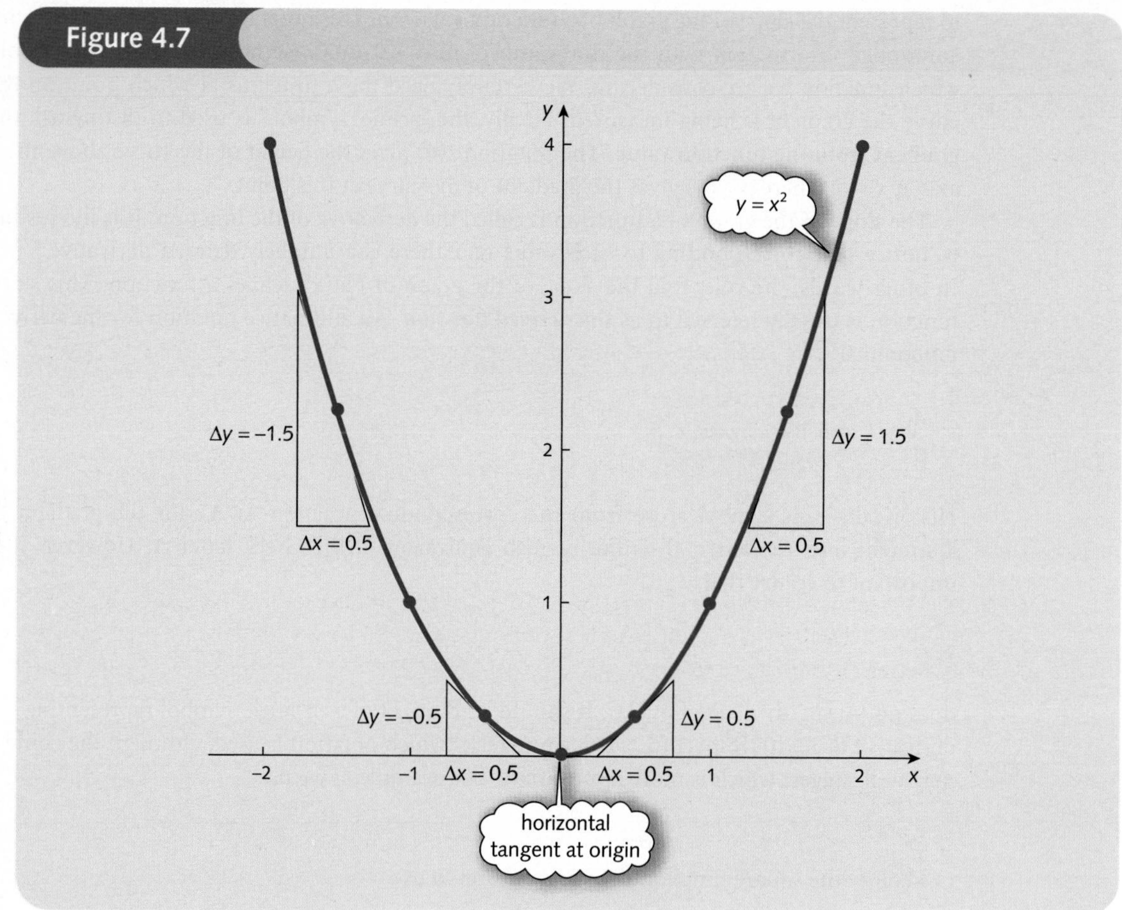

The corresponding graph of the square function is sketched in Figure 4.7. From the graph we see that the slopes of the tangents are

$$f'(-1.5) = \frac{-1.5}{0.5} = -3$$

$$f'(-0.5) = \frac{-1.5}{0.5} = -1$$

$$f'(0) = 0$$

$$f'(0.5) = \frac{0.5}{0.5} = 1$$

$$f'(1.5) = \frac{1.5}{0.5} = 3$$

The value of $f'(0)$ is zero because the tangent is horizontal at $x = 0$. Notice that

$$f'(-1.5) = -f'(1.5) \quad \text{and} \quad f'(-0.5) = -f'(0.5)$$

This is to be expected because the graph is symmetric about the y axis. The slopes of the tangents to the left of the y axis have the same size as those of the corresponding tangents to the right. However, they have opposite signs since the curve slopes downhill on one side and uphill on the other.

Practice Problem

2 Complete the following table of function values and hence sketch an accurate graph of $f(x) = x^3$.

x	−1.50	−1.25	−1.00	−0.75	−0.50	−0.25	0.00
$f(x)$		−1.95			−0.13		

x	0.25	0.50	0.75	1.00	1.25	1.50
$f(x)$		0.13			1.95	

Draw the tangents to the graph at $x = -1$, 0 and 1. Hence estimate the values of $f'(-1)$, $f'(0)$ and $f'(1)$.

Problem 2 should convince you how hard it is in practice to calculate $f'(a)$ exactly using graphs. It is impossible to sketch a perfectly smooth curve using graph paper and pencil, and it is equally difficult to judge, by eye, precisely where the tangent should be. There is also the problem of measuring the vertical and horizontal distances required for the slope of the tangent. These inherent errors may compound to produce quite inaccurate values for $f'(a)$. Fortunately, there is a really simple formula that can be used to find $f'(a)$ when f is a power function. It can be proved that

$$\text{if } f(x) = x^n \text{ then } f'(x) = nx^{n-1}$$

or, equivalently,

$$\text{if } y = x^n \text{ then } \frac{dy}{dx} = nx^{n-1}$$

The process of finding the derived function symbolically (rather than using graphs) is known as ***differentiation***. In order to differentiate x^n all that needs to be done is to bring the power down to the front and then to subtract 1 from the power:

$$x^n \text{ differentiates to } nx^{n-1}$$

To differentiate the square function we set $n = 2$ in this formula to deduce that

$$f(x) = x^2 \text{ differentiates to } f'(x) = 2x^{2-1}$$

that is,

$$f'(x) = 2x^1 = 2x$$

Using this result we see that

$$f'(-1.5) = 2 \times (-1.5) = -3$$

$$f'(-0.5) = 2 \times (-0.5) = -1$$

$$f'(0) = 2 \times (0) = 0$$

$$f'(0.5) = 2 \times (0.5) = 1$$

$$f'(1.5) = 2 \times (1.5) = 3$$

which are in agreement with the results obtained graphically in the preceding example.

Practice Problem

3 If $f(x) = x^3$ write down a formula for $f'(x)$. Calculate $f'(-1)$, $f'(0)$ and $f'(1)$. Confirm that these are in agreement with your rough estimates obtained in Problem 2.

Example

Differentiate

(a) $y = x^4$ **(b)** $y = x^{10}$ **(c)** $y = x$ **(d)** $y = 1$ **(e)** $y = 1/x^4$ **(f)** $y = \sqrt{x}$

Solution

(a) To differentiate $y = x^4$ we bring down the power (that is, 4) to the front and then subtract 1 from the power (that is, $4 - 1 = 3$) to deduce that

$$\frac{dy}{dx} = 4x^3$$

(b) Similarly,

$$\text{if} \quad y = x^{10} \quad \text{then} \quad \frac{dy}{dx} = 10x^9$$

(c) To use the general formula to differentiate x we first need to express $y = x$ in the form $y = x^n$ for some number n. In this case $n = 1$ because $x^1 = x$, so

$$\frac{dy}{dx} = 1x^0 = 1 \quad \text{since} \quad x^0 = 1$$

This result is also obvious from the graph of $y = x$ sketched in Figure 4.8.

(d) Again, to differentiate 1 we need to express $y = 1$ in the form $y = x^n$. In this case $n = 0$ because $x^0 = 1$, so

$$\frac{dy}{dx} = 0x^{-1} = 0$$

This result is also obvious from the graph of $y = 1$ sketched in Figure 4.9.

(e) Noting that $1/x^4 = x^{-4}$ it follows that

$$\text{if} \quad y = \frac{1}{x^4} \quad \text{then} \quad \frac{dy}{dx} = -4x^{-5} = -\frac{4}{x^5}$$

The power has decreased to -5 because $-4 - 1 = -5$.

Figure 4.8

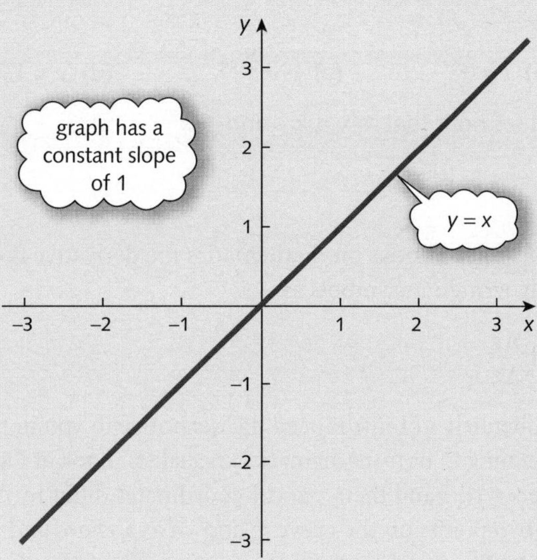

Figure 4.9

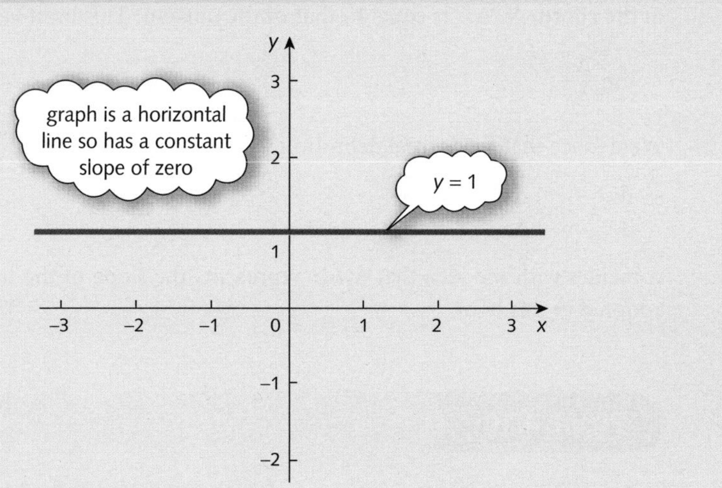

(f) Noting that $\sqrt{x} = x^{1/2}$ it follows that if

$$y = \sqrt{x} \quad \text{then} \quad \frac{dy}{dx} = \frac{1}{2}x^{-1/2}$$

$$= \frac{1}{2x^{1/2}}$$ negative powers denote reciprocals

$$= \frac{1}{2\sqrt{x}}$$ fractional powers denote roots

The power has decreased to $-^1/_2$ because $^1/_2 - 1 = -^1/_2$.

Practice Problem

4 Differentiate

(a) $y = x^5$ **(b)** $y = x^6$ **(c)** $y = x^{100}$ **(d)** $y = 1/x$ **(e)** $y = 1/x^2$

[Hint: in parts (d) and (e) note that $1/x = x^{-1}$ and $1/x^2 = x^{-2}$.]

In more advanced books on mathematics the derivative is defined via the concept of a limit and is usually written in symbols as

$$\frac{dy}{dx} = \lim_{\Delta x \to 0} \frac{\Delta y}{\Delta x}$$

We have deliberately not introduced the derivative to you in this way because the notation can appear frightening to non-mathematics specialists. Look at Figure 4.10. Points A and B both lie on the curve $y = f(x)$ and their x and y coordinates differ by Δx and Δy respectively. A line AB which joins two points on the curve is known as a ***chord*** and it has slope $\Delta y/\Delta x$.

Now look at Figure 4.11, which shows a variety of chords, AB_1, AB_2, AB_3, . . . , corresponding to smaller and smaller 'widths' Δx. As the right-hand end points, B_1, B_2, B_3, . . . , get closer to A, the 'width', Δx, tends to zero. More significantly, the slope of the chord gets closer to that of the tangent at A. We describe this by saying that in the limit, as Δx tends to zero, the slope of the chord, $\Delta y/\Delta x$, is equal to that of the tangent. This limit is written

$$\lim_{\Delta x \to 0} \frac{\Delta y}{\Delta x}$$

We deduce that the formal definition

$$\frac{dy}{dx} = \lim_{\Delta x \to 0} \frac{\Delta y}{\Delta x}$$

coincides with the idea that dy/dx represents the slope of the tangent, which is the approach adopted in this book.

Figure 4.10

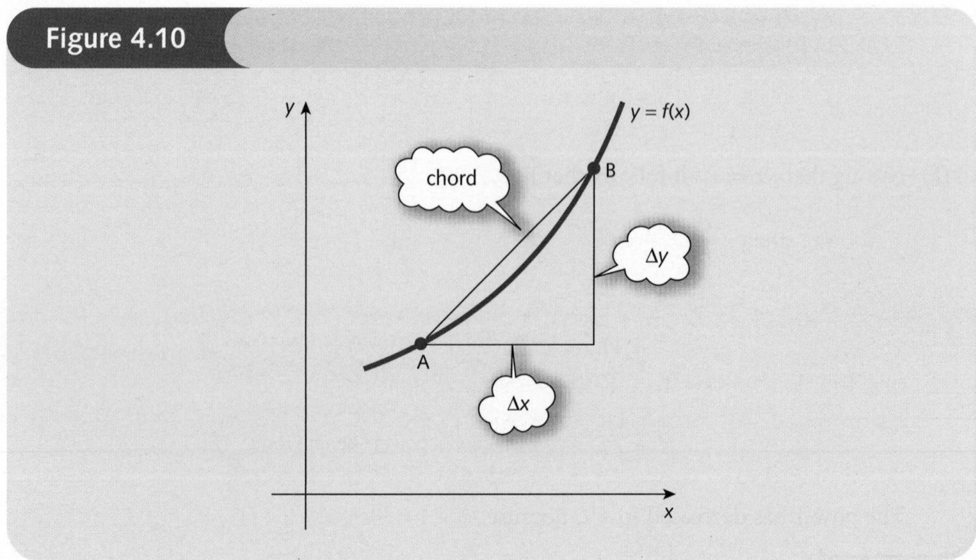

Figure 4.11

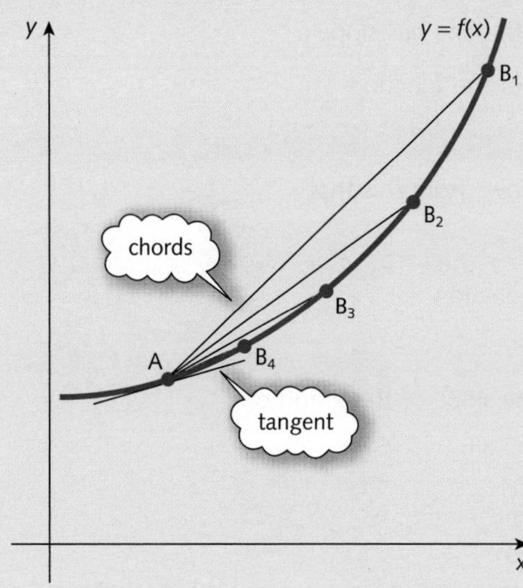

Advice

If you have met differentiation before, you might be interested in using this definition to prove results. You are advised to consult Appendix 1 at the end of this book.

Key Terms

Chord A line joining two points on a curve.

Derivative The gradient of the tangent to a curve at a point. The derivative at $x = a$ is written $f'(a)$.

Derived function The rule, f', which gives the gradient of a function, f, at a general point.

Differentiation The process or operation of determining the first derivative of a function.

Gradient The gradient of a line measures steepness and is the vertical change divided by the horizontal change between any two points on the line. The gradient of a curve at a point is that of the tangent at that point.

Slope An alternative word for gradient.

Tangent A line that just touches a curve at a point.

Practice Problems

5 Find the slope of the straight line passing through

(a) (2, 5) and (4, 9) **(b)** (3, −1) and (7, −5) **(c)** (7, 19) and (4, 19)

6 Verify that the points (0, 2) and (3, 0) lie on the line

$$2x + 3y = 6$$

Hence find the slope of this line. Is the line uphill, downhill or horizontal?

➔

7 Verify that the points (0, b) and (1, $a + b$) lie on the line

$$y = ax + b$$

Hence show that this line has slope a.

8 Sketch the graph of the function

$$f(x) = 5$$

Explain why it follows from this that

$$f'(x) = 0$$

9 Differentiate the function

$$f(x) = x^7$$

Hence calculate the slope of the graph of

$$y = x^7$$

at the point $x = 2$.

10 Differentiate

(a) $y = x^8$ **(b)** $y = x^{50}$ **(c)** $y = x^{19}$ **(d)** $y = x^{999}$

11 Differentiate the following functions, giving your answer in a similar form, without negative or fractional indices.

(a) $f(x) = \frac{1}{x^3}$ **(b)** $f(x) = \sqrt[3]{x}$ **(c)** $f(x) = \frac{1}{\sqrt[4]{x}}$ **(d)** $y = x\sqrt{x}$

12 Complete the following table of function values for the function, $f(x) = x^2 - 2x$:

x	-1	-0.5	0	0.5	1	1.5	2	2.5
$x^2 - 2x$								

Sketch the graph of this function and, by measuring the slope of this graph, estimate

(a) $f'(-0.5)$ **(b)** $f'(1)$ **(c)** $f'(1.5)$

13 For each of the graphs

(a) $y = \sqrt{x}$ **(b)** $y = x\sqrt{x}$ **(c)** $y = \frac{1}{\sqrt{x}}$

A is the point where $x = 4$, and B is the point where $x = 4.1$. In each case find

(i) the y coordinates of A and B.

(ii) the gradient of the chord AB

(iii) the value of $\frac{dy}{dx}$ at A.

Compare your answers to parts (ii) and (iii).

14 Find the coordinates of the point(s) at which the curve has the specified gradient.

(a) $y = x^{2/3}$, gradient = 1/3 **(b)** $y = x^5$, gradient = 405

(c) $y = \frac{1}{x^2}$, gradient = 16 **(d)** $y = \frac{1}{x\sqrt{x}}$, gradient = $-\frac{3}{64}$

section 4.2

Rules of differentiation

Objectives

At the end of this section you should be able to:

- Use the constant rule to differentiate a function of the form $cf(x)$.
- Use the sum rule to differentiate a function of the form $f(x) + g(x)$.
- Use the difference rule to differentiate a function of the form $f(x) - g(x)$.
- Evaluate and interpret second-order derivatives.

Advice

In this section we consider three elementary rules of differentiation. Subsequent sections of this chapter describe various applications to economics. However, before you can tackle these successfully, you must have a thorough grasp of the basic techniques involved. The problems in this section are repetitive in nature. This is deliberate. Although the rules themselves are straightforward, it is necessary for you to practise them over and over again before you can become proficient in using them. In fact, you will not be able to get much further with the rest of this book until you have mastered the rules of this section.

Rule 1 The constant rule

If $h(x) = cf(x)$ then $h'(x) = cf'(x)$

for any constant c.

This rule tells you how to find the derivative of a constant multiple of a function:

differentiate the function and multiply by the constant

Example

Differentiate

(a) $y = 2x^4$ **(b)** $y = 10x$

Solution

(a) To differentiate $2x^4$ we first differentiate x^4 to get $4x^3$ and then multiply by 2. Hence

$$\text{if} \quad y = 2x^4 \quad \text{then} \quad \frac{dy}{dx} = 2(4x^3) = 8x^3$$

(b) To differentiate $10x$ we first differentiate x to get 1 and then multiply by 10. Hence

$$\text{if} \quad y = 10x \quad \text{then} \quad \frac{dy}{dx} = 10(1) = 10$$

Practice Problem

1 Differentiate

(a) $y = 4x^3$ **(b)** $y = 2/x$

The constant rule can be used to show that

constants differentiate to zero

To see this, note that the equation

$$y = c$$

is the same as

$$y = cx^0$$

because $x^0 = 1$. By the constant rule we first differentiate x^0 to get $0x^{-1}$ and then multiply by c. Hence

$$\text{if} \quad y = c \quad \text{then} \quad \frac{dy}{dx} = c(0x^{-1}) = 0$$

This result is also apparent from the graph of $y = c$, sketched in Figure 4.12, which is a horizontal line c units away from the x axis. It is an important result and explains why lone constants lurking in mathematical expressions disappear when differentiated.

Rule 2 The sum rule

$$\text{If} \quad h(x) = f(x) + g(x) \quad \text{then} \quad h'(x) = f'(x) + g'(x)$$

This rule tells you how to find the derivative of the sum of two functions:

differentiate each function separately and add

Figure 4.12

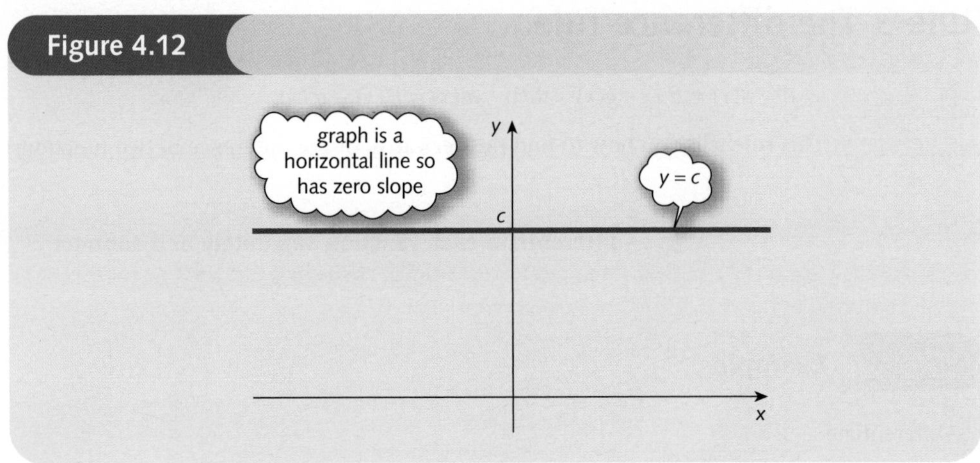

Example

Differentiate

(a) $y = x^2 + x^{50}$ **(b)** $y = x^3 + 3$

Solution

(a) To differentiate $x^2 + x^{50}$ we need to differentiate x^2 and x^{50} separately and to add. Now

x^2 differentiates to $2x$

and

x^{50} differentiates to $50x^{49}$

so

$$\text{if} \quad y = x^2 + x^{50} \quad \text{then} \quad \frac{dy}{dx} = 2x + 50x^{49}$$

(b) To differentiate $x^3 + 3$ we need to differentiate x^3 and 3 separately and to add. Now

x^3 differentiates to $3x^2$

and

3 differentiates to 0

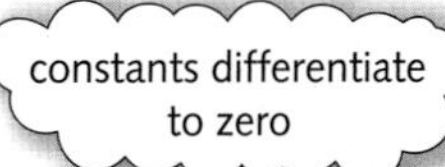

so

$$\text{if} \quad y = x^3 + 3 \quad \text{then} \quad \frac{dy}{dx} = 3x^2 + 0 = 3x^2$$

Practice Problem

2 Differentiate

(a) $y = x^5 + x$ **(b)** $y = x^2 + 5$

Rule 3 The difference rule

$$\text{If} \quad h(x) = f(x) - g(x) \quad \text{then} \quad h'(x) = f'(x) - g'(x)$$

This rule tells you how to find the derivative of the difference of two functions:

differentiate each function separately and subtract

Example

Differentiate

(a) $y = x^5 - x^2$ **(b)** $y = x - \frac{1}{x^2}$

Solution

(a) To differentiate $x^5 - x^2$ we need to differentiate x^5 and x^2 separately and to subtract. Now

x^5 differentiates to $5x^4$

and

x^2 differentiates to $2x$

so

$$\text{if} \quad y = x^5 - x^2 \quad \text{then} \quad \frac{dy}{dx} = 5x^4 - 2x$$

(b) To differentiate $x - \frac{1}{x^2}$ we need to differentiate x and $\frac{1}{x^2}$ separately and subtract. Now

x differentiates to 1

and

$\frac{1}{x^2}$ differentiates to $-\frac{2}{x^3}$

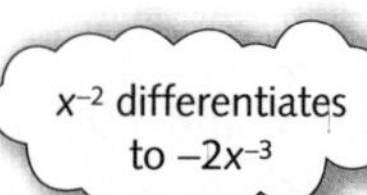

so

$$\text{if} \quad y = x - \frac{1}{x^2} \quad \text{then} \quad \frac{dy}{dx} = 1 - \left(-\frac{2}{x^3}\right) = 1 + \frac{2}{x^3}$$

Practice Problem

3 Differentiate

(a) $y = x^2 - x^3$ **(b)** $y = 50 - \frac{1}{x^3}$

It is possible to combine these three rules and so to find the derivative of more involved functions, as the following example demonstrates.

Example

Differentiate

(a) $y = 3x^5 + 2x^3$ **(b)** $y = x^3 + 7x^2 - 2x + 10$ **(c)** $y = 2\sqrt{x} + \frac{3}{x}$

Solution

(a) The sum rule shows that to differentiate $3x^5 + 2x^3$ we need to differentiate $3x^5$ and $2x^3$ separately and to add. By the constant rule

$3x^5$ differentiates to $3(5x^4) = 15x^4$

and

$2x^3$ differentiates to $2(3x^2) = 6x^2$

so

$$\text{if} \quad y = 3x^5 + 2x^3 \quad \text{then} \quad \frac{dy}{dx} = 15x^4 + 6x^2$$

With practice you will soon find that you can just write the derivative down in a single line of working by differentiating term by term. For the function

$$y = 3x^5 + 2x^3$$

we could just write

$$\frac{dy}{dx} = 3(5x^4) + 2(3x^2) = 15x^4 + 6x^2$$

(b) So far we have only considered expressions comprising at most two terms. However, the sum and difference rules still apply to lengthier expressions, so we can differentiate term by term as before. For the function

$$y = x^3 + 7x^2 - 2x + 10$$

we get

$$\frac{dy}{dx} = 3x^2 + 7(2x) - 2(1) + 0 = 3x^2 + 14x - 2$$

(c) To differentiate

$$y = 2\sqrt{x} + \frac{3}{x}$$

we first rewrite it using the notation of indices as

$$y = 2x^{1/2} + 3x^{-1}$$

Differentiating term by term then gives

$$\frac{dy}{dx} = 2\left(\frac{1}{2}\right)x^{-1/2} + 3(-1)x^{-2} = x^{-1/2} - 3x^{-2}$$

which can be written in the more familiar form

$$= \frac{1}{\sqrt{x}} - \frac{3}{x^2}$$

Practice Problem

4 Differentiate

(a) $y = 9x^5 + 2x^2$

(b) $y = 5x^8 - \dfrac{3}{x}$

(c) $y = x^2 + 6x + 3$

(d) $y = 2x^4 + 12x^3 - 4x^2 + 7x - 400$

Whenever a function is differentiated, the thing that you end up with is itself a function. This suggests the possibility of differentiating a second time to get the 'slope of the slope function'. This is written as

$f''(x)$ read '*f* double dashed of *x*'

or

$\dfrac{d^2y}{dx^2}$ read 'dee two *y* by dee *x* squared'

For example, if

$$f(x) = 5x^2 - 7x + 12$$

then differentiating once gives

$$f'(x) = 10x - 7$$

and if we now differentiate $f'(x)$ we get

$$f''(x) = 10$$

The function $f'(x)$ is called the ***first-order derivative*** and $f''(x)$ is called the ***second-order derivative.***

Example

Evaluate $f''(1)$ where

$$f(x) = x^7 + \frac{1}{x}$$

Solution

To find $f''(1)$ we need to differentiate

$$f(x) = x^7 + x^{-1}$$

twice and to put $x = 1$ into the end result. Differentiating once gives

$$f'(x) = 7x^6 + (-1)x^{-2} = 7x^6 - x^{-2}$$

and differentiating a second time gives

$$f''(x) = 7(6x^5) - (-2)x^{-3} = 42x^5 + 2x^{-3}$$

Finally, substituting $x = 1$ into

$$f''(x) = 42x^5 + \frac{2}{x^3}$$

gives

$$f''(1) = 42 + 2 = 44$$

Practice Problem

5 Evaluate $f''(6)$ where

$$f(x) = 4x^3 - 5x^2$$

It is possible to give a graphical interpretation of the sign of the second-order derivative. Remember that the first-order derivative, $f'(x)$, measures the gradient of a curve. If the derivative of $f'(x)$ is positive (that is, if $f''(x) > 0$) then $f'(x)$ is increasing. This means that the graph gets steeper as you move from left to right and so the curve bends upwards. On the other hand, if $f''(x) < 0$, the gradient, $f'(x)$ must be decreasing, so the curve bends downwards. These two cases are illustrated in Figure 4.13. For this function, $f''(x) < 0$ to the left of $x = a$, and $f''(x) > 0$ to the right of $x = a$. At $x = a$ itself, the curve changes from bending downwards to bending upwards and at this point, $f''(a) = 0$.

Figure 4.13

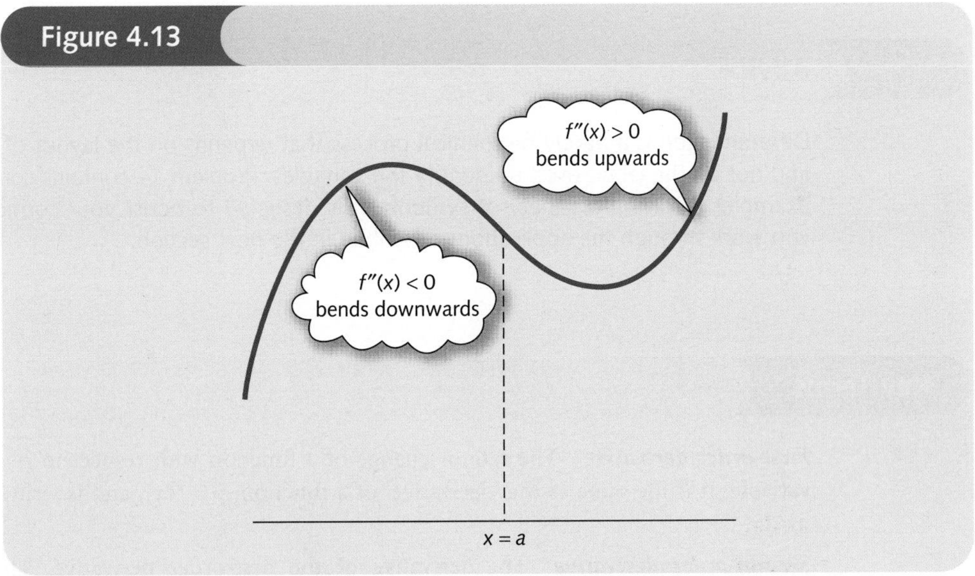

Example

Use the second-order derivative to show that the graph of the quadratic

$$y = ax^2 + bx + c$$

bends upwards when $a > 0$ and bends downwards when $a < 0$.

➔

Solution

If $y = ax^2 + bx + c$ then $\dfrac{dy}{dx} = 2ax + b$ and $\dfrac{d^2y}{dx^2} = 2a$

If $a > 0$ then $\dfrac{d^2y}{dx^2} = 2a > 0$ so the parabola bends upwards

If $a < 0$ then $\dfrac{d^2y}{dx^2} = 2a < 0$ so the parabola bends downwards

Of course, if $a = 0$, the equation reduces to $y = bx + c$, which is the equation of a straight line, so the graph bends neither upwards nor downwards.

Throughout this section the functions have all been of the form $y = f(x)$, where the letters x and y denote the variables involved. In economic functions, different symbols are used. It should be obvious, however, that we can still differentiate such functions by applying the rules of this section. For example, if a supply function is given by

$$Q = P^2 + 3P + 1$$

and we need to find the derivative of Q with respect to P then we can apply the sum and difference rules to obtain

$$\frac{dQ}{dP} = 2P + 3$$

Advice

Differentiation is a purely mechanical process that depends on the layout of the function and not on the labels used to identify the variables. Problem 14 contains some additional examples involving a variety of symbols. It is designed to boost your confidence before you work through the applications described in the next section.

Key Terms

First-order derivative The rate of change of a function with respect to its independent variable. It is the same as the 'derivative' of a function, $y = f(x)$, and is written as $f'(x)$ or dy/dx.

Second-order derivative The derivative of the first-order derivative. The expression obtained when the original function, $y = f(x)$, is differentiated twice in succession and is written as $f''(x)$ or d^2y/dx^2.

Practice Problems

6 Differentiate

(a) $y = 5x^2$ **(b)** $y = \frac{3}{x}$ **(c)** $y = 2x + 3$

(d) $y = x^2 + x + 1$ **(e)** $y = x^2 - 3x + 2$ **(f)** $y = 3x - \frac{7}{x}$

(g) $y = 2x^3 - 6x^2 + 49x - 54$ **(h)** $y = ax + b$ **(i)** $y = ax^2 + bx + c$

(j) $y = 4\sqrt{x} - \frac{3}{x} + \frac{7}{x^2}$

7 Evaluate $f'(x)$ for each of the following functions at the given point

(a) $f(x) = 3x^9$ at $x = 1$

(b) $f(x) = x^2 - 2x$ at $x = 3$

(c) $f(x) = x^3 - 4x^2 + 2x - 8$ at $x = 0$

(d) $f(x) = 5x^4 - \frac{4}{x^4}$ at $x = -1$

(e) $f(x) = \sqrt{x} - \frac{2}{x}$ at $x = 4$

8 By writing $x^2\left(x^2 + 2x - \frac{5}{x^2}\right) = x^4 + 2x^3 - 5$ differentiate $x^2\left(x^2 + 2x - \frac{5}{x^2}\right)$

Use a similar approach to differentiate

(a) $x^2(3x - 4)$ **(b)** $x(3x^3 - 2x^2 + 6x - 7)$ **(c)** $(x + 1)(x - 6)$

(d) $\frac{x^2 - 3}{x}$ **(e)** $\frac{x - 4x^2}{x^3}$ **(f)** $\frac{x^2 - 3x + 5}{x^2}$

9 Find expressions for d^2y/dx^2 in the case when

(a) $y = 7x^2 - x$

(b) $y = \frac{1}{x^2}$

(c) $y = ax + b$

10 Evaluate $f''(2)$ for the function

$$f(x) = x^3 - 4x^2 + 10x - 7$$

11 Use the second-order derivative to show that the graph of the cubic,

$$f(x) = ax^3 + bx^2 + cx + d \quad (a > 0)$$

bends upwards when $x > -b/3a$ and bends downwards when $x < -b/3a$.

12 If $f(x) = x^2 - 6x + 8$, evaluate $f'(3)$. What information does this provide about the graph of $y = f(x)$ at $x = 3$?

13 By writing $\sqrt{4x} = \sqrt{4} \times \sqrt{x} = 2\sqrt{x}$, differentiate $\sqrt{4x}$.

Use a similar approach to differentiate

(a) $\sqrt{25x}$ **(b)** $\sqrt[3]{27x}$ **(c)** $\sqrt[4]{16x^3}$ **(d)** $\sqrt{\frac{25}{x}}$

➔

14 Find expressions for

(a) $\frac{dQ}{dP}$ for the supply function $Q = P^2 + P + 1$

(b) $\frac{d(TR)}{dQ}$ for the total revenue function $TR = 50Q - 3Q^2$

(c) $\frac{d(AC)}{dQ}$ for the average cost function $AC = \frac{30}{Q} + 10$

(d) $\frac{dC}{dY}$ for the consumption function $C = 3Y + 7$

(e) $\frac{dQ}{dL}$ for the production function $Q = 10\sqrt{L}$

(f) $\frac{d\pi}{dQ}$ for the profit function $\pi = -2Q^3 + 15Q^2 - 24Q - 3$

section 4.3

Marginal functions

Objectives

At the end of this section you should be able to:

- Calculate marginal revenue and marginal cost.
- Derive the relationship between marginal and average revenue for both a monopoly and perfect competition.
- Calculate marginal product of labour.
- State the law of diminishing marginal productivity using the notation of calculus.
- Calculate marginal propensity to consume and marginal propensity to save.

At this stage you may be wondering what on earth differentiation has got to do with economics. In fact, we cannot get very far with economic theory without making use of calculus. In this section we concentrate on three main areas that illustrate its applicability:

- revenue and cost
- production
- consumption and savings.

We consider each of these in turn.

4.3.1 Revenue and cost

In Chapter 2 we investigated the basic properties of the revenue function, TR. It is defined to be PQ, where P denotes the price of a good and Q denotes the quantity demanded. In practice, we usually know the demand equation, which provides a relationship between P and Q. This enables a formula for TR to be written down solely in terms of Q. For example, if

$$P = 100 - 2Q$$

then

$$\begin{aligned} \text{TR} &= PQ \\ &= (100 - 2Q)Q \\ &= 100Q - 2Q^2 \end{aligned}$$

The formula can be used to calculate the value of TR corresponding to any value of Q. Not content with this, we are also interested in the effect on TR of a change in the value of Q from some existing level. To do this we introduce the concept of marginal revenue. The *marginal revenue*, MR, of a good is defined by

$$\text{MR} = \frac{\text{d(TR)}}{\text{d}Q}$$

marginal revenue is the derivative of total revenue with respect to demand

For example, the marginal revenue function corresponding to

$$\text{TR} = 100Q - 2Q^2$$

is given by

$$\frac{\text{d(TR)}}{\text{d}Q} = 100 - 4Q$$

If the current demand is 15, say, then

$$\text{MR} = 100 - 4(15) = 40$$

You may be familiar with an alternative definition often quoted in elementary economics textbooks. Marginal revenue is sometimes taken to be the change in TR brought about by a 1 unit increase in Q. It is easy to check that this gives an acceptable approximation to MR, although it is not quite the same as the exact value obtained by differentiation. For example, substituting $Q = 15$ into the total revenue function considered previously gives

$$\text{TR} = 100(15) - 2(15)^2 = 1050$$

An increase of 1 unit in the value of Q produces a total revenue

$$\text{TR} = 100(16) - 2(16)^2 = 1088$$

This is an increase of 38, which, according to the non-calculus definition, is the value of MR when Q is 15. This compares with the exact value of 40 obtained by differentiation.

It is instructive to give a graphical interpretation of these two approaches. In Figure 4.14 the point A lies on the TR curve corresponding to a quantity Q_0. The exact value of MR at this point is equal to the derivative

$$\frac{\text{d(TR)}}{\text{d}Q}$$

and so is given by the slope of the tangent at A. The point B also lies on the curve but corresponds to a 1 unit increase in Q. The vertical distance from A to B therefore equals the change in TR when Q increases by 1 unit. The slope of the chord joining A and B is

$$\frac{\Delta(\text{TR})}{\Delta Q} = \frac{\Delta(\text{TR})}{1} = \Delta(\text{TR})$$

Figure 4.14

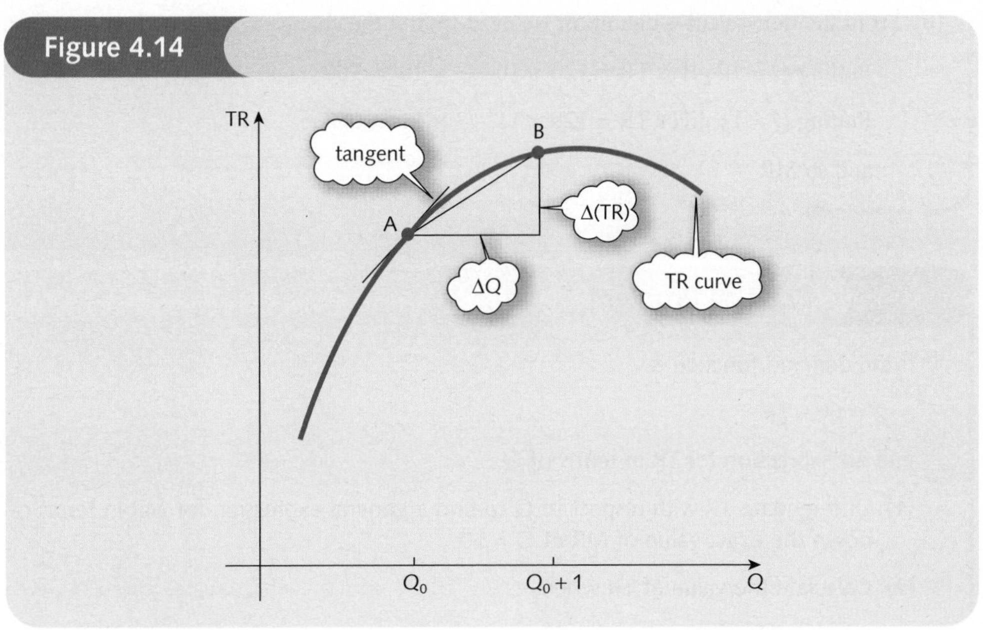

In other words, the slope of the chord is equal to the value of MR obtained from the non-calculus definition. Inspection of the diagram reveals that the slope of the tangent is approximately the same as that of the chord joining A and B. In this case the slope of the tangent is slightly the larger of the two, but there is not much in it. We therefore see that the 1 unit increase approach produces a reasonable approximation to the exact value of MR given by

$$\frac{d(TR)}{dQ}$$

Example

If the demand function is

$$P = 120 - 3Q$$

find an expression for TR in terms of Q.
Find the value of MR at $Q = 10$ using

(a) differentiation

(b) the 1 unit increase approach

Solution

$$TR = PQ = (120 - 3Q)Q = 120Q - 3Q^2$$

(a) The general expression for MR is given by

$$\frac{d(TR)}{dQ} = 120 - 6Q$$

so at $Q = 10$,

$$MR = 120 - 6 \times 10 = 60$$

➔

(b) From the non-calculus definition we need to find the change in TR as Q increases from 10 to 11.

Putting $Q = 10$ gives $\text{TR} = 120 \times 10 - 3 \times 10^2 = 900$

Putting $Q = 11$ gives $\text{TR} = 120 \times 11 - 3 \times 11^2 = 957$

and so $\text{MR} \simeq 57$

Practice Problem

1 If the demand function is

$$P = 60 - Q$$

find an expression for TR in terms of Q.

(1) Differentiate TR with respect to Q to find a general expression for MR in terms of Q. Hence write down the exact value of MR at Q = 50.

(2) Calculate the value of TR when

(a) $Q = 50$ **(b)** $Q = 51$

and hence confirm that the 1 unit increase approach gives a reasonable approximation to the exact value of MR obtained in part (1).

The approximation indicated by Figure 4.14 holds for any value of ΔQ. The slope of the tangent at A is the marginal revenue, MR. The slope of the chord joining A and B is Δ(TR)/ΔQ. It follows that

$$\text{MR} \simeq \frac{\Delta(\text{TR})}{\Delta Q}$$

This equation can be transposed to give

$$\Delta(\text{TR}) \simeq \text{MR} \times \Delta Q$$

multiply both sides by ΔQ

that is,

change in total revenue	$\simeq$	marginal revenue	$\times$	change in demand

Moreover, Figure 4.14 shows that the smaller the value of ΔQ, the better the approximation becomes. This, of course, is similar to the argument used at the end of Section 4.1 when we discussed the formal definition of a derivative as a limit.

Example

If the total revenue function of a good is given by

$$100Q - Q^2$$

write down an expression for the marginal revenue function. If the current demand is 60, estimate the change in the value of TR due to a 2 unit increase in Q.

Solution

If

$$\text{TR} = 100Q - Q^2$$

then

$$\text{MR} = \frac{\text{d(TR)}}{\text{d}Q} = 100 - 2Q$$

When Q = 60

$$\text{MR} = 100 - 2(60) = -20$$

If Q increases by 2 units, $\Delta Q = 2$ and the formula

$$\Delta(\text{TR}) \simeq \text{MR} \times \Delta Q$$

shows that the change in total revenue is approximately

$$(-20) \times 2 = -40$$

A 2 unit increase in Q therefore leads to a decrease in TR of about 40.

Practice Problem

2 If the total revenue function of a good is given by

$$1000Q - 4Q^2$$

write down an expression for the marginal revenue function. If the current demand is 30, find the approximate change in the value of TR due to a

(a) 3 unit increase in Q

(b) 2 unit decrease in Q

The simple model of demand, originally introduced in Section 1.3, assumed that price, P, and quantity, Q, are linearly related according to an equation

$$P = aQ + b$$

where the slope, a, is negative and the intercept, b, is positive. A downward-sloping demand curve such as this corresponds to the case of a ***monopolist***. A single firm, or possibly a group of firms forming a cartel, is assumed to be the only supplier of a particular product and so has control over the market price. As the firm raises the price, so demand falls. The associated total revenue function is given by

$$\begin{aligned}\text{TR} &= PQ\\ &= (aQ + b)Q\\ &= aQ^2 + bQ\end{aligned}$$

An expression for marginal revenue is obtained by differentiating TR with respect to Q to get

$$\text{MR} = 2aQ + b$$

Figure 4.15

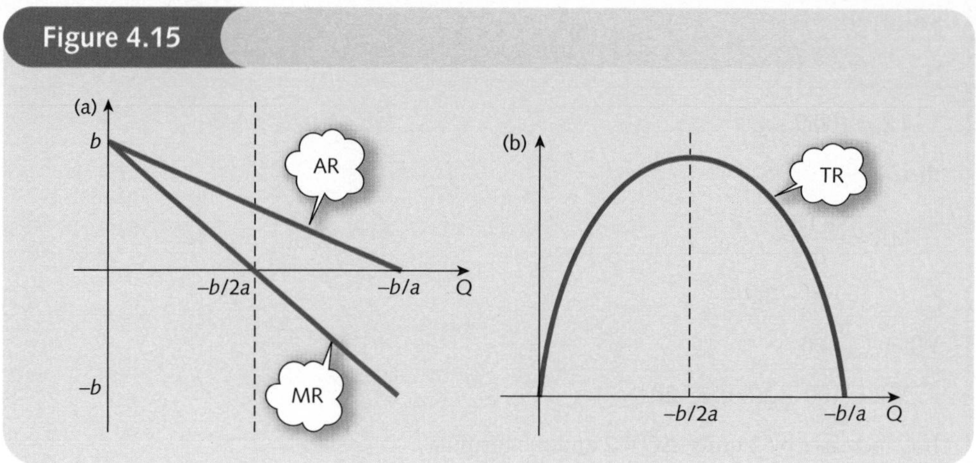

It is interesting to notice that, on the assumption of a linear demand equation, the marginal revenue is also linear with the same intercept, b, but with slope $2a$. The marginal revenue curve slopes downhill exactly twice as fast as the demand curve. This is illustrated in Figure 4.15(a).

The *average revenue*, AR, is defined by

$$\text{AR} = \frac{\text{TR}}{Q}$$

and, since $\text{TR} = PQ$, we have

$$\text{AR} = \frac{PQ}{Q} = P$$

For this reason the demand curve is labelled average revenue in Figure 4.15(a). The above derivation of the result $\text{AR} = P$ is independent of the particular demand function. Consequently, the terms 'average revenue curve' and 'demand curve' are synonymous.

Figure 4.15(a) shows that the marginal revenue takes both positive and negative values. This is to be expected. The total revenue function is a quadratic and its graph has the familiar parabolic shape indicated in Figure 4.15(b). To the left of $-b/2a$ the graph is uphill, corresponding to a positive value of marginal revenue, whereas to the right of this point it is downhill, giving a negative value of marginal revenue. More significantly, at the maximum point of the TR curve, the tangent is horizontal with zero slope and so MR is zero.

At the other extreme from a monopolist is the case of *perfect competition*. For this model we assume that there are a large number of firms all selling an identical product and that there are no barriers to entry into the industry. Since any individual firm produces a tiny proportion of the total output, it has no control over price. The firm can sell only at the prevailing market price and, because the firm is relatively small, it can sell any number of goods at this price. If the fixed price is denoted by b then the demand function is

$$P = b$$

and the associated total revenue function is

$$\text{TR} = PQ = bQ$$

An expression for marginal revenue is obtained by differentiating TR with respect to Q and, since b is just a constant, we see that

$$\text{MR} = b$$

In the case of perfect competition, the average and marginal revenue curves are the same. They are horizontal straight lines, b units above the Q axis as shown in Figure 4.16.

Figure 4.16

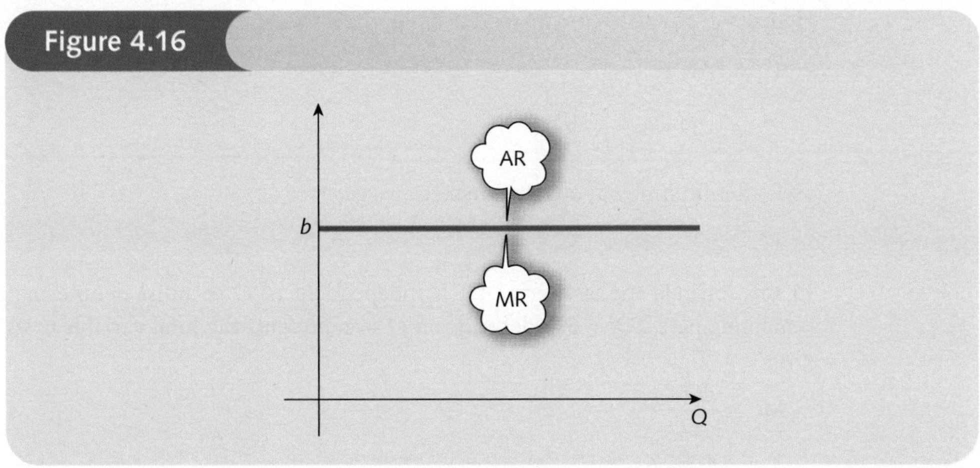

So far we have concentrated on the total revenue function. Exactly the same principle can be used for other economic functions. For instance, we define the ***marginal cost***, MC, by

$$\text{MC} = \frac{\text{d(TC)}}{\text{d}Q}$$

marginal cost is the derivative of total cost with respect to output

Again, using a simple geometrical argument, it is easy to see that if Q changes by a small amount ΔQ then the corresponding change in TC is given by

$$\Delta(\text{TC}) \simeq \text{MC} \times \Delta Q$$

$$\boxed{\begin{array}{c}\text{change in}\\ \text{total cost}\end{array}} \simeq \boxed{\begin{array}{c}\text{marginal}\\ \text{cost}\end{array}} \times \boxed{\begin{array}{c}\text{change in}\\ \text{output}\end{array}}$$

In particular, putting $\Delta Q = 1$ gives

$$\Delta(\text{TC}) \simeq \text{MC}$$

so that MC gives the approximate change in TC when Q increases by 1 unit.

Example

If the average cost function of a good is

$$\text{AC} = 2Q + 6 + \frac{13}{Q}$$

find an expression for MC. If the current output is 15, estimate the effect on TC of a 3 unit decrease in Q.

Solution

We first need to find an expression for TC using the given formula for AC. Now we know that the average cost is just the total cost divided by Q: that is,

$$\text{AC} = \frac{\text{TC}}{Q}$$

Hence

$$TC = (AC)Q$$
$$= \left(2Q + 6 + \frac{13}{Q}\right)Q$$

and, after multiplying out the brackets, we get

$$TC = 2Q^2 + 6Q + 13$$

In this formula the last term, 13, is independent of Q so must denote the fixed costs. The remaining part, $2Q^2 + 6Q$, depends on Q so represents the total variable costs. Differentiating gives

$$MC = \frac{d(TC)}{dQ}$$
$$= 4Q + 6$$

Notice that because the fixed costs are constant they differentiate to zero and so have no effect on the marginal cost. When $Q = 15$,

$$MC = 4(15) + 6 = 66$$

Also, if Q decreases by 3 units then $\Delta Q = -3$. Hence the change in TC is given by

$$\Delta(TC) \simeq MC \times \Delta Q$$
$$= 66 \times (-3)$$
$$= -198$$

so TC decreases by 198 units approximately.

Practice Problem

3 Find the marginal cost given the average cost function

$$AC = \frac{100}{Q} + 2$$

Deduce that a 1 unit increase in Q will always result in a 2 unit increase in TC, irrespective of the current level of output.

4.3.2 Production

Production functions were introduced in Section 2.3. In the simplest case output, Q, is assumed to be a function of labour, L, and capital, K. Moreover, in the short run the input K can be assumed to be fixed, so Q is then only a function of one input L. (This is not a valid assumption in the long run and in general Q must be regarded as a function of at least two inputs. Methods for handling this situation are considered in the next chapter.) The variable L is usually measured in terms of the number of workers or possibly in terms of the number of worker hours. Motivated by our previous work, we define the ***marginal product of labour***, MP_L, by

$$MP_L = \frac{dQ}{dL}$$

marginal product of labour is the derivative of output with respect to labour

As before, this gives the approximate change in Q that results from using 1 more unit of L.

Example

If the production function is

$$Q = 300\sqrt{L} - 4L$$

where Q denotes output and L denotes the size of the workforce, calculate the value of MP_L when

(a) $L = 1$

(b) $L = 9$

(c) $L = 100$

(d) $L = 2500$

and discuss the implications of these results.

Solution

If

$$Q = 300\sqrt{L} - 4L = 300L^{1/2} - 4L$$

then

$$\begin{aligned} MP_L &= \frac{dQ}{dL} \\ &= 300(\tfrac{1}{2}L^{-1/2}) - 4 \\ &= 150L^{-1/2} - 4 \\ &= \frac{150}{\sqrt{L}} - 4 \end{aligned}$$

(a) When $L = 1$

$$MP_L = \frac{150}{\sqrt{1}} - 4 = 146$$

(b) When $L = 9$

$$MP_L = \frac{150}{\sqrt{9}} - 4 = 46$$

(c) When $L = 100$

$$MP_L = \frac{150}{\sqrt{100}} - 4 = 11$$

(d) When $L = 2500$

$$MP_L = \frac{150}{\sqrt{2500}} - 4 = -1$$

→

Notice that the values of MP_L decline with increasing L. Part (a) shows that if the workforce consists of only one person then to employ two people would increase output by approximately 146. In part (b) we see that to increase the number of workers from 9 to 10 would result in about 46 additional units of output. In part (c) we see that a 1 unit increase in labour from a level of 100 increases output by only 11. In part (d) the situation is even worse. This indicates that to increase staff actually reduces output! The latter is a rather surprising result, but it is borne out by what occurs in real production processes. This may be due to problems of overcrowding on the shopfloor or to the need to create an elaborate administration to organize the larger workforce.

This example illustrates the *law of diminishing marginal productivity* (sometimes called the *law of diminishing returns*). It states that the increase in output due to a 1 unit increase in labour will eventually decline. In other words, once the size of the workforce has reached a certain threshold level, the marginal product of labour will get smaller. In the previous example, the value of MP_L continually goes down with rising L. This is not always so. It is possible for the marginal product of labour to remain constant or to go up to begin with for small values of L. However, if it is to satisfy the law of diminishing marginal productivity then there must be some value of L above which MP_L decreases.

A typical product curve is sketched in Figure 4.17, which has slope

$$\frac{dQ}{dL} = MP_L$$

Between 0 and L_0 the curve bends upwards, becoming progressively steeper, and so the slope function, MP_L, increases. Mathematically, this means that the slope of MP_L is positive: that is,

$$\frac{d(MP_L)}{dQ} > 0$$

Now MP_L is itself the derivative of Q with respect to L, so we can use the notation for the second derivative and write this as

$$\frac{d^2Q}{dL^2} > 0$$

Figure 4.17

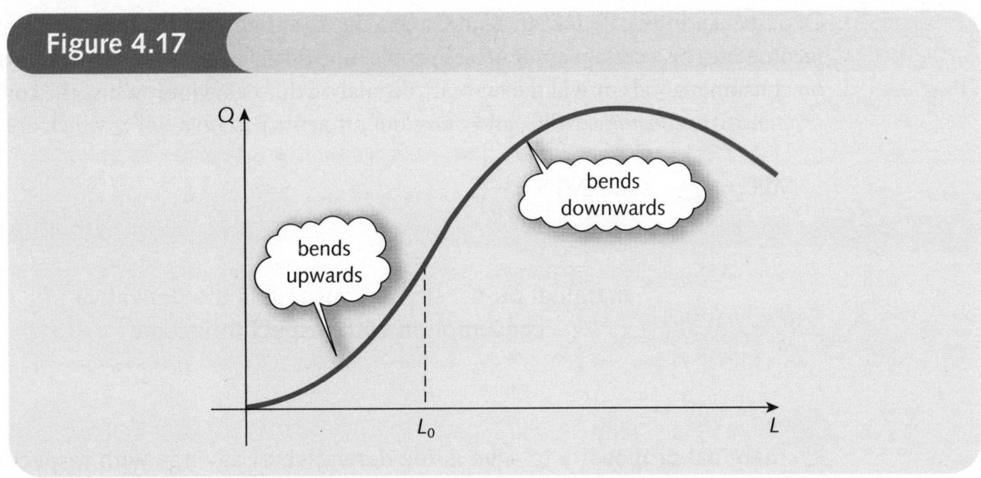

Similarly, if L exceeds the threshold value of L_0, then Figure 4.17 shows that the product curve bends downwards and the slope decreases. In this region, the slope of the slope function is negative, so that

$$\frac{d^2Q}{dL^2} < 0$$

The law of diminishing returns states that this must happen eventually: that is,

$$\frac{d^2Q}{dL^2} < 0$$

for sufficiently large L.

Practice Problem

4 A Cobb–Douglas production function is given by

$$Q = 5L^{1/2}K^{1/2}$$

Assuming that capital, K, is fixed at 100, write down a formula for Q in terms of L only. Calculate the marginal product of labour when

(a) $L = 1$ **(b)** $L = 9$ **(c)** $L = 10\,000$

Verify that the law of diminishing marginal productivity holds in this case.

4.3.3 Consumption and savings

In Chapter 1 the relationship between consumption, C, savings, S, and national income, Y, was investigated. If we assume that national income is only used up in consumption and savings then

$$Y = C + S$$

Of particular interest is the effect on C and S due to variations in Y. Expressed simply, if national income rises by a certain amount, are people more likely to go out and spend their extra income on consumer goods or will they save it? To analyse this behaviour we use the concepts *marginal propensity to consume*, MPC, and *marginal propensity to save*, MPS, which are defined by

$$\text{MPC} = \frac{dC}{dY} \quad \text{and} \quad \text{MPS} = \frac{dS}{dY}$$

marginal propensity to consume is the derivative of consumption with respect to income

marginal propensity to save is the derivative of savings with respect to income

These definitions are consistent with those given in Section 1.6, where MPC and MPS were taken to be the slopes of the linear consumption and savings curves, respectively. At first sight it appears that, in general, we need to work out two derivatives in order to evaluate MPC and MPS. However, this is not strictly necessary. Recall that we can do whatever we like to an equation provided we do the same thing to both sides. Consequently, we can differentiate both sides of the equation

$$Y = C + S$$

with respect to Y to deduce that

$$\frac{dY}{dY} = \frac{dC}{dY} + \frac{dS}{dY} = \text{MPC} + \text{MPS}$$

Now we are already familiar with the result that when we differentiate x with respect to x the answer is 1. In this case Y plays the role of x, so

$$\frac{dY}{dY} = 1$$

Hence

$$1 = \text{MPC} + \text{MPS}$$

This formula is identical to the result given in Section 1.6 for simple linear functions. In practice, it means that we need only work out one of the derivatives. The remaining derivative can then be calculated directly from this equation.

Example

If the consumption function is

$$C = 0.01Y^2 + 0.2Y + 50$$

calculate MPC and MPS when $Y = 30$.

Solution

In this example the consumption function is given, so we begin by finding MPC. To do this we differentiate C with respect to Y. If

$$C = 0.01Y^2 + 0.2Y + 50$$

then

$$\frac{dC}{dY} = 0.02Y + 0.2$$

so, when $Y = 30$,

$$MPC = 0.02(30) + 0.2 = 0.8$$

To find the corresponding value of MPS we use the formula

$$MPC + MPS = 1$$

which gives

$$MPS = 1 - MPC = 1 - 0.8 = 0.2$$

This indicates that when national income increases by 1 unit (from its current level of 30) consumption rises by approximately 0.8 units, whereas savings rise by only about 0.2 units. At this level of income the nation has a greater propensity to consume than it has to save.

Practice Problem

5 If the savings function is given by

$$S = 0.02Y^2 - Y + 100$$

calculate the values of MPS and MPC when $Y = 40$. Give a brief interpretation of these results.

Key Terms

Average revenue Total revenue per unit of output: $AR = TR/Q = P$.

Law of diminishing marginal productivity or law of diminishing returns Once the size of the workforce exceeds a particular value, the increase in output due to a 1 unit increase in labour will decline: $d^2Q/dL^2 < 0$ for sufficiently large L.

Marginal cost The cost of producing 1 more unit of output: $MC = d(TC)/dQ$.

Marginal product of labour The extra output produced by 1 more unit of labour: $MP_L = dQ/dL$.

Marginal propensity to consume The fraction of a rise in national income which goes on consumption: $MPC = dC/dY$.

Marginal propensity to save The fraction of a rise in national income which goes into savings: $MPS = dS/dY$.

Marginal revenue The extra revenue gained by selling 1 more unit of a good: $MR = d(TR)/dQ$.

Monopolist The only firm in the industry.

Perfect competition A situation in which there are no barriers to entry in an industry where there are many firms selling an identical product at the market price.

Practice Problems

6 If the demand function is

$$P = 100 - 4Q$$

find expressions for TR and MR in terms of Q. Hence estimate the change in TR brought about by a 0.3 unit increase in output from a current level of 12 units.

7 If the demand function is

$$P = 80 - 3Q$$

show that

$$\text{MR} = 2P - 80$$

8 A monopolist's demand function is given by

$$P + Q = 100$$

Write down expressions for TR and MR in terms of Q and sketch their graphs. Find the value of Q which gives a marginal revenue of zero and comment on the significance of this value.

9 The fixed costs of producing a good are 100 and the variable costs are 2 + Q/10 per unit.

(a) Find expressions for TC and MC.

(b) Evaluate MC at Q = 30 and hence estimate the change in TC brought about by a 2 unit increase in output from a current level of 30 units.

(c) At what level of output does MC = 22?

10 If the average cost function of a good is

$$\text{AC} = \frac{15}{Q} + 2Q + 9$$

find an expression for TC. What are the fixed costs in this case? Write down an expression for the marginal cost function.

11 A firm's production function is

$$Q = 50L - 0.01L^2$$

where L denotes the size of the workforce. Find the value of MP_L in the case when

(a) $L = 1$ **(b)** $L = 10$ **(c)** $L = 100$ **(d)** $L = 1000$

Does the law of diminishing marginal productivity apply to this particular function?

12 Show that the law of diminishing marginal productivity holds for the production function

$$Q = 6L^2 - 0.2L^3$$

13 If the consumption function is

$$C = 50 + 2\sqrt{Y}$$

calculate MPC and MPS when $Y = 36$ and give an interpretation of these results.

14 The consumption function is

$$C = 0.01Y^2 + 0.8Y + 100$$

(a) Calculate the values of MPC and MPS when $Y = 8$.

(b) Use the fact that $C + S = Y$ to obtain a formula for S in terms of Y. By differentiating this expression find the value of MPS at $Y = 8$ and verify that this agrees with your answer to part (a).

section 4.4

Further rules of differentiation

Objectives

At the end of this section you should be able to:

- Use the chain rule to differentiate a function of a function.
- Use the product rule to differentiate the product of two functions.
- Use the quotient rule to differentiate the quotient of two functions.
- Differentiate complicated functions using a combination of rules.

Section 4.2 introduced you to the basic rules of differentiation. Unfortunately, not all functions can be differentiated using these rules alone. For example, we are unable to differentiate the functions

$$x\sqrt{(2x-3)} \quad \text{and} \quad \frac{x}{x^2+1}$$

using just the constant, sum or difference rules. The aim of the present section is to describe three further rules which allow you to find the derivative of more complicated expressions. Indeed, the totality of all six rules will enable you to differentiate any mathematical function. Although you may find that the rules described in this section take you slightly longer to grasp than before, they are vital to any understanding of economic theory.

The first rule that we investigate is called the chain rule and it can be used to differentiate functions such as

$$y = (2x+3)^{10} \quad \text{and} \quad y = \sqrt{(1+x^2)}$$

The distinguishing feature of these expressions is that they represent a 'function of a function'. To understand what we mean by this, consider how you might evaluate

$$y = (2x+3)^{10}$$

Figure 4.18

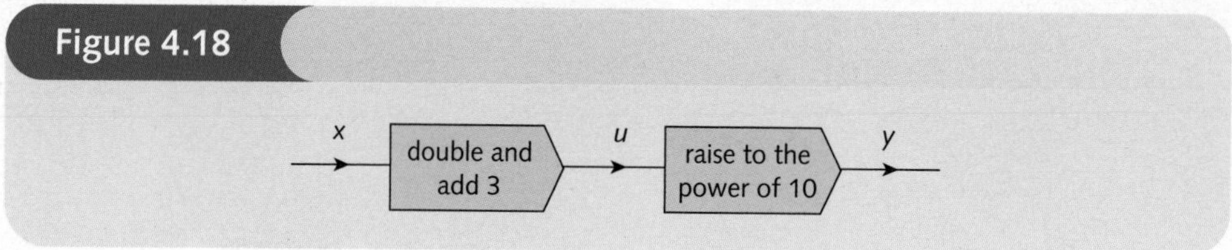

on a calculator. You would first work out an intermediate number u, say, given by

$$u = 2x + 3$$

and then raise it to the power of 10 to get

$$y = u^{10}$$

This process is illustrated using the flow chart in Figure 4.18. Note how the incoming number x is first processed by the inner function, 'double and add 3'. The output u from this is then passed on to the outer function, 'raise to the power of 10', to produce the final outgoing number y.

The function

$$y = \sqrt{(1 + x^2)}$$

can be viewed in the same way. To calculate y you perform the inner function, 'square and add 1', followed by the outer function, 'take square roots'.

The chain rule for differentiating a function of a function may now be stated.

Rule 4 The chain rule

If y is a function of u, which is itself a function of x, then

$$\frac{dy}{dx} = \frac{dy}{du} \times \frac{du}{dx}$$

differentiate the outer function and multiply by the derivative of the inner function

To illustrate this rule, let us return to the function

$$y = (2x + 3)^{10}$$

in which

$$y = u^{10} \quad \text{and} \quad u = 2x + 3$$

Now

$$\frac{dy}{du} = 10u^9 = 10(2x + 3)^9$$

$$\frac{du}{dx} = 2$$

The chain rule then gives

$$\frac{dy}{dx} = \frac{dy}{du} \times \frac{du}{dx} = 10(2x + 3)^9(2) = 20(2x + 3)^9$$

With practice it is possible to perform the differentiation without explicitly introducing the variable u. To differentiate

$$y = (2x + 3)^{10}$$

we first differentiate the outer power function to get

$$10(2x + 3)^9$$

and then multiply by the derivative of the inner function, $2x + 3$, which is 2, so

$$\frac{dy}{dx} = 20(2x + 3)^9$$

Example

Differentiate

(a) $y = (3x^2 - 5x + 2)^4$

(b) $y = \frac{1}{3x + 7}$

(c) $y = \sqrt{(1 + x^2)}$

Solution

(a) The chain rule shows that to differentiate $(3x^2 - 5x + 2)^4$ we first differentiate the outer power function to get

$$4(3x^2 - 5x + 2)^3$$

and then multiply by the derivative of the inner function, $3x^2 - 5x + 2$, which is $6x - 5$. Hence if

$$y = (3x^2 - 5x + 2)^4 \quad \text{then} \quad \frac{dy}{dx} = 4(3x^2 - 5x + 2)^3(6x - 5)$$

(b) To use the chain rule to differentiate

$$y = \frac{1}{3x + 7}$$

recall that reciprocals are denoted by negative powers, so that

$$y = (3x + 7)^{-1}$$

The outer power function differentiates to get

$$-(3x + 7)^{-2}$$

and the inner function, $3x + 7$, differentiates to get 3. By the chain rule we just multiply these together to deduce that

$$\text{if} \quad y = \frac{1}{3x + 7} \quad \text{then} \quad \frac{dy}{dx} = -(3x + 7)^{-2}(3) = \frac{-3}{(3x + 7)^2}$$

(c) To use the chain rule to differentiate

$$y = \sqrt{(1 + x^2)}$$

recall that roots are denoted by fractional powers, so that

$$y = (1 + x^2)^{1/2}$$

→

The outer power function differentiates to get

$$\frac{1}{2}(1+x^2)^{-1/2}$$

and the inner function, $1 + x^2$, differentiates to get $2x$. By the chain rule we just multiply these together to deduce that

$$\text{if} \quad y = \sqrt{(1+x^2)} \quad \text{then} \quad \frac{dy}{dx} = \frac{1}{2}(1+x^2)^{-1/2}(2x) = \frac{x}{\sqrt{(1+x^2)}}$$

Practice Problem

1 Differentiate

(a) $y = (3x - 4)^5$ **(b)** $y = (x^2 + 3x + 5)^3$ **(c)** $y = \dfrac{1}{2x - 3}$ **(d)** $y = \sqrt{(4x - 3)}$

The next rule is used to differentiate the product of two functions, $f(x)g(x)$. In order to give a clear statement of this rule, we write

$$u = f(x) \quad \text{and} \quad v = g(x)$$

Rule 5 The product rule

$$\text{If} \quad y = uv \quad \text{then} \quad \frac{dy}{dx} = u\frac{dv}{dx} \times v\frac{du}{dx}$$

This rule tells you how to differentiate the product of two functions:

multiply each function by the derivative of the other and add

Example

Differentiate

(a) $y = x^2(2x + 1)^3$ (b) $x\sqrt{(6x + 1)}$ (c) $y = \dfrac{x}{1 + x}$

Solution

(a) The function $x^2(2x + 1)^3$ involves the product of two simpler functions, namely x^2 and $(2x + 1)^3$, which we denote by u and v respectively. (It does not matter which function we label u and which we label v. The same answer is obtained if u is $(2x + 1)^3$ and v is x^2. You might like to check this for yourself later.) Now if

$$u = x^2 \quad \text{and} \quad v = (2x + 1)^3$$

then

$$\frac{du}{dx} = 2x \quad \text{and} \quad \frac{dv}{dx} = 6(2x + 1)^2$$

where we have used the chain rule to find $\mathrm{d}v/\mathrm{d}x$. By the product rule,

$$\frac{\mathrm{d}y}{\mathrm{d}x} = u\frac{\mathrm{d}v}{\mathrm{d}x} + v\frac{\mathrm{d}u}{\mathrm{d}x}$$
$$= x^2[6(2x+1)^2] + (2x+1)^3(2x)$$

The first term is obtained by leaving u alone and multiplying it by the derivative of v. Similarly, the second term is obtained by leaving v alone and multiplying it by the derivative of u.

If desired, the final answer may be simplified by taking out a common factor of $2x(2x+1)^2$. This factor goes into the first term $3x$ times and into the second $2x + 1$ times. Hence

$$\frac{\mathrm{d}y}{\mathrm{d}x} = 2x(2x+1)^2[3x + (2x+1)]$$
$$= 2x(2x+1)^2(5x+1)$$

(b) The function $x\sqrt{(6x+1)}$ involves the product of the simpler functions

$$u = x \quad \text{and} \quad v = \sqrt{(6x+1)} = (6x+1)^{1/2}$$

for which

$$\frac{\mathrm{d}u}{\mathrm{d}x} = 1 \quad \text{and} \quad \frac{\mathrm{d}v}{\mathrm{d}x} = \frac{1}{2}(6x+1)^{-1/2} \times 6 = 3(6x+1)^{-1/2}$$

where we have used the chain rule to find $\mathrm{d}v/\mathrm{d}x$. By the product rule,

$$\frac{\mathrm{d}y}{\mathrm{d}x} = u\frac{\mathrm{d}v}{\mathrm{d}x} + v\frac{\mathrm{d}u}{\mathrm{d}x}$$
$$= x[3(6x+1)^{-1/2}] + (6x+1)^{1/2}(1)$$
$$= \frac{3x}{\sqrt{(6x+1)}} + \sqrt{(6x+1)}$$

If desired, this can be simplified by putting the second term over a common denominator

$$\sqrt{(6x+1)}$$

To do this we multiply the top and bottom of the second term by $\sqrt{(6x+1)}$ to get

$$\frac{6x+1}{\sqrt{(6x+1)}}$$

$\sqrt{(6x+1)} \times \sqrt{(6x+1)} = 6x+1$

Hence

$$\frac{\mathrm{d}y}{\mathrm{d}x} = \frac{3x + (6x+1)}{\sqrt{(6x+1)}} = \frac{9x+1}{\sqrt{(6x+1)}}$$

(c) At first sight it is hard to see how we can use the product rule to differentiate

$$\frac{x}{1+x}$$

since it appears to be the quotient and not the product of two functions. However, if we recall that reciprocals are equivalent to negative powers, we may rewrite it as

$$x(1+x)^{-1}$$

It follows that we can put

$$u = x \quad \text{and} \quad v = (1+x)^{-1}$$

➔

which gives

$$\frac{du}{dx} = 1 \quad \text{and} \quad \frac{dv}{dx} = -(1 + x)^{-2}$$

where we have used the chain rule to find dv/dx. By the product rule

$$\frac{dy}{dx} = u\frac{dv}{dx} + v\frac{du}{dx}$$

$$\frac{dy}{dx} = x[-(1 + x)^{-2}] + (1 + x)^{-1}(1)$$

$$= \frac{-x}{(1 + x)^2} + \frac{1}{1 + x}$$

If desired, this can be simplified by putting the second term over a common denominator

$$(1 + x)^2$$

To do this we multiply the top and bottom of the second term by $1 + x$ to get

$$\frac{1 + x}{(1 + x)^2}$$

Hence

$$\frac{dy}{dx} = \frac{-x}{(1 + x)^2} + \frac{1 + x}{(1 + x)^2} = \frac{-x + (1 + x)}{(1 + x)^2} = \frac{1}{(1 + x)^2}$$

Practice Problem

2 Differentiate

(a) $y = x(3x - 1)^6$ **(b)** $y = x^3\sqrt{(2x + 3)}$ **(c)** $y = \frac{x}{x - 2}$

Advice

You may have found the product rule the hardest of the rules so far. This may have been due to the algebraic manipulation that is required to simplify the final expression. If this is the case, do not worry about it at this stage. The important thing is that you can use the product rule to obtain some sort of an answer even if you cannot tidy it up at the end. This is not to say that the simplification of an expression is pointless. If the result of differentiation is to be used in a subsequent piece of theory, it may well save time in the long run if it is simplified first.

One of the most difficult parts of Practice Problem 2 is part (c), since this involves algebraic fractions. For this function, it is necessary to manipulate negative indices and to put two individual fractions over a common denominator. You may feel that you are unable to do either of these processes with confidence. For this reason we conclude this section with a rule that is specifically designed to differentiate this type of function. The rule itself is quite complicated. However, as will become apparent, it does the algebra for you, so you may prefer to use it rather than the product rule when differentiating algebraic fractions.

Rule 6 The quotient rule

$$\text{If} \quad y = \frac{u}{v} \quad \text{then} \quad \frac{\mathrm{d}y}{\mathrm{d}x} = \frac{v\mathrm{d}u/\mathrm{d}x - u\mathrm{d}v/\mathrm{d}x}{v^2}$$

This rule tells you how to differentiate the quotient of two functions:

bottom times derivative of top, minus top times derivative of bottom, all over bottom squared

Example

Differentiate

(a) $y = \dfrac{x}{1 + x}$ (b) $y = \dfrac{1 + x^2}{2 - x^3}$

Solution

(a) In the quotient rule, u is used as the label for the numerator and v is used for the denominator, so to differentiate

$$\frac{x}{1 + x}$$

we must take

$$u = x \quad \text{and} \quad v = 1 + x$$

for which

$$\frac{\mathrm{d}u}{\mathrm{d}x} = 1 \quad \text{and} \quad \frac{\mathrm{d}v}{\mathrm{d}x} = 1$$

By the quotient rule

$$\begin{aligned}\frac{\mathrm{d}y}{\mathrm{d}x} &= \frac{v\mathrm{d}u/\mathrm{d}x - u\mathrm{d}v/\mathrm{d}x}{v^2} \\ &= \frac{(1 + x)(1) - x(1)}{(1 + x^2)} \\ &= \frac{1 + x - x}{(1 + x)^2} \\ &= \frac{1}{(1 + x)^2}\end{aligned}$$

Notice how the quotient rule automatically puts the final expression over a common denominator. Compare this with the algebra required to obtain the same answer using the product rule in part (c) of the previous example.

(b) The numerator of the algebraic fraction

$$\frac{1 + x^2}{2 - x^3}$$

is $1 + x^2$ and the denominator is $2 - x^3$, so we take

$$u = 1 + x^2 \quad \text{and} \quad v = 2 - x^3$$

for which

$$\frac{du}{dx} = 2x \quad \text{and} \quad \frac{dv}{dx} = -3x^2$$

By the quotient rule

$$\frac{dy}{dx} = \frac{v du/dx - u dv/dx}{v^2}$$

$$= \frac{(2 - x^3)(2x) - (1 + x^2)(-3x^2)}{(2 - x^3)^2}$$

$$= \frac{4x - 2x^4 + 3x^2 + 3x^4}{(2 - x^3)^2}$$

$$= \frac{x^4 + 3x^2 + 4x}{(2 - x^3)^2}$$

Practice Problem

3 Differentiate

(a) $y = \dfrac{x}{x - 2}$ **(b)** $y = \dfrac{x - 1}{x + 1}$

[You might like to check that your answer to part (a) is the same as that obtained in Practice Problem 2(c).]

Advice

The product and quotient rules give alternative methods for the differentiation of algebraic fractions. It does not matter which rule you go for; use whichever rule is easiest for you. Practice Problems 4, 5, 6 and 7 at the end of this section should give you further practice at 'technique bashing' if you feel you need it.

Practice Problems

4 Use the chain rule to differentiate

(a) $y = (2x + 1)^{10}$ **(b)** $y = (x^2 + 3x - 5)^3$ **(c)** $y = \dfrac{1}{7x - 3}$

(d) $y = \dfrac{1}{x^2 + 1}$ **(e)** $y = \sqrt{(8x - 1)}$

5 Use the product rule to differentiate

(a) $y = x^2(x + 5)^3$ **(b)** $y = x^5(4x + 5)^2$ **(c)** $y = x^4\sqrt{(x + 1)}$

6 Use the quotient rule to differentiate

(a) $y = \frac{x^2}{x + 4}$ **(b)** $y = \frac{2x - 1}{x + 1}$ **(c)** $y = \frac{x^3}{\sqrt{(x - 1)}}$

7 Differentiate

(a) $y = x(x - 3)^4$ **(b)** $y = x\sqrt{(2x - 3)}$ **(c)** $y = \frac{x}{x + 5}$ **(d)** $y = \frac{x}{x^2 + 1}$

8 Differentiate

$$y = (5x + 7)^2$$

(a) by using the chain rule

(b) by first multiplying out the brackets and then differentiating term by term.

9 Differentiate

$$y = x^5(x + 2)^2$$

(a) by using the product rule

(b) by first multiplying out the brackets and then differentiating term by term.

10 Find expressions for marginal revenue in the case when the demand equation is given by

(a) $P = \sqrt{(100 - 2Q)}$ **(b)** $P = \frac{1000}{\sqrt{(2 + Q)}}$

11 If the consumption function is

$$C = \frac{300 + 2Y^2}{1 + Y}$$

calculate MPC and MPS when $Y = 36$ and give an interpretation of these results.

section 4.5

Elasticity

Objectives

At the end of this section you should be able to:

- Calculate price elasticity averaged along an arc.
- Calculate price elasticity evaluated at a point.
- Decide whether supply and demand are inelastic, unit elastic or elastic.
- Understand the relationship between price elasticity of demand and revenue.
- Determine the price elasticity for general linear demand functions.

One important problem in business is to determine the effect on revenue of a change in the price of a good. Let us suppose that a firm's demand curve is downward-sloping. If the firm lowers the price then it will receive less for each item, but the number of items sold increases. The formula for total revenue, TR, is

$$\text{TR} = PQ$$

and it is not immediately obvious what the net effect on TR will be as P decreases and Q increases. The crucial factor here is not the absolute changes in P and Q but rather the proportional or percentage changes. Intuitively, we expect that if the percentage rise in Q is greater than the percentage fall in P then the firm experiences an increase in revenue. Under these circumstances we say that demand is *elastic*, since the demand is relatively sensitive to changes in price. Similarly, demand is said to be *inelastic* if demand is relatively insensitive to price changes. In this case, the percentage change in quantity is less than the percentage change in price. A firm can then increase revenue by raising the price of the good. Although demand falls as a result, the increase in price more than compensates for the reduced volume of sales and revenue rises. Of course, it could happen that the percentage changes in price and quantity are equal, leaving revenue unchanged. We use the term ***unit elastic*** to describe this situation.

We quantify the responsiveness of demand to price change by defining the ***price elasticity of demand*** to be

$$E = \frac{\text{percentage change in demand}}{\text{percentage change in price}}$$

Notice that because the demand curve slopes downwards, a positive change in price leads to a negative change in quantity and vice versa. Consequently, the value of E is always negative. It is conventional to avoid this by deliberately changing the sign and taking

$$E = -\frac{\text{percentage change in demand}}{\text{percentage change in price}}$$

which makes E positive. The previous classification of demand functions can now be restated more succinctly in terms of E.

Demand is said to be

- **inelastic if $E < 1$**
- **unit elastic if $E = 1$**
- **elastic if $E > 1$.**

Advice

You should note that not all economists adopt the convention of ignoring the sign to make E positive. If the negative sign is left in, the demand will be inelastic if $E > -1$, unit elastic if $E = -1$ and elastic if $E < -1$. You should check with your lecturer the particular convention that you need to adopt.

As usual, we denote the changes in P and Q by ΔP and ΔQ respectively, and seek a formula for E in terms of these symbols. To motivate this, suppose that the price of a good is \$12 and that it rises to \$18. A moment's thought should convince you that the percentage change in price is then 50%. You can probably work this out in your head without thinking too hard. However, it is worthwhile identifying the mathematical process involved. To obtain this figure we first express the change

$$18 - 12 = 6$$

as a fraction of the original to get

$$\frac{6}{12} = 0.5$$

and then multiply by 100 to express it as a percentage. This simple example gives us a clue as to how we might find a formula for E. In general, the percentage change in price is

$$\frac{\Delta P}{P} \times 100$$

change in price expressed as a fraction of the original price

multiply by 100 to convert fractions into percentages

Similarly, the percentage change in quantity is

$$\frac{\Delta Q}{Q} \times 100$$

Figure 4.19

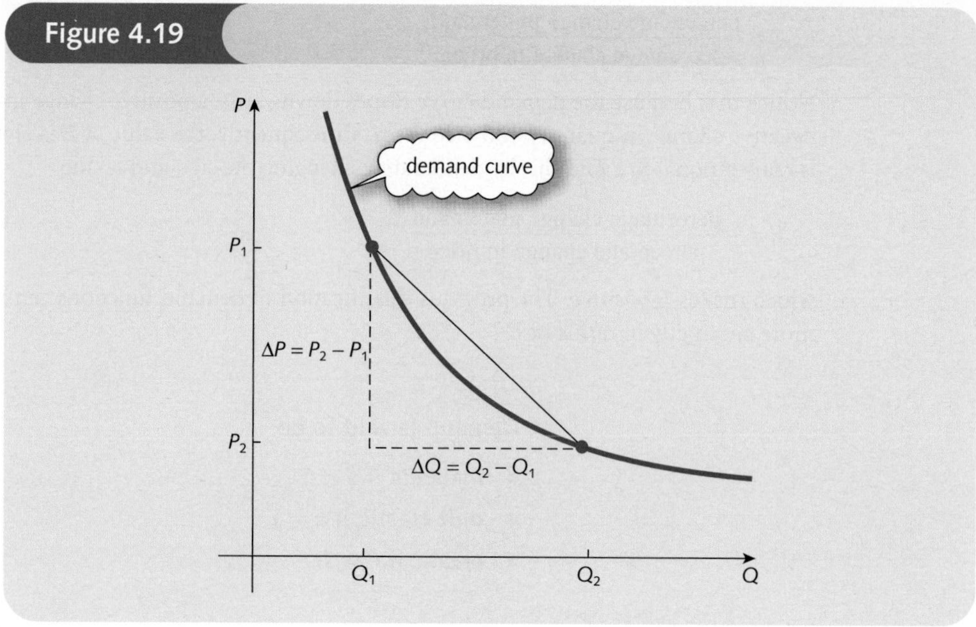

Hence

$$E = -\left(\frac{\Delta Q}{Q} \times 100\right) \div \left(\frac{\Delta P}{P} \times 100\right)$$

Now, when we divide two fractions we turn the denominator upside down and multiply, so

$$E = -\left(\frac{\Delta Q}{Q} \times \cancel{100}\right) \times \left(\frac{P}{\cancel{100} \times \Delta P}\right)$$

$$= -\frac{P}{Q} \times \frac{\Delta Q}{\Delta P}$$

A typical demand curve is illustrated in Figure 4.19, in which a price fall from P_1 to P_2 causes an increase in demand from Q_1 to Q_2.

Example

Determine the elasticity of demand when the price falls from 136 to 119, given the demand function

$$P = 200 - Q^2$$

Solution

In the notation of Figure 4.19 we are given that

$$P_1 = 136 \quad \text{and} \quad P_2 = 119$$

The corresponding values of Q_1 and Q_2 are obtained from the demand equation

$$P = 200 - Q^2$$

by substituting $P = 136$ and 119 respectively and solving for Q. For example, if $P = 136$ then

$$136 = 200 - Q^2$$

which rearranges to give

$$Q^2 = 200 - 136 = 64$$

This has solution $Q = \pm 8$ and, since we can obviously ignore the negative quantity, we have $Q_1 = 8$. Similarly, setting $P = 119$ gives $Q_2 = 9$. The elasticity formula is

$$E = -\frac{P}{Q} \times \frac{\Delta Q}{\Delta P}$$

and the values of ΔP and ΔQ are easily worked out to be

$$\Delta P = 119 - 136 = -17$$

$$\Delta Q = 9 - 8 = 1$$

However, it is not at all clear what to take for P and Q. Do we take P to be 136 or 119? Clearly we are going to get two different answers depending on our choice. A sensible compromise is to use their average and take

$$P = \tfrac{1}{2}(136 + 119) = 127.5$$

Similarly, averaging the Q values gives

$$Q = \tfrac{1}{2}(8 + 9) = 8.5$$

Hence

$$E = -\frac{127.5}{8.5} \times \left(\frac{1}{-17}\right) = 0.88$$

The particular application of the general formula considered in the previous example provides an estimate of elasticity averaged over a section of the demand curve between (Q_1, P_1) and (Q_2, P_2). For this reason it is called ***arc elasticity*** and is obtained by replacing P by $\tfrac{1}{2}(P_1 + P_2)$ and Q by $\tfrac{1}{2}(Q_1 + Q_2)$ in the general formula.

Practice Problem

1 Given the demand function

$$P = 1000 - 2Q$$

calculate the arc elasticity as P falls from 210 to 200.

A disappointing feature of the previous example is the need to compromise and calculate the elasticity averaged along an arc rather than calculate the exact value at a point. A formula for the latter can easily be deduced from

$$E = -\frac{P}{Q} \times \frac{\Delta Q}{\Delta P}$$

by considering the limit as ΔQ and ΔP tend to zero in Figure 4.19. All that happens is that the arc shrinks to a point and the ratio $\Delta Q/\Delta P$ tends to dQ/dP. The price elasticity at a point may therefore be found from

$$E = -\frac{P}{Q} \times \frac{dQ}{dP}$$

Example

Given the demand function

$$P = 50 - 2Q$$

find the elasticity when the price is 30. Is demand inelastic, unit elastic or elastic at this price?

Solution

To find dQ/dP we need to differentiate Q with respect to P. However, we are actually given a formula for P in terms of Q, so we need to transpose

$$P = 50 - 2Q$$

for Q. Adding $2Q$ to both sides gives

$$P + 2Q = 50$$

and if we subtract P then

$$2Q = 50 - P$$

Finally, dividing through by 2 gives

$$Q = 25 - {}^1\!/_2P$$

Hence

$$\frac{dQ}{dP} = -{}^1\!/_2$$

We are given that $P = 30$ so, at this price, demand is

$$Q = 25 - {}^1\!/_2(30) = 10$$

These values can now be substituted into

$$E = -\frac{P}{Q} \times \frac{dQ}{dP}$$

to get

$$E = -\frac{30}{10} \times \left(-\frac{1}{2}\right) = 1.5$$

Moreover, since $1.5 > 1$, demand is elastic at this price.

Practice Problem

2 Given the demand function

$$P = 100 - Q$$

calculate the price elasticity of demand when the price is

(a) 10 **(b)** 50 **(c)** 90

Is the demand inelastic, unit elastic or elastic at these prices?

It is quite common in economics to be given the demand function in the form

$$P = f(Q)$$

where P is a function of Q. In order to evaluate elasticity it is necessary to find

$$\frac{dQ}{dP}$$

which assumes that Q is actually given as a function of P. Consequently, we may have to transpose the demand equation and find an expression for Q in terms of P before we perform the differentiation. This was the approach taken in the previous example. Unfortunately, if $f(Q)$ is a complicated expression, it may be difficult, if not impossible, to carry out the initial rearrangement to extract Q. An alternative approach is based on the fact that

$$\frac{dQ}{dP} = \frac{1}{dP/dQ}$$

A proof of this can be obtained via the chain rule, although we omit the details. This result shows that we can find dQ/dP by just differentiating the original demand function to get dP/dQ and reciprocating.

Example

Given the demand function

$$P = -Q^2 - 4Q + 96$$

find the price elasticity of demand when $P = 51$. If this price rises by 2%, calculate the corresponding percentage change in demand.

Solution

We are given that $P = 51$, so to find the corresponding demand we need to solve the quadratic equation

$$-Q^2 - 4Q + 96 = 51$$

that is,

$$-Q^2 - 4Q + 45 = 0$$

To do this we use the standard formula

$$\frac{-b \pm \sqrt{(b^2 - 4ac)}}{2a}$$

discussed in Section 2.1, which gives

$$Q = \frac{-(-4) \pm \sqrt{((-4)^2 - 4(-1)(45))}}{2(-1)}$$

$$= \frac{4 \pm \sqrt{196}}{-2}$$

$$= \frac{4 \pm 14}{-2}$$

The two solutions are −9 and 5. As usual, the negative value can be ignored, since it does not make sense to have a negative quantity, so $Q = 5$.

➔

To find the value of E we also need to calculate

$$\frac{dQ}{dP}$$

from the demand equation, $P = -Q^2 - 4Q + 96$. It is not at all easy to transpose this for Q. Indeed, we would have to use the formula for solving a quadratic, as above, replacing the number 51 by the letter P. Unfortunately this expression involves square roots and the subsequent differentiation is quite messy. (You might like to have a go at this yourself!) However, it is easy to differentiate the given expression with respect to Q to get

$$\frac{dP}{dQ} = -2Q - 4$$

and so

$$\frac{dQ}{dP} = \frac{1}{dP/dQ} = \frac{1}{-2Q - 4}$$

Finally, putting $Q = 5$ gives

$$\frac{dQ}{dP} = -\frac{1}{14}$$

The price elasticity of demand is given by

$$E = -\frac{P}{Q} \times \frac{dQ}{dP}$$

and if we substitute $P = 51$, $Q = 5$ and $dQ/dP = -1/14$ we get

$$E = -\frac{51}{5} \times \left(-\frac{1}{14}\right) = 0.73$$

To discover the effect on Q due to a 2% rise in P we return to the original definition

$$E = -\frac{\text{percentage change in demand}}{\text{percentage change in price}}$$

We know that $E = 0.73$ and that the percentage change in price is 2, so

$$0.73 = -\frac{\text{percentage change in demand}}{2}$$

which shows that demand changes by

$$-0.73 \times 2 = -1.46\%$$

A 2% rise in price therefore leads to a fall in demand of 1.46%.

Practice Problem

3 Given the demand equation

$$P = -Q^2 - 10Q + 150$$

find the price elasticity of demand when $Q = 4$. Estimate the percentage change in price needed to increase demand by 10%.

The *price elasticity of supply* is defined in an analogous way to that of demand. We define

$$E = \frac{\text{percentage change in supply}}{\text{percentage change in price}}$$

This time, however, there is no need to fiddle the sign. An increase in price leads to an increase in supply, so E is automatically positive. In symbols,

$$E = \frac{P}{Q} \times \frac{\Delta Q}{\Delta P}$$

If (Q_1, P_1) and (Q_2, P_2) denote two points on the supply curve then arc elasticity is obtained, as before, by setting

$$\Delta P = P_2 - P_1$$

$$\Delta Q = Q_2 - Q_1$$

$$P = \tfrac{1}{2}(P_1 + P_2)$$

$$Q = \tfrac{1}{2}(Q_1 + Q_2)$$

The corresponding formula for point elasticity is

$$E = \frac{P}{Q} \times \frac{dQ}{dP}$$

Example

Given the supply function

$$P = 10 + \sqrt{Q}$$

find the price elasticity of supply

(a) averaged along an arc between $Q = 100$ and $Q = 105$

(b) at the point $Q = 100$

Solution

(a) We are given that

$$Q_1 = 100, Q_2 = 105$$

so that

$$P_1 = 10 + \sqrt{100} = 20 \quad \text{and} \quad P_2 = 10 + \sqrt{105} = 20.247$$

Hence

$$\Delta P = 20.247 - 20 = 0.247, \qquad \Delta Q = 105 - 100 = 5$$

$$P = \frac{1}{2}(20 + 20.247) = 20.123, \qquad Q = \frac{1}{2}(100 + 105) = 102.5$$

The formula for arc elasticity gives

$$E = \frac{P}{Q} \times \frac{\Delta Q}{\Delta P} = \frac{20.123}{102.5} \times \frac{5}{0.247} = 3.97$$

→

(b) To evaluate the elasticity at the point $Q = 100$, we need to find the derivative, $\frac{dQ}{dP}$. The supply equation

$$P = 10 + Q^{1/2}$$

differentiates to give

$$\frac{dP}{dQ} = \frac{1}{2}Q^{-1/2} = \frac{1}{2\sqrt{Q}}$$

so that

$$\frac{dQ}{dP} = 2\sqrt{Q}$$

At the point $Q = 100$, we get

$$\frac{dQ}{dP} = 2\sqrt{100} = 20$$

The formula for point elasticity gives

$$E = \frac{P}{Q} \times \frac{dQ}{dP} = \frac{20}{100} \times 20 = 4$$

Notice that, as expected, the answers to parts (a) and (b) are nearly the same.

Practice Problem

4 If the supply equation is

$$Q = 150 + 5P + 0.1P^2$$

calculate the price elasticity of supply

(a) averaged along an arc between $P = 9$ and $P = 11$

(b) at the point $P = 10$

Advice

The concept of elasticity can be applied to more general functions and we consider some of these in the next chapter. For the moment we investigate the theoretical properties of demand elasticity. The following material is more difficult to understand than the foregoing, so you may prefer just to concentrate on the conclusions and skip the intermediate derivations.

We begin by analysing the relationship between elasticity and marginal revenue. Marginal revenue, MR, is given by

$$\text{MR} = \frac{d(\text{TR})}{dQ}$$

Now TR is equal to the product PQ, so we can apply the product rule to differentiate it. If

$$u = P \quad \text{and} \quad v = Q$$

then

$$\frac{du}{dQ} = \frac{dP}{dQ} \quad \text{and} \quad \frac{dv}{dQ} = \frac{dQ}{dQ} = 1$$

By the product rule

$$\begin{aligned} \text{MR} &= u\frac{dv}{dQ} + v\frac{du}{dQ} \\ &= P + Q \times \frac{dP}{dQ} \\ &= P\left(1 + \frac{Q}{P} \times \frac{dP}{dQ}\right) \end{aligned}$$

check this by multiplying out the brackets

Now

$$-\frac{P}{Q} \times \frac{dQ}{dP} = E$$

so

$$\frac{Q}{P} \times \frac{dP}{dQ} = -\frac{1}{E}$$

turn both sides upside down and multiply by −1

This can be substituted into the expression for MR to get

$$\text{MR} = P\left(1 - \frac{1}{E}\right)$$

The connection between marginal revenue and demand elasticity is now complete, and this formula can be used to justify the intuitive argument that we gave at the beginning of this section concerning revenue and elasticity. Observe that if $E < 1$ then $1/E > 1$, so MR is negative for any value of P. It follows that the revenue function is decreasing in regions where demand is inelastic, because MR determines the slope of the revenue curve. Similarly, if $E > 1$ then $1/E < 1$, so MR is positive for any price, P, and the revenue curve is uphill. In other words, the revenue function is increasing in regions where demand is elastic. Finally, if $E = 1$ then MR is 0, and so the slope of the revenue curve is horizontal at points where demand is unit elastic.

Throughout this section we have taken specific functions and evaluated the elasticity at particular points. It is more instructive to consider general functions and to deduce general expressions for elasticity. Consider the standard linear downward-sloping demand function

$$P = aQ + b$$

when $a < 0$ and $b > 0$. As noted in Section 4.3, this typifies the demand function faced by a monopolist. To transpose this equation for Q, we subtract b from both sides to get

$$aQ = P - b$$

and then divide through by a to get

$$Q = \frac{1}{a}(P - b)$$

Hence

$$\frac{dQ}{dP} = \frac{1}{a}$$

The formula for elasticity of demand is

$$E = -\frac{P}{Q} \times \frac{dQ}{dP}$$

so replacing Q by $(1/a)(P - b)$ and dQ/dP by $1/a$ gives

$$E = \frac{-P}{(1/a)(P - b)} \times \frac{1}{a}$$

$$= \frac{-P}{P - b}$$

$$= \frac{P}{b - P}$$

multiply top and bottom by −1

Notice that this formula involves P and b but not a. Elasticity is therefore independent of the slope of linear demand curves. In particular, this shows that, corresponding to any price P, the elasticities of the two demand functions sketched in Figure 4.20 are identical. This is perhaps a rather surprising result. We might have expected demand to be more elastic at point A than at point B, since A is on the steeper curve. However, the mathematics shows that this is not the case. (Can you explain, in economic terms, why this is so?)

Another interesting feature of the result

$$E = \frac{P}{b - P}$$

is the fact that b occurs in the denominator of this fraction, so that corresponding to any price, P, the larger the value of the intercept, b, the smaller the elasticity. In Figure 4.21, elasticity at C is smaller than that at D because C lies on the curve with the larger intercept.

The dependence of E on P is also worthy of note. It shows that elasticity varies along a linear demand curve. This is illustrated in Figure 4.22. At the left-hand end, $P = b$, so

Figure 4.20

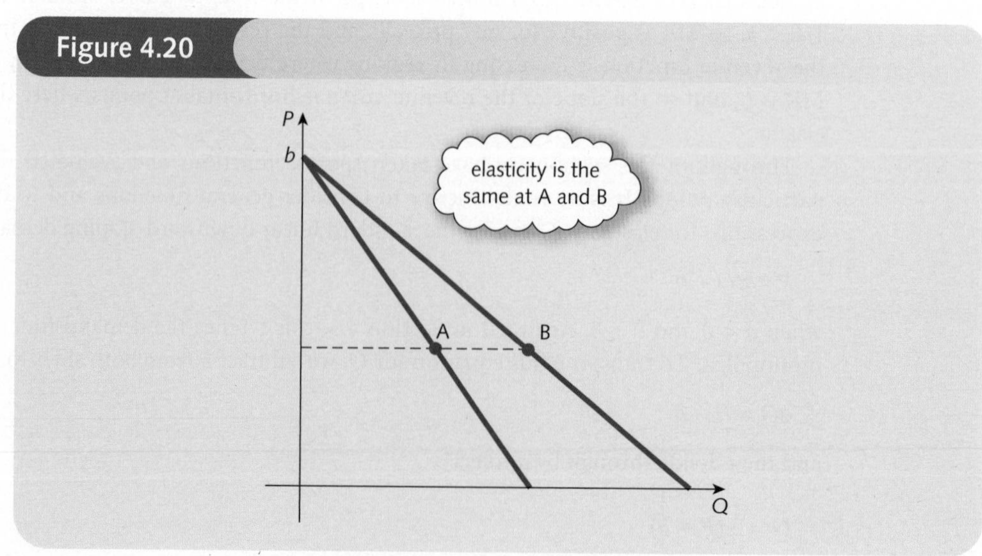

Figure 4.21

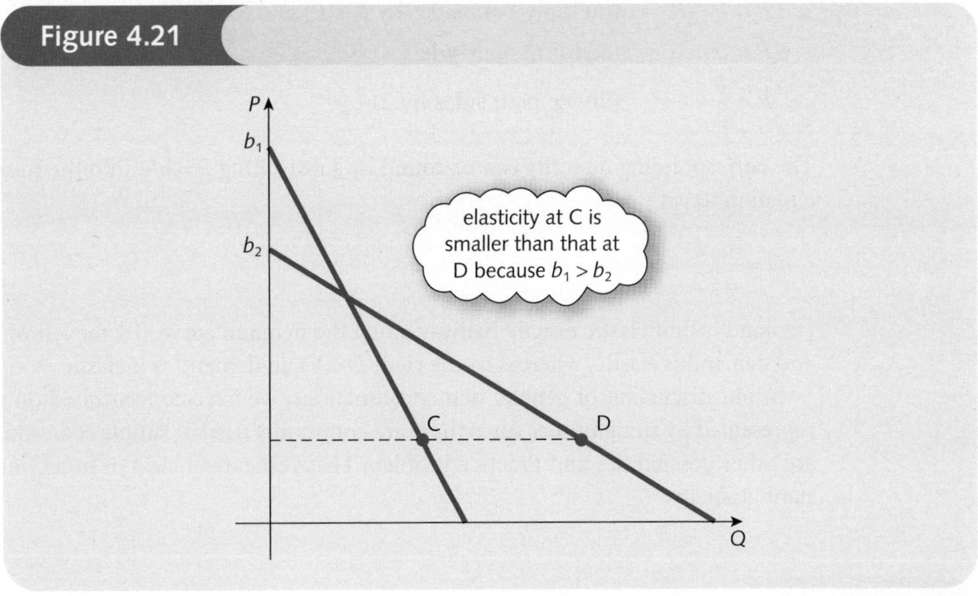

Figure 4.22

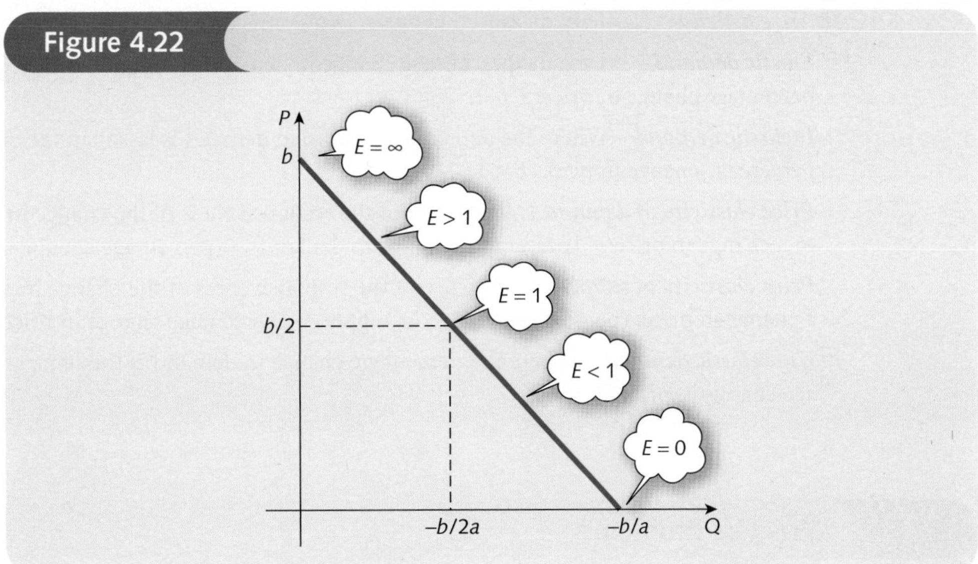

$$E = \frac{b}{b - b} = \frac{b}{0} = \infty$$

read 'infinity'

At the right-hand end, $P = 0$, so

$$E = \frac{0}{b - 0} = \frac{0}{b} = 0$$

As you move down the demand curve, the elasticity decreases from ∞ to 0, taking all possible values. Demand is unit elastic when $E = 1$ and the price at which this occurs can be found by solving

$$\frac{P}{b - P} = 1 \text{ for } P$$

$$P = b - P \quad \text{(multiply both sides by } b - P\text{)}$$
$$2P = b \quad \text{(add } P \text{ to both sides)}$$
$$P = \frac{b}{2} \quad \text{(divide both sides by 2)}$$

The corresponding quantity can be found by substituting $P = b/2$ into the transposed demand equation to get

$$Q = \frac{1}{a}\left(\frac{b}{2} - b\right) = -\frac{b}{2a}$$

Demand is unit elastic exactly halfway along the demand curve. To the left of this point $E > 1$ and demand is elastic, whereas to the right $E < 1$ and demand is inelastic.

In our discussion of general demand functions, we have concentrated on those which are represented by straight lines since these are commonly used in simple economic models. There are other possibilities and Practice Problem 11 investigates a class of functions that have constant elasticity.

Key Terms

Arc elasticity Elasticity measured between two points on a curve.

Elastic demand Where the percentage change in demand is more than the corresponding percentage change in price: $E > 1$.

Inelastic demand Where the percentage change in demand is less than the corresponding percentage change in price: $E < 1$.

Price elasticity of demand A measure of the responsiveness of the change in demand due to a change in price: – (percentage change in demand) ÷ (percentage change in price).

Price elasticity of supply A measure of the responsiveness of the change in supply due to a change in price: (percentage change in supply) ÷ (percentage change in price).

Unit elastic demand Where the percentage change in demand is the same as the percentage change in price: $E = 1$.

Practice Problems

5 Given the demand function

$$P = 500 - 4Q^2$$

calculate the price elasticity of demand averaged along an arc joining $Q = 8$ and $Q = 10$.

6 Find the price elasticity of demand at the point $Q = 9$ for the demand function

$$P = 500 - 4Q^2$$

and compare your answer with that of Practice Problem 5.

7 Find the price elasticity of demand at $P = 6$ for each of the following demand functions:

(a) $P = 30 - 2Q$

(b) $P = 30 - 12Q$

(c) $P = \sqrt{(100 - 2Q)}$

8 If the demand equation is

$$Q + 4P = 60$$

find a general expression for the price elasticity of demand in terms of P. For what value of P is demand unit elastic?

9 Consider the supply equation

$$Q = 4 + 0.1P^2$$

(a) Write down an expression for dQ/dP.

(b) Show that the supply equation can be rearranged as

$$P = \sqrt{(10Q - 40)}$$

Differentiate this to find an expression for dP/dQ.

(c) Use your answers to parts (a) and (b) to verify that

$$\frac{dQ}{dP} = \frac{1}{dP/dQ}$$

(d) Calculate the elasticity of supply at the point $Q = 14$.

10 If the supply equation is

$$Q = 7 + 0.1P + 0.004P^2$$

find the price elasticity of supply if the current price is 80.

(a) Is supply elastic, inelastic or unit elastic at this price?

(b) Estimate the percentage change in supply if the price rises by 5%.

11 Show that the price elasticity of demand is constant for the demand functions

$$P = \frac{A}{Q^n}$$

where A and n are positive constants.

12 Find a general expression for the point elasticity of supply for the function,

$$Q = aP + b \qquad (a > 0)$$

Deduce that the supply function is

(a) unit elastic when $b = 0$

(b) inelastic when $b > 0$

Give a brief geometrical interpretation of these results.

section 4.6

Optimization of economic functions

Objectives

At the end of this section you should be able to:

- Use the first-order derivative to find the stationary points of a function.
- Use the second-order derivative to classify the stationary points of a function.
- Find the maximum and minimum points of an economic function.
- Use stationary points to sketch graphs of economic functions.

In Section 2.1 a simple three-step strategy was described for sketching graphs of quadratic functions of the form

$$f(x) = ax^2 + bx + c$$

The basic idea is to solve the corresponding equation

$$ax^2 + bx + c = 0$$

to find where the graph crosses the x axis. Provided that the quadratic equation has at least one solution, it is then possible to deduce the coordinates of the maximum or minimum point of the parabola. For example, if there are two solutions, then by symmetry the graph turns round at the point exactly halfway between these solutions. Unfortunately, if the quadratic equation has no solution then only a limited sketch can be obtained using this approach.

In this section we show how the techniques of calculus can be used to find the coordinates of the turning point of a parabola. The beauty of this approach is that it can be used to locate the maximum and minimum points of any economic function, not just those represented by quadratics. Look at the graph in Figure 4.23 (overleaf). Points B, C, D, E, F and G are referred to as the ***stationary points*** (sometimes called ***critical points***, ***turning points*** or ***extrema***) of the function. At a stationary point the tangent to the graph is horizontal and so has zero slope.

Figure 4.23

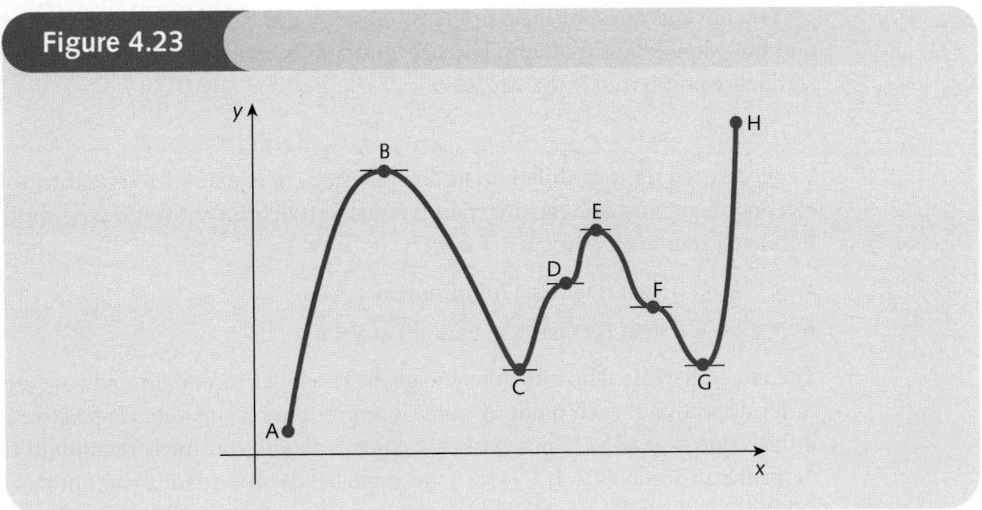

Consequently, at a stationary point of a function $f(x)$,

$$f'(x) = 0$$

The reason for using the word 'stationary' is historical. Calculus was originally used by astronomers to predict planetary motion. If a graph of the distance travelled by an object is sketched against time then the speed of the object is given by the slope, since this represents the rate of change of distance with respect to time. It follows that if the graph is horizontal at some point then the speed is zero and the object is instantaneously at rest: that is, stationary.

Stationary points are classified into one of three types: local maxima, local minima and stationary points of inflection.

At a ***local maximum*** (sometimes called a ***relative maximum***) the graph falls away on both sides. Points B and E are the local maxima for the function sketched in Figure 4.23. The word 'local' is used to highlight the fact that, although these are the maximum points relative to their locality or neighbourhood, they may not be the overall or global maximum. In Figure 4.23 the highest point on the graph actually occurs at the right-hand end, H, which is not a stationary point, since the slope is not zero at H.

At a ***local minimum*** (sometimes called a ***relative minimum***) the graph rises on both sides. Points C and G are the local minima in Figure 4.23. Again, it is not necessary for the global minimum to be one of the local minima. In Figure 4.23 the lowest point on the graph occurs at the left-hand end, A, which is not a stationary point.

At a ***stationary point of inflection*** the graph rises on one side and falls on the other. The stationary points of inflection in Figure 4.23 are labelled D and F. These points are of little value in economics, although they do sometimes assist in sketching graphs of economic functions. Maxima and minima, on the other hand, are important. The calculation of the maximum points of the revenue and profit functions is clearly worthwhile. Likewise, it is useful to be able to find the minimum points of average cost functions.

For most examples in economics, the local maximum and minimum points coincide with the global maximum and minimum. For this reason we shall drop the word 'local' when describing stationary points. However, it should always be borne in mind that the global maximum and minimum could actually be attained at an end point and this possibility may need to be checked. This can be done by comparing the function values at the end points with those of the stationary points and then deciding which of them gives rise to the largest or smallest values.

Two obvious questions remain. How do we find the stationary points of any given function and how do we classify them? The first question is easily answered. As we mentioned earlier, stationary points satisfy the equation

$$f'(x) = 0$$

so all we need do is to differentiate the function, to equate to zero and to solve the resulting algebraic equation. The classification is equally straightforward. It can be shown that if a function has a stationary point at $x = a$ then

- if $f''(a) > 0$ then $f(x)$ has a minimum at $x = a$
- if $f''(a) < 0$ then $f(x)$ has a maximum at $x = a$

Therefore, all we need do is to differentiate the function a second time and to evaluate this second-order derivative at each point. A point is a minimum if this value is positive and a maximum if this value is negative. These facts are consistent with our interpretation of the second-order derivative in Section 4.2. If $f''(a) > 0$ the graph bends upwards at $x = a$ (points C and G in Figure 4.23). If $f''(a) < 0$ the graph bends downwards at $x = a$ (points B and E in Figure 4.23). There is, of course, a third possibility, namely $f''(a) = 0$. Sadly, when this happens it provides no information whatsoever about the stationary point. The point $x = a$ could be a maximum, minimum or inflection. This situation is illustrated in Practice Problem 7 at the end of this section.

Advice

If you are unlucky enough to encounter this case, you can always classify the point by tabulating the function values in the vicinity and use these to produce a local sketch.

To summarize, the method for finding and classifying stationary points of a function, $f(x)$, is as follows:

Step 1

Solve the equation $f'(x) = 0$ to find the stationary points, $x = a$.

Step 2

If

- $f''(a) > 0$ then the function has a minimum at $x = a$
- $f''(a) < 0$ then the function has a maximum at $x = a$
- $f''(a) = 0$ then the point cannot be classified using the available information

Example

Find and classify the stationary points of the following functions. Hence sketch their graphs.

(a) $f(x) = x^2 - 4x + 5$ **(b)** $f(x) = 2x^3 + 3x^2 - 12x + 4$

Solution

(a) In order to use steps 1 and 2 we need to find the first- and second-order derivatives of the function

$$f(x) = x^2 - 4x + 5$$

Differentiating once gives

$$f'(x) = 2x - 4$$

and differentiating a second time gives

$$f''(x) = 2$$

Step 1

The stationary points are the solutions of the equation

$$f'(x) = 0$$

so we need to solve

$$2x - 4 = 0$$

This is a linear equation so has just one solution. Adding 4 to both sides gives

$$2x = 4$$

and dividing through by 2 shows that the stationary point occurs at

$$x = 2$$

Step 2

To classify this point we need to evaluate

$$f''(2)$$

In this case

$$f''(x) = 2$$

for all values of x, so in particular

$$f''(2) = 2$$

This number is positive, so the function has a minimum at $x = 2$.

We have shown that the minimum point occurs at $x = 2$. The corresponding value of y is easily found by substituting this number into the function to get

$$y = (2)^2 - 4(2) + 5 = 1$$

so the minimum point has coordinates (2, 1). A graph of $f(x)$ is shown in Figure 4.24 (overleaf).

(b) In order to use steps 1 and 2 we need to find the first- and second-order derivatives of the function

$$f(x) = 2x^3 + 3x^2 - 12x + 4$$

Differentiating once gives

$$f'(x) = 6x^2 + 6x - 12$$

and differentiating a second time gives

$$f''(x) = 12x + 6$$

Step 1

The stationary points are the solutions of the equation

$$f'(x) = 0$$

→

Figure 4.24

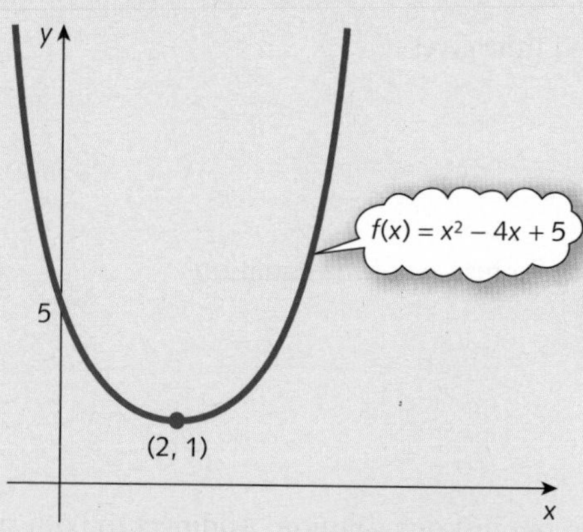

so we need to solve

$$6x^2 + 6x - 12 = 0$$

This is a quadratic equation and so can be solved using 'the formula'. However, before doing so, it is a good idea to divide both sides by 6 to avoid large numbers. The resulting equation

$$x^2 + x - 2 = 0$$

has solution

$$x = \frac{-1 \pm \sqrt{(1^2 - 4(1)(-2))}}{2(1)} = \frac{-1 \pm \sqrt{9}}{2} = \frac{-1 \pm 3}{2} = -2,\ 1$$

In general, whenever $f(x)$ is a cubic function the stationary points are the solutions of a quadratic equation, $f'(x) = 0$. Moreover, we know from Section 2.1 that such an equation can have two, one or no solutions. It follows that a cubic equation can have two, one or no stationary points. In this particular example we have seen that there are two stationary points, at $x = -2$ and $x = 1$.

Step 2

To classify these points we need to evaluate $f''(-2)$ and $f''(1)$. Now

$$f''(-2) = 12(-2) + 6 = -18$$

This is negative, so there is a maximum at $x = -2$. When $x = -2$,

$$y = 2(-2)^3 + 3(-2)^2 - 12(-2) + 4 = 24$$

so the maximum point has coordinates (−2, 24). Now

$$f''(1) = 12(1) + 6 = 18$$

This is positive, so there is a minimum at $x = 1$. When $x = 1$,

$$y = 2(1)^3 + 3(1)^2 - 12(1) + 4 = -3$$

so the minimum point has coordinates (1, −3).

This information enables a partial sketch to be drawn as shown in Figure 4.25. Before we can be confident about the complete picture it is useful to plot a few more points such as those below.

x	−10	0	10
y	−1816	4	2184

Figure 4.25

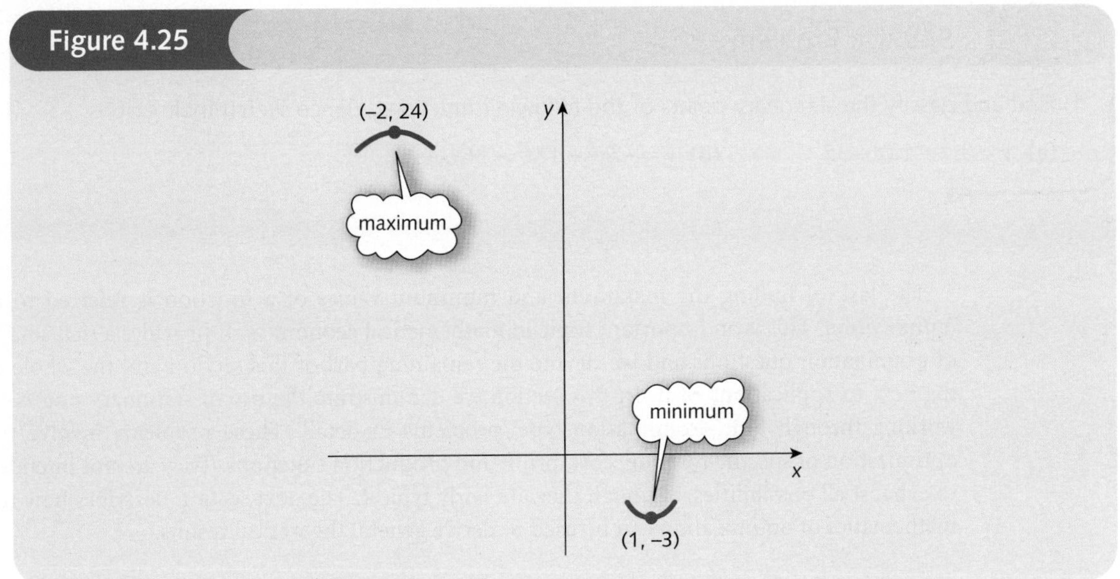

This table indicates that when x is positive the graph falls steeply downwards from a great height. Similarly, when x is negative the graph quickly disappears off the bottom of the page. The curve cannot wiggle and turn round except at the two stationary points already plotted (otherwise it would have more stationary points, which we know is not the case). We now have enough information to join up the pieces and so sketch a complete picture as shown in Figure 4.26.

In an ideal world it would be nice to calculate the three points at which the graph crosses the x axis. These are the solutions of

$$2x^3 + 3x^2 - 12x + 4 = 0$$

There is a formula for solving cubic equations, just as there is for quadratic equations, but it is extremely complicated and is beyond the scope of this book.

Figure 4.26

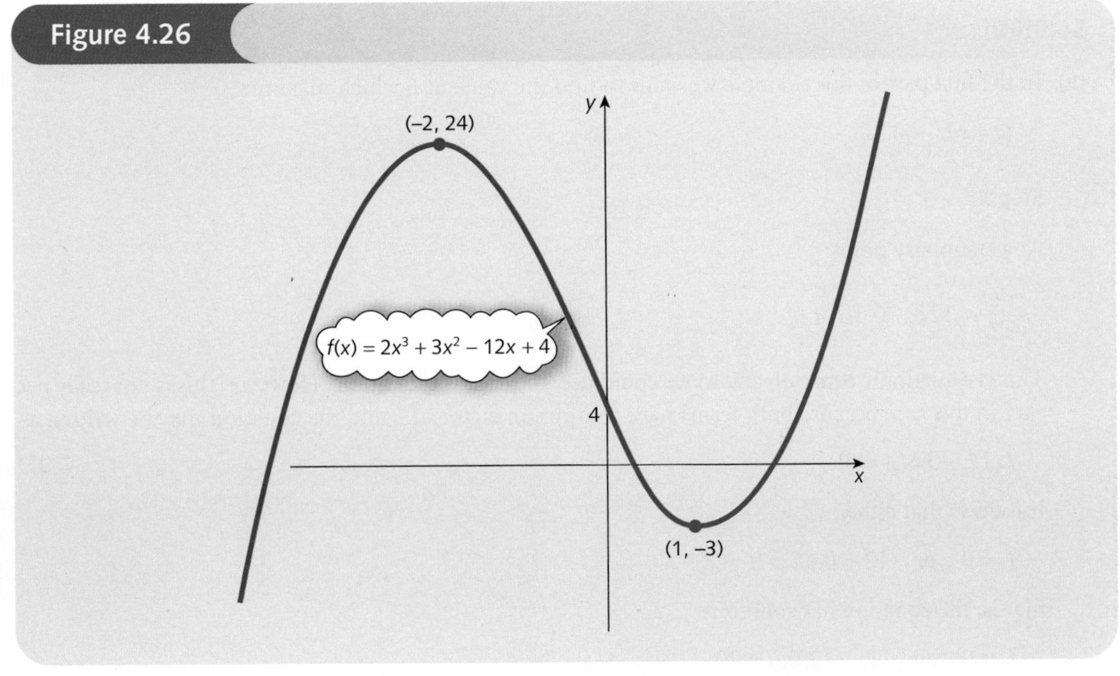

Practice Problem

1 Find and classify the stationary points of the following functions. Hence sketch their graphs.

(a) $y = 3x^2 + 12x - 35$ **(b)** $y = -2x^3 + 15x^2 - 36x + 27$

The task of finding the maximum and minimum values of a function is referred to as *optimization*. This is an important topic in mathematical economics. It provides a rich source of examination questions and we devote the remaining part of this section and the whole of the next to applications of it. In this section we demonstrate the use of stationary points by working through four 'examination-type' problems in detail. These problems involve the optimization of specific revenue, cost, profit and production functions. They are not intended to exhaust all possibilities, although they are fairly typical. The next section describes how the mathematics of optimization can be used to derive general theoretical results.

Example

A firm's short-run production function is given by

$$Q = 6L^2 - 0.2L^3$$

where L denotes the number of workers.

(a) Find the size of the workforce that maximizes output and hence sketch a graph of this production function.

(b) Find the size of the workforce that maximizes the average product of labour. Calculate MP_L and AP_L at this value of L. What do you observe?

Solution

(a) In the first part of this example we want to find the value of L which maximizes

$$Q = 6L^2 - 0.2L^3$$

Step 1

At a stationary point

$$\frac{dQ}{dL} = 12L - 0.6L^2 = 0$$

This is a quadratic equation and so we could use 'the formula' to find L. However, this is not really necessary in this case because both terms have a common factor of L and the equation may be written as

$$L(12 - 0.6L) = 0$$

It follows that either

$$L = 0 \quad \text{or} \quad 12 - 0.6L = 0$$

that is, the equation has solutions

$$L = 0 \quad \text{and} \quad L = 12/0.6 = 20$$

Step 2

It is obvious on economic grounds that $L = 0$ is a minimum and presumably $L = 20$ is the maximum. We can, of course, check this by differentiating a second time to get

$$\frac{d^2Q}{dL^2} = 12 - 1.2L$$

When $L = 0$,

$$\frac{d^2Q}{dL^2} = 12 > 0$$

which confirms that $L = 0$ is a minimum. The corresponding output is given by

$$Q = 6(0)^2 - 0.2(0)^3 = 0$$

as expected. When $L = 20$,

$$\frac{d^2Q}{dL^2} = -12 < 0$$

which confirms that $L = 20$ is a maximum.

The firm should therefore employ 20 workers to achieve a maximum output

$$Q = 6(20)^2 - 0.2(20)^3 = 800$$

We have shown that the minimum point on the graph has coordinates (0, 0) and the maximum point has coordinates (20, 800). There are no further turning points, so the graph of the production function has the shape sketched in Figure 4.27.

It is possible to find the precise values of L at which the graph crosses the horizontal axis. The production function is given by

$$Q = 6L^2 - 0.2L^3$$

so we need to solve

$$6L^2 - 0.2L^3 = 0$$

Figure 4.27

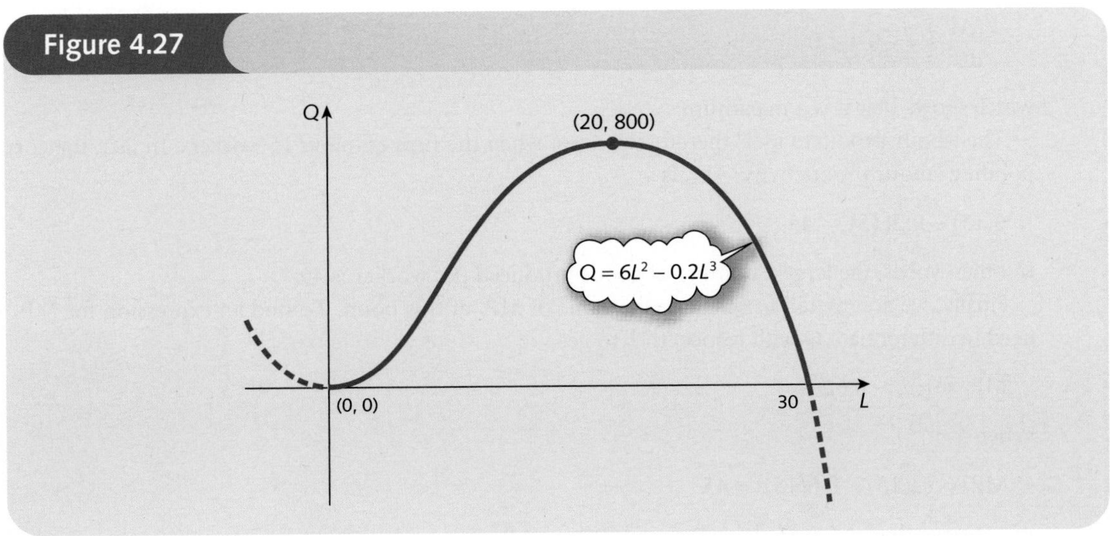

We can take out a factor of L^2 to get

$$L^2(6 - 0.2L) = 0$$

Hence, either

$$L^2 = 0 \quad \text{or} \quad 6 - 0.2L = 0$$

The first of these merely confirms the fact that the curve passes through the origin, whereas the second shows that the curve intersects the L axis at $L = 6/0.2 = 30$.

(b) In the second part of this example we want to find the value of L which maximizes the average product of labour. This is a concept that we have not met before in this book, although it is not difficult to guess how it might be defined.

The ***average product of labour***, AP_L, is taken to be total output divided by labour, so that in symbols

$$AP_L = \frac{Q}{L}$$

This is sometimes called ***labour productivity***, since it measures the average output per worker.

In this example,

$$AP_L = \frac{6L^2 - 0.2L^3}{L} = 6L - 0.2L^2$$

Step 1

At a stationary point

$$\frac{d(AP_L)}{dL} = 0$$

so

$$6 - 0.4L = 0$$

which has solution $L = 6/0.4 = 15$.

Step 2

To classify this stationary point we differentiate a second time to get

$$\frac{d^2(AP_L)}{dL^2} = -0.4 < 0$$

which shows that it is a maximum.

The labour productivity is therefore greatest when the firm employs 15 workers. In fact, the corresponding labour productivity, AP_L, is

$$6(15) - 0.2(15)^2 = 45$$

In other words, the largest number of goods produced per worker is 45.

Finally, we are invited to calculate the value of MP_L at this point. To find an expression for MP_L we need to differentiate Q with respect to L to get

$$MP_L = 12L - 0.6L^2$$

When $L = 15$,

$$MP_L = 12(15) - 0.6(15)^2 = 45$$

We observe that at $L = 15$ the values of MP_L and AP_L are equal.

In this particular example we discovered that at the point of maximum average product of labour

marginal product of labour	=	average product of labour

There is nothing special about this example and in the next section we show that this result holds for any production function.

Practice Problem

2 A firm's short-run production function is given by

$$Q = 300L^2 - L^4$$

where L denotes the number of workers. Find the size of the workforce that maximizes the average product of labour and verify that at this value of L

$$\text{MP}_L = \text{AP}_L$$

Example

The demand equation of a good is

$$P + Q = 30$$

and the total cost function is

$$\text{TC} = \tfrac{1}{2}Q^2 + 6Q + 7$$

(a) Find the level of output that maximizes total revenue.

(b) Find the level of output that maximizes profit. Calculate MR and MC at this value of Q. What do you observe?

Solution

(a) In the first part of this example we want to find the value of Q which maximizes total revenue. To do this we use the given demand equation to find an expression for TR and then apply the theory of stationary points in the usual way.

The total revenue is defined by

$$\text{TR} = PQ$$

We seek the value of Q which maximizes TR, so we express TR in terms of the variable Q only. The demand equation

$$P + Q = 30$$

can be rearranged to get

$$P = 30 - Q$$

Hence

$$\begin{aligned}\text{TR} &= (30 - Q)Q \\ &= 30Q - Q^2\end{aligned}$$

→

Step 1

At a stationary point

$$\frac{\mathrm{d(TR)}}{\mathrm{d}Q} = 0$$

so

$$30 - 2Q = 0$$

which has solution $Q = 30/2 = 15$.

Step 2

To classify this point we differentiate a second time to get

$$\frac{\mathrm{d^2(TR)}}{\mathrm{d}Q^2} = -2$$

This is negative, so TR has a maximum at $Q = 15$.

(b) In the second part of this example we want to find the value of Q which maximizes profit. To do this we begin by determining an expression for profit in terms of Q. Once this has been done, it is then a simple matter to work out the first- and second-order derivatives and so to find and classify the stationary points of the profit function.

The profit function is defined by

$$\pi = \mathrm{TR} - \mathrm{TC}$$

From part (a)

$$\mathrm{TR} = 30Q - Q^2$$

We are given the total cost function

$$\mathrm{TC} = \tfrac{1}{2}Q^2 + 6Q + 7$$

Hence

$$\begin{aligned}\pi &= (30Q - Q^2) - (\tfrac{1}{2}Q^2 + 6Q + 7)\\ &= 30Q - Q^2 - \tfrac{1}{2}Q^2 - 6Q - 7\\ &= -\tfrac{3}{2}Q^2 + 24Q - 7\end{aligned}$$

Step 1

At a stationary point

$$\frac{\mathrm{d}\pi}{\mathrm{d}Q} = 0$$

so

$$-3Q + 24 = 0$$

which has solution $Q = 24/3 = 8$.

Step 2

To classify this point we differentiate a second time to get

$$\frac{\mathrm{d}^2\pi}{\mathrm{d}Q^2} = -3$$

This is negative, so π has a maximum at $Q = 8$. In fact, the corresponding maximum profit is

$$\pi = -\tfrac{3}{2}(8)^2 + 24(8) - 7 = 89$$

Finally, we are invited to calculate the marginal revenue and marginal cost at this particular value of Q. To find expressions for MR and MC we need only differentiate TR and TC, respectively. If

$$TR = 30Q - Q^2$$

then

$$MR = \frac{d(TR)}{dQ} = 30 - 2Q$$

so when $Q = 8$

$$MR = 30 - 2(8) = 14$$

If

$$TC = \tfrac{1}{2}Q^2 + 6Q + 7$$

then

$$MC = \frac{d(TC)}{dQ} = Q + 6$$

so when $Q = 8$

$$MC = 8 + 6 = 14$$

We observe that at $Q = 8$, the values of MR and MC are equal.

In this particular example we discovered that at the point of maximum profit,

$$\boxed{\text{marginal revenue}} = \boxed{\text{marginal cost}}$$

There is nothing special about this example and in the next section we show that this result holds for any profit function.

Practice Problem

3 The demand equation of a good is given by

$$P + 2Q = 20$$

and the total cost function is

$$Q^3 - 8Q^2 + 20Q + 2$$

(a) Find the level of output that maximizes total revenue.

(b) Find the maximum profit and the value of Q at which it is achieved. Verify that, at this value of Q, MR = MC.

Example

The cost of building an office block, x floors high, is made up of three components:

(1) \$10 million for the land

(2) \$$^1\!/_4$ million per floor

(3) specialized costs of \$10 000x per floor.

How many floors should the block contain if the average cost per floor is to be minimized?

Solution

The \$10 million for the land is a fixed cost because it is independent of the number of floors. Each floor costs \$$^1\!/_4$ million, so if the building has x floors altogether then the cost will be 250 000x.

In addition there are specialized costs of 10 000x per floor, so if there are x floors this will be

$$(10\,000x)x = 10\,000x^2$$

Notice the square term here, which means that the specialized costs rise dramatically with increasing x. This is to be expected, since a tall building requires a more complicated design. It may also be necessary to use more expensive materials.

The total cost, TC, is the sum of the three components: that is,

$$TC = 10\,000\,000 + 250\,000x + 10\,000x^2$$

The average cost per floor, AC, is found by dividing the total cost by the number of floors: that is,

$$\begin{aligned} AC &= \frac{TC}{x} \\ &= \frac{10\,000\,000 + 250\,000x + 10\,000x^2}{x} \\ &= \frac{10\,000\,000}{x} + 250\,000 + 10\,000x \\ &= 10\,000\,000x^{-1} + 250\,000 + 10\,000x \end{aligned}$$

Step 1

At a stationary point

$$\frac{d(AC)}{dx} = 0$$

In this case

$$\frac{d(AC)}{dx} = -10\,000\,000x^{-2} + 10\,000 = \frac{-10\,000\,000}{x^2} + 10\,000$$

so we need to solve

$$10\,000 = \frac{10\,000\,000}{x^2} \text{ or equivalently } 10\,000x^2 = 10\,000\,000$$

Hence

$$x^2 = \frac{10\,000\,000}{10\,000} = 1000$$

This has solution

$$x = \pm\sqrt{1000} = \pm 31.6$$

We can obviously ignore the negative value because it does not make sense to build an office block with a negative number of floors, so we can deduce that $x = 31.6$.

Step 2

To confirm that this is a minimum we need to differentiate a second time. Now

$$\frac{d(AC)}{dx} = -10\,000\,000x^{-2} + 10\,000$$

so

$$\frac{d^2(AC)}{dx^2} = -2(-10\,000\,000)x^{-3} = \frac{20\,000\,000}{x^3}$$

When $x = 31.6$ we see that

$$\frac{d^2(AC)}{dx^2} = \frac{20\,000\,000}{(31.6)^3} = 633.8$$

It follows that $x = 31.6$ is indeed a minimum because the second-order derivative is a positive number. 

At this stage it is tempting to state that the answer is 31.6. This is mathematically correct but is a physical impossibility since x must be a whole number. To decide whether to take x to be 31 or 32 we simply evaluate AC for these two values of x and choose the one that produces the lower average cost.

When $x = 31$,

$$\text{AC} = \frac{10\ 000\ 000}{31} + 250\ 000 + 10\ 000(31) = \$882\ 581$$

When $x = 32$,

$$\text{AC} = \frac{10\ 000\ 000}{32} + 250\ 000 + 10\ 000(32) = \$882\ 500$$

Therefore an office block 32 floors high produces the lowest average cost per floor.

Practice Problem

4 The total cost function of a good is given by

$$\text{TC} = Q^2 + 3Q + 36$$

Calculate the level of output that minimizes average cost. Find AC and MC at this value of Q. What do you observe?

Example

The supply and demand equations of a good are given by

$$P = Q_S + 8$$

and

$$P = -3Q_D + 80$$

respectively.

The government decides to impose a tax, t, per unit. Find the value of t which maximizes the government's total tax revenue on the assumption that equilibrium conditions prevail in the market.

Solution

The idea of taxation was first introduced in Chapter 1. In Section 1.3 the equilibrium price and quantity were calculated from a given value of t. In this example t is unknown but the analysis is exactly the same. All we need to do is to carry the letter t through the usual calculations and then to choose t at the end so as to maximize the total tax revenue.

To take account of the tax we replace P by $P - t$ in the supply equation. This is because the price that the supplier actually receives is the price, P, that the consumer pays less the tax, t, deducted by the government. The new supply equation is then

$$P - t = Q_S + 8$$

so that

$$P = Q_S + 8 + t$$

In equilibrium

$$Q_S = Q_D$$

If this common value is denoted by Q then the supply and demand equations become

$$P = Q + 8 + t$$
$$P = -3Q + 80$$

Hence

$$Q + 8 + t = -3Q + 80$$

since both sides are equal to P. This can be rearranged to give

$$Q = -3Q + 72 - t \quad \text{(subtract } 8 + t \text{ from both sides)}$$
$$4Q = 72 - t \quad \text{(add } 3Q \text{ to both sides)}$$
$$Q = 18 - \tfrac{1}{4}t \quad \text{(divide both sides by 4)}$$

Now, if the number of goods sold is Q and the government raises t per good then the total tax revenue, T, is given by

$$T = tQ$$
$$= t(18 - \tfrac{1}{4}t)$$
$$= 18t - \tfrac{1}{4}t^2$$

This then is the expression that we wish to maximize.

Step 1

At a stationary point

$$\frac{dT}{dt} = 0$$

so

$$18 - \tfrac{1}{2}t = 0$$

which has solution

$$t = 36$$

Step 2

To classify this point we differentiate a second time to get

$$\frac{d^2T}{dt^2} = \tfrac{1}{2} < 0$$

which confirms that it is a maximum.

Hence the government should impose a tax of \$36 on each good.

Practice Problem

5 The supply and demand equations of a good are given by

$$P = {}^1/_2 Q_S + 25$$

and

$$P = -2Q_D + 50$$

respectively.

The government decides to impose a tax, t, per unit. Find the value of t which maximizes the government's total tax revenue on the assumption that equilibrium conditions prevail in the market.

We conclude this section by describing the use of a computer package to solve optimization problems. Although a spreadsheet could be used to do this, by tabulating the values of a function, it cannot handle the associated mathematics. A symbolic computation system such as Maple, Matlab, Mathcad or Derive can not only sketch the graphs of functions, but also differentiate and solve algebraic equations. Consequently, it is possible to obtain the exact solution using one of these packages. In this book we have chosen to use Maple.

Advice

A simple introduction to this package is described in the Getting Started section at the very beginning of this book. If you have not used Maple before, go back and read through this section now.

The following example makes use of three basic Maple instructions: `plot`, `diff` and `solve`. As the name suggests, `plot` produces a graph of a function by joining together points which are accurately plotted over a specified range of values. The instruction `diff`, not surprisingly, differentiates a given expression with respect to any stated variable, and `solve` finds the exact solution of an equation.

Example MAPLE

The price, P, of a good varies over time, t, during a 15-year period according to

$$P = 0.064t^3 - 1.44t^2 + 9.6t + 10 \quad (0 \le t \le 15)$$

(a) Sketch a graph of this function and use it to estimate the local maximum and minimum points.

(b) Find the exact coordinates of these points using calculus.

Solution

It is convenient to give the cubic expression the name `price`, and to do this in Maple, we type

```
>price:=0.064*t^3-1.44*t^2+9.6*t+10;
```

(a) To plot a graph of this function for values of t between 0 and 15 we type

```
>plot(price,t=0..15);
```

Maple responds by producing a graph of price over the specified range (see Figure 4.28). The graph shows that there is one local maximum and one local minimum. (It also shows very clearly that the overall, or global, minimum and maximum occur at the ends, 0 and 15 respectively.) If you now move the cursor to some point on the plot and click, you will discover that two things happen. You will first notice that the graph is now surrounded by a box. More significantly, if you look carefully at the top of the screen, you will see that a graphics toolbar has appeared. In the left-hand corner of this is a small window containing the coordinates of the position of the cursor. To estimate the local maximum and minimum all you need do is to move the cursor to the relevant points, click, and read off the answer from the screen. Looking carefully at Figure 4.28, in which the cursor is positioned over the local maximum, we see that the coordinates of this point are approximately (5.02, 30.01). A similar estimate could be found for the local minimum point.

(b) To find the exact coordinates we need to use calculus. The simple instruction

```
>diff(price,t);
```

will produce the first derivative of price with respect to t. However, since we want to equate this to zero and solve the associated equation, it makes sense to give this a name. You can use whatever combination of symbols you like for a name in Maple, provided it does not begin with a number and it has not already been reserved by Maple. So, you are not allowed to use `1deriv`, say (because it starts with the digit 1), or `subs` (which Maple recognizes as one of its own in-house instructions for substituting numbers for letters in an expression). If we choose to call it `deriv1` we type:

```
>deriv1:=diff(price,t);
```

and Maple responds with

$deriv1 := .192t^2 - 2.88t + 9.6$

To find the stationary points, we need to equate this to zero and solve for t. This is achieved in Maple by typing:

```
>solve(deriv1=0,t);
```

and Maple responds with:

5. , 10.

These are the values of t at the stationary points. It is clear from the graph in Figure 4.28 that $t = 5$ is a local maximum and $t = 10$ is a local minimum. To find the price at the maximum we substitute $t = 5$ into the expression for price, so we type:

```
>subs(t=5,price);
```

and Maple responds with

30.000

To find the price at the local minimum we edit the instruction to create

```
>subs(t=10,price);
```

and Maple responds with

26.000

The local maximum and minimum have coordinates (5, 30) and (10, 26) respectively. ➔

Figure 4.28

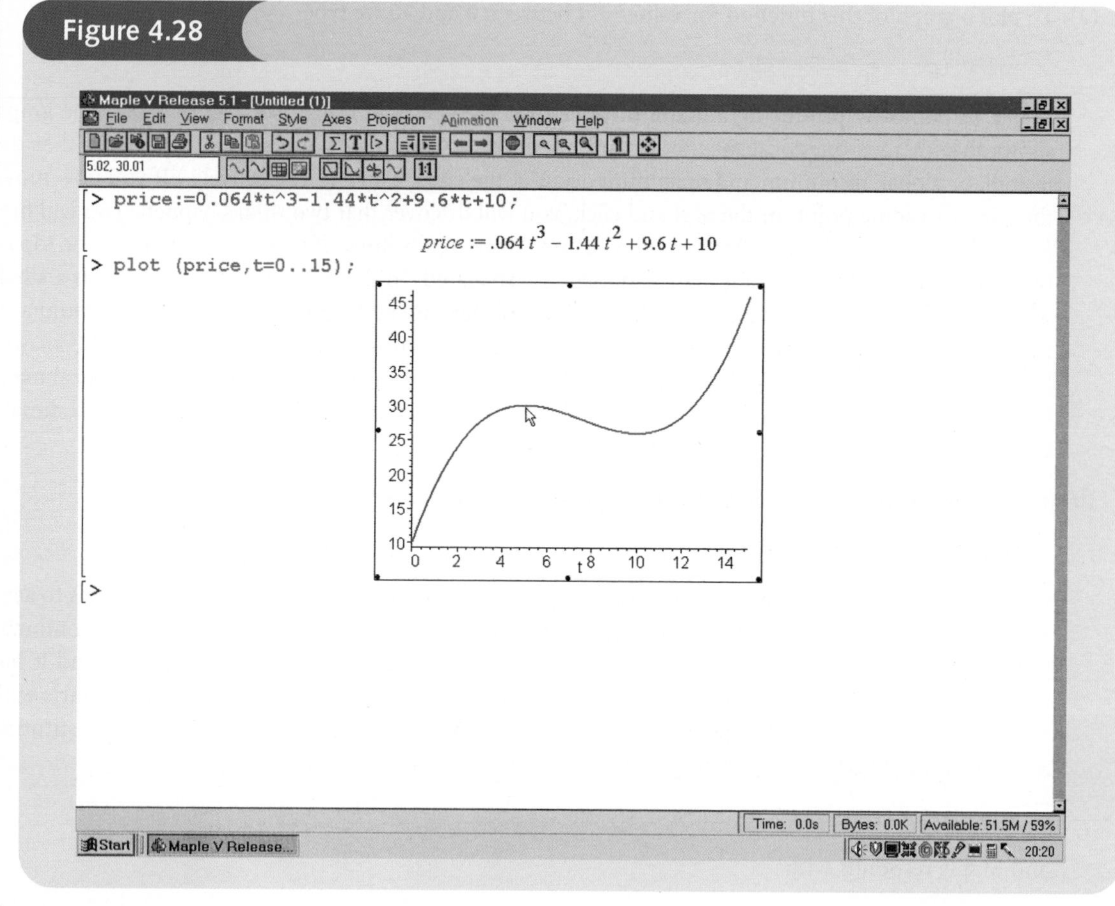

Key Terms

Average product of labour (or labour productivity) Output per worker: $AP_L = Q/L$.

Local (or relative) maximum A point on a curve which has the highest function value in comparison with other values in its neighbourhood; at such a point the first-order derivative is zero and the second-order derivative is either zero or negative.

Local (or relative) minimum A point on a curve which has the lowest function value in comparison with other values in its neighbourhood; at such a point the first-order derivative is zero and the second-order derivative is either zero or positive.

Optimization The determination of the optimal (usually stationary) points of a function.

Stationary points (critical points, turning points, extrema) Points on a graph at which the tangent is horizontal; at a stationary point the first-order derivative is zero.

Stationary point of inflection A stationary point that is neither a maximum nor a minimum; at such a point both the first- and second-order derivatives are zero.

Practice Problems

6 Find and classify the stationary points of the following functions. Hence give a rough sketch of their graphs.

(a) $y = -x^2 + x + 1$ **(b)** $y = x^2 - 4x + 4$

(c) $y = x^2 - 20x + 105$ **(d)** $y = -x^3 + 3x$

7 Show that all of the following functions have a stationary point at $x = 0$. Verify in each case that $f''(0) = 0$. Classify these points by producing a rough sketch of each function.

(a) $f(x) = x^3$ **(b)** $f(x) = x^4$ **(c)** $f(x) = -x^6$

8 If the demand equation of a good is

$$P = 40 - 2Q$$

find the level of output that maximizes total revenue.

9 If fixed costs are 15 and the variable costs are 2Q per unit, write down expressions for TC, AC and MC. Find the value of Q which minimizes AC and verify that AC = MC at this point.

10 A firm's short-run production function is given by

$$Q = 30L^2 - 0.5L^3$$

Find the value of L which maximizes AP_L and verify that $MP_L = AP_L$ at this point.

11 If the fixed costs are 13 and the variable costs are Q + 2 per unit, show that the average cost function is

$$AC = \frac{13}{Q} + Q + 2$$

(a) Calculate the values of AC when Q = 1, 2, 3, . . . , 6. Plot these points on graph paper and hence produce an accurate graph of AC against Q.

(b) Use your graph to estimate the minimum average cost.

(c) Use differentiation to confirm your estimate obtained in part (b).

12 An electronic components firm launches a new product on 1 January. During the following year a rough estimate of the number of orders, S, received t days after the launch is given by

$$S = t^2 - 0.002t^3$$

(a) What is the maximum number of orders received on any one day of the year?

(b) After how many days does the firm experience the greatest increase in orders?

13 If the demand equation of a good is

$$P = \sqrt{(1000 - 4Q)}$$

find the value of Q which maximizes total revenue.

14 The demand and total cost functions of a good are

$$4P + Q - 16 = 0$$

and

$$TC = 4 + 2Q - \frac{3Q^2}{10} + \frac{Q^3}{20}$$

respectively.

➔

(a) Find expressions for TR, π, MR and MC in terms of Q.

(b) Solve the equation

$$\frac{d\pi}{dQ} = 0$$

and hence determine the value of Q which maximizes profit.

(c) Verify that, at the point of maximum profit,

$$MR = MC$$

15 The supply and demand equations of a good are given by

$$3P - Q_S = 3$$

and

$$2P + Q_D = 14$$

respectively.

The government decides to impose a tax, t, per unit. Find the value of t (in dollars) which maximizes the government's total tax revenue on the assumption that equilibrium conditions prevail in the market.

16 **(Maple)** Plot a graph of each of the following functions over the specified range of values and use these graphs to estimate the coordinates of all of the stationary points. Use calculus to find the exact coordinates of these points.

(a) $y = 3x^4 - 28x^3 + 84x^2 - 96x + 30 \quad (0 \le x \le 5)$

(b) $y = x^4 - 8x^3 + 18x^2 - 10 \quad (-1 \le x \le 4)$

(c) $y = \dfrac{x}{x^2 + 1} \quad (-4 \le x \le 4)$

17 **(Maple)**

(a) Attempt to use Maple to plot a graph of the function $y = 1/x$ over the range $-4 \le x \le 4$. What difficulty do you encounter? Explain briefly why this has occurred for this particular function.

(b) One way of avoiding the difficulty in part (a) is to restrict the range of the y values. Produce a plot by typing

```
plot(1/x,x=-4..4,y=-3..3);
```

(c) Use the approach suggested in part (b) to plot a graph of the curve

$$y = \frac{x - 3}{(x + 1)(x - 2)}$$

on the interval $-2 \le x \le 6$. Use calculus to find all of the stationary points.

18 **(Maple)** The total cost, TC, and total revenue, TR, functions of a good are given by

$$TC = 80Q - \frac{15}{2}Q^2 + \frac{1}{3}Q^3 \quad \text{and} \quad TR = 50Q - Q^2$$

Obtain Maple expressions for π, MC and MR, naming them profit, MC and MR respectively. Plot all three functions on the same diagram using the instruction:

```
plot({profit,MC,MR},Q=0..14);
```

Use this diagram to show that

(a) when the profit is a minimum, MR = MC and the MC curve cuts the MR curve from above

(b) when the profit is a maximum, MR = MC and the MC curve cuts the MR curve from below.

19 **(Maple)** A firm's short-run production function is given by

$$Q = 300L^{0.8}(240 - 5L)^{0.5} \quad (0 \leq L \leq 48)$$

where L is the size of the workforce. Plot a graph of this function and hence estimate the level of employment needed to maximize output. Confirm this by using differentiation.

section 4.7

Further optimization of economic functions

Objectives

At the end of this section you should be able to:

- Show that, at the point of maximum profit, marginal revenue equals marginal cost.
- Show that, at the point of maximum profit, the slope of the marginal revenue curve is less than that of marginal cost.
- Maximize profits of a firm with and without price discrimination in different markets.
- Show that, at the point of maximum average product of labour, average product of labour equals marginal product of labour.

The previous section demonstrated how mathematics can be used to optimize particular economic functions. Those examples suggested two important results:

(a) If a firm maximizes profit then MR = MC.

(b) If a firm maximizes average product of labour then $\text{AP}_L = \text{MP}_L$.

Although these results were found to hold for all of the examples considered in Section 4.6, it does not necessarily follow that the results are always true. The aim of this section is to prove these assertions without reference to specific functions and hence to demonstrate their generality.

Advice

You may prefer to skip these proofs at a first reading and just concentrate on the worked example (and Practice Problems 1 and 8) on price discrimination.

Justification of result (a) turns out to be really quite easy. Profit, π, is defined to be the difference between total revenue, TR, and total cost, TC: that is,

$$\pi = \text{TR} - \text{TC}$$

To find the stationary points of π we differentiate with respect to Q and equate to zero: that is,

$$\frac{d\pi}{dQ} = \frac{d(\text{TR})}{dQ} - \frac{d(\text{TC})}{dQ} = 0$$

where we have used the difference rule to differentiate the right-hand side. In Section 4.3 we defined

$$\text{MR} = \frac{d(\text{TR})}{dQ} \quad \text{and} \quad \text{MC} = \frac{d(\text{TC})}{dQ}$$

so the above equation is equivalent to

$$\text{MR} - \text{MC} = 0$$

and so MR = MC as required.

The stationary points of the profit function can therefore be found by sketching the MR and MC curves on the same diagram and inspecting the points of intersection. Figure 4.29 shows typical marginal revenue and marginal cost curves. The result

$$\text{MR} = \text{MC}$$

holds for any stationary point. Consequently, if this equation has more than one solution then we need some further information before we can decide on the profit-maximizing level of output. In Figure 4.29 there are two points of intersection, Q_1 and Q_2, and it turns out (as you discovered in Practice Problems 3 and 14 in the previous section) that one of these is a maximum while the other is a minimum. Obviously, in any actual example, we can classify these points by evaluating second-order derivatives. However, it would be nice to make this decision just by inspecting the graphs of marginal revenue and marginal cost. To see how this can be done let us return to the equation

Figure 4.29

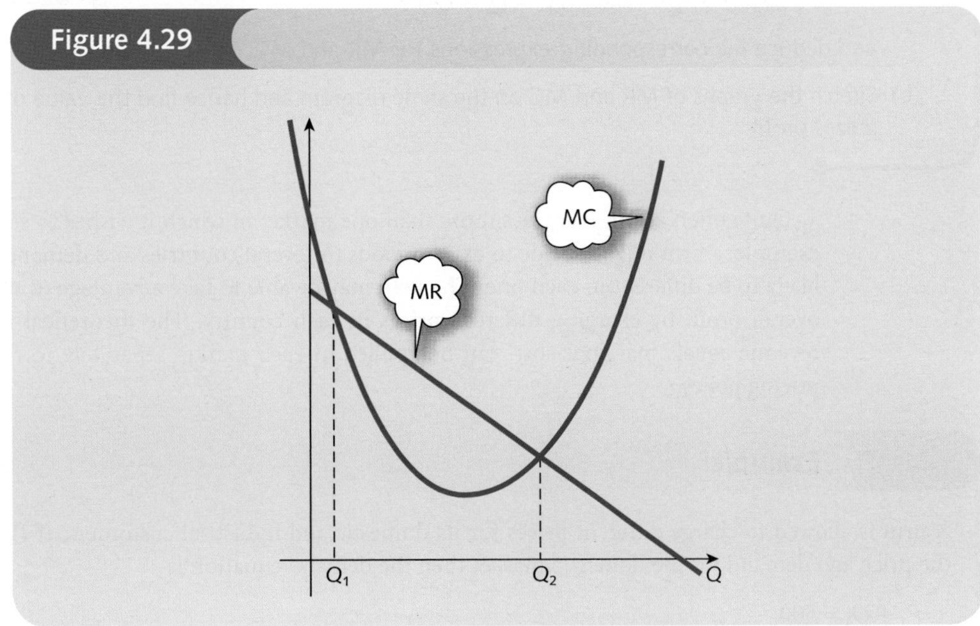

$$\frac{d\pi}{dQ} = \text{MR} - \text{MC}$$

and differentiate again with respect to Q to get

$$\frac{d^2\pi}{dQ^2} = \frac{d(\text{MR})}{dQ} - \frac{d(\text{MC})}{dQ}$$

Now if $d^2\pi/dQ^2 < 0$ then the profit is a maximum. This will be so when

$$\frac{d(\text{MR})}{dQ} < \frac{d(\text{MC})}{dQ}$$

that is, when the slope of the marginal revenue curve is less than the slope of the marginal cost curve.

Looking at Figure 4.29, we deduce that this criterion is true at Q_2, so this must be the desired level of output needed to maximize profit. Note also from Figure 4.29 that the statement 'the slope of the marginal revenue curve is less than the slope of the marginal cost curve' is equivalent to saying that 'the marginal cost curve cuts the marginal revenue curve from below'. It is this latter form that is often quoted in economics textbooks. A similar argument shows that, at a minimum point, the marginal cost curve cuts the marginal revenue curve from above and so we can deduce that profit is a minimum at Q_1 in Figure 4.29. In practice, the task of sketching the graphs of MR and MC and reading off the coordinates of the points of intersection is not an attractive one, particularly if MR and MC are complicated functions. However, it might turn out that MR and MC are both linear, in which case a graphical approach is feasible.

Practice Problem

1 A monopolist's demand function is

$$P = 25 - 0.5Q$$

The fixed costs of production are 7 and the variable costs are $Q + 1$ per unit.

(a) Show that

$$\text{TR} = 25Q - 0.5Q^2 \quad \text{and} \quad \text{TC} = Q^2 + Q + 7$$

and deduce the corresponding expressions for MR and MC.

(b) Sketch the graphs of MR and MC on the same diagram and hence find the value of Q which maximizes profit.

Quite often a firm identifies more than one market in which it wishes to sell its goods. For example, a firm might decide to export goods to several countries and demand conditions are likely to be different in each one. The firm may be able to take advantage of this and increase overall profit by charging different prices in each country. The theoretical result 'marginal revenue equals marginal cost' can be applied in each market separately to find the optimal pricing policy.

Example

A firm is allowed to charge different prices for its domestic and industrial customers. If P_1 and Q_1 denote the price and demand for the domestic market then the demand equation is

$$P_1 + Q_1 = 500$$

If P_2 and Q_2 denote the price and demand for the industrial market then the demand equation is

$$2P_2 + 3Q_2 = 720$$

The total cost function is

$$\text{TC} = 50\,000 + 20Q$$

where $Q = Q_1 + Q_2$. Determine the prices (in dollars) that the firm should charge to maximize profits

(a) with price discrimination

(b) without price discrimination.

Compare the profits obtained in parts (a) and (b).

Solution

(a) The important thing to notice is that the total cost function is independent of the market and so marginal costs are the same in each case. In fact, since

$$\text{TC} = 50\,000 + 20Q$$

we have MC = 20. All we have to do to maximize profits is to find an expression for the marginal revenue for each market and to equate this to the constant value of marginal cost.

Domestic market

The demand equation

$$P_1 + Q_1 = 500$$

rearranges to give

$$P_1 = 500 - Q_1$$

so the total revenue function for this market is

$$\text{TR}_1 = (500 - Q_1)Q_1 = 500Q_1 - Q_1^2$$

Hence

$$\text{MR}_1 = \frac{\text{d}(\text{TR}_1)}{\text{d}Q_1} = 500 - 2Q_1$$

For maximum profit

$$\text{MR}_1 = \text{MC}$$

so

$$500 - 2Q_1 = 20$$

which has solution $Q_1 = 240$. The corresponding price is found by substituting this value into the demand equation to get

$$P_1 = 500 - 240 = \$260$$

To maximize profit the firm should charge its domestic customers \$260 per good.

Industrial market

The demand equation

$$2P_2 + 3Q_2 = 720$$

rearranges to give

$$P_2 = 360 - {}^3\!/_2Q_2$$

➔

so the total revenue function for this market is

$$TR_2 = (360 - \tfrac{3}{2}Q_2)Q_2 = 360Q_2 - \tfrac{3}{2}Q_2^2$$

Hence

$$MR_2 = \frac{d(TR_2)}{dQ_2}$$
$$= 360 - 3Q_2$$

For maximum profit

$$MR_2 = MC$$

so

$$360 - 3Q_2 = 20$$

which has solution $Q_2 = 340/3$. The corresponding price is obtained by substituting this value into the demand equation to get

$$P_2 = 360 - \frac{3}{2}\left(\frac{340}{3}\right) = \$190$$

To maximize profits the firm should charge its industrial customers \$190 per good, which is lower than the price charged to its domestic customers.

(b) If there is no price discrimination then $P_1 = P_2 = P$, say, and the demand functions for the domestic and industrial markets become

$$P + Q_1 = 500$$

and

$$2P + 3Q_2 = 720$$

respectively. We can use these to deduce a single demand equation for the combined market. We need to relate the price, P, of each good to the total demand, $Q = Q_1 + Q_2$.

This can be done by rearranging the given demand equations for Q_1 and Q_2 and then adding. For the domestic market

$$Q_1 = 500 - P$$

and for the industrial market

$$Q_2 = 240 - \tfrac{2}{3}P$$

Hence

$$Q = Q_1 + Q_2 = 740 - \tfrac{5}{3}P$$

The demand equation for the combined market is therefore

$$Q + \tfrac{5}{3}P = 740$$

The usual procedure for profit maximization can now be applied. This demand equation rearranges to give

$$P = 444 - \tfrac{3}{5}Q$$

enabling the total revenue function to be written down as

$$TR = \left(444 - \frac{3}{5}Q\right)Q = 444Q - \frac{3Q^2}{5}$$

Hence

$$\text{MR} = \frac{\text{d(TR)}}{\text{d}Q}$$

$$= 444 - \frac{6}{5}Q$$

For maximum profit

$$\text{MR} = \text{MC}$$

so

$$444 - {}^6\!/_5Q = 20$$

which has solution $Q = 1060/3$. The corresponding price is found by substituting this value into the demand equation to get

$$P = 444 - \frac{3}{5}\left(\frac{1060}{3}\right) = \$232$$

To maximize profit without discrimination the firm needs to charge a uniform price of \$232 for each good. Notice that this price lies between the prices charged to its domestic and industrial customers with discrimination.

To evaluate the profit under each policy we need to work out the total revenue and subtract the total cost. In part (a) the firm sells 240 goods at \$260 each in the domestic market and sells 340/3 goods at \$190 each in the industrial market, so the total revenue received is

$$240 \times 260 + \frac{340}{3} \times 190 = \$83\,933.33$$

The total number of goods produced is

$$240 + \frac{340}{3} = \frac{1060}{3}$$

so the total cost is

$$50\,000 + 20 \times \frac{1060}{3} = \$57\,066.67$$

Therefore the profit with price discrimination is

$$83\,933.33 - 57\,066.67 = \$26\,866.67$$

In part (b) the firm sells 1060/3 goods at \$232 each, so total revenue is

$$\frac{1060}{3} \times 232 = \$81\,973.33$$

Now the total number of goods produced under both pricing policies is the same: that is, 1060/3. Consequently, the total cost of production in part (b) must be the same as part (a): that is,

$$\text{TC} = \$57\,066.67$$

The profit without price discrimination is

$$81\,973.33 - 57\,066.67 = \$24\,906.66$$

As expected, the profits are higher with discrimination than without.

Practice Problem

2 A firm has the possibility of charging different prices in its domestic and foreign markets. The corresponding demand equations are given by

$$Q_1 = 300 - P_1$$

$$Q_2 = 400 - 2P_2$$

The total cost function is

$$\text{TC} = 5000 + 100Q$$

where $Q = Q_1 + Q_2$.

Determine the prices (in dollars) that the firm should charge to maximize profits

(a) with price discrimination

(b) without price discrimination

Compare the profits obtained in parts (a) and (b).

In the previous example and in Practice Problem 2 we assumed that the marginal costs were the same in each market. The level of output that maximizes profit with price discrimination was found by equating marginal revenue to this common value of marginal cost. It follows that the marginal revenue must be the same in each market. In symbols

$$\text{MR}_1 = \text{MC} \quad \text{and} \quad \text{MR}_2 = \text{MC}$$

so

$$\text{MR}_1 = \text{MR}_2$$

This fact is obvious on economic grounds. If it were not true then the firm's policy would be to increase sales in the market where marginal revenue is higher and to decrease sales by the same amount in the market where the marginal revenue is lower. The effect would be to increase revenue while keeping costs fixed, thereby raising profit. This property leads to an interesting result connecting price, P, with elasticity of demand, E. In Section 4.5 we derived the formula

$$\text{MR} = P\left(1 - \frac{1}{E}\right)$$

If we let the price elasticity of demand in two markets be denoted by E_1 and E_2 corresponding to prices P_1 and P_2 then the equation

$$\text{MR}_1 = \text{MR}_2$$

becomes

$$P_1\left(1 - \frac{1}{E_1}\right) = P_2\left(1 - \frac{1}{E_2}\right)$$

This equation holds whenever a firm chooses its prices P_1 and P_2 to maximize profits in each market. Note that if $E_1 < E_2$ then this equation can only be true if $P_1 > P_2$. In other words, the firm charges the higher price in the market with the lower elasticity of demand.

Practice Problem

3 Calculate the price elasticity of demand at the point of maximum profit for each of the demand functions given in Practice Problem 2 with price discrimination. Verify that the firm charges the higher price in the market with the lower elasticity of demand.

The previous discussion concentrated on profit. We now turn our attention to average product of labour and prove result (2) stated at the beginning of this section. This concept is defined by

$$\text{AP}_L = \frac{Q}{L}$$

where Q is output and L is labour. The maximization of AP_L is a little more complicated than before, since it is necessary to use the quotient rule to differentiate this function. In the notation of Section 4.4 we write

$$u = Q \quad \text{and} \quad v = L$$

so

$$\frac{\text{d}u}{\text{d}L} = \frac{\text{d}Q}{\text{d}L} = \text{MP}_L \quad \text{and} \quad \frac{\text{d}v}{\text{d}L} = \frac{\text{d}L}{\text{d}L} = 1$$

where we have used the fact that the derivative of output with respect to labour is the marginal product of labour.

The quotient rule gives

$$\frac{\text{d}(\text{AP}_L)}{\text{d}L} = \frac{v\,\text{d}u/\text{d}L - u\,\text{d}v/\text{d}L}{v^2}$$

$$= \frac{L(\text{MP}_L) - Q(1)}{L^2}$$

$$= \frac{\text{MP}_L - Q/L}{L}$$

divide top and bottom by L

$$= \frac{\text{MP}_L - \text{AP}_L}{L}$$

by definition, $\text{AP}_L = \frac{Q}{L}$

At a stationary point

$$\frac{\text{d}(\text{AP}_L)}{\text{d}L} = 0$$

so

$$\frac{\text{MP}_L - \text{AP}_L}{L} = 0$$

Hence

$$\text{MP}_L = \text{AP}_L$$

as required.

This analysis shows that, at a stationary point of the average product of labour function, the marginal product of labour equals the average product of labour. The above argument provides

Figure 4.30

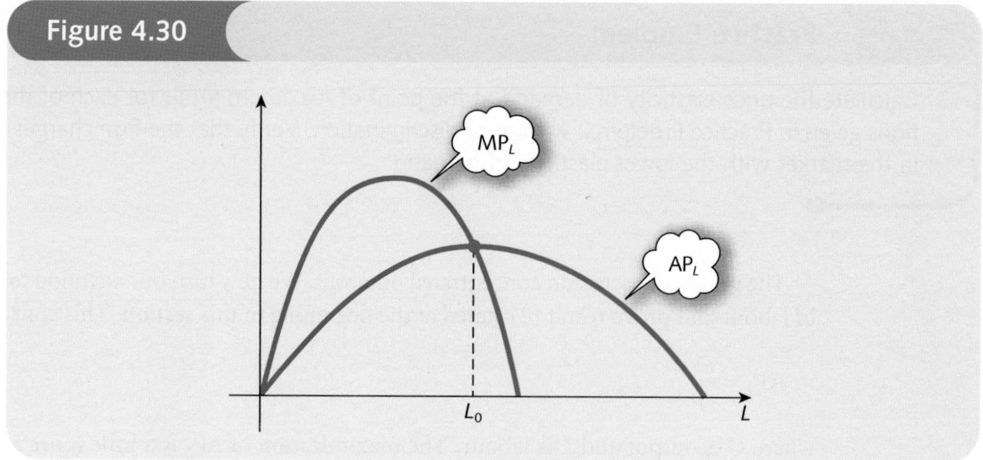

a formal proof that this result is true for any average product of labour function. Figure 4.30 shows typical average and marginal product functions. Note that the two curves intersect at the peak of the AP_L curve. To the left of this point the AP_L function is increasing, so that

$$\frac{\text{d}(\text{AP}_L)}{\text{d}L} > 0$$

Now we have just seen that

$$\frac{\text{d}(\text{AP}_L)}{\text{d}L} = \frac{\text{MP}_L - \text{AP}_L}{L}$$

so we deduce that, to the left of the maximum, $\text{MP}_L > \text{AP}_L$. In other words, in this region the graph of marginal product of labour lies above that of average product of labour. Similarly, to the right of the maximum, AP_L is decreasing, so that

$$\frac{\text{d}(\text{AP}_L)}{\text{d}L} < 0$$

and hence $\text{MP}_L < \text{AP}_L$. The graph of marginal product of labour therefore lies below that of average product of labour in this region.

We deduce that if the stationary point is a maximum then the MP_L curve cuts the AP_L curve from above. A similar argument can be used for any average function. The particular case of the average cost function is investigated in Practice Problem 9.

Practice Problems

4 Show that if the marginal cost curve cuts the marginal revenue curve from above then profit is a minimum.

5 A firm's demand function is

$$P = aQ + b \quad (a < 0, b > 0)$$

Fixed costs are c and variable costs per unit are d.

(a) Write down general expressions for TR and TC.

(b) By differentiating the expressions in part (a), deduce MR and MC.

(c) Use your answers to (b) to show that profit, π, is maximized when

$$Q = \frac{d - b}{2a}$$

6 (a) In Section 4.5 the following relationship between marginal revenue, MR, and price elasticity of demand, E, was derived:

$$\text{MR} = P\left(1 - \frac{1}{E}\right)$$

Use this result to show that at the point of maximum total revenue, $E = 1$.

(b) Verify the result of part (a) for the demand function

$$2P + 3Q = 60$$

7 The economic order quantity, EOQ, is used in cost accounting to minimize the total cost, TC, to order and carry a firm's stock over the period of a year.

The annual cost of placing orders, ACO, is given by

$$\text{ACO} = \frac{(\text{ARU})(\text{CO})}{\text{EOQ}}$$

where

$$\text{ARU} = \text{annual required units}$$

$$\text{CO} = \text{cost per order}$$

The annual carrying cost, ACC, is given by

$$\text{ACC} = (\text{CU})(\text{CC})\frac{(\text{EOQ})}{2}$$

where

$$\text{CU} = \text{cost per unit}$$

$$\text{CC} = \text{carrying cost}$$

and (EOQ)/2 provides an estimate of the average number of units in stock at any given time of the year. Assuming that ARU, CO, CU and CC are all constant, show that the total cost

$$\text{TC} = \text{ACO} + \text{ACC}$$

is minimized when

$$\text{EOQ} = \sqrt{\frac{2(\text{ARU})(\text{CO})}{(\text{CU})(\text{CC})}}$$

8 The demand functions for a firm's domestic and foreign markets are

$$P_1 = 50 - 5Q_1$$

$$P_2 = 30 - 4Q_2$$

and the total cost function is

$$\text{TC} = 10 + 10Q$$

where $Q = Q_1 + Q_2$. Determine the prices needed to maximize profit

(a) with price discrimination

(b) without price discrimination

Compare the profits obtained in parts (a) and (b).

➔

9 **(a)** Show that, at a stationary point of an average cost function, average cost equals marginal cost.

(b) Show that if the marginal cost curve cuts the average cost curve from below then average cost is a minimum.

10 In a competitive market the equilibrium price, P, and quantity, Q, are found by setting $Q_S = Q_D = Q$ in the supply and demand equations

$$P = aQ_S + b \quad (a > 0, b > 0)$$

$$P = -cQ_D + d \quad (c > 0, d > 0)$$

If the government levies an excise tax, t, per unit, show that

$$Q = \frac{d - b - t}{a + c}$$

Deduce that the government's tax revenue, $T = tQ$, is maximized by taking

$$t = \frac{d - b}{2}$$

section 4.8

The derivative of the exponential and natural logarithm functions

Objectives

At the end of this section you should be able to:

- Differentiate the exponential function.
- Differentiate the natural logarithm function.
- Use the chain, product and quotient rules to differentiate combinations of these functions.
- Appreciate the use of the exponential function in economic modelling.

In this section we investigate the derived functions associated with the exponential and natural logarithm functions, e^x and $\ln x$. The approach that we adopt is similar to that used in Section 4.1. The derivative of a function determines the slope of the graph of a function. Consequently, to discover how to differentiate an unfamiliar function we first produce an accurate sketch and then measure the slopes of the tangents at selected points.

Advice

The functions, e^x and $\ln x$ were first introduced in Section 2.4. You might find it useful to remind yourself how these functions are defined before working through the rest of the current section.

Example

Complete the following table of function values and hence sketch a graph of $f(x) = e^x$.

x	−2.0	−1.5	−1.0	−0.5	0.0	0.5	1.0	1.5
$f(x)$								

Draw tangents to the graph at $x = -1$, 0 and 1. Hence estimate the values of $f'(-1)$, $f'(0)$ and $f'(1)$. Suggest a general formula for the derived function $f'(x)$.

Solution

Using a calculator we obtain

x	−2.0	−1.5	−1.0	−0.5	0.0	0.5	1.0	1.5
$f(x)$	0.14	0.22	0.37	0.61	1.00	1.65	2.72	4.48

The corresponding graph of the exponential function is sketched in Figure 4.31. From the graph we see that the slopes of the tangents are

Figure 4.31

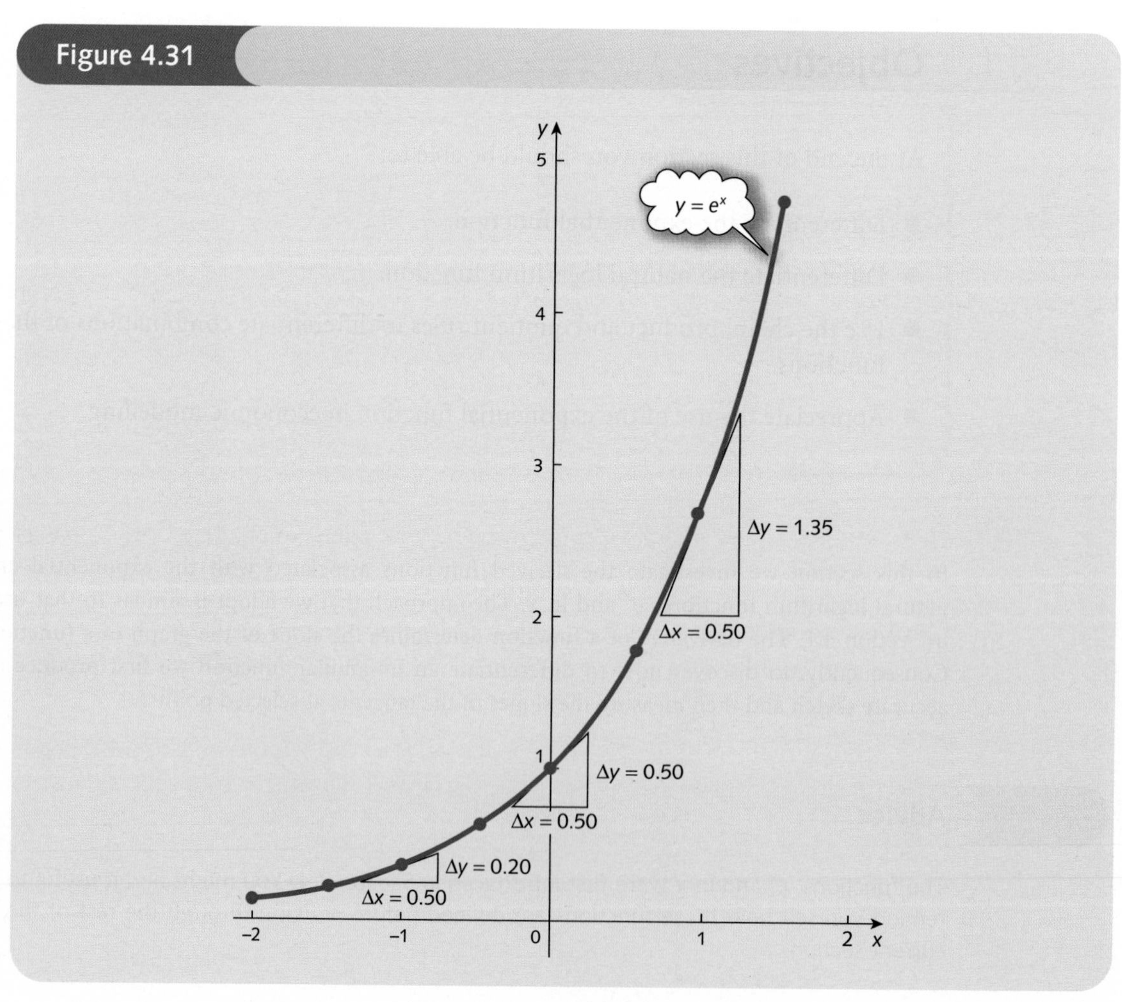

$$f'(-1) = \frac{0.20}{0.50} = 0.4$$

$$f'(0) = \frac{0.50}{0.50} = 1.0$$

$$f'(1) = \frac{1.35}{0.50} = 2.7$$

These results are obtained by measurement and so are quoted to only 1 decimal place. We cannot really expect to achieve any greater accuracy using this approach.

The values of x, $f(x)$ and $f'(x)$ are summarized in the following table. The values of $f(x)$ are rounded to 1 decimal place in order to compare with the graphical estimates of $f'(x)$.

x	−1	0	1
$f(x)$	0.4	1.0	2.7
$f'(x)$	0.4	1.0	2.7

Notice that the values of $f(x)$ and $f'(x)$ are identical to within the accuracy quoted.

These results suggest that the slope of the graph at each point is the same as the function value at that point: that is, e^x differentiates to itself. Symbolically,

$$\text{if} \quad f(x) = e^x \quad \text{then} \quad f'(x) = e^x$$

or, equivalently,

$$\text{if} \quad y = e^x \quad \text{then} \quad \frac{dy}{dx} = e^x$$

Practice Problem

1 Use your calculator to complete the following table of function values and hence sketch an accurate graph of $f(x) = \ln x$.

x	0.50	1.00	1.50	2.00	2.50	3.00	3.50	4.00
$f(x)$			0.41				1.25	

Draw the tangents to the graph at $x = 1$, 2 and 3. Hence estimate the values of $f'(1)$, $f'(2)$ and $f'(3)$. Suggest a general formula for the derived function $f'(x)$.

[Hint: for the last part you may find it helpful to rewrite your estimates of $f'(x)$ as simple fractions.]

In fact, it is possible to prove that, for any value of the constant m,

$$\text{if} \quad y = e^{mx} \quad \text{then} \quad \frac{dy}{dx} = me^{mx}$$

and

$$\text{if} \quad y = \ln mx \quad \text{then} \quad \frac{dy}{dx} = \frac{1}{x}$$

In particular, we see by setting $m = 1$ that

$$e^x \quad \text{differentiates to} \quad e^x$$

and that

$$\ln x \quad \text{differentiates to} \quad \frac{1}{x}$$

which agree with our practical investigations.

Example

Differentiate

(a) $y = e^{2x}$ **(b)** $y = e^{-7x}$ **(c)** $y = \ln 5x \quad (x > 0)$ **(d)** $y = \ln 559x \quad (x > 0)$

Solution

(a) Setting $m = 2$ in the general formula shows that

$$\text{if} \quad y = e^{2x} \quad \text{then} \quad \frac{dy}{dx} = 2e^{2x}$$

Notice that when exponential functions are differentiated the power itself does not change. All that happens is that the coefficient of x comes down to the front.

(b) Setting $m = -7$ in the general formula shows that

$$\text{if} \quad y = e^{-7x} \quad \text{then} \quad \frac{dy}{dx} = -7e^{-7x}$$

(c) Setting $m = 5$ in the general formula shows that

$$\text{if} \quad y = \ln 5x \quad \text{then} \quad \frac{dy}{dx} = \frac{1}{x}$$

Notice the restriction $x > 0$ stated in the question. This is needed to ensure that we do not attempt to take the logarithm of a negative number, which is impossible.

(d) Setting $m = 559$ in the general formula shows that

$$\text{if} \quad y = \ln 559x \quad \text{then} \quad \frac{dy}{dx} = \frac{1}{x}$$

Notice that we get the same answer as part (c). The derivative of the natural logarithm function does not depend on the coefficient of x. This fact may seem rather strange but it is easily accounted for. The third rule of logarithms shows that $\ln 559x$ is the same as

$$\ln 559 + \ln x$$

The first term is merely a constant, so differentiates to zero, and the second term differentiates to $1/x$.

Practice Problem

2 Differentiate

(a) $y = e^{3x}$ **(b)** $y = e^{-x}$ **(c)** $y = \ln 3x \ (x > 0)$ **(d)** $y = \ln 51\,234x \ (x > 0)$

The chain rule can be used to explain what happens to the m when differentiating e^{mx}. The outer function is the exponential, which differentiates to itself, and the inner function is mx, which differentiates to m. Hence, by the chain rule,

$$\text{if} \quad y = e^{mx} \quad \text{then} \quad \frac{dy}{dx} = e^{mx} \times m = me^{mx}$$

Similarly, noting that the natural logarithm function differentiates to the reciprocal function,

$$\text{if} \quad y = \ln mx \quad \text{then} \quad \frac{dy}{dx} = \frac{1}{mx} \times m = \frac{1}{x}$$

The chain, product and quotient rules can be used to differentiate more complicated functions involving e^x and $\ln x$.

Example

Differentiate

(a) $y = x^3e^{2x}$ **(b)** $y = \ln(x^2 + 2x + 1)$ **(c)** $y = \dfrac{e^{3x}}{x^2 + 2}$

Solution

(a) The function x^3e^{2x} involves the product of two simpler functions, x^3 and e^{2x}, so we need to use the product rule to differentiate it. Putting

$$u = x^3 \quad \text{and} \quad v = e^{2x}$$

gives

$$\frac{du}{dx} = 3x^2 \quad \text{and} \quad \frac{dv}{dx} = 2e^{2x}$$

By the product rule

$$\begin{aligned}\frac{dy}{dx} &= u\frac{dv}{dx} + v\frac{du}{dx} \\ &= x^3[2e^{2x}] + e^{2x}[3x^2] \\ &= 2x^3e^{2x} + 3x^2e^{2x}\end{aligned}$$

There is a common factor of x^2e^{2x}, which goes into the first term $2x$ times and into the second term 3 times. Hence

$$\frac{dy}{dx} = x^2e^{2x}(2x + 3)$$

(b) The expression $\ln(x^2 + 2x + 1)$ can be regarded as a function of a function, so we can use the chain rule to differentiate it. We first differentiate the outer log function to get

$$\frac{1}{x^2 + 2x + 1}$$

and then multiply by the derivative of the inner function, $x^2 + 2x + 1$, which is $2x + 2$. Hence

$$\frac{dy}{dx} = \frac{2x + 2}{x^2 + 2x + 1}$$

(c) The function

$$\frac{e^{3x}}{x^2+2}$$

is the quotient of the simpler functions

$$u = e^{3x} \quad \text{and} \quad v = x^2 + 2$$

for which

$$\frac{du}{dx} = 3e^{3x} \quad \text{and} \quad \frac{dv}{dx} = 2x$$

By the quotient rule

$$\frac{dy}{dx} = \frac{v\dfrac{du}{dx} - u\dfrac{dv}{dx}}{v^2}$$

$$= \frac{(x^2+2)(3e^{3x}) - e^{3x}(2x)}{(x^2+2)^2}$$

$$= \frac{e^{3x}[3(x^2+2) - 2x]}{(x^2+2)^2}$$

$$= \frac{e^{3x}(3x^2 - 2x + 6)}{(x^2+2)^2}$$

Practice Problem

3 Differentiate

(a) $y = x^4 \ln x$

(b) $y = e^{x^2}$

(c) $y = \dfrac{\ln x}{x+2}$

Advice

If you ever need to differentiate a function of the form:

ln(an inner function involving products, quotients or powers of x)

then it is usually quicker to use the rules of logs to expand the expression before you begin. The three rules are

Rule 1 $\ln(x \times y) = \ln x + \ln y$

Rule 2 $\ln(x \div y) = \ln x - \ln y$

Rule 3 $\ln x^m = m \ln x$

The following example shows how to apply this 'trick' in practice.

Example

Differentiate

(a) $y = \ln(x(x+1)^4)$ (b) $y = \ln\left(\frac{x}{\sqrt{(x+5)}}\right)$

Solution

(a) From rule 1

$$\ln(x(x+1)^4) = \ln x + \ln(x+1)^4$$

which can be simplified further using rule 3 to give

$$y = \ln x + 4\ln(x+1)$$

Differentiation of this new expression is trivial. We see immediately that

$$\frac{dy}{dx} = \frac{1}{x} + \frac{4}{x+1}$$

If desired the final answer can be put over a common denominator

$$\frac{1}{x} + \frac{4}{x+1} = \frac{(x+1)+4x}{x(x+1)} = \frac{5x+1}{x(x+1)}$$

(b) The quickest way to differentiate

$$y = \ln\left(\frac{x}{\sqrt{(x+5)}}\right)$$

is to expand first to get

$$y = \ln x - \ln(x+5)^{1/2} \quad \text{(rule 2)}$$

$$= \ln x - \tfrac{1}{2}\ln(x+5) \quad \text{(rule 3)}$$

Again this expression is easy to differentiate:

$$\frac{dy}{dx} = \frac{1}{x} - \frac{1}{2(x+5)}$$

If desired, this can be written as a single fraction:

$$\frac{1}{x} - \frac{1}{2(x+5)} = \frac{2(x+5)-x}{2x(x+5)} = \frac{x+10}{2x(x+5)}$$

Practice Problem

4 Differentiate the following functions by first expanding each expression using the rules of logs:

(a) $y = \ln(x^3(x+2)^4)$ **(b)** $y = \ln\left(\frac{x^2}{2x+3}\right)$

Exponential and natural logarithm functions provide good mathematical models in many areas of economics and we conclude this chapter with some illustrative examples.

Example

A firm's short-run production function is given by

$$Q = L^2e^{-0.01L}$$

Find the value of L which maximizes the average product of labour.

Solution

The average product of labour is given by

$$AP_L = \frac{Q}{L} = \frac{L^2e^{-0.01L}}{L} = Le^{-0.01L}$$

To maximize this function we adopt the strategy described in Section 4.6.

Step 1

At a stationary point

$$\frac{d(AP_L)}{dL} = 0$$

To differentiate $Le^{-0.01L}$, we use the product rule. If

$$u = L \quad \text{and} \quad v = e^{-0.01L}$$

then

$$\frac{du}{dL} = 1 \quad \text{and} \quad \frac{dv}{dL} = -0.01e^{-0.01L}$$

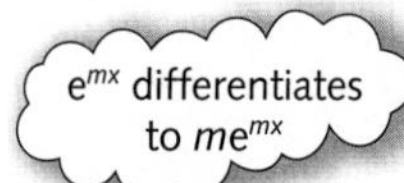

By the product rule

$$\frac{d(AP_L)}{dL} = u\frac{dv}{dL} + v\frac{du}{dL}$$

$$= L(-0.01e^{-0.01L}) + e^{-0.01L}$$

$$= (1 - 0.01L)e^{-0.01L}$$

We know that a negative exponential is never equal to zero. (Although $e^{-0.01L}$ gets ever closer to zero as L increases, it never actually reaches it for finite values of L.) Hence the only way that

$$(1 - 0.01L)e^{-0.01L}$$

can equal zero is when

$$1 - 0.01L = 0$$

which has solution $L = 100$.

Step 2

To show that this is a maximum we need to differentiate a second time. To do this we apply the product rule to

$$(1 - 0.01L)e^{-0.01L}$$

taking

$$u = 1 - 0.01L \quad \text{and} \quad v = e^{-0.01L}$$

for which

$$\frac{du}{dL} = -0.01 \quad \text{and} \quad \frac{dv}{dL} = -0.01e^{-0.01L}$$

Hence

$$\frac{d^2(AP_L)}{dL^2} = u\frac{dv}{dL} + v\frac{du}{dL}$$

$$= (1 - 0.01L)(-0.01e^{-0.01L}) + e^{-0.01L}(-0.01) = (-0.02 + 0.0001L)e^{-0.01L}$$

Finally, putting $L = 100$ into this gives

$$\frac{d^2(AP_L)}{dL^2} = -0.0037$$

The fact that this is negative shows that the stationary point, $L = 100$, is indeed a maximum.

Practice Problem

5 The demand function of a good is given by

$$Q = 1000e^{-0.2P}$$

If fixed costs are 100 and the variable costs are 2 per unit, show that the profit function is given by

$$\pi = 1000Pe^{-0.2P} - 2000e^{-0.2P} - 100$$

Find the price needed to maximize profit.

Example

A firm estimates that the total revenue received from the sale of Q goods is given by

$$TR = \ln(1 + 1000Q^2)$$

Calculate the marginal revenue when $Q = 10$.

Solution

The marginal revenue function is obtained by differentiating the total revenue function. To differentiate $\ln(1 + 1000Q^2)$ we use the chain rule. We first differentiate the outer log function to get

$$\frac{1}{1 + 1000Q^2}$$

natural logs differentiate to reciprocals

and then multiply by the derivative of the inner function, $1 + 1000Q^2$, to get $2000Q$. Hence

$$MR = \frac{d(TR)}{dQ} = \frac{2000Q}{1 + 1000Q^2}$$

At $Q = 10$,

$$MR = \frac{2000(10)}{1 + 1000(10)^2} = 0.2$$

Practice Problem

6 If the demand equation is

$$P = 200 - 40 \ln(Q + 1)$$

calculate the price elasticity of demand when $Q = 20$.

Practice Problems

7 Write down the derivative of

(a) $y = e^{6x}$ **(b)** $y = e^{-342x}$ **(c)** $y = 2e^{-x} + 4e^{x}$ **(d)** $y = 10e^{4x} - 2x^2 + 7$

8 Write down the derivative of

(a) $y = \ln(3x) \quad (x > 0)$ **(b)** $y = \ln(-13x) \quad (x < 0)$

9 Use the chain rule to differentiate

(a) $y = e^{x^3}$ **(b)** $y = \ln(x^4 + 3x^2)$

10 Use the product rule to differentiate

(a) $y = x^4e^{2x}$ **(b)** $y = x \ln x$

11 Use the quotient rule to differentiate

(a) $y = \dfrac{e^{4x}}{x^2 + 2}$ **(b)** $y = \dfrac{e^x}{\ln x}$

12 Use the rules of logarithms to expand each of the following functions. Hence write their derivatives.

(a) $y = \ln\left(\dfrac{x}{x-1}\right)$ **(b)** $y = \ln(x\sqrt{(3x - 1)})$ **(c)** $y = \ln\sqrt{\dfrac{x+1}{x-1}}$

13 Find and classify the stationary points of

(a) $y = xe^{-x}$ **(b)** $y = \ln x - x$

Hence sketch their graphs.

14 Find the output needed to maximize profit given that the total cost and total revenue functions are

$$\text{TC} = 2Q \quad \text{and} \quad \text{TR} = 100 \ln(Q + 1)$$

respectively.

15 If a firm's production function is given by

$$Q = 700Le^{-0.02L}$$

find the value of L which maximizes output.

16 The demand function of a good is given by

$$P = 100e^{-0.1Q}$$

Show that demand is unit elastic when $Q = 10$.

17 The growth rate of an economic variable, y, is defined to be $\dfrac{dy}{dt} \div y$.

Use this definition to find the growth rate of the variable, $y = Ae^{kt}$.

chapter 5

Partial Differentiation

This chapter continues the topic of calculus by describing how to differentiate functions of more than one variable. In many ways this chapter can be regarded as the climax of the whole book. It is the summit of the mathematical mountain that we have been merrily climbing. Not only are the associated mathematical ideas and techniques quite sophisticated, but also partial differentiation provides a rich source of applications. In one sense there is no new material presented here. If you know how to differentiate a function of one variable then you also know how to partially differentiate a function of several variables because the rules are the same. Similarly, if you can optimize a function of one variable then you need have no fear of unconstrained and constrained optimization. Of course, if you cannot use the elementary rules of differentiation or cannot find the maximum and minimum values of a function as described in Chapter 4 then you really are fighting a lost cause. Under these circumstances you are best advised to omit this chapter entirely. There is no harm in doing this, because it does not form the prerequisite for any of the later topics. However, you will miss out on one of the most elegant and useful branches of mathematics.

There are six sections to this chapter. It is important that Sections 5.1 and 5.2 are read first, but the remaining sections can be studied in any order. Sections 5.1 and 5.2 follow the familiar pattern. We begin by looking at the mathematical techniques and then use them to determine marginal functions and elasticities. Section 5.3 describes the multiplier concept and completes the topic of statics which you studied in Chapter 1.

The final three sections are devoted to optimization. For functions of several variables, optimization problems are split into two groups, unconstrained and constrained.

Unconstrained problems, tackled in Section 5.4, involve the maximization and minimization of functions in which the variables are free to take any values whatsoever. In a constrained problem only certain combinations of the variables are examined. For example, a firm might wish to minimize costs but is constrained by the need to satisfy production quotas, or an individual might want to maximize utility but is subject to a budgetary constraint, and so on. There are two ways of solving constrained problems: the method of substitution and the method of Lagrange multipliers, described in Sections 5.5 and 5.6 respectively.

section 5.1

Functions of several variables

Objectives

At the end of this section you should be able to:

- Use the function notation, $z = f(x, y)$.
- Determine the first-order partial derivatives, f_x and f_y.
- Determine the second-order partial derivatives, f_{xx}, f_{xy}, f_{yx} and f_{yy}.
- Appreciate that, for most functions, $f_{xy} = f_{yx}$.
- Use the small increments formula.
- Perform implicit differentiation.

Most relationships in economics involve more than two variables. The demand for a good depends not only on its own price but also on the price of substitutable and complementary goods, incomes of consumers, advertising expenditure and so on. Likewise, the output from a production process depends on a variety of inputs, including land, capital and labour. To analyse general economic behaviour we must extend the concept of a function, and particularly the differential calculus, to functions of several variables.

A *function*, f, *of two variables* is a rule that assigns to each incoming pair of numbers, (x, y), a uniquely defined outgoing number, z. This is illustrated in Figure 5.1. The 'black box'

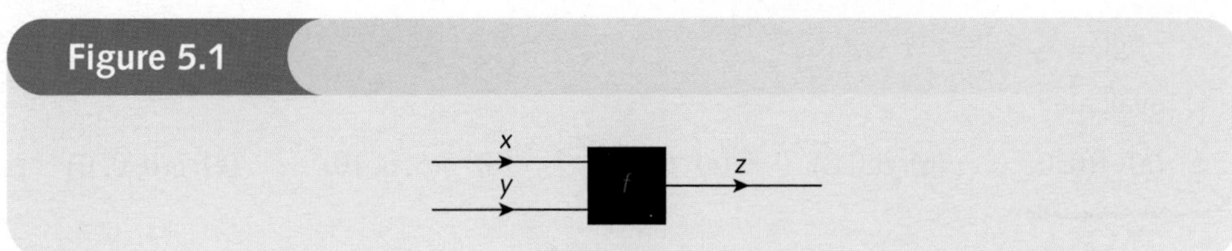

f performs some arithmetic operation on x and y to produce z. For example, the rule might be 'multiply the two numbers together and add twice the second number'. In symbols we write this either as

$$f(x, y) = xy + 2y$$

or as

$$z = xy + 2y$$

In order to be able to evaluate the function we have to specify the numerical values of both x and y.

Example

If $f(x, y) = xy + 2y$ evaluate

(a) $f(3, 4)$ **(b)** $f(4, 3)$

Solution

(a) Substituting $x = 3$ and $y = 4$ gives

$$f(3, 4) = 3(4) + 2(4) = 20$$

(b) Substituting $x = 4$ and $y = 3$ gives

$$f(4, 3) = 4(3) + 2(3) = 18$$

Note that, for this function, $f(3, 4)$ is not the same as $f(4, 3)$, so in general we must be careful to write down the correct ordering of the variables.

We have used the labels x and y for the two incoming numbers (called the ***independent*** variables) and z for the outgoing number (called the ***dependent*** variable). We could equally well have written the above function as

$$y = x_1x_2 + 2x_2$$

say, using x_1 and x_2 to denote the independent variables and using y this time to denote the dependent variable. The use of subscripts may seem rather cumbersome, but it does provide an obvious extension to functions of more than two variables. In general, a function of n variables can be written

$$y = f(x_1, x_2, \ldots, x_n)$$

Practice Problem

1 If

$$f(x, y) = 5x + xy^2 - 10$$

and

$$g(x_1, x_2, x_3) = x_1 + x_2 + x_3$$

evaluate

(a) $f(0, 0)$ **(b)** $f(1, 2)$ **(c)** $f(2, 1)$ **(d)** $g(5, 6, 10)$ **(e)** $g(0, 0, 0)$ **(f)** $g(10, 5, 6)$

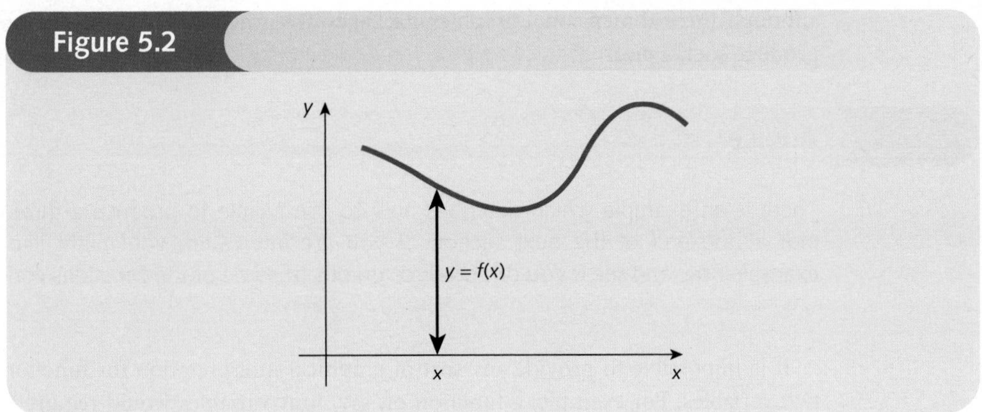

Figure 5.2

A function of one variable can be given a pictorial description using graphs, which help to give an intuitive feel for its behaviour. Figure 5.2 shows the graph of a typical function

$$y = f(x)$$

in which the horizontal axis determines the incoming number, x, and the vertical axis determines the corresponding outgoing number, y. The height of the curve directly above any point on the x axis represents the value of the function at this point.

An obvious question to ask is whether there is a pictorial representation of functions of several variables. The answer is yes in the case of functions of two variables, although it is not particularly easy to construct. A function

$$z = f(x, y)$$

can be thought of as a surface, rather like a mountain range, in three-dimensional space as shown in Figure 5.3. If you visualize the incoming point with coordinates (x, y) as lying in a horizontal plane then the height of the surface, z, directly above it represents the value of the function at this point. As you can probably imagine, it is not an easy task to sketch the surface by hand from an equation such as

$$f(x, y) = xy^3 + 4x$$

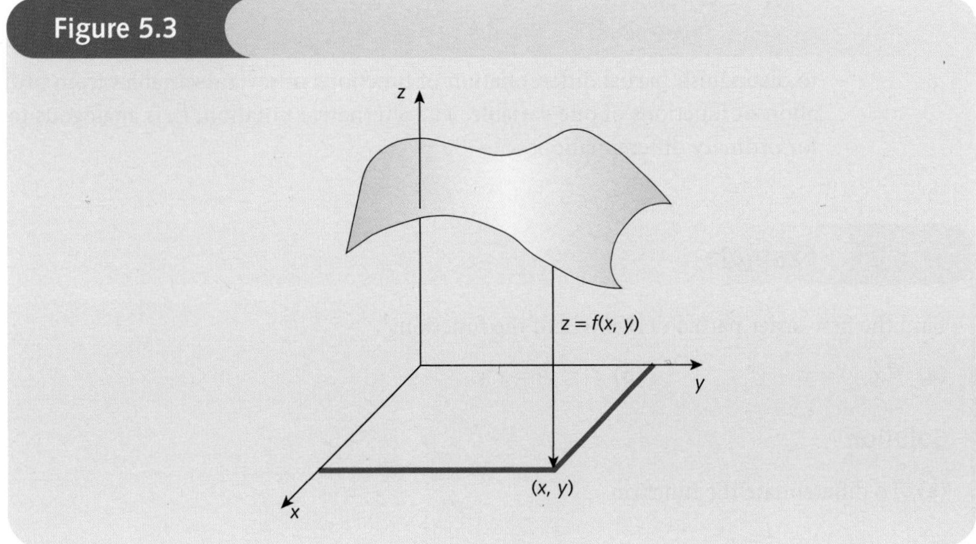

Figure 5.3

although three-dimensional graphics packages are available for most computers which can produce such a plot.

Advice

There is an example which describes how to use Maple to produce a three-dimensional plot at the end of the next section. If you are interested, you might like to read the example now and see if you can produce graphs of some of the functions considered here.

It is impossible to provide any sort of graphical interpretation for functions of more than two variables. For example, a function of, say, four variables would require five dimensions, one for each of the incoming variables and a further one for the outgoing variable! In spite of this setback we can still perform the task of differentiating functions of several variables and, as we shall see in the remaining sections of this chapter, such derivatives play a vital role in analysing economic behaviour.

Given a function of two variables,

$$z = f(x, y)$$

we can determine two first-order derivatives. The *partial derivative of* f *with respect to* x is written as

$$\frac{\partial z}{\partial x} \quad \text{or} \quad \frac{\partial f}{\partial x} \quad \text{or} \quad f_x$$

and is found by differentiating f with respect to x, with y held constant. Similarly, the *partial derivative of* f *with respect to* y is written as

$$\frac{\partial z}{\partial y} \quad \text{or} \quad \frac{\partial f}{\partial y} \quad \text{or} \quad f_y$$

and is found by differentiating f with respect to y, with x held constant. We use curly dees in the notation

$$\frac{\partial f}{\partial x}$$

read 'partial dee *f* by dee *x*'

to distinguish partial differentiation of functions of several variables from ordinary differentiation of functions of one variable. The alternative notation, f_x, is analogous to the f' notation for ordinary differentiation.

Example

Find the first-order partial derivatives of the functions

(a) $f(x, y) = x^2 + y^3$ **(b)** $f(x, y) = x^2y$

Solution

(a) To differentiate the function

$$f(x, y) = x^2 + y^3$$

with respect to x we work as follows. By the sum rule we know that we can differentiate each part separately and add. Now, when we differentiate x^2 with respect to x we get $2x$. However, when we differentiate y^3 with respect to x we get 0. To see this, note from the definition of partial differentiation with respect to x that the variable y is held constant. Of course, if y is a constant then so is y^3 and, as we discovered in Chapter 4, constants differentiate to zero. Hence

$$\frac{\partial f}{\partial x} = 2x + 0 = 2x$$

In the same way

$$\frac{\partial f}{\partial y} = 0 + 3y^2 = 3y^2$$

This time x is held constant, so x^2 goes to zero, and when we differentiate y^3 with respect to y we get $3y^2$.

(b) To differentiate the function

$$f(x, y) = x^2y$$

with respect to x, we differentiate in the normal way, taking x as the variable while pretending that y is a constant. Now, when we differentiate a constant multiple of x^2 we differentiate x^2 to get $2x$ and then multiply by the constant. For example,

$7x^2$ differentiates to $7(2x) = 14x$

$-100x^2$ differentiates to $-100(2x) = -200x$

and

cx^2 differentiates to $c(2x) = 2cx$

for any constant c. In our case, y, plays the role of a constant, so

x^2y differentiates to $(2x)y = 2xy$

Hence

$$f_x = 2xy$$

Similarly, to find f_y we treat y as the variable and x as a constant in the expression

$$f(x, y) = x^2y$$

Now, when we differentiate a constant multiple of y we just get the constant, so cy differentiates to c. In our case, x^2 plays the role of c, so x^2y differentiates to x^2. Hence

$$f_y = x^2$$

Practice Problem

2 Find expressions for the first-order partial derivatives for the functions

(a) $f(x, y) = 5x^4 - y^2$ **(b)** $f(x, y) = x^2y^3 - 10x$

In general, when we differentiate a function of two variables, the thing we end up with is itself a function of two variables. This suggests the possibility of differentiating a second time. In fact there are four ***second-order partial derivatives***. We write

$$\frac{\partial^2 z}{\partial x^2} \quad \text{or} \quad \frac{\partial^2 f}{\partial x^2} \quad \text{or} \quad f_{xx}$$

for the function obtained by differentiating twice with respect to x,

$$\frac{\partial^2 z}{\partial y^2} \quad \text{or} \quad \frac{\partial^2 f}{\partial y^2} \quad \text{or} \quad f_{yy}$$

for the function obtained by differentiating twice with respect to y,

$$\frac{\partial^2 z}{\partial y \partial x} \quad \text{or} \quad \frac{\partial^2 f}{\partial y \partial x} \quad \text{or} \quad f_{yx}$$

for the function obtained by differentiating first with respect to x and then with respect to y, and

$$\frac{\partial^2 z}{\partial x \partial y} \quad \text{or} \quad \frac{\partial^2 f}{\partial x \partial y} \quad \text{or} \quad f_{xy}$$

for the function obtained by differentiating first with respect to y and then with respect to x.

Example

Find expressions for the second-order partial derivatives f_{xx}, f_{yy}, f_{yx} and f_{xy} for the functions

(a) $f(x, y) = x^2 + y^3$ **(b)** $f(x, y) = x^2y$

Solution

(a) The first-order partial derivatives of the function

$$f(x, y) = x^2 + y^3$$

have already been found and are given by

$$f_x = 2x, f_y = 3y^2$$

To find f_{xx} we differentiate f_x with respect to x to get

$$f_{xx} = 2$$

To find f_{yy} we differentiate f_y with respect to y to get

$$f_{yy} = 6y$$

To find f_{yx} we differentiate f_x with respect to y to get

$$f_{yx} = 0$$

Note how f_{yx} is obtained. Starting with the original function

$$f(x, y) = x^2 + y^3$$

we first differentiate with respect to x to get $2x$ and when we differentiate this with respect to y we keep x constant, so it goes to zero. Finally, to find f_{xy} we differentiate f_y with respect to x to get

$$f_{xy} = 0$$

Note how f_{xy} is obtained. Starting with the original function

$$f(x, y) = x^2 + y^3$$

we first differentiate with respect to y to get $3y^2$ and when we differentiate this with respect to x we keep y constant, so it goes to zero.

(b) The first-order partial derivatives of the function

$$f(x, y) = x^2y$$

have already been found and are given by

$$f_x = 2xy, \quad f_y = x^2$$

Hence

$$f_{xx} = 2y, \quad f_{yy} = 0, \quad f_{yx} = 2x, \quad f_{xy} = 2x$$

Practice Problem

3 Find expressions for the second-order partial derivatives of the functions

(a) $f(x, y) = 5x^4 - y^2$ **(b)** $f(x, y) = x^2y^3 - 10x$

[Hint: you might find your answer to Practice Problem 2 useful.]

Looking back at the expressions obtained in the previous example and Practice Problem 3, notice that in all cases

$$\boxed{\frac{\partial^2 f}{\partial y \partial x} = \frac{\partial^2 f}{\partial x \partial y}}$$

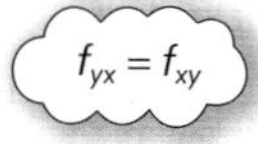

It can be shown that this result holds for all functions that arise in economics. It is immaterial in which order the partial differentiation is performed. Differentiating with respect to x then y gives the same expression as differentiating with respect to y then x. (In fact, there are some weird mathematical functions for which this result is not true, although they need not concern us.)

Although we have concentrated exclusively on functions of two variables, it should be obvious how to work out partial derivatives of functions of more than two variables. For the general function

$$y = f(x_1, x_2, \ldots, x_n)$$

there are n first-order partial derivatives, written as

$$\frac{\partial f}{\partial x_i} \quad \text{or} \quad f_i \quad (i = 1, 2, \ldots, n)$$

which are found by differentiating with respect to one variable at a time, keeping the remaining $n - 1$ variables fixed. The second-order partial derivatives are determined in an analogous way.

Example

Find the derivative, f_{31}, for the function

$$f(x_1, x_2, x_3) = x_1^3 + x_1x_3^2 + 5x_2^4$$

➔

Solution

We need to find

$$f_{31} = \frac{\partial^2 f}{\partial x_3 \partial x_1}$$

which denotes the function obtained by differentiating first with respect to x_1 and then with respect to x_3. Differentiating with respect to x_1 gives

$$f_1 = \frac{\partial f}{\partial x_1} = 3x_1^2 + x_3^2$$

and if we further differentiate this with respect to x_3 we get

$$f_{31} = \frac{\partial^2 f}{\partial x_3 \partial x_1} = 2x_3$$

In fact, as we have just noted for functions of two variables, we get the same answer if we differentiate in reverse order. You might like to check this for yourself.

Practice Problem

4 Find expressions for the partial derivatives f_1, f_{11} and f_{21} in the case when

$$f(x_1, x_2, x_3) = x_1x_2 + x_1^5 - x_2^2x_3$$

We have seen how to work out partial derivatives but have yet to give any meaning to them. To provide an interpretation of a partial derivative, let us take one step back for a moment and recall the corresponding situation for functions of one variable of the form

$$y = f(x)$$

The derivative, dy/dx, gives the rate of change of y with respect to x. In other words, if x changes by a small amount Δx then the corresponding change in y satisfies

$$\Delta y \simeq \frac{dy}{dx}\Delta x$$

Moreover, the accuracy of the approximation improves as Δx becomes smaller and smaller.

Advice

You might like to remind yourself of the reasoning behind this approximation, which was explained graphically in Section 4.3.1.

Given the way in which a partial derivative is found, we can deduce that for a function of two variables

$$z = f(x, y)$$

if x changes by a small amount Δx and y is held fixed then the corresponding change in z satisfies

$$\Delta z \simeq \frac{\partial z}{\partial x}\Delta x$$

Similarly, if y changes by Δy and x is fixed then z changes by

$$\Delta z \simeq \frac{\partial z}{\partial y}\Delta y$$

In practice, of course, x and y may both change simultaneously. If this is the case then the net change in z will be the sum of the individual changes brought about by changes in x and y separately, so that

$$\boxed{\Delta z \simeq \frac{\partial z}{\partial x}\Delta x + \frac{\partial z}{\partial y}\Delta y}$$

This is referred to as the *small increments formula*. Although this is only an approximation, it can be shown that for most functions the corresponding error tends to zero as Δx and Δy both tend to zero. For this reason the formula is sometimes quoted with an equality sign and written as

$$dz = \frac{\partial z}{\partial x}dx + \frac{\partial z}{\partial y}dy$$

where the symbols dx, dy and dz are called *differentials* and represent limiting values of Δx, Δy and Δz, respectively.

Example

If

$$z = x^3y - y^3x$$

evaluate

$$\frac{\partial z}{\partial x} \quad \text{and} \quad \frac{\partial z}{\partial y}$$

at the point (1, 3). Hence estimate the change in z when x increases from 1 to 1.1 and y decreases from 3 to 2.8 simultaneously.

Solution

If

$$z = x^3y - y^3x$$

then

$$\partial z/\partial x = 3x^2y - y^3 \quad \text{and} \quad \partial z/\partial y = x^3 - 3y^2x$$

so at the point (1, 3)

$$\frac{\partial z}{\partial x} = 3(1)^2(3) - 3^3 = -18$$

$$\frac{\partial z}{\partial y} = 1^3 - 3(3)^2(1) = -26$$

Now, since x increases from 1 to 1.1, the change in x is

$$\Delta x = 0.1$$

positive numbers denote increases

→

and, since y decreases from 3 to 2.8, the change in y is

$$\Delta y = -0.2$$

negative numbers denote decreases

The small increments formula states that

$$\Delta z \simeq \frac{\partial z}{\partial x}\Delta x + \frac{\partial z}{\partial y}\Delta y$$

The change in z is therefore

$$\Delta z \simeq (-18)(0.1) + (-26)(-0.2) = 3.4$$

so z increases by approximately 3.4.

Practice Problem

5 If

$$z = xy - 5x + 2y$$

evaluate

$$\frac{\partial z}{\partial x} \text{ and } \frac{\partial z}{\partial y}$$

at the point (2, 6).

(a) Use the small increments formula to estimate the change in z as x decreases from 2 to 1.9 and y increases from 6 to 6.1.

(b) Confirm your estimate of part (a) by evaluating z at (2, 6) and (1.9, 6.1).

One important application of the small increments formula is to implicit differentiation. We hope by now that you are entirely happy differentiating functions of one variable such as

$$y = x^3 + 2x^2 + 5$$

$\frac{dy}{dx} = 3x^2 + 4x$

Suppose, however, that you are asked to find dy/dx given the equation

$$y^3 + 2xy^2 - x = 5$$

This is much more difficult. The reason for the difference is that in the first case y is given explicitly in terms of x whereas in the second case the functional dependence of y on x is only given implicitly. You would need to somehow rearrange this equation and to write y in terms of x before you could differentiate it. Unfortunately, this is an impossible task because of the presence of the y^3 term. The trick here is to regard the expression on the left-hand side of the equation as a function of the two variables x and y, so that

$$f(x, y) = y^3 + 2xy^2 - x$$

or equivalently

$$z = y^3 + 2xy^2 - x$$

The equation

$$y^3 + 2xy^2 - x = 5$$

then reads

$$z = 5$$

In general, the differential form of the small increments formula states that

$$dz = \frac{\partial z}{\partial x}dx + \frac{\partial z}{\partial y}dy$$

In our particular case, z takes the constant value of 5, so does not change. Hence $dz = 0$ and the formula reduces to

$$0 = \frac{\partial z}{\partial x}dx + \frac{\partial z}{\partial y}dy$$

which rearranges as

$$\frac{\partial z}{\partial y}dy = -\frac{\partial z}{\partial x}dx$$

that is,

$$\frac{dy}{dx} = -\frac{\partial z/\partial x}{\partial z/\partial y}$$

This formula can be used to find dy/dx given any implicit function

$$f(x, y) = \text{constant}$$

that is,

$$\boxed{\text{if } f(x, y) = \text{constant} \quad \text{then} \quad \frac{dy}{dx} = -\frac{f_x}{f_y}}$$

The technique of finding dy/dx from $-f_x/f_y$ is called ***implicit differentiation*** and can be used whenever it is difficult or impossible to obtain an explicit representation for y in terms of x.

Example

Use implicit differentiation to find an expression for dy/dx given that

$$y^3 + 2xy^2 - x = 5$$

Solution

For the function

$$f(x, y) = y^3 + 2xy^2 - x$$

we have

$$f_x = 2y^2 - 1 \quad \text{and} \quad f_y = 3y^2 + 4xy$$

so that

$$\frac{dy}{dx} = -\frac{f_x}{f_y} = -\left(\frac{2y^2 - 1}{3y^2 + 4xy}\right) = \frac{-2y^2 + 1}{3y^2 + 4xy}$$

Advice

There is an alternative way of thinking about implicit differentiation which is based on the chain rule and does not depend on partial differentiation at all. This is described in Appendix 2. You might find it easier to use than the method described above.

Practice Problem

6 Use implicit differentiation to find expressions for dy/dx given that

(a) $xy - y^3 + y = 0$ **(b)** $y^5 - xy^2 = 10$

Key Terms

Dependent variable A variable whose value is determined by that taken by the independent variables; in $z = f(x, y)$, the dependent variable is z.

Differentials Limiting values of incremental changes. In the limit, the approximation $\Delta z \simeq \frac{\partial z}{\partial x} \times \Delta x$ becomes $dz \simeq \frac{\partial z}{\partial x} \times dx$ where dz and dx are the differentials.

Function of two variables A rule that assigns to each pair of incoming numbers, x and y, a uniquely defined outgoing number, z.

Implicit differentiation The process of obtaining dy/dx where the function is not given explicitly as an expression for y in terms of x.

Independent variable Variables whose values determine that of the dependent variable; in $z = f(x, y)$, the independent variables are x and y.

Partial derivative The derivative of a function of two or more variables with respect to one of these variables, the others being regarded as constant.

Second-order partial derivative The partial derivative of a first-order partial derivative. For example, f_{xy} is the second-order partial derivative when f is differentiated first with respect to y and then with respect to x.

Small increments formula The result $\Delta z \simeq \frac{\partial z}{\partial x}\Delta x + \frac{\partial z}{\partial y}\Delta y$

Symmetric function A function of two or more variables which is unchanged by any permutation of the variables. A function of two variables is symmetric when $f(x, y) = f(y, x)$

Practice Problems

7 If

$$f(x, y) = 3x^2y^3$$

evaluate $f(2, 3)$, $f(5, 1)$ and $f(0, 7)$.

8 If

$$f(x, y) = 2xy + 3x$$

verify that $f(5, 7) \neq f(7, 5)$. Are there any pairs of numbers, (x, y) for which $f(x, y) = f(y, x)$?

9 Write down expressions for the first-order partial derivatives, $\frac{\partial z}{\partial x}$ and $\frac{\partial z}{\partial y}$ for

(a) $z = x^2 + 4y^5$ **(b)** $z = 3x^3 - 2e^y$ **(c)** $z = xy + 6y$ **(d)** $z = x^6y^2 + 5y^3$

10 If

$$f(x, y) = x^4y^5 - x^2 + y^2$$

write down expressions for the first-order partial derivatives, f_x and f_y. Hence evaluate $f_x(1, 0)$ and $f_y(1, 1)$.

11 Find expressions for all first- and second-order partial derivatives of the following functions. In each case verify that

$$\frac{\partial^2 z}{\partial y \partial x} = \frac{\partial^2 z}{\partial x \partial y}$$

(a) $z = xy$ **(b)** $z = e^x y$ **(c)** $z = x^2 + 2x + y$ **(d)** $z = 16x^{1/4}y^{3/4}$ **(e)** $z = \frac{y}{x^2} + \frac{x}{y}$

12 Use the small increments formula to estimate the change in

$$z = x^2y^4 - x^6 + 4y$$

when

(a) x increases from 1 to 1.1 and y remains fixed at 0

(b) x remains fixed at 1 and y decreases from 0 to –0.5

(c) x increases from 1 to 1.1 and y decreases from 0 to –0.5

13 If

$$z = x^2y^3 - 10xy + y^2$$

evaluate z_x and z_y at the point (2, 3). Hence estimate the change in z as x increases by 0.2 and y decreases by 0.1.

14 **(a)** If

$$f(x, y) = y - x^3 + 2x$$

write down expressions for f_x and f_y. Hence use implicit differentiation to find dy/dx given that

$$y - x^3 + 2x = 1$$

(b) Confirm your answer to part (a) by rearranging the equation

$$y - x^3 + 2x = 1$$

to give y explicitly in terms of x and using ordinary differentiation.

15 Verify that $x = 1$, $y = -1$ satisfy the equation $x^2 - 2y^3 = 3$. Use implicit differentiation to find the value of dy/dx at this point.

16 A function of three variables is given by

$$f(x_1, x_2, x_3) = \frac{x_1 x_3^2}{x_2} + \ln(x_2 x_3)$$

Find all of the first- and second-order derivatives of this function and verify that

$$f_{12} = f_{21}, \quad f_{13} = f_{31} \quad \text{and} \quad f_{23} = f_{32}$$

section 5.2

Partial elasticity and marginal functions

Objectives

At the end of this section you should be able to:

- Calculate partial elasticities.
- Calculate marginal utilities.
- Calculate the marginal rate of commodity substitution along an indifference curve.
- Calculate marginal products.
- Calculate the marginal rate of technical substitution along an isoquant.
- State Euler's theorem for homogeneous production functions.

The first section of this chapter described the technique of partial differentiation. Hopefully, you have discovered that partial differentiation is no more difficult than ordinary differentiation. The only difference is that for functions of several variables you have to be clear at the outset which letter in a mathematical expression is to be the variable, and to bear in mind that all remaining letters are then just constants in disguise! Once you have done this, the actual differentiation itself obeys the usual rules. In Sections 4.3 and 4.5 we considered various microeconomic applications. Given the intimate relationship between ordinary and partial differentiation, you should not be too surprised to learn that we can extend these applications to functions of several variables. We concentrate on three main areas:

- elasticity of demand
- utility
- production.

We consider each of these in turn.

5.2.1 Elasticity of demand

Suppose that the demand, Q, for a certain good depends on its price, P, the price of an alternative good, P_A, and the income of consumers, Y, so that

$$Q = f(P, P_A, Y)$$

for some demand function, f.

Of particular interest is the responsiveness of demand to changes in any one of these three variables. This can be measured quantitatively using elasticity. The *(own) price elasticity of demand* is defined to be

$$E_P = -\frac{\text{percentage change in } Q}{\text{percentage change in } P}$$

with P_A and Y held constant. This definition is identical to the one given in Section 4.5, so following the same mathematical argument presented there we deduce that

$$E_P = -\frac{P}{Q} \times \frac{\partial Q}{\partial P}$$

The partial derivative notation is used here because Q is now a function of several variables, and P_A and Y are held constant.

Advice

You may recall that the introduction of the minus sign is an artificial device designed to make E_P positive. This policy is not universal and you are advised to check which convention your own tutor uses.

In an analogous way we can measure the responsiveness of demand to changes in the price of the alternative good. The *cross-price elasticity of demand* is defined to be

$$E_{P_A} = \frac{\text{percentage change in } Q}{\text{percentage change in } P_A}$$

with P and Y held constant. Again, the usual mathematical argument shows that

$$E_{P_A} = \frac{P_A}{Q} \times \frac{\partial Q}{\partial P_A}$$

The sign of E_{P_A} could turn out to be positive or negative depending on the nature of the alternative good. If the alternative good is substitutable then Q increases as P_A rises, because consumers buy more of the given good as it becomes relatively less expensive. Consequently,

$$\frac{\partial Q}{\partial P_A} > 0$$

and so $E_{P_A} > 0$. If the alternative good is complementary then Q decreases as P_A rises, because the bundle of goods as a whole becomes more expensive. Consequently,

$$\frac{\partial Q}{\partial P_A} < 0$$

and so $E_{P_A} < 0$.

Finally, the *income elasticity of demand* is defined to be

$$E_Y = \frac{\text{percentage change in } Q}{\text{percentage change in } Y}$$

and can be found from

$$E_Y = \frac{Y}{Q} \times \frac{\partial Q}{\partial Y}$$

Again, E_Y can be positive or negative. If the good is superior then demand rises as income rises and E_Y is positive. However, if the good is inferior then demand falls as income rises and E_Y is negative.

Example

Given the demand function

$$Q = 100 - 2P + P_A + 0.1Y$$

where $P = 10$, $P_A = 12$ and $Y = 1000$, find the

(a) price elasticity of demand

(b) cross-price elasticity of demand

(c) income elasticity of demand

Is the alternative good substitutable or complementary?

Solution

We begin by calculating the value of Q when $P = 10$, $P_A = 12$ and $Y = 1000$. The demand equation gives

$$Q = 100 - 2(10) + 12 + 0.1(1000) = 192$$

(a) To find the price elasticity of demand we partially differentiate

$$Q = 100 - 2P + P_A + 0.1Y$$

with respect to P to get

$$\frac{\partial Q}{\partial Y} = -2$$

Hence

$$E_P = -\frac{P}{Q} \times \frac{\partial Q}{\partial P} = -\frac{10}{192} \times (-2) = 0.10$$

(b) To find the cross-price elasticity of demand we partially differentiate

$$Q = 100 - 2P + P_A + 0.1Y$$

with respect to P_A to get

$$\frac{\partial Q}{\partial P_A} = 1$$

Hence

$$E_{P_A} = \frac{P_A}{Q} \times \frac{\partial Q}{\partial P_A} = \frac{12}{192} \times 1 = 0.06$$

The fact that this is positive shows that the two goods are substitutable.

(c) To find the income elasticity of demand we partially differentiate

$$Q = 100 - 2P + P_A + 0.1Y$$

with respect to Y to get

$$\frac{\partial Q}{\partial Y} = 0.1$$

Hence

$$E_Y = \frac{Y}{Q} \times \frac{\partial Q}{\partial Y} = \frac{1000}{192} \times 0.1 = 0.52$$

Practice Problem

1 Given the demand function

$$Q = 500 - 3P - 2P_A + 0.01Y$$

where $P = 20$, $P_A = 30$ and $Y = 5000$, find

(a) the price elasticity of demand

(b) the cross-price elasticity of demand

(c) the income elasticity of demand

If income rises by 5%, calculate the corresponding percentage change in demand. Is the good inferior or superior?

5.2.2 Utility

So far in this book we have concentrated almost exclusively on the behaviour of producers. In this case it is straightforward to identify the primary objective, which is to maximize profit. We now turn our attention to consumers. Unfortunately, it is not so easy to identify the motivation for their behaviour. One tentative suggestion is that consumers try to maximize earned income. However, if this were the case then individuals would try to work 24 hours a day for 7 days a week, which is not so. In practice, people like to allocate a reasonable proportion of time to leisure activities.

Consumers are faced with a choice of how many hours each week to spend working and how many to devote to leisure. In the same way, a consumer needs to decide how many items of various goods to buy and has a preference between the options available. To analyse the behaviour of consumers quantitatively we associate with each set of options a number, U, called *utility*, which indicates the level of satisfaction. Suppose that there are two goods, G1 and G2, and that the consumer buys x_1 items of G1 and x_2 items of G2. The variable U is then a function of x_1 and x_2, which we write as

$$U = U(x_1, x_2)$$

If

$$U(3, 7) = 20 \quad \text{and} \quad U(4, 5) = 25$$

for example, then the consumer derives greater satisfaction from buying 4 items of G1 and 5 items of G2 than from buying 3 items of G1 and 7 items of G2.

Utility is a function of two variables, so we can work out two first-order partial derivatives,

$$\frac{\partial U}{\partial x_1} \quad \text{and} \quad \frac{\partial U}{\partial x_2}$$

The derivative

$$\frac{\partial U}{\partial x_1}$$

gives the rate of change of U with respect to x_i and is called the *marginal utility of* x_i. If x_i changes by a small amount Δx_i and the other variable is held fixed then the change in U satisfies

$$\Delta U \simeq \frac{\partial U}{\partial x_i}\Delta x_i$$

If x_1 and x_2 both change then the net change in U can be found from the small increments formula

$$\Delta U \simeq \frac{\partial U}{\partial x_1}\Delta x_1 + \frac{\partial U}{\partial x_2}\Delta x_2$$

Example

Given the utility function

$$U = x_1^{1/4}x_2^{3/4}$$

determine the value of the marginal utilities

$$\frac{\partial U}{\partial x_1} \quad \text{and} \quad \frac{\partial U}{\partial x_2}$$

when $x_1 = 100$ and $x_2 = 200$. Hence estimate the change in utility when x_1 decreases from 100 to 99 and x_2 increases from 200 to 201.

Solution

If

$$U = x_1^{1/4}x_2^{3/4}$$

then

$$\frac{\partial U}{\partial x_1} = \tfrac{1}{4}x_1^{-3/4}x_2^{3/4} \quad \text{and} \quad \frac{\partial U}{\partial x_2} = \tfrac{3}{4}x_1^{1/4}x_2^{-1/4}$$

so when $x_1 = 100$ and $x_2 = 200$

$$\frac{\partial U}{\partial x_1} = \tfrac{1}{4}(100)^{-3/4}(200)^{3/4} = 0.42$$

$$\frac{\partial U}{\partial x_2} = \tfrac{3}{4}(100)^{1/4}(200)^{-1/4} = 0.63$$

Now x_1 decreases by 1 unit, so

$$\Delta x_1 = -1$$

and x_2 increases by 1 unit, so

$$\Delta x_2 = 1$$

The small increments formula states that

$$\Delta U \simeq \frac{\partial U}{\partial x_1}\Delta x_1 + \frac{\partial U}{\partial x_2}\Delta x_2$$

The change in utility is therefore

$$\Delta U \simeq (0.42)(-1) + (0.63)(1) = 0.21$$

Note that for the particular utility function

$$U = x_1^{1/4}x_2^{3/4}$$

given in the above example, the second-order derivatives

$$\frac{\partial^2 U}{\partial x_1^2} = \frac{-3}{16}x_1^{-7/4}x_2^{3/4} \quad \text{and} \quad \frac{\partial^2 U}{\partial x_2^2} = \frac{-3}{16}x_1^{1/4}x_2^{-5/4}$$

are both negative. Now $\partial^2 U/\partial x_1^2$ is the partial derivative of marginal utility $\partial U/\partial x_1$ with respect to x_1. The fact that this is negative means that marginal utility of x_1 decreases as x_1 rises. In other words, as the consumption of good G1 increases, each additional item of G1 bought confers less utility than the previous item. A similar property holds for G2. This is known as the ***law of diminishing marginal utility***.

Advice

You might like to compare this with the law of diminishing marginal productivity discussed in Section 4.3.2.

Practice Problem

2 An individual's utility function is given by

$$U = 1000x_1 + 450x_2 + 5x_1x_2 - 2x_1^2 - x_2^2$$

where x_1 is the amount of leisure measured in hours per week and x_2 is earned income measured in dollars per week.

Determine the value of the marginal utilities

$$\frac{\partial U}{\partial x_1} \text{ and } \frac{\partial U}{\partial x_2}$$

when $x_1 = 138$ and $x_2 = 500$.

Hence estimate the change in U if the individual works for an extra hour, which increases earned income by \$15 per week.

Does the law of diminishing marginal utility hold for this function?

It was pointed out in Section 5.1 that functions of two variables could be represented by surfaces in three dimensions. This is all very well in theory, but in practice the task of sketching such a surface by hand is virtually impossible. This difficulty has been faced by geographers for years and the way they circumvent the problem is to produce a two-dimensional contour map. A contour is a curve joining all points at the same height above sea level. Exactly the same device can be used for utility functions. Rather than attempt to sketch the surface, we draw an *indifference map*. This consists of *indifference curves* joining points (x_1, x_2) which give the same value of utility. Mathematically, an indifference curve is defined by an equation

$$U(x_1, x_2) = U_0$$

for some fixed value of U_0. A typical indifference map is sketched in Figure 5.4.

Points A and B both lie on the lower indifference curve, $U_0 = 20$. Point A corresponds to the case when the consumer buys a_1 units of G1 and a_2 units of G2. Likewise, point B corresponds

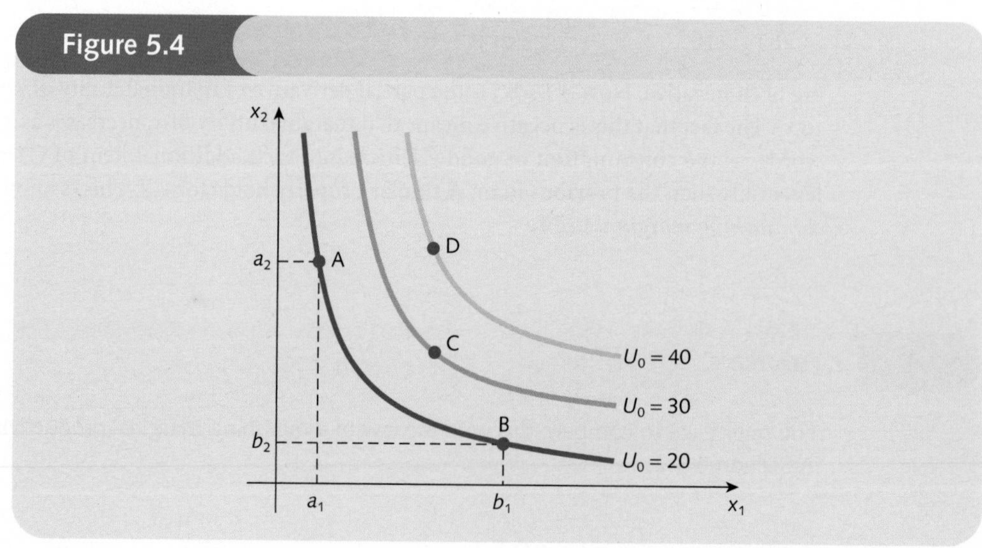

to the case when the consumer buys b_1 units of G1 and b_2 units of G2. Both of these combinations yield the same level of satisfaction and the consumer is indifferent to choosing between them. In symbols we have

$$U(a_1, a_2) = 20 \quad \text{and} \quad U(b_1, b_2) = 20$$

Points C and D lie on indifference curves that are further away from the origin. The combinations of goods that these points represent yield higher levels of utility and so are ranked above those of A and B.

Indifference curves are usually downward-sloping. If fewer purchases are made of G1 then the consumer has to compensate for this by buying more of type G2 to maintain the same level of satisfaction. Note also from Figure 5.4 that the slope of an indifference curve varies along its length, taking large negative values close to the vertical axis and becoming almost zero as the curve approaches the horizontal axis. Again this is to be expected for any function that obeys the law of diminishing marginal utility. A consumer who currently owns a large number of items of G2 and relatively few of G1 is likely to value G1 more highly. Consequently, he or she might be satisfied in sacrificing a large number of items of G2 to gain just one or two extra items of G1. In this region the marginal utility of x_1 is much greater than that of x_2, which accounts for the steepness of the curve close to the vertical axis. Similarly, as the curve approaches the horizontal axis, the situation is reversed and the curve flattens off. We quantify this exchange of goods by introducing the ***marginal rate of commodity substitution***, MRCS. This is defined to be the increase in x_2 necessary to maintain a constant value of utility when x_1 decreases by 1 unit. This is illustrated in Figure 5.5.

Starting at point E, we move 1 unit to the left. The value of MRCS is then the vertical distance that we need to travel if we are to remain on the indifference curve passing through E. Now this sort of '1 unit change' definition is precisely the approach that we took in Section 4.3 when discussing marginal functions. In that section we actually defined the marginal function to be the derived function and we showed that the '1 unit change' definition gave a good approximation to it. If we do the same here then we can define

$$\text{MRCS} = -\frac{dx_2}{dx_1}$$

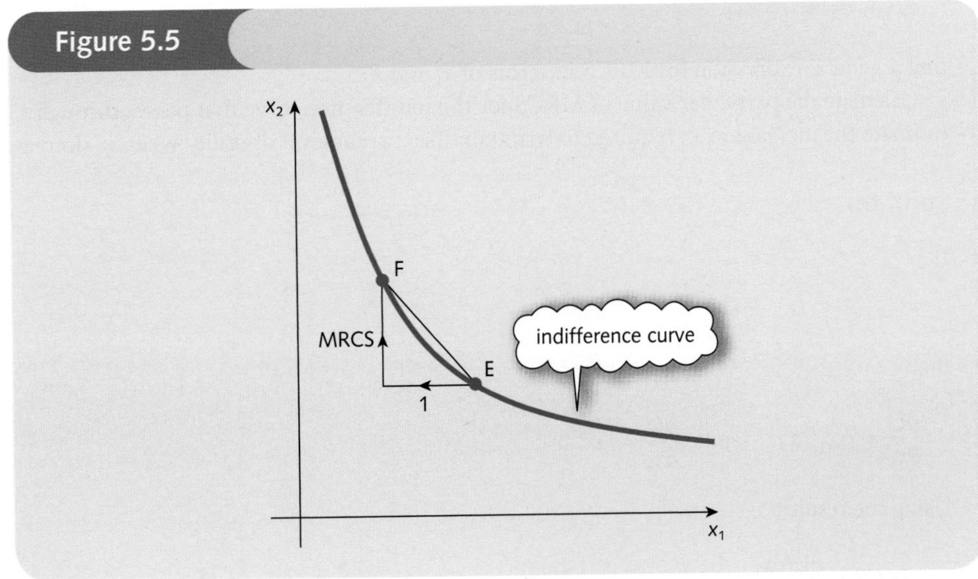

The derivative, dx_2/dx_1, determines the slope of an indifference curve when x_1 is plotted on the horizontal axis and x_2 is plotted on the vertical axis. This is negative, so we deliberately put a minus sign in front to make MRCS positive. This definition is useful only if we can find the equation of an indifference curve with x_2 given explicitly in terms of x_1. However, we may only know the utility function

$$U(x_1, x_2)$$

so that the indifference curve is determined implicitly from an equation

$$U(x_1, x_2) = U_0$$

This is precisely the situation that we discussed at the end of Section 5.1. The formula for implicit differentiation gives

$$\frac{dx_2}{dx_1} = -\frac{\partial U/\partial x_1}{\partial U/\partial x_2}$$

Hence

$$\text{MRCS} = -\frac{dx_2}{dx_1} = \frac{\partial U/\partial x_1}{\partial U/\partial x_2}$$

marginal rate of commodity substitution is the marginal utility of x_1 divided by the marginal utility of x_2

Example

Given the utility function

$$U = x_1^{1/2}x_2^{1/2}$$

find a general expression for MRCS in terms of x_1 and x_2.

Calculate the particular value of MRCS for the indifference curve that passes through (300, 500). Hence estimate the increase in x_2 required to maintain the current level of utility when x_1 decreases by 3 units.

Solution

If

$$U = x_1^{1/2}x_2^{1/2}$$

then

$$\frac{\partial U}{\partial x_1} = \tfrac{1}{2}x_1^{-1/2}x_2^{1/2} \quad \text{and} \quad \frac{\partial U}{\partial x_2} = \tfrac{1}{2}x_1^{1/2}x_2^{-1/2}$$

Using the result

$$\text{MRCS} = \frac{\partial U/\partial x_1}{\partial U/\partial x_2}$$

we see that

$$\text{MRCS} = \frac{\frac{1}{2}x_1^{-1/2}x_2^{1/2}}{\frac{1}{2}x_1^{1/2}x_2^{-1/2}}$$

$$= x_1^{-1}x_2^{1}$$

rule 2 of indices;
$b^m \div b^n = b^{m-n}$

$$= \frac{x_2}{x_1}$$

$b^1 = b$,
$b^{-1} = \frac{1}{b}$

At the point (300, 500)

$$\text{MRCS} = \frac{500}{300} = \frac{5}{3}$$

Now MRCS approximates the increase in x_2 required to maintain a constant level of utility when x_1 decreases by 1 unit. In this example x_1 decreases by 3 units, so we multiply MRCS by 3. The approximate increase in x_2 is

$$\frac{5}{3} \times 3 = 5$$

We can check the accuracy of this approximation by evaluating U at the old point (300, 500) and the new point (297, 505). We get

$$U(300, 500) = (300)^{1/2}(500)^{1/2} = 387.30$$

$$U(297, 505) = (297)^{1/2}(505)^{1/2} = 387.28$$

This shows that, to all intents and purposes, the two points do indeed lie on the same indifference curve.

Practice Problem

3 Calculate the value of MRCS for the utility function given in Practice Problem 2 at the point (138, 500). Hence estimate the increase in earned income required to maintain the current level of utility if leisure time falls by 2 hours per week.

5.2.3 Production

Production functions were first introduced in Section 2.3. We assume that output, Q, depends on capital, K, and labour, L, so we can write

$$Q = f(K, L)$$

Such functions can be analysed in a similar way to utility functions. The partial derivative

$$\frac{\partial Q}{\partial K}$$

gives the rate of change of output with respect to capital and is called the ***marginal product of capital***, MP_K. If capital changes by a small amount ΔK, with labour held constant, then the corresponding change in Q is given by

$$\Delta Q \simeq \frac{\partial Q}{\partial K}\Delta K$$

Similarly,

$$\frac{\partial Q}{\partial L}$$

gives the rate of change of output with respect to labour and is called the ***marginal product of labour***, MP_L. If labour changes by a small amount ΔL, with capital held constant, then the corresponding change in Q is given by

$$\Delta Q \simeq \frac{\partial Q}{\partial L}\Delta L$$

If K and L both change simultaneously, then the net change in Q can be found from the small increments formula

$$\Delta Q \simeq \frac{\partial Q}{\partial K}\Delta K + \frac{\partial Q}{\partial L}\Delta L$$

The contours of a production function are called ***isoquants***. In Greek 'iso' means 'equal', so the word 'isoquant' literally translates as 'equal quantity'. Points on an isoquant represent all possible combinations of inputs (K, L) which produce a constant level of output, Q_0. A typical isoquant map is sketched in Figure 5.6. Notice that we have adopted the standard convention of plotting labour on the horizontal axis and capital on the vertical axis.

The lower curve determines the input pairs needed to output 100 units. Higher levels of output correspond to isoquants further away from the origin. Again, the general shape of the curves is to be expected. For example, as capital is reduced it is necessary to increase labour to compensate and so maintain production levels. Moreover, if capital continues to decrease, the rate of substitution of labour for capital goes up. We quantify this exchange of inputs by defining the ***marginal rate of technical substitution***, MRTS, to be

$$-\frac{dK}{dL}$$

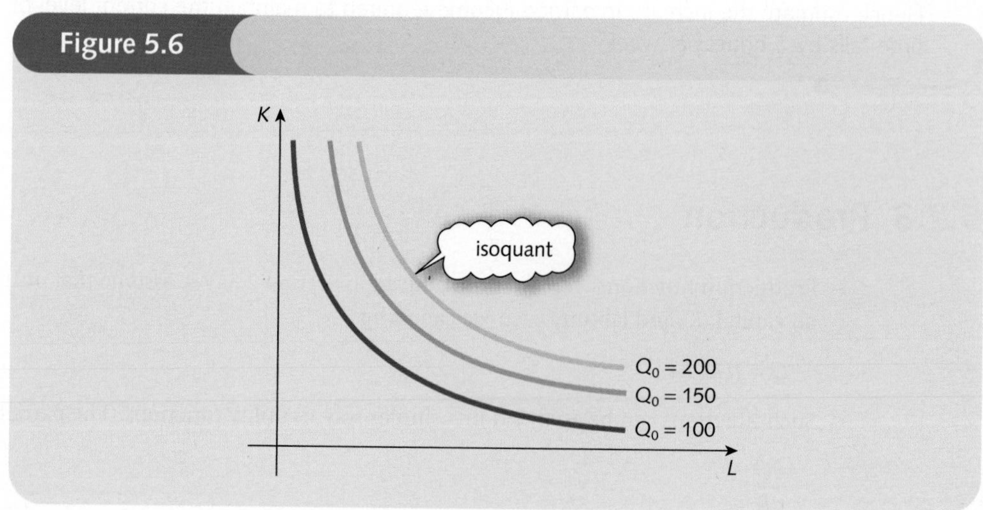

so that MRTS is the positive value of the slope of an isoquant. As in the case of a utility function, the formula for implicit differentiation shows that

$$\text{MRTS} = \frac{\partial Q/\partial L}{\partial Q/\partial K} = \frac{\text{MP}_L}{\text{MP}_K}$$

marginal rate of technical substitution is the marginal product of labour divided by the marginal product of capital

Example

Find an expression for MRTS for the general Cobb–Douglas production function

$$Q = AK^{\alpha}L^{\beta}$$

where A, α and β are positive constants.

Solution

We begin by finding the marginal products. Partial differentiation of

$$Q = AK^{\alpha}L^{\beta}$$

with respect to K and L gives

$$\text{MP}_K = \alpha AK^{\alpha-1}L^{\beta} \quad \text{and} \quad \text{MP}_L = \beta AK^{\alpha}L^{\beta-1}$$

Hence

$$\text{MRTS} = \frac{\text{MP}_L}{\text{MP}_K} = \frac{\beta AK^{\alpha}L^{\beta-1}}{\alpha AK^{\alpha-1}L^{\beta}} = \frac{\beta K}{\alpha L}$$

Practice Problem

4 Given the production function

$$Q = K^2 + 2L^2$$

write down expressions for the marginal products

$$\frac{\partial Q}{\partial K} \quad \text{and} \quad \frac{\partial Q}{\partial L}$$

Hence show that

(a) $\text{MRTS} = \dfrac{2L}{K}$

(b) $K\dfrac{\partial Q}{\partial K} + L\dfrac{\partial Q}{\partial L} = 2Q$

Advice

Production functions and the concept of homogeneity were covered in Section 2.3. You might find it useful to revise this before reading the next paragraph.

Recall that a production function is described as being homogeneous of degree n if, for any number λ,

$$f(\lambda K, \lambda L) = \lambda^n f(K, L)$$

A production function is then said to display decreasing returns to scale, constant returns to scale or increasing returns to scale, depending on whether $n < 1$, $n = 1$ or $n > 1$, respectively. One useful result concerning homogeneous functions is known as *Euler's theorem*, which states that

$$\boxed{K\frac{\partial f}{\partial K} + L\frac{\partial f}{\partial L} = nf(K, L)}$$

In fact, you have already verified this in Practice Problem 4(b) for the particular production function

$$Q = K^2 + 2L^2$$

which is easily shown to be homogeneous of degree 2. We have no intention of proving this theorem, although you are invited to confirm its validity for general Cobb–Douglas production functions in Practice Problem 10 at the end of this section.

The special case $n = 1$ is worthy of note because the right-hand side is then simply $f(K, L)$, which is equal to the output, Q. Euler's theorem for homogeneous production functions of degree 1 states that

capital times marginal product of capital	+	labour times marginal product of labour	=	total output

If each input factor is paid an amount equal to its marginal product then each term on the left-hand side gives the total bill for that factor. For example, if each unit of labour is paid MP_L then the cost of L units of labour is $L(MP_L)$. Provided that the production function displays constant returns to scale, Euler's theorem shows that the sum of the factor payments is equal to the total output.

We conclude this section with an example that shows how the computer package Maple can be used to handle functions of two variables.

Example MAPLE

Consider the production function

$$Q = 2K^{0.2}L^{0.8} \quad (0 \le K \le 1000, 0 \le L \le 1000)$$

(a) Draw a three-dimensional plot of this function together with its isoquant map.

(b) Use the instruction `diff` to find an expression for MRTS.

Solution

(a) Let us name this function prod. To do this, we type

```
>prod:=2*K^0.2*L^0.8;
```

The instruction for a three-dimensional plot is plot3d. This is used in the same way as ordinary plot. The only difference is that we must specify the range of both K and L, so we type

```
>plot3d(prod,K=0..1000,L=0..1000);
```

If you do this, you get a most uninspiring picture of the surface. Most of the surface is 'coming straight towards you', so you cannot see it properly. Maple does, however, allow you to rotate the surface to get a better perspective. To do this, click on the surface to make the graphics toolbar appear. This is shown in Figure 5.7.

Figure 5.7

Advice

It is well worth playing around with some of the buttons on the toolbar to investigate some of the useful features of the package. For example, click on the first indicated button. This creates a cuboid on the screen. To rotate the axes, simply hold the mouse button down and drag the cursor around. Figure 5.8 (overleaf) shows one such perspective with the origin at the front. It shows clearly how the output rises with increasing capital and labour and that this effect is more pronounced with increasing values of L than K.

To obtain an isoquant map we need to call up the more sophisticated plotting routines, from which we select the one called contourplot. You type

```
>with(plots):
```

you can end this command with a colon ':'

followed by

```
>contourplot(prod,K=0..1000,L=0..1000);
```

The response from Maple is the isoquant map shown in Figure 5.9 (overleaf).

(b) Partial differentiation is performed, as usual, via the instruction diff. Typing

```
>diff(output,K);
```

performs the differentiation with respect to K and shows that MP_K is given by

$$\frac{0.4L^{0.8}}{K^{0.8}}$$

→

Figure 5.8

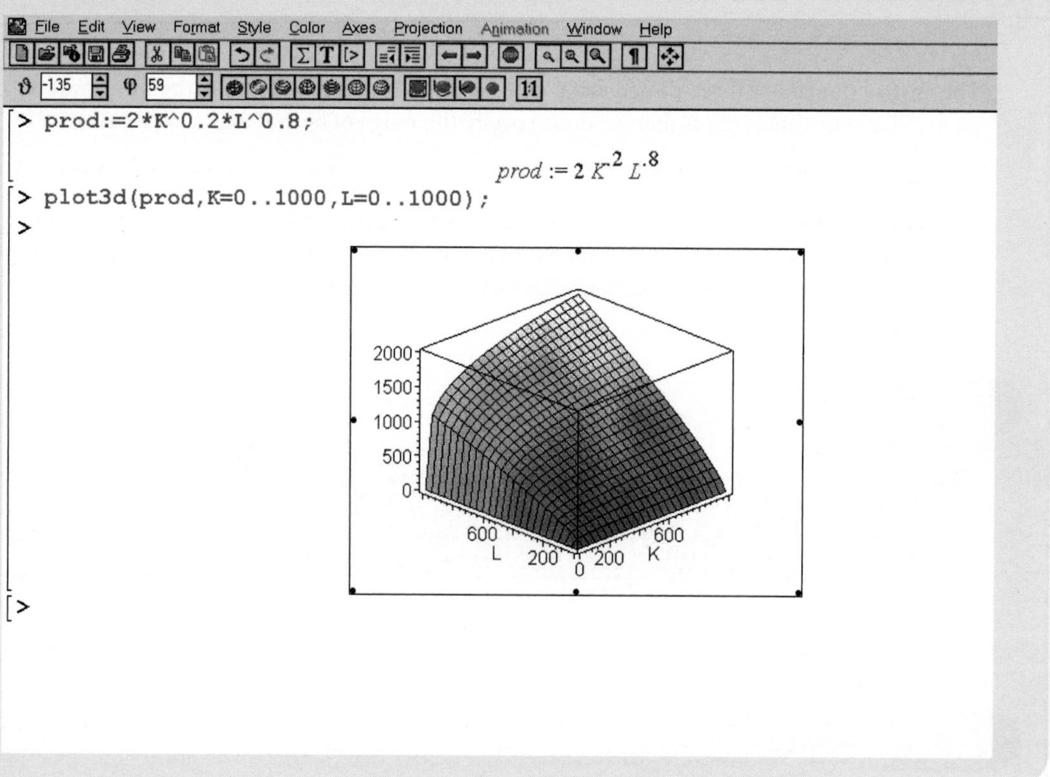

Figure 5.9

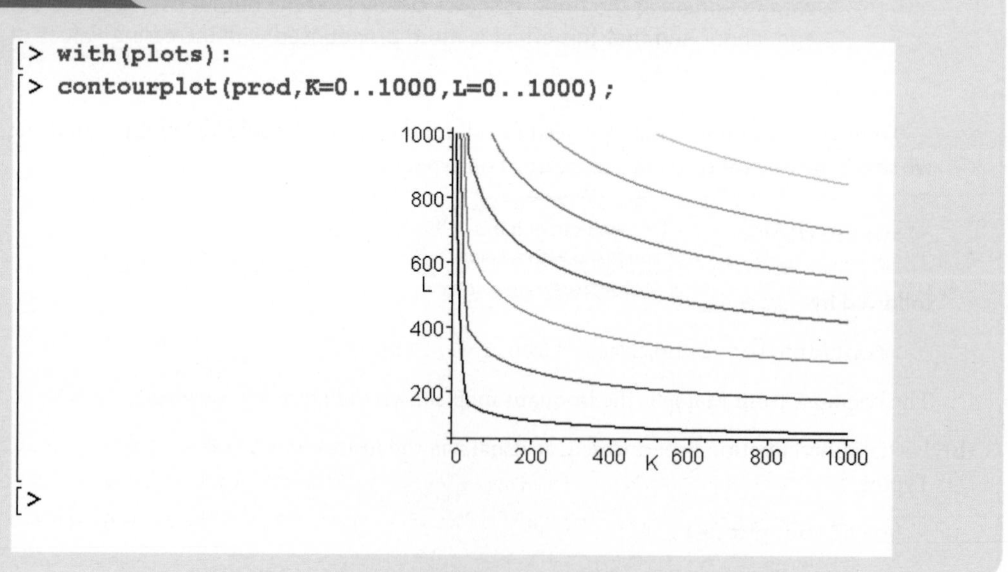

Typing

```
>diff(output,L);
```

performs the differentiation with respect to L and shows that MP_L is given by

$$\frac{1.6K^{0.2}}{L^{0.2}}$$

The MRTS is found by dividing one by the other. In fact, this can be done in a single line:

```
>MRTS:=diff(output,L)/diff(output,K);
```

which gives

$$\text{MRTS} = \frac{4K}{L}$$

Key Terms

Cross-price elasticity of demand The responsiveness of demand for one good to a change in the price of another: (percentage change in quantity) ÷ (percentage change in the price of the alternative good).

Euler's theorem If each input is paid the value of its marginal product, the total costs of these inputs is equal to total output, provided there are constant returns to scale.

Income elasticity of demand The responsiveness of demand for one good to a change in income: (percentage change in quantity) ÷ (percentage change in income).

Indifference curves A curve indicating all combinations of two goods which give the same level of utility.

Indifference map A diagram showing the graphs of a set of indifference curves. The further the curve is from the origin, the greater the level of utility.

Isoquant A curve indicating all combinations of two factors which give the same level of output.

Law of diminishing marginal utility The law which states that the increase in utility due to the consumption of an additional good will eventually decline: $\partial^2U/\partial x_i^2 < 0$ for sufficiently large x_i.

Marginal product of capital The additional output produced by a 1 unit increase in capital: $MP_K = \partial Q/\partial K$.

Marginal product of labour The additional output produced by a 1 unit increase in labour: $MP_L = \partial Q/\partial L$.

Marginal rate of commodity substitution The amount by which one input needs to increase to maintain a constant value of utility when the other input decreases by 1 unit: $\text{MRTS} = \partial U/\partial x_1 \div \partial U/\partial x_2$.

Marginal rate of technical substitution The amount by which capital needs to rise to maintain a constant level of output when labour decreases by 1 unit: $\text{MRTS} = MP_L/MP_K$.

Marginal utility The extra satisfaction gained by consuming 1 extra unit of a good: $\partial U/\partial x_i$.

Price elasticity of demand The responsiveness of demand for one good to a change in its own price: – (percentage change in quantity) ÷ (percentage change in the price).

Utility The satisfaction gained from the consumption of a good.

Practice Problems

5 Given the demand function

$$Q = 200 - 2P - P_A + 0.1Y^2$$

where $P = 10$, $P_A = 15$ and $Y = 100$, find

(a) the price elasticity of demand

(b) the cross-price elasticity of demand

(c) the income elasticity of demand

Estimate the percentage change in demand if P_A rises by 3%. Is the alternative good substitutable or complementary?

6 Given the demand function

$$Q = \frac{P_A Y^2}{P}$$

where $P_A = 10$, $Y = 2$ and $P = 4$, find the income elasticity of demand. If P_A and P are fixed, estimate the percentage change in Y needed to raise Q by 2%.

7 Given the utility function

$$U = x_1^{1/2} x_2^{1/3}$$

determine the value of the marginal utilities

$$\frac{\partial U}{\partial x_1} \text{ and } \frac{\partial U}{\partial x_2}$$

at the point (25, 8). Hence

(a) estimate the change in utility when x_1 and x_2 both increase by 1 unit

(b) find the marginal rate of commodity substitution at this point

8 Evaluate MP_K and MP_L for the production function

$$Q = 2LK + \sqrt{L}$$

given that the current levels of K and L are 7 and 4, respectively. Hence

(a) write down the value of MRTS

(b) estimate the increase in capital needed to maintain the current level of output given a 1 unit decrease in labour

9 If $Q = 2K^3 + 3L^2K$ show that $K(MP_K) + L(MP_L) = 3Q$.

10 Verify Euler's theorem for the Cobb–Douglas production function

$$Q = AK^{\alpha}L^{\beta}$$

[Hint: this function was shown to be homogeneous of degree $\alpha + \beta$ in Section 2.3.]

11 If a firm's production function is given by

$$Q = 5L + 7K$$

sketch the isoquant corresponding an output level, Q = 700. Use your graph to find the value of MRTS and confirm this using partial differentiation.

12 **(Maple)** Consider the production function

$$Q = L(5\sqrt{K} + \sqrt{L}) \quad (0 \le K \le 3, 0 \le L \le 5)$$

(a) Draw a three-dimensional plot of this function. Rotate the axes to give a clear view of the surface. Draw the corresponding isoquant map.

(b) Find an expression for MRTS.

(c) Given that $L = 4$, find the value of K for which MRTS = 2.

13 **(Maple)** Consider the production function

$$Q = (0.3K^{-3} + 0.7L^{-3})^{-1/3} \quad (1 \le K \le 10,\ 1 \le L \le 10)$$

(a) Draw a three-dimensional plot of this function. Rotate the axes to give a clear view of the surface.

(b) Draw the corresponding isoquant map. Deduce that the marginal rate of technical substitution diminishes with increasing L.

(c) Find an expression for MRTS.

(d) Find the slope of the isoquant $Q = 4$ at the point $L = 8$.

section 5.3

Comparative statics

Objectives

At the end of this section you should be able to:

- Use structural equations to derive the reduced form of macroeconomic models.
- Calculate national income multipliers.
- Use multipliers to give a qualitative description of economic models.
- Use multipliers to give a quantitative description of economic models.
- Calculate multipliers for the linear one-commodity market model.

Advice

The content of this section is quite difficult since it depends on ideas covered earlier in this book. You might find it helpful to read quickly through Section 1.6 now before tackling the new material.

The simplest macroeconomic model, discussed in Section 1.6, assumes that there are two sectors, households and firms, and that household consumption, C, is modelled by a linear relationship of the form

$$C = aY + b \tag{1}$$

In this equation Y denotes national income and a and b are parameters. The parameter a is the marginal propensity to consume and lies in the range $0 < a < 1$. The parameter b is the autonomous consumption and satisfies $b > 0$. In equilibrium

$$Y = C + I \tag{2}$$

where I denotes investment, which is assumed to be given by

$$I = I^* \tag{3}$$

for some constant I^*. Equations (1), (2) and (3) describe the structure of the model and as such are called *structural equations*. Substituting equations (1) and (3) into equation (2) gives

$$Y = aY + b + I^*$$

$$Y - aY = b + I^* \quad \text{(subtract } aY \text{ from both sides)}$$

$$(1 - a)Y = b + I^* \quad \text{(take out a common factor of } Y\text{)}$$

$$Y = \frac{b + I^*}{1 - a} \quad \text{(divide both sides by } 1 - a\text{)}$$

This is known as the *reduced form* because it compresses the model into a single equation in which the endogenous variable, Y, is expressed in terms of the exogenous variable, I^*, and parameters, a and b. The process of analysing the equilibrium level of income in this way is referred to as *statics* because it assumes that the equilibrium state is attained instantaneously. The branch of mathematical economics which investigates time dependence is known as *dynamics* and is considered in Chapter 9.

We should like to do rather more than just to calculate the equilibrium values here. In particular, we are interested in the effect on the endogenous variables in a model brought about by changes in the exogenous variables and parameters. This is known as *comparative statics*, since we seek to compare the effects obtained by varying each variable and parameter in turn. The actual mechanism for change will be ignored and it will be assumed that the system returns to equilibrium instantaneously. The equation

$$Y = \frac{b + I^*}{1 - a}$$

shows that Y is a function of three variables, a, b and I^*, so we can write down three partial derivatives

$$\frac{\partial Y}{\partial a}, \quad \frac{\partial Y}{\partial b}, \quad \frac{\partial Y}{\partial I^*}$$

The only hard one to work out is the first, and this is found using the chain rule by writing

$$Y = (b + I^*)(1 - a)^{-1}$$

which gives

$$\frac{\partial Y}{\partial a} = (b + I^*)(-1)(1 - a)^{-2}(-1) = \frac{b + I^*}{(1 - a)^2}$$

To interpret this derivative let us suppose that the marginal propensity to consume, a, changes by Δa with b and I^* held constant. The corresponding change in Y is given by

$$\Delta Y = \frac{\partial Y}{\partial a}\Delta a$$

Strictly speaking, the '$=$' sign should really be '$\simeq$'. However, as we have seen in the previous two sections, provided that Δa is small the approximation is reasonably accurate. In any case we could argue that the model itself is only a first approximation to what is really happening in the economy and so any further small inaccuracies that are introduced are unlikely to have any significant effect on our conclusions. The above equation shows that the change in national income is found by multiplying the change in the marginal propensity to consume by the partial derivative $\partial Y/\partial a$. For this reason the partial derivative is called the *marginal propensity to consume multiplier* for Y. In the same way, $\partial Y/\partial b$ and $\partial Y/\partial I^*$ are called the *autonomous consumption multiplier* and the *investment multiplier*, respectively.

Multipliers enable us to explain the behaviour of the model both qualitatively and quantitatively. The qualitative behaviour can be described simply by inspecting the multipliers as they stand, before any numerical values are assigned to the variables and parameters. It is usually possible to state whether the multipliers are positive or negative and hence whether an increase in an exogenous variable or parameter leads to an increase or decrease in the corresponding endogenous variable. In the present model it is apparent that the marginal propensity to consume multiplier for Y is positive because it is known that b and I^* are both positive, and the denominator $(1 - a)^2$ is clearly positive. Therefore, national income rises whenever a rises.

Once the exogenous variables and parameters have been assigned specific numerical values, the behaviour of the model can be explained quantitatively. For example, if $b = 10$, $I^* = 30$ and $a = 0.5$ then the marginal propensity to consume multiplier is

$$\frac{b + I^*}{(1 - a)^2} = \frac{10 + 30}{(1 - 0.5)^2} = 160$$

This means that when the marginal propensity to consume rises by, say, 0.02 units the change in national income is

$$160 \times 0.02 = 3.2$$

Of course, if a, b and I^* change by amounts Δa, Δb and ΔI^* simultaneously then the small increments formula shows that the change in Y can be found from

$$\Delta Y = \frac{\partial Y}{\partial a}\Delta a + \frac{\partial Y}{\partial b}\Delta b + \frac{\partial Y}{\partial I^*}\Delta I^*$$

Example

Use the equation

$$Y = \frac{b + I^*}{1 - a}$$

to find the investment multiplier.

Deduce that an increase in investment always leads to an increase in national income.

Calculate the change in national income when investment rises by 4 units and the marginal propensity to consume is 0.6.

Solution

Writing

$$Y = \frac{b}{1 - a} + \frac{I^*}{1 - a}$$

we see that

$$\frac{\partial Y}{\partial I^*} = \frac{1}{1 - a}$$

which is positive because $a < 1$. Therefore national income rises whenever I^* rises.

When $a = 0.6$ the investment multiplier is

$$\frac{1}{1 - a} = \frac{1}{1 - 0.6} = \frac{1}{0.4} = 2.5$$

so that when investment rises by 4 units the change in national income is

$$2.5 \times 4 = 10$$

Practice Problem

1 By substituting

$$Y = \frac{b + I^*}{1 - a}$$

into

$$C = aY + b$$

write down the reduced equation for C in terms of a, b and I^*. Hence show that the investment multiplier for C is

$$\frac{a}{1 - a}$$

Deduce that an increase in investment always leads to an increase in consumption. Calculate the change in consumption when investment rises by 2 units if the marginal propensity to consume is 1/2.

The following example is more difficult because it involves three sectors: households, firms and government. However, the basic strategy for analysing the model is the same. We first obtain the reduced form, which is differentiated to determine the relevant multipliers. These can then be used to discuss the behaviour of national income both qualitatively and quantitatively.

Example

Consider the three-sector model

$$Y = C + I + G \qquad (1)$$

$$C = aY_d + b \qquad (0 < a < 1, b > 0) \qquad (2)$$

$$Y_d = Y - T \qquad (3)$$

$$T = tY + T^* \qquad (0 < t < 1, T^* > 0) \qquad (4)$$

$$I = I^* \qquad (I^* > 0) \qquad (5)$$

$$G = G^* \qquad (G^* > 0) \qquad (6)$$

where G denotes government expenditure and T denotes taxation.

(a) Show that

$$Y = \frac{-aT^* + b + I^* + G^*}{1 - a + at}$$

(b) Write down the government expenditure multiplier and autonomous taxation multiplier. Deduce the direction of change in Y due to increases in G^* and T^*.

(c) If it is government policy to finance any increase in expenditure, ΔG^*, by an increase in autonomous taxation, ΔT^*, so that

$$\Delta G^* = \Delta T^*$$

show that national income rises by an amount that is less than the rise in expenditure.

(d) If $a = 0.7$, $b = 50$, $T^* = 200$, $t = 0.2$, $I^* = 100$ and $G^* = 300$, calculate the equilibrium level of national income, Y, and the change in Y due to a 10 unit increase in government expenditure.

➔

Solution

(a) We need to 'solve' equations (1)–(6) for Y. An obvious first move is to substitute equations (2), (5) and (6) into equation (1) to get

$$Y = aY_d + b + I^* + G^* \tag{7}$$

Now from equations (3) and (4)

$$\begin{aligned} Y_d &= Y - T \\ &= Y - (tY + T^*) \\ &= Y - tY - T^* \end{aligned}$$

so this can be put into equation (7) to get

$$\begin{aligned} Y &= a(Y - tY - T^*) + b + I^* + G^* \\ &= aY - atY - aT^* + b + I^* + G^* \end{aligned}$$

Collecting terms in Y on the left-hand side gives

$$(1 - a + at)Y = -aT^* + b + I^* + G^*$$

which produces the desired equation

$$Y = \frac{-aT^* + b + I^* + G^*}{1 - a + at}$$

(b) The government expenditure multiplier is

$$\frac{\partial Y}{\partial G^*} = \frac{1}{1 - a + at}$$

and the autonomous taxation multiplier is

$$\frac{\partial Y}{\partial T^*} = \frac{-a}{1 - a + at}$$

We are given that $a < 1$, so $1 - a > 0$. Also, we know that a and t are both positive, so their product, at, must be positive. The expression $(1 - a) + at$ is therefore positive, being the sum of two positive terms. The government expenditure multiplier is therefore positive, which shows that any increase in G^* leads to an increase in Y. The autonomous taxation multiplier is negative because its numerator is negative and its denominator is positive. This shows that any increase in T^* leads to a decrease in Y.

(c) Government policy is to finance a rise in expenditure out of autonomous taxation, so that

$$\Delta G^* = \Delta T^*$$

From the small increments formula

$$\Delta Y = \frac{\partial Y}{\partial G^*}\Delta G^* + \frac{\partial Y}{\partial T^*}\Delta T^*$$

we deduce that

$$\Delta Y = \left(\frac{\partial Y}{\partial G^*} + \frac{\partial Y}{\partial T^*}\right)\Delta G^* = \left(\frac{1}{1 - a + at} + \frac{-a}{1 - a + at}\right)\Delta G^* = \left(\frac{1 - a}{1 - a + at}\right)\Delta G^*$$

The multiplier

$$\frac{1 - a}{1 - a + at}$$

is called the *balanced budget multiplier* and is positive because the numerator and denominator are both positive. An increase in government expenditure leads to an increase in national income. However, the denominator is greater than the numerator by an amount at, so that

$$\frac{1-a}{1-a+at} < 1$$

and $\Delta Y < \Delta G^*$, showing that the rise in national income is less than the rise in expenditure.

(d) To solve this part of the problem we simply substitute the numerical values $a = 0.7$, $b = 50$, $T^* = 200$, $t = 0.2$, $I^* = 100$ and $G^* = 300$ into the results of parts (a) and (b). From part (a)

$$Y = \frac{-aT^* + b + I^* + G^*}{1-a+at} = \frac{-0.7(200) + 50 + 100 + 300}{1 - 0.7 + 0.7(0.2)} = 704.5$$

From part (b) the government expenditure multiplier is

$$\frac{1}{1-a+at} = \frac{1}{0.44} = 2.27$$

and we are given that $\Delta G^* = 10$, so the change in national income is

$$2.27 \times 10 = 22.7$$

Practice Problem

2 Consider the four-sector model

$$\begin{aligned}
Y &= C + I + G + X - M \\
C &= aY + b && (0 < a < 1, b > 0) \\
I &= I^* && (I^* > 0) \\
G &= G^* && (G^* > 0) \\
X &= X^* && (X^* > 0) \\
M &= mY + M^* && (0 < m < 1, M^* > 0)
\end{aligned}$$

where X and M denote exports and imports respectively and m is the marginal propensity to import.

(a) Show that

$$Y = \frac{b + I^* + G^* + X^* - M^*}{1 - a + m}$$

(b) Write down the autonomous export multiplier

$$\frac{\partial Y}{\partial X^*}$$

and the marginal propensity to import multiplier

$$\frac{\partial Y}{\partial m}$$

Deduce the direction of change in Y due to increases in X^* and m.

(c) If $a = 0.8$, $b = 120$, $I^* = 100$, $G^* = 300$, $X^* = 150$, $m = 0.1$ and $M^* = 40$, calculate the equilibrium level of national income, Y, and the change in Y due to a 10 unit increase in autonomous exports.

So far, all of the examples of comparative statics that we have considered have been taken from macroeconomics. The same approach can be used in microeconomics. For example, let us analyse the equilibrium price and quantity in supply and demand theory.

Figure 5.10 illustrates the simple linear one-commodity market model described in Section 1.3. The equilibrium values of price and quantity are determined from the point of intersection of the supply and demand curves. The supply curve is a straight line with a positive slope and intercept, so its equation may be written as

$$P = aQ_S + b \quad (a > 0, b > 0)$$

The demand equation is also linear but has a negative slope and a positive intercept, so its equation may be written as

$$P = -cQ_D + d \quad (c > 0, d > 0)$$

It is apparent from Figure 5.10 that in order for these two lines to intersect in the positive quadrant, it is necessary for the intercept on the demand curve to lie above that on the supply curve, so we require

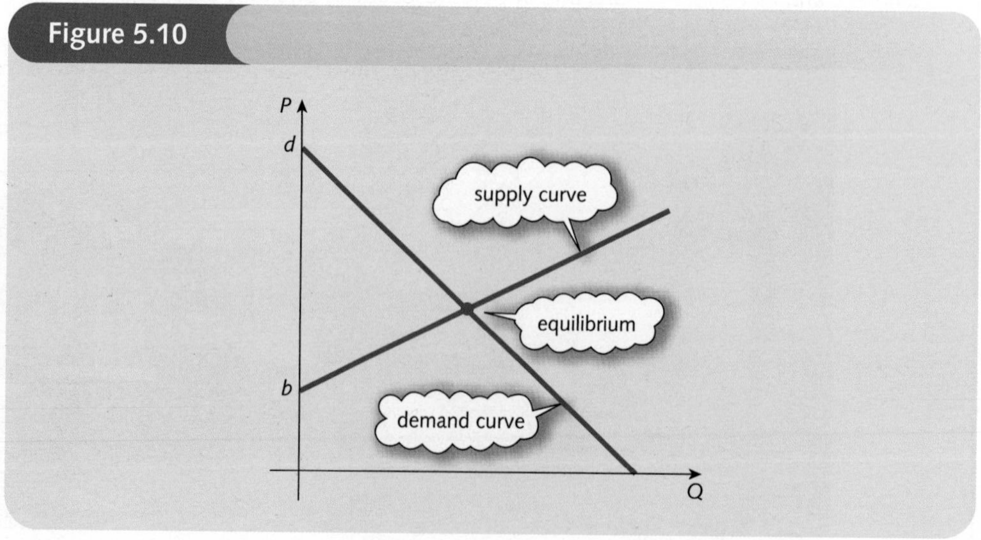

$$d > b$$

or equivalently

$$d - b > 0$$

In equilibrium Q_S and Q_D are equal. If we let their common value be denoted by Q then the supply and demand equations become

$$P = aQ + b$$

$$P = -cQ + d$$

and so

$$aQ + b = -cQ + d$$

since both sides are equal to P.

To solve for Q we first collect like terms together, which gives

$$(a + c)Q = d - b$$

and then divide by the coefficient of Q to get

$$Q = \frac{d - b}{a + c}$$

(Incidentally, this confirms the restriction $d - b > 0$. If this were not true then Q would be either zero or negative, which does not make economic sense.)

Equilibrium quantity is a function of the four parameters a, b, c and d, so there are four multipliers

$$\frac{\partial Q}{\partial a} = -\frac{d - b}{(a + c)^2}$$

$$\frac{\partial Q}{\partial b} = -\frac{1}{a + c}$$

$$\frac{\partial Q}{\partial c} = -\frac{d - b}{(a + c)^2}$$

$$\frac{\partial Q}{\partial d} = \frac{1}{a + c}$$

where the chain rule is used to find $\partial Q/\partial a$ and $\partial Q/\partial c$.

We noted previously that all of the parameters are positive and that $d - b > 0$, so

$$\frac{\partial Q}{\partial a} < 0, \frac{\partial Q}{\partial b} < 0, \frac{\partial Q}{\partial c} < 0 \quad \text{and} \quad \frac{\partial Q}{\partial d} > 0$$

This shows that an increase in a, b or c causes a decrease in Q, whereas an increase in d causes an increase in Q.

Example

Give a graphical confirmation of the sign of the multiplier

$$\frac{\partial Q}{\partial a}$$

Figure 5.11

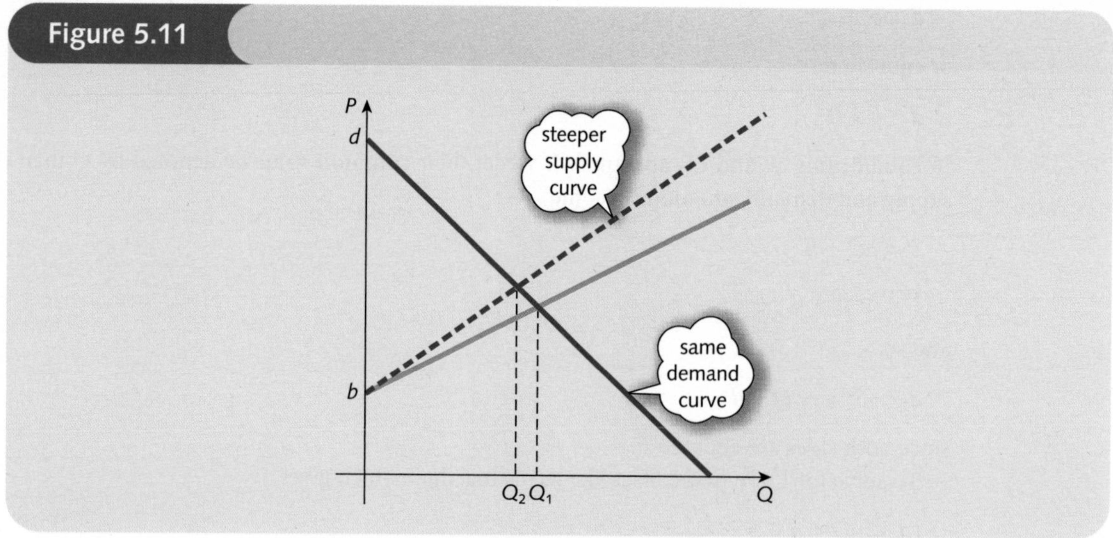

Solution

From the supply equation

$$P = aQ_S + b$$

we see that a small increase in the value of the parameter a causes the supply curve to become slightly steeper, as indicated by the dashed line in Figure 5.11. The effect is to shift the point of intersection to the left and so the equilibrium quantity decreases from Q_1 to Q_2 which is consistent with a negative value of the multiplier, $\partial Q/\partial a$.

Given any pair of supply and demand equations, we can easily calculate the effect on the equilibrium quantity. For example, consider the equations

$$P = Q_S + 1$$

$$P = -2Q_D + 5$$

and let us suppose that we need to calculate the change in equilibrium quantity when the coefficient of Q_S increases from 1 to 1.1. In this case we have

$$a = 1, b = 1, c = 2, d = 5$$

To find ΔQ we first evaluate the multiplier

$$\frac{\partial Q}{\partial a} = -\frac{d - b}{(a + c)^2} = -\frac{5 - 1}{(1 + 2)^2} = -0.44$$

and then multiply by 0.1 to get

$$\Delta Q = (-0.44) \times 0.1 = -0.044$$

An increase of 0.1 in the slope of the supply curve therefore produces a decrease of 0.044 in the equilibrium quantity.

Practice Problem

3 Give a graphical confirmation of the sign of the multiplier

$$\frac{\partial Q}{\partial d}$$

for the linear one-commodity market model

$$P = aQ_S + b \qquad (a > 0, b > 0)$$

$$P = -cQ_D + d \quad (c > 0, d > 0)$$

Throughout this section all of the relations in each model have been assumed to be linear. It is possible to analyse non-linear relations in a similar way, although this is beyond the scope of this book.

Advice

We shall return to this topic again in Chapter 7 when we use Cramer's rule to solve the structural equations of a linear model.

Key Terms

Autonomous consumption multiplier The number by which you multiply the change in autonomous consumption to deduce the corresponding change in, say, national income: $\partial Y/\partial b$.

Balanced budget multiplier The number by which you multiply the change in government expenditure to deduce the corresponding change in, say, national income: $\partial Y/\partial G^*$, assuming that this change is financed entirely by a change in taxation.

Comparative statics Examination of the effect on equilibrium values due to changes in the parameters of an economic model.

Dynamics Analysis of how equilibrium values vary over time.

Investment multiplier The number by which you multiply the change in investment to deduce the corresponding change in, say, national income: $\partial Y/\partial I^*$.

Marginal propensity to consume multiplier The number by which you multiply the change in MPC to deduce the corresponding change in, say, national income: $\partial Y/\partial a$.

Reduced form The final equation obtained when exogenous variables are eliminated in the course of solving a set of structural equations in a macroeconomic model.

Statics The determination of the equilibrium values of variables in an economic model which do not change over time.

Structural equations A collection of equations that describe the equilibrium conditions of a macroeconomic model.

Practice Problems

4 Consider the three-sector model

$$Y = C + I + G \quad (1)$$

$$C = aY_d + b \qquad (0 < a < 1, b > 0) \quad (2)$$

$$Y_d = Y - T \quad (3)$$

$$T = T^* \qquad (T^* > 0) \quad (4)$$

$$I = I^* \qquad (I^* > 0) \quad (5)$$

$$G = G^* \qquad (G^* > 0) \quad (6)$$

(a) Show that

$$C = \frac{aI^* + aG^* - aT^* + b}{1 - a}$$

(b) Write down the investment multiplier for C. Decide the direction of change in C due to an increase in I^*.

(c) If $a = 0.9$, $b = 80$, $I^* = 60$, $G^* = 40$, $T^* = 20$, calculate the equilibrium level of consumption, C, and also the change in C due to a 2 unit change in investment.

5 Consider the four-sector macroeconomic model

$$Y = C + I + G + X - M$$

$$C = aY_d + b \qquad (0 < a < 1, b > 0)$$

$$Y_d = Y - T$$

$$T = tY + T^* \qquad (0 < t < 1, T^* > 0)$$

$$I = I^* \qquad (I^* > 0)$$

$$G = G^* \qquad (G^* > 0)$$

$$X = X^* \qquad (X^* > 0)$$

$$M = mY_d + M^* \qquad (0 < m < 1, M^* > 0)$$

(1) Show that

$$Y = \frac{b + (m - a)T^* + I^* + G^* + X^* - M^*}{1 - a + at + m - mt}$$

(2) (a) Write down the autonomous taxation multiplier. Deduce that an increase in T^* causes a decrease in Y on the assumption that a country's marginal propensity to import, m, is less than its marginal propensity to consume, a.

(b) Write down the government expenditure multiplier. Deduce that an increase in G^* causes an increase in Y.

(3) Let $a = 0.7$, $b = 150$, $t = 0.25$, $m = 0.1$, $T^* = 100$, $I^* = 100$, $G^* = 500$, $M^* = 300$ and $X^* = 160$.

(a) Calculate the equilibrium level of national income.

(b) Calculate the change in Y due to an 11 unit increase in G^*.

(c) Find the increase in autonomous taxation required to restore Y to its level calculated in part (a).

6 Show that the equilibrium price for a linear one-commodity market model

$$P = aQ_S + b \quad (a > 0, b > 0)$$

$$P = -cQ_D + d \qquad (c > 0, d > 0)$$

where $d - b > 0$, is given by

$$P = \frac{ad + bc}{a + c}$$

Find expressions for the multipliers

$$\frac{\partial P}{\partial a}, \quad \frac{\partial P}{\partial b}, \quad \frac{\partial P}{\partial c}, \quad \frac{\partial P}{\partial d}$$

and deduce the direction of change in P due to an increase in a, b, c or d.

7 **(1)** For the commodity market

$$Y = C + I$$

$$C = aY + b \quad (0 < a < 1, b > 0)$$

$$I = cr + d \quad (c < 0, d > 0)$$

where r is the interest rate.

Show that, when the commodity market is in equilibrium,

$$(1 - a)Y - cr = b + d$$

(2) For the money market

(money supply)	$M_S = M_S^*$	$(M_S^* > 0)$
(total demand for money)	$M_D = k_1Y + k_2r + k_3$	$(k_1 > 0, k_2 < 0, k_3 > 0)$
(equilibrium)	$M_D = M_S$	

Show that when the money market is in equilibrium

$$k_1Y + k_2r = M_S^* - k_3$$

(3) **(a)** By solving the simultaneous equations derived in parts (1) and (2) show that when the commodity and money markets are both in equilibrium

$$Y = \frac{k_2(b + d) + c(M_S^* - k_3)}{(1 - a)k_2 + ck_1}$$

(b) Write down the money supply multiplier, $\partial Y/\partial M_S^*$ and deduce that an increase in M_S^* causes an increase in Y.

section 5.4

Unconstrained optimization

Objectives

At the end of this section you should be able to:

- Use the first-order partial derivatives to find the stationary points of a function of two variables.
- Use the second-order partial derivatives to classify the stationary points of a function of two variables.
- Find the maximum profit of a firm that produces two goods.
- Find the maximum profit of a firm that sells a single good in different markets with price discrimination.

As you might expect, methods for finding the maximum and minimum points of a function of two variables are similar to those used for functions of one variable. However, the nature of economic functions of several variables forces us to subdivide optimization problems into two types, unconstrained and constrained. To understand the distinction, consider the utility function

$$U(x_1, x_2) = x_1^{1/4}x_2^{3/4}$$

The value of U measures the satisfaction gained from buying x_1 items of a good G1 and x_2 items of a good G2. The natural thing to do here is to try to pick x_1 and x_2 to make U as large as possible, thereby maximizing utility. However, a moment's thought should convince you that, as it stands, this problem does not have a finite solution. The factor $x_1^{1/4}$ can be made as large as we please by taking ever-increasing values of x_1 and likewise for the factor $x_2^{3/4}$. In other words, utility increases without bound as more and more items of goods G1 and G2 are bought. In practice, of course, this does not occur, since there is a limit to the amount of money that an individual has to spend on these goods. For example, suppose that the cost of each item of G1 and G2 is \$2 and \$3, respectively, and that we allocate \$100 for the purchase of these goods. The total cost of buying x_1 items of G1 and x_2 items of G2 is

$$2x_1 + 3x_2$$

so we require

$$2x_1 + 3x_2 = 100$$

The problem now is to maximize the utility function

$$U = x_1^{1/4}x_2^{3/4}$$

subject to the budgetary constraint

$$2x_1 + 3x_2 = 100$$

The constraint prevents us from taking ever-increasing values of x_1 and x_2 and leads to a finite solution.

We describe how to solve constrained optimization problems in the following two sections. For the moment we concentrate on the simple case of optimizing functions

$$z = f(x, y)$$

without any constraints. This is typified by the problem of profit maximization, which usually has a finite solution without the need to impose constraints. In a sense the constraints are built into the profit function, which is defined by

$$\pi = \text{TR} - \text{TC}$$

because there is a conflict between trying to make total revenue, TR, as large as possible while trying to make total cost, TC, as small as possible.

Let us begin by recalling how to find and classify stationary points of functions of one variable

$$y = f(x)$$

In Section 4.6 we used the following strategy:

Step 1

Solve the equation

$$f'(x) = 0$$

to find the stationary points, $x = a$.

Step 2

If

- $f''(a) > 0$ then the function has a minimum at $x = a$
- $f''(a) < 0$ then the function has a maximum at $x = a$
- $f''(a) = 0$ then the point cannot be classified using the available information

For functions of two variables

$$z = f(x, y)$$

the stationary points are found by solving the simultaneous equations

$$\frac{\partial z}{\partial x} = 0$$

$$\frac{\partial z}{\partial y} = 0$$

Figure 5.12

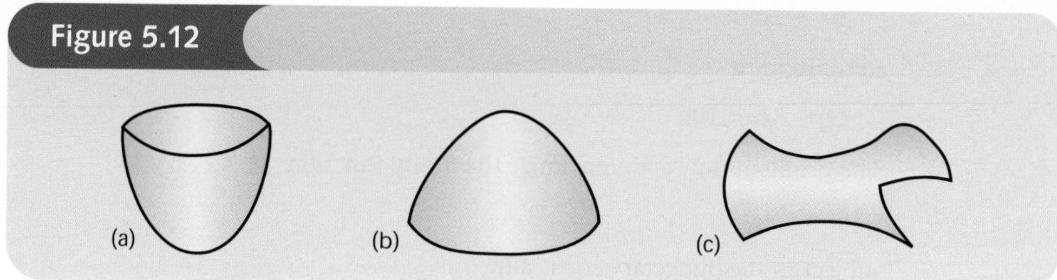

that is,

$$f_x(x, y) = 0$$

$$f_y(x, y) = 0$$

This is a natural extension of the one-variable case. We first write down expressions for the first-order partial derivatives and then equate to zero. This represents a system of two equations for the two unknowns x and y, which we hope can be solved. Stationary points obtained in this way can be classified into one of three types: *minimum*, *maximum* and *saddle point*.

Figure 5.12(a) shows the shape of a surface in the neighbourhood of a minimum. It can be thought of as the bottom of a bowl-shaped valley. If you stand at the minimum point and walk in any direction then you are certain to start moving upwards. Mathematically, we can classify a stationary point (a, b) as a minimum provided that all three of the following conditions hold:

$$\frac{\partial^2 z}{\partial x^2} > 0, \quad \frac{\partial^2 z}{\partial y^2} > 0, \quad \left(\frac{\partial^2 z}{\partial x^2}\right)\left(\frac{\partial^2 z}{\partial y^2}\right) - \left(\frac{\partial^2 z}{\partial x \partial y}\right)^2 > 0$$

when $x = a$ and $y = b$: that is,

$$f_{xx}(a, b) > 0, \quad f_{yy}(a, b) > 0, \quad f_{xx}(a, b)f_{yy}(a, b) - [f_{xy}(a, b)]^2 > 0$$

This triple requirement is obviously more complicated than the single condition needed in the case of a function of one variable. However, once the second-order partial derivatives have been evaluated at the stationary point, the three conditions are easily checked.

Figure 5.12(b) shows the shape of a surface in the neighbourhood of a maximum. It can be thought of as the summit of a mountain. If you stand at the maximum point and walk in any direction then you are certain to start moving downwards. Mathematically, we can classify a stationary point (a, b) as a maximum provided that all three of the following conditions hold:

$$\frac{\partial^2 z}{\partial x^2} < 0, \quad \frac{\partial^2 z}{\partial y^2} < 0, \quad \left(\frac{\partial^2 z}{\partial x^2}\right)\left(\frac{\partial^2 z}{\partial y^2}\right) - \left(\frac{\partial^2 z}{\partial x \partial y}\right)^2 > 0$$

when $x = a$ and $y = b$: that is,

$$f_{xx}(a, b) < 0, \quad f_{yy}(a, b) < 0, \quad f_{xx}(a, b)f_{yy}(a, b) - [f_{xy}(a, b)]^2 > 0$$

Of course, any particular mountain range may well have lots of valleys and summits. Likewise, a function of two variables can have more than one minimum or maximum.

Figure 5.12(c) shows the shape of a surface in the neighbourhood of a saddle point. As its name suggests, it can be thought of as the middle of a horse's saddle. If you sit at this point and edge towards the head or tail then you start moving upwards. On the other hand, if you edge sideways then you start moving downwards (and will probably fall off!). Mathematically, we can classify a stationary point (a, b) as a saddle point provided that the following single condition holds:

$$\left(\frac{\partial^2 z}{\partial x^2}\right)\left(\frac{\partial^2 z}{\partial y^2}\right) - \left(\frac{\partial^2 z}{\partial x \partial y}\right)^2 < 0$$

when $x = a$ and $y = b$: that is,

$$f_{xx}(a, b)f_{yy}(a, b) - [f_{xy}(a, b)]^2 < 0$$

To summarize, the method for finding and classifying stationary points of a function $f(x, y)$ is as follows:

Step 1

Solve the simultaneous equations

$$f_x(x, y) = 0$$

$$f_y(x, y) = 0$$

to find the stationary points, (a, b).

Step 2

If

- $f_{xx} > 0, f_{yy} > 0$ and $f_{xx}f_{yy} - f_{xy}^2 > 0$ at (a, b) then the function has a minimum at (a, b)
- $f_{xx} < 0, f_{yy} < 0$ and $f_{xx}f_{yy} - f_{xy}^2 > 0$ at (a, b) then the function has a maximum at (a, b)
- $f_{xx}f_{yy} - f_{xy}^2 < 0$ at (a, b) then the function has a saddle point at (a, b)

Advice

The second-order conditions needed to classify a stationary point can be expressed more succinctly using Hessians. Details are given in Appendix 3 at the end of this book, although you will need to be familiar with determinants of 2×2 matrices, which are covered later in Section 7.2.

Example

Find and classify the stationary points of the function

$$f(x, y) = x^3 - 3x + xy^2$$

Solution

In order to use steps 1 and 2 we need to find all first- and second-order partial derivatives of the function

$$f(x, y) = x^3 - 3x + xy^2$$

These are easily worked out as

$$f_x = 3x^2 - 3 + y^2$$

$$f_y = 2xy$$

$$f_{xx} = 6x$$

$$f_{xy} = 2y$$

$$f_{yy} = 2x$$

→

Step 1

The stationary points are the solutions of the simultaneous equations

$$f_x(x, y) = 0$$
$$f_y(x, y) = 0$$

so we need to solve

$$3x^2 - 3 + y^2 = 0$$
$$2xy = 0$$

There have been many occasions throughout this book when we have solved simultaneous equations. So far these have been linear. This time, however, we need to solve a pair of non-linear equations. Unfortunately, there is no standard method for solving such systems. We have to rely on our wits in any particular instance. The trick here is to begin with the second equation

$$2xy = 0$$

The only way that the product of three numbers can be equal to zero is when one or more of the individual numbers forming the product are zero. We know that $2 \neq 0$, so either $x = 0$ or $y = 0$. We investigate these two possibilities separately.

- Case 1: $x = 0$. Substituting $x = 0$ into the first equation

  $$3x^2 - 3 + y^2 = 0$$

  gives

  $$-3 + y^2 = 0$$

  that is,

  $$y^2 = 3$$

  There are therefore two possibilities for y to go with $x = 0$, namely $y = -\sqrt{3}$ and $y = \sqrt{3}$. Hence $(0, -\sqrt{3})$ and $(0, \sqrt{3})$ are stationary points.

- Case 2: $y = 0$. Substituting $y = 0$ into the first equation

  $$3x^2 - 3 + y^2 = 0$$

  gives

  $$3x^2 - 3 = 0$$

  that is,

  $$x^2 = 1$$

  There are therefore two possibilities for x to go with $y = 0$, namely $x = -1$ and $x = 1$. Hence $(-1, 0)$ and $(1, 0)$ are stationary points.

These two cases indicate that there are precisely four stationary points, $(0, -\sqrt{3})$, $(0, \sqrt{3})$, $(-1, 0)$, $(1, 0)$.

Step 2

To classify these points we need to evaluate the second-order partial derivatives

$$f_{xx} = 6x, \quad f_{yy} = 2x, \quad f_{xy} = 2y$$

at each point and check the signs of

$$f_{xx}, \; f_{yy}, \; f_{xx}f_{yy} - f_{xy}^2$$

- Point $(0, -\sqrt{3})$

$$f_{xx} = 6(0) = 0, \quad f_{yy} = 2(0) = 0, \quad f_{xy} = -2\sqrt{3}$$

Hence

$$f_{xx}f_{yy} - f_{xy}^2 = 0(0) - (-2\sqrt{3})^2 = -12 < 0$$

and so $(0, -\sqrt{3})$ is a saddle point.

- Point $(0, \sqrt{3})$

$$f_{xx} = 6(0) = 0, \qquad f_{yy} = 2(0) = 0, \qquad f_{xy} = 2\sqrt{3}$$

Hence

$$f_{xx}f_{yy} - f_{xy}^2 = 0(0) - (2\sqrt{3})^2 = -12 < 0$$

and so $(0, \sqrt{3})$ is a saddle point.

- Point $(-1, 0)$

$$f_{xx} = 6(-1) = -6, \; f_{yy} = 2(-1) = -2, \; f_{xy} = 2(0) = 0$$

Hence

$$f_{xx}f_{yy} - f_{xy}^2 = (-6)(-2) - 0^2 = 12 > 0$$

and so $(-1, 0)$ is not a saddle point. Moreover, since

$$f_{xx} < 0 \quad \text{and} \quad f_{yy} < 0$$

we deduce that $(-1, 0)$ is a maximum.

- Point $(1, 0)$

$$f_{xx} = 6(1) = 6, \quad f_{yy} = 2(1) = 2, \quad f_{xy} = 2(0) = 0$$

Hence

$$f_{xx}f_{yy} - f_{xy}^2 = 6(2) - 0^2 = 12 > 0$$

and so $(1, 0)$ is not a saddle point. Moreover, since

$$f_{xx} > 0 \quad \text{and} \quad f_{yy} > 0$$

we deduce that $(1, 0)$ is a minimum.

Practice Problem

1 Find and classify the stationary points of the function

$$f(x, y) = x^2 + 6y - 3y^2 + 10$$

We now consider two examples from economics, both involving the maximization of profit. The first considers the case of a firm producing two different goods, whereas the second involves a single good sold in two different markets.

Example

A firm is a perfectly competitive producer and sells two goods G1 and G2 at \$1000 and \$800, respectively, each. The total cost of producing these goods is given by

$$TC = 2Q_1^2 + 2Q_1Q_2 + Q_2^2$$

where Q_1 and Q_2 denote the output levels of G1 and G2, respectively. Find the maximum profit and the values of Q_1 and Q_2 at which this is achieved.

Solution

The fact that the firm is perfectly competitive tells us that the price of each good is fixed by the market and does not depend on Q_1 and Q_2. The actual prices are stated in the question as \$1000 and \$800. If the firm sells Q_1 items of G1 priced at \$1000 then the revenue is

$$TR_1 = 1000Q_1$$

Similarly, if the firm sells Q_2 items of G2 priced at \$800 then the revenue is

$$TR_2 = 800Q_2$$

The total revenue from the sale of both goods is then

$$TR = TR_1 + TR_2 = 1000Q_1 + 800Q_2$$

We are given that the total cost is

$$TC = 2Q_1^2 + 2Q_1Q_2 + Q_2^2$$

so the profit function is

$$\begin{aligned}\pi &= TR - TC \\ &= (1000Q_1 + 800Q_2) - (2Q_1^2 + 2Q_1Q_2 + Q_2^2) \\ &= 1000Q_1 + 800Q_2 - 2Q_1^2 - 2Q_1Q_2 - Q_2^2\end{aligned}$$

This is a function of the two variables, Q_1 and Q_2, that we wish to optimize. The first- and second-order partial derivatives are

$$\frac{\partial \pi}{\partial Q_1} = 1000 - 4Q_1 - 2Q_2$$

$$\frac{\partial \pi}{\partial Q_2} = 800 - 2Q_1 - 2Q_2$$

$$\frac{\partial^2 \pi}{\partial Q_1^2} = -4$$

$$\frac{\partial^2 \pi}{\partial Q_1 \partial Q_2} = -2$$

$$\frac{\partial^2 \pi}{\partial Q_2^2} = -2$$

The two-step strategy then gives the following:

Step 1

At a stationary point

$$\frac{\partial \pi}{\partial Q_1} = 0$$

$$\frac{\partial \pi}{\partial Q_2} = 0$$

so we need to solve the simultaneous equations

$$1000 - 4Q_1 - 2Q_2 = 0$$

$$800 - 2Q_1 - 2Q_2 = 0$$

that is,

$$4Q_1 + 2Q_2 = 1000 \quad (1)$$

$$2Q_1 + 2Q_2 = 800 \quad (2)$$

The variable Q_2 can be eliminated by subtracting equation (2) from (1) to get

$$2Q_1 = 200$$

and so $Q_1 = 100$. Substituting this into either equation (1) or (2) gives $Q_2 = 300$. The profit function therefore has one stationary point at (100, 300).

Step 2

To show that the point really is a maximum we need to check that

$$\frac{\partial^2 \pi}{\partial Q_1^2} < 0, \quad \frac{\partial^2 \pi}{\partial Q_2^2} < 0, \quad \left(\frac{\partial^2 \pi}{\partial Q_1^2}\right)\left(\frac{\partial^2 \pi}{\partial Q_2^2}\right) - \left(\frac{\partial^2 \pi}{\partial Q_1 \partial Q_2}\right)^2 > 0$$

at this point. In this example the second-order partial derivatives are all constant. We have

$$\frac{\partial^2 \pi}{\partial Q_1^2} = -4 < 0 \quad ✓$$

$$\frac{\partial^2 \pi}{\partial Q_2^2} = -2 < 0 \quad ✓$$

$$\left(\frac{\partial^2 \pi}{\partial Q_1^2}\right)\left(\frac{\partial^2 \pi}{\partial Q_2^2}\right) - \left(\frac{\partial^2 \pi}{\partial Q_1 \partial Q_2}\right)^2 = (-4)(-2) - (-2)^2 = 4 > 0 \quad ✓$$

confirming that the firm's profit is maximized by producing 100 items of G1 and 300 items of G2.

The actual value of this profit is obtained by substituting $Q_1 = 100$ and $Q_2 = 300$ into the expression

$$\pi = 1000Q_1 + 800Q_2 - 2Q_1^2 - 2Q_1Q_2 - Q_2^2$$

to get

$$\pi = 1000(100) + 800(300) - 2(100)^2 - 2(100)(300) - (300)^2 = \$170\,000$$

Practice Problem

2 A firm is a monopolistic producer of two goods G1 and G2. The prices are related to quantities Q_1 and Q_2 according to the demand equations

$$P_1 = 50 - Q_1$$

$$P_2 = 95 - 3Q_2$$

If the total cost function is

$$TC = Q_1^2 + 3Q_1Q_2 + Q_2^2$$

show that the firm's profit function is

$$\pi = 50Q_1 - 2Q_1^2 + 95Q_2 - 4Q_2^2 - 3Q_1Q_2$$

Hence find the values of Q_1 and Q_2 which maximize π and deduce the corresponding prices.

Example

A firm is allowed to charge different prices for its domestic and industrial customers. If P_1 and Q_1 denote the price and demand for the domestic market then the demand equation is

$$P_1 + Q_1 = 500$$

If P_2 and Q_2 denote the price and demand for the industrial market then the demand equation is

$$2P_2 + 3Q_2 = 720$$

The total cost function is

$$TC = 50\,000 + 20Q$$

where $Q = Q_1 + Q_2$. Determine the firm's pricing policy that maximizes profit with price discrimination and calculate the value of the maximum profit.

Solution

The topic of price discrimination has already been discussed in Section 4.7. This particular problem is identical to the worked example solved in that section using ordinary differentiation. You might like to compare the details of the two approaches.

Our current aim is to find an expression for profit in terms of Q_1 and Q_2 which can then be optimized using partial differentiation. For the domestic market the demand equation is

$$P_1 + Q_1 = 500$$

which rearranges as

$$P_1 = 500 - Q_1$$

The total revenue function for this market is then

$$TR_1 = P_1Q_1 = (500 - Q_1)Q_1 = 500Q_1 - Q_1^2$$

For the industrial market the demand equation is

$$2P_2 + 3Q_2 = 720$$

which rearranges as

$$P_2 = 360 - \tfrac{3}{2}Q_2$$

The total revenue function for this market is then

$$TR_2 = P_2Q_2 = (360 - \tfrac{3}{2}Q_2)Q_2 = 360Q_2 - \tfrac{3}{2}Q_2^2$$

The total revenue received from sales in both markets is

$$TR = TR_1 + TR_2 = 500Q_1 - Q_1^2 + 360Q_2 - \tfrac{3}{2}Q_2^2$$

The total cost of producing these goods is given by

$$TC = 50\,000 + 20Q$$

and, since $Q = Q_1 + Q_2$, we can write this as

$$\begin{aligned} TC &= 50\,000 + 20(Q_1 + Q_2) \\ &= 50\,000 + 20Q_1 + 20Q_2 \end{aligned}$$

The firm's profit function is therefore

$$\begin{aligned} \pi &= TR - TC \\ &= (500Q_1 - Q_1^2 + 360Q_2 - {}^3\!/\!_2Q_2^2) - (50\,000 + 20Q_1 + 20Q_2) \\ &= 480Q_1 - Q_1^2 + 340Q_2 - {}^3\!/\!_2Q_2^2 - 50\,000 \end{aligned}$$

This is a function of the two variables, Q_1 and Q_2, that we wish to optimize. The first- and second-order partial derivatives are

$$\frac{\partial \pi}{\partial Q_1} = 480 - 2Q_1$$

$$\frac{\partial \pi}{\partial Q_2} = 340 - 3Q_2$$

$$\frac{\partial^2 \pi}{\partial Q_1^2} = -2$$

$$\frac{\partial^2 \pi}{\partial Q_1 \partial Q_2} = 0$$

$$\frac{\partial^2 \pi}{\partial Q_2^2} = -3$$

The two-step strategy gives the following:

Step 1

At a stationary point

$$\frac{\partial \pi}{\partial Q_1} = 0$$

$$\frac{\partial \pi}{\partial Q_2} = 0$$

so we need to solve the simultaneous equations

$$480 - 2Q_1 = 0$$

$$340 - 3Q_2 = 0$$

These are easily solved because they are 'uncoupled'. The first equation immediately gives

$$Q_1 = \frac{480}{2} = 240$$

while the second gives

$$Q_2 = \frac{340}{3}$$

Step 2

It is easy to check that the conditions for a maximum are satisfied:

$$\frac{\partial^2 \pi}{\partial Q_1^2} = -2 < 0$$

$$\frac{\partial^2 \pi}{\partial Q_2^2} = -3 < 0$$

$$\left(\frac{\partial^2 \pi}{\partial Q_1^2}\right)\left(\frac{\partial^2 \pi}{\partial Q_2^2}\right) - \left(\frac{\partial^2 \pi}{\partial Q_1 \partial Q_2}\right)^2 = (-2)(-3) - 0^2 = 6 > 0$$

The question actually asks for the optimum prices rather than the quantities. These are found by substituting

$$Q_1 = 240 \quad \text{and} \quad Q_2 = \frac{340}{3}$$

into the corresponding demand equations. For the domestic market

$$P_1 = 500 - Q_1 = 500 - 240 = \$260$$

For the industrial market

$$P_2 = 360 - \frac{3}{2}Q_1 = 360 - \frac{3}{2}\left(\frac{340}{3}\right) = \$190$$

Finally, we substitute the values of Q_1 and Q_2 into the profit function

$$\pi = 480Q_1 - Q_1^2 + 340Q_2 - {}^3\!/_2Q_2^2 - 50\,000$$

to deduce that the maximum profit is \$26 866.67.

Practice Problem

3 A firm has the possibility of charging different prices in its domestic and foreign markets. The corresponding demand equations are given by

$$Q_1 = 300 - P_1$$

$$Q_2 = 400 - 2P_2$$

The total cost function is

$$TC = 5000 + 100Q$$

where $Q = Q_1 + Q_2$. Determine the prices that the firm should charge to maximize profit with price discrimination and calculate the value of this profit.

[You have already solved this particular example in Practice Problem 2(a) of Section 4.7.]

Example

MAPLE

A utility function is given by

$$U = 4x_1 + 2x_2 - x_1^2 - x_1^2 + x_1x_2 \quad (0 \le x_1 \le 5, 0 \le x_2 \le 5)$$

where x_1 and x_2 denote the number of units of goods 1 and 2 that are consumed.

(a) Draw a three-dimensional plot of this function and hence estimate the values of x_1 and x_2 at the stationary point. Is this a maximum, minimum or saddle point?

(b) Find the exact values of x_1 and x_2 at the stationary point using calculus.

Solution

(a) We can name this function `utility` by typing

```
>utility:=4*x1+2*x2-x1^2-x2^2+x1*x2;
```

and then plot the surface by typing

```
>plot3d(utility,x1=0..5,x2=0..5);
```

This surface, rotated so that the origin is to the front of the picture, is drawn in Figure 5.13, which shows that there is just one stationary point. This is clearly a maximum with approximate coordinates (3, 3).

(b) The partial derivatives are worked out by typing

```
>derivx1:=diff(utility,x1);
```

which gives

derivx1:=4−2x1+x2

and then typing

```
>derivx2:=diff(utility,x2);
```

which gives

derivx1:=2−2x2+x1

Figure 5.13

```
> utility:=4*x1+2*x2-x1^2-x2^2+x1*x2;
```

$$utility := 4\,x1 + 2\,x2 - x1^2 - x2^2 + x1\,x2$$

```
> plot3d(utility,x1=0..5,x2=0..5);
```

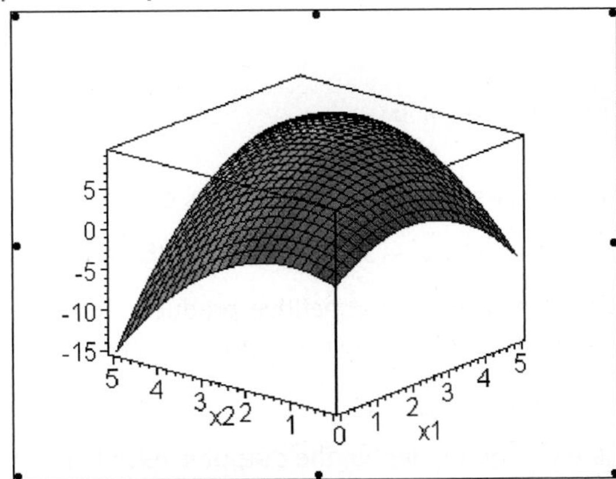


```
>
```

The stationary point is then found by setting each of these derivatives to zero and solving the resulting simultaneous equations. This is achieved by typing

```
>solve({derivx1=0,derivx2=0},{x1,x2});
```

which generates the response

$$\left\{x_2 = \frac{8}{3}, x_1 = \frac{10}{3}\right\}$$

so we conclude that the utility function has a maximum point when

$$x_1 = \frac{10}{3} \quad \text{and} \quad x_2 = \frac{8}{3}$$

Key Terms

Maximum point A point on a surface which has the highest function value in comparison with other values in its neighbourhood; at such a point the surface looks like the top of a mountain.

Minimum point A point on a surface which has the lowest function value in comparison with other values in its neighbourhood; at such a point the surface looks like the bottom of a valley or bowl.

Saddle point A stationary point which is neither a maximum or minimum and at which the surface looks like the middle of a horse's saddle.

Advice

As usual we conclude this section with some additional problems for you to try. Practice Problems 4, 5 and 6 are similar to those given in the text. Practice Problems 7, 8 and 9 are slightly different, so you may prefer to concentrate on these if you find that you do not have the time to attempt all of them. Practice Problems 10 and 11 provide practice in using Maple.

Practice Problems

4 Find and classify the stationary points of the following functions:

(a) $f(x, y) = x^3 + y^3 - 3x - 3y$ **(b)** $f(x, y) = x^3 + 3xy^2 - 3x^2 - 3y^2 + 10$

5 A firm is a perfectly competitive producer and sells two goods G1 and G2 at \$70 and \$50, respectively, each. The total cost of producing these goods is given by

$$TC = Q_1^2 + Q_1Q_2 + Q_2^2$$

where Q_1 and Q_2 denote the output levels of G1 and G2. Find the maximum profit and the values of Q_1 and Q_2 at which this is achieved.

6 The demand functions for a firm's domestic and foreign markets are

$$P_1 = 50 - 5Q_1$$

$$P_2 = 30 - 4Q_2$$

and the total cost function is

$$TC = 10 + 10Q$$

where $Q = Q_1 + Q_2$. Determine the prices needed to maximize profit with price discrimination and calculate the value of the maximum profit.

[You have already solved this particular example in Practice Problem 8(a) of Section 4.7.]

7 A firm's production function is given by

$$Q = 2L^{1/2} + 3K^{1/2}$$

where Q, L and K denote the number of units of output, labour and capital. Labour costs are \$2 per unit, capital costs are \$1 per unit and output sells at \$8 per unit. Show that the profit function is

$$\pi = 16L^{1/2} + 24K^{1/2} - 2L - K$$

and hence find the maximum profit and the values of L and K at which it is achieved.

8 An individual's utility function is given by

$$U = 260x_1 + 310x_2 + 5x_1x_2 - 10x_1^2 - x_2^2$$

where x_1 is the amount of leisure measured in hours per week and x_2 is earned income measured in dollars per week. Find the values of x_1 and x_2 which maximize U. What is the corresponding hourly rate of pay?

9 A monopolist produces the same product at two factories. The cost functions for each factory are as follows

$$TC_1 = 8Q_1 \quad \text{and} \quad TC_2 = Q_2^2$$

The demand function for the good is

$$P = 100 - 2Q$$

where $Q = Q_1 + Q_2$. Find the values of Q_1 and Q_2 which maximize profit.

10 **(Maple)** Draw the surface representing each of the following functions. By rotating the surface, estimate the x and y values of the stationary points and state whether they are maxima, minima or saddle points. Use calculus to find the exact location of these points.

(a) $z = x^2 + y^2 - 2x - 4y + 15 \quad (0 \le x \le 4, 0 \le y \le 4)$

(b) $z = 39 - x^2 - y^2 + 2y \quad (-2 \le x \le 2, -2 \le y \le 2)$

(c) $z = x^2 - y^2 - 4x + 4y \quad (0 \le x \le 4, 0 \le y \le 4)$

11 **(Maple)** A monopolistic producer charges different prices at home and abroad. The demand functions of the domestic and foreign markets are given by

$$P_1 + Q_1 = 100 \quad \text{and} \quad P_2 + 2Q_2 = 80$$

respectively. The firm's total cost function is

$$TC = (Q_1 + Q_2)^2$$

(a) Show that the firm's profit function is given by

$$\pi = 100Q_1 + 80Q_2 - 2Q_1^2 - 3Q_2^2 - 2Q_1Q_2$$

Use calculus to show that profit is maximized when $Q_1 = 22$ and $Q_2 = 6$, and find the corresponding prices.

(b) The foreign country believes that the firm is guilty of dumping because the good sells at a higher price in the home market, so decides to restrict the sales to a maximum of 2, so that $Q_2 \le 2$. By plotting π in the region $0 \le Q_1 \le 30$, $0 \le Q_2 \le 2$, explain why the profit is maximized when $Q_2 = 2$. Use calculus to find value of Q_1, and compare the corresponding profit with that of the free market in part (a).

section 5.5

Constrained optimization

Objectives

At the end of this section you should be able to:

- Give a graphical interpretation of constrained optimization.
- Show that when a firm maximizes output subject to a cost constraint, the ratio of marginal product to price is the same for all inputs.
- Show that when a consumer maximizes utility subject to a budgetary constraint, the ratio of marginal utility to price is the same for all goods.
- Use the method of substitution to solve constrained optimization problems in economics.

Advice

In this section we begin by proving some theoretical results before describing the method of substitution. You might prefer to skip the theory at a first reading, and begin with the two worked examples.

In Section 5.4 we described how to find the optimum (that is, maximum or minimum) of a function of two variables

$$z = f(x, y)$$

where the variables x and y are free to take any values. As we pointed out at the beginning of that section, this assumption is unrealistic in many economic situations. An individual wishing to maximize utility is subject to an income constraint and a firm wishing to maximize output is subject to a cost constraint.

In general, we want to optimize a function,

$$z = f(x, y)$$

called the *objective* function subject to a constraint

$$\varphi(x, y) = M$$

Here φ, the Greek letter phi, is a known function of two variables and M is a known constant. The problem is to pick the pair of numbers (x, y) which maximizes or minimizes $f(x, y)$ as before. This time, however, we limit the choice of pairs to those which satisfy

$$\varphi(x, y) = M$$

A graphical interpretation should make this clear. To be specific, let us suppose that a firm wants to maximize output and that the production function is of the form

$$Q = f(K, L)$$

Let the costs of each unit of capital and labour be P_K and P_L respectively. The cost to the firm of using as input K units of capital and L units of labour is

$$P_K K + P_L L$$

so if the firm has a fixed amount, M, to spend on these inputs then

$$P_K K + P_L L = M$$

The problem is one of trying to maximize the objective function

$$Q = f(K, L)$$

subject to the cost constraint

$$P_K K + P_L L = M$$

Sketched in Figure 5.14 (overleaf) is a typical isoquant map. As usual, points on any one isoquant yield the same level of output and as output rises the isoquants themselves move further away from the origin. Also sketched in Figure 5.14 is the cost constraint. This is called an *isocost curve* because it gives all combinations of K and L which can be bought for a fixed cost, M.

The fact that

$$P_K K + P_L L = M$$

is represented by a straight line should come as no surprise to you by now. We can even rewrite it in the more familiar '$y = ax + b$' form and so identify its slope and intercept. In Figure 5.14, L is plotted on the horizontal axis and K is plotted on the vertical axis, so we need to rearrange

$$P_K K + P_L L = M$$

to express K in terms of L. Subtracting $P_L L$ from both sides and dividing through by P_K gives

$$K = \left(-\frac{P_L}{P_K}\right)L + \frac{M}{P_K}$$

The isocost curve is therefore a straight line with slope $-P_L/P_K$ and intercept M/P_K. Graphically, our constrained problem is to choose that point on the isocost line which maximizes output. This is given by the point labelled A in Figure 5.14. Point A certainly lies on the isocost line and it maximizes output because it also lies on the highest isoquant. Other points, such as B and C, also satisfy the constraint but they lie on lower isoquants and so yield smaller levels of output than A. Point A is characterized by the fact that the isocost line is tangential to an isoquant. In other words, the slope of the isocost line is the same as that of the isoquant at A.

Figure 5.14

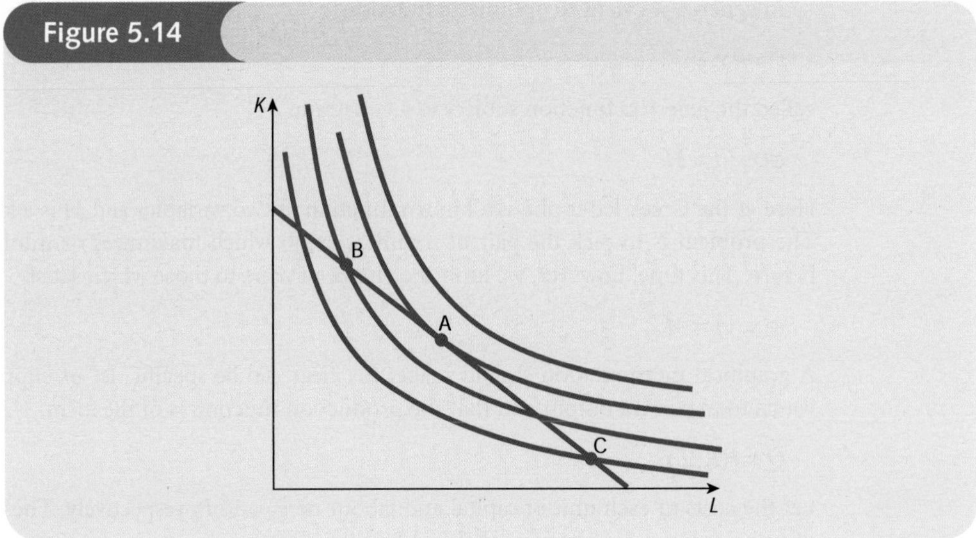

Now we have already shown that the isocost line has slope $-P_L/P_K$. In Section 5.2 we defined the marginal rate of technical substitution, MRTS, to be minus the slope of an isoquant, so at point A we must have

$$\frac{P_L}{P_K} = \text{MRTS}$$

We also showed that

$$\text{MRTS} = \frac{\text{MP}_L}{\text{MP}_K}$$

so

$$\frac{P_L}{P_K} = \frac{\text{MP}_L}{\text{MP}_K}$$

the ratio of the input prices is equal to the ratio of their marginal products

This relationship can be rearranged as

$$\frac{\text{MP}_L}{P_L} = \frac{\text{MP}_K}{P_K}$$

so when output is maximized subject to a cost constraint

the ratio of marginal product to price is the same for all inputs

The marginal product determines the change in output due to a 1 unit increase in input. This optimization condition therefore states that the last dollar spent on labour yields the same addition to output as the last dollar spent on capital.

The above discussion has concentrated on production functions. An analogous situation arises when we maximize utility functions

$$U = U(x_1, x_2)$$

where x_1, x_2 denote the number of items of goods G1, G2 that an individual buys. If the prices of these goods are denoted by P_1 and P_2 and the individual has a fixed budget, M, to spend on these goods then the corresponding constraint is

$$P_1x_1 + P_2x_2 = M$$

This budgetary constraint plays the role of the cost constraint, and indifference curves are analogous to isoquants. Consequently, we analyse the problem by superimposing the budget line on an indifference map. The corresponding diagram is virtually indistinguishable from that of Figure 5.14. The only change is that the axes would be labelled x_1 and x_2 rather than L and K. Once again, the maximum point of the constrained problem occurs at the point of tangency, so that at this point the slope of the budget line is that of an indifference curve. Hence

$$\frac{P_1}{P_2} = \text{MRCS}$$

In Section 5.2 we derived the result

$$\text{MRCS} = \frac{\partial U/\partial x_1}{\partial U/\partial x_2}$$

Writing the partial derivatives $\partial U/\partial x_i$ more concisely as U_i we can deduce that

$$\frac{P_1}{P_2} = \frac{U_1}{U_2}$$

that is,

the ratio of the prices of the goods is equal to the ratio of their marginal utilities

Again, this relationship can be rearranged into the more familiar form

$$\frac{U_1}{P_1} = \frac{U_2}{P_2}$$

so when utility is maximized subject to a budgetary constraint,

the ratio of marginal utility to price is the same for all goods consumed

If individuals allocate their budgets between goods in this way then utility is maximized when the last dollar spent on each good yields the same addition to total utility. Under these circumstances, the consumer has achieved maximum satisfaction within the constraint of a fixed budget, so there is no tendency to reallocate income between these goods. Obviously, the consumer's equilibrium will be affected if there is a change in external conditions such as income or the price of any good. For example, suppose that P_1 suddenly increases, while P_2 and M remain fixed. If this happens then the equation

$$\frac{U_1}{P_1} = \frac{U_2}{P_2}$$

turns into an inequality

$$\frac{U_1}{P_1} < \frac{U_2}{P_2}$$

so equilibrium no longer holds. Given that P_1 has increased, consumers find that the last dollar spent no longer buys as many items of G1, so utility can be increased by purchasing more of G2 and less of G1. By the law of diminishing marginal utility, the effect is to increase U_1 and to decrease U_2. The process of reallocation continues until the ratio of marginal utilities to prices is again equal and equilibrium is again established.

The graphical approach provides a useful interpretation of constrained optimization. It has also enabled us to justify some familiar results in microeconomics. However, it does not give us a practical way of actually solving such problems. It is very difficult to produce an accurate isoquant or indifference map from any given production or utility function. We now describe an alternative approach, known as the *method of substitution*. To illustrate the method we begin with an easy example.

Example

Find the minimum value of the objective function

$$z = -2x^2 + y^2$$

subject to the constraint $y = 2x - 1$.

Solution

In this example we need to optimize the function

$$z = -2x^2 + y^2$$

given that x and y are related by

$$y = 2x - 1$$

The obvious thing to do is to substitute the expression for y given by the constraint directly into the function that we are trying to optimize to get

$$\begin{aligned} z &= -2x^2 + (2x - 1)^2 \\ &= -2x^2 + 4x^2 - 4x + 1 \\ &= 2x^2 - 4x + 1 \end{aligned}$$

Note the wonderful effect that this has on z. Instead of z being a function of two variables, x and y, it is now just a function of the one variable, x. Consequently, the minimum value of z can be found using the theory of stationary points discussed in Chapter 4.

At a stationary point

$$\frac{dz}{dx} = 0$$

that is,

$$4x - 4 = 0$$

which has solution $x = 1$. Differentiating a second time we see that

$$\frac{d^2z}{dx^2} = 4 > 0$$

confirming that the stationary point is a minimum. The value of z can be found by substituting $x = 1$ into

$$z = 2x^2 - 4x + 1$$

to get

$$z = 2(1)^2 - 4(1) + 1 = -1$$

It is also possible to find the value of y at the minimum. To do this we substitute $x = 1$ into the constraint

$$y = 2x - 1$$

to get

$$y = 2(1) - 1 = 1$$

The constrained function therefore has a minimum value of -1 at the point (1, 1).

The method of substitution for optimizing

$$z = f(x, y)$$

subject to

$$\varphi(x, y) = M$$

may be summarized as follows.

Step 1

Use the constraint

$$\varphi(x, y) = M$$

to express y in terms of x.

Step 2

Substitute this expression for y into the objective function

$$z = f(x, y)$$

to write z as a function of x only.

Step 3

Use the theory of stationary points of functions of one variable to optimize z.

Practice Problem

1 Find the maximum value of the objective function

$$z = 2x^2 - 3xy + 2y + 10$$

subject to the constraint $y = x$.

Advice

The most difficult part of the three-step strategy is step 1, where we rearrange the given constraint to write y in terms of x. In the previous example and in Practice Problem 1 this step was exceptionally easy because the constraint was linear. In both cases the constraint was even presented in the appropriate form to begin with, so no extra work was required. In general, if the constraint is non-linear, it may be difficult or impossible to perform the initial rearrangement. If this happens then you could try working the other way round and expressing x in terms of y, although there is no guarantee that this will be possible either. However, when step 1 can be tackled successfully, the method does provide a really quick way of solving constrained optimization problems.

To illustrate this we now use the method of substitution to solve two economic problems that both involve production functions. In the first example output is maximized subject to cost constraint and in the second example cost is minimized subject to an output constraint.

Example

A firm's unit capital and labour costs are \$1 and \$2 respectively. If the production function is given by

$$Q = 4LK + L^2$$

find the maximum output and the levels of K and L at which it is achieved when the total input costs are fixed at \$105. Verify that the ratio of marginal product to price is the same for both inputs at the optimum.

Solution

We are told that 1 unit of capital costs \$1 and that 1 unit of labour costs \$2. If the firm uses K units of capital and L units of labour then the total input costs are

$$K + 2L$$

This is fixed at \$105, so

$$K + 2L = 105$$

The mathematical problem is to maximize the objective function

$$Q = 4LK + L^2$$

subject to the constraint

$$K + 2L = 105$$

The three-step strategy is as follows:

Step 1

Rearranging the constraint to express K in terms of L gives

$$K = 105 - 2L$$

Step 2

Substituting this into the objective function

$$Q = 4LK + L^2$$

gives

$$Q = 4L(105 - 2L) + L^2 = 420L - 7L^2$$

and so output is now a function of the one variable, L.

Step 3

At a stationary point

$$\frac{dQ}{dL} = 0$$

that is,

$$420 - 14L = 0$$

which has solution $L = 30$. Differentiating a second time gives

$$\frac{d^2Q}{dL^2} = -14 < 0$$

confirming that the stationary point is a maximum.

The maximum output is found by substituting $L = 30$ into the objective function

$$Q = 420L - 7L^2$$

to get

$$Q = 420(30) - 7(30)^2 = 6300$$

The corresponding level of capital is found by substituting $L = 30$ into the constraint

$$K = 105 - 2L$$

to get

$$K = 105 - 2(30) = 45$$

The firm should therefore use 30 units of labour and 45 units of capital to produce a maximum output of 6300.

Finally, we are asked to check that the ratio of marginal product to price is the same for both inputs. From the formula

$$Q = 4LK + L^2$$

we see that the marginal products are given by

$$\text{MP}_L = \frac{\partial Q}{\partial L} = 4K + 2L \quad \text{and} \quad \text{MP}_K = \frac{\partial Q}{\partial K} = 4L$$

so at the optimum

$$\text{MP}_L = 4(45) + 2(30) = 240$$

and

$$\text{MP}_K = 4(30) = 120$$

The ratios of marginal products to prices are then

$$\frac{\text{MP}_L}{P_L} = \frac{240}{2} = 120$$

and

$$\frac{\text{MP}_K}{P_K} = \frac{120}{1} = 120$$

which are seen to be the same.

Practice Problem

2 An individual's utility function is given by

$$U = x_1x_2$$

where x_1 and x_2 denote the number of items of two goods, G1 and G2. The prices of the goods are \$2 and \$10 respectively. Assuming that the individual has \$400 available to spend on these goods, find the utility-maximizing values of x_1 and x_2. Verify that the ratio of marginal utility to price is the same for both goods at the optimum.

Example

A firm's production function is given by

$$Q = 2K^{1/2}L^{1/2}$$

Unit capital and labour costs are \$4 and \$3 respectively. Find the values of K and L which minimize total input costs if the firm is contracted to provide 160 units of output.

Solution

Given that capital and labour costs are \$4 and \$3 per unit, the total cost of using K units of capital and L units of labour is

$$\text{TC} = 4K + 3L$$

The firm's production quota is 160, so

$$2K^{1/2}L^{1/2} = 160$$

The mathematical problem is to minimize the objective function

$$\text{TC} = 4K + 3L$$

subject to the constraint

$$2K^{1/2}L^{1/2} = 160$$

Step 1

Rearranging the constraint to express L in terms of K gives

$$L^{1/2} = \frac{80}{K^{1/2}} \quad \text{(divide both sides by } 2K^{1/2}\text{)}$$

$$L = \frac{6400}{K} \quad \text{(square both sides)}$$

Step 2

Substituting this into the objective function

$$\text{TC} = 4K + 3L$$

gives

$$TC = 4K + \frac{19\,200}{K}$$

and so total cost is now a function of the one variable, K.

Step 3

At a stationary point

$$\frac{d(TC)}{dK} = 0$$

that is,

$$4 - \frac{19\,200}{K^2} = 0$$

This can be written as

$$4 = \frac{19\,200}{K^2}$$

so that

$$K^2 = \frac{19\,200}{4} = 4800$$

Hence

$$K = \sqrt{4800} = 69.28$$

Differentiating a second time gives

$$\frac{d^2(TC)}{dK^2} = \frac{38\,400}{K^3} > 0 \quad \text{because} \quad K > 0$$

confirming that the stationary point is a minimum.

Finally, the value of L can be found by substituting $K = 69.28$ into the constraint

$$L = \frac{6400}{K}$$

to get

$$L = \frac{6400}{69.28} = 92.38$$

We are not asked for the minimum cost, although this could easily be found by substituting the values of K and L into the objective function.

Practice Problem

3 A firm's total cost function is given by

$$TC = 3x_1^2 + 2x_1x_2 + 7x_2^2$$

where x_1 and x_2 denote the number of items of goods G1 and G2, respectively, that are produced. Find the values of x_1 and x_2 which minimize costs if the firm is committed to providing 40 goods of either type in total.

Key Terms

Isocost curve A line showing all combinations of two factors which can be bought for a fixed cost.

Method of substitution The method of solving constrained optimization problems whereby the constraint is used to eliminate one of the variables in the objective function.

Objective function A function that one seeks to optimize (usually) subject to constraints.

Practice Problems

4 Find the maximum value of

$$z = 6x - 3x^2 + 2y$$

subject to the constraint

$$y - x^2 = 2$$

5 Find the maximum value of

$$z = 80x - 0.1x^2 + 100y - 0.2y^2$$

subject to the constraint

$$x + y = 500$$

6 A firm's production function is given by

$$Q = 10K^{1/2}L^{1/4}$$

Unit capital and labour costs are \$4 and \$5 respectively and the firm spends a total of \$60 on these inputs. Find the values of *K* and *L* which maximize output.

7 A firm's production function is given by

$$Q = 50KL$$

Unit capital and labour costs are \$2 and \$3 respectively. Find the values of *K* and *L* which minimize total input costs if the production quota is 1200.

8 A firm's production function is given by

$$Q = 2L^{1/2} + 3K^{1/2}$$

where Q, *L* and *K* denote the number of units of output, labour and capital respectively. Labour costs are \$2 per unit, capital costs are \$1 per unit and output sells at \$8 per unit. If the firm is prepared to spend \$99 on input costs, find the maximum profit and the values of *K* and *L* at which it is achieved.

[You might like to compare your answer with the corresponding unconstrained problem that you solved in Practice Problem 7 of Section 5.4.]

9 A consumer's utility function is

$$U = \ln x_1 + 2 \ln x_2$$

Find the values of x_1 and x_2 which maximize *U* subject to the budgetary constraint

$$2x_1 + 3x_2 = 18$$

section 5.6

Lagrange multipliers

Objectives

At the end of this section you should be able to:

- Use the method of Lagrange multipliers to solve constrained optimization problems.
- Give an economic interpretation of Lagrange multipliers.
- Use Lagrange multipliers to maximize a Cobb–Douglas production function subject to a cost constraint.
- Use Lagrange multipliers to show that when a firm maximizes output subject to a cost constraint, the ratio of marginal product to price is the same for all inputs.

We now describe the method of Lagrange multipliers for solving constrained optimization problems. This is the preferred method, since it handles non-linear constraints and problems involving more than two variables with ease. It also provides some additional information that is useful when solving economic problems.

To optimize an objective function

$$f(x, y)$$

subject to a constraint

$$\varphi(x, y) = M$$

we work as follows.

Step 1

Define a new function

$$g(x, y, \lambda) = f(x, y) + \lambda[M - \varphi(x, y)]$$

Step 2

Solve the simultaneous equations

$$\frac{\partial g}{\partial x} = 0$$

$$\frac{\partial g}{\partial y} = 0$$

$$\frac{\partial g}{\partial \lambda} = 0$$

for the three unknowns, x, y and λ.

The basic steps of the method are straightforward. In step 1 we combine the objective function and constraint into a single function. To do this we first rearrange the constraint as

$$M - \varphi(x, y)$$

and multiply by the scalar (i.e. number) λ (the Greek letter 'lambda'). This scalar is called the *Lagrange multiplier*. Finally, we add on the objective function to produce the new function

$$g(x, y, \lambda) = f(x, y) + \lambda[M - \varphi(x, y)]$$

This is called the *Lagrangian* function. The right-hand side involves the three letters x, y and λ, so g is a function of three variables.

In step 2 we work out the three first-order partial derivatives

$$\frac{\partial g}{\partial x}, \frac{\partial g}{\partial y}, \frac{\partial g}{\partial \lambda}$$

and equate these to zero to produce a system of three simultaneous equations for the three unknowns x, y and λ. The point (x, y) is then the optimal solution of the constrained problem. The number λ can also be given a meaning and we consider this later. For the moment we consider a simple example to get us started.

Example

Use Lagrange multipliers to find the optimal value of

$$x^2 - 3xy + 12x$$

subject to the constraint

$$2x + 3y = 6$$

Solution

Step 1

In this example

$$f(x, y) = x^2 - 3xy + 12x$$

$$\varphi(x, y) = 2x + 3y$$

$$M = 6$$

so the Lagrangian function is given by

$$g(x, y, \lambda) = x^2 - 3xy + 12x + \lambda(6 - 2x - 3y)$$

Step 2

Working out the three partial derivatives of g gives

$$\frac{\partial g}{\partial x} = 2x - 3y + 12 - 2\lambda$$

$$\frac{\partial g}{\partial y} = -3x - 3\lambda$$

$$\frac{\partial g}{\partial \lambda} = 6 - 2x - 3y$$

so we need to solve the simultaneous equations

$$\begin{aligned} 2x - 3y + 12 - 2\lambda &= 0 \\ -3x - 3\lambda &= 0 \\ 6 - 2x - 3y &= 0 \end{aligned}$$

that is,

$$2x - 3y - 2\lambda = -12 \quad (1)$$

$$-3x - 3\lambda = 0 \quad (2)$$

$$2x + 3y = 6 \quad (3)$$

We can eliminate x from equation (2) by multiplying equation (1) by 3, multiplying equation (2) by 2 and adding. Similarly, x can be eliminated from equation (3) by subtracting equation (3) from (1). These operations give

$$-9y - 12\lambda = -36 \quad (4)$$

$$-6y - 2\lambda = -18 \quad (5)$$

The variable y can be eliminated by multiplying equation (4) by 6 and equation (5) by 9, and subtracting to get

$$-54\lambda = -54 \quad (6)$$

so $\lambda = 1$. Substituting this into equations (5) and (2) gives $y = 8/3$ and $x = -1$ respectively.

The optimal solution is therefore $(-1, 8/3)$ and the corresponding value of the objective function

$$x^2 - 3xy + 12x$$

is

$$(-1)^2 - 3(-1)(8/3) + 12(-1) = -3$$

Practice Problem

1 Use Lagrange multipliers to optimize

$$2x^2 - xy$$

subject to

$$x + y = 12$$

Looking back at the worked example and your own solution to Practice Problem 1, notice that the third equation in step 2 is just a restatement of the original constraint. It is easy to see that this is always the case because if

$$g(x, y, \lambda) = f(x, y) + \lambda[M - \varphi(x, y)]$$

then

$$\frac{\partial g}{\partial \lambda} = M - \varphi(x,y)$$

The equation

$$\frac{\partial g}{\partial \lambda} = 0$$

then implies the constraint

$$\varphi(x, y) = M$$

It is possible to make use of second-order partial derivatives to classify the optimal point. Unfortunately, these conditions are quite complicated and are considered in Appendix 3. In all problems that we consider there is only a single optimum and it is usually obvious on economic grounds whether it is a maximum or a minimum.

Example

A monopolistic producer of two goods, G1 and G2, has a joint total cost function

$$\text{TC} = 10Q_1 + Q_1Q_2 + 10Q_2$$

where Q_1 and Q_2 denote the quantities of G1 and G2 respectively. If P_1 and P_2 denote the corresponding prices then the demand equations are

$$P_1 = 50 - Q_1 + Q_2$$

$$P_2 = 30 + 2Q_1 - Q_2$$

Find the maximum profit if the firm is contracted to produce a total of 15 goods of either type. Estimate the new optimal profit if the production quota rises by 1 unit.

Solution

The first thing that we need to do is to write down expressions for the objective function and constraint. The objective function is profit and is given by

$$\pi = \text{TR} - \text{TC}$$

The total cost function is given to be

$$\text{TC} = 10Q_1 + Q_1Q_2 + 10Q_2$$

However, we need to use the demand equations to obtain an expression for TR. Total revenue from the sale of G1 is

$$\text{TR}_1 = P_1Q_1 = (50 - Q_1 + Q_2)Q_1 = 50Q_1 - Q_1^2 + Q_2Q_1$$

and total revenue from the sale of G2 is

$$\text{TR}_2 = P_2Q_2 = (30 + 2Q_1 - Q_2)Q_2 = 30Q_2 + 2Q_1Q_2 - Q_2^2$$

so

$$\begin{aligned}\text{TR} &= \text{TR}_1 + \text{TR}_2\\ &= 50Q_1 - Q_1^2 + Q_2Q_1 + 30Q_2 + 2Q_1Q_2 - Q_2^2\\ &= 50Q_1 - Q_1^2 + 3Q_1Q_2 + 30Q_2 - Q_2^2\end{aligned}$$

Hence

$$\begin{aligned}\pi &= \text{TR} - \text{TC}\\ &= (50Q_1 - Q_1^2 + 3Q_1Q_2 + 30Q_2 - Q_2^2) - (10Q_1 + Q_1Q_2 + 10Q_2)\\ &= 40Q_1 - Q_1^2 + 2Q_1Q_2 + 20Q_2 - Q_2^2\end{aligned}$$

The constraint is more easily determined. We are told that the firm produces 15 goods in total, so

$$Q_1 + Q_2 = 15$$

The mathematical problem is to maximize the objective function

$$\pi = 40Q_1 - Q_1^2 + 2Q_1Q_2 + 20Q_2 - Q_2^2$$

subject to the constraint

$$Q_1 + Q_2 = 15$$

Step 1

The Lagrangian function is

$$g(Q_1, Q_2, \lambda) = 40Q_1 - Q_1^2 + 2Q_1Q_2 + 20Q_2 - Q_2^2 + \lambda(15 - Q_1 - Q_2)$$

Step 2

The simultaneous equations

$$\frac{\partial g}{\partial Q_1} = 0, \quad \frac{\partial g}{\partial Q_2} = 0, \quad \frac{\partial g}{\partial \lambda} = 0$$

are

$$\begin{aligned}40 - 2Q_1 + 2Q_2 - \lambda &= 0\\ 2Q_1 + 20 - 2Q_2 - \lambda &= 0\\ 15 - Q_1 - Q_2 &= 0\end{aligned}$$

that is,

$$-2Q_1 + 2Q_2 - \lambda = -40 \qquad (1)$$

$$2Q_1 - 2Q_2 - \lambda = -20 \qquad (2)$$

$$Q_1 + Q_2 = 15 \qquad (3)$$

The obvious way of solving this system is to add equations (1) and (2) to get

$$-2\lambda = -60$$

so $\lambda = 30$. Putting this into equation (1) gives

$$-2Q_1 + 2Q_2 = -10 \qquad (4)$$

Equations (3) and (4) constitute a system of two equations for the two unknowns Q_1 and Q_2. We can eliminate Q_1 by multiplying equation (3) by 2 and adding equation (4) to get

$$4Q_2 = 20$$

➔

so $Q_2 = 5$. Substituting this into equation (3) gives

$$Q_1 = 15 - 5 = 10$$

The maximum profit is found by substituting $Q_1 = 10$ and $Q_2 = 5$ into the formula for π to get

$$\pi = 40(10) - (10)^2 + 2(10)(5) + 20(5) - 5^2 = 475$$

The final part of this example wants us to find the new optimal profit when the production quota rises by 1 unit. One way of doing this is just to repeat the calculations replacing the previous quota of 15 by 16, although this is extremely tedious and not strictly necessary. There is a convenient shortcut based on the value of the Lagrange multiplier λ. To understand this, let us replace the production quota, 15, by the variable M, so that the Lagrangian function is

$$g(Q_1, Q_2, \lambda, M) = 40Q_1 - Q_1^2 + 2Q_1Q_2 + 20Q_2 - Q_2^2 + \lambda(M - Q_1 - Q_2)$$

The expression on the right-hand side involves Q_1, Q_2, λ and M, so g is now a function of four variables. If we partially differentiate with respect to M then

$$\frac{\partial g}{\partial M} = \lambda$$

We see that λ is a multiplier not only in the mathematical but also in the economic sense. It represents the (approximate) change in g due to a 1 unit increase in M. Moreover, if the constraint is satisfied, then

$$Q_1 + Q_2 = M$$

and the expression for g reduces to

$$40Q_1 - Q_1^2 + 2Q_1Q_2 + 20Q_2 - Q_2^2$$

which is equal to profit. The value of the Lagrange multiplier represents the change in optimal profit brought about by a 1 unit increase in the production quota. In this case, $\lambda = 30$, so profit rises by 30 to become 505.

The interpretation placed on the value of λ in this example applies quite generally. Given an objective function

$$f(x, y)$$

and constraint

$$\varphi(x, y) = M$$

the value of λ gives the approximate change in the optimal value of f due to a 1 unit increase in M.

Practice Problem

2 A consumer's utility function is given by

$$U(x_1, x_2) = 2x_1x_2 + 3x_1$$

where x_1 and x_2 denote the number of items of two goods G1 and G2 that are bought. Each item costs \$1 for G1 and \$2 for G2. Use Lagrange multipliers to find the maximum value of U if the consumer's income is \$83. Estimate the new optimal utility if the consumer's income rises by \$1.

Example

Use Lagrange multipliers to find expressions for K and L which maximize output given by a Cobb–Douglas production function

$$Q = AK^{\alpha}L^{\beta} \quad (A, \alpha \text{ and } \beta \text{ are positive constants})$$

subject to a cost constraint

$$P_K K + P_L L = M$$

Solution

This example appears very hard at first sight because it does not involve specific numbers. However, it is easy to handle such generalized problems provided that we do not panic.

Step 1

The Lagrangian function is

$$g(K, L, \lambda) = AK^{\alpha}L^{\beta} + \lambda(M - P_K K - P_L L)$$

Step 2

The simultaneous equations

$$\frac{\partial g}{\partial K} = 0, \quad \frac{\partial g}{\partial L} = 0, \quad \frac{\partial g}{\partial \lambda} = 0$$

are

$$A\alpha K^{\alpha-1}L^{\beta} - \lambda P_K = 0 \qquad (1)$$

$$A\beta K^{\alpha}L^{\beta-1} - \lambda P_L = 0 \qquad (2)$$

$$M - P_K K - P_L L = 0 \qquad (3)$$

These equations look rather forbidding. Before we begin to solve them it pays to simplify equations (1) and (2) slightly by introducing $Q = AK^{\alpha}L^{\beta}$. Notice that

$$A\alpha K^{\alpha-1}L^{\beta} = \frac{\alpha(AK^{\alpha}L^{\beta})}{K} = \frac{\alpha Q}{K}$$

$$A\beta K^{\alpha}L^{\beta-1} = \frac{\beta(AK^{\alpha}L^{\beta})}{L} = \frac{\beta Q}{L}$$

so equations (1), (2) and (3) can be written

$$\frac{\alpha Q}{K} - \lambda P_K = 0 \qquad (4)$$

$$\frac{\beta Q}{L} - \lambda P_L = 0 \qquad (5)$$

$$P_K K + P_L L = M \qquad (6)$$

Equations (4) and (5) can be rearranged to give

$$\lambda = \frac{\alpha Q}{P_K K} \quad \text{and} \quad \lambda = \frac{\beta Q}{P_L L}$$

➔

so that

$$\frac{\alpha Q}{P_K K} = \frac{\beta Q}{P_L L}$$

and hence

$$\frac{P_K K}{\alpha} = \frac{P_L L}{\beta} \quad \text{(divide both sides by } Q \text{ and turn both sides upside down)}$$

that is,

$$P_K K = \frac{\alpha}{\beta} P_L L \quad \text{(multiply through by } \alpha\text{)} \tag{7}$$

Substituting this into equation (6) gives

$$\frac{\alpha}{\beta} P_L L + P_L L = M$$

$$\alpha L + \beta L = \frac{\beta M}{P_L} \quad \text{(multiply through by } \beta/P_L\text{)}$$

$$(\alpha + \beta)L = \frac{\beta M}{P_L} \quad \text{(factorize)}$$

$$L = \frac{\beta M}{(\alpha + \beta)P_L} \quad \text{(divide through by } \alpha + \beta\text{)}$$

Finally, we can put this into equation (7) to get

$$P_K K = \frac{\alpha M}{\alpha + \beta}$$

so

$$K = \frac{\alpha M}{(\alpha + \beta)P_K}$$

The values of K and L which optimize Q are therefore

$$\frac{\alpha M}{(\alpha + \beta)P_K} \quad \text{and} \quad \frac{\beta M}{(\alpha + \beta)P_L}$$

Practice Problem

3 Use Lagrange multipliers to find expressions for x_1 and x_2 which maximize the utility function

$$U = x_1^{1/2} + x_2^{1/2}$$

subject to the general budgetary constraint

$$P_1 x_1 + P_2 x_2 = M$$

The previous example illustrates the power of mathematics when solving economics problems. The main advantage of using algebra and calculus rather than just graphs and tables of numbers is their generality. In future, if we need to maximize any particular Cobb–Douglas production function subject to any particular cost constraint, then all we have to do is to quote the result of the previous example. By substituting specific values of M, α, β, P_K and P_L into the general formulas for K and L, we can write down the solution in a matter of seconds. In fact, we can use mathematics to generalize still further. Rather than work with production functions of a prescribed form such as

$$Q = AK^{\alpha}L^{\beta}$$

we can obtain results pertaining to any production function

$$Q = f(K, L)$$

For instance, we can use Lagrange multipliers to justify a result that we derived graphically in Section 5.5. At the beginning of that section we showed that when output is maximized subject to a cost constraint, the ratio of marginal product to price is the same for all inputs. To obtain this result using Lagrange multipliers we simply write down the Lagrangian function

$$g(K, L, \lambda) = f(K, L) + \lambda(M - P_K K - P_L L)$$

which corresponds to a production function

$$f(K, L)$$

and cost constraint

$$P_K K + P_L L = M$$

The simultaneous equations

$$\frac{\partial g}{\partial K} = 0, \quad \frac{\partial g}{\partial L} = 0, \quad \frac{\partial g}{\partial \lambda} = 0$$

are

$$\text{MP}_K - \lambda P_K = 0 \qquad (1)$$

$$\text{MP}_L - \lambda P_L = 0 \qquad (2)$$

$$M - P_K K - P_L L = 0 \qquad (3)$$

because

$$\frac{\partial f}{\partial K} = \text{MP}_K \quad \text{and} \quad \frac{\partial f}{\partial L} = \text{MP}_L$$

Equations (1) and (2) can be rearranged to give

$$\lambda = \frac{\text{MP}_K}{P_K} \quad \text{and} \quad \lambda = \frac{\text{MP}_L}{P_L}$$

so

$$\frac{\text{MP}_K}{P_K} = \frac{\text{MP}_L}{P_L}$$

as required.

Key Terms

Lagrange multiplier The number λ which is used in the Lagrangian function. In economics this gives the change in the value of the objective function when the value of the constraint is increased by 1 unit.

Lagrangian The function $f(x, y) + \lambda[M - \varphi(x, y)]$, where $f(x, y)$ is the objective function and $\phi(x, y) = M$ is the constraint. The stationary point of this function is the solution of the associated constrained optimization problem.

Advice

There now follow five new Practice Problems for you to attempt. If you feel that you need even more practice then you are advised to rework the nine Practice Problems given in Section 5.5 using Lagrange multipliers.

Practice Problems

4 Use Lagrange multipliers to maximize

$$z = x + 2xy$$

subject to the constraint

$$x + 2y = 5$$

5 A firm that manufactures speciality bicycles has a profit function

$$\pi = 5x^2 - 10xy + 3y^2 + 240x$$

where x denotes the number of frames and y denotes the number of wheels. Find the maximum profit assuming that the firm does not want any spare frames or wheels left over at the end of the production run.

6 A monopolistic producer of two goods, G1 and G2, has a total cost function

$$\text{TC} = 5Q_1 + 10Q_2$$

where Q_1 and Q_2 denote the quantities of G1 and G2 respectively. If P_1 and P_2 denote the corresponding prices then the demand equations are

$$P_1 = 50 - Q_1 - Q_2$$

$$P_2 = 100 - Q_1 - 4Q_2$$

Find the maximum profit if the firm's total costs are fixed at \$100. Estimate the new optimal profit if total costs rise to \$101.

7 Find the maximum value of

$$Q = 10\sqrt{(KL)}$$

subject to the cost constraint

$$K + 4L = 16$$

Estimate the change in the optimal value of Q if the cost constraint is changed to

$$K + 4L = 17$$

8 A consumer's utility function is given by

$$U = \alpha \ln x_1 + \beta \ln x_2$$

Find the values of x_1 and x_2 which maximize U subject to the budgetary constraint

$$P_1x_1 + P_2x_2 = M$$

chapter 6
Integration

This chapter concludes the topic of calculus by considering the integration of functions of one variable. It is in two sections, which should be read in the order that they appear.

Section 6.1 introduces the idea of integration as the opposite process to that of differentiation. It enables you to recover an expression for the total revenue function from any given marginal revenue function, to recover the total cost function from any marginal cost function and so on. You will no doubt be pleased to discover that no new mathematical techniques are needed for this. All that is required is for you to put your brain into reverse gear. Of course, driving backwards is a little harder to master than going forwards. However, with practice you should find integration almost as easy as differentiation.

Section 6.2 shows how integration can be used to find the area under the graph of a function. This process is called definite integration. We can apply the technique to supply and demand curves and so calculate producer's and consumer's surpluses. Definite integration can also be used to determine capital stock and to discount a continuous revenue stream.

section 6.1

Indefinite integration

Objectives

At the end of this section you should be able to:

- Recognize the notation for indefinite integration.
- Write down the integrals of simple power and exponential functions.
- Integrate functions of the form $af(x) + bg(x)$.
- Find the total cost function given any marginal cost function.
- Find the total revenue function given any marginal revenue function.
- Find the consumption and savings functions given either the marginal propensity to consume or the marginal propensity to save.

Throughout mathematics there are many pairs of operations which cancel each other out and take you back to where you started. Perhaps the most obvious pair is multiplication and division. If you multiply a number by a non-zero constant, k, and then divide by k you end up with the number you first thought of. This situation is described by saying that the two operations are *inverses* of each other. In calculus, the inverse of differentiation is called *integration*.

Suppose that you are required to find a function, $F(x)$, which differentiates to

$$f(x) = 3x^2$$

Can you guess what $F(x)$ is in this case? Given such a simple function it is straightforward to write down the answer by inspection. It is

$$F(x) = x^3$$

because

$$F'(x) = 3x^2 = f(x) \quad ✓$$

as required.

As a second example, consider

$$f(x) = x^7$$

Can you think of a function, $F(x)$, which differentiates to this? Recall that when power functions are differentiated the power decreases by 1, so it makes sense to do the opposite here and to try

$$F(x) = x^8$$

Unfortunately, this does not quite work out, because it differentiates to

$$8x^7$$

which is eight times too big. This suggests that we try

$$F(x) = \tfrac{1}{8}x^8$$

which does work because

$$F'(x) = \tfrac{8}{8}x^7 = x^7 = f(x) \quad ✓$$

In general, if $F'(x) = f(x)$ then $F(x)$ is said to be the *integral* (sometimes called the *anti-derivative* or *primitive*) of $f(x)$ and is written

$$F(x) = \int f(x)\mathrm{d}x$$

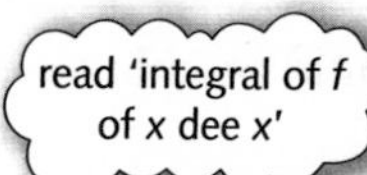

In this notation

$$\int 3x^2\mathrm{d}x = x^3$$

and

$$\int x^7\mathrm{d}x = \tfrac{1}{8}x^8$$

Example

Find

$$\int \frac{1}{x^4}\mathrm{d}x$$

Solution

Writing $\frac{1}{x^4}$ in the form x^{-4} suggests that we try

$$F(x) = x^{-3}$$

which gives

$$F'(x) = -3x^{-4}.$$

This is (-3) times too big, so

$$\int \frac{1}{x^4}\mathrm{d}x = -\frac{1}{3}x^{-3} = -\frac{1}{3x^3}$$

Here is a problem for you to try. Do not let the notation

$$\int \mathrm{d}x$$

put you off. It is merely an instruction for you to think of a function that differentiates to whatever is squashed between the integral sign '∫' and dx. If you get stuck, try adding 1 on to the power. Differentiate your guess and if it does not quite work out then go back and try again, adjusting the coefficient accordingly.

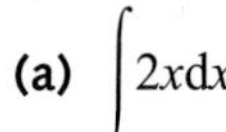

Practice Problem

1 Find

(a) $\int 2x\mathrm{d}x$ **(b)** $\int 4x^3\mathrm{d}x$ **(c)** $\int 100x^{99}\mathrm{d}x$ **(d)** $\int x^3\mathrm{d}x$ **(e)** $\int x^{18}\mathrm{d}x$

In Problem 1(a) you probably wrote

$$\int 2x\mathrm{d}x = x^2$$

However, there are other possibilities. For example, both of the functions

$$x^2 + 6 \quad \text{and} \quad x^2 - 59$$

differentiate to $2x$, because constants differentiate to zero. In fact, we can add any constant, c, to x^2 to obtain a function that differentiates to $2x$. Hence

$$\int 2x\mathrm{d}x = x^2 + c$$

The arbitrary constant, c, is called the *constant of integration*. In general, if $F(x)$ is any function that differentiates to $f(x)$ then so does

$$F(x) + c$$

Hence

$$\boxed{\int f(x)\mathrm{d}x = F(x) + c}$$

In Problem 1 you used guesswork to find various integrals. In theory most integrals can be worked out in this way. However, considerable ingenuity (and luck!) may be required when integrating complicated functions. It is possible to develop various rules similar to those of differentiation, which we discussed in Chapter 4, although even then we sometimes have to resort to sheer trickery. It is not our intention to plod through each rule as we did in Chapter 4, for the simple reason that they are rarely needed in economics. However, it is worthwhile showing you a direct way of integrating simple functions such as

$$2x - 3x^2 + 10x^3 \quad \text{and} \quad x - \mathrm{e}^{2x} + 5$$

We begin by finding general formulae for

$$\int x^n\mathrm{d}x \quad \text{and} \quad \int \mathrm{e}^{mx}\mathrm{d}x$$

To integrate $f(x) = x^n$ an obvious first guess is

$$F(x) = x^{n+1}$$

This gives

$$F'(x) = (n + 1)x^n$$

which is $n + 1$ times too big. This suggests that we try again with

$$F(x) = \frac{1}{n+1}x^{n+1}$$

which checks out because

$$F'(x) = \frac{n+1}{n+1}x^n = x^n = f(x) \quad ✓$$

Hence

$$\boxed{\int x^n \mathrm{d}x = \frac{1}{n+1}x^{n+1} + c}$$

To integrate a power function you simply add 1 to the power and divide by the number you get. This formula holds whenever n is positive, negative, a whole number or a fraction. There is just one exception to the rule, when $n = -1$. The formula cannot be used to integrate

$$\frac{1}{x}$$

because it is impossible to divide by zero. An alternative result is therefore required in this case. We know from Chapter 4 that the natural logarithm function

$$\ln x$$

differentiates to give

$$\frac{1}{x}$$

and so

$$\boxed{\int \frac{1}{x}\mathrm{d}x = \ln x + c}$$

The last basic integral that we wish to determine is

$$\int \mathrm{e}^{mx}\mathrm{d}x$$

In Section 4.8 we showed that to differentiate an exponential function all we need to do is to multiply by the coefficient of x. To integrate we do exactly the opposite and divide by the coefficient of x, so

$$\boxed{\int \mathrm{e}^{mx}\mathrm{d}x = \frac{1}{m}\mathrm{e}^{mx} + c}$$

It is easy to check that this is correct, because if

$$F(x) = \frac{1}{m}\mathrm{e}^{mx}$$

then

$$F'(x) = \frac{m}{m}\mathrm{e}^{mx} = \mathrm{e}^{mx} \quad ✓$$

Example

Find

(a) $\int x^6 \mathrm{d}x$ **(b)** $\int \frac{1}{x^2}\mathrm{d}x$ **(c)** $\int \sqrt{x}\,\mathrm{d}x$ **(d)** $\int \mathrm{e}^{2x}\mathrm{d}x$

Solution

The formula

$$\int x^n \mathrm{d}x = \frac{1}{n+1}x^{n+1} + c$$

can be used to find the first three integrals by substituting particular values for n.

(a) Putting $n = 6$ gives

$$\int x^6 \mathrm{d}x = \frac{1}{7}x^7 + c$$

(b) Putting $n = -2$ gives

$$\int \frac{1}{x^2}\mathrm{d}x = \int x^{-2}\mathrm{d}x = \frac{1}{-1}x^{-1} + c = -\frac{1}{x} + c$$

(c) Putting $n = 1/2$ gives

$$\int \sqrt{x}\,\mathrm{d}x = \int x^{1/2}\mathrm{d}x = \frac{1}{3/2}x^{3/2} + c = \frac{2}{3}x^{3/2} + c$$

(d) To find

$$\int \mathrm{e}^{2x}\mathrm{d}x$$

we put $m = 2$ into the formula

$$\int \mathrm{e}^{mx}\mathrm{d}x = \frac{1}{m}\mathrm{e}^{mx} + c$$

to get

$$\int \mathrm{e}^{2x}\mathrm{d}x = \frac{1}{2}\mathrm{e}^{2x} + c$$

Practice Problem

2 Find

(a) $\int x^4 \mathrm{d}x$ **(b)** $\int \frac{1}{x^3}\mathrm{d}x$ **(c)** $\int x^{1/3}\mathrm{d}x$ **(d)** $\int \mathrm{e}^{3x}\mathrm{d}x$ **(e)** $\int 1\mathrm{d}x$

(f) $\int x\mathrm{d}x$ **(g)** $\int \frac{1}{x}\mathrm{d}x$

[Hint: in parts (b), (e) and (f) note that $1/x^3 = x^{-3}$, $1 = x^0$ and $x = x^1$ respectively.]

In Section 4.2 we described three rules of differentiation known as the constant, sum and difference rules. Given that integration is the inverse operation, these three rules also apply whenever we integrate a function. The integral of a constant multiple of a function is obtained by integrating the function and multiplying by the constant. The integral of the sum (or difference) of two functions is obtained by integrating the functions separately and adding (or subtracting). These three rules can be combined into the single rule:

$$\int [af(x) + bg(x)]\mathrm{d}x = a\int f(x)\mathrm{d}x + b\int g(x)\mathrm{d}x$$

This enables us to integrate an expression 'term by term', as the following example demonstrates.

Example

Find

(a) $\int (2x^2 - 4x^6)\mathrm{d}x$ **(b)** $\int \left(7\mathrm{e}^{-x} + \frac{2}{x}\right)\mathrm{d}x$ **(c)** $\int (5x^2 + 3x + 2)\mathrm{d}x$

Solution

(a) $$\int (2x^2 - 4x^6)\mathrm{d}x = 2\int x^2\mathrm{d}x - 4\int x^6\mathrm{d}x$$

Putting $n = 2$ and $n = 6$ into

$$\int x^n\mathrm{d}x = \frac{1}{n+1}x^{n+1}$$

gives

$$\int x^2\mathrm{d}x = \frac{1}{3}x^3 \quad \text{and} \quad \int x^6\mathrm{d}x = \frac{1}{7}x^7$$

Hence

$$\int (2x^2 - 4x^6)\mathrm{d}x = \frac{2}{3}x^3 - \frac{4}{7}x^7$$

Finally, we add an arbitrary constant to get

$$\int (2x^2 - 4x^6)\mathrm{d}x = \frac{2}{3}x^3 - \frac{4}{7}x^7 + c$$

As a check:

$$\text{if} \quad F(x) = \frac{2}{3}x^3 - \frac{4}{7}x^7 + c \quad \text{then} \quad F'(x) = 2x^2 - 4x^6 \quad \checkmark$$

(b) $$\int \left(7\mathrm{e}^{-x} + \frac{2}{x}\right)\mathrm{d}x = 7\int \mathrm{e}^{-x}\mathrm{d}x + 2\int \frac{1}{x}\mathrm{d}x$$

Now

$$\int \mathrm{e}^{mx}\mathrm{d}x = \frac{1}{m}\mathrm{e}^{mx}$$

so putting $m = -1$ gives

$$\int \mathrm{e}^{-x}\mathrm{d}x = \frac{1}{-1}\mathrm{e}^{-x} = -\mathrm{e}^{-x}$$

Also, we know that the reciprocal function integrates to the natural logarithm function, so

$$\int \frac{1}{x}\,\mathrm{d}x = \ln x$$

Hence

$$\int \left(7\mathrm{e}^{-x} + \frac{2}{x}\right)\mathrm{d}x = -7\mathrm{e}^{-x} + 2\ln x$$

Finally, we add an arbitrary constant to get

$$\int \left(7\mathrm{e}^{-x} + \frac{2}{x}\right)\mathrm{d}x = -7\mathrm{e}^{-x} + 2\ln x + c$$

As a check:

if $F(x) = -7\mathrm{e}^{-x} + 2\ln x + c$ then $F'(x) = 7\mathrm{e}^{-x} + \frac{2}{x}$ ✓

(c) $$\int (5x^2 + 3x + 2)\mathrm{d}x = 5\int x^2\mathrm{d}x + 3\int x\mathrm{d}x + 2\int 1\mathrm{d}x$$

Putting $n = 2$, 1 and 0 into

$$\int x^n\mathrm{d}x = \frac{1}{n+1}x^{n+1}$$

gives

$$\int x^2\mathrm{d}x = \frac{1}{3}x^3, \quad \int x\mathrm{d}x = \frac{1}{2}x^2, \quad \int 1\mathrm{d}x = x$$

Hence

$$\int (5x^2 + 3x + 2)\mathrm{d}x = \frac{5}{3}x^3 + \frac{3}{2}x^2 + 2x$$

Finally, we add an arbitrary constant to get

$$\int (5x^2 + 3x + 2)\mathrm{d}x = \frac{5}{3}x^3 + \frac{3}{2}x^2 + 2x + c$$

As a check:

if $F(x) = \frac{5}{3}x^3 + \frac{3}{2}x^2 + 2x + c$ then $F'(x) = 5x^2 + 3x + 2$ ✓

Advice

We have written out the solution to this example in detail to show you exactly how integration is performed. With practice you will probably find that you can just write the answer down in a single line of working, although it is always a good idea to check (at least in your head, if not on paper), by differentiating your answer, that you have not made any mistakes.

The technique of integration that we have investigated produces a function of x. In the next section a different type of integration is discussed which produces a single number as the end result. For this reason we use the word ***indefinite*** to describe the type of integration considered here to distinguish it from the ***definite*** integration in Section 6.2.

Practice Problem

3 Find the indefinite integrals

(a) $\int (2x - 4x^3)\mathrm{d}x$ **(b)** $\int \left(10x^4 + \frac{5}{x^2}\right)\mathrm{d}x$ **(c)** $\int (7x^2 - 3x + 2)\mathrm{d}x$

In Section 4.3 we described several applications of differentiation to economics. Starting with any basic economic function, we can differentiate to obtain the corresponding marginal function. Integration allows us to work backwards and to recover the original function from any marginal function. For example, by integrating a marginal cost function the total cost function is found. Likewise, given a marginal revenue function, integration enables us to determine the total revenue function, which in turn can be used to find the demand function. These ideas are illustrated in the following example, which also shows how the constant of integration can be given a specific numerical value in economic problems.

Example

(a) A firm's marginal cost function is

$$\mathrm{MC} = Q^2 + 2Q + 4$$

Find the total cost function if the fixed costs are 100.

(b) The marginal revenue function of a monopolistic producer is

$$\mathrm{MR} = 10 - 4Q$$

Find the total revenue function and deduce the corresponding demand equation.

(c) Find an expression for the consumption function if the marginal propensity to consume is given by

$$\mathrm{MPC} = 0.5 + \frac{0.1}{\sqrt{Y}}$$

and consumption is 85 when income is 100.

Solution

(a) We need to find the total cost from the marginal cost function

$$\mathrm{MC} = Q^2 + 2Q + 4$$

Now

$$\mathrm{MC} = \frac{\mathrm{d(TC)}}{\mathrm{d}Q}$$

so

$$\begin{aligned}\mathrm{TC} &= \int \mathrm{MC}\,\mathrm{d}Q \\ &= \int (Q^2 + 2Q + 4)\mathrm{d}Q \\ &= \frac{Q^3}{3} + Q^2 + 4Q + c\end{aligned}$$

The fixed costs are given to be 100. These are independent of the number of goods produced and represent the costs incurred when the firm does not produce any goods whatsoever. Putting $Q = 0$ into the TC function gives

$$\text{TC} = \frac{0^3}{3} + 0^2 + 4(0) + c = c$$

The constant of integration is therefore equal to the fixed costs of production, so $c = 100$. Hence

$$\text{TC} = \frac{Q^3}{3} + Q^2 + 4Q + 100$$

(b) We need to find the total revenue from the marginal revenue function

$$\text{MR} = 10 - 4Q$$

Now

$$\text{MR} = \frac{\text{d(TR)}}{\text{d}Q}$$

so

$$\begin{aligned}\text{TR} &= \int \text{MR}\text{d}Q \\ &= \int (10 - 4Q)\text{d}Q \\ &= 10 - 2Q^2 + c\end{aligned}$$

Unlike in part (a) of this example we have not been given any additional information to help us to pin down the value of c. We do know, however, that when the firm produces no goods the revenue is zero, so that $\text{TR} = 0$ when $Q = 0$. Putting this condition into

$$\text{TR} = 10Q - 2Q^2 + c$$

gives

$$0 = 10(0) - 2(0)^2 + c = c$$

The constant of integration is therefore equal to zero. Hence

$$\text{TR} = 10Q - 2Q^2$$

Finally, we can deduce the demand equation from this. To find an expression for total revenue from any given demand equation we normally multiply by Q, because $\text{TR} = PQ$. This time we work backwards, so we divide by Q to get

$$P = \frac{\text{TR}}{Q} = \frac{10Q - 2Q^2}{Q} = 10 - 2Q$$

so the demand equation is

$$P = 10 - 2Q$$

(c) We need to find consumption given that the marginal propensity to consume is

$$\text{MPC} = 0.5 + \frac{0.1}{\sqrt{Y}}$$

Now

$$\text{MPC} = \frac{\text{d}C}{\text{d}Y}$$

➔

so

$$C = \int \text{MRC}\,dY$$

$$= \int \left(0.5 + \frac{0.1}{\sqrt{Y}}\right) dY$$

$$= 0.5Y + 0.2\sqrt{Y} + c$$

where the second term is found from

$$\int \frac{0.1}{\sqrt{Y}}\,dY = 0.1\int Y^{-1/2}dY = 0.1\left(\frac{1}{1/2}Y^{1/2}\right) = 0.2\sqrt{Y}$$

The constant of integration can be calculated from the additional information that $C = 85$ when $Y = 100$. Putting $Y = 100$ into the expression for C gives

$$85 = 0.5(100) + 0.2\sqrt{100} + c = 52 + c$$

and so

$$c = 85 - 52 = 33$$

Hence

$$C = 0.5Y + 0.2\sqrt{Y} + 33$$

Practice Problem

4 **(a)** A firm's marginal cost function is

$$\text{MC} = 2$$

Find an expression for the total cost function if the fixed costs are 500. Hence find the total cost of producing 40 goods.

(b) The marginal revenue function of a monopolistic producer is

$$\text{MR} = 100 - 6Q$$

Find the total revenue function and deduce the corresponding demand equation.

(c) Find an expression for the savings function if the marginal propensity to save is given by

$$\text{MPS} = 0.4 - 0.1Y^{-1/2}$$

and savings are zero when income is 100.

Example MAPLE

A firm's marginal revenue and marginal cost functions are given by

$$\text{MR} = 500 - 0.5Q^2 \quad \text{and} \quad \text{MC} = 140 + 0.4Q^2$$

Given that total fixed costs are 1000, plot a graph of the profit function. Find the range of values of Q for which the firm makes a profit and find its maximum value.

Solution

Let us begin by naming these two expressions as MR and MC by typing

```
>MR:=500-0.5*Q^2;
```

and

```
>MC:=140+0.4*Q^2;
```

Total revenue is the integral of marginal revenue and this is achieved by typing

```
>TR:=int(MR,Q);
```

which gives

$TR:=500.Q-0.1666666667Q^3$

Notice that Maple forgets to include the constant of integration! As it happens, the constant of integration is zero in this case because TR = 0 when $Q = 0$. However, when we integrate MC we need to add on the fixed costs of 1000. We type

```
>TC:=int(MC,Q)+1000;
```

which gives

$TC:=140.Q+0.1333333333Q^2+1000$

Profit is then found by subtracting TC from TR so we type:

```
>profit:=TR-TC;
```

which gives

$profit:=360.Q-.3000000000Q^3-1000$

In order to sketch a graph of the profit function we need to specify the range of values of Q. Now total revenue is defined as PQ, so the demand equation can be found by dividing TR by Q to get

$$P = 500 - \frac{1}{6}Q^2$$

This equation is valid only when $P \geq 0$, i.e. when

$$500 - \frac{1}{6}Q^2 \geq 0$$

This will be so provided

$$Q^2 \leq 500 \times 6 = 3000$$

so we require $Q \leq \sqrt{3000} = 54.8$.

If we now choose to sketch the graph between 0 and 50, we type

```
>plot(profit,Q=0..50);
```

which produces the diagram shown in Figure 6.1 (overleaf).

The firm makes a profit when the graph lies above the horizontal axis. Figure 6.1 shows that this happens between $Q = 3$ and $Q = 33$ approximately. The diagram also shows that the maximum profit is roughly \$4000, which occurs when $Q = 20$.

If more precise values are required then these can easily be obtained by typing

```
>solve(profit=0,Q);
```

Figure 6.1

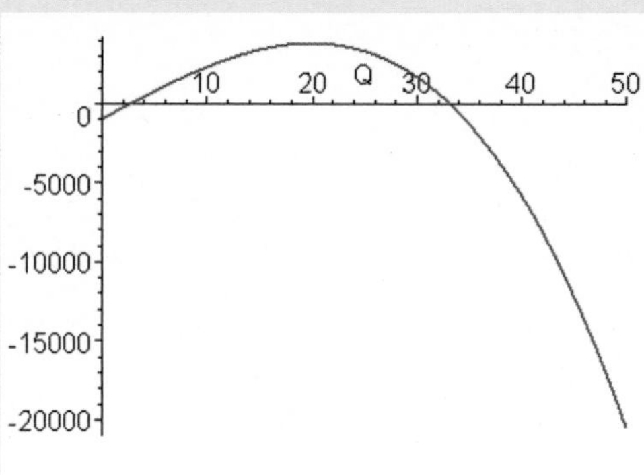

which gives

```
-35.95428113, 2.795992680, 33.15828845
```

Only the positive solutions are relevant, so we deduce that the firm makes a profit provided $2.8 \le Q \le 33.2$.

Maximum profit is achieved when MR = MC, so we type

```
solve(MR=MC,Q);
```

which gives

```
20., -20.
```

Again, the negative solution can be disregarded. The profit itself is found by substituting $Q = 20$ into the profit function

```
>subs(Q=20,profit);
```

which gives $3800.

Key Terms

Anti-derivative A function whose derivative is a given function.

Constant of integration The arbitrary constant that appears in an expression when finding an indefinite integral.

Definite integration The process of finding the area under a graph by subtracting the values obtained when the limits are substituted into the anti-derivative.

Indefinite integration The process of obtaining an anti-derivative.

Integral The number $\int_a^b f(x)\mathrm{d}x$ (definite integral) or the function $\int f(x)\mathrm{d}x$ (indefinite integral).

Integration The generic name for the evaluation of definite or indefinite integrals.

Inverse The operation that reverses the effect of a given operation and takes you back to the original. For example, the inverse of halving is doubling.

Primitive An alternative word for an anti-derivative.

Practice Problems

5 Find

(a) $\int 6x^5 dx$ **(b)** $\int x^4 dx$ **(c)** $\int 10e^{10x} dx$

(d) $\int \frac{1}{x} dx$ **(e)** $\int x^{3/2} dx$ **(f)** $\int (2x^3 - 6x) dx$

(g) $\int (x^2 - 8x + 3) dx$ **(h)** $\int (ax + b) dx$ **(i)** $\int \left(7x^3 + 4e^{-2x} - \frac{3}{x^2}\right) dx$

6 **(1)** Differentiate

$$F(x) = (2x + 1)^5$$

Hence find

$$\int (2x + 1)^4 dx$$

(2) Use the approach suggested in part (1) to find

(a) $\int (3x - 2)^7 dx$ **(b)** $\int (2 - 4x)^9 dx$

(c) $\int (ax + b)^n dx \ (n \neq -1)$ **(d)** $\int \frac{1}{7x + 3} dx$

7 **(a)** Find the total cost if the marginal cost is

$$MC = Q + 5$$

and fixed costs are 20.

(b) Find the total cost if the marginal cost is

$$MC = 3e^{0.5Q}$$

and fixed costs are 10.

8 Find the total revenue and demand functions corresponding to each of the following marginal revenue functions:

(a) $MR = 20 - 2Q$ **(b)** $MR = \frac{6}{\sqrt{Q}}$

9 Find the consumption function if the marginal propensity to consume is 0.6 and consumption is 10 when income is 5. Deduce the corresponding savings function.

10 Find the short-run production functions corresponding to each of the following marginal product of labour functions:

(a) $1000 - 3L^2$ **(b)** $\frac{6}{\sqrt{L}} - 0.01$

11 **(a)** Show that

$$\sqrt{x}(\sqrt{x} + x^2) = x + x^{5/2}$$

Hence find

$$\int \sqrt{x}(\sqrt{x} + x^2) dx$$

➔

(b) Use the approach suggested in part (a) to integrate each of the following functions:

$$x^4\left(x^6 + \frac{1}{x^2}\right), \quad e^{2x}(e^{3x} + e^{-x} + 3), \quad x^{3/2}\left(\sqrt{x} - \frac{1}{\sqrt{x}}\right)$$

12 (a) Show that

$$\frac{x^4 - x^2 + \sqrt{x}}{x} = x^3 - x + x^{-1/2}$$

Hence find

$$\int \frac{x^4 - x^2 + \sqrt{x}}{x}\,dx$$

(b) Use the approach suggested in part (a) to integrate each of the following functions:

$$\frac{x^2 - x}{x^3}, \quad \frac{e^x - e^{-x}}{e^{2x}}, \quad \frac{\sqrt{x} - x\sqrt{x} + x^2}{x\sqrt{x}}$$

13 (Maple) If the marginal propensity to consume is given by

$$\text{MPC} = \frac{1}{3} + \frac{1}{2\sqrt{Y}}$$

and $C = 19/3$ when $Y = 4$, find an expression for the consumption function. Plot both the consumption function and the line $C = Y$ on the same diagram. Use your graph to show that consumption exceeds income for all values of Y in the range $0 < Y < k$ for some number k which should be stated.

14 (Maple)

(a) Find an expression for a firm's short-run production function given that

$$\text{MP}_L = 10e^{-0.02L}(50 - L)$$

(b) Find an expression for the total revenue function given that

$$\text{MR} = \frac{100Q}{1 + 50Q^2}$$

section 6.2

Definite integration

Objectives

At the end of this section you should be able to:

- Recognize the notation for definite integration.
- Evaluate definite integrals in simple cases.
- Calculate the consumer's surplus.
- Calculate the producer's surplus.
- Calculate the capital stock formation.
- Calculate the present value of a continuous revenue stream.

One rather tedious task that you may remember from school is that of finding areas. Sketched in Figure 6.2 (overleaf) is a region bounded by the curve $y = x^2$, the lines $x = 1$, $x = 2$, and the x axis. At school you may well have been asked to find the area of this region by 'counting' squares on graph paper. A much quicker and more accurate way of calculating this area is to use integration. We begin by integrating the function

$$f(x) = x^2$$

to get

$$F(x) = \frac{1}{3}x^3$$

In our case we want to find the area under the curve between $x = 1$ and $x = 2$, so we evaluate

$$F(1) = \frac{1}{3}(1)^3 = \frac{1}{3}$$

$$F(2) = \frac{1}{3}(2)^3 = \frac{8}{3}$$

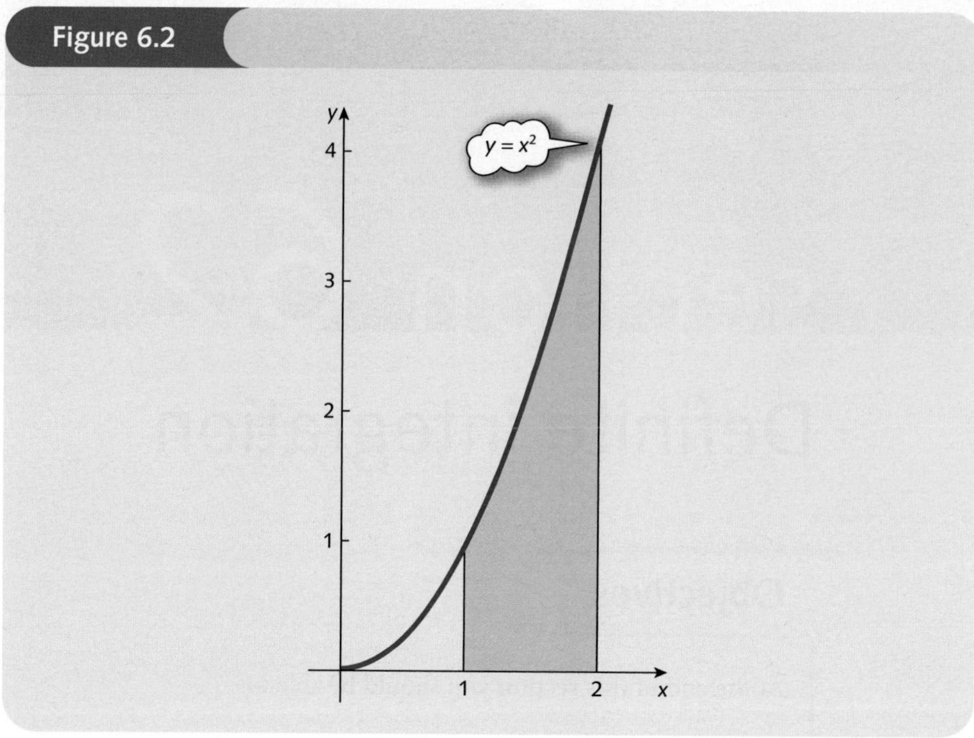

Finally, we subtract $F(1)$ from $F(2)$ to get

$$F(2) - F(1) = \frac{8}{3} - \frac{1}{3} = \frac{7}{3}$$

This number is the exact value of the area of the region sketched in Figure 6.2. Given the connection with integration, we write this area as

$$\int_1^2 x^2 \mathrm{d}x$$

In general, the *definite integral*

$$\int_a^b f(x)\mathrm{d}x$$

denotes the area under the graph of $f(x)$ between $x = a$ and $x = b$ as shown in Figure 6.3. The numbers a and b are called the *limits of integration*, and it is assumed throughout this section that $a < b$ and that $f(x) \geq 0$ as indicated in Figure 6.3.

The technique of evaluating definite integrals is as follows. A function $F(x)$ is found which differentiates to $f(x)$. Methods of obtaining $F(x)$ have already been described in Section 6.1. The new function, $F(x)$, is then evaluated at the limits $x = a$ and $x = b$ to get $F(a)$ and $F(b)$. Finally, the second number is subtracted from the first to get the answer

$$F(b) - F(a)$$

In symbols,

$$\int_a^b f(x)\mathrm{d}x = F(b) - F(a)$$

Figure 6.3

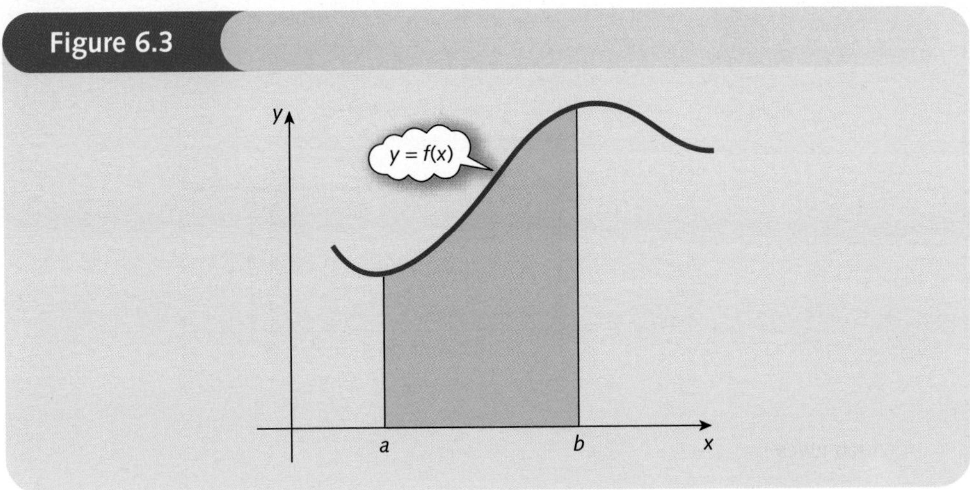

The process of evaluating a function at two distinct values of x and subtracting one from the other occurs sufficiently frequently in mathematics to warrant a special notation. We write

$$[F(x)]_a^b$$

as an abbreviation for $F(b) - F(a)$, so that definite integrals are evaluated as

$$\int_a^b f(x)\mathrm{d}x = [F(x)]_a^b = F(b) - F(a)$$

where $F(x)$ is the indefinite integral of $f(x)$. Using this notation, the evaluation of

$$\int_1^2 x^2\mathrm{d}x$$

would be written as

$$\int_1^2 x^2\mathrm{d}x = \left[\frac{1}{3}x^3\right]_1^2 = \frac{1}{3}(2)^3 - \frac{1}{3}(1)^3 = \frac{7}{3}$$

Note that it is not necessary to include the constant of integration, because it cancels out when we subtract $F(a)$ from $F(b)$.

Example

Evaluate the definite integrals

(a) $\int_2^6 3\mathrm{d}x$ **(b)** $\int_0^2 (x + 1)\mathrm{d}x$

Solution

(a) $\int_2^6 3\mathrm{d}x = [3x]_2^6 = 3(6) - 3(2) = 12$

This value can be confirmed graphically. Figure 6.4 (overleaf) shows the region under the graph of $y = 3$ between $x = 2$ and $x = 6$. This is a rectangle, so its area can be found from the formula

area = base × height

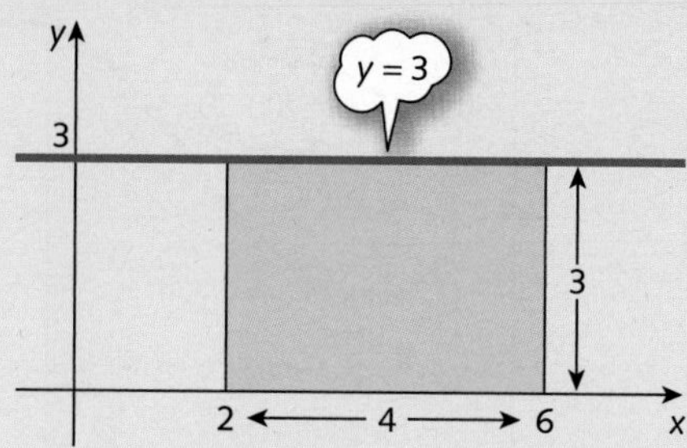

which gives

area = 4 × 3 = 12 ✓

(b) $\int_0^2 (x+1)\mathrm{d}x = \left[\frac{x^2}{2} + x\right]_0^2 = \left(\frac{2^2}{2} + 2\right) - \left(\frac{0^2}{2} + 0\right) = 4$

Again this value can be confirmed graphically. Figure 6.5(a) shows the region under the graph of $y = x + 1$ between $x = 0$ and $x = 2$. This can also be regarded as one-half of the rectangle illustrated in Figure 6.5(b). This rectangle has a base of 2 units and a height of 4 units, so has area

2 × 4 = 8

The area of the region shown in Figure 6.5(a) is therefore

½ × 8 = 4 ✓

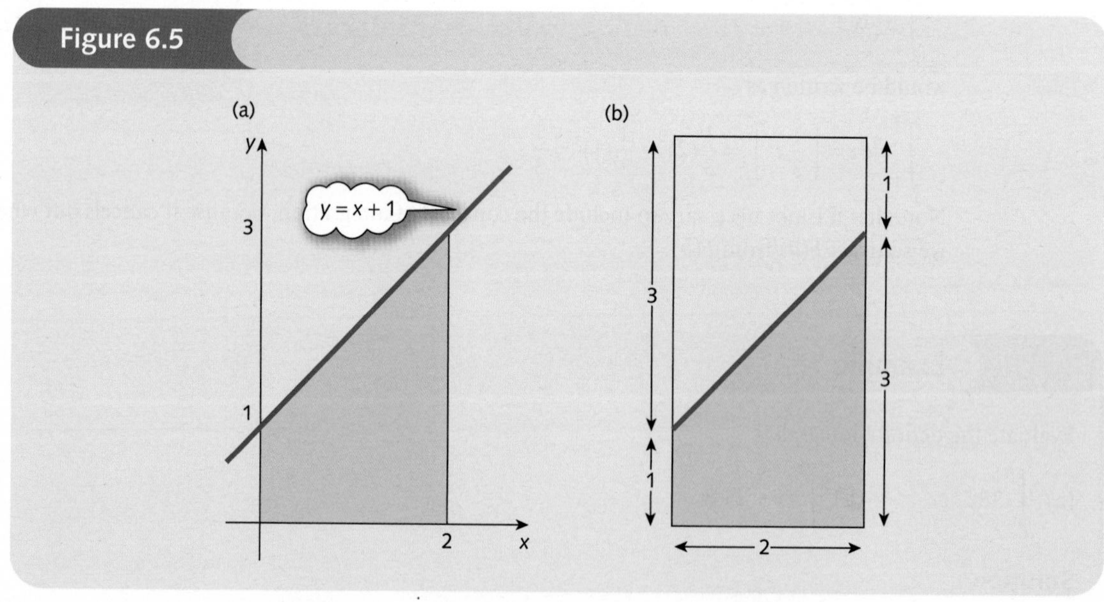

In this example we deliberately chose two very simple functions so that we could demonstrate the fact that definite integrals really do give the areas under graphs. The beauty of the integration technique, however, is that it can be used to calculate areas under quite complicated functions for which alternative methods would fail to produce the exact value.

Practice Problem

1 Evaluate the following definite integrals:

(a) $\int_0^1 x^3 dx$ **(b)** $\int_2^5 (2x - 1)dx$ **(c)** $\int_1^4 (x^2 - x + 1)dx$ **(d)** $\int_0^1 e^x dx$

To illustrate the applicability of definite integration we concentrate on four topics:

- consumer's surplus
- producer's surplus
- investment flow
- discounting.

We consider each of these in turn.

6.2.1 Consumer's surplus

The demand function, $P = f(Q)$, sketched in Figure 6.6, gives the different prices that consumers are prepared to pay for various quantities of a good. At $Q = Q_0$ the price $P = P_0$. The total amount of money spent on Q_0 goods is then Q_0P_0, which is given by the area of the rectangle OABC. Now, P_0 is the price that consumers are prepared to pay for the last unit that they buy, which is the Q_0th good. For quantities up to Q_0 they would actually be willing to pay the higher price given by the demand curve. The shaded area BCD therefore represents the benefit to the consumer of paying the fixed price of P_0 and is called the *consumer's surplus*, CS. The value of CS can be found by observing that

area BCD = area OABD − area OABC

Figure 6.6

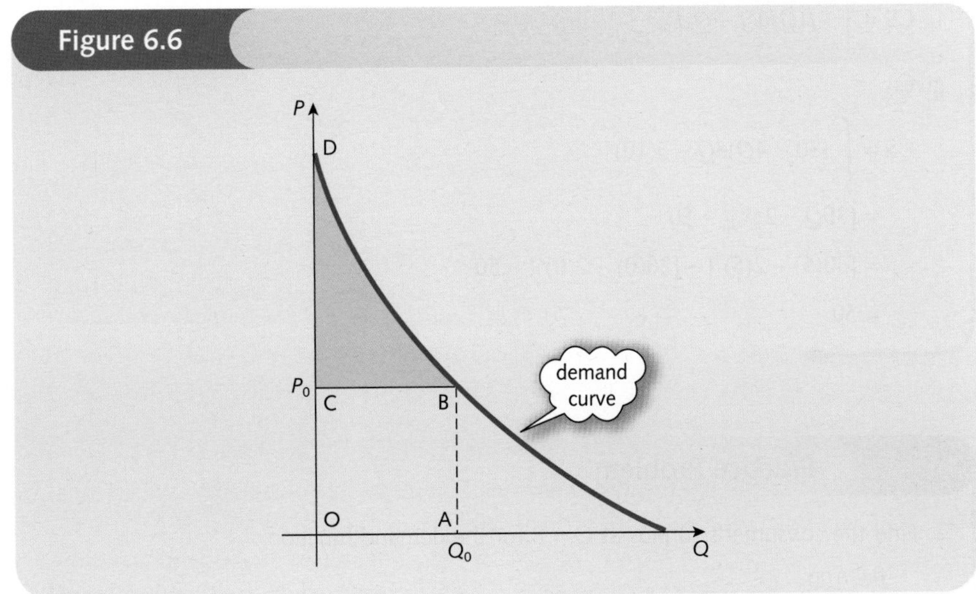

The area OABD is the area under the demand curve $P = f(Q)$, between $Q = 0$ and $Q = Q_0$, and so is equal to

$$\int_0^{Q_0} f(Q)\mathrm{d}Q$$

The region OABC is a rectangle with base Q_0 and height P_0 so

$$\text{area OABC} = Q_0 P_0$$

Hence

$$\boxed{\text{CS} = \int_0^{Q_0} f(Q)\mathrm{d}Q - Q_0 P_0}$$

Example

Find the consumer's surplus at $Q = 5$ for the demand function

$$P = 30 - 4Q$$

Solution

In this case

$$f(Q) = 30 - 4Q$$

and $Q_0 = 5$. The price is easily found by substituting $Q = 5$ into

$$P = 30 - 4Q$$

to get

$$P_0 = 30 - 4(5) = 10$$

The formula for consumer's surplus

$$\text{CS} = \int_0^{Q_0} f(Q)\mathrm{d}Q - Q_0 P_0$$

gives

$$\begin{aligned}\text{CS} &= \int_0^{5} (30 - 4Q)\mathrm{d}Q - 5(10)\\ &= [30Q - 2Q^2]_0^5 - 50\\ &= [30(5) - 2(5)^2] - [30(0) - 2(0)^2] - 50\\ &= 50\end{aligned}$$

Practice Problem

2 Find the consumer's surplus at $Q = 8$ for the demand function

$$P = 100 - Q^2$$

Figure 6.7

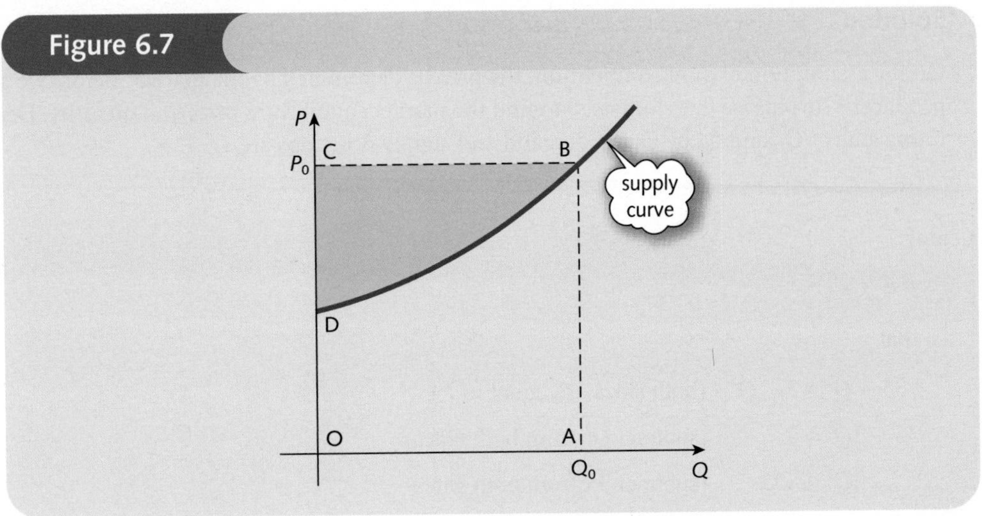

6.2.2 Producer's surplus

The supply function, $P = g(Q)$, sketched in Figure 6.7 gives the different prices at which producers are prepared to supply various quantities of a good. At $Q = Q_0$ the price $P = P_0$. Assuming that all goods are sold, the total amount of money received is then Q_0P_0, which is given by the area of the rectangle OABC.

Now, P_0 is the price at which the producer is prepared to supply the last unit, which is the Q_0th good. For quantities up to Q_0 they would actually be willing to accept the lower price given by the supply curve. The shaded area BCD therefore represents the benefit to the producer of selling at the fixed price of P_0 and is called the ***producer's surplus***, PS. The value of PS is found by observing that

area BCD = area OABC − area OABD

The region OABC is a rectangle with base Q_0 and height P_0, so

area OABC = Q_0P_0

The area OABD is the area under the supply curve $P = g(Q)$, between $Q = 0$ and $Q = Q_0$, and so is equal to

$$\int_0^{Q_0} g(Q)\mathrm{d}Q$$

Hence

$$\boxed{\text{PS} = Q_0P_0 - \int_0^{Q_0} g(Q)\mathrm{d}Q}$$

Example

Given the demand function

$$P = 35 - Q_D^2$$

and supply function

$$P = 3 + Q_S^2$$

find the producer's surplus assuming pure competition.

Solution

On the assumption of pure competition, the price is determined by the market. Before we can calculate the producer's surplus we therefore need to find the market equilibrium price and quantity. Denoting the common value of Q_D and Q_S by Q, the demand and supply functions are

$$P = 35 - Q^2$$

and

$$P = 3 + Q^2$$

so that

$$35 - Q^2 = 3 + Q^2 \quad \text{(both sides are equal to } P\text{)}$$

$$35 - 2Q^2 = 3 \quad \text{(subtract } Q^2 \text{ from both sides)}$$

$$-2Q^2 = -32 \quad \text{(subtract 35 from both sides)}$$

$$Q^2 = 16 \quad \text{(divide both sides by } -2\text{)}$$

which has solution $Q = \pm 4$. We can obviously ignore the negative solution because it does not make economic sense. The equilibrium quantity is therefore equal to 4. The corresponding price can be found by substituting this into either the demand or the supply equation. From the demand equation we have

$$P_0 = 35 - (4)^2 = 19$$

The formula for the producer's surplus,

$$\text{PS} = Q_0P_0 - \int_0^{Q_0} g(Q)\text{d}Q$$

gives

$$\text{PS} = 4(19) - \int_0^4 (3 + Q^2)\text{d}Q$$

$$= 76 - \left[3Q + \frac{Q^3}{3}\right]_0^4$$

$$= 76 - \{[3(4) + \tfrac{1}{3}(4)^3] - [3(0) - \tfrac{1}{3}(0)^3]\}$$

$$= 42\tfrac{2}{3}$$

Practice Problem

3 Given the demand equation

$$P = 50 - 2Q_D$$

and supply equation

$$P = 10 + 2Q_S$$

calculate

(a) the consumer's surplus

(b) the producer's surplus

assuming pure competition.

6.2.3 Investment flow

Net investment, I, is defined to be the rate of change of capital stock, K, so that

$$I = \frac{\mathrm{d}K}{\mathrm{d}t}$$

Here $I(t)$ denotes the flow of money, measured in dollars per year, and $K(t)$ is the amount of capital accumulated at time t as a result of this investment flow and is measured in dollars.

Given a formula for capital stock in terms of time, we simply differentiate to find net investment. Conversely, if we know the net investment function then we integrate to find the capital stock. In particular, to calculate the capital formation during the time period from $t = t_1$ to $t = t_2$ we evaluate the definite integral

$$\int_{t_1}^{t_2} I(t)\mathrm{d}t$$

Example

If the investment flow is

$$I(t) = 9000\sqrt{t}$$

calculate

(a) the capital formation from the end of the first year to the end of the fourth year

(b) the number of years required before the capital stock exceeds \$100 000.

Solution

(a) In this part we need to calculate the capital formation from $t = 1$ to $t = 4$, so we evaluate the definite integral

$$\begin{aligned}\int_1^4 9000\sqrt{t}\,\mathrm{d}t &= 9000\int_1^4 t^{1/2}\mathrm{d}t\\ &= 9000\left[\frac{2}{3}t^{3/2}\right]_1^4\\ &= 9000\left[\frac{2}{3}(4)^{3/2} - \frac{2}{3}(1)^{3/2}\right]\\ &= 9000\left(\frac{16}{3} - \frac{2}{3}\right)\\ &= \$42\,000\end{aligned}$$

(b) In this part we need to calculate the number of years required to accumulate a total of \$100 000. After T years the capital stock is

$$\int_0^T 9000\sqrt{t}\,\mathrm{d}t = 9000\int_0^T t^{1/2}\mathrm{d}t$$

We want to find the value of T so that

$$9000\int_0^T t^{1/2}\mathrm{d}t = 100\,000$$

The integral is easily evaluated as

$$9000\left[\frac{2}{3}t^{3/2}\right]_0^T = 9000\left(\frac{2}{3}T^{3/2} - \frac{2}{3}(0)^{3/2}\right) = 6000T^{3/2}$$

so T satisfies

$$6000T^{3/2} = 100\ 000$$

This non-linear equation can be solved by dividing both sides by 6000 to get

$$T^{3/2} = 16.67$$

and then raising both sides to the power of $^2/_3$, which gives

$$T = 6.5$$

The capital stock reaches the $100 000 level about halfway through the seventh year.

Practice Problem

4 If the net investment function is given by

$$I(t) = 800t^{1/3}$$

calculate

(a) the capital formation from the end of the first year to the end of the eighth year

(b) the number of years required before the capital stock exceeds $48 600.

Example

MAPLE

During the next year, a firm's revenue and cost flows, measured in thousands of dollars per month, are to be modelled by

$$r(t) = 0.5t^3 - 6t^2 + 18t + 30 \quad \text{and} \quad c(t) = 6.5t + 12$$

respectively.

(1) Evaluate the definite integrals

(a) $\int_0^4 (r - c)\mathrm{d}t$ **(b)** $\int_4^9 (r - c)\mathrm{d}t$

(2) Plot a graph of $r(t)$ and $c(t)$ on the same diagram and hence interpret the values obtained in part (1).

Solution

(1) We begin by naming these expressions `r` and `c` by typing

```
>r:=0.5*t^3-6*t^2+18*t+30;
```

and

```
>c:=6.5*t+12;
```

(a) To evaluate the definite integral $\int_0^4 (r - c)dt$, we use the instruction `int` as before, except that we now specify the limits, 0 and 4. We type

```
>int(r-c,t=0..4);
```

which gives the answer

```
68
```

(b) Similarly, the second integral is evaluated by editing the instruction as:

```
>int(r-c,t=4..9);
```

which gives

```
-78.125
```

(2) To plot both r and c on the same diagram, type

```
>plot({r,c},t=0..12);
```

The result is sketched in Figure 6.8.

Now, $r(t)$ represents the rate of change of revenue with respect to time, so the revenue itself is the area under the graph of $r(t)$. Similarly, the area under the graph of $c(t)$ determines the total cost. Hence the area between the two curves gives the predicted profit during that time period. Figure 6.8 shows that between $t = 0$ and $t = 4$ revenue exceeds costs, so the firm is likely to make a positive profit of \$68 000. On the other hand, between $t = 4$ and $t = 9$ the graph of $c(t)$ lies above that of $r(t)$, which explains why the value of the integral is negative. During this period the firm is expected to make a loss of \$78 125.

Figure 6.8

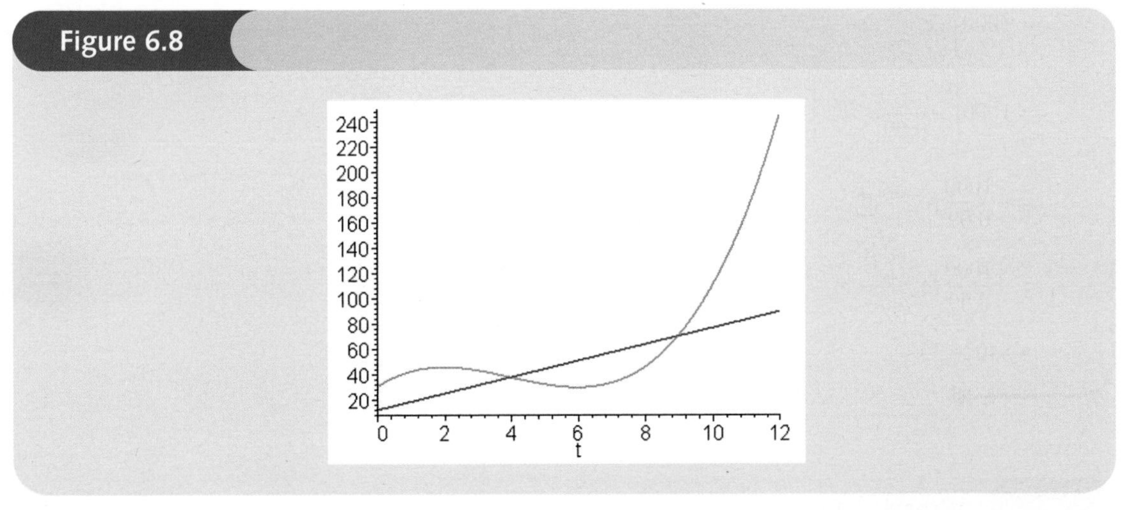

6.2.4 Discounting

In Chapter 3 the formula

$$P = Se^{-rt/100}$$

was used to calculate the present value, P, when a single future value, S, is discounted at r% interest continuously for t years. We also discussed the idea of an annuity. This is a fund that provides a series of discrete regular payments and we showed how to calculate the original lump sum needed to secure these payments for a prescribed number of years. This amount is called the present value of the annuity. If the fund is to provide a continuous revenue stream for n years at an annual rate of S dollars per year then the present value can be found by evaluating the definite integral

$$P = \int_0^n S e^{-rt/100} dt$$

Example

Calculate the present value of a continuous revenue stream for 5 years at a constant rate of \$1000 per year if the discount rate is 9%.

Solution

The present value is found from the formula

$$P = \int_0^n S e^{-rt/100} dt$$

with $S = 1000$, $r = 9$ and $n = 5$, so

$$P = \int_0^5 1000 e^{-0.09t} dt$$

$$= 1000 \int_0^5 e^{-0.09t} dt$$

$$= 1000 \left[-\frac{1}{0.09} e^{-0.09t} \right]_0^5$$

$$= -\frac{1000}{0.09} [e^{-0.09t}]_0^5$$

$$= -\frac{1000}{0.09} (e^{-0.45} - 1)$$

$e^0 = 1$

$$= \$4026.35$$

Practice Problem

5 Calculate the present value of a continuous revenue stream for 10 years at a constant rate of \$5000 per year if the discount rate is 6%.

Key Terms

Consumer's surplus The excess cost that a person would have been prepared to pay for goods over and above what is actually paid.

Definite integral The number $\int_a^b f(x)\mathrm{d}x$ which represents the area under the graph of $f(x)$ between $x = a$ and $x = b$.

Limits of integration The numbers a and b which appear in the definite integral, $\int_a^b f(x)\mathrm{d}x$.

Net investment Rate of change of capital stock over time: $I = \mathrm{d}K/\mathrm{d}t$.

Producer's surplus The excess revenue that a producer has actually received over and above the lower revenue that it was prepared to accept for the supply of its goods.

Practice Problems

6 Find the area under the graph of

$$f(x) = 4x^3 - 3x^2 + 4x + 2$$

between $x = 1$ and $x = 2$.

7 Evaluate each of the following definite integrals:

(a) $\int_0^2 x^3\mathrm{d}x$ **(b)** $\int_{-2}^2 x^3\mathrm{d}x$

By sketching a rough graph of the cube function between and $x = -2$ and 2, suggest a reason for your answer to part (b). What is the actual area between the x axis and the graph of $y = x^3$ over this range?

8 (Maple)

(a) Plot a graph of $y = \frac{1}{x^2}$ on $1 \le x \le 30$.

Evaluate $\int_1^N \frac{1}{x^2}\mathrm{d}x$ in the cases when N is 2, 20 and 200.

What do these results suggest about the value of $\int_1^\infty \frac{1}{x^2}\mathrm{d}x$?

(b) Repeat part (a) for the function $y = \frac{1}{\sqrt{x}}$.

9 (Maple)

(a) Plot a graph of $y = \frac{1}{\sqrt{x}}$ on $0.01 \le x \le 1$.

Evaluate $\int_N^1 \frac{1}{\sqrt{x}}\mathrm{d}x$ in the cases when N is 0.1, 0.01 and 0.001.

What do these results suggest about the value of $\int_0^1 \frac{1}{\sqrt{x}}\mathrm{d}x$?

(b) Repeat part (a) for the function $y = \frac{1}{x^2}$.

➔

10 Find the consumer's surplus at $P = 5$ for the following demand functions:

(a) $P = 25 - 2Q$ **(b)** $P = \dfrac{10}{\sqrt{Q}}$

11 Find the producer's surplus at $Q = 9$ for the following supply functions:

(a) $P = 12 + 2Q$ **(b)** $P = 20\sqrt{Q} + 15$

12 Given the demand function

$$P = -Q_D^2 - 4Q_D + 68$$

and the supply function

$$P = Q_S^2 - 2Q_S + 12$$

find

(a) the consumer's surplus

(b) the producer's surplus

assuming pure competition.

13 If the net investment function is given by

$$I(t) = 100e^{0.1t}$$

calculate

(a) the capital formation from the end of the second year to the end of the fifth year

(b) the number of years required before the capital stock exceeds $100 000.

14 Find the expression for capital formation between $t = 0$ and $t = T$ for the following net investment functions:

(a) $I(t) = At^{\alpha}$ **(b)** $I(t) = Ae^{\alpha t}$

where A and α are positive constants.

15 Calculate the present value of a continuous revenue stream of $1000 per year if the discount rate is 5% and the money is paid

(a) for 3 years **(b)** for 10 years **(c)** for 100 years **(d)** in perpetuity

16 The present value of a continuous revenue stream of $5000 per year with a discount rate of 10% over n years is $25 000. Find the value of n correct to 1 decimal place.

chapter 7 Matrices

The impression that you may have gained from reading this book is that mathematics consists of one main topic, calculus, and that every other topic is just a variation on this theme. This is far from the truth and in this chapter and the next we look at two refreshingly different branches of mathematics. It would be useful for you to have studied Chapter 1, although even this is not essential. There are four sections. It is essential that Sections 7.1 and 7.2 are read first, but the remaining sections can be read in either order.

Section 7.1 introduces the concept of a matrix, which is a convenient mathematical way of representing information displayed in a table. By defining the matrix operations of addition, subtraction and multiplication it is possible to develop an algebra of matrices. Simple economic examples are used to illustrate these definitions and it is shown that the rules of matrix manipulation are almost identical to those of ordinary arithmetic. In Section 7.2 you are shown how to calculate the inverse of a matrix. This is analogous to the reciprocal of a number and enables matrix equations to be solved. In particular, inverses provide an alternative way of solving systems of simultaneous linear equations and so can be used to solve problems in statics. Section 7.3 describes Cramer's rule for solving systems of linear equations. This method is a particularly useful way of solving economic models where only a selection of endogenous variables need to be determined.

In Section 7.4 we discuss a new economic topic known as input–output analysis and show how the matrix operations described in the first two sections can be used to determine the flow of money between firms.

section 7.1

Basic matrix operations

Objectives

At the end of this section you should be able to:

- Understand the notation and terminology of matrix algebra.
- Find the transpose of a matrix.
- Add and subtract matrices.
- Multiply a matrix by a scalar.
- Multiply matrices together.
- Represent a system of linear equations in matrix notation.

Suppose that a firm produces three types of good, G1, G2 and G3, which it sells to two customers, C1 and C2. The monthly sales for these goods are given in Table 7.1. During the month the firm sells 3 items of G2 to customer C1, 6 items of G3 to customer C2, and so on. It may well be obvious from the context exactly what these numbers represent. Under these circumstances it makes sense to ignore the table headings and to write this information more concisely as

$$\mathbf{A} = \begin{bmatrix} 7 & 3 & 4 \\ 1 & 5 & 6 \end{bmatrix}$$

Table 7.1

		Monthly sales for goods		
		G1	**G2**	**G3**
Sold to	*C1*	7	3	4
customer	*C2*	1	5	6

which is an example of a matrix. Quite generally, any rectangular array of numbers surrounded by a pair of brackets is called a *matrix* (plural *matrices*) and the individual numbers constituting the array are called *entries* or *elements*. In this book we use square brackets, although it is equally correct to use parentheses (that is, round brackets) instead. It helps to think of a matrix as being made up of rows and columns. The matrix **A** has two rows and three columns and is said to have order 2×3. In general, a matrix of *order* $m \times n$ has m rows and n columns.

We denote matrices by capital letters in bold type (that is, **A**, **B**, **C**, . . .) and their elements by the corresponding lower-case letter in ordinary type. In fact, we use a rather clever double subscript notation so that a_{ij} stands for the element of **A** which occurs in row i and column j. Referring to the matrix **A** above, we see that

$$a_{12} = 3 \quad \text{(row 1 and column 2 of } \mathbf{A}\text{)}$$

A general matrix **D** of order 3×2 would be written

$$\begin{bmatrix} d_{11} & d_{12} \\ d_{21} & d_{22} \\ d_{31} & d_{32} \end{bmatrix}$$

Similarly, a 3×3 matrix labelled **E** would be written

$$\begin{bmatrix} e_{11} & e_{12} & e_{13} \\ e_{21} & e_{22} & e_{23} \\ e_{31} & e_{32} & e_{33} \end{bmatrix}$$

Example

Let

$$\mathbf{B} = \begin{bmatrix} 2 & 1 \\ -1 & 6 \end{bmatrix} \qquad \mathbf{C} = \begin{bmatrix} 1 & 3 & 4 & 0 \\ 1 & 2 & 1 & 1 \\ 1 & 4 & 5 & 7 \end{bmatrix}$$

(a) State the orders of the matrices **B** and **C**.

(b) Write down the values of b_{22} and c_{34}.

Solution

(a) Matrices **B** and **C** have orders 2×2 and 3×4 respectively.

(b) $b_{22} = 6$ (row 2 and column 2 of **B**)
$c_{34} = 7$ (row 3 and column 4 of **C**)

Practice Problem

1 Let

$$\mathbf{A} = \begin{bmatrix} 1 & 2 \\ 3 & 4 \end{bmatrix} \qquad \mathbf{B} = [1 \quad -1 \quad 0 \quad 6 \quad 2] \qquad \mathbf{C} = \begin{bmatrix} 1 & 0 & 2 & 3 & 1 \\ 5 & 7 & 9 & 0 & 2 \\ 3 & 4 & 6 & 7 & 8 \end{bmatrix} \qquad \mathbf{D} = [6]$$

(a) State the orders of the matrices **A**, **B**, **C** and **D**.

(b) Write down the values of

$$a_{11}, \quad a_{22}, \quad b_{14}, \quad c_{25}, \quad c_{33}, \quad c_{43}, \quad d_{11}$$

All we have done so far is to explain what matrices are and to provide some notation for handling them. A matrix certainly gives us a convenient shorthand to describe information presented in a table. However, we would like to go further than this and to use matrices to solve problems in economics. To do this we describe several mathematical operations that can be performed on matrices, namely

- transposition
- addition and subtraction
- scalar multiplication
- matrix multiplication.

One obvious omission from the list is matrix division. Strictly speaking, it is impossible to divide one matrix by another, although we can get fairly close to the idea of division by defining something called an inverse, which we consider in Section 7.2.

Advice

If you have not met matrices before, you might like to split this section into two separate parts. You are advised to work through the material as far as 7.1.4 now, leaving matrix multiplication for another session.

7.1.1 Transposition

In Table 7.1 the rows correspond to the two customers and the columns correspond to the three goods. The matrix representation of the table is then

$$\mathbf{A} = \begin{bmatrix} 7 & 3 & 4 \\ 1 & 5 & 6 \end{bmatrix}$$

The same information about monthly sales could easily have been presented the other way round, as shown in Table 7.2. The matrix representation would then be

$$\mathbf{B} = \begin{bmatrix} 7 & 1 \\ 3 & 5 \\ 4 & 6 \end{bmatrix}$$

We describe this situation by saying that **A** and **B** are transposes of each other and write

$$\mathbf{A}^T = \mathbf{B}$$

read 'A transpose equals B'

Table 7.2

		Sold to customer	
		C1	C2
Monthly sales for goods	G1	7	1
	G2	3	5
	G3	4	6

or equivalently

$\mathbf{B}^{\mathrm{T}} = \mathbf{A}$ read 'B transpose equals A'

The *transpose* of a matrix is found by replacing rows by columns, so that the first row becomes the first column, the second row becomes the second column, and so on. The number of rows of **A** is then the same as the number of columns of $\mathbf{A}^{\mathrm{T}}$ and vice versa. Consequently, if **A** has order $m \times n$ then $\mathbf{A}^{\mathrm{T}}$ has order $n \times m$.

Example

Write down the transpose of the matrices

$$\mathbf{D} = \begin{bmatrix} 1 & 7 & 0 & 3 \\ 2 & 4 & 6 & 0 \\ 5 & 1 & 9 & 2 \end{bmatrix} \qquad \mathbf{E} = \begin{bmatrix} -6 \\ 3 \end{bmatrix}$$

Solution

The transpose of the 3×4 matrix **D** is the 4×3 matrix

$$\mathbf{D}^{\mathrm{T}} = \begin{bmatrix} 1 & 2 & 5 \\ 7 & 4 & 1 \\ 0 & 6 & 9 \\ 3 & 0 & 2 \end{bmatrix}$$

The transpose of the 2×1 matrix **E** is the 1×2 matrix

$$\mathbf{E}^{\mathrm{T}} = [-6 \quad 3]$$

Practice Problem

2 Write down the transpose of the following matrices:

$$\mathbf{A} = \begin{bmatrix} 1 & 4 & 0 & 1 & 2 \\ 3 & 7 & 6 & 1 & 4 \\ 2 & 1 & 3 & 5 & -1 \\ 2 & -5 & 1 & 8 & 0 \end{bmatrix}$$

$$\mathbf{B} = [1 \quad 5 \quad 7 \quad 9]$$

$$\mathbf{C} = \begin{bmatrix} 1 & 2 & 3 \\ 2 & 4 & 5 \\ 3 & 5 & 6 \end{bmatrix}$$

There are two particular shapes of matrices which are given special names. A matrix that has only one row, such as

$$\mathbf{c} = [5 \quad 2 \quad 1 \quad -4]$$

is called a *row vector*, and a matrix that has only one column, such as

$$\mathbf{d} = \begin{bmatrix} -3 \\ 10 \\ 6 \\ -7 \\ 1 \\ 9 \\ 2 \end{bmatrix}$$

is called a *column vector*. It is standard practice to identify vectors using lower-case rather than upper-case letters. In books they are set in bold type. If you are writing them down by hand then you should underline the letters and put

$\underline{c}$ (or possibly $\underset{\sim}{c}$) and $\underline{d}$ (or possibly $\underset{\sim}{d}$)

This is a useful convention since it helps to distinguish scalar quantities such as x, y, a, b, which denote single numbers, from vector quantities such as $\mathbf{x}$, $\mathbf{y}$, $\mathbf{a}$, $\mathbf{b}$, which denote matrices with one row or column. Incidentally, it is actually quite expensive to print column vectors in books and journals since it is wasteful on space, particularly if the number of elements is large. It is then more convenient to use the transpose notation and write the vector horizontally. For example, the 7×1 matrix $\mathbf{d}$ given above would be printed as

$$\mathbf{d} = [-3 \quad 10 \quad 6 \quad -7 \quad 1 \quad 9 \quad 2]^T$$

where the superscript T tells us that it is the column vector that is intended.

7.1.2 Addition and subtraction

Let us suppose that, for the two-customer three-product example, the matrix

$$\mathbf{A} = \begin{bmatrix} 7 & 3 & 4 \\ 1 & 5 & 6 \end{bmatrix}$$

gives the sales for the month of January. Similarly, the monthly sales for February might be given by

$$\mathbf{B} = \begin{bmatrix} 6 & 2 & 1 \\ 0 & 4 & 4 \end{bmatrix}$$

This means, for example, that customer C1 buys 7 items of G1 in January and 6 items of G1 in February. Customer C1 therefore buys a total of

$$7 + 6 = 13$$

items of G1 during the two months. A similar process can be applied to the remaining goods and customers, so that the matrix giving the sales for the two months is

$$\mathbf{C} = \begin{bmatrix} 7+6 & 3+2 & 4+1 \\ 1+0 & 5+4 & 6+4 \end{bmatrix}$$

$$= \begin{bmatrix} 13 & 5 & 5 \\ 1 & 9 & 10 \end{bmatrix}$$

We describe this by saying that **C** is the **sum** of the two matrices **A** and **B** and we write

$$\mathbf{C} = \mathbf{A} + \mathbf{B}$$

In general, to add (or subtract) two matrices of the same size, we simply add (or subtract) their corresponding elements. It is obvious from this definition that, for any two $m \times n$ matrices, **A** and **B**,

$$\mathbf{A} + \mathbf{B} = \mathbf{B} + \mathbf{A}$$

because it is immaterial which way round two numbers are added. Note that in order to combine matrices in this way it is necessary for them to have the same order. For example, it is impossible to add the matrices

$$\mathbf{D} = \begin{bmatrix} 1 & -7 \\ 1 & 3 \end{bmatrix} \quad \text{and} \quad \mathbf{E} = \begin{bmatrix} 1 & 2 \\ 1 & 1 \\ 3 & 5 \end{bmatrix}$$

because **D** has order 2×2 and **E** has order 3×2.

Example

Let

$$\mathbf{A} = \begin{bmatrix} 9 & -3 \\ 4 & 1 \\ 2 & 0 \end{bmatrix} \quad \text{and} \quad \mathbf{B} = \begin{bmatrix} 5 & 2 \\ -1 & 6 \\ 3 & 4 \end{bmatrix}$$

Find

(a) $\mathbf{A} + \mathbf{B}$ (b) $\mathbf{A} - \mathbf{B}$ (c) $\mathbf{A} - \mathbf{A}$

Solution

(a) $$\mathbf{A} + \mathbf{B} = \begin{bmatrix} 9 & -3 \\ 4 & 1 \\ 2 & 0 \end{bmatrix} + \begin{bmatrix} 5 & 2 \\ -1 & 6 \\ 3 & 4 \end{bmatrix} = \begin{bmatrix} 14 & -1 \\ 3 & 7 \\ 5 & 4 \end{bmatrix}$$

(b) $$\mathbf{A} - \mathbf{B} = \begin{bmatrix} 9 & -3 \\ 4 & 1 \\ 2 & 0 \end{bmatrix} - \begin{bmatrix} 5 & 2 \\ -1 & 6 \\ 3 & 4 \end{bmatrix} = \begin{bmatrix} 4 & -5 \\ 5 & -5 \\ -1 & -4 \end{bmatrix}$$

(c) $$\mathbf{A} - \mathbf{A} = \begin{bmatrix} 9 & -3 \\ 4 & 1 \\ 2 & 0 \end{bmatrix} - \begin{bmatrix} 9 & -3 \\ 4 & 1 \\ 2 & 0 \end{bmatrix} = \begin{bmatrix} 0 & 0 \\ 0 & 0 \\ 0 & 0 \end{bmatrix}$$

The result of part (c) of this example is a 3×2 matrix in which every entry is zero. Such a matrix is called a *zero matrix* and is written **0**. In fact, there are lots of zero matrices, each corresponding to a particular order. For example,

$$[0] \quad \begin{bmatrix} 0 & 0 \\ 0 & 0 \end{bmatrix} \quad \begin{bmatrix} 0 \\ 0 \\ 0 \\ 0 \end{bmatrix} \quad \begin{bmatrix} 0 & 0 & 0 & 0 & 0 & 0 \\ 0 & 0 & 0 & 0 & 0 & 0 \\ 0 & 0 & 0 & 0 & 0 & 0 \\ 0 & 0 & 0 & 0 & 0 & 0 \end{bmatrix}$$

are the 1×1, 2×2, 4×1 and 4×6 zero matrices respectively. However, despite this, we shall use the single symbol **0** for all of these since it is usually clear in any actual example what the order is and hence which particular zero matrix is being used. It follows from the definition of addition and subtraction that, for any matrix **A**,

$$\mathbf{A} - \mathbf{A} = \mathbf{0}$$

$$\mathbf{A} + \mathbf{0} = \mathbf{A}$$

The role played by the matrix **0** in matrix algebra is therefore similar to that of the number 0 in ordinary arithmetic.

Practice Problem

3 Let

$$\mathbf{A} = \begin{bmatrix} 7 & 5 \\ 2 & 1 \end{bmatrix} \quad \mathbf{B} = \begin{bmatrix} 5 \\ 4 \end{bmatrix} \quad \mathbf{C} = \begin{bmatrix} 2 \\ 2 \end{bmatrix} \quad \mathbf{D} = \begin{bmatrix} -6 & 2 \\ 1 & -9 \end{bmatrix} \quad \mathbf{0} = \begin{bmatrix} 0 \\ 0 \end{bmatrix}$$

Find (where possible)

(a) $\mathbf{A} + \mathbf{D}$ (b) $\mathbf{A} + \mathbf{C}$ (c) $\mathbf{B} - \mathbf{C}$ (d) $\mathbf{C} - \mathbf{0}$ (e) $\mathbf{D} - \mathbf{D}$

7.1.3 Scalar multiplication

Returning to the two-customer three-product example, let us suppose that the sales are the same each month and are given by

$$\mathbf{A} = \begin{bmatrix} 7 & 3 & 4 \\ 1 & 5 & 6 \end{bmatrix}$$

This means, for example, that customer C1 buys 7 items of G1 every month, so in a whole year C1 buys

$$12 \times 7 = 84$$

items of G1. A similar process applies to the remaining goods and customers, and the matrix giving the annual sales is

$$\mathbf{B} = \begin{bmatrix} 12 \times 7 & 12 \times 3 & 12 \times 4 \\ 12 \times 1 & 12 \times 5 & 12 \times 6 \end{bmatrix} = \begin{bmatrix} 84 & 36 & 48 \\ 12 & 60 & 72 \end{bmatrix}$$

Matrix **B** is found by scaling each element in **A** by 12 and we write

$$\mathbf{B} = 12\mathbf{A}$$

In general, to multiply a matrix **A** by a scalar k we simply multiply each element of **A** by k.

Example

If

$$\mathbf{A} = \begin{bmatrix} 1 & 2 & 3 \\ 4 & 5 & 6 \\ 7 & 8 & 9 \end{bmatrix}$$

find

(a) $2\mathbf{A}$ (b) $-\mathbf{A}$ (c) $0\mathbf{A}$

➔

Solution

(a) $$2\mathbf{A} = \begin{bmatrix} 2 & 4 & 6 \\ 8 & 10 & 12 \\ 14 & 16 & 18 \end{bmatrix}$$

(b) $$-\mathbf{A} = (-1)\mathbf{A} = \begin{bmatrix} -1 & -2 & -3 \\ -4 & -5 & -6 \\ -7 & -8 & -9 \end{bmatrix}$$

(c) $$0\mathbf{A} = \begin{bmatrix} 0 & 0 & 0 \\ 0 & 0 & 0 \\ 0 & 0 & 0 \end{bmatrix} = \mathbf{0}$$

In ordinary arithmetic we know that

$$a(b + c) = ab + ac$$

for any three numbers a, b and c. It follows from our definitions of matrix addition and scalar multiplication that

$$k(\mathbf{A} + \mathbf{B}) = k\mathbf{A} + k\mathbf{B}$$

for any $m \times n$ matrices **A** and **B**, and scalar k.

Another property of matrices is

$$k(l\mathbf{A}) = (kl)\mathbf{A}$$

for scalars k and l. Again this follows from the comparable property

$$a(bc) = (ab)c$$

for ordinary numbers.

You are invited to check these two matrix properties for yourself in the following problem.

Practice Problem

4 Let

$$\mathbf{A} = \begin{bmatrix} 1 & -2 \\ 3 & 5 \\ 0 & 4 \end{bmatrix} \quad \text{and} \quad \mathbf{B} = \begin{bmatrix} 0 & -1 \\ 2 & 7 \\ 1 & 6 \end{bmatrix}$$

(1) Find

(a) $2\mathbf{A}$ **(b)** $2\mathbf{B}$ **(c)** $\mathbf{A} + \mathbf{B}$ **(d)** $2(\mathbf{A} + \mathbf{B})$

Hence verify that

$$2(\mathbf{A} + \mathbf{B}) = 2\mathbf{A} + 2\mathbf{B}$$

(2) Find

(a) $3\mathbf{A}$ **(b)** $-6\mathbf{A}$

Hence verify that

$$-2(3\mathbf{A}) = -6\mathbf{A}$$

7.1.4 Matrix multiplication

Advice

Hopefully, you have found the matrix operations considered so far in this section easy to understand. We now turn our attention to matrix multiplication. If you have never multiplied matrices before, you may find that it requires a bit more effort to grasp and you should allow yourself extra time to work through the problems. There is no need to worry. Once you have performed a dozen or so matrix multiplications, you will find that the technique becomes second nature, although the process may appear rather strange and complicated at first sight.

We begin by showing you how to multiply a row vector by a column vector. To illustrate this let us suppose that goods G1, G2 and G3 sell at \$50, \$30 and \$20, respectively, and let us introduce the row vector

$$\mathbf{p} = [50 \quad 30 \quad 20]$$

If the firm sells a total of 100, 200 and 175 goods of type G1, G2 and G3, respectively, then we can write this information as the column vector

$$\mathbf{q} = \begin{bmatrix} 100 \\ 200 \\ 175 \end{bmatrix}$$

The total revenue received from the sale of G1 is found by multiplying the price, \$50, by the quantity, 100, to get

$$\$50 \times 100 = \$5000$$

Similarly, the revenue from G2 and G3 is

$$\$30 \times 200 = \$6000$$

and

$$20 \times 175 = \$3500$$

respectively. The total revenue of the firm is therefore

$$\text{TR} = \$5000 + \$6000 + \$3500 = \$14\ 500$$

The value of TR is a single number and can be regarded as a 1×1 matrix: that is,

$$[14\ 500]$$

This 1×1 matrix is obtained by multiplying together the price vector, **p**, and the quantity vector, **q**, to get

$$[50 \quad 30 \quad 20] \begin{bmatrix} 100 \\ 200 \\ 175 \end{bmatrix} = [14\ 500]$$

The value 14 500 is found by multiplying the corresponding elements of **p** and **q** and then adding together: that is,

$$[50 \quad 30 \quad 20]\begin{bmatrix}100\\200\\175\end{bmatrix} = [5000 + 6000 + 3500] = [14\,500]$$

In general, if **a** is the row vector

$$[a_{11} \quad a_{12} \quad a_{13} \quad \ldots \quad a_{1s}]$$

and **b** is the column vector

$$\begin{bmatrix}b_{11}\\b_{21}\\b_{31}\\\vdots\\b_{s1}\end{bmatrix}$$

then we define the matrix product

$$\mathbf{ab} = [a_{11} \quad a_{12} \quad a_{13} \quad \ldots \quad a_{1s}]\begin{bmatrix}b_{11}\\b_{21}\\b_{31}\\\vdots\\b_{s1}\end{bmatrix}$$

to be the 1×1 matrix

$$[a_{11}b_{11} + a_{12}b_{21} + a_{13}b_{31} + \ldots + a_{1s}b_{s1}]$$

It is important to notice that the single element in the 1×1 matrix **ab** is found by multiplying each element of **a** by the corresponding element of **b**. Consequently, it is essential that both vectors have the same number of elements. In other words, if **a** has order $1 \times s$ and **b** has order $t \times 1$ then it is only possible to form the product **ab** when $s = t$.

Example

If

$$\mathbf{a} = [1 \quad 2 \quad 3 \quad 4], \quad \mathbf{b} = \begin{bmatrix}2\\5\\-1\\0\end{bmatrix} \quad \text{and} \quad \mathbf{c} = \begin{bmatrix}6\\9\\2\end{bmatrix}$$

find **ab** and **ac**.

Solution

Using the definition of the multiplication of a row vector, by a column vector, we have

$$\mathbf{ab} = [1 \quad 2 \quad 3 \quad 4]\begin{bmatrix}2\\5\\-1\\0\end{bmatrix} = [1(2) + 2(5) + 3(-1) + 4(0)] = [9]$$

We have set out the calculations in this way so that you can see how the value, 9, is obtained. There is no need for you to indicate this in your own answers and you may simply write

$$[1 \quad 2 \quad 3 \quad 4]\begin{bmatrix} 2 \\ 5 \\ -1 \\ 0 \end{bmatrix} = [9]$$

without bothering to insert any intermediate steps.

It is impossible to multiply **a** and **c** because **a** has four elements and **c** has only three elements. You can see the problem if you actually try to perform the calculations, since there is no entry in **c** with which to multiply the 4 in **a**.

$$[1 \quad 2 \quad 3 \quad 4]\begin{bmatrix} 6 \\ 9 \\ 2 \end{bmatrix} = [1(6) + 2(9) + 3(2) + 4(?)]$$

Practice Problem

5 Let

$$\mathbf{a} = [1 \quad -1 \quad 0 \quad 3 \quad 2], \quad \mathbf{b} = [1 \quad 2 \quad 9], \quad \mathbf{c} = \begin{bmatrix} 0 \\ -1 \\ 1 \\ 1 \\ 2 \end{bmatrix} \quad \text{and} \quad \mathbf{d} = \begin{bmatrix} -2 \\ 1 \\ 0 \end{bmatrix}$$

Find (where possible)

(a) **ac** **(b)** **bd** **(c)** **ad**

We now turn our attention to general matrix multiplication, which is defined as follows. If **A** is an $m \times s$ matrix and **B** is an $s \times n$ matrix then

$$\mathbf{C} = \mathbf{AB}$$

is an $m \times n$ matrix and c_{ij} is found by multiplying the *i*th row of **A** into the *j*th column of **B**.

There are three things to notice about this definition. Firstly, the number of columns of **A** is the same as the number of rows of **B**. Unless this condition is satisfied it is impossible to form the product **AB**. Secondly, the matrix **C** has order $m \times n$, where m is the number of rows of **A** and n is the number of columns of **B**. Finally, the elements of **C** are found by multiplying row vectors by column vectors. The best way of understanding this definition is to consider an example.

Example

Find **AB** in the case when

$$\mathbf{A} = \begin{bmatrix} 2 & 1 & 0 \\ 1 & 0 & 4 \end{bmatrix} \quad \text{and} \quad \mathbf{B} = \begin{bmatrix} 3 & 1 & 2 & 1 \\ 1 & 0 & 1 & 2 \\ 5 & 4 & 1 & 1 \end{bmatrix}$$

Solution

It is a good idea to check before you begin any detailed calculations that it is possible to multiply these matrices and also to identify the order of the resulting matrix. In this case

A is a 2 × 3 matrix and **B** is a 3 × 4 matrix

Matrix **A** has three columns and **B** has the same number of rows, so it *is* possible to find **AB**. Moreover, **AB** must have order 2 × 4 because **A** has two rows and **B** has four columns. Hence

$$\begin{bmatrix} 2 & 1 & 0 \\ 1 & 0 & 4 \end{bmatrix}\begin{bmatrix} 3 & 1 & 2 & 1 \\ 1 & 0 & 1 & 2 \\ 5 & 4 & 1 & 1 \end{bmatrix} = \begin{bmatrix} c_{11} & c_{12} & c_{13} & c_{14} \\ c_{21} & c_{22} & c_{23} & c_{24} \end{bmatrix}$$

All that remains for us to do is to calculate the eight numbers c_{ij}.

The number c_{11} in the top left-hand corner lies in the first row and first column, so to find its value we multiply the first row of **A** into the first column of **B** to get

$$\begin{bmatrix} \boxed{2 \quad 1 \quad 0} \\ 1 \quad 0 \quad 4 \end{bmatrix}\begin{bmatrix} \boxed{3} & 1 & 2 & 1 \\ \boxed{1} & 0 & 1 & 2 \\ \boxed{5} & 4 & 1 & 1 \end{bmatrix} = \begin{bmatrix} \boxed{7} & c_{12} & c_{13} & c_{14} \\ c_{21} & c_{22} & c_{23} & c_{24} \end{bmatrix}$$

because 2(3) + 1(1) + 0(5) = 7.

The number c_{12} lies in the first row and second column, so to find its value we multiply the first row of **A** into the second column of **B** to get

$$\begin{bmatrix} \boxed{2 \quad 1 \quad 0} \\ 1 \quad 0 \quad 4 \end{bmatrix}\begin{bmatrix} 3 & \boxed{1} & 2 & 1 \\ 1 & \boxed{0} & 1 & 2 \\ 5 & \boxed{4} & 1 & 1 \end{bmatrix} = \begin{bmatrix} 7 & \boxed{2} & c_{13} & c_{14} \\ c_{21} & c_{22} & c_{23} & c_{24} \end{bmatrix}$$

because 2(1) + 1(0) + 0(4) = 2.

The values of c_{13} and c_{14} are then found in a similar way by multiplying the first row of **A** into the third and fourth columns of **B**, respectively, to get

$$\begin{bmatrix} \boxed{2 \quad 1 \quad 0} \\ 1 \quad 0 \quad 4 \end{bmatrix}\begin{bmatrix} 3 & 1 & \boxed{2} & \boxed{1} \\ 1 & 0 & \boxed{1} & \boxed{2} \\ 5 & 4 & \boxed{1} & \boxed{1} \end{bmatrix} = \begin{bmatrix} 7 & 2 & \boxed{5} & \boxed{4} \\ c_{21} & c_{22} & c_{23} & c_{24} \end{bmatrix}$$

because 2(2) + 1(1) + 0(1) = 5 and 2(1) + 1(2) + 0(1) = 4.

Finally, we repeat the whole procedure along the second row of **C**. The elements c_{21}, c_{22}, c_{23} and c_{24} are calculated by multiplying the second row of **A** into the four columns of **B** in succession to get

$$\begin{bmatrix} 2 \quad 1 \quad 0 \\ \boxed{1 \quad 0 \quad 4} \end{bmatrix}\begin{bmatrix} \boxed{3} & \boxed{1} & \boxed{2} & \boxed{1} \\ \boxed{1} & \boxed{0} & \boxed{1} & \boxed{2} \\ \boxed{5} & \boxed{4} & \boxed{1} & \boxed{1} \end{bmatrix} = \begin{bmatrix} 7 & 2 & 5 & 4 \\ \boxed{23} & \boxed{17} & \boxed{6} & \boxed{5} \end{bmatrix}$$

because

1(3) + 0(1) + 4(5) = 23

1(1) + 0(0) + 4(4) = 17

1(2) + 0(1) + 4(1) = 6

1(1) + 0(2) + 4(1) = 5

In this example we have indicated how to build up the matrix **C** in a step-by-step manner and have used boxes to show you how the calculations are performed. This approach has been adopted merely as a teaching device. There is no need for you to set your calculations out in this way and you are encouraged to write down your answer in a single line of working.

Advice

Take the trouble to check before you begin that it is possible to form the matrix product and to anticipate the order of the end result. This can be done by jotting down the orders of the original matrices side by side. The product exists if the inner numbers are the same and the order of the answer is given by the outer numbers: that is,

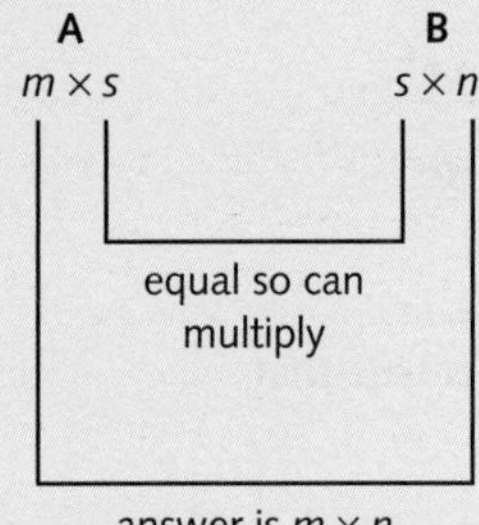

For example, if **A**, **B** and **C** have orders 3×5, 5×2 and 3×4 respectively, then **AB** exists and has order 3×2 because

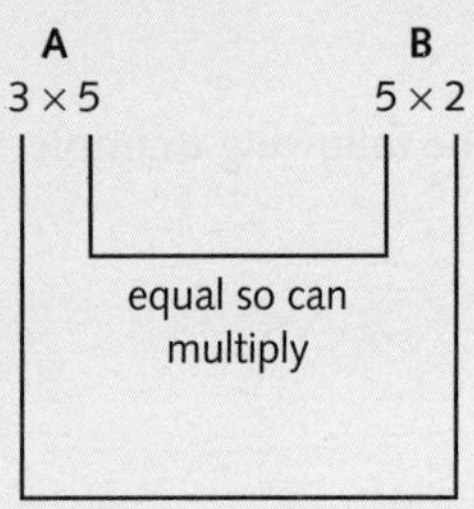

but it is impossible to form **AC** because

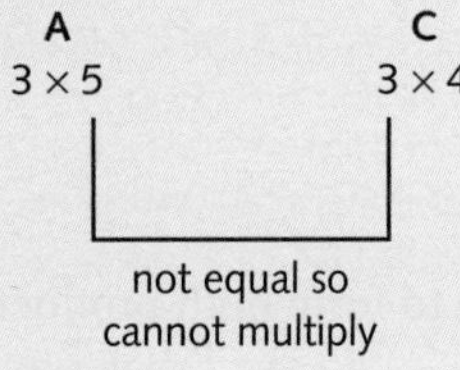

Practice Problem

6 Write down the order of the matrices

$$\mathbf{A} = \begin{bmatrix} 1 & 2 \\ 0 & 1 \\ 3 & 1 \end{bmatrix} \quad \text{and} \quad \mathbf{B} = \begin{bmatrix} 1 & 2 \\ 3 & 4 \end{bmatrix}$$

Hence verify that it is possible to form the matrix product

$$\mathbf{C} = \mathbf{AB}$$

and write down the order of **C**. Calculate all of the elements of **C**.

We have already noted that matrix operations have similar properties to those of ordinary arithmetic. Some particular rules of arithmetic are:

$a(b + c) = ab + ac$ (distributive law)

$(a + b)c = ac + bc$ (distributive law)

$a(bc) = (ab)c$ (associative law)

$ab = ba$ (commutative law)

An obvious question to ask is whether they have a counterpart in matrix algebra. It turns out that provided the matrices **A**, **B** and **C** have the correct orders for the appropriate sums and products to exist then

$$\mathbf{A}(\mathbf{B} + \mathbf{C}) = \mathbf{AB} + \mathbf{AC}$$

$$(\mathbf{A} + \mathbf{B})\mathbf{C} = \mathbf{AC} + \mathbf{BC}$$

$$\mathbf{A}(\mathbf{BC}) = (\mathbf{AB})\mathbf{C}$$

However, although it is true that

$$ab = ba$$

for numbers, this result does **not** extend to matrices. Even if **AB** and **BA** both exist it is not necessarily true that

$$\mathbf{AB} = \mathbf{BA}$$

This is illustrated in the following example.

Example

If

$$\mathbf{A} = \begin{bmatrix} 1 & -1 \\ 2 & 1 \end{bmatrix} \quad \text{and} \quad \mathbf{B} = \begin{bmatrix} 1 & 3 \\ 1 & 2 \end{bmatrix}$$

evaluate **AB** and **BA**.

Solution

It is easy to check that it is possible to form both products **AB** and **BA** and that they both have order 2×2. In fact

$$\mathbf{AB} = \begin{bmatrix} 1 & -1 \\ 2 & 1 \end{bmatrix}\begin{bmatrix} 1 & 3 \\ 1 & 2 \end{bmatrix} = \begin{bmatrix} 0 & 1 \\ 3 & 8 \end{bmatrix}$$

$$\mathbf{BA} = \begin{bmatrix} 1 & 3 \\ 1 & 2 \end{bmatrix}\begin{bmatrix} 1 & -1 \\ 2 & 1 \end{bmatrix} = \begin{bmatrix} 7 & 2 \\ 5 & 1 \end{bmatrix}$$

so $\mathbf{AB} \neq \mathbf{BA}$.

There are certain pairs of matrices which do commute (that is, for which **AB** = **BA**) and we shall investigate some of these in the next section. However, these are very much the exception. We therefore have the 'non-property' that, in general,

$$\mathbf{AB} \neq \mathbf{BA}$$

Practice Problems

7 Let

$$\mathbf{A}=\begin{bmatrix}2 & 1 & 1\\5 & 1 & 0\\-1 & 1 & 4\end{bmatrix}\quad \mathbf{B}=\begin{bmatrix}1\\2\\1\end{bmatrix}\quad \mathbf{C}=\begin{bmatrix}1 & 2\\3 & 1\end{bmatrix}\quad \mathbf{D}=\begin{bmatrix}1 & 1\\-1 & 1\\2 & 1\end{bmatrix}\quad \text{and}\quad \mathbf{E}=\begin{bmatrix}1 & 2 & 3\\4 & 5 & 6\end{bmatrix}$$

Find (where possible)

(a) $\mathbf{AB}$ **(b)** $\mathbf{BA}$ **(c)** $\mathbf{CD}$ **(d)** $\mathbf{DC}$

(e) $\mathbf{AE}$ **(f)** $\mathbf{EA}$ **(g)** $\mathbf{DE}$ **(h)** $\mathbf{ED}$

8 Evaluate the matrix product **Ax**, where

$$\mathbf{A}=\begin{bmatrix}1 & 4 & 7\\2 & 6 & 5\\8 & 9 & 5\end{bmatrix}\quad \text{and}\quad \mathbf{x}=\begin{bmatrix}x\\y\\z\end{bmatrix}$$

Hence show that the system of linear equations

$$x + 4y + 7z = -3$$
$$2x + 6y + 5z = 10$$
$$8x + 9y + 5z = 1$$

can be written as $\mathbf{Ax} = \mathbf{b}$ where

$$\mathbf{b}=\begin{bmatrix}-3\\10\\1\end{bmatrix}$$

We conclude this section by showing you how to express a familiar problem in matrix notation. Section 1.2 described the method of elimination for solving systems of simultaneous linear equations. For example, we might want to find values of x and y which satisfy

$$2x - 5y = 6$$
$$7x + 8y = -1$$

Motivated by the result of Problem 8 we write this as

$$\mathbf{Ax} = \mathbf{b}$$

where

$$\mathbf{A}=\begin{bmatrix}2 & -5\\7 & 8\end{bmatrix}\quad \mathbf{x}=\begin{bmatrix}x\\y\end{bmatrix}\quad \mathbf{b}=\begin{bmatrix}6\\-1\end{bmatrix}$$

It is easy to check that this is correct simply by multiplying out **Ax** to get

$$\begin{bmatrix}2 & -5\\7 & 8\end{bmatrix}\begin{bmatrix}x\\y\end{bmatrix}=\begin{bmatrix}2x-5y\\7x+8y\end{bmatrix}$$

and so the matrix equation $\mathbf{Ax} = \mathbf{b}$ reads

$$\begin{bmatrix}2x-5y\\7x+8y\end{bmatrix}=\begin{bmatrix}6\\-1\end{bmatrix}$$

that is,

$$2x - 5y = 6$$

$$7x + 8y = -1$$

Quite generally, any system of n linear equations in n unknowns can be written as

$$\mathbf{Ax} = \mathbf{b}$$

where **A**, **x** and **b** are $n \times n$, $n \times 1$ and $n \times 1$ matrices respectively. The matrix **A** consists of the coefficients, the vector **x** consists of the unknowns and the vector **b** consists of the right-hand sides. The definition of matrix multiplication allows us to write a linear system in terms of matrices, although it is not immediately obvious that there is any advantage in doing so. In the next section we introduce the concept of a matrix inverse and show you how to use this to solve systems of equations expressed in matrix form.

Throughout this section we have noted various properties that matrices satisfy. For convenience these are summarized in the next subsection.

7.1.5 Summary

Provided that the indicated sums and products make sense,

$$\mathbf{A} + \mathbf{B} = \mathbf{B} + \mathbf{A}$$

$$\mathbf{A} - \mathbf{A} = \mathbf{0}$$

$$\mathbf{A} + \mathbf{0} = \mathbf{A}$$

$$k(\mathbf{A} + \mathbf{B}) = k\mathbf{A} + k\mathbf{B}$$

$$k(l\mathbf{A}) = (kl)\mathbf{A}$$

$$\mathbf{A}(\mathbf{B} + \mathbf{C}) = \mathbf{AB} + \mathbf{AC}$$

$$(\mathbf{A} + \mathbf{B})\mathbf{C} = \mathbf{AC} + \mathbf{BC}$$

$$\mathbf{A}(\mathbf{BC}) = (\mathbf{AB})\mathbf{C}$$

We also have the non-property that, in general,

$$\mathbf{AB} \neq \mathbf{BA}$$

Key Terms

Column vector A matrix with one column.

Elements The individual numbers inside a matrix. (Also called entries.)

Matrix A rectangular array of numbers, set out in rows and columns, surrounded by a pair of brackets. (Plural matrices.)

Order The dimensions of a matrix. A matrix with m rows and n columns has order $m \times n$.

Row vector A matrix with one row.

Transpose of a matrix The matrix obtained from a given matrix by interchanging rows and columns. The transpose of a matrix **A** is written $\mathbf{A}^T$.

Zero matrix A matrix in which every element is zero.

Advice

There now follow some Practice Problems. If you find that you do not have time to do all of them, you should at least attempt Practice Problem 15 because the result of this problem will be used in the next section.

Practice Problems

9 The monthly sales (in thousands) of burgers (B1) and bites (B2) in three fast-food restaurants (R1, R2, R3) are as follows:

	R1	R2	R3
B1	35	27	13
B2	42	39	24

January

	R1	R2	R3
B1	31	17	3
B2	25	29	16

February

(a) Write down two 2×3 matrices **J** and **F** representing sales in January and February respectively.

(b) By finding **J** + **F**, write down the matrix for the total sales over the two months.

(c) By finding **J** − **F**, write down the matrix for the difference in sales for the two months.

10 If

$$\mathbf{A} = \begin{bmatrix} 2 & 3 & 1 & 9 \\ 1 & 0 & 5 & 0 \\ 6 & 7 & 8 & 4 \end{bmatrix} \qquad \mathbf{B} = \begin{bmatrix} 1 & 7 & 9 & 6 \\ 2 & 1 & 0 & 5 \\ 6 & 4 & 5 & 3 \end{bmatrix}$$

work out

(a) $2\mathbf{A}$ **(b)** $2\mathbf{B}$ **(c)** $2\mathbf{A} + 2\mathbf{B}$ **(d)** $2(\mathbf{A} + \mathbf{B})$

Do you notice any connection between your answers to parts (c) and (d)?

11 A firm manufactures three products, P1, P2 and P3, which it sells to two customers, C1 and C2. The number of items of each product that are sold to these customers is given by

$$\mathbf{A} = \begin{matrix} \\ \text{C1} \\ \text{C2} \end{matrix} \begin{matrix} \text{P1} & \text{P2} & \text{P3} \\ \left[\begin{matrix} 6 \\ 2 \end{matrix}\right. & \begin{matrix} 7 \\ 1 \end{matrix} & \left.\begin{matrix} 9 \\ 2 \end{matrix}\right] \end{matrix}$$

➔

The firm charges both customers the same price for each product according to

$$\mathbf{B} = \begin{matrix} & \text{P1} & \text{P2} & \text{P3} \\ [& 100 & 500 & 200 &]^{\mathrm{T}} \end{matrix}$$

To make each item of type P1, P2 and P3, the firm uses four raw materials, R1, R2, R3 and R4. The number of tonnes required per item is given by

$$\mathbf{C} = \begin{matrix} \\ \text{P1} \\ \text{P2} \\ \text{P3} \end{matrix} \begin{matrix} \begin{matrix} \text{R1} & \text{R2} & \text{R3} & \text{R4} \end{matrix} \\ \begin{bmatrix} 1 & 0 & 0 & 1 \\ 1 & 1 & 2 & 1 \\ 0 & 0 & 1 & 1 \end{bmatrix} \end{matrix}$$

The cost per tonne of raw materials is

$$\mathbf{D} = \begin{matrix} & \text{R1} & \text{R2} & \text{R3} & \text{R4} \\ [& 20 & 10 & 15 & 15 &]^{\mathrm{T}} \end{matrix}$$

In addition, let

$$\mathbf{E} = [1 \quad 1]$$

Find the following matrix products and give an interpretation of each one.

(a) $\mathbf{AB}$ **(b)** $\mathbf{AC}$ **(c)** $\mathbf{CD}$ **(d)** $\mathbf{ACD}$ **(e)** $\mathbf{EAB}$

(f) $\mathbf{EACD}$ **(g)** $\mathbf{EAB} - \mathbf{EACD}$

12 **(1)** Let

$$\mathbf{A} = \begin{bmatrix} 1 & 2 \\ 3 & 4 \\ 5 & 6 \end{bmatrix} \quad \text{and} \quad \mathbf{B} = \begin{bmatrix} 1 & -1 \\ 2 & 1 \\ -3 & 4 \end{bmatrix}$$

Find

(a) $\mathbf{A}^{\mathrm{T}}$ **(b)** $\mathbf{B}^{\mathrm{T}}$ **(c)** $\mathbf{A} + \mathbf{B}$ **(d)** $(\mathbf{A} + \mathbf{B})^{\mathrm{T}}$

Do you notice any connection between $(\mathbf{A} + \mathbf{B})^{\mathrm{T}}$, $\mathbf{A}^{\mathrm{T}}$ and $\mathbf{B}^{\mathrm{T}}$?

(2) Let

$$\mathbf{C} = \begin{bmatrix} 1 & 4 \\ 5 & 9 \end{bmatrix} \quad \text{and} \quad \mathbf{D} = \begin{bmatrix} 2 & 1 & 0 \\ -1 & 0 & 1 \end{bmatrix}$$

Find

(a) $\mathbf{C}^{\mathrm{T}}$ **(b)** $\mathbf{D}^{\mathrm{T}}$ **(c)** $\mathbf{CD}$ **(d)** $(\mathbf{CD})^{\mathrm{T}}$

Do you notice any connection between $(\mathbf{CD})^{\mathrm{T}}$, $\mathbf{C}^{\mathrm{T}}$ and $\mathbf{D}^{\mathrm{T}}$?

13 Verify the equations

(a) $\mathbf{A}(\mathbf{B} + \mathbf{C}) = \mathbf{AB} + \mathbf{AC}$ **(b)** $(\mathbf{AB})\mathbf{C} = \mathbf{A}(\mathbf{BC})$

in the case when

$$\mathbf{A} = \begin{bmatrix} 5 & -3 \\ 2 & 1 \end{bmatrix}, \quad \mathbf{B} = \begin{bmatrix} 1 & 5 \\ 4 & 0 \end{bmatrix} \quad \text{and} \quad \mathbf{C} = \begin{bmatrix} -1 & 1 \\ 1 & 2 \end{bmatrix}$$

14 If

$$\mathbf{A} = [1 \quad 2 \quad -4 \quad 3] \quad \text{and} \quad \mathbf{B} = \begin{bmatrix} 1 \\ 7 \\ 3 \\ -2 \end{bmatrix}$$

find **AB** and **BA**.

15 Let

$$\mathbf{A} = \begin{bmatrix} a & b \\ c & d \end{bmatrix}, \quad \mathbf{A}^{-1} = \frac{1}{ad - bc}\begin{bmatrix} d & -b \\ -c & a \end{bmatrix} \quad (ad - bc \neq 0)$$

$$\mathbf{I} = \begin{bmatrix} 1 & 0 \\ 0 & 1 \end{bmatrix} \quad \text{and} \quad \mathbf{x} = \begin{bmatrix} x \\ y \end{bmatrix}$$

Show that

(a) $\mathbf{AI} = \mathbf{A}$ and $\mathbf{IA} = \mathbf{A}$ **(b)** $\mathbf{A}^{-1}\mathbf{A} = \mathbf{I}$ and $\mathbf{AA}^{-1} = \mathbf{I}$ **(c)** $\mathbf{Ix} = \mathbf{x}$

16 **(a)** Evaluate the matrix product, **Ax**, where

$$\mathbf{A} = \begin{bmatrix} 7 & 5 \\ 1 & 3 \end{bmatrix} \quad \text{and} \quad \mathbf{x} = \begin{bmatrix} x \\ y \end{bmatrix}$$

Hence show that the system of linear equations

$$7x + 5y = 3$$
$$x + 3y = 2$$

can be written as **Ax** = **b** where $\mathbf{b} = \begin{bmatrix} 3 \\ 2 \end{bmatrix}$.

(b) The system of equations

$$2x + 3y - 2z = 6$$
$$x - y + 2z = 3$$
$$4x + 2y + 5z = 1$$

can be expressed in the form **Ax** = **b**. Write down the matrices **A**, **x** and **b**.

section 7.2

Matrix inversion

Objectives

At the end of this section you should be able to:

- Write down the 2×2 and 3×3 identity matrices.
- Detect whether a matrix is singular or non-singular.
- Calculate the determinant and inverse of a 2×2 matrix.
- Calculate the cofactors of a 3×3 matrix.
- Use cofactors to find the determinant and inverse of a 3×3 matrix.
- Use matrix inverses to solve systems of linear equations arising in economics.

In this and the following section we consider *square* matrices, in which the number of rows and columns are equal. For simplicity we concentrate on 2×2 and 3×3 matrices, although the ideas and techniques apply more generally to $n \times n$ matrices of any size. We have already seen that, with one notable exception, the algebra of matrices is virtually the same as the algebra of numbers. There are, however, two important properties of numbers which we have yet to consider. The first is the existence of a number, 1, which satisfies

$$a1 = a \quad \text{and} \quad 1a = a$$

for any number, a. The second is the fact that corresponding to any non-zero number, a, we can find another number, a^{-1}, with the property that

$$a^{-1}a = 1 \quad \text{and} \quad aa^{-1} = 1$$

$a^{-1} = \frac{1}{a}$

If you have worked through Practice Problem 15 of Section 7.1 you will know how to extend these to 2×2 matrices. In part (a) you showed that, for any 2×2 matrix, $\mathbf{A}$,

$$\mathbf{AI} = \mathbf{A} \quad \text{and} \quad \mathbf{IA} = \mathbf{A}$$

where

$$\mathbf{I} = \begin{bmatrix} 1 & 0 \\ 0 & 1 \end{bmatrix}$$

The matrix **I** is called the *identity* matrix and is analogous to the number 1 in ordinary arithmetic. You also showed in part (b) of Practice Problem 15 that corresponding to the 2×2 matrix

$$\mathbf{A} = \begin{bmatrix} a & b \\ c & d \end{bmatrix}$$

there is another matrix

$$\mathbf{A}^{-1} = \frac{1}{ad - bc}\begin{bmatrix} d & -b \\ -c & a \end{bmatrix}$$

with the property that

$$\mathbf{A}^{-1}\mathbf{A} = \mathbf{I} \quad \text{and} \quad \mathbf{A}\mathbf{A}^{-1} = \mathbf{I}$$

The matrix $\mathbf{A}^{-1}$ is said to be the *inverse* of **A** and is analogous to the reciprocal of a number. The formula for $\mathbf{A}^{-1}$ looks rather complicated but the construction of $\mathbf{A}^{-1}$ is in fact very easy. Starting with some matrix

$$\mathbf{A} = \begin{bmatrix} a & b \\ c & d \end{bmatrix}$$

we first swap the two numbers on the leading diagonal (that is, the elements along the line joining the top left-hand corner to the bottom right-hand corner of **A**) to get

$$\begin{bmatrix} d & b \\ c & a \end{bmatrix}$$

swap
a and *d*

Secondly, we change the sign of the 'off-diagonal' elements to get

$$\begin{bmatrix} d & -b \\ -c & a \end{bmatrix}$$

change signs
of *b* and *c*

Finally, we multiply the matrix by the scalar

$$\frac{1}{ad - bc}$$

to get

$$\frac{1}{ad - bc}\begin{bmatrix} d & -b \\ -c & a \end{bmatrix}$$

divide each element
by $ad - bc$

The number $ad - bc$ is called the *determinant* of **A** and is written as

$$\det(\mathbf{A}) \quad \text{or} \quad |\mathbf{A}| \quad \text{or} \quad \begin{vmatrix} a & b \\ c & d \end{vmatrix}$$

Notice that the last step in the calculation is impossible if

$$|\mathbf{A}| = 0$$

because we cannot divide by zero. We deduce that the inverse of a matrix exists only if the matrix has a non-zero determinant. This is comparable to the situation in arithmetic where a reciprocal of a number exists provided the number is non-zero. If the matrix has a non-zero determinant, it is said to be *non-singular*; otherwise it is said to be *singular*.

Example

Find the inverse of the following matrices. Are these matrices singular or non-singular?

$$\mathbf{A} = \begin{bmatrix} 1 & 2 \\ 3 & 4 \end{bmatrix} \quad \text{and} \quad \mathbf{B} = \begin{bmatrix} 2 & 5 \\ 4 & 10 \end{bmatrix}$$

Solution

We begin by calculating the determinant of

$$\mathbf{A} = \begin{bmatrix} 1 & 2 \\ 3 & 4 \end{bmatrix}$$

to see whether or not the inverse exists.

$$\det(\mathbf{A}) = \begin{vmatrix} 1 & 2 \\ 3 & 4 \end{vmatrix} = 1(4) - 2(3) = 4 - 6 = -2$$

We see that $\det(\mathbf{A}) \neq 0$, so the matrix is non-singular and the inverse exists. To find $\mathbf{A}^{-1}$ we swap the diagonal elements, 1 and 4, change the sign of the off-diagonal elements, 2 and 3, and divide by the determinant, −2. Hence

$$\mathbf{A}^{-1} = -\frac{1}{2}\begin{bmatrix} 4 & -2 \\ -3 & 1 \end{bmatrix} = \begin{bmatrix} -2 & 1 \\ 3/2 & -1/2 \end{bmatrix}$$

Of course, if $\mathbf{A}^{-1}$ really is the inverse of $\mathbf{A}$, then $\mathbf{A}^{-1}\mathbf{A}$ and $\mathbf{A}\mathbf{A}^{-1}$ should multiply out to give $\mathbf{I}$. As a check:

$$\mathbf{A}^{-1}\mathbf{A} = \begin{bmatrix} -2 & 1 \\ 3/2 & -1/2 \end{bmatrix}\begin{bmatrix} 1 & 2 \\ 3 & 4 \end{bmatrix} = \begin{bmatrix} 1 & 0 \\ 0 & 1 \end{bmatrix} \quad ✓$$

$$\mathbf{A}\mathbf{A}^{-1} = \begin{bmatrix} 1 & 2 \\ 3 & 4 \end{bmatrix}\begin{bmatrix} -2 & 1 \\ 3/2 & -1/2 \end{bmatrix} = \begin{bmatrix} 1 & 0 \\ 0 & 1 \end{bmatrix} \quad ✓$$

To discover whether or not the matrix

$$\mathbf{B} = \begin{bmatrix} 2 & 5 \\ 4 & 10 \end{bmatrix}$$

has an inverse we need to find its determinant.

$$\det(\mathbf{B}) = \begin{bmatrix} 2 & 5 \\ 4 & 10 \end{bmatrix} = 2(10) - 5(4) = 20 - 20 = 0$$

We see that $\det(\mathbf{B}) = 0$, so this matrix is singular and the inverse does not exist.

Practice Problem

1 Find (where possible) the inverse of the following matrices. Are these matrices singular or non-singular?

$$\mathbf{A} = \begin{bmatrix} 6 & 4 \\ 1 & 2 \end{bmatrix} \quad \mathbf{B} = \begin{bmatrix} 6 & 4 \\ 3 & 2 \end{bmatrix}$$

One reason for calculating the inverse of a matrix is that it helps us to solve matrix equations in the same way that the reciprocal of a number is used to solve algebraic equations. We have already seen in Section 7.1 how to express a system of linear equations in matrix form. Any 2×2 system

$$ax + by = e$$

$$cx + dy = f$$

can be written as

$$\mathbf{Ax} = \mathbf{b}$$

where

$$\mathbf{A} = \begin{bmatrix} a & b \\ c & d \end{bmatrix} \quad \mathbf{x} = \begin{bmatrix} x \\ y \end{bmatrix} \quad \mathbf{b} = \begin{bmatrix} e \\ f \end{bmatrix}$$

The coefficient matrix, **A**, and right-hand-side vector, **b**, are assumed to be given and the problem is to determine the vector of unknowns, **x**. Multiplying both sides of

$$\mathbf{Ax} = \mathbf{b}$$

by $\mathbf{A}^{-1}$ gives

$$\mathbf{A}^{-1}(\mathbf{Ax}) = \mathbf{A}^{-1}\mathbf{b}$$

$$(\mathbf{A}^{-1}\mathbf{A})\mathbf{x} = \mathbf{A}^{-1}\mathbf{b} \quad \text{(associative property)}$$

$$\mathbf{Ix} = \mathbf{A}^{-1}\mathbf{b} \quad \text{(definition of an inverse)}$$

$$\mathbf{x} = \mathbf{A}^{-1}\mathbf{b} \quad \text{(Practice Problem 15(c) in Section 7.1)}$$

The solution vector **x** can therefore be found simply by multiplying $\mathbf{A}^{-1}$ by **b**. We are assuming here that $\mathbf{A}^{-1}$ exists. If the coefficient matrix is singular then the inverse cannot be found and the system of linear equations does not possess a unique solution; there are either infinitely many solutions or no solution.

Advice

These special cases are dealt with using the elimination method described in Section 1.2. You might find it instructive to revise both Sections 1.2 and 1.3.

The following two examples illustrate the use of inverses to solve systems of linear equations. The first is taken from microeconomics and the second from macroeconomics.

Example

The equilibrium prices P_1 and P_2 for two goods satisfy the equations

$$-4P_1 + P_2 = -13$$

$$2P_1 - 5P_2 = -7$$

Express this system in matrix form and hence find the values of P_1 and P_2.

➔

Solution

Using the notation of matrices, the simultaneous equations

$$-4P_1 + P_2 = -13$$
$$2P_1 - 5P_2 = -7$$

can be written as

$$\begin{bmatrix} -4 & 1 \\ 2 & -5 \end{bmatrix}\begin{bmatrix} P_1 \\ P_2 \end{bmatrix} = \begin{bmatrix} -13 \\ -7 \end{bmatrix}$$

that is, as

$$\mathbf{Ax} = \mathbf{b}$$

where

$$\mathbf{A} = \begin{bmatrix} -4 & 1 \\ 2 & -5 \end{bmatrix} \quad \mathbf{x} = \begin{bmatrix} P_1 \\ P_2 \end{bmatrix} \quad \mathbf{b} = \begin{bmatrix} -13 \\ -7 \end{bmatrix}$$

The matrix **A** has determinant

$$\begin{vmatrix} -4 & 1 \\ 2 & -5 \end{vmatrix} = (-4)(-5) - (1)(2) = 20 - 2 = 18$$

To find $\mathbf{A}^{-1}$ we swap the diagonal elements, −4 and −5, change the sign of the off-diagonal elements, 1 and 2, and divide by the determinant, 18, to get

$$\mathbf{A}^{-1} = \frac{1}{18}\begin{bmatrix} -5 & -1 \\ -2 & -4 \end{bmatrix}$$

Finally, to calculate **x** we multiply $\mathbf{A}^{-1}$ by **b** to get

$$\mathbf{x} = \mathbf{A}^{-1}\mathbf{b}$$
$$= \frac{1}{18}\begin{bmatrix} -5 & -1 \\ -2 & -4 \end{bmatrix}\begin{bmatrix} -13 \\ -7 \end{bmatrix}$$
$$= \frac{1}{18}\begin{bmatrix} 72 \\ 54 \end{bmatrix} = \begin{bmatrix} 4 \\ 3 \end{bmatrix}$$

Hence $P_1 = 4$ and $P_2 = 3$.

Practice Problem

2 The equilibrium prices P_1 and P_2 for two goods satisfy the equations

$$9P_1 + P_2 = 43$$
$$2P_1 + 7P_2 = 57$$

Express this system in matrix form and hence find the values of P_1 and P_2.

[You have already solved this particular system in Practice Problem 4 of Section 1.3. You might like to compare the work involved in solving this system using the method of elimination described in Chapter 1 and the method based on matrix inverses considered here.]

Example

The equilibrium levels of consumption, C, and income, Y, for the simple two-sector macro-economic model satisfy the structural equations

$$Y = C + I^*$$

$$C = aY + b$$

where a and b are parameters in the range $0 < a < 1$ and $b > 0$, and I^* denotes investment. Express this system in matrix form and hence express Y and C in terms of a, b and I^*. Give an economic interpretation of the inverse matrix.

Solution

The reduced form of the structural equations for this simple model has already been found in Section 5.3. It is instructive to reconsider this problem using matrices. The objective is to express the endogenous variables, Y and C, in terms of the exogenous variable I^* and parameters a and b. The 'unknowns' of this problem are therefore Y and C, and we begin by rearranging the structural equations so that these variables appear on the left-hand sides. Subtracting C from both sides of

$$Y = C + I^*$$

gives

$$Y - C = I^* \qquad (1)$$

and if we subtract aY from both sides of

$$C = aY + b$$

we get

$$-aY + C = b \qquad (2)$$

(It is convenient to put the term involving Y first so that the variables align with those of equation (1).)

In matrix form, equations (1) and (2) become

$$\begin{bmatrix} 1 & -1 \\ -a & 1 \end{bmatrix}\begin{bmatrix} Y \\ C \end{bmatrix} = \begin{bmatrix} I^* \\ b \end{bmatrix}$$

that is,

$$\mathbf{Ax} = \mathbf{b}$$

where

$$\mathbf{A} = \begin{bmatrix} 1 & -1 \\ -a & 1 \end{bmatrix} \quad \mathbf{x} = \begin{bmatrix} Y \\ C \end{bmatrix} \quad \mathbf{b} = \begin{bmatrix} I^* \\ b \end{bmatrix}$$

The matrix $\mathbf{A}$ has determinant

$$\begin{vmatrix} 1 & -1 \\ -a & 1 \end{vmatrix} = 1(1) - (-1)(-a) = 1 - a$$

which is non-zero because $a < 1$.

To find $\mathbf{A}^{-1}$, we swap the diagonal elements, 1 and 1, change the sign of the off-diagonal elements, -1 and $-a$, and divide by the determinant, $1 - a$, to get

$$\mathbf{A}^{-1} = \frac{1}{1-a}\begin{bmatrix} 1 & 1 \\ a & 1 \end{bmatrix}$$

Finally, to determine $\mathbf{x}$ we multiply $\mathbf{A}^{-1}$ by $\mathbf{b}$ to get

$$\mathbf{x} = \mathbf{A}^{-1}\mathbf{b}$$
$$= \frac{1}{1-a}\begin{bmatrix} 1 & 1 \\ a & 1 \end{bmatrix}\begin{bmatrix} I^* \\ b \end{bmatrix} = \frac{1}{1-a}\begin{bmatrix} I^* + b \\ aI^* + b \end{bmatrix}$$

Hence

$$Y = \frac{I^* + b}{1-a} \quad \text{and} \quad C = \frac{aI^* + b}{1-a}$$

The inverse matrix obviously provides a useful way of solving the structural equations of a macroeconomic model. In addition, the elements of the inverse matrix can be given an important economic interpretation. To see this, let us suppose that the investment I^* changes by an amount ΔI^* to become $I^* + \Delta I^*$, with the parameter b held fixed. The new values of Y and C are obtained by replacing I^* by $I^* + \Delta I^*$ in the expressions for Y and C, and are given by

$$\frac{I^* + \Delta I^* + b}{1-a} \quad \text{and} \quad \frac{a(I^* + \Delta I^*) + b}{1-a}$$

respectively. The change in the value of Y is therefore

$$\Delta Y = \frac{I^* + \Delta I^* + b}{1-a} - \frac{I^* + b}{1-a} = \left(\frac{1}{1-a}\right)\Delta I^*$$

and the change in the value of C is

$$\Delta C = \frac{a(I^* + \Delta I^*) + b}{1-a} - \frac{aI^* + b}{1-a} = \left(\frac{a}{1-a}\right)\Delta I^*$$

In other words, the changes to Y and C are found by multiplying the change in I^* by

$$\frac{1}{1-a} \quad \text{and} \quad \frac{a}{1-a}$$

respectively. For this reason we call

$$\frac{1}{1-a}$$

the investment multiplier for Y and

$$\frac{a}{1-a}$$

the investment multiplier for C.

Now the inverse matrix is

$$\mathbf{A}^{-1} = \begin{bmatrix} \dfrac{1}{1-a} & \dfrac{1}{1-a} \\ \dfrac{a}{1-a} & \dfrac{1}{1-a} \end{bmatrix}$$

and we see that these multipliers are precisely the elements that appear in the first column. It is easy to show, using a similar argument, that the second column contains the multipliers for Y and C due to changes in the autonomous consumption, b. The four elements in the inverse matrix can thus be interpreted as follows:

	I^*	b
Y	investment multiplier for Y	autonomous consumption multiplier for Y
C	investment multiplier for C	autonomous consumption multiplier for C

Practice Problem

3 The general linear supply and demand equations for a one-commodity market model are given by

$$P = aQ_S + b \quad (a > 0, b > 0)$$

$$P = -cQ_D + d \quad (c > 0, d > 0)$$

Show that in matrix notation the equilibrium price, *P*, and quantity, *Q*, satisfy

$$\begin{bmatrix} 1 & -a \\ 1 & c \end{bmatrix} \begin{bmatrix} P \\ Q \end{bmatrix} = \begin{bmatrix} b \\ d \end{bmatrix}$$

Solve this system to express *P* and *Q* in terms of *a*, *b*, *c* and *d*. Write down the multiplier for *Q* due to changes in *b* and deduce that an increase in *b* leads to a decrease in *Q*.

The concepts of determinant, inverse and identity matrices apply equally well to 3 × 3 matrices. The identity matrix is easily dealt with. It can be shown that the 3 × 3 identity matrix is

$$\mathbf{I} = \begin{bmatrix} 1 & 0 & 0 \\ 0 & 1 & 0 \\ 0 & 0 & 1 \end{bmatrix}$$

You are invited to check that, for any 3 × 3 matrix **A**,

$$\mathbf{AI} = \mathbf{A} \quad \text{and} \quad \mathbf{IA} = \mathbf{A}$$

Before we can discuss the determinant and inverse of a 3 × 3 matrix we need to introduce an additional concept known as a *cofactor*. Corresponding to each element a_{ij} of a matrix **A**, there is a cofactor, A_{ij}. A 3 × 3 matrix has nine elements, so there are nine cofactors to be computed. The cofactor, A_{ij}, is defined to be the determinant of the 2 × 2 matrix obtained by deleting row *i* and column *j* of **A**, prefixed by a '+' or '−' sign according to the following pattern

$$\begin{bmatrix} + & - & + \\ - & + & - \\ + & - & + \end{bmatrix}$$

For example, suppose we wish to calculate A_{23}, which is the cofactor associated with a_{23} in the matrix

$$\mathbf{A} = \begin{bmatrix} a_{11} & a_{12} & a_{13} \\ a_{21} & a_{22} & a_{23} \\ a_{31} & a_{32} & a_{33} \end{bmatrix}$$

The element a_{23} lies in the second row and third column. Consequently, we delete the second row and third column to produce the 2 × 2 matrix

$$\begin{bmatrix} a_{11} & a_{12} & \not{a}_{13} \\ \not{a}_{21} & \not{a}_{22} & \not{a}_{23} \\ a_{31} & a_{32} & \not{a}_{33} \end{bmatrix}$$

The cofactor, A_{23}, is the determinant of this 2 × 2 matrix prefixed by a '−' sign because from the pattern

$$\begin{bmatrix} + & - & + \\ - & + & \boxed{-} \\ + & - & + \end{bmatrix}$$

we see that a_{23} is in a minus position. In other words,

$$\begin{aligned}\mathbf{A}_{23} &= -\begin{vmatrix} a_{11} & a_{12} \\ a_{31} & a_{32} \end{vmatrix} \\ &= -(a_{11}a_{32} - a_{12}a_{31}) \\ &= -a_{11}a_{32} + a_{12}a_{31}\end{aligned}$$

Example

Find all the cofactors of the matrix

$$\mathbf{A} = \begin{bmatrix} 2 & 4 & 1 \\ 4 & 3 & 7 \\ 2 & 1 & 3 \end{bmatrix}$$

Solution

Let us start in the top left-hand corner and work row by row. For cofactor A_{11}, the element $a_{11} = 2$ lies in the first row and first column, so we delete this row and column to produce the 2×2 matrix

$$\begin{bmatrix} \not{2} & \not{4} & \not{1} \\ \not{4} & 3 & 7 \\ \not{2} & 1 & 3 \end{bmatrix}$$

Cofactor A_{11} is the determinant of this 2×2 matrix, prefixed by a '+' sign because from the pattern

$$\begin{bmatrix} \boxed{+} & - & + \\ - & + & - \\ + & - & + \end{bmatrix}$$

we see that a_{11} is in a plus position. Hence

$$\begin{aligned}A_{11} &= +\begin{vmatrix} 3 & 7 \\ 1 & 3 \end{vmatrix} \\ &= +(3(3) - 7(1)) \\ &= 9 - 7 \\ &= 2\end{aligned}$$

For cofactor A_{12}, the element $a_{12} = 4$ lies in the first row and second column, so we delete this row and column to produce the 2×2 matrix

$$\begin{bmatrix} \not{2} & \not{4} & \not{1} \\ 4 & \not{3} & 7 \\ 2 & \not{1} & 3 \end{bmatrix}$$

Cofactor A_{12} is the determinant of this 2×2 matrix, prefixed by a '–' sign because from the pattern

$$\begin{bmatrix} + & \boxed{-} & + \\ - & + & - \\ + & - & + \end{bmatrix}$$

we see that a_{12} is in a minus position. Hence

$$A_{12} = +\begin{vmatrix} 4 & 7 \\ 2 & 3 \end{vmatrix}$$
$$= -(4(3) - 7(2))$$
$$= -(12 - 14)$$
$$= 2$$

We can continue in this way to find the remaining cofactors

$$A_{13} = +\begin{vmatrix} 4 & 3 \\ 2 & 1 \end{vmatrix} = -2$$

$$A_{21} = -\begin{vmatrix} 4 & 1 \\ 1 & 3 \end{vmatrix} = -11$$

$$A_{22} = +\begin{vmatrix} 2 & 1 \\ 2 & 3 \end{vmatrix} = 4$$

$$A_{23} = -\begin{vmatrix} 2 & 4 \\ 2 & 1 \end{vmatrix} = 6$$

$$A_{31} = +\begin{vmatrix} 4 & 1 \\ 3 & 7 \end{vmatrix} = 25$$

$$A_{32} = -\begin{vmatrix} 2 & 1 \\ 4 & 7 \end{vmatrix} = -10$$

$$A_{33} = +\begin{vmatrix} 2 & 4 \\ 4 & 3 \end{vmatrix} = -10$$

Practice Problem

4 Find all the cofactors of the matrix

$$\mathbf{A} = \begin{bmatrix} 1 & 3 & 3 \\ 1 & 4 & 3 \\ 1 & 3 & 4 \end{bmatrix}$$

We are now in a position to describe how to calculate the determinant and inverse of a 3×3 matrix. The determinant is found by multiplying the elements in any one row or column by their corresponding cofactors and adding together. It does not matter which row or column is chosen; exactly the same answer is obtained in each case. If we expand along the first row of the matrix

$$\mathbf{A} = \begin{bmatrix} a_{11} & a_{12} & a_{13} \\ a_{21} & a_{22} & a_{23} \\ a_{31} & a_{32} & a_{33} \end{bmatrix}$$

we get

$$\det(\mathbf{A}) = a_{11}A_{11} + a_{12}A_{12} + a_{13}A_{13}$$

Similarly, if we expand down the second column, we get

$$\det(\mathbf{A}) = a_{12}A_{12} + a_{22}A_{22} + a_{32}A_{32}$$

The fact that we get the same answer irrespective of the row and column that we use for expansion is an extremely useful property. It provides us with an obvious check on our calculations. Also, there are occasions when it is more convenient to expand along certain rows or columns than others.

Example

Find the determinants of the following matrices:

$$\mathbf{A} = \begin{bmatrix} 2 & 4 & 1 \\ 4 & 3 & 7 \\ 2 & 1 & 3 \end{bmatrix} \quad \text{and} \quad \mathbf{B} = \begin{bmatrix} 10 & 7 & 5 \\ 0 & 2 & 0 \\ 2 & 7 & 3 \end{bmatrix}$$

Solution

We have already calculated all nine cofactors of the matrix

$$\mathbf{A} = \begin{bmatrix} 2 & 4 & 1 \\ 4 & 3 & 7 \\ 2 & 1 & 3 \end{bmatrix}$$

in the previous example. It is immaterial which row or column we use. Let us choose the second row. The cofactors corresponding to the three elements 4, 3, 7 in the second row were found to be −11, 4, 6, respectively. Consequently, if we expand along this row, we get

$$\begin{vmatrix} 2 & 4 & 1 \\ 4 & 3 & 7 \\ 2 & 1 & 3 \end{vmatrix} = 4(-11) + 3(4) + 7(6) = 10$$

As a check, let us also expand down the third column. The elements in this column are 1, 7, 3 with cofactors −2, 6, −10, respectively. Hence, if we multiply each element by its cofactor and add, we get

$$1(-2) + 7(6) + 3(-10) = 10$$

which is the same as before. If you are interested, you might like to confirm for yourself that the value of 10 is also obtained when expanding along rows 1 and 3, and down columns 1 and 2.

The matrix

$$\mathbf{B} = \begin{bmatrix} 10 & 7 & 5 \\ 0 & 2 & 0 \\ 2 & 7 & 3 \end{bmatrix}$$

is entirely new to us, so we have no prior knowledge about its cofactors. In general, we need to evaluate all three cofactors in any one row or column to find the determinant of a 3 × 3 matrix. In this case, however, we can be much lazier. Observe that all but one of the elements in the second row are zero, so when we expand along this row we get

$$\begin{aligned} \det(\mathbf{B}) &= b_{21}B_{21} + b_{22}B_{22} + b_{23}B_{23} \\ &= 0B_{21} + 2B_{22} + 0B_{23} \\ &= 2B_{22} \end{aligned}$$

Hence B_{22} is the only cofactor that we need to find. This corresponds to the element in the second row and second column, so we delete this row and column to produce the 2 × 2 matrix

$$\begin{bmatrix} 10 & 7 & 5 \\ 0 & 2 & 0 \\ 2 & 7 & 3 \end{bmatrix}$$

The element b_{22} is in a plus position, so

$$B_{22} = +\begin{vmatrix} 10 & 5 \\ 2 & 3 \end{vmatrix} = 20$$

Hence,

$$\det(\mathbf{B}) = 2B_{22} = 2 \times 20 = 40$$

Practice Problem

5 Find the determinants of

$$\mathbf{A} = \begin{bmatrix} 1 & 3 & 3 \\ 1 & 4 & 3 \\ 1 & 3 & 4 \end{bmatrix} \quad \text{and} \quad \mathbf{B} = \begin{bmatrix} 270 & -372 & 0 \\ 552 & 201 & 0 \\ 999 & 413 & 0 \end{bmatrix}$$

[Hint: you might find your answer to Practice Problem 4 useful when calculating the determinant of **A**.]

The inverse of the 3 × 3 matrix

$$\mathbf{A} = \begin{bmatrix} a_{11} & a_{12} & a_{13} \\ a_{21} & a_{22} & a_{23} \\ a_{31} & a_{32} & a_{33} \end{bmatrix}$$

is given by

$$\mathbf{A}^{-1} = \frac{1}{|\mathbf{A}|}\begin{bmatrix} A_{11} & A_{21} & A_{31} \\ A_{12} & A_{22} & A_{32} \\ A_{13} & A_{23} & A_{33} \end{bmatrix}$$

Once the cofactors of **A** have been found, it is easy to construct $\mathbf{A}^{-1}$. We first stack the cofactors in their natural positions

$$\begin{bmatrix} A_{11} & A_{12} & A_{13} \\ A_{21} & A_{22} & A_{23} \\ A_{31} & A_{32} & A_{33} \end{bmatrix}$$

called the adjugate matrix

Secondly, we take the transpose to get

$$\begin{bmatrix} A_{11} & A_{21} & A_{31} \\ A_{12} & A_{22} & A_{32} \\ A_{13} & A_{23} & A_{33} \end{bmatrix}$$

called the adjoint matrix

Finally, we multiply by the scalar

$$\frac{1}{|\mathbf{A}|}$$

to get

$$\mathbf{A}^{-1} = \frac{1}{|\mathbf{A}|}\begin{bmatrix} A_{11} & A_{21} & A_{31} \\ A_{12} & A_{22} & A_{32} \\ A_{13} & A_{23} & A_{33} \end{bmatrix}$$

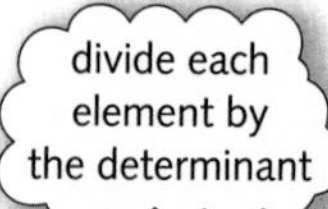

The last step is impossible if

$$|\mathbf{A}| = 0$$

because we cannot divide by zero. Under these circumstances the inverse does not exist and the matrix is singular.

Advice

It is a good idea to check that no mistakes have been made by verifying that

$$\mathbf{A}^{-1}\mathbf{A} = \mathbf{I} \quad \text{and} \quad \mathbf{A}\mathbf{A}^{-1} = \mathbf{I}$$

Example

Find the inverse of

$$\mathbf{A} = \begin{bmatrix} 2 & 4 & 1 \\ 4 & 3 & 7 \\ 2 & 1 & 3 \end{bmatrix}$$

Solution

The cofactors of this particular matrix have already been calculated as

$$\begin{array}{lll} A_{11} = 2, & A_{12} = 2, & A_{13} = -2 \\ A_{21} = -11, & A_{22} = 4, & A_{23} = 6 \\ A_{31} = 25, & A_{32} = -10, & A_{33} = -10 \end{array}$$

Stacking these numbers in their natural positions gives the adjugate matrix

$$\begin{bmatrix} 2 & 2 & -2 \\ -11 & 4 & 6 \\ 25 & -10 & -10 \end{bmatrix}$$

The adjoint matrix is found by transposing this to get

$$\begin{bmatrix} 2 & -11 & 25 \\ 2 & 4 & -10 \\ -2 & 6 & -10 \end{bmatrix}$$

In the previous example the determinant was found to be 10, so

$$\mathbf{A}^{-1} = \frac{1}{10}\begin{bmatrix} 2 & -11 & 25 \\ 2 & 4 & -10 \\ -2 & 6 & -10 \end{bmatrix} = \begin{bmatrix} 1/5 & -11/10 & 5/2 \\ 1/5 & 2/5 & -1 \\ -1/5 & 3/5 & -1 \end{bmatrix}$$

As a check

$$\mathbf{A}^{-1}\mathbf{A} = \begin{bmatrix} 1/5 & -11/10 & 5/2 \\ 1/5 & 2/5 & -1 \\ -1/5 & 3/5 & -1 \end{bmatrix}\begin{bmatrix} 2 & 4 & 1 \\ 4 & 3 & 7 \\ 2 & 1 & 3 \end{bmatrix} = \begin{bmatrix} 1 & 0 & 0 \\ 0 & 1 & 0 \\ 0 & 0 & 1 \end{bmatrix} = \mathbf{I} \quad ✓$$

$$\mathbf{A}\mathbf{A}^{-1} = \begin{bmatrix} 2 & 4 & 1 \\ 4 & 3 & 7 \\ 2 & 1 & 3 \end{bmatrix}\begin{bmatrix} 1/5 & -11/10 & 5/2 \\ 1/5 & 2/5 & -1 \\ -1/5 & 3/5 & -1 \end{bmatrix} = \begin{bmatrix} 1 & 0 & 0 \\ 0 & 1 & 0 \\ 0 & 0 & 1 \end{bmatrix} = \mathbf{I} \quad ✓$$

Practice Problem

6 Find (where possible) the inverses of

$$\mathbf{A} = \begin{bmatrix} 1 & 3 & 3 \\ 1 & 4 & 3 \\ 1 & 3 & 4 \end{bmatrix} \quad \text{and} \quad \mathbf{B} = \begin{bmatrix} 270 & -372 & 0 \\ 552 & 201 & 0 \\ 999 & 413 & 0 \end{bmatrix}$$

[Hint: you might find your answers to Practice Problems 4 and 5 useful.]

Inverses of 3×3 matrices can be used to solve systems of three linear equations in three unknowns. The general system

$$a_{11}x + a_{12}y + a_{13}z = b_1$$

$$a_{21}x + a_{22}y + a_{23}z = b_2$$

$$a_{31}x + a_{32}y + a_{33}z = b_3$$

can be written as

$$\mathbf{Ax} = \mathbf{b}$$

where

$$\mathbf{A} = \begin{bmatrix} a_{11} & a_{12} & a_{13} \\ a_{21} & a_{22} & a_{23} \\ a_{31} & a_{32} & a_{33} \end{bmatrix} \quad \mathbf{x} = \begin{bmatrix} x \\ y \\ z \end{bmatrix} \quad \mathbf{b} = \begin{bmatrix} b_1 \\ b_2 \\ b_3 \end{bmatrix}$$

The vector of unknowns, **x**, can be found by inverting the coefficient matrix, **A**, and multiplying by the right-hand-side vector, **b**, to get

$$\mathbf{x} = \mathbf{A}^{-1}\mathbf{b}$$

Example

Determine the equilibrium prices of three interdependent commodities that satisfy

$$2P_1 + 4P_2 + P_3 = 77$$

$$4P_1 + 3P_2 + 7P_3 = 114$$

$$2P_1 + P_2 + 3P_3 = 48$$

Solution

In matrix notation this system of equations can be written as

$$\mathbf{Ax} = \mathbf{b}$$

where

$$\mathbf{A} = \begin{bmatrix} 2 & 4 & 1 \\ 4 & 3 & 7 \\ 2 & 1 & 3 \end{bmatrix} \quad \mathbf{x} = \begin{bmatrix} P_1 \\ P_2 \\ P_3 \end{bmatrix} \quad \mathbf{b} = \begin{bmatrix} 77 \\ 114 \\ 48 \end{bmatrix}$$

The inverse of the coefficient matrix has already been found in the previous example and is

$$\mathbf{A}^{-1} = \begin{bmatrix} 1/5 & -11/10 & 5/2 \\ 1/5 & 2/5 & -1 \\ -1/5 & 3/5 & -1 \end{bmatrix}$$

so

$$\begin{bmatrix} P_1 \\ P_2 \\ P_3 \end{bmatrix} = \begin{bmatrix} 1/5 & -11/10 & 5/2 \\ 1/5 & 2/5 & -1 \\ -1/5 & 3/5 & -1 \end{bmatrix} \begin{bmatrix} 77 \\ 114 \\ 48 \end{bmatrix} = \begin{bmatrix} 10 \\ 13 \\ 5 \end{bmatrix}$$

The equilibrium prices are therefore given by

$$P_1 = 10, \quad P_2 = 13, \quad P_3 = 5$$

Practice Problem

7 Determine the equilibrium prices of three interdependent commodities that satisfy

$$P_1 + 3P_2 + 3P_3 = 32$$

$$P_1 + 4P_2 + 3P_3 = 37$$

$$P_1 + 3P_2 + 4P_3 = 35$$

[Hint: you might find your answer to Practice Problem 6 useful.]

Throughout this section, we have concentrated on 2×2 and 3×3 matrices. The method described can be extended to larger matrices of order $n \times n$. However, the cofactor approach is very inefficient. The amount of working rises dramatically as n increases, making this method inappropriate for large matrices. The preferred method for solving simultaneous equations is based on the elimination idea that we described in Section 1.2. This is easily programmed, and a computer can solve large systems of equations in a matter of seconds. We shall not pursue this approach any further, since the practical implementation of this method requires an understanding of the build-up of rounding errors on a computer. (The interested reader is referred to any textbook on numerical analysis.) Instead, we describe how to use Maple to get results, and we illustrate this with a simple 4×4 example.

Example MAPLE

Solve the equations

$$
\begin{aligned}
2x_1 - x_2 + 4x_3 + 5x_4 &= -17 \\
x_1 + 6x_2 + 2x_3 - 4x_4 &= 24 \\
5x_1 + 2x_2 - x_3 - 7x_4 &= 40 \\
3x_1 + 2x_2 + 2x_3 + 6x_4 &= -16
\end{aligned}
$$

Solution

Before you can use Maple to solve problems involving matrices, it is necessary first to load the specialist linear algebra routines. You do this by typing:

```
with(linalg):
```

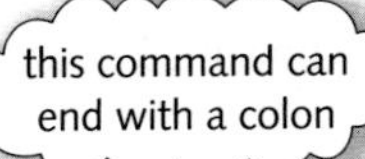

In matrix notation, the system can be written as

$$\mathbf{Ax} = \mathbf{b}$$

where

$$
\mathbf{A} = \begin{bmatrix} 2 & -1 & 4 & 5 \\ 1 & 6 & 2 & -4 \\ 5 & 2 & -1 & -7 \\ 3 & 2 & 2 & 6 \end{bmatrix} \quad
\mathbf{x} = \begin{bmatrix} x_1 \\ x_2 \\ x_3 \\ x_4 \end{bmatrix} \quad
\mathbf{b} = \begin{bmatrix} -17 \\ 24 \\ 40 \\ -16 \end{bmatrix}
$$

To input the matrix **A**, you type

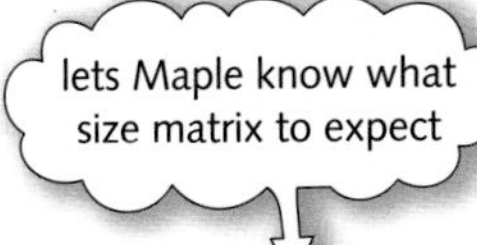

the elements are entered row by row using square brackets

```
>A:=array(1..4,1..4,[[2,-1,4,5],[1,6,2,-4],[5,2,-1,-7],[3,2,2,6]]);
```

To solve the linear equations all you have to do is to type

the right-hand side is written horizontally in square brackets

```
>linsolve(A,[-17,24,40,-16]);
```

→

Maple responds with the solution

```
[2,1,0,-4]
```

so $x_1 = 2$, $x_2 = 1$, $x_3 = 0$, $x_4 = -4$.

Notice that Maple has solved the equations without explicitly showing the inverse on the screen. If you need to see this, then just type

```
>C:=inverse(A);
```

which gives

$$\begin{bmatrix} \frac{14}{533} & \frac{-1}{13} & \frac{6}{41} & \frac{4}{41} \\ \frac{-73}{533} & \frac{3}{26} & \frac{-2}{41} & \frac{11}{82} \\ \frac{148}{533} & \frac{3}{26} & \frac{-1}{41} & \frac{-15}{82} \\ \frac{-32}{533} & \frac{-1}{26} & \frac{-2}{41} & \frac{11}{82} \end{bmatrix}$$

The solution can also be found from this by working out $\mathbf{A}^{-1}\mathbf{b}$. This is achieved by first creating the 4×1 right-hand side column matrix **b** by typing:

```
>b:=array(1..4,1..1,[[-17],[24],[40],[-16]]);
```

In Maple, the symbol for matrix multiplication is &* so we can find **x** by typing

```
>x:=C&*b;
```

Unfortunately, Maple rather unimaginatively responds with

```
x:=C&*b
```

To actually perform the calculation you need to evaluate this matrix product by typing

```
>evalm(x);
```

which produces the desired effect and yields the solution

$$\begin{bmatrix} 2 \\ 1 \\ 0 \\ -4 \end{bmatrix}$$

Key Terms

Cofactor (of an element) The cofactor of the element, a_{ij}, is the determinant of the matrix left when row i and column j are deleted, multiplied by +1 or −1, depending on whether $i + j$ is even or odd, respectively.

Determinant A determinant can be expanded as the sum of the products of the elements in any one row or column and their respective cofactors.

Identity matrix An $n \times n$ matrix, $\mathbf{I}$, in which every element on the main diagonal is 1 and the other elements are all 0. If $\mathbf{A}$ is any $n \times n$ matrix then $\mathbf{AI} = \mathbf{I} = \mathbf{IA}$.

Inverse matrix A matrix, $\mathbf{A}^{-1}$ with the property that $\mathbf{A}^{-1}\mathbf{A} = \mathbf{I} = \mathbf{AA}^{-1}$.

Non-singular matrix A square matrix with a non-zero determinant.

Singular matrix A square matrix with a zero determinant. A singular matrix fails to possess an inverse.

Square matrix A matrix with the same number of rows as columns.

Practice Problems

8 Let

$$\mathbf{A} = \begin{bmatrix} 2 & 1 \\ 5 & 1 \end{bmatrix} \quad \text{and} \quad \mathbf{B} = \begin{bmatrix} 1 & 0 \\ 2 & 4 \end{bmatrix}$$

(1) Find

(a) $|\mathbf{A}|$ **(b)** $|\mathbf{B}|$ **(c)** $|\mathbf{AB}|$

Do you notice any connection between $|\mathbf{A}|$, $|\mathbf{B}|$ and $|\mathbf{AB}|$?

(2) Find

(a) $\mathbf{A}^{-1}$ **(b)** $\mathbf{B}^{-1}$ **(c)** $(\mathbf{AB})^{-1}$

Do you notice any connection between $\mathbf{A}^{-1}$, $\mathbf{B}^{-1}$ and $(\mathbf{AB})^{-1}$?

9 If the matrices

$$\begin{bmatrix} 2 & -1 \\ 3 & a \end{bmatrix} \quad \text{and} \quad \begin{bmatrix} 2 & b \\ 3 & -4 \end{bmatrix}$$

are singular, find the values of a and b.

10 Use matrices to solve the following pairs of simultaneous equations:

(a) $3x + 4y = -1$
$5x - y = 6$

(b) $x + 3y = 8$
$4x - y = 6$

11 Calculate the inverse of

$$\begin{bmatrix} -3 & 1 \\ 2 & -9 \end{bmatrix}$$

Hence find the equilibrium prices in the two-commodity market model given in Practice Problem 12 of Section 1.3.

12 For the commodity market

$$C = aY + b \quad (0 < a < 1, b > 0)$$

$$I = cr + d \quad (c < 0, d > 0)$$

➔

For the money market

$$M_S = M_S^*$$

$$M_D = k_1 Y + k_2 r + k_3 \quad (k_1, k_3 > 0, k_2 < 0)$$

Show that when the commodity and money markets are both in equilibrium, the income, Y, and interest rate, r, satisfy the matrix equation

$$\begin{bmatrix} 1-a & -c \\ k_1 & k_2 \end{bmatrix}\begin{bmatrix} Y \\ r \end{bmatrix} = \begin{bmatrix} b+d \\ M_S^* - k_3 \end{bmatrix}$$

and solve this system for Y and r. Write down the multiplier for r due to changes in M_S^* and deduce that interest rates fall as the money supply grows.

13 Find (where possible) the inverse of the matrices

$$\mathbf{A} = \begin{bmatrix} 2 & 1 & -1 \\ 1 & 3 & 2 \\ -1 & 2 & 1 \end{bmatrix} \quad \mathbf{B} = \begin{bmatrix} 1 & 4 & 5 \\ 2 & 1 & 3 \\ -1 & 3 & 2 \end{bmatrix}$$

Are these matrices singular or non-singular?

14 Find the determinant of the matrix

$$\mathbf{A} = \begin{bmatrix} 2 & 1 & 3 \\ 1 & 0 & a \\ 3 & 1 & 4 \end{bmatrix}$$

in terms of a. Deduce that this matrix is non-singular provided $a \neq 1$ and find $\mathbf{A}^{-1}$ in this case.

15 Find the inverse of

$$\begin{bmatrix} -2 & 2 & 1 \\ 1 & -5 & -1 \\ 2 & -1 & -6 \end{bmatrix}$$

Hence find the equilibrium prices of the three-commodity market model given in Practice Problem 13 of Section 1.3.

16 **(Maple)** Use the Maple instruction `det(A)` to find the determinant of the 4×4 matrix

$$\begin{bmatrix} a & 2 & 1 & -1 \\ 2 & 3 & 4 & 1 \\ 0 & -1 & 5 & 6 \\ 1 & 2 & 4 & -3 \end{bmatrix}$$

For what value of a does this matrix fail to possess an inverse? [Do not forget to load the linear algebra package before you begin. You do this by typing `with(linalg):`]

17 **(Maple)**

(a) Find the determinant of the general 3×3 matrix, **A**

$$\begin{bmatrix} a & b & c \\ d & e & f \\ g & h & i \end{bmatrix}$$

Deduce that this matrix is singular whenever the second row is a multiple of the first row.

(b) Find the inverse, $\mathbf{A}^{-1}$, of the general 3×3 matrix given in part (a) and verify that $\mathbf{AA}^{-1} = \mathbf{I}$.

18 **(Maple)** The equilibrium levels of income, Y, consumption, C, and taxation, T, satisfy the equations

$$Y = C + I^* + G^*$$

$$C = a(Y - T) + b$$

$$T = tY$$

Show that this system can be written in the form **Ax** = **b** where

$$\mathbf{A} = \begin{bmatrix} 1 & -1 & 0 \\ -a & 1 & a \\ t & 0 & -1 \end{bmatrix} \quad \mathbf{x} = \begin{bmatrix} Y \\ C \\ T \end{bmatrix} \quad \mathbf{b} = \begin{bmatrix} I^* + G^* \\ b \\ 0 \end{bmatrix}$$

(a) Find $\mathbf{A}^{-1}$ and give an interpretation of the element in the first row and second column of $\mathbf{A}^{-1}$.

(b) By using $\mathbf{A}^{-1}$, or otherwise, solve this system of equations.

section 7.3

Cramer's rule

Objectives

At the end of this section you should be able to:

- Appreciate the limitations of using inverses to solve systems of linear equations.
- Use Cramer's rule to solve systems of linear equations.
- Apply Cramer's rule to analyse static macroeconomic models.
- Apply Cramer's rule to solve two-country trading models.

In Section 7.2 we described the mechanics of calculating the determinant and inverse of 2×2 and 3×3 matrices. These concepts can be extended to larger systems in an obvious way, although the amount of effort needed rises dramatically as the size of the matrix increases. For example, consider the work involved in solving the system

$$\begin{bmatrix} 1 & 0 & 2 & 3 \\ -1 & 5 & 4 & 1 \\ 0 & 7 & -3 & 6 \\ 2 & 4 & 5 & 1 \end{bmatrix} \begin{bmatrix} x_1 \\ x_2 \\ x_3 \\ x_4 \end{bmatrix} = \begin{bmatrix} -1 \\ 1 \\ -24 \\ 15 \end{bmatrix}$$

using the method of matrix inversion. In this case the coefficient matrix has order 4×4 and so has 16 elements. Corresponding to each of these elements there is a cofactor. This is defined to be the 3×3 determinant obtained by deleting the row and column containing the element, prefixed by a '+' or '–' according to the following pattern:

$$\begin{bmatrix} + & - & + & - \\ - & + & - & + \\ + & - & + & - \\ - & + & - & + \end{bmatrix}$$

Determinants are found by expanding along any one row or column and inverses are found by stacking cofactors as before. However, given that there are 16 cofactors to be calculated, even the most enthusiastic student is likely to view the prospect with some trepidation. To make matters worse, it frequently happens in economics that only a few of the variables x_i are actually needed. For instance, it could be that the variable x_3 is the only one of interest. Under these circumstances it is clearly wasteful expending a large amount of effort calculating the inverse matrix, particularly since the values of the remaining variables, x_1, x_2 and x_4, are not required.

In this section we describe an alternative method that finds the value of one variable at a time. This new method requires less effort if only a selection of the variables is required. It is known as Cramer's rule and makes use of matrix determinants. *Cramer's rule* for solving any $n \times n$ system, $\mathbf{Ax} = \mathbf{b}$, states that the ith variable, x_i, can be found from

$$x_i = \frac{\det(\mathbf{A}_i)}{\det(\mathbf{A})}$$

where $\mathbf{A}_i$ is the $n \times n$ matrix found by replacing the ith column of $\mathbf{A}$ by the right-hand-side vector $\mathbf{b}$. To understand this, consider the simple 2×2 system

$$\begin{bmatrix} 7 & 2 \\ 4 & 5 \end{bmatrix} \begin{bmatrix} x_1 \\ x_2 \end{bmatrix} = \begin{bmatrix} -6 \\ 12 \end{bmatrix}$$

and suppose that we need to find the value of the second variable, x_2, say. According to Cramer's rule, this is given by

$$x_2 = \frac{\det(\mathbf{A}_2)}{\det(\mathbf{A})}$$

where

$$\mathbf{A} = \begin{bmatrix} 7 & 2 \\ 4 & 5 \end{bmatrix} \quad \text{and} \quad \mathbf{A}_2 = \begin{bmatrix} 7 & -6 \\ 4 & 12 \end{bmatrix}$$

Notice that x_2 is given by the quotient of two determinants. The one on the bottom is that of the original coefficient matrix $\mathbf{A}$. The one on the top is that of the matrix found from $\mathbf{A}$ by replacing the second column (since we are trying to find the second variable) by the right-hand-side vector

$$\begin{bmatrix} -6 \\ 12 \end{bmatrix}$$

In this case the determinants are easily worked out to get

$$\det(\mathbf{A}_2) = \begin{vmatrix} 7 & -6 \\ 4 & 12 \end{vmatrix} = 7(12) - (-6)(4) = 108$$

$$\det(\mathbf{A}) = \begin{vmatrix} 7 & 2 \\ 4 & 5 \end{vmatrix} = 7(5) - 2(4) = 27$$

Hence

$$x_2 = \frac{108}{27} = 4$$

Example

Solve the system of equations

$$\begin{bmatrix} 1 & 2 & 3 \\ -4 & 1 & 6 \\ 2 & 7 & 5 \end{bmatrix} \begin{bmatrix} x_1 \\ x_2 \\ x_3 \end{bmatrix} = \begin{bmatrix} 9 \\ -9 \\ 13 \end{bmatrix}$$

using Cramer's rule to find x_1.

→

Solution

Cramer's rule gives

$$x_1 = \frac{\det(\mathbf{A}_1)}{\det(\mathbf{A})}$$

where **A** is the coefficient matrix

$$\begin{bmatrix} 1 & 2 & 3 \\ -4 & 1 & 6 \\ 2 & 7 & 5 \end{bmatrix}$$

and $\mathbf{A}_1$ is constructed by replacing the first column of **A** by the right-hand-side vector

$$\begin{bmatrix} 9 \\ -9 \\ 13 \end{bmatrix}$$

which gives

$$\mathbf{A}_1 = \begin{bmatrix} 9 & 2 & 3 \\ -9 & 1 & 6 \\ 13 & 7 & 5 \end{bmatrix}$$

If we expand each of these determinants along the top row, we get

$$\begin{aligned} \det(\mathbf{A}_1) &= \begin{vmatrix} 9 & 2 & 3 \\ -9 & 1 & 6 \\ 13 & 7 & 5 \end{vmatrix} \\ &= 9\begin{vmatrix} 1 & 6 \\ 7 & 5 \end{vmatrix} - 2\begin{vmatrix} -9 & 6 \\ 13 & 5 \end{vmatrix} + 3\begin{vmatrix} -9 & 1 \\ 13 & 7 \end{vmatrix} \\ &= 9(-37) - 2(-123) + 3(-76) \\ &= -315 \end{aligned}$$

and

$$\begin{aligned} \det(\mathbf{A}) &= \begin{vmatrix} 1 & 2 & 3 \\ -4 & 1 & 6 \\ 2 & 7 & 5 \end{vmatrix} \\ &= 1\begin{vmatrix} 1 & 6 \\ 7 & 5 \end{vmatrix} - 2\begin{vmatrix} -4 & 6 \\ 2 & 5 \end{vmatrix} + 3\begin{vmatrix} -4 & 1 \\ 2 & 7 \end{vmatrix} \\ &= 1(-37) - 2(-32) + 3(-30) \\ &= -63 \end{aligned}$$

Hence

$$x_1 = \frac{\det(\mathbf{A}_1)}{\det(\mathbf{A})} = \frac{-315}{-63} = 5$$

Practice Problem

1 **(a)** Solve the system of equations

$$2x_1 + 4x_2 = 16$$
$$3x_1 - 5x_2 = -9$$

using Cramer's rule to find x_2.

(b) Solve the system of equations

$$4x_1 + x_2 + 3x_3 = 8$$
$$-2x_1 + 5x_2 + x_3 = 4$$
$$3x_1 + 2x_2 + 4x_3 = 9$$

using Cramer's rule to find x_3.

We now illustrate the use of Cramer's rule to analyse economic models. We begin by considering the three-sector macroeconomic model involving government activity.

Advice

The incorporation of government expenditure and taxation into the model has already been considered in Section 5.3, and you might like to compare the working involved in the two approaches.

Example

The equilibrium levels of income, Y, disposable income, Y_d, and taxation, T, for a three-sector macroeconomic model satisfy the structural equations

$$Y = C + I^* + G^*$$
$$C = aY_d + b \qquad (0 < a < 1,\ b > 0)$$
$$Y_d = Y - T$$
$$T = tY + T^* \qquad (0 < t < 1,\ T^* > 0)$$

Show that this system can be written as $\mathbf{Ax} = \mathbf{b}$, where

$$\mathbf{A} = \begin{bmatrix} 1 & -1 & 0 & 0 \\ 0 & 1 & -a & 0 \\ -1 & 0 & 1 & 1 \\ -t & 0 & 0 & 1 \end{bmatrix} \quad \mathbf{x} = \begin{bmatrix} Y \\ C \\ Y_d \\ T \end{bmatrix} \quad \mathbf{b} = \begin{bmatrix} I^* + G^* \\ b \\ 0 \\ T^* \end{bmatrix}$$

Use Cramer's rule to solve this system for Y.

Solution

In this model the endogenous variables are Y, C, Y_d and T, so we begin by manipulating the equations so that these variables appear on the left-hand sides. Moreover, since the vector of 'unknowns', $\mathbf{x}$, is given to be

$$\begin{bmatrix} Y \\ C \\ Y_d \\ T \end{bmatrix}$$

we need to arrange the equations so that the variables appear in the order Y, C, Y_d, T. For example, in the case of the third equation

$$Y_d = Y - T$$

we first subtract Y and add T to both sides to get

$$Y_d - Y + T = 0$$

but then reorder the terms to obtain

$$-Y + Y_d + T = 0$$

Performing a similar process with the remaining equations gives

$$\begin{aligned} Y - C &= I^* + G^* \\ C - aY_d &= b \\ -Y + Y_d + T &= 0 \\ -tY + T &= T^* \end{aligned}$$

so that in matrix form they become

$$\begin{bmatrix} 1 & -1 & 0 & 0 \\ 0 & 1 & -a & 0 \\ -1 & 0 & 1 & 1 \\ -t & 0 & 0 & 1 \end{bmatrix} \begin{bmatrix} Y \\ C \\ Y_d \\ T \end{bmatrix} = \begin{bmatrix} I^* + G^* \\ b \\ 0 \\ T^* \end{bmatrix}$$

The variable Y is the first, so Cramer's rule gives

$$Y = \frac{\det(\mathbf{A}_1)}{\det(\mathbf{A})}$$

where

$$\mathbf{A}_1 = \begin{bmatrix} I^* + G^* & -1 & 0 & 0 \\ b & 1 & -a & 0 \\ 0 & 0 & 1 & 1 \\ T^* & 0 & 0 & 1 \end{bmatrix}$$

and

$$\mathbf{A} = \begin{bmatrix} 1 & -1 & 0 & 0 \\ 0 & 1 & -a & 0 \\ -1 & 0 & 1 & 1 \\ -t & 0 & 0 & 1 \end{bmatrix}$$

The calculations are fairly easy to perform, in spite of the fact that both matrices are 4×4, because they contain a high proportion of zeros. Expanding $\mathbf{A}_1$ along the first row gives

$$\det(\mathbf{A}_1) = (I^* + G^*)\begin{vmatrix} 1 & -a & 0 \\ 0 & 1 & 1 \\ 0 & 0 & 1 \end{vmatrix} - (-1)\begin{vmatrix} b & -a & 0 \\ 0 & 1 & 1 \\ T^* & 0 & 1 \end{vmatrix}$$

along the first row the pattern is + − + −

Notice that there is no point in evaluating the last two cofactors in the first row, since the corresponding elements are both zero.

For the first of these 3 × 3 determinants we choose to expand down the first column, since this column has only one non-zero element. This gives

$$\begin{vmatrix} 1 & -a & 0 \\ 0 & 1 & 1 \\ 0 & 0 & 1 \end{vmatrix} = (1)\begin{vmatrix} 1 & 1 \\ 0 & 1 \end{vmatrix} = 1$$

It is immaterial which row or column we choose for the second 3 × 3 determinant, since they all contain two non-zero elements. Working along the first row gives

$$\begin{vmatrix} b & -a & 0 \\ 0 & 1 & 1 \\ T^* & 0 & 1 \end{vmatrix} = b\begin{vmatrix} 1 & 1 \\ 0 & 1 \end{vmatrix} - (-a)\begin{vmatrix} 0 & 1 \\ T^* & 1 \end{vmatrix} = b - aT^*$$

Hence

$$\det(\mathbf{A}_1) = (I^* + G^*)(1) - (-1)(b - aT^*) = I^* + G^* + b - aT^*$$

A similar process can be applied to matrix **A**. Expanding along the top row gives

$$\det(\mathbf{A}) = (1)\begin{vmatrix} 1 & -a & 0 \\ 0 & 1 & 1 \\ 0 & 0 & 1 \end{vmatrix} - (-1)\begin{vmatrix} 0 & -a & 0 \\ -1 & 1 & 1 \\ -t & 0 & 1 \end{vmatrix}$$

The first of these 3 × 3 determinants has already been found to be 1 in our previous calculations. The second 3 × 3 determinant is new and if we expand this along the first row we get

$$\begin{vmatrix} 0 & -a & 0 \\ -1 & 1 & 1 \\ -t & 0 & 1 \end{vmatrix} = -(-a)\begin{vmatrix} -1 & 1 \\ -t & 1 \end{vmatrix} = a(-1 + t)$$

Hence

$$\det(\mathbf{A}) = (1)(1) - (-1)a(-1 + t) = 1 - a + at$$

Finally we use Cramer's rule to deduce that

$$Y = \frac{I^* + G^* + b - aT^*}{1 - a + at}$$

Practice Problem

2 Use Cramer's rule to solve the following system of equations for Y_d.

$$\begin{bmatrix} 1 & -1 & 0 & 0 \\ 0 & 1 & -a & 0 \\ -1 & 0 & 1 & 1 \\ -t & 0 & 0 & 1 \end{bmatrix}\begin{bmatrix} Y \\ C \\ Y_d \\ T \end{bmatrix} = \begin{bmatrix} I^* + G^* \\ b \\ 0 \\ T^* \end{bmatrix}$$

[Hint: the determinant of the coefficient matrix has already been evaluated in the previous worked example.]

We conclude this section by introducing foreign trade into our model. In all of our previous macroeconomic models we have implicitly assumed that the behaviour of different countries has no effect on the national income of the other countries. In reality this is clearly not the case and we now investigate how the economies of trading nations interact. To simplify the situation we shall ignore all government activity and suppose that there are just two countries, labelled 1 and 2, trading with each other but not with any other country. We shall use an obvious subscript notation so that Y_1 denotes the national income of country 1, C_2 denotes the consumption of country 2 and so on. In the absence of government activity the equation defining equilibrium in country i is

$$Y_i = C_i + I_i + X_i - M_i$$

where I_i is the investment of country i, X_i is the exports of country i and M_i is the imports of country i. As usual, we shall assume that I_i is determined exogenously and takes a known value I_i^*.

Given that there are only two countries, which trade between themselves, the exports of one country must be the same as the imports of the other. In symbols we write

$$X_1 = M_2 \quad \text{and} \quad X_2 = M_1$$

We shall assume that imports are a fraction of national income, so that

$$M_i = m_i Y_i$$

where the marginal propensity to import, m_i, satisfies $0 < m_i < 1$.

Once expressions for C_i and M_i are given, we can derive a system of two simultaneous equations for the two unknowns, Y_1 and Y_2, which can be solved either by using Cramer's rule or by using matrix inverses.

Example

The equations defining a model of two trading nations are given by

$$Y_1 = C_1 + I_1^* + X_1 - M_1 \qquad Y_2 = C_2 + I_2^* + X_2 - M_2$$

$$C_1 = 0.8Y_1 + 200 \qquad C_2 = 0.9Y_2 + 100$$

$$M_1 = 0.2Y_1 \qquad M_2 = 0.1Y_2$$

Express this system in matrix form and hence write Y_1 in terms of I_1^* and I_2^*.

Write down the multiplier for Y_1 due to changes in I_2^* and hence describe the effect on the national income of country 1 due to changes in the investment in country 2.

Solution

In this problem there are six equations for six endogenous variables, Y_1, C_1, M_1 and Y_2, C_2, M_2. However, rather than working with a 6×6 matrix, we perform some preliminary algebra to reduce it to only two equations in two unknowns. To do this we substitute the expressions for C_1 and M_1 into the first equation to get

$$Y_1 = 0.8Y_1 + 200 + I_1^* + X_1 - 0.2Y_1$$

Also, since $X_1 = M_2 = 0.1Y_2$, this becomes

$$Y_1 = 0.8Y_1 + 200 + I_1^* + 0.1Y_2 - 0.2Y_1$$

which rearranges as

$$0.4Y_1 - 0.1Y_2 = 200 + I_1^*$$

A similar procedure applied to the second set of equations for country 2 gives

$$-0.2Y_1 + 0.2Y_2 = 100 + I_2^*$$

In matrix form this pair of equations can be written as

$$\begin{bmatrix} 0.4 & -0.1 \\ -0.2 & 0.2 \end{bmatrix} \begin{bmatrix} Y_1 \\ Y_2 \end{bmatrix} = \begin{bmatrix} 200 + I_1^* \\ 100 + I_2^* \end{bmatrix}$$

Cramer's rule gives

$$Y_1 = \frac{\begin{vmatrix} 200 + I_1^* & -0.1 \\ 200 + I_2^* & 0.2 \end{vmatrix}}{\begin{vmatrix} 0.4 & -0.1 \\ -0.2 & 0.2 \end{vmatrix}} = \frac{50 + 0.2I_1^* + 0.1I_2^*}{0.06}$$

To find the multiplier for Y_1 due to changes in I_2^* we consider what happens to Y_1 when I_2^* changes by an amount ΔI_2^*. The new value of Y_1 is obtained by replacing I_2^* by $I_2^* + \Delta I_2^*$ to get

$$\frac{50 + 0.2I_1^* + 0.1(I_2^* + \Delta I_2^*)}{0.06}$$

so the corresponding change in Y_1 is

$$\Delta Y_1 = \frac{50 + 0.2I_1^* + 0.1(I_2^* + \Delta I_2^*)}{0.06} - \frac{50 + 0.2I_1^* + 0.1I_2^*}{0.06}$$

$$= \frac{0.1}{0.06}\Delta I_2^* = \frac{5}{3}\Delta I_2^*$$

We deduce that any increase in investment in country 2 leads to an increase in the national income in country 1. Moreover, because $5/3 > 1$, the increase in national income is greater than the increase in investment.

Practice Problem

3 The equations defining a model of two trading nations are given by

$$Y_1 = C_1 + I_1^* + X_1 - M_1 \qquad Y_2 = C_2 + I_2^* + X_2 - M_2$$

$$C_1 = 0.7Y_1 + 50 \qquad C_2 = 0.8Y_2 + 100$$

$$I_1^* = 200 \qquad I_2^* = 300$$

$$M_1 = 0.3Y_1 \qquad M_2 = 0.1Y_2$$

Express this system in matrix form and hence find the values of Y_1 and Y_2. Calculate the balance of payments between these countries.

Key Terms

Cramer's rule A method of solving simultaneous equations, $\mathbf{Ax} = \mathbf{b}$, by the use of determinants. The *i*th variable x_i can be computed using $\det(\mathbf{A}_i)/\det(\mathbf{A})$ where $\mathbf{A}_i$ is the determinant of the matrix obtained from $\mathbf{A}$ by replacing the *i*th column by $\mathbf{b}$.

Practice Problems

4 Use Cramer's rule to solve

(a) $\begin{bmatrix} 4 & -1 \\ -2 & 5 \end{bmatrix}\begin{bmatrix} x_1 \\ x_2 \end{bmatrix} = \begin{bmatrix} 13 \\ 7 \end{bmatrix}$

for x_1

(b) $\begin{bmatrix} 3 & 2 & -2 \\ 4 & 3 & 3 \\ 2 & -1 & 1 \end{bmatrix}\begin{bmatrix} x_1 \\ x_2 \\ x_3 \end{bmatrix} = \begin{bmatrix} -15 \\ 17 \\ -1 \end{bmatrix}$

for x_2

(c) $\begin{bmatrix} 1 & 0 & 2 & 3 \\ -1 & 5 & 4 & 1 \\ 0 & 7 & -3 & 6 \\ 2 & 4 & 5 & 1 \end{bmatrix}\begin{bmatrix} x_1 \\ x_2 \\ x_3 \\ x_4 \end{bmatrix} = \begin{bmatrix} -1 \\ 1 \\ -24 \\ 15 \end{bmatrix}$

for x_4

5 Consider the macroeconomic model defined by

$$Y = C + I^* + G^* + X^* - M$$

$$C = aY + b \qquad (0 < a < 1,\ b > 0)$$

$$M = mY + M^* \qquad (0 < m < 1,\ M^* > 0)$$

Show that this system can be written as **Ax** = **b**, where

$$\mathbf{A} = \begin{bmatrix} 1 & -1 & 1 \\ -a & 1 & 0 \\ -m & 0 & 1 \end{bmatrix} \quad \mathbf{x} = \begin{bmatrix} Y \\ C \\ M \end{bmatrix} \quad \mathbf{b} = \begin{bmatrix} I^* + G^* + X^* \\ b \\ M^* \end{bmatrix}$$

Use Cramer's rule to show that

$$Y = \frac{b + I^* + G^* + X^* - M^*}{1 - a + m}$$

Write down the autonomous investment multiplier for Y and deduce that Y increases as as I^* increases.

6 Consider the macroeconomic model defined by

national income: $Y = C + I + G^* \quad (G^* > 0)$

consumption: $C = aY + b \quad (0 < a < 1,\ b > 0)$

investment: $I = cr + d \quad (c < 0,\ d > 0)$

money supply: $M_S^* = k_1 Y + k_2 r \quad (k_1 > 0,\ k_2 < 0,\ M_S^* > 0)$

Show that this system can be written as **Ax** = **b**, where

$$\mathbf{A} = \begin{bmatrix} 1 & -1 & -1 & 0 \\ -a & 1 & 0 & 0 \\ 0 & 0 & 1 & -c \\ k_1 & 0 & 0 & k_2 \end{bmatrix} \quad \mathbf{x} = \begin{bmatrix} Y \\ C \\ I \\ r \end{bmatrix} \quad \mathbf{b} = \begin{bmatrix} G^* \\ b \\ d \\ M_S^* \end{bmatrix}$$

Use Cramer's rule to show that

$$r = \frac{M_S^*(1 - a) - k_1(b + d + G^*)}{k_2(1 - a) + ck_1}$$

Write down the government expenditure multiplier for r and deduce that the interest rate, r, increases as government expenditure, G^*, increases.

7 The equations defining a model of two trading nations are given by

$$\begin{aligned} Y_1 &= C_1 + I_1^* + X_1 - M_1 & \quad Y_2 &= C_2 + I_2^* + X_2 - M_2 \\ C_1 &= 0.6Y_1 + 50 & \quad C_2 &= 0.8Y_2 + 80 \\ M_1 &= 0.2Y_1 & \quad M_2 &= 0.1Y_2 \end{aligned}$$

If $I_2^* = 70$, find the value of I_1^* if the balance of payments is zero.
[Hint: construct a system of three equations for the three unknowns, Y_1, Y_2 and I_1^*.]

8 The equations defining a general model of two trading countries are given by

$$\begin{aligned} Y_1 &= C_1 + I_1^* + X_1 - M_1 & \quad Y_2 &= C_2 + I_2^* + X_2 - M_2 \\ C_1 &= a_1Y_1 + b_1 & \quad C_2 &= a_2Y_2 + b_2 \\ M_1 &= m_1Y_1 & \quad M_2 &= m_2Y_2 \end{aligned}$$

where $0 < a_i < 1$, $b_i > 0$ and $0 < m_i < 1$ ($i = 1, 2$). Express this system in matrix form and use Cramer's rule to solve this system for Y_1. Write down the multiplier for Y_1 due to changes in I_2^* and hence give a general description of the effect on the national income of one country due to a change in investment in the other.

section 7.4

Input–output analysis

Objectives

At the end of this section you should be able to:

- Understand what is meant by a matrix of technical coefficients.
- Calculate the final demand vector given the total output vector.
- Calculate the total output vector given the final demand vector.
- Calculate the multipliers in simple input–output models.

The simplest model of the macroeconomy assumes that there are only two sectors: households and firms. The flow of money between these sectors is illustrated in Figure 7.1.

The 'black box' labelled 'firms' belies a considerable amount of economic activity. Firms exchange goods and services between themselves as well as providing them for external consumption by households. For example, the steel industry uses raw materials such as iron ore and coal to produce steel. This, in turn, is bought by mechanical engineering firms to produce machine tools. These tools are then used by other firms, including those in the steel industry. It is even possible for some businesses to use as input some of their own output. For example, in the agricultural sector, a farm might use arable land to produce grain, some of which is

Figure 7.1

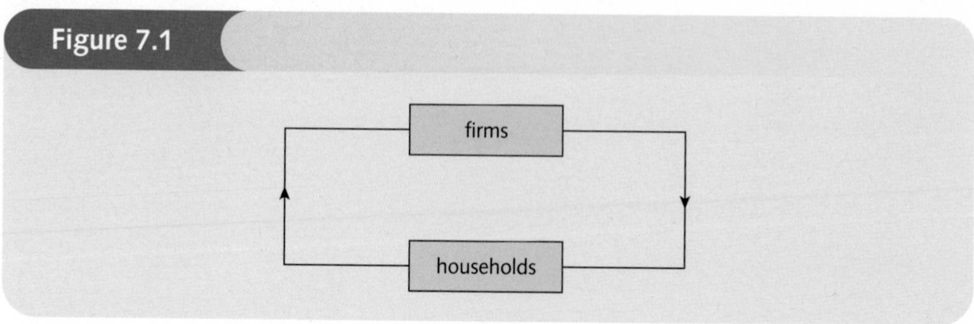

Table 7.3

		Output	
		I1	I2
Input	*I1*	0.1	0.5
	I2	0.3	0.2

recycled as animal foodstuffs. Output destined for households is called *final* (or *external*) *demand*. Output that is used as input by another (or the same) firm is called *intermediate output*. The problem of identifying individual firms and goods, and of tracking down the flow of money between firms for these goods, is known as *input–output analysis*.

Suppose that there are just two industries, I1 and I2, and that \$1 worth of output of I1 requires as input 10 cents worth of I1 and 30 cents worth of I2. The corresponding figures for I2 are 50 cents and 20 cents respectively. This information can be displayed in tabular form as shown in Table 7.3.

The matrix obtained by stripping away the headings in Table 7.3 is

$$\mathbf{A} = \begin{bmatrix} 0.1 & 0.5 \\ 0.3 & 0.2 \end{bmatrix}$$

and is called the *matrix of technical coefficients* (sometimes called the *technology matrix*). The columns of **A** give the inputs needed to produce \$1 worth of output. In general, if there are n industries then the matrix of technical coefficients has order $n \times n$. Element a_{ij} gives the input needed from the ith industry to produces \$1 worth of output for the jth industry.

We shall make the important assumption that the production functions for each industry in the model exhibit constant returns to scale. This means that the technical coefficients can be thought of as proportions that are independent of the level of output. For example, suppose that we wish to produce 500 monetary units of output of I1 instead of just 1 unit. The first column of **A** shows that the input requirements are

$0.1 \times 500 = 50$ units of I1

$0.3 \times 500 = 150$ units of I2

Similarly, if we produce 400 units of I2 then the second column of **A** shows that we use

$0.5 \times 400 = 200$ units of I1

$0.2 \times 400 = 80$ units of I2

In this situation, of the 500 units of I1 that are produced, 50 go back into I1 and 200 are used in I2. This means that there are 250 units left which are available for external demand. Similarly, of the 400 units of I2 that are produced, 230 are used as intermediate output, leaving 170 units to satisfy external demand. The flow of money for this simple input–output model is illustrated in Figure 7.2.

For the general case of n industries we would like to be able to use the matrix of technical coefficients to provide answers to the following questions.

Question 1

How much output is available for final demand given the total output level?

Figure 7.2

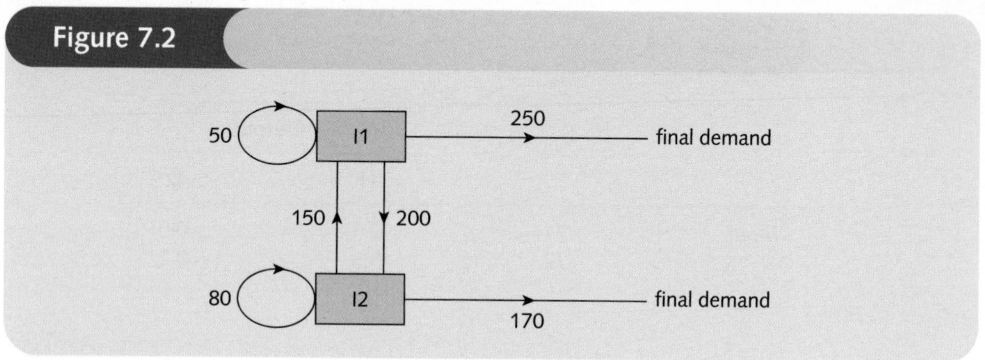

Question 2

How much total output is required to satisfy a given level of final demand?

Question 3

What changes need to be made to total output when final demand changes by a given amount?

It turns out that all three questions can be answered using one basic matrix equation, which we now derive. We begin by returning to the simple two-industry model with matrix of technical coefficients

$$\mathbf{A} = \begin{bmatrix} 0.1 & 0.5 \\ 0.3 & 0.2 \end{bmatrix}$$

Let us denote the final demand for I1 and I2 by d_1 and d_2 respectively, and denote the total outputs by x_1 and x_2. Total output from I1 gets used up in three different ways. Firstly, some of the output from I1 gets used up as input to I1. The precise proportion is given by the element

$$a_{11} = 0.1$$

so I1 uses $0.1x_1$ units of its own output. Secondly, some of the output from I1 gets used as input to I2. The element

$$a_{12} = 0.5$$

gives the amount of I1 that is used to make 1 unit of I2. We make a total of x_2 units of I2, so we use up $0.5x_2$ units of I1 in this way. Finally, some of the output of I1 satisfies final demand, which we denote by d_1. The total amount of I1 that is used is therefore

$$0.1x_1 + 0.5x_2 + d_1$$

If we assume that the total output from I1 is just sufficient to meet these requirements then

$$x_1 = 0.1x_1 + 0.5x_2 + d_1$$

Similarly, if I2 produces output to satisfy the input requirements of the two industries as well as final demand then

$$x_2 = 0.3x_1 + 0.2x_2 + d_2$$

In matrix notation these two equations can be written as

$$\begin{bmatrix} x_1 \\ x_2 \end{bmatrix} = \begin{bmatrix} 0.1 & 0.5 \\ 0.3 & 0.2 \end{bmatrix} \begin{bmatrix} x_1 \\ x_2 \end{bmatrix} + \begin{bmatrix} d_1 \\ d_2 \end{bmatrix}$$

that is, $\mathbf{x} = \mathbf{Ax} + \mathbf{d}$

where $\mathbf{x}$ is the total output vector

$$\begin{bmatrix} x_1 \\ x_2 \end{bmatrix}$$

and $\mathbf{d}$ is the final demand vector

$$\begin{bmatrix} d_1 \\ d_2 \end{bmatrix}$$

For the general case of n industries, we write x_i and d_i for the total output and final demand for the ith industry. Of the x_i units of output of industry i that are produced,

$a_{i1}x_1$ is used as input for industry 1

$a_{i2}x_1$ is used as input for industry 2

⋮

$a_{in}x_n$ is used as input for industry n

and

d_i is used for external demand

Hence

$$x_i = a_{i1}x_1 + a_{i2}x_2 + \ldots + a_{in}x_n + d_i$$

In matrix form, the totality of equations obtained by setting $i = 1, 2, \ldots, n$, in turn, can be written as

$$\begin{bmatrix} x_1 \\ x_2 \\ \vdots \\ x_n \end{bmatrix} = \begin{bmatrix} a_{11} & a_{12} & \ldots & a_{1n} \\ a_{21} & a_{22} & \ldots & a_{2n} \\ \vdots & \vdots & \vdots & \vdots \\ a_{n1} & a_{n2} & \ldots & a_{nn} \end{bmatrix} \begin{bmatrix} x_1 \\ x_2 \\ \vdots \\ x_n \end{bmatrix} = \begin{bmatrix} d_1 \\ d_2 \\ \vdots \\ d_n \end{bmatrix}$$

that is, as

$$\mathbf{x} = \mathbf{Ax} + \mathbf{d}$$

where $\mathbf{A}$ is the $n \times n$ matrix of technical coefficients, $\mathbf{x}$ is the $n \times 1$ total output vector and $\mathbf{d}$ is the $n \times 1$ final demand vector.

The three questions posed can now be answered.

Question 1

How much output is available for final demand given the total output level?

Answer 1

In this case the vector $\mathbf{x}$ is assumed to be known and we need to calculate the unknown vector $\mathbf{d}$. The matrix equation

$$\mathbf{x} = \mathbf{Ax} + \mathbf{d}$$

immediately gives $\mathbf{d} = \mathbf{x} - \mathbf{Ax}$

and the right-hand side is easily evaluated to get $\mathbf{d}$.

Example

The output levels of machinery, electricity and oil of a small country are 3000, 5000 and 2000 respectively.

- Each unit of machinery requires inputs of 0.3 units of electricity and 0.3 units of oil.
- Each unit of electricity requires inputs of 0.1 units of machinery and 0.2 units of oil.
- Each unit of oil requires inputs of 0.2 units of machinery and 0.1 units of electricity.

Determine the machinery, electricity and oil available for export.

Solution

Let us denote the total output for machinery, electricity and oil by x_1, x_2 and x_3 respectively, so that

$$x_1 = 3000, \quad x_2 = 5000, \quad x_3 = 2000$$

The first bullet point of the problem statement provides details of the input requirements for machinery. To produce 1 unit of machinery we use 0 units of machinery, 0.3 units of electricity and 0.3 units of oil. The first column of the matrix of technical coefficients is therefore

$$\begin{bmatrix} 0 \\ 0.3 \\ 0.3 \end{bmatrix}$$

Likewise, the third and fourth sentences give the input requirements for electricity and oil, so the complete matrix is

$$\mathbf{A} = \begin{bmatrix} 0 & 0.1 & 0.2 \\ 0.3 & 0 & 0.1 \\ 0.3 & 0.2 & 0 \end{bmatrix}$$

From the equation

$$\mathbf{d} = \mathbf{x} - \mathbf{A}\mathbf{x}$$

we see that the final demand vector is

$$\begin{bmatrix} d_1 \\ d_2 \\ d_3 \end{bmatrix} = \begin{bmatrix} 3000 \\ 5000 \\ 2000 \end{bmatrix} - \begin{bmatrix} 0 & 0.1 & 0.2 \\ 0.3 & 0 & 0.1 \\ 0.3 & 0.2 & 0 \end{bmatrix} \begin{bmatrix} 3000 \\ 5000 \\ 2000 \end{bmatrix}$$

$$= \begin{bmatrix} 3000 \\ 5000 \\ 2000 \end{bmatrix} - \begin{bmatrix} 900 \\ 1100 \\ 1900 \end{bmatrix}$$

$$= \begin{bmatrix} 2100 \\ 3900 \\ 100 \end{bmatrix}$$

The country therefore has 2100, 3900 and 100 units of machinery, electricity and oil, respectively, available for export.

Practice Problem

1 Determine the final demand vector for three firms given the matrix of technical coefficients

$$\mathbf{A} = \begin{bmatrix} 0.2 & 0.4 & 0.2 \\ 0.1 & 0.2 & 0.1 \\ 0.1 & 0.1 & 0 \end{bmatrix}$$

and the total output vector

$$\mathbf{x} = \begin{bmatrix} 1000 \\ 300 \\ 700 \end{bmatrix}$$

Question 2

How much total output is required to satisfy a given level of final demand?

Answer 2

In this case the vector **d** is assumed to be known and we need to calculate the unknown vector **x**. The matrix equation

$$\mathbf{x} = \mathbf{Ax} + \mathbf{d}$$

rearranges to give

$$\mathbf{x} - \mathbf{Ax} = \mathbf{d}$$

or equivalently

$$(\mathbf{I} - \mathbf{A})\mathbf{x} = \mathbf{d}$$

because

$$(\mathbf{I} - \mathbf{A})\mathbf{x} = \mathbf{Ix} - \mathbf{Ax} = \mathbf{x} - \mathbf{Ax}$$

This represents a system of linear equations in which the coefficient matrix is $\mathbf{I} - \mathbf{A}$ and the right-hand-side vector is **d**. From Section 7.2 we know that we can solve this by multiplying the inverse of the coefficient matrix by the right-hand-side vector to get

$$\mathbf{x} = (\mathbf{I} - \mathbf{A})^{-1}\mathbf{d}$$

In the context of input–output analysis the matrix $(\mathbf{I} - \mathbf{A})^{-1}$ is called the *Leontief inverse*.

Example

Given the matrix of technical coefficients

$$\mathbf{A} = \begin{bmatrix} 0.3 & 0.1 & 0.1 \\ 0.2 & 0.2 & 0.2 \\ 0.4 & 0.2 & 0.3 \end{bmatrix}$$

for three industries, I1, I2 and I3, determine the total outputs required to satisfy final demands of 49, 106 and 17 respectively.

➔

Solution

To solve this problem we need to find the inverse of **I** – **A** and then to multiply by the final demand vector. The matrix **I** – **A** is

$$\begin{bmatrix} 1 & 0 & 0 \\ 0 & 1 & 0 \\ 0 & 0 & 1 \end{bmatrix} - \begin{bmatrix} 0.3 & 0.1 & 0.1 \\ 0.2 & 0.2 & 0.2 \\ 0.4 & 0.2 & 0.3 \end{bmatrix} = \begin{bmatrix} 0.7 & -0.1 & -0.1 \\ -0.2 & 0.8 & -0.2 \\ -0.4 & -0.2 & 0.7 \end{bmatrix}$$

The inverse of this matrix is then found by calculating its cofactors. If we call this matrix **B** then the cofactors, B_{ij}, corresponding to elements b_{ij} are given by

$$B_{11} = 0.52, \quad B_{12} = 0.22, \quad B_{13} = 0.36$$

$$B_{21} = 0.09, \quad B_{22} = 0.45, \quad B_{23} = 0.18$$

$$B_{31} = 0.10, \quad B_{32} = 0.16, \quad B_{33} = 0.54$$

By expanding along the first row we see that

$$|\mathbf{B}| = 0.7(0.52) + (-0.1)(0.22) + (-0.1)(0.36) = 0.306$$

Hence

$$\mathbf{B}^{-1} = (\mathbf{I} - \mathbf{A})^{-1} = \frac{1}{0.306}\begin{bmatrix} 0.52 & 0.09 & 0.10 \\ 0.22 & 0.45 & 0.16 \\ 0.36 & 0.18 & 0.54 \end{bmatrix}$$

We are given that

$$\mathbf{d} = \begin{bmatrix} 49 \\ 106 \\ 17 \end{bmatrix}$$

so the equation

$$\mathbf{x} = (\mathbf{I} - \mathbf{A})^{-1}\mathbf{d}$$

gives

$$\begin{bmatrix} x_1 \\ x_2 \\ x_3 \end{bmatrix} = \frac{1}{0.306}\begin{bmatrix} 0.52 & 0.09 & 0.10 \\ 0.22 & 0.45 & 0.16 \\ 0.36 & 0.18 & 0.54 \end{bmatrix}\begin{bmatrix} 49 \\ 106 \\ 17 \end{bmatrix} = \begin{bmatrix} 120 \\ 200 \\ 150 \end{bmatrix}$$

Practice Problem

2 Each unit of engineering output requires as input 0.2 units of engineering and 0.4 units of transport. Each unit of transport output requires as input 0.2 units of engineering and 0.1 units of transport. Determine the level of total output needed to satisfy a final demand of 760 units of engineering and 420 units of transport.

Question 3

What changes need to be made to total output when final demand changes by a given amount?

Answer 3

In this case we assume that the current total output vector, $\mathbf{x}$, is chosen to satisfy some existing final demand vector, $\mathbf{d}$, so that

$$\mathbf{x} = \mathbf{Ax} + \mathbf{d}$$

or equivalently

$$\mathbf{x} = (\mathbf{I} - \mathbf{A})^{-1}\mathbf{d} \tag{1}$$

Suppose that the final demand vector changes by an amount $\Delta\mathbf{d}$, so that the new final demand vector is $\mathbf{d} + \Delta\mathbf{d}$. In order to satisfy the new requirements, the total output vector, $\mathbf{x} + \Delta\mathbf{x}$, is then given by

$$\begin{aligned}\mathbf{x} + \Delta\mathbf{x} &= (\mathbf{I} - \mathbf{A})^{-1}(\mathbf{d} + \Delta\mathbf{d}) \\ &= (\mathbf{I} - \mathbf{A})^{-1}\mathbf{d} + (\mathbf{I} - \mathbf{A})^{-1}\Delta\mathbf{d} \qquad (2)\end{aligned}$$

where we have used the distributive law to multiply out the brackets. However, from equation (1) we know that

$$(\mathbf{I} - \mathbf{A})^{-1}\mathbf{d} = \mathbf{x}$$

so equation (2) becomes

$$\mathbf{x} + \Delta\mathbf{x} = \mathbf{x} + (\mathbf{I} - \mathbf{A})^{-1}\Delta\mathbf{d}$$

and if $\mathbf{x}$ is subtracted from both sides then

$$\Delta\mathbf{x} = (\mathbf{I} - \mathbf{A})^{-1}\Delta\mathbf{d}$$

Notice that this equation does not have $\mathbf{d}$ or $\mathbf{x}$ in it. This shows that the change in output, $\Delta\mathbf{x}$, does not depend on the existing final demand or existing total output. It depends only on the change, $\Delta\mathbf{d}$. It is also interesting to observe that the mathematics needed to solve this is the same as that for Question 2. Both require the calculation of the Leontief inverse followed by a simple matrix multiplication.

Example

Consider the following inter-industrial flow table for two industries, I1 and I2.

		Output		
		I1	I2	Final demand
Input	*I1*	200	300	500
	I2	100	100	300

Assuming that the total output is just sufficient to meet the input and final demand requirements, write down

(a) the total output vector

(b) the matrix of technical coefficients

Hence calculate the new total output vector needed when the final demand for I1 rises by 100 units. 

Solution

(a) To calculate the current total outputs for I1 and I2 all we have to do is to add together the numbers along each row of the table. The first row shows that I1 uses 200 units of I1 as input, I2 uses 300 units of I1 as input and 500 units of I1 are used in final demand. The total number of units of I1 is then

$$200 + 300 + 500 = 1000$$

Assuming that the total output of I1 exactly matches these requirements, we can deduce that

$$x_1 = 1000$$

Similarly, from the second row of the table,

$$x_2 = 100 + 100 + 300 = 500$$

Hence the total output vector is

$$\mathbf{x} = \begin{bmatrix} 1000 \\ 500 \end{bmatrix}$$

(b) The first column of the matrix of technical coefficients represents the inputs needed to produce 1 unit of I1. The first column of the inter-industrial flow table gives the inputs needed to produce the current total output of I1, which we found in part (a) to be 1000. In all input–output models we assume that production is subject to constant returns to scale, so we divide the first column of the inter-industrial flow table by 1000 to find the inputs needed to produce just 1 unit of output. In part (a) the total output for I2 was found to be 500, so the second column of the matrix of technical coefficients is calculated by dividing the second column of the inter-industrial flow table by 500. Hence

$$\mathbf{A} = \begin{bmatrix} \frac{200}{1000} & \frac{300}{500} \\ \frac{100}{1000} & \frac{100}{500} \end{bmatrix}$$

$$= \begin{bmatrix} 0.2 & 0.6 \\ 0.1 & 0.2 \end{bmatrix}$$

If the demand for I1 rises by 100 units and the demand for I2 remains constant, the vector giving the change in final demand is

$$\Delta\mathbf{d} = \begin{bmatrix} 100 \\ 0 \end{bmatrix}$$

To determine the corresponding change in output we use the equation

$$\Delta\mathbf{x} = (\mathbf{I} - \mathbf{A})^{-1}\Delta\mathbf{d}$$

Subtracting **A** from the identity matrix gives

$$\mathbf{I} - \mathbf{A} = \begin{bmatrix} 1 & 0 \\ 0 & 1 \end{bmatrix} - \begin{bmatrix} 0.2 & 0.6 \\ 0.1 & 0.2 \end{bmatrix} = \begin{bmatrix} 0.8 & -0.6 \\ -0.1 & 0.8 \end{bmatrix}$$

which has determinant

$$|\mathbf{I} - \mathbf{A}| = (0.8)(0.8) - (-0.6)(-0.1) = 0.58$$

The inverse of $\mathbf{I} - \mathbf{A}$ is then

$$(\mathbf{I} - \mathbf{A})^{-1} = \frac{1}{0.58}\begin{bmatrix} 0.8 & 0.6 \\ 0.1 & 0.8 \end{bmatrix}$$

so

$$\Delta\mathbf{x} = \frac{1}{0.58}\begin{bmatrix} 0.8 & 0.6 \\ 0.1 & 0.8 \end{bmatrix}\begin{bmatrix} 100 \\ 0 \end{bmatrix} = \begin{bmatrix} 138 \\ 17 \end{bmatrix}$$

to the nearest unit. There is an increase in total output of I2 despite the fact that the final demand for I2 remains unchanged. This is to be expected because in order to meet the increase in final demand for I1 it is necessary to raise output of I1, which in turn requires more inputs of both I1 and I2. Any change to just one industry has a knock-on effect throughout all of the industries in the model.

From part (a) the current total output vector is

$$\mathbf{x} = \begin{bmatrix} 1000 \\ 500 \end{bmatrix}$$

so the new total output vector is

$$\mathbf{x} + \Delta\mathbf{x} = \begin{bmatrix} 1000 \\ 500 \end{bmatrix} + \begin{bmatrix} 138 \\ 17 \end{bmatrix} = \begin{bmatrix} 1138 \\ 517 \end{bmatrix}$$

Practice Problems

3 Write down the 4×4 matrix of technical coefficients using the information provided in the following inter-industrial flow table. You may assume that the total outputs are just sufficient to satisfy the input requirements and final demands.

		Output				
		I1	I2	I3	I4	Final demand
Input	*I1*	0	300	100	100	500
	I2	100	0	200	100	100
	I3	200	100	0	400	1300
	I4	300	0	100	0	600

4 Given the matrix of technical coefficients

$$\mathbf{A} = \begin{matrix} & \begin{matrix} \text{I1} & \text{I2} & \text{I3} \end{matrix} \\ \begin{matrix} \text{I1} \\ \text{I2} \\ \text{I3} \end{matrix} & \begin{bmatrix} 0.1 & 0.2 & 0.2 \\ 0.1 & 0.1 & 0.1 \\ 0.1 & 0.3 & 0.1 \end{bmatrix} \end{matrix}$$

determine the changes in total output for the three industries when the final demand for I1 rises by 1000 units and the final demand for I3 falls by 800 units simultaneously.

We conclude this section with a postscript highlighting again the connection between the multiplier concept and the matrix inverse. Suppose that we have a three-industry model and that the Leontief inverse, $(\mathbf{I} - \mathbf{A})^{-1}$, is given by

$$\begin{bmatrix} b_{11} & b_{12} & b_{13} \\ b_{21} & b_{22} & b_{23} \\ b_{31} & b_{32} & b_{33} \end{bmatrix}$$

The equation

$$\mathbf{x} = (\mathbf{I} - \mathbf{A})^{-1}\mathbf{d}$$

is then

$$\begin{bmatrix} x_1 \\ x_2 \\ x_3 \end{bmatrix} = \begin{bmatrix} b_{11} & b_{12} & b_{13} \\ b_{21} & b_{22} & b_{23} \\ b_{31} & b_{32} & b_{33} \end{bmatrix} \begin{bmatrix} d_1 \\ d_2 \\ d_3 \end{bmatrix}$$

so that

$$x_1 = b_{11}d_1 + b_{12}d_2 + b_{13}d_3$$

$$x_2 = b_{21}d_1 + b_{22}d_2 + b_{23}d_3$$

$$x_3 = b_{31}d_1 + b_{32}d_2 + b_{33}d_3$$

The first equation shows that x_1 is a function of the three variables d_1, d_2 and d_3. Consequently, we can write down three partial derivatives

$$\frac{\partial x_1}{\partial d_1} = b_{11}, \quad \frac{\partial x_1}{\partial d_2} = b_{12}, \quad \frac{\partial x_1}{\partial d_3} = b_{13}$$

In the same way, the second and third equations give

$$\frac{\partial x_2}{\partial d_1} = b_{21}, \quad \frac{\partial x_2}{\partial d_2} = b_{22}, \quad \frac{\partial x_2}{\partial d_3} = b_{23}$$

$$\frac{\partial x_3}{\partial d_1} = b_{31}, \quad \frac{\partial x_3}{\partial d_2} = b_{32}, \quad \frac{\partial x_3}{\partial d_3} = b_{33}$$

Recall from Chapter 5 that partial derivatives determine the multipliers in economic models. These nine partial derivatives show that if we regard the final demands as exogenous variables and the total outputs as endogenous variables then the multipliers are the elements in the matrix $(\mathbf{I} - \mathbf{A})^{-1}$. More precisely, the multiplier of the variable x_i due to changes in d_j is the element b_{ij}, which lies in the ith row and jth column of $(\mathbf{I} - \mathbf{A})^{-1}$. This result can also be seen more directly from the equation

$$\Delta\mathbf{x} = (\mathbf{I} - \mathbf{A})^{-1}\Delta\mathbf{d}$$

If we put $\Delta\mathbf{x} = [\Delta x_1 \quad \Delta x_2 \quad \Delta x_3]^{\mathrm{T}}$ and $\Delta\mathbf{d} = [\Delta d_1 \quad \Delta d_2 \quad \Delta d_3]^{\mathrm{T}}$ then this matrix equation leads to

$$\Delta x_1 = b_{11}\Delta d_1 + b_{12}\Delta d_2 + b_{13}\Delta d_3$$

$$\Delta x_2 = b_{21}\Delta d_1 + b_{22}\Delta d_2 + b_{23}\Delta d_3$$

$$\Delta x_3 = b_{31}\Delta d_1 + b_{32}\Delta d_2 + b_{33}\Delta d_3$$

We see from the ith equation that the contribution to the change Δx_i due to the change Δd_j is $b_{ij}\Delta d_j$. In other words, if d_j changes by Δd_j and all other final demands are fixed, then we can calculate the corresponding change in x_i by multiplying Δd_j by b_{ij}.

Key Terms

External demand Output that is used by households.

Final demand An alternative to 'external demand'.

Input–output analysis Examination of how inputs and outputs from various sectors of the economy are matched to the total resources available.

Intermediate output Output from one sector which is used as input by another (or the same) sector.

Leontief inverse The inverse of **I** – **A**, where **A** is the matrix of technical coefficients.

Matrix of technical coefficients (or technology matrix) A square matrix in which element a_{ij} is the input required from the ith sector to produce 1 unit of output for the jth sector.

Practice Problems

5 Calculate the available final demand for five firms if the matrix of technical coefficients is

$$\begin{array}{c} \\ \text{F1} \\ \text{F2} \\ \text{F3} \\ \text{F4} \\ \text{F5} \end{array}
\begin{array}{c} \begin{array}{ccccc} \text{F1} & \text{F2} & \text{F3} & \text{F4} & \text{F5} \end{array} \\
\begin{bmatrix} 0 & 0.1 & 0.1 & 0 & 0.2 \\ 0.1 & 0 & 0.2 & 0 & 0.1 \\ 0 & 0 & 0 & 0.3 & 0.1 \\ 0.2 & 0.1 & 0.1 & 0 & 0.1 \\ 0 & 0.3 & 0 & 0.1 & 0 \end{bmatrix} \end{array}$$

and the total output vector is

$$[1000 \quad 1500 \quad 2000 \quad 5000 \quad 1000]^{\text{T}}$$

6 Each unit of water output requires inputs of 0.1 units of steel and 0.2 units of electricity.

Each unit of steel output requires inputs of 0.1 units of water and 0.2 units of electricity.

Each unit of electricity output requires inputs of 0.2 units of water and 0.1 units of steel.

(a) Determine the level of total output needed to satisfy a final demand of 750 units of water, 300 units of steel and 700 units of electricity.

(b) Write down the multiplier for water output due to changes in final demand for electricity. Hence calculate the change in water output due to a 100 unit increase in final demand for electricity.

7 Consider the following inter-industrial flow table for two industries I1 and I2.

		Output		
		I1	**I2**	**Final demand**
Input	*I1*	100	100	300
	I2	200	500	300

➔

Assuming that the total output is just sufficient to meet the input and final demand requirements, find

(a) the current total output vector, $\mathbf{x}$

(b) the matrix of technical coefficients, $\mathbf{A}$

(c) the matrix of multipliers, $(\mathbf{I} - \mathbf{A})^{-1}$

(d) the future total output vector, $\mathbf{x} + \Delta\mathbf{x}$, if final demand for I1 rises by 150 units and final demand for I2 falls by 50 units simultaneously

8 An economy consists of three industries: agriculture, mining and manufacturing.

One unit of agricultural output requires 0.2 units of its own output, 0.3 units of mining output and 0.4 units of manufacturing output.

One unit of mining output requires 0.2 units of agricultural output, 0.4 units of its own output and 0.2 units of manufacturing output.

One unit of manufacturing output requires 0.3 units of agricultural output, 0.3 units of mining output and 0.1 units of its own output.

(a) Write down the matrix of technical coefficients and find the Leontief inverse.

(b) Determine the levels of total output needed to satisfy a final demand of 10 000 units of agricultural output, 30 000 units of mining output and 40 000 units of manufacturing output.

(c) If the final demand for agricultural output rises by 1000 units and the final demand for manufacturing output falls by 1000 units, calculate the change in mining output.

chapter 8

Linear Programming

Several methods were described in Chapter 5 for optimizing functions of two variables subject to constraints. In economics not all relationships between variables are represented by equations and we now consider the case when the constraints are given by inequalities. Provided the function to be optimized is linear and the inequalities are all linear, the problem is said to be one of linear programming. For simplicity we concentrate on problems involving just two unknowns and describe a graphical method of solution.

There are two sections, which should be read in the order that they appear. Section 8.1 describes the basic mathematical techniques and considers special cases when problems have either no solution or infinitely many solutions. Section 8.2 shows how an economic problem, initially given in words, can be expressed as a linear programming problem and hence solved.

The material in this chapter can be read at any stage, since it requires only an understanding of how to sketch a straight line on graph paper.

section 8.1

Graphical solution of linear programming problems

Objectives

At the end of this section you should be able to:

- Identify the region defined by a linear inequality.
- Sketch the feasible region defined by simultaneous linear inequalities.
- Solve linear programming problems graphically.
- Appreciate that a linear programming problem may have infinitely many solutions.
- Appreciate that a linear programming problem may have no finite solution.

In this and the following section we show you how to set up and solve linear programming problems. This process falls naturally into two separate phases. The first phase concerns problem formulation; a problem, initially given in words, is expressed in mathematical symbols. The second phase involves the actual solution of such a problem. Experience indicates that students usually find the first phase the more difficult. For this reason, we postpone consideration of problem formulation until Section 8.2 and begin by investigating techniques for their mathematical solution.

Advice

You may like to glance at one or two of the examples given in Section 8.2 now to get a feel for the type of problem that can be solved using these techniques.

Before you can consider linear programming it is essential that you know how to sketch linear inequalities. In Section 1.1 we discovered that a linear equation of the form

$$dx + ey = f$$

Figure 8.1

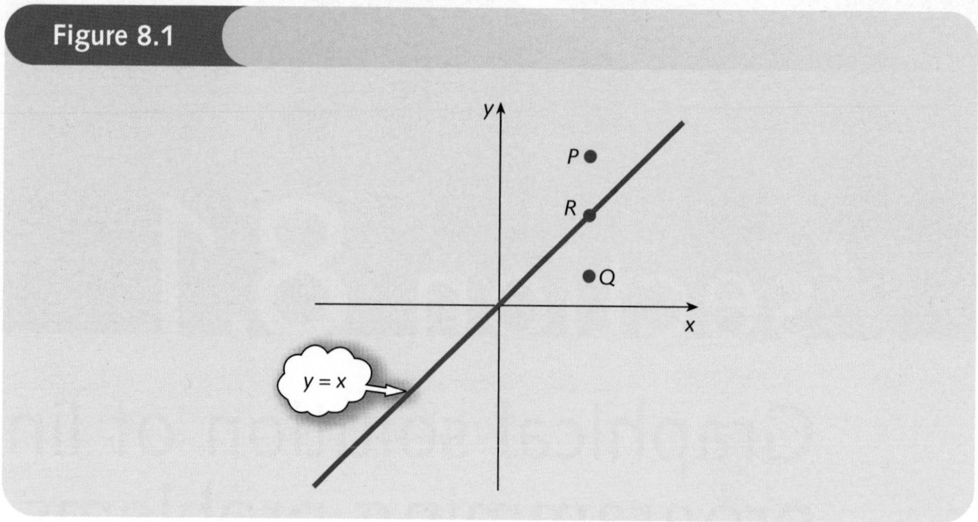

can be represented by a straight line on graph paper. We can give a similar graphical interpretation for linear inequalities involving two variables when the equals sign is replaced by one of

$<$ (less than)

$\leq$ (less than or equal to)

$>$ (greater than)

$\geq$ (greater than or equal to)

To illustrate this consider the simple inequality

$$y \geq x$$

We would like to identify those points with coordinates (x, y) for which this inequality is true. Clearly this has something to do with the straight line

$$y = x$$

sketched in Figure 8.1.

If a point P lies above the line then the y coordinate is greater than the x coordinate, so that

$$y > x$$

Similarly, if a point Q lies below the line then the y coordinate is less than the x coordinate, so that

$$y < x$$

Of course, the coordinates of a point R which actually lies on the line satisfy

$$y = x$$

Hence we see that the inequality

$$y \geq x$$

holds for any point that lies on or above the line $y = x$.

It is useful to be able to indicate this region pictorially. We do this by shading one half of the coordinate plane. There are actually two schools of thought here. Some people like to shade the region containing the points for which the inequality is true. Others prefer to shade the region

Figure 8.2

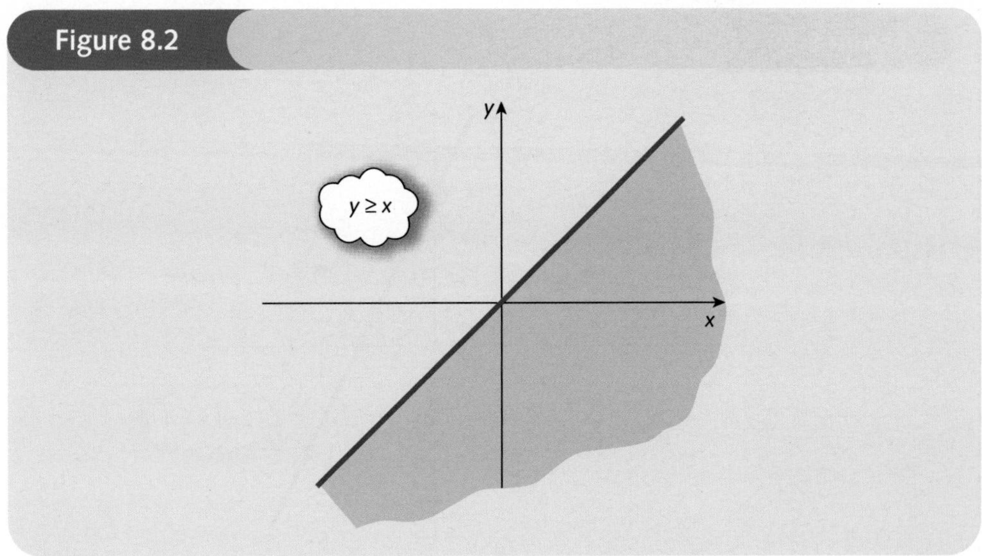

for which it is false. In this book we adopt the latter approach and always shade the region that we are *not* interested in, as shown in Figure 8.2. This may seem a strange choice, but the reason for making it will soon become apparent.

In general, to sketch an inequality of the form

$$dx + ey < f$$

$$dx + ey \le f$$

$$dx + ey > f$$

$$dx + ey \ge f$$

we first sketch the corresponding line

$$dx + ey = f$$

and then decide which side of the line to deal with. An easy way of doing this is to pick a 'test point', (x, y). It does not matter what point is chosen, provided it does not actually lie on the line itself. The numbers x and y are then substituted into the original inequality. If the inequality is satisfied then the side containing the test point is the region of interest. If not, then we go for the region on the other side of the line.

Example

Sketch the region

$$2x + y < 4$$

Solution

We first sketch the line

$$2x + y = 4$$

When $x = 0$ we get

$$y = 4$$

Figure 8.3

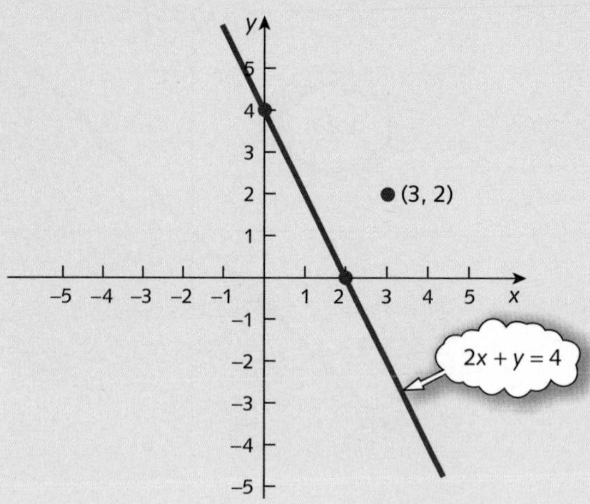

When $y = 0$ we get

$$2x = 4$$

and so $x = 4/2 = 2$.

The line passes through (0, 4) and (2, 0) and is shown in Figure 8.3. For a test point let us take (3, 2), which lies above the line. Substituting $x = 3$ and $y = 2$ into the expression $2x + y$ gives

$$2(3) + 2 = 8$$

This is *not* less than 4, so the test point does not satisfy the inequality. It follows that the region of interest lies below the line. This is illustrated in Figure 8.4. In this example the symbol $<$ is used rather than $\leq$. Hence the points on the line itself are not included in the region of interest. We have chosen to indicate this by using a broken line for the boundary.

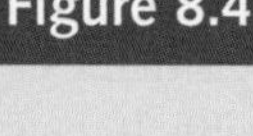

Figure 8.4

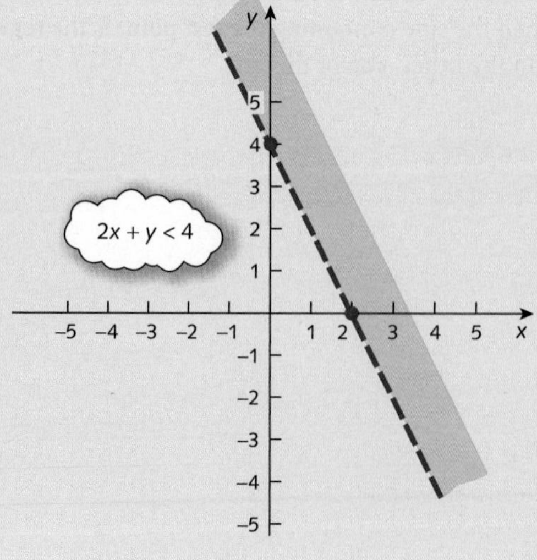

Practice Problem

1 Sketch the straight line

$$-x + 3y = 6$$

on graph paper. By considering the test point (1, 4) indicate the region

$$-x + 3y > 6$$

We now consider the region defined by simultaneous linear inequalities. This is known as a *feasible region*. It consists of those points (x, y) which satisfy several inequalities at the same time. We find it by sketching the regions defined by each inequality in turn. The feasible region is then the unshaded part of the plane corresponding to the intersection of all of the individual regions.

Example

Sketch the feasible region

$$x + 2y \leq 12$$

$$-x + y \leq 3$$

$$x \geq 0$$

$$y \geq 0$$

Solution

In this problem the easiest inequalities to handle are the last two. These merely indicate that x and y are non-negative and so we need only consider points in the top right-hand quadrant of the plane, as shown in Figure 8.5.

Figure 8.5

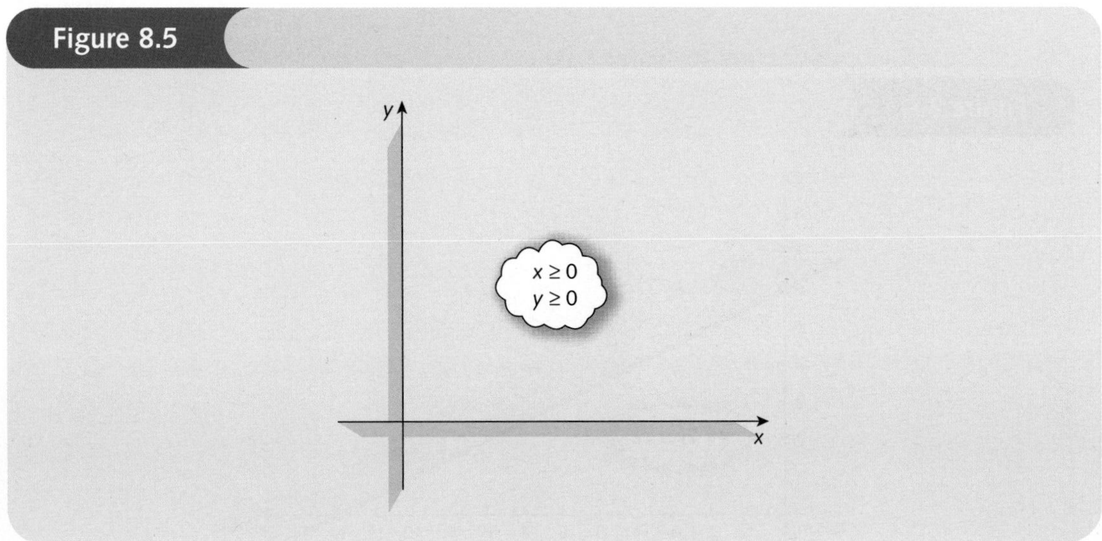

For the inequality

$$x + 2y \leq 12$$

we need to sketch the line

$$x + 2y = 12$$

When $x = 0$ we get

$$2y = 12$$

and so $y = 12/2 = 6$.

When $y = 0$ we get

$$x = 12$$

The line passes through (0, 6) and (12, 0).

For a test point let us take (0, 0), since such a choice minimizes the amount of arithmetic that we have to do. Substituting $x = 0$ and $y = 0$ into the inequality gives

$$0 + 2(0) \leq 12$$

which is obviously true. Now the region containing the origin lies below the line, so we shade the region that lies above it. This is indicated in Figure 8.6.

For the inequality

$$-x + y \leq 3$$

we need to sketch the line

$$-x + y = 3$$

When $x = 0$ we get

$$y = 3$$

When $y = 0$ we get

$$-x = 3$$

and so $x = 3/(-1) = -3$.

Figure 8.6

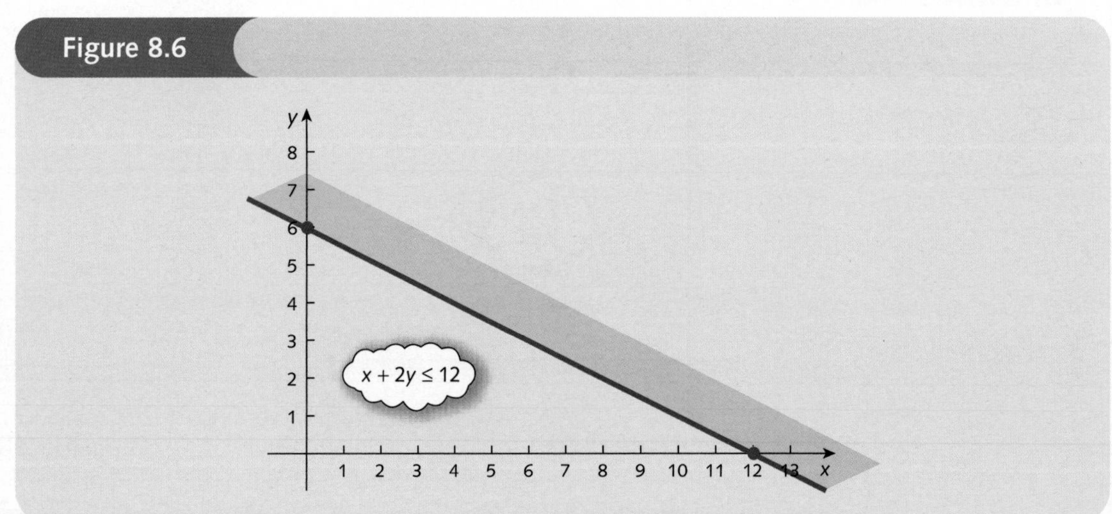

Figure 8.7

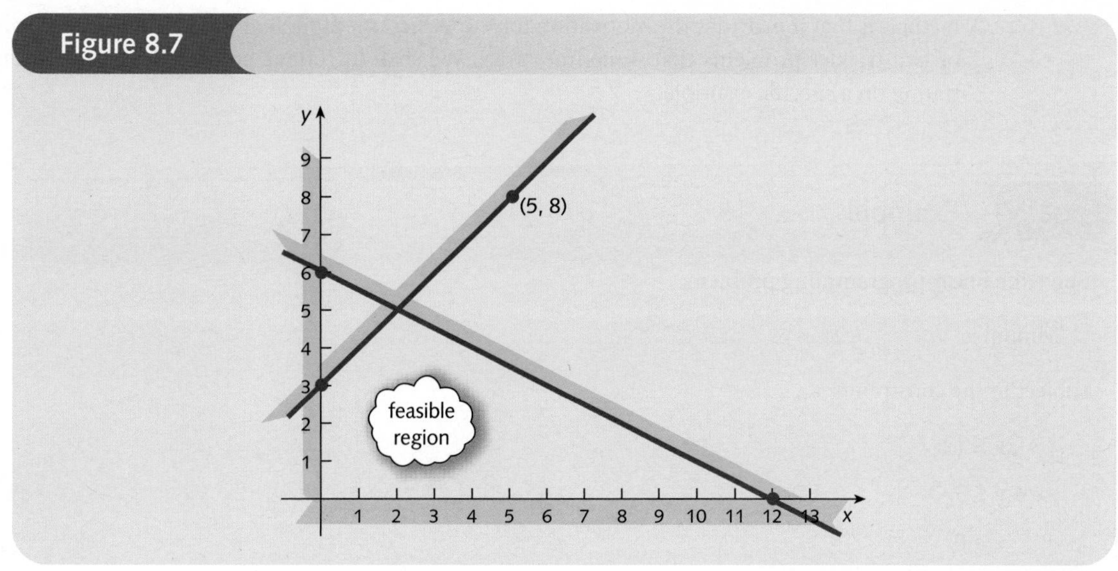

The line passes through (0, 3) and (−3, 0). Unfortunately, the second point does not lie on the diagram as we have drawn it. At this stage we can either redraw the x axis to include −3 or we can try finding another point on the line which does fit on the graph. For example, putting $x = 5$ gives

$$-5 + y = 3$$

so $y = 3 + 5 = 8$. Hence the line passes through (5, 8), which can now be plotted along with (0, 3) to sketch the line. At the test point (0, 0) the inequality reads

$$-0 + 0 \leq 3$$

which is obviously true. We are therefore interested in the region below the line, since this contains the origin. As usual we indicate this by shading the region on the other side. The complete picture is shown in Figure 8.7.

Points (x, y) which satisfy all four inequalities must lie in the unshaded 'hole' in the middle. Incidentally, this explains why we did not adopt the convention of shading the region of interest. Had we done so, our task would have been to identify the most heavily shaded part of the diagram, which is not so easy.

Practice Problem

2 Sketch the feasible region

$$x + 2y \leq 10$$
$$-3x + y \leq 10$$
$$x \geq 0$$
$$y \geq 0$$

We are now in a position to explain exactly what we mean by a linear programming problem and how such a problem can be solved graphically. We actually intend to describe two slightly different methods of solution. One of these is fairly sophisticated and difficult to use, while the other is more straightforward. The justification for bothering with the 'harder' method is that it provides the motivation for the 'easier' method. It also helps us to handle one or two trickier problems that sometimes arise. We shall introduce both methods by concentrating on a specific example.

Example

Solve the linear programming problem:

Minimize $-2x + y$

subject to the constraints

$$x + 2y \leq 12$$

$$-x + y \leq 3$$

$$x \geq 0$$

$$y \geq 0$$

Solution

In general, there are three ingredients making up a linear programming problem. Firstly, there are several unknowns to be determined. In this example there are just two unknowns, x and y. Secondly, there is a mathematical expression of the form

$$ax + by$$

which we want to either maximize or minimize. Such an expression is called an *objective function*. In this example, $a = -2$, $b = 1$ and the problem is one of minimization. Finally, the unknowns x and y are subject to a collection of linear inequalities. Quite often (but not always) two of the inequalities are $x \geq 0$ and $y \geq 0$. These are referred to as *non-negativity constraints*. In this example there are a total of four constraints including the non-negativity constraints.

Geometrically, points (x, y) which satisfy simultaneous linear inequalities define a feasible region in the coordinate plane. In fact, for this particular problem, the feasible region has already been sketched in Figure 8.7.

The problem now is to try to identify that point inside the feasible region which minimizes the value of the objective function. One naïve way of doing this might be to use trial and error: that is, we could evaluate the objective function at every point within the region and choose the point which produces the smallest value. For instance, (1, 1) lies in the region and when the values $x = 1$ and $y = 1$ are substituted into

$$-2x + y$$

we get

$$-2(1) + 1 = -1$$

Similarly we might try (3.4, 2.1), which produces

$$-2(3.4) + 2.1 = -4.7$$

which is an improvement, since $-4.7 < -1$.

The drawback of this approach is that there are infinitely many points inside the region, so it is going to take a very long time before we can be certain of the solution! A more systematic approach is to superimpose, on top of the feasible region, the family of straight lines,

$$-2x + y = c$$

Figure 8.8

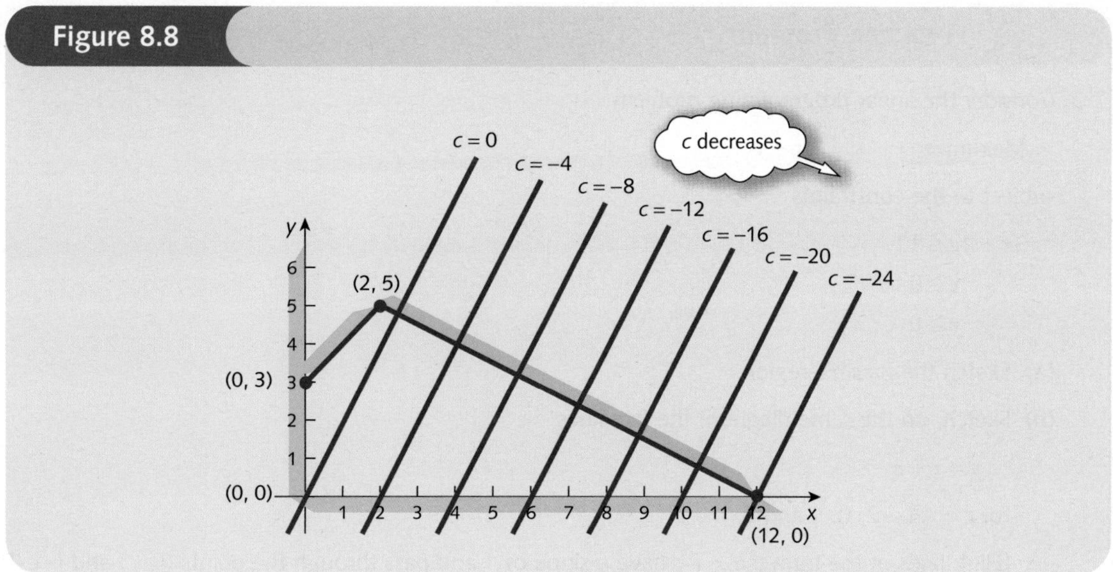

for various values of the constant c. Looking back at the objective function, you will notice that the number c is precisely the thing that we want to minimize. That such an equation represents a straight line should come as no surprise to you by now. Indeed, we know from the rearrangement

$$y = 2x + c$$

that the line has a slope of 2 with a y intercept of c. Consequently, all of these lines are parallel to each other, their precise location being determined from the number c.

Now when $y = 0$ the equation reads

$$0 = 2x + c$$

and so has solution $x = -c/2$. Hence the line passes through the point $(-c/2, 0)$. A selection of lines is sketched in Figure 8.8 for values of c in the range 0 to −24. These have been sketched using the information that they pass through $(-c/2, 0)$ and have a slope of 2. Note that as c decreases from 0 to −24, the lines sweep across the feasible region from left to right. Also, once c goes below −24 the lines no longer intersect this region. The minimum value of c (which, you may remember, is just the value of the objective function) is therefore −24. Moreover, when $c = -24$, the line

$$-2x + y = c$$

intersects the feasible region in exactly one point, namely (12, 0). This then must be the solution of our problem. The point (12, 0) lies in the feasible region as required and because it also lies on the line

$$-2x + y = -24$$

we know that the corresponding value of the objective function is −24, which is the minimum value. Other points in the feasible region also lie on lines

$$-2x + y = c$$

but with larger values of c.

Practice Problem

3 Consider the linear programming problem

Maximize $-x + y$

subject to the constraints

$$3x + 4y \leq 12$$
$$x \geq 0$$
$$y \geq 0$$

(a) Sketch the feasible region.

(b) Sketch, on the same diagram, the five lines

$$y = x + c$$

for $c = -4, -2, 0, 1$ and 3.

[Hint: lines of the form $y = x + c$ have a slope of 1 and pass through the points $(0, c)$ and $(-c, 0)$.]

(c) Use your answers to part (b) to solve the given linear programming problem.

In the previous example, and in Practice Problem 3, the optimal value of the objective function is attained at one of the corners of the feasible region. This is not simply a coincidence. It can be shown that the solution of any linear programming problem always occurs at one of the corners. Consequently, the trial-and-error approach suggested earlier is not so naïve after all. The only possible candidates for the answer are the corners and so only a finite number of points need ever be examined. This method may be summarized:

Step 1

Sketch the feasible region.

Step 2

Identify the corners of the feasible region and find their coordinates.

Step 3

Evaluate the objective function at the corners and choose the one which has the maximum or minimum value.

Returning to the previous example, we work as follows:

Step 1

The feasible region has already been sketched in Figure 8.7.

Step 2

There are four corners with coordinates (0, 0), (0, 3), (2, 5) and (12, 0).

Step 3

Corner	Objective function
(0, 0)	$-2(0) + 0 = 0$
(0, 3)	$-2(0) + 3 = 3$
(2, 5)	$-2(2) + 5 = 1$
(12, 0)	$-2(12) + 0 = -24$

From this we see that the minimum occurs at (12, 0), at which the objective function is –24. Incidentally, if we also require the maximum then this can be deduced without further effort. From the table the maximum is 3, which occurs at (0, 3).

Example

Solve the linear programming problem

Maximize $5x + 3y$

subject to

$$2x + 4y \leq 8$$
$$x \geq 0$$
$$y \geq 0$$

Solution

Step 1

The non-negativity constraints $x \geq 0$ and $y \geq 0$ indicate that the region is bounded by the coordinate axes in the positive quadrant.

The line $2x + 4y = 8$ passes through (0, 2) and (4, 0). Also, at the test point (0, 0) the inequality

$$2x + 4y \leq 8$$

reads

$$0 \leq 8$$

which is true. We are therefore interested in the region below the line, since this region contains the test point, (0, 0). The feasible region is sketched in Figure 8.9 (overleaf).

Step 2

The feasible region is a triangle with three corners, (0, 0), (0, 2) and (4, 0).

Step 3

Corner	Objective function
(0, 0)	$5(0) + 3(0) = 0$
(0, 2)	$5(0) + 3(2) = 6$
(4, 0)	$5(4) + 3(0) = 20$

The maximum value of the objective function is 20, which occurs when $x = 4$ and $y = 0$.

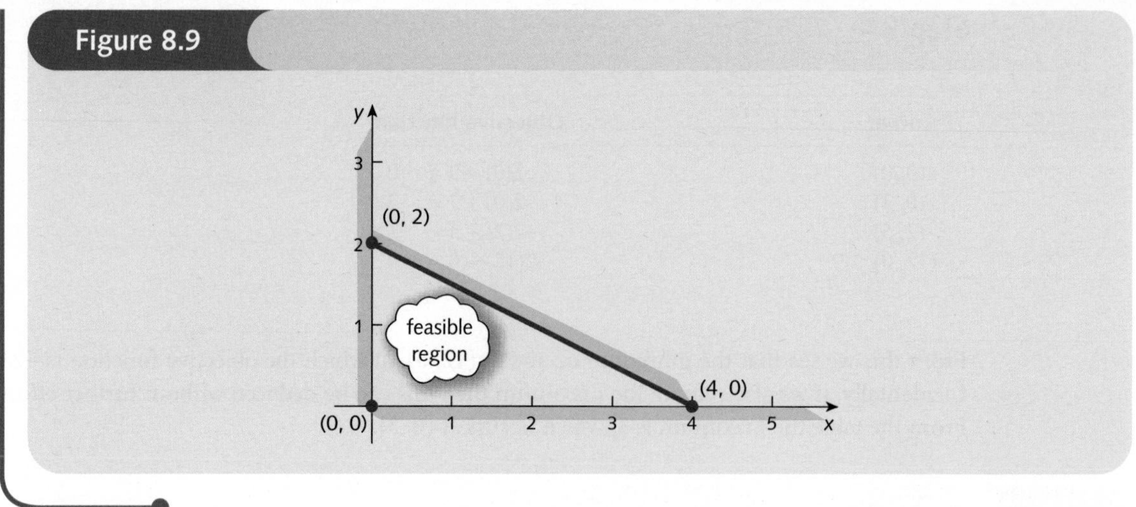

Practice Problems

4 Solve the linear programming problem

Minimize $x - y$

subject to

$$2x + y \leq 2$$
$$x \geq 0$$
$$y \geq 0$$

5 Solve the linear programming problem

Maximize $3x + 5y$

subject to

$$x + 2y \leq 10$$
$$3x + y \leq 10$$
$$x \geq 0$$
$$y \geq 0$$

[Hint: you might find your answer to Practice Problem 2 useful.]

In Section 1.2 we showed you how to solve a system of simultaneous linear equations. We discovered that such a system does not always have a unique solution. It is possible for a problem to have either no solution or infinitely many solutions. An analogous situation arises in linear programming. We conclude this section by considering two examples that illustrate these special cases.

Example

Solve the linear programming problem

Maximize $x + 2y$

subject to

$$2x + 4y \leq 8$$
$$x \geq 0$$
$$y \geq 0$$

Solution

Step 1

The feasible region is identical to the one sketched in Figure 8.9 for the previous worked example.

Step 2

As before, the feasible region has three corners, (0, 0), (0, 2) and (4, 0).

Step 3

Corner	Objective function
(0, 0)	$0 + 2(0) = 0$
(0, 2)	$0 + 2(2) = 4$
(4, 0)	$4 + 2(0) = 4$

This time, however, the maximum value is 4, which actually occurs at two corners, (0, 2) and (4, 0). This shows that the problem does not have a unique solution. To explain what is going on here we return to the method introduced at the beginning of this section. We superimpose the family of lines obtained by setting the objective function equal to some constant, c. The parallel lines

$$x + 2y = c$$

pass through the points $(0, c/2)$ and $(c, 0)$.

A selection of lines is sketched in Figure 8.10 (overleaf) for values of c between 0 and 4. These particular values are chosen since they produce lines that cross the feasible region. As c increases, the lines sweep across the region from left to right. Moreover, when c goes above 4 the lines no longer intersect the region. The maximum value that c (that is, the objective function) can take is therefore 4. However, instead of the line

$$x + 2y = 4$$

intersecting the region at only one point, it intersects along a whole line segment of points. Any point on the line joining the two corners (0, 2) and (4, 0) will be a solution. This follows because any point on this line segment lies in the feasible region and the corresponding value of the objective function on this line is 4, which is the maximum value.

Figure 8.10

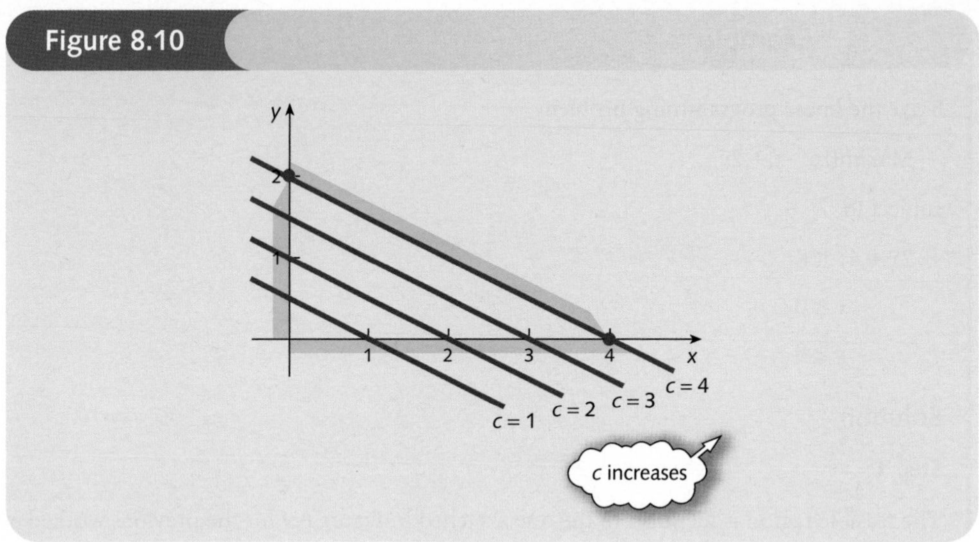

This example suggests the general result. If, in step 3, the maximum (or minimum) occurs at *two* corners then the problem has infinitely many solutions. Any point on the line segment joining these corners, including the two corners themselves, is also a solution.

Example

Solve the linear programming problem

Maximize $3x + 2y$

subject to

$$x + 4y \geq 8$$

$$x + y \geq 5$$

$$2x + y \geq 6$$

$$x \geq 0$$

$$y \geq 0$$

What can you say about the solution if this problem is one of minimization rather than maximization?

Solution

Step 1

As usual the non-negativity constraints indicate that we need only consider the positive quadrant.

The line $x + 4y = 8$ passes through (0, 2) and (8, 0).
The line $x + y = 5$ passes through (0, 5) and (5, 0).
The line $2x + y = 6$ passes through (0, 6) and (3, 0).

Also, the test point (0, 0) does not satisfy any of the corresponding constraints because the three inequality signs are all '≥'. We are therefore interested in the region *above* all of these lines, as shown in Figure 8.11.

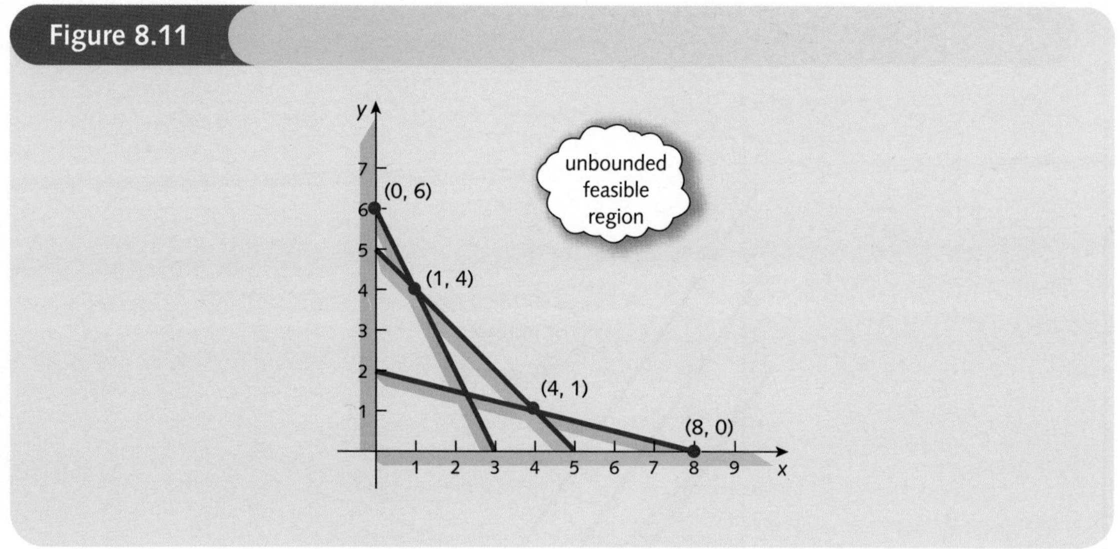

Step 2

The feasible region has four corners, (0, 6), (1, 4), (4, 1) and (8, 0).

Step 3

Corner	Objective function
(0, 6)	$3(0) + 2(6) = 12$
(1, 4)	$3(1) + 2(4) = 11$
(4, 1)	$3(4) + 2(1) = 14$
(8, 0)	$3(8) + 2(0) = 24$

From the table, the minimum and maximum values of the objective function are 11 and 24, which occur at (1, 4) and (8, 0) respectively. However, we do have a slightly unusual situation in that the feasible region is not enclosed on all sides. We describe this by saying that the feasible region is *unbounded*. It is open at the top and, strictly speaking, it does not make sense to talk about the corners of such a region. Are we therefore justified in applying the 'easy' method in this case? To answer this question we superimpose the family of lines

$$3x + 2y = c$$

representing the objective function, as shown in Figure 8.12 (overleaf).

When $c = 11$ the line intersects the region at only one point (1, 4). However, as c increases from this value, the lines sweep across the feasible region and never leave it, no matter how large c becomes. Consequently, if the problem is one of maximization we conclude that it does not have a finite solution. We can substitute huge values of x and y into $3x + 2y$ and get an ever-increasing result. On the other hand, if the problem is one of minimization then it does have a solution at the corner (1, 4). This, of course, is the answer obtained previously using the 'easy' method.

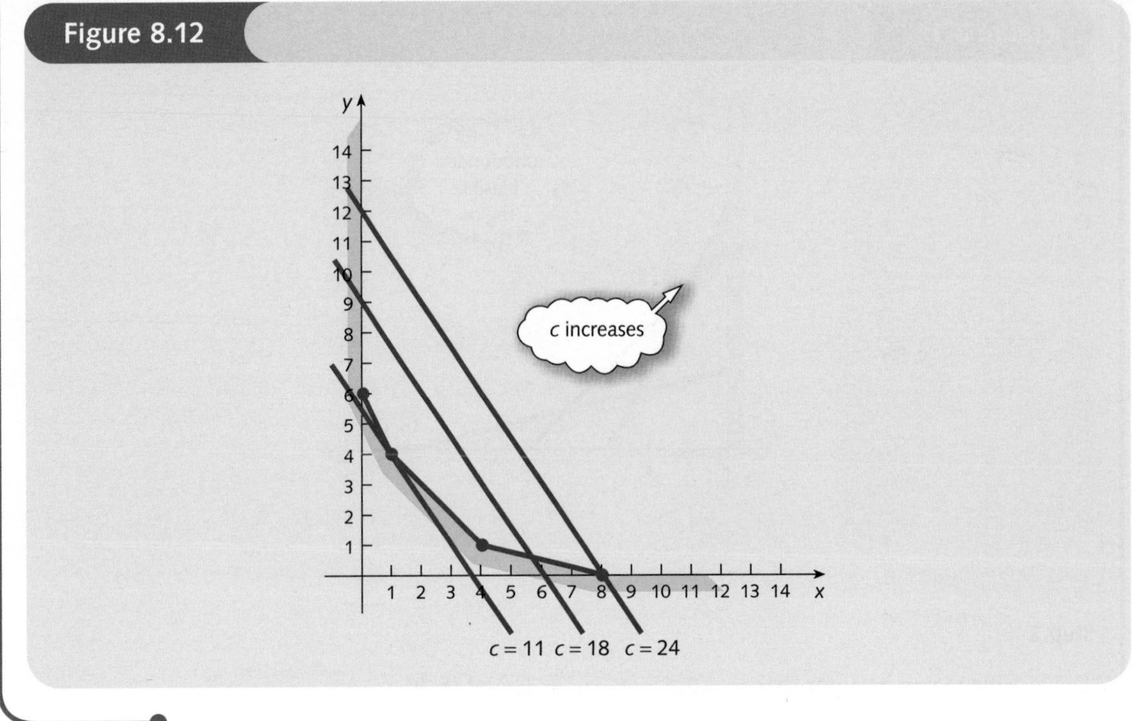

This example shows that a linear programming problem may not have a finite solution when the feasible region is unbounded. However, when a solution does exist, it may be found simply by inspecting the corners in the normal way. In practice, linear programming problems arise from realistic economic situations. We would therefore expect the problem to possess a sensible (that is, finite) answer and so the difficulty of the non-existence of a solution rarely occurs.

Key Terms

Feasible region The set of points that satisfy all of the constraints in a linear programming problem.

Non-negativity constraints The constraints, $x \geq 0$, $y \geq 0$, etc.

Objective function The function that is optimized in a linear programming problem.

Unbounded region A feasible region that is not completely enclosed by a polygon. The associated linear programming problem may not have a finite solution.

Practice Problems

6 Sketch the feasible regions defined by the following sets of inequalities:

(a) $5x + 3y \leq 30$	**(b)** $2x + 5y \leq 20$	**(c)** $x - 2y \leq 3$
$7x + 2y \leq 28$	$x + y \leq 5$	$x - y \leq 4$
$x \geq 0$	$x \geq 0$	$x \geq 1$
$y \geq 0$	$y \geq 0$	$y \geq 0$

7 Use your answers to Practice Problem 6 to solve the following linear programming problems.

(a) Maximize $4x + 9y$

subject to

$$5x + 3y \le 30$$
$$7x + 2y \le 28$$
$$x \ge 0$$
$$y \ge 0$$

(b) Maximize $3x + 6y$

subject to

$$2x + 5y \le 20$$
$$x + y \le 5$$
$$x \ge 0$$
$$y \ge 0$$

(c) Maximize $x + y$

subject to

$$x - 2y \le 3$$
$$x - y \le 4$$
$$x \ge 1$$
$$y \ge 0$$

8 What can you say about the solution to Practice Problem 7(c) if the problem is one of maximization rather than minimization? Explain your answer by superimposing the family of lines

$$x + y = c$$

on the feasible region.

9 Solve the following linear programming problems.

(a) Maximize $2x + 3y$

subject to

$$2x + y \le 8$$
$$x + y \le 6$$
$$x + 2y \le 10$$
$$x \ge 0$$
$$y \ge 0$$

(b) Maximize $-8x + 4y$

subject to

$$x - y \le 2$$
$$2x - y \ge -3$$
$$x - y \ge -4$$
$$x \ge 0$$
$$y \ge 0$$

10 Explain why each of the following problems fails to possess a solution.

(a) Maximize $x + y$

subject to

$$y \ge 2$$
$$x \le 2$$
$$x - y \le 1$$
$$x \ge 0$$
$$y \ge 0$$

(b) Maximize $x + y$

subject to

$$2x - y \ge -1$$
$$x - 2y \le 2$$
$$x \ge 0$$
$$y \ge 0$$

➔

11 Solve the linear programming problem

Maximize $6x + 2y$

subject to

$$\begin{aligned} x - y &\geq 0 \\ 3x + y &\geq 8 \\ x &\geq 0 \\ y &\geq 0 \end{aligned}$$

12 Show that the linear programming problem given in Practice Problem 7(a) can be expressed in matrix notation as

Maximize $\mathbf{c}^{\mathrm{T}}\mathbf{x}$

subject to

$$\begin{aligned} \mathbf{Ax} &\leq \mathbf{b} \\ \mathbf{x} &\geq \mathbf{0} \end{aligned}$$

where **c**, **x**, **b** and **0** are 2×1 matrices and **A** is a 2×2 matrix, which should be stated.

section 8.2

Applications of linear programming

Objectives

At the end of this section you should be able to:

- Identify the unknowns in a linear programming problem.
- Find an expression for the objective function and decide whether it should be maximized or minimized.
- Write down all of the constraints, including any obvious ones not mentioned explicitly in the problem specification.
- Solve linear programming problems expressed in words, remembering to check that the answer makes sense.

The impression possibly given so far is that linear programming is a mathematical technique designed to solve rather abstract problems. This is misleading since linear programming problems do arise from concrete situations. We now put the record straight by considering three realistic examples that lead naturally to such problems. In doing so we shall develop an important skill that can loosely be called problem formulation. Here we start with information, perhaps only vaguely given in words, and try to express it using the more precise language of mathematics. Once this has been done, it is a simple matter of applying mathematical techniques to produce the solution.

Example

A small manufacturer produces two kinds of good, A and B, for which demand exceeds capacity. The production costs for A and B are \$6 and \$3, respectively, each, and the corresponding selling prices are \$7 and \$4. In addition, the transport costs are 20 cents and 30 cents for each good of type A and B, respectively. The conditions of a bank loan limit the manufacturer to maximum weekly production costs of \$2700 and maximum weekly transport costs of \$120. How should the manufacturer arrange production to maximize profit?

Solution

As mentioned in Section 8.1, there are three things constituting a linear programming problem: a pair of unknowns x and y, an objective function that needs maximizing or minimizing, and some constraints. We consider each of these in turn.

The manufacturer has to decide exactly how many goods of types A and B to produce each week. These are therefore the unknowns of this problem and we denote these unknowns by the letters x and y: that is, we let

x = number of goods of type A to be made each week

y = number of goods of type B to be made each week

The final sentence of the problem states that the manufacturer should choose these quantities to maximize profit. Hence we need to find a formula for profit in terms of x and y. Now for each good of type A the production costs are \$6 and the transport costs are 20 cents. The total cost is therefore \$6.20. If the selling price is \$7, it follows that the profit made on a single item is 80 cents. Consequently, when x goods of type A are made the total profit is x times this amount, \$0.8x. Notice that the question states that 'demand exceeds capacity', so all goods are guaranteed to be sold. Exactly the same reasoning can be applied to B. The profit is 70 cents each, so when y goods of type B are made the total profit is \$0.7y. Hence the profit resulting from the production of both A and B is

$$0.8x + 0.7y$$

This then is the objective function that we want to maximize.

The next thing to do is to read through the original specification to see what restrictions are to be imposed on the production levels. We see that the total weekly production costs must not exceed \$2700. The production costs are \$6 for A and \$3 for B. Hence if x goods are made of type A and y goods are made of type B the total cost is

$$6x + 3y$$

so we require

$$6x + 3y \leq 2700$$

Similarly, the total cost of transporting the goods is

$$0.2x + 0.3y$$

and since this must not exceed \$120 we need

$$0.2x + 0.3y \leq 120$$

On the face of it there appear to be no further constraints given in the problem. However, a moment's thought should convince you that we are missing two important constraints, namely

$$x \geq 0 \quad \text{and} \quad y \geq 0$$

Although these are not mentioned explicitly, it is obvious that it is impossible to manufacture a negative number of goods.

Collecting all of the ingredients together, the linear programming problem may be stated:

Maximize $0.8x + 0.7y$

subject to

$$6x + 3y \leq 2700$$

$$0.2x + 0.3y \leq 120$$

$$x \geq 0$$

$$y \geq 0$$

The problem can now be solved using the method described in Section 8.1.

Step 1

As usual the non-negativity constraints indicate that we need only consider points in the positive quadrant.

The line $6x + 3y = 2700$ passes through (0, 900) and (450, 0).

The line $0.2x + 0.3y = 120$ passes through (0, 400) and (600, 0).

Also, using the origin as the test point reveals that the region of interest lies below both lines. It is sketched in Figure 8.13.

Step 2

The feasible region has four corners altogether, three of which have obvious coordinates (0, 0), (0, 400) and (450, 0). Unfortunately it is not at all easy to read off from the diagram the exact coordinates of the remaining corner. This is formed by the intersection of the two lines

$$6x + \quad 3y = 2700 \qquad (1)$$

$$0.2x + 0.3y = 120 \qquad (2)$$

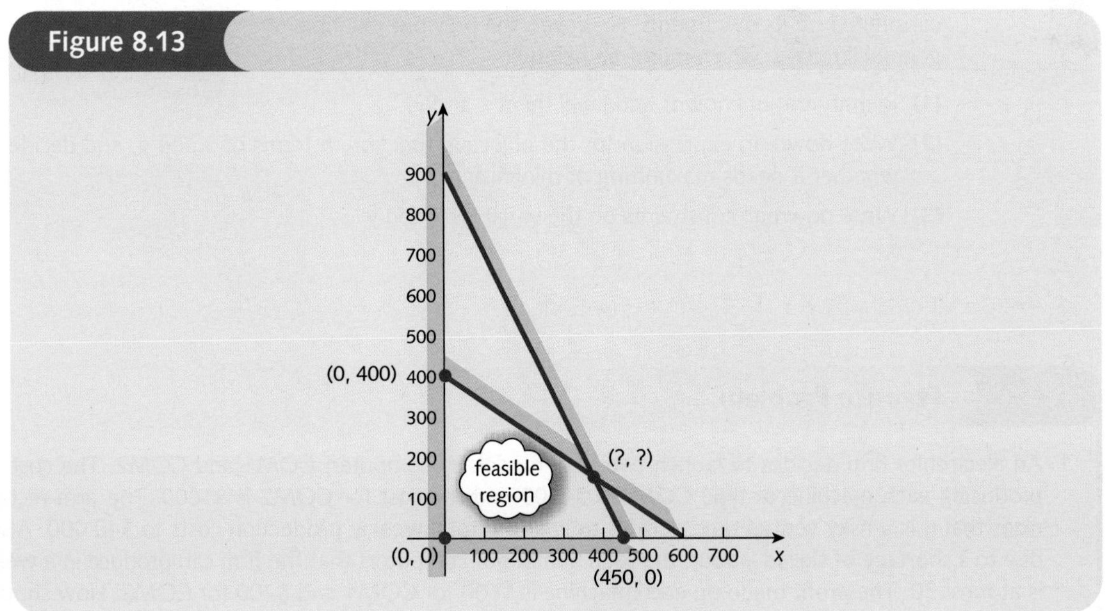

You may have encountered this difficulty when solving Practice Problem 8 in Section 8.1. If desired, we can always find the exact coordinates by treating the corresponding equations as a pair of simultaneous equations and solving them algebraically. The variable x can be eliminated by multiplying equation (2) by 30 and subtracting from (1) to get

$$\begin{aligned} 6x + 3y &= 2700 \\ 6x + 9y &= 3600 \\ \hline -6y &= -900 \end{aligned} \tag{3}$$

Equation (3) gives $y = 150$ and if this value is substituted into either of the original equations, it is easy to see that $x = 375$. The fourth corner therefore has coordinates (375, 150).

Step 3

Corner	Objective function
(0, 0)	0
(0, 400)	280
(450, 0)	360
(375, 150)	405

The maximum weekly profit is \$405, which occurs when 375 goods of type A and 150 goods of type B are manufactured.

Advice

It is impossible to give a precise description of problem formulation. Each case has to be considered on its own merits. However, the previous example does suggest the following general strategy, which might be helpful:

(1) Identify the unknowns and label them x and y.

(2) Write down an expression for the objective function in terms of x and y, and decide whether it needs maximizing or minimizing.

(3) Write down all constraints on the variables x and y.

Practice Problem

1 An electronics firm decides to launch two new models of computer, COM1 and COM2. The cost of producing each machine of type COM1 is \$1200 and the cost for COM2 is \$1600. The firm recognizes that it is a risky venture and decides to limit the total weekly production costs to \$40 000. Also, due to a shortage of skilled labour, the total number of computers that the firm can produce in a week is at most 30. The profit made on each machine is \$600 for COM1 and \$700 for COM2. How should the firm arrange production to maximize profit?

Example

A food producer uses two processing plants, P1 and P2 that operates 7 days a week. After processing, beef is graded into high-, medium- and low-quality foodstuffs. High-quality beef is sold to butchers, medium-quality beef is used in super-market ready-meals and the low-quality beef is used in dog food. The producer has contracted to provide 120 kg of high-, 80 kg of medium- and 240 kg of low-quality beef each week. It costs \$4000 per day to run plant P1 and \$3200 per day to run plant P2. Each day P1 processes 60 kg of high-quality beef, 20 kg of medium-quality beef and 40 kg of low-quality beef. The corresponding quantities for P2 are 20 kg, 20 kg and 120 kg, respectively. How many days each week should the plants be operated to fulfil the beef contract most economically?

Solution

The clue to the unknowns of this problem can be found in the final sentence 'How many days each week . . .'. We let

x = number of days each week that plant P1 is operated

y = number of days each week that plant P2 is operated

The objective function is harder to ascertain. The phrase 'to fulfil the contract most economically' is rather vague. It could mean that we want to maximize profit, as in the previous example. Unfortunately, insufficient information is given in the problem to determine profit, since we do not know the selling prices of the three grades of meat. We are, however, given the operating cost, so we take this to be the objective function, which then needs to be minimized. The daily costs for plants P1 and P2 are \$4000 and \$3200, respectively. Consequently, if plant P1 is operated for x days and plant P2 for y days then the total weekly cost is

$$4000x + 3200y$$

The remaining information is used to determine the constraints. The producer has contracted to provide 120 kg of high-quality beef each week. This means that *at least* this amount must be processed to fulfil

the contract. High-quality beef comes from two sources. Plant P1 processes 60 kg per day, whereas plant P2 processes 20 kg per day. Hence the total weekly output of high-quality beef is

$$60x + 20y$$

The contract is therefore satisfied provided

$$60x + 20y \geq 120$$

A similar argument holds for both medium- and low-quality beef. The corresponding constraints are

$$20x + 20y \geq 80$$

$$40x + 120y \geq 240$$

Looking back, it is easy to check that every piece of numerical information has been used in the formulation so far. However, there are still four further constraints to write down! These are based on common sense, but they do need to be built into the statement of the linear programming problem. The number of days in a week is 7, so the values of x and y must range between 0 and 7. Thus we have

$$x \geq 0 \quad \text{and} \quad y \geq 0$$

$$x \leq 7 \quad \text{and} \quad y \leq 7$$

The complete problem may now be stated:

$$\text{Minimize} \quad 4000x + 3200y$$

subject to

$$60x + 20y \geq 120$$

$$20x + 20y \geq 80$$

$$40x + 120y \geq 240$$

$$x \leq 7$$

$$y \leq 7$$

$$x \geq 0$$

$$y \geq 0$$

It can now be solved by applying the method described in Section 8.1.

Step 1

The feasible region can be sketched in the usual way. Note that the last four constraints merely indicate that the region is boxed in by the vertical and horizontal lines $x = 0$, $x = 7$, $y = 0$ and $y = 7$.

The line $60x + 20y = 120$ passes through (0, 6) and (2, 0).
The line $20x + 20y = 80$ passes through (0, 4) and (4, 0).
The line $40x + 120y = 240$ passes through (0, 2) and (6, 0).

Also the test point (0, 0) does not satisfy any of the first three constraints, so the feasible region lies above these lines, as shown in Figure 8.14.

Step 2

The feasible region has corners (7, 0), (7, 7), (0, 7), (1, 3), (3, 1), (0, 6) and (6, 0).

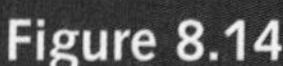

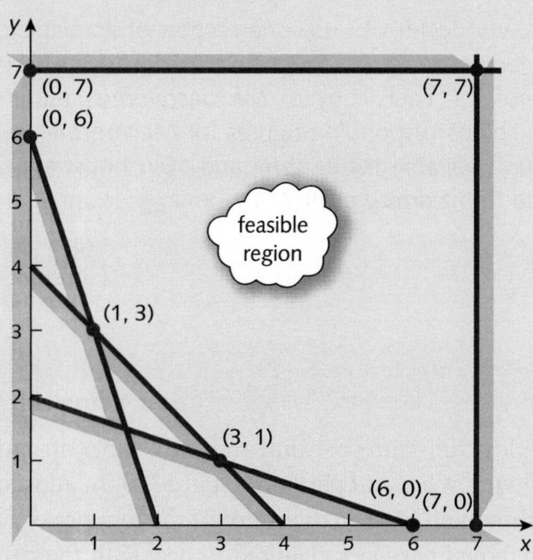

Step 3

Corner	Objective function
(7, 0)	28 000
(7, 7)	50 400
(0, 7)	22 400
(0, 6)	19 200
(3, 1)	15 200
(1, 3)	13 600
(6, 0)	24 000

The minimum cost is \$13 600 and is obtained by operating plant P1 for 1 day a week and plant P2 for 3 days a week.

Advice

Perhaps one of the more difficult aspects of formulating linear programming problems is remembering to include the obvious constraints arising from common sense. In the previous example, the last four constraints are so obvious that it is all too easy to omit them, particularly because they are not mentioned explicitly in the original specification. We might therefore include the following point in our general strategy:

(4) Write down any obvious constraints, such as the non-negativity constraints, that you may have forgotten about in (3).

Practice Problem

2 A small publishing company decides to use one section of its plant to produce two textbooks called *Microeconomics* and *Macroeconomics*. The profit made on each copy is \$12 for *Microeconomics* and \$18 for *Macroeconomics*. Each copy of *Microeconomics* requires 12 minutes for printing and 18 minutes for binding. The corresponding figures for *Macroeconomics* are 15 and 9 minutes respectively. There are 10 hours available for printing and $10\frac{1}{2}$ hours available for binding. How many of each should be produced to maximize profit?

Example

An insurance company employs full- and part-time staff, who work 40 and 20 hours per week respectively. Full-time staff are paid \$800 per week and part-time staff \$320. In addition, it is company policy that the number of part-time staff should not exceed one-third of the number of full-time staff.

If the number of worker-hours per week required to deal with the company's work is 900, how many workers of each type should be employed in order to complete the workload at minimum cost?

Solution

If the company employs x full-time staff and y part-time staff then the company would like to choose x and y to minimize its weekly salary costs. Also, since full- and part-time staff are paid \$800 and \$320 per week, respectively, the total wage bill is then

$$800x + 320y$$

which is the objective function that needs to be minimized.

Full- and part-time staff work 40 and 20 hours per week, respectively, so the total number of worker-hours available is

$$40x + 20y$$

It is required that this is at least 900, so we obtain the constraint

$$40x + 20y \geq 900$$

A further constraint on the company arises from the fact that the number of part-time staff cannot exceed one-third of the number of full-time staff. This means, for example, that if the company employs 30 full-time staff then it is not allowed to employ more than 10 part-time staff because

$$1/3 \times 30 = 10$$

In general, if x denotes the number of full-time staff then the number of part-time staff, y, cannot exceed $x/3$: that is,

$$y \leq x/3$$

In addition, we have the obvious non-negativity constraints

$$x \geq 0 \quad \text{and} \quad y \geq 0$$

The complete problem may now be stated:

$$\text{Minimize} \quad 800x + 320y$$

subject to

$$40x + 20y \geq 900$$

$$y \leq x/3$$

$$x \geq 0$$

$$y \geq 0$$

It can now be solved by applying the method described in Section 8.1.

Step 1

The feasible region can be sketched in the usual way. The line $y = x/3$ passes through (0, 0), (3, 1), (6, 2) and so on. Unfortunately, because the origin actually lies on the line, it is necessary to use some other point as a test point. For example, substituting $x = 30$, $y = 5$ into the inequality

$$y \leq x/3$$

gives

$$5 \leq (30)/3$$

This inequality is clearly true, indicating that (30, 5), which lies below the line, is in the region of interest. The constraint

$$40x + 20y \geq 900$$

is easier to handle. The corresponding line passes through (0, 45) and ($22\frac{1}{2}$, 0), and using the origin as a test point shows that we need to shade the region below the line. The feasible region is sketched in Figure 8.15.

Step 2

The feasible region has two corners. One of these is obviously ($22\frac{1}{2}$, 0). However, it is not possible to write down directly from the diagram the coordinates of the other corner. This is formed by the intersection of the two lines

Figure 8.15

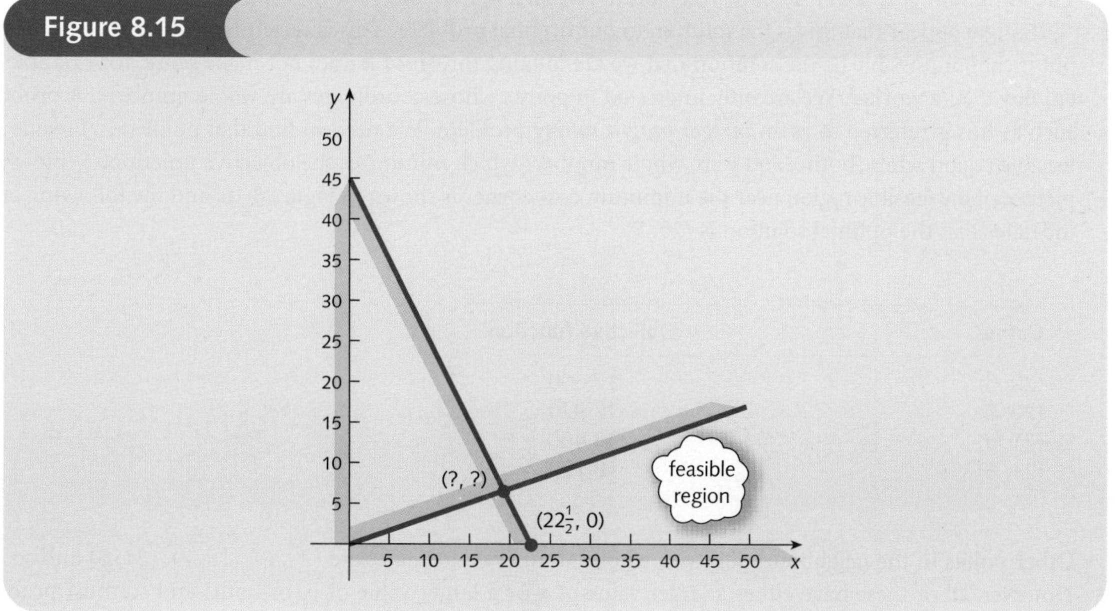

$$y = x/3 \quad (1)$$

$$40x + 20y = 900 \quad (2)$$

and so we must solve this system algebraically. In this case the easiest thing to do is to substitute equation (1) into equation (2) to eliminate y immediately. This gives

$$40x + \frac{20}{3}x = 900$$

that is,

$$\frac{140}{3}x = 900$$

which has solution

$$x = \frac{2700}{140} = \frac{135}{7} = 19^2/_7$$

Finally, from equation (1),

$$y = \frac{1}{3}x = \frac{1}{3} \times \frac{135}{7} = \frac{45}{7} = 6^3/_7$$

The feasible region therefore has coordinates ($19^2/_7$, $6^3/_7$) and ($22^1/_2$, 0).

Step 3

Corner	Objective function
($19^2/_7$, $6^3/_7$)	17 485$^5/_7$
($22^1/_2$, 0)	18 000

The minimum cost is \$17 485.71, which occurs when $x = 19^2/_7$ and $y = 6^3/_7$.

It might appear that this is the solution to our original problem. This is certainly mathematically correct, but it cannot possibly be the solution that we are looking for, since it does not make sense, for example, to employ $^2/_7$ of a worker. We are only interested in points whose coordinates are whole numbers. A problem such as this is referred to as an ***integer programming*** problem. We need to find that point (x, y) inside the feasible region where both x and y are whole numbers which minimizes the objective function. A 'blow-up' picture of the feasible region near the minimum cost corner is shown in Figure 8.16, and the following table indicates that the optimal solution is (20, 5).

Corner	Objective function
(20, 5)	17 600
(20, 6)	17 920
(21, 5)	18 400
(21, 6)	18 720

Other points in the neighbourhood with whole-number coordinates are (20, 6), (20, 7), (21, 5) and so on. However, all of these have either a larger value of x or a larger value of y (or both) and so must produce a larger total cost. The company should therefore employ 20 full- and 5 part-time staff to minimize its salary bill.

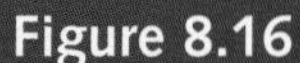

Figure 8.16

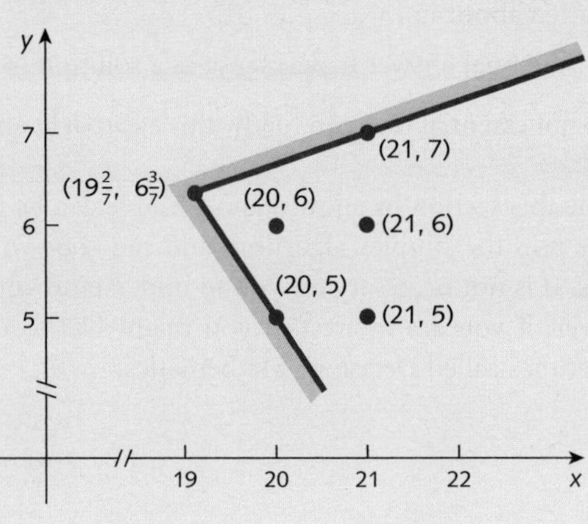

Advice

This example highlights the need to look back at the original problem to make sure that the final answer makes sense. It is very tempting when solving linear programming problems just to write down the solution without thinking, neatly underline it and then go on to another problem. Unfortunately, it is all to easy to make mistakes, both in the problem formulation and in sketching the feasible region. Spending a few moments checking the validity of your solution may well help you to discover any blunders that you have made, as well as suggesting possible modifications to the solution procedure, as in the previous example. We might, therefore, conclude the general strategy with the following step:

(5) Check that the final answer makes sense as a solution to the original problem.

Practice Problem

3 An individual spends 95% of earned income on essential goods and services, leaving only 5% to be spent on luxury goods, which is subdivided between trendy clothes and visits to the theatre. The cost of each item of clothing is \$150 and a trip to the theatre costs \$70. The corresponding utility function is

$$U = 3x + 7y$$

where x and y denote the number of trendy clothes and theatre visits per year, respectively. In order to maintain a reasonable appearance throughout the year, it is vital that at least nine new items of clothing are purchased each year. Given that annual earned income is \$42 000, find the values of x and y which maximize utility.

In this section we have described how to formulate linear programming problems. The general strategy may be summarized:

(1) Identify the unknowns and label them x and y.

(2) Write down an expression for the objective function in terms of x and y, and decide whether it needs maximizing or minimizing.

(3) Write down all constraints on the variables x and y.

(4) Write down any obvious constraints, such as the non-negativity constraints, that you may have forgotten about in (3).

(5) Check that the final answer makes sense as a solution to the original problem.

Obviously it is not essential that you follow this approach, although you may wish to refer to it if you get stuck.

We conclude this section by showing how Maple can be used to solve linear programming problems. This uses the simplex algorithm and can cope with problems involving more than two unknowns. It is not necessary for you to understand what the simplex algorithm is to use Maple. However, if you are interested, you might like to consult an academic textbook on Discrete (sometimes called Decision) Mathematics.

Example

MAPLE

A manufacturer produces three types of camera, X, Y and Z. Camera X requires 4 hours of staff time, 5 units of raw materials and 1 machine-hour to produce. The corresponding figures for camera Y are 3, 9 and 1, respectively. For Z, the figures are 3, 6 and 4, respectively. The profits earned from the sale of each camera of type X, Y and Z are \$400, \$300 and \$250, respectively. The total number of staff-hours available is 800, the number of units of raw materials available is 1100, and the number of machine-hours available is 400. In addition, because of limited demand, it is decided that no more than 100 cameras of any one type should be produced. How should the manufacturer arrange production to maximize profit?

Solution

In this problem a decision needs to be made about the number of each type of camera to produce. We shall denote these unknowns by x, y and z. The objective is to maximize profit, so we type

```
>profit:=400*x+300*y+250*z;
```

The next thing to do is to input the constraints. Production levels are limited by staff-hours, raw materials, machine-hours and demand. We consider each of these in turn and label the constraints `c1`, `c2`, `c3` and so on:

```
>c1:=4*x+3*y+3*z<=800;

>c2:=5*x+9*y+6*z<=1100;

>c3:=x+y+4*z<=400;

>c4:=x<=100;

>c5:=y<=100;

>c6:=z<=100;
```

in Maple the inequality ≤ is typed as <=

Maple has many routines based on the simplex method, which we load by typing

```
>with(simplex):
```

this command can end with a colon ':'

We are finally in a position to solve the problem, which we do by typing:

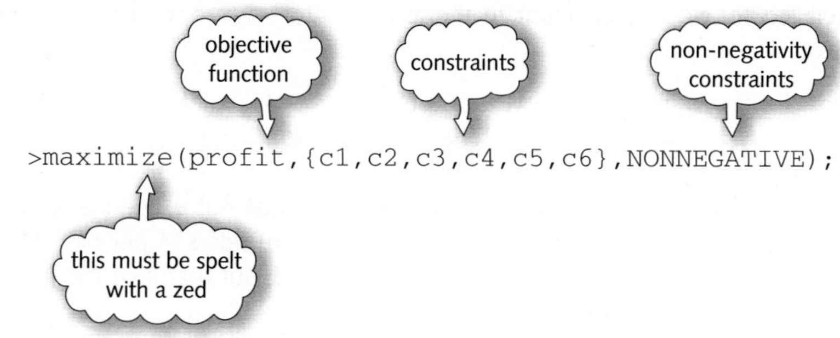

which gives

```
{x=52, y=100, z=62}
```

Profit is maximized by making 52 cameras of type X, 100 of type Y and 62 of type Z. The value of the profit itself can be found by typing

```
>subs({x=52,y=100,z=62},profit);
```

which gives $66 300.

Key Terms

Integer programming A linear programming problem in which the search for solution is restricted to points in the feasible region with whole-number coordinates.

Practice Problems

4 A manufacturer produces two models of racing bike, B and C, each of which must be processed through two machine shops. Machine shop 1 is available for 120 hours per month and machine shop 2 for 180 hours per month. The manufacture of each bike of type B takes 6 hours in shop 1 and 3 hours in shop 2. The corresponding times for C are 4 and 10 hours, respectively. If the profit is $180 and $220 per bike of type B and C respectively, how should the manufacturer arrange production to maximize total profit?

5 A small firm manufactures and sells litre cartons of non-alcoholic cocktails, 'The Caribbean' and 'Mr Fruity', which sell for $1 and $1.25, respectively. Each is made by mixing fresh orange, pineapple and apple juices in different proportions. The Caribbean consists of 1 part orange, 6 parts pineapple and 1 part apple. Mr Fruity consists of 2 parts orange, 3 parts pineapple and 1 part apple. The firm can buy up to 300 litres of orange juice, up to 1125 litres of pineapple juice and up to 195 litres of apple juice each week at a cost of $0.72, $0.64 and $0.48 per litre, respectively.

Find the number of cartons of 'The Caribbean' and 'Mr Fruity' that the firm should produce to maximize profits. You may assume that non-alcoholic cocktails are so popular that the firm can sell all that it produces.

6 In a student's diet a meal consists of beefburgers and chips. Beefburgers have 1 unit of nutrient N1, 4 units of N2 and 125 calories per ounce. The figures for chips are $\frac{1}{2}$ unit of N1, 1 unit of N2 and 60 calories per ounce. In the interests of the student's health it is essential for the meal to contain at least 7 units of N1 and 22 units of N2.

→

What should the student ask for on the next visit to the refectory to satisfy the nutrient requirements and minimize the number of calories?

7 A manufacturer of outdoor clothing makes wax jackets and trousers. Each jacket requires 1 hour to make, whereas each pair of trousers takes 40 minutes. The materials for a jacket cost \$32 and those for a pair of trousers cost \$40. The company can devote only 34 hours per week to the production of jackets and trousers, and the firm's total weekly cost for materials must not exceed \$1200. The company sells the jackets at a profit of \$12 each and the trousers at a profit of \$14 per pair. Market research indicates that the firm can sell all of the jackets that are produced, but that it can sell at most half as many pairs of trousers as jackets.

(a) How many jackets and trousers should the firm produce each week to maximize profit?

(b) Due to the changes in demand, the company has to change its profit margin on a pair of trousers. Assuming that the profit margin on a jacket remains at \$12 and the manufacturing constraints are unchanged, find the minimum and maximum profit margins on a pair of trousers which the company can allow before it should change its strategy for optimum output.

8 A farmer wishes to feed pigs with minimum cost but needs to ensure that each receives at least 1.6 kg of protein, at least 0.3 kg of amino acid and no more than 0.3 kg of calcium per day. Foods available are fish meal and meat scraps, which contain protein, calcium and amino acid according to the following table:

	Kg protein per kg feed	**Kg calcium per kg feed**	**Kg amino acid per kg feed**
Fish meal	0.60	0.05	0.18
Meat scraps	0.50	0.11	0.05

Fish meal costs \$0.65 per kg, whereas meat scraps cost \$0.52 per kg. Determine a minimum-cost feeding programme.

9 **(Maple)** A firm allocates resources L, K and R to the production of three goods X, Y and Z. The following table gives the amount of each input required to produce 1 unit of the given output:

			Output	
		X	***Y***	***Z***
	L	2	3	4
Input	*K*	1	5	2
	R	2	1	1

The value of each unit of output is \$10, \$5 and \$8 for X, Y and Z.

The resources available are 480, 300 and 180 for L, K and R.

Find the optimal resource allocation that maximizes the value of the total output.

10 **(Maple)** A farmer has two fields, the more fertile of which is 1400 hectares, the area of the other being 800 hectares. She grazes cattle and grows barley by fencing off areas in both fields. Some of the barley is kept for feed in the winter months. The profit per hectare is given by this table:

Year 1

	Large field	Small field
Barley	\$400	\$220
Cattle	\$350	\$200

Year 2

	Large field	Small field
Barley	\$420	\$240
Cattle	\$540	\$320

For winter feed alone she must allocate 1 hectare to barley for every 2 hectares allocated to cattle. Also the farmer finds that she cannot increase the area allocated to cattle by more than 20% from one year to the next.

Show that this can be formulated as a linear programming problem involving 8 unknowns and 15 constraints. Use the simplex method to solve this problem.

chapter 9
Dynamics

This chapter provides a simple introduction to the dynamics of economic systems. So far, all of our models have been static. We have implicitly assumed that equilibrium prices and incomes are somehow attained instantaneously, as if by magic. In practice, economic variables take time to vary. The incorporation of time into economic models is known as *dynamics* and it enables us to decide whether equilibrium values are actually achieved and, if so, exactly how individual variables approach these values. There are two sections that can be read in either order.

In Section 9.1 we consider the case when time, t, is a discrete variable taking whole-number values. This describes the situation in which variables change only after fixed periods. For example, the prices of certain agricultural products change from one season to the next but are fixed throughout each season. We express this time dependence using an obvious subscript notation. The price during the first period is denoted by P_1, the price during the second period is P_2, and so on. Equations that relate the price in one period, P_t, to that of the previous period, P_{t-1}, are called difference equations and a method is described for the solution of such equations.

In Section 9.2 we consider the case when time is a continuous variable taking all possible values in a certain interval. This describes the situation in which variables change from one moment to the next. For example, the prices of certain commodities, such as oil, effectively change instantaneously and are not fixed on a seasonal basis. We express this time dependence using the usual function notation, $P(t)$. It is appropriate to model the rate of change of P with respect to t using the derivative, $P'(t)$. Equations that involve the derivatives of an unknown function are called differential equations and a method is described for solving such equations.

It is not possible in an introductory book such as this to give you more than a flavour of the mathematics of dynamics. However, in spite of this, we show you how to solve dynamic systems in both macroeconomics and microeconomics. Also, we hope that this chapter will encourage you to read other books that describe more advanced methods and models.

section 9.1

Difference equations

Objectives

At the end of this section you should be able to:

- Find the complementary function of a difference equation.
- Find the particular solution of a difference equation.
- Analyse the stability of economic systems.
- Solve lagged national income determination models.
- Solve single-commodity-market models with lagged supply.

A *difference equation* (sometimes called a *recurrence relation*) is an equation that relates consecutive terms of a sequence of numbers. For example, the equation

$$Y_t = 2Y_{t-1}$$

describes sequences in which one number is twice its predecessor. There are obviously many sequences that satisfy this requirement, including

2, 4, 8, 16, . . .

5, 10, 20, 40, . . .

−1, −2, −4, −8, . . .

In order to determine the sequence uniquely, we need to be given some additional information, such as the first term. It is conventional to write the first term as Y_0, and once this is given a specific value all remaining terms are known.

Example

Write down the first four terms of the sequence defined by

$$Y_0 = 3 \quad \text{and} \quad Y_t = 2Y_{t-1}$$

and obtain a formula for the general term Y_t in terms of t.

Solution

If $Y_0 = 3$ then

$$Y_1 = 2Y_0 = 2 \times 3 = 6$$

$$Y_2 = 2Y_1 = 2 \times 6 = 12$$

$$Y_3 = 2Y_2 = 2 \times 12 = 24$$

In order to produce a formula for the general term, we write these as

$$Y_1 = 2Y_0 = 2^1 \times 3$$

$$Y_2 = 2Y_1 = 2^2 \times 3$$

$$Y_3 = 2Y_2 = 2^3 \times 3$$

It is now obvious from this pattern that the general term, given by

$$Y_t = 3(2^t)$$

is the solution of the difference equation

$$Y_t = 2Y_{t-1}$$

with *initial condition*

$$Y_0 = 3$$

The following problem gives you an opportunity to solve difference equations for yourself.

Practice Problem

1 Starting with the given initial conditions, write down the first four terms of each of the following sequences. By expressing these as an appropriate power, write down a formula for the general term, Y_t in terms of t.

(1) (a) $Y_t = 3Y_{t-1}$; $Y_0 = 1$ **(b)** $Y_t = 3Y_{t-1}$; $Y_0 = 7$ **(c)** $Y_t = 3Y_{t-1}$; $Y_0 = A$

(2) (a) $Y_t = \frac{1}{2}Y_{t-1}$; $Y_0 = 1$ **(b)** $Y_t = \frac{1}{2}Y_{t-1}$; $Y_0 = 7$ **(c)** $Y_t = \frac{1}{2}Y_{t-1}$; $Y_0 = A$

(3) $Y_t = bY_{t-1}$; $Y_0 = A$

The result of the last part of Practice Problem 1 shows that the solution of the general equation

$$Y_t = bY_{t-1} \tag{1}$$

with initial condition

$$Y_0 = A$$

is given by

$$Y_t = A(b^t)$$

Before we can consider the use of difference equations in economic models, we must examine the solution of more general equations of the form

$$Y_t = bY_{t-1} + c \tag{2}$$

where the right-hand side now includes a non-zero constant, c. We begin by defining some terminology. The *general solution* of equation (2) can be written as the sum of two separate expressions known as the complementary function (CF) and the particular solution (PS). The *complementary function* is the name that we give to the solution of equation (2) when the constant, c, is zero. In this case, equation (2) reduces to equation (1), and so

$$\text{CF} = A(b^t)$$

The *particular solution* is the name that we give to any solution of equation (2) that we are clever enough to 'spot'. This turns out to be rather easier to do than might at first appear and we will see how this can be done in a moment. Finally, once CF and PS have been found, we can write down the general solution of equation (2) as

$$Y_t = \text{CF} + \text{PS} = A(b^t) + \text{PS}$$

A proof of this result can be found in Practice Problem 8 at the end of this section. The letter A is no longer equal to the first term, Y_0, although it can easily be calculated, as the following example demonstrates.

Example

Solve the following difference equations with the specified initial conditions. Comment on the qualitative behaviour of the solution in each case.

(a) $Y_t = 4Y_{t-1} + 21$; $Y_0 = 1$

(b) $Y_t = \frac{1}{3}Y_{t-1} + 8$; $Y_0 = 2$

Solution

(a) The difference equation

$$Y_t = 4Y_{t-1} + 21$$

is of the standard form

$$Y_t = bY_{t-1} + c$$

and so can be solved using the complementary function and particular solution. The complementary function is the general solution of the equation when the constant term on the right-hand side is replaced by zero: that is, it is the solution of

$$Y_t = 4Y_{t-1}$$

which is $A(4^t)$.

The particular solution is any solution of the original equation

$$Y_t = 4Y_{t-1} + 21$$

that we are able to find. In effect, we need to think of a sequence of numbers, Y_t, such that when this is substituted into

$$Y_t - 4Y_{t-1}$$

we obtain the constant value of 21. One obvious sequence likely to work is a constant sequence,

$$Y_t = D$$

for some number, D. If this is substituted into

$$Y_t = 4Y_{t-1} + 21$$

we obtain

$$D = 4D + 21$$

(Note that $Y_t = D$ whatever the value of t so Y_{t-1} is also equal to D.) This algebraic equation can be re-arranged to get

$$-3D = 21$$

and so $D = -7$.

We have therefore shown that the complementary function is given by

$$\text{CF} = A(4^t)$$

and that the particular solution is

$$\text{PS} = -7$$

Hence

$$Y_t = \text{CF} + \text{PS} = A(4^t) - 7$$

which is the general solution of the difference equation

$$Y_t = 4Y_{t-1} + 21$$

To find the specific solution that satisfies the initial condition

$$Y_0 = 1$$

we simply put $t = 0$ in the general solution to get

$$Y_0 = A(4^0) - 7 = 1$$

that is,

$$A - 7 = 1$$

which gives

$$A = 8$$

The solution is

$$Y_t = 8(4^t) - 7$$

A graphical interpretation of this solution is shown in Figure 9.1 where Y_t is plotted against t. It is tempting to join the points up with a smooth curve. However, this does not make sense because t is allowed to take only whole-number values. Consequently, we join up the points with horizontal lines to create the 'staircase' which more properly reflects the fact that t is discrete. Figure 9.1 shows that the values of Y_t increase without bound as t increases. This is also apparent from the formula for Y_t because

Figure 9.1

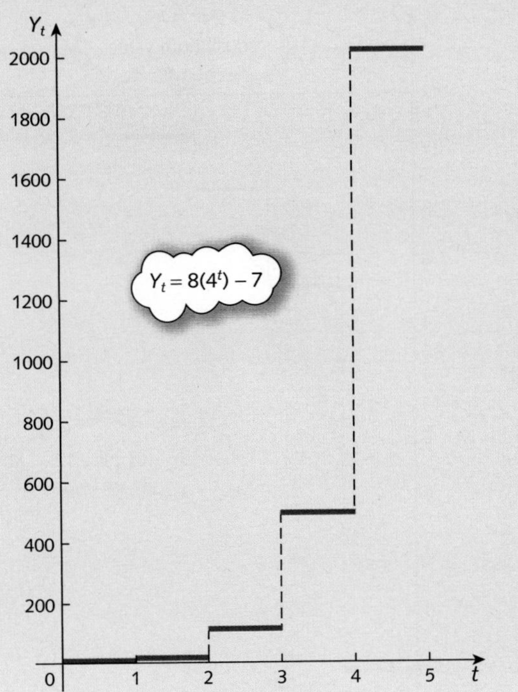

the numbers 4^t get ever larger as t increases. We describe this by saying that the time path *diverges uniformly* or *explodes*. This sort of behaviour can be expected to occur for any solution

$$Y_t = A(b^t) + \text{PS}$$

where $b > 1$.

(b) The difference equation

$$Y_t = \frac{1}{3}Y_{t-1} + 8$$

can be solved in a similar way to that of part (a). The complementary function is given by

$$\text{CF} = A\left(\frac{1}{3}\right)^t$$

and for a particular solution we try

$$Y_t = D$$

for some constant D. Substituting this into the difference equation gives

$$D = \frac{1}{3}D + 8$$

which has solution $D = 12$, so

$$\text{PS} = 12$$

The general solution is therefore

$$Y_t = \text{CF} + \text{PS} = A\left(\frac{1}{3}\right)^t + 12$$

➔

Figure 9.2

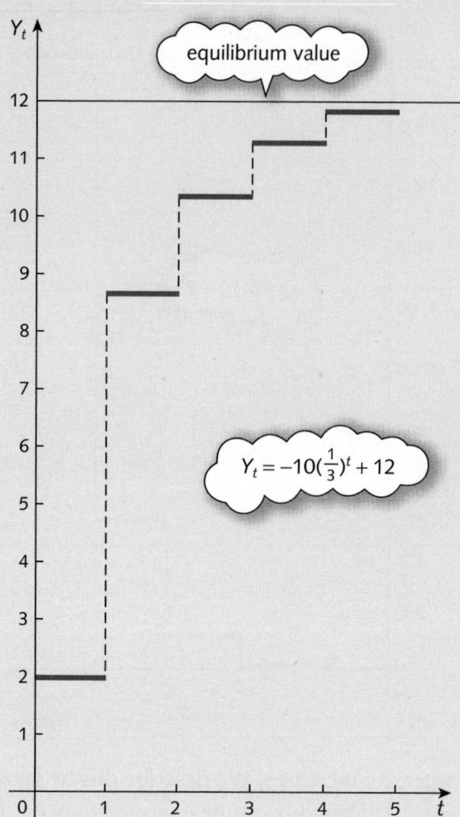

Finally, the specific value of A can be found from the initial condition

$$Y_0 = 2$$

Setting $t = 0$ in the general solution gives

$$2 = A\left(\frac{1}{3}\right)^0 + 12 = A + 12$$

and so A is -10. The solution is

$$Y_t = -10\left(\frac{1}{3}\right)^t + 12$$

This solution is sketched in Figure 9.2, which shows that the values of Y_t increase but eventually settle down at 12. We describe this by saying that the time path *converges uniformly* to the value of 12, which is referred to as the *equilibrium value*. This behaviour is also apparent from the formula for Y_t because the number $(1/3)^t$ gets ever smaller as t increases. In effect, the complementary function disappears, leaving just the particular solution. The particular solution is the equilibrium value of Y_t whereas the complementary function measures the deviation from the equilibrium which, in this case, converges to zero as t increases. This sort of behaviour can be expected to occur for any solution

$$Y_t = A(b^t) + \text{PS}$$

when $0 < b < 1$.

Practice Problem

2 Solve the following difference equations with the specified initial conditions:

(a) $Y_t = -\frac{1}{2}Y_{t-1} + 6;\ Y_0 = 0$ (b) $Y_t = -2Y_t + 9;\ Y_0 = 4$

In each case, sketch the corresponding 'staircase' diagram and comment on the qualitative behaviour of the solution as t increases.

The results of the previous example and Practice Problem 2 can be summarized:

If $b > 1$ then Y_t displays uniform divergence.
If $0 < b < 1$ then Y_t displays uniform convergence.
If $-1 < b < 0$ then Y_t displays oscillatory convergence.
If $b < -1$ then Y_t displays oscillatory divergence.

The remaining possibilities, $b = 1$, $b = -1$ and $b = 0$, are considered in Practice Problem 6 at the end of this section, which shows that Y_t converges when b is 0 but diverges when b is 1 or -1. We conclude that the solution of the difference equation eventually settles down to an equilibrium state only when b lies in the range $-1 < b < 1$.

If convergence does occur in an economic model, the model is said to be *stable*. If the variables diverge, it is said to be *unstable*.

We now investigate two applications of difference equations taken from macroeconomics and microeconomics, respectively:

- national income determination
- supply and demand analysis.

We consider each of these in turn.

9.1.1 National income determination

In Section 1.6 we introduced a simple two-sector model with structural equations

$$Y = C + I$$

$$C = aY + b$$

$$I = I^*$$

where b and I^* denote autonomous consumption and investment, and a is the marginal propensity to consume, which lies in the range $0 < a < 1$. In writing down the equations in this form, we are implicitly assuming that only one time period is involved, that consumption depends on national income within this time period and that equilibrium values are attained instantaneously. In practice, there is a time lag between consumption and national income. Consumption, C_t, in period t depends on national income, Y_{t-1}, in the previous period, $t - 1$. The corresponding consumption function is given by

$$C_t = aY_{t-1} + b$$

If we assume that investment is the same in all time periods then

$$I_t = I^*$$

for some constant, I^*. Finally, if the flow of money is in balance in each time period, we also have

$$Y_t = C_t + I_t$$

Substituting the expressions for C_t and I_t into this gives

$$Y_t = aY_{t-1} + b + I^*$$

which we recognize as a difference equation of the standard form given in this section. This equation can therefore be solved and the time path analysed.

Example

Consider a two-sector model:

$$Y_t = C_t + I_t$$

$$C_t = 0.8Y_{t-1} + 100$$

$$I_t = 200$$

Find an expression for Y_t when $Y_0 = 1700$. Is this system stable or unstable?

Solution

Substituting the expressions for C_t and I_t into

$$Y_t = C_t + I_t$$

gives

$$\begin{aligned} Y_t &= (0.8Y_{t-1} + 100) + 200 \\ &= 0.8Y_{t-1} + 300 \end{aligned}$$

The complementary function is given by

$$\text{CF} = A(0.8)^t$$

and for a particular solution we try

$$Y_t = D$$

for some constant D. Substituting this into the difference equation gives

$$D = 0.8D + 300$$

which has solution $D = 1500$. The general solution is therefore

$$Y_t = A(0.8)^t + 1500$$

The initial condition,

$$Y_0 = 1700$$

gives

$$1700 = A(0.8)^0 + 1500 = A + 1500$$

and so A is 200. The solution is

$$Y_t = 200(0.8)^t + 1500$$

As t increases, $(0.8)^t$ converges to zero and so Y_t eventually settles down at the equilibrium level of 1500. The system is therefore stable. Note also that because 0.8 lies between 0 and 1, the time path displays uniform convergence.

Practice Problem

3 Consider the two-sector model:

$$Y_t = C_t + I_t$$

$$C_t = 0.9Y_{t-1} + 250$$

$$I_t = 350$$

Find an expression for Y_t when $Y_0 = 6500$. Is this system stable or unstable?

In the previous example, and again in Practice Problem 3, we noted that the model is stable and that it displays uniform convergence. If we return to the general equation

$$Y_t = aY_{t-1} + b + I^*$$

it is easy to see that this is always the case for the simple two-sector model because the coefficient of Y_{t-1} is the marginal propensity to consume, which is known to lie between 0 and 1.

9.1.2 Supply and demand analysis

In Section 1.3 we introduced a simple model of supply and demand for a single good in an isolated market. If we assume that the supply and demand functions are both linear then we have the relations

$$Q_S = aP - b$$

$$Q_D = -cP + d$$

for some positive constants, a, b, c and d. (Previously, we have written P in terms of Q and have sketched the supply and demand curves with Q on the horizontal axis and P on the vertical axis. It turns out that it is more convenient in the present context to work the other way round and to write Q as a function of P.) In writing down these equations, we are implicitly assuming that only one time period is involved, that supply and demand are dependent only on the price in this time period, and that equilibrium values are attained instantaneously. However, for certain goods, there is a time lag between supply and price. For example, a farmer needs to decide precisely how much of any crop to sow well in advance of the time of sale. This decision is made on the basis of the price at the time of planting and not on the price prevailing at harvest time, which is unknown. In other words, the supply, Q_{St}, in period t depends on the price, P_{t-1}, in the preceding period $t - 1$. The corresponding time-dependent supply and demand equations are

$$Q_{St} = aP_{t-1} - b$$

$$Q_{Dt} = -cP_t + d$$

If we assume that, within each time period, demand and supply are equal, so that all goods are sold, then

$$Q_{Dt} = Q_{St}$$

that is,

$$-cP_t + d = aP_{t-1} - b$$

This equation can be rearranged as

$$-cP_t = aP_{t-1} - b - d \qquad \text{(subtract } d \text{ from both sides)}$$

$$P_t = \left(-\frac{a}{c}\right)P_{t-1} + \frac{b+d}{c} \qquad \text{(divide both sides by } -c\text{)}$$

which is a difference equation of the standard form. The equation can therefore be solved in the usual way and the time path analysed. Once a formula for P_t is obtained, we can use the demand equation

$$Q_t = -cP_t + d$$

to deduce a corresponding formula for Q_t by substituting the expression for P_t into the right-hand side.

Example

Consider the supply and demand equations

$$Q_{St} = 4P_{t-1} - 10$$

$$Q_{Dt} = -5P_t + 35$$

Assuming that the market is in equilibrium, find expressions for P_t and Q_t when $P_0 = 6$. Is the system stable or unstable?

Solution

If

$$Q_{Dt} = Q_{St}$$

then

$$-5P_t + 35 = 4P_{t-1} - 10$$

which rearranges to give

$$-5P_t = 4P_{t-1} - 45 \qquad \text{(subtract 35 from both sides)}$$

$$P_t = -0.8P_{t-1} + 9 \qquad \text{(divide both sides by } -5\text{)}$$

The complementary function is given by

$$\text{CF} = A(-0.8)^t$$

and for a particular solution we try

$$P_t = D$$

for some constant D. Substituting this into the difference equation gives

$$D = -0.8D + 9$$

which has solution $D = 5$. The general solution is therefore

$$P_t = A(-0.8)^t + 5$$

The initial condition, $P_0 = 6$, gives

$$6 = A(-0.8)^0 + 5 = A + 5$$

and so A is 1. The solution is

$$P_t = (-0.8)^t + 5$$

An expression for Q_t can be found by substituting this into the demand equation

$$Q_t = -5P_t + 35$$

to get

$$\begin{aligned} Q_t &= -5[(-0.8)^t + 5] + 35 \\ &= -5(-0.8)^t + 10 \end{aligned}$$

As t increases, $(-0.8)^t$ converges to zero and so P_t and Q_t eventually settle down at the equilibrium levels of 5 and 10 respectively. The system is therefore stable. Note also that because -0.8 lies between -1 and 0, the time paths display oscillatory convergence.

Practice Problems

4 Consider the supply and demand equations

$$Q_{St} = P_{t-1} - 8$$

$$Q_{Dt} = -2P_t + 22$$

Assuming equilibrium, find expressions for P_t and Q_t when $P_0 = 11$. Is the system stable or unstable?

5 Consider the supply and demand equations

$$Q_{St} = 3P_{t-1} - 20$$

$$Q_{Dt} = -2P_t + 80$$

Assuming equilibrium, find expressions for P_t and Q_t when $P_0 = 8$. Is the system stable or unstable?

Two features emerge from the previous example and Practice Problems 4 and 5. Firstly, the time paths are always oscillatory. Secondly, the system is not necessarily stable and so equilibrium might not be attained. These properties can be explained if we return to the general equation

$$P_t = \left(-\frac{a}{c}\right)P_{t-1} + \frac{b+d}{c}$$

The coefficient of P_{t-1} is $-a/c$. Given that a and c are both positive, it follows that $-a/c$ is negative and so oscillations will always be present. Moreover,

- if $a > c$ then $-a/c < -1$ and P_t diverges
- if $a < c$ then $-1 < -a/c < 0$ and P_t converges.

We conclude that stability depends on the relative sizes of a and c, which govern the slopes of the supply and demand curves. Bearing in mind that we have chosen to consider supply and demand equations in which Q is expressed in terms of P, namely

$$Q_S = aP - b$$

$$Q_D = -cP + d$$

we deduce that the system is stable whenever the supply curve is flatter than the demand curve when P is plotted on the horizontal axis.

Throughout this section we have concentrated on linear models. An obvious question to ask is whether we can extend these to cover the case of non-linear relationships. Unfortunately, the associated mathematics gets hard very quickly, even for mildly non-linear problems. It is usually impossible to find an explicit formula for the solution of such difference equations. Under these circumstances, we fall back on the tried and trusted approach of actually calculating the first few values until we can identify its behaviour. A spreadsheet provides an ideal way of doing this, since the parameters in the model can be easily changed.

Example — EXCEL

Consider the supply and demand equations

$$Q_{St} = P_{t-1}^{0.8}$$

$$Q_{Dt} = 12 - P_t$$

(a) Assuming that the market is in equilibrium, write down a difference equation for price.

(b) Given that $P_0 = 1$, find the values of the price, P_t for $t = 1, 2, \ldots, 10$ and plot a graph of P_t against t. Describe the qualitative behaviour of the time path.

Solution

(a) If

$$Q_{Dt} = Q_{St}$$

then

$$12 - P_t = P_{t-1}^{0.8}$$

which rearranges to give

$$P_t = 12 - P_{t-1}^{0.8}$$

Notice that this difference equation is not of the form considered in this section, so we cannot obtain an explicit formula for P_t in terms of t.

(b) We are given that $P_0 = 1$, so we can compute the values of $P_1, P_2, \ldots$ in turn. Setting $t = 1$ in the difference equation gives

$$\begin{aligned} P_1 &= 12 - P_0^{0.8} \\ &= 12 - 1^{0.8} \\ &= 11 \end{aligned}$$

This number can now be substituted into the difference equation, with $t = 2$, to get

$$P_2 = 12 - P_1^{0.8} = 12 - 11^{0.8} = 5.190\,52$$

and so on.

Figure 9.3

	A	B	C
1	Market Equilibrium		
2			
3			
4	Time	Price	
5	0	1	
6	1		
7	2		
8	3		
9	4		
10	5		
11	6		
12	7		
13	8		
14	9		
15	10		
16			
17			

Excel provides an easy way to perform the calculations. We type in the values, 0, 1, 2, . . . , 10 for each time period down the first column, type in the value of the initial price, P_0, in the second column, and then copy the relevant formula down the second column to generate successive values of P_t. Once this has been done, Chart Wizard can be used to draw the time path.

The most appropriate way of representing the results graphically is to use a bar chart. We would like the numbers in the first column of the spreadsheet to act as labels for the bars on the horizontal axis. Unfortunately, unless we tell Excel that we want to do this, it will actually produce two sets of bars on the same diagram, using the numbers in column A as heights for the first set of bars, and the numbers in column B as heights for the second set. This can be avoided by entering the values down the first column as text. This is done by first highlighting column A and then selecting **Format: Cells** from the menu bar. We choose the **Number** tab, and click on **Text** and **OK**. We can now finally enter the values of 0, 1, 2, . . . , 10 in the first column, together with the headings and numerical value of P_0 in column 2 as shown in Figure 9.3.

The remaining entries in column B are worked out using the difference equation

$$P_t = 12 - P_{t-1}^{0.8}$$

For example, the entry in cell B6 is P_1 which is calculated from

$$P_1 = 12 - P_0^{0.8}$$

The value of P_0 is located in cell B5, so we need to type the formula

=12–B5^0.8

into B6. Subsequent values are worked out in the same way, so all we need do is drag the formula down to B15 to complete the table.

Chart Wizard can now be used to plot this time path. We first highlight both columns and then click on Chart Wizard. The bar chart that we want is the one that is automatically displayed, so we just press the Finish button. To close the gaps between the bars, click on any one bar. A square dot will appear on each bar to indicate that all bars have been selected. We then choose **Format: Selected data series** for the menu bar. Finally, click on the **Options** tab, reduce the **Gap Width** to zero and click **OK**. The final spreadsheet is shown in Figure 9.4. It illustrates the oscillatory convergence and shows that the price eventually settles down to a value just greater than 7.

➔

Figure 9.4

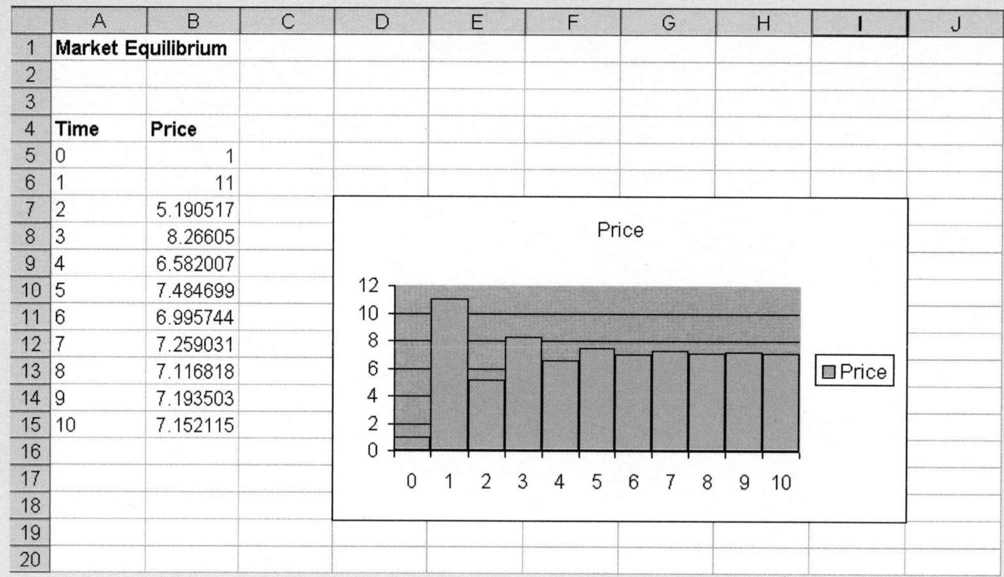

	A	B
1	**Market Equilibrium**	
2		
3		
4	**Time**	**Price**
5	0	1
6	1	11
7	2	5.190517
8	3	8.26605
9	4	6.582007
10	5	7.484699
11	6	6.995744
12	7	7.259031
13	8	7.116818
14	9	7.193503
15	10	7.152115
16		
17		
18		
19		
20		

Key Terms

Complementary function of a difference equation The solution of the difference equation, $Y_t = bY_{t-1} + c$ when the constant c is replaced by zero.

Difference equation An equation that relates consecutive terms of a sequence of numbers.

Dynamics Analysis of how equilibrium values vary over time.

Equilibrium value of a difference equation A solution of a difference equation that does not vary over time; it is the limiting value of Y_n as n tends to infinity.

General solution of a difference equation The solution of a difference equation that contains an arbitary constant. It is the sum of the complementary function and particular solution.

Initial condition The value of Y_0 that needs to be specified to obtain a unique solution of a difference equation.

Particular solution of a difference equation Any one solution of a difference equation such as $Y_t = bY_{t-1} + c$.

Recurrence relation An alternative phrase for a difference equation. It is an expression for Y_n in terms of Y_{n-1} (and possibly Y_{n-2}, Y_{n-3} etc).

Stable (unstable) equilibrium An economic model in which the solution of the associated difference equation converges (diverges).

Uniformly convergent sequence A sequence of numbers which progressively increases (or decreases) to a finite limit.

Uniformly divergent sequence A sequence of numbers which progressively increases (or decreases) without a finite limit.

Practice Problems

6 Calculate the first four terms of the sequences defined by the following difference equations. Hence write down a formula for Y_t in terms of t. Comment on the qualitative behaviour of the solution in each case.

(a) $Y_t = Y_{t-1} + 2;\ Y_0 = 0$ **(b)** $Y_t = -Y_{t-1} + 6;\ Y_0 = 4$ **(c)** $Y_t = 0Y_{t-1} + 3;\ Y_0 = 3$

7 Solve the following difference equations with the specified initial conditions:

(a) $Y_t = \frac{1}{4}Y_{t-1} + 6;\ Y_0 = 1$ **(b)** $Y_t = -4Y_{t-1} + 5;\ Y_0 = 2$

Comment on the qualitative behaviour of the solution as t increases.

8 Show, by substituting into the difference equation, that

$$Y_t = A(b^t) + D \quad \text{where} \quad D = \frac{c}{1-b}$$

is a solution of

$$Y_t = bY_{t-1} + c \quad (b \neq 1)$$

9 Consider the two-sector model:

$$Y_t = C_t + I_t$$

$$C_t = 0.7Y_{t-1} + 400$$

$$I_t = 0.1Y_{t-1} + 100$$

Given that $Y_0 = 3000$, find an expression for Y_t. Is this system stable or unstable?

10 Consider the supply and demand equations

$$Q_{St} = 0.4P_{t-1} - 12$$

$$Q_{Dt} = -0.8P_t + 60$$

Assuming that equilibrium conditions prevail, find an expression for P_t when $P_0 = 70$. Is the system stable or unstable?

11 The Harrod–Domar model of the growth of an economy is based on three assumptions.

(1) Savings, S_t, in any time period are proportional to income, Y_t, in that period, so that

$$S_t = \alpha Y_t \quad (\alpha > 0)$$

(2) Investment, I_t, in any time period is proportional to the change in income from the previous period to the current period so that

$$I_t = \beta(Y_t - Y_{t-1}) \quad (\beta > 0)$$

(3) Investment and savings are equal in any period so that

$$I_t = S_t$$

Use these assumptions to show that

$$Y_t = \left(\frac{\beta}{\beta - \alpha}\right)Y_{t-1}$$

and hence write down a formula for Y_t in terms of Y_0. Comment on the stability of the system in the case when $\alpha = 0.1$ and $\beta = 1.4$, and write down expressions for S_t and I_t in terms of Y_0. ➔

12 Consider the difference equation

$$Y_t = 0.1Y_{t-1} + 5(0.6)^t$$

(a) Write down the complementary function.

(b) By substituting $Y_t = D(0.6)^t$ into this equation, find a particular solution.

(c) Use your answers to parts (a) and (b) to write down the general solution and hence find the specific solution that satisfies the initial condition, $Y_0 = 9$.

(d) Is the solution in part (c) stable or unstable?

13 Consider the difference equation

$$Y_t = 0.2Y_{t-1} + 0.8t + 5$$

(a) Write down the complementary function.

(b) By substituting $Y_t = Dt + E$ into this equation, find a particular solution.

(c) Use your answers to parts (a) and (b) to write down the general solution and hence find the specific solution that satisfies the initial condition, $Y_0 = 10$.

(d) Is the solution in part (c) stable or unstable?

14 **(Excel)** Consider the two-sector model

$$C_t = 100 + 0.6Y_{t-1}^{0.8}$$

$$Y_t = C_t + 60$$

(a) Write down a difference equation for Y_t.

(b) Given that $Y_0 = 10$, calculate the values of Y_t for $t = 1, 2, \ldots 8$ and plot these values on a diagram. Is this system stable or unstable?

(c) Does the qualitative behaviour of the system depend on the initial value of Y_0?

15 **(Excel)** An economic growth model is based on three assumptions:

(1) Aggregate output, y_t, in time period t depends on capital stock, k_t, according to

$$y_t = k_t^{0.6}$$

(2) Capital stock in time period $t + 1$ is given by

$$k_{t+1} = 0.99k_t + s_t$$

where the first term reflects the fact that capital stock has depreciated by 1%, and the second term denotes the output that is saved during period t.

(3) Savings during period t are one-fifth of income so that

$$s_t = 0.2y_t$$

(a) Use these assumptions to write down a difference equation for k_t.

(b) Given that $k_0 = 7000$, find the equilibrium level of capital stock and state whether k_t displays uniform or oscillatory convergence. Do you get the same behaviour for other initial values of capital stock?

section 9.2

Differential equations

Objectives

At the end of this section you should be able to:

- Find the complementary function of a differential equation.
- Find the particular solution of a differential equation.
- Analyse the stability of economic systems.
- Solve continuous time national income determination models.
- Solve continuous time supply and demand models.

A *differential equation* is an equation that involves the derivative of an unknown function. Several examples have already been considered in Chapter 6. For instance, in Section 6.2 we noted the relationship

$$\frac{dK}{dt} = I$$

where K and I denote capital stock and net investment, respectively. Given any expression for $I(t)$, this represents a differential equation for the unknown function, $K(t)$. In such a simple case as this, we can solve the differential equation by integrating both sides with respect to t. For example, if $I(t) = t$ the equation becomes

$$\frac{dK}{dt} = t$$

and so

$$K(t) = \int t \, dt = \frac{t^2}{2} + c$$

where c is a constant of integration. The function $K(t)$ is said to be the *general solution* of the differential equation and c is referred to as an *arbitrary constant*. Some additional information is needed if the solution is to be pinned down uniquely. This is usually provided in the form of an *initial condition* in which we specify the value of K at $t = 0$. For example, the capital stock may be known to be 500 initially. Substituting $t = 0$ into the general solution

$$K(t) = \frac{t^2}{2} + c$$

gives

$$K(0) = \frac{0^2}{2} + c = 500$$

and so c is 500. Hence the solution is

$$K(t) = \frac{t^2}{2} + 500$$

In this section we investigate more complicated differential equations such as

$$\frac{dy}{dt} = 5y \quad \text{and} \quad \frac{dy}{dt} = -y + 3$$

The right-hand sides of these equations are given in terms of y rather than t and cannot be solved by direct integration. It turns out that the solution of such equations involves the exponential function, e^t which was first introduced in Section 2.4. The letter e denotes the number

2.718 28 . . .

and so e^t simply means 2.718 28 . . . raised to the power of t. A graph of e^t against t is sketched in Figure 9.5, which shows that e^t grows rapidly with rising t. In fact, the basic shape is the same for all exponential functions of the form

$$y = e^{mt}$$

when m is a positive constant. Figure 9.6 shows the graph of the negative exponential function, e^{-t}. The general behaviour of this function is to be expected because

$$e^{-t} = \frac{1}{e^t}$$

negative powers denote reciprocals

Figure 9.5

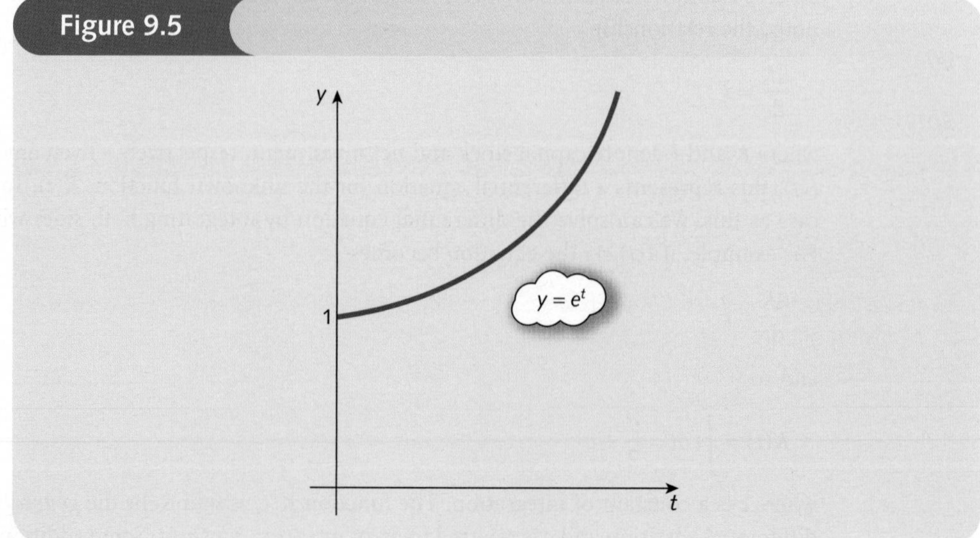

Figure 9.6

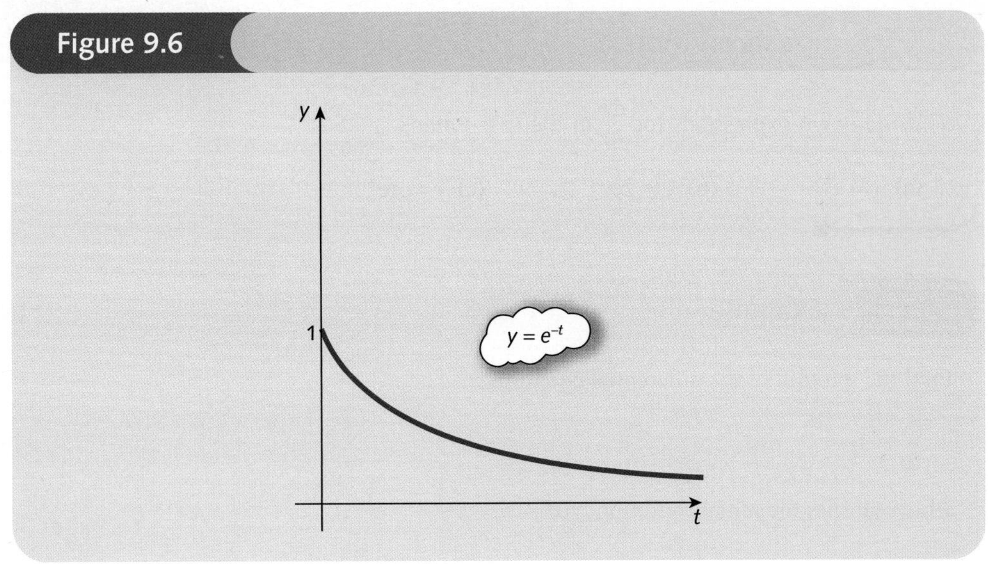

and, as t increases, the denominator increases, causing e^{-t} to converge to zero. The basic shape illustrated in Figure 9.6 is the same for all exponential functions of the form

$$y = e^{mt}$$

when m is a negative constant.

The most important property of the exponential function is that it differentiates to itself: that is,

$$\text{if} \quad y = e^t \quad \text{then} \quad \frac{dy}{dt} = e^t$$

Example

Write down expressions for $\frac{dy}{dt}$ in the case when

(a) $y = e^{3t}$ **(b)** $y = 7e^{-4t} - 9$

Solution

In Section 4.8 we showed that:

$$\text{if } y = e^{mt} \quad \text{then} \quad \frac{dy}{dt} = me^{mt}$$

so that e^{mt} differentiates to m times itself.

Applying this result to

(a) $y = e^{3t}$ gives $\frac{dy}{dt} = 3e^{3t}$

(b) $y = 7e^{-4t} - 9$ gives $\frac{dy}{dt} = -28e^{-4t}$

Practice Problem

1 Write down expressions for $\frac{dy}{dt}$ in the case when

(a) $y = e^{5t}$ **(b)** $y = 2e^{-3t}$ **(c)** $y = Ae^{mt}$

Example

Find the solution of the differential equation

$$\frac{dy}{dt} = 3y$$

which satisfies the initial condition, $y(0) = 5$.

Solution

The solution of the equation

$$\frac{dy}{dt} = 3y$$

is any function, $y(t)$, which differentiates to three times itself. We have noted that e^{mt} differentiates to m times itself, so an obvious candidate for the solution is

$$y = e^{3t}$$

However, there are many functions with the same property, including

$$y = 2e^{3t},\ y = 5e^{3t} \quad \text{and} \quad y = -7.52e^{3t}$$

Indeed, any function of the form

$$y = Ae^{3t}$$

satisfies this differential equation because

$$\frac{dy}{dt} = 3(Ae^{3t}) = 3y$$

The precise value of the constant A is determined from the initial condition

$$y(0) = 5$$

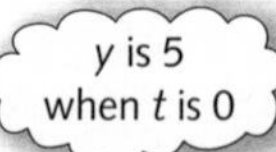

If we substitute $t = 0$ into the general solution

$$y(t) = Ae^{3t}$$

we get

$$y(0) = Ae^0 = A$$

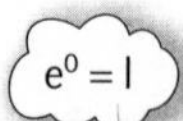

and so A is 5. The solution is

$$y(t) = 5e^{3t}$$

Practice Problem

2 **(a)** Find the solution of the differential equation

$$\frac{dy}{dt} = 4y$$

which satisfies the initial condition, $y(0) = 6$.

(b) Find the solution of the differential equation

$$\frac{dy}{dt} = -5y$$

which satisfies the initial condition, $y(0) = 2$.

Consider the differential equation

$$\frac{dy}{dt} = my + c \qquad (1)$$

where m and c are constants. The general solution of equation (1) is the sum of two separate functions, known as the complementary function (CF) and particular solution (PS). These are defined in much the same way as their counterparts for difference equations discussed in the previous section. The ***complementary function*** is the solution of equation (1) when the constant term on the right-hand side is replaced by zero. In other words, the complementary function is the solution of

$$\frac{dy}{dt} = my$$

The results of Practice Problem 2 show that this is given by

$$\text{CF} = Ae^{mt}$$

The ***particular solution*** is any solution that we are able to find of the original equation (1). This can be done by 'guesswork', just as we did in Section 9.1. Finally, once CF and PS have been determined, the general solution of equation (1) can be written down as

$$y = \text{CF} + \text{PS} = Ae^{mt} + \text{PS}$$

As usual, the specific value of A can be worked out at the very end of the calculations via an initial condition.

Example

Solve the differential equation

$$\frac{dy}{dt} = -2y + 100$$

in the case when the initial condition is

(a) $y(0) = 10$ **(b)** $y(0) = 90$ **(c)** $y(0) = 50$

Comment on the qualitative behaviour of the solution in each case.

Solution

The differential equation

$$\frac{dy}{dt} = -2y + 100$$

is of the standard form

$$\frac{dy}{dt} = my + c$$

and so can be solved using the complementary function and particular solution.

The complementary function is the general solution of the equation when the constant term is taken to be zero: that is, it is the solution of

$$\frac{dy}{dt} = -2y$$

which is Ae^{-2t}. The particular solution is any solution of the original equation

$$\frac{dy}{dt} = -2y + 100$$

that we are able to find. In effect, we need to think of a function, $y(t)$, such that when it is substituted into

$$\frac{dy}{dt} + 2y$$

we obtain the constant value of 100. One obvious function likely to work is a constant function,

$$y(t) = D$$

for some constant D. If this is substituted into

$$\frac{dy}{dt} = -2y + 100$$

we obtain

$$0 = -2D + 100$$

(Note that $dy/dt = 0$ because constants differentiate to zero.) This algebraic equation can be rearranged to get

$$2D = 100$$

and so $D = 50$.

We have therefore shown that the complementary function is given by

$$\text{CF} = Ae^{-2t}$$

and that the particular solution is

$$\text{PS} = 50$$

Hence

$$y(t) = \text{CF} + \text{PS} = Ae^{-2t} + 50$$

This is the general solution of the differential equation

$$\frac{dy}{dt} = -2y + 100$$

Figure 9.7

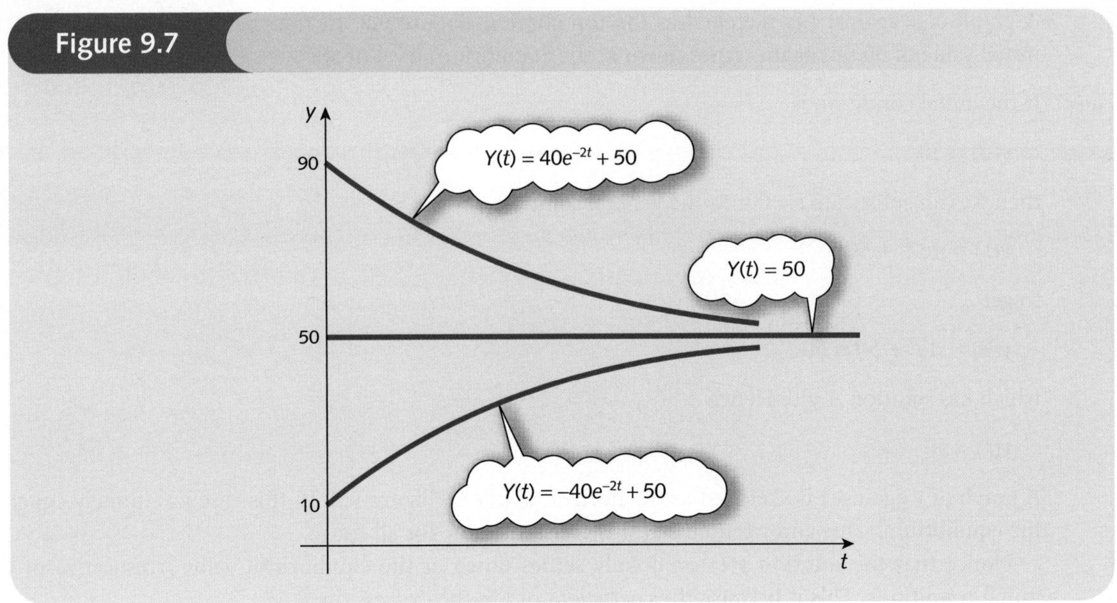

(a) To find the specific solution that satisfies the initial condition

$$y(0) = 10$$

we simply put $t = 0$ into the general solution to get

$$y(0) = Ae^0 + 50 = 10$$

that is

$$A + 50 = 10$$

which gives

$$A = -40$$

The solution is

$$y(t) = -40e^{-2t} + 50$$

A graph of y against t is sketched as the bottom graph in Figure 9.7. This shows that $y(t)$ increases from its initial value of 10 and settles down at the value of 50 for sufficiently large t. As usual, this limit is called the *equilibrium value* and is equal to the particular solution. The complementary function measures the deviation from the equilibrium.

(b) If the initial condition is

$$y(0) = 90$$

then we can substitute $t = 0$ into the general solution

$$y(t) = Ae^{-2t} + 50$$

to get

$$y(0) = Ae^0 + 50 = 90$$

which has solution $A = 40$. Hence

$$y(t) = 40e^{-2t} + 50$$

➔

A graph of y against t is sketched as the top curve in Figure 9.7. In this case $y(t)$ decreases from its initial value of 90 but again settles down at the equilibrium level of 50.

(c) If the initial condition is

$$y(0) = 50$$

then we can substitute $t = 0$ into the general solution

$$y(t) = Ae^{-2t} + 50$$

to get

$$y(0) = Ae^{0} + 50 = 50$$

which has solution $A = 0$. Hence

$$y(t) = 50$$

A graph of y against t is sketched as the horizontal line in Figure 9.7. In this case y is initially equal to the equilibrium value and y remains at this constant value for all time.

Notice that the solution $y(t)$ eventually settles down at the equilibrium value irrespective of the initial conditions. This is because the coefficient of t in the expression

$$\text{CF} = Ae^{-2t}$$

is negative, causing CF to converge to zero as t increases. We would expect convergence to occur for any solution

$$y(t) = Ae^{mt} + D$$

when $m < 0$.

Practice Problem

3 Solve the following differential equation subject to the given initial condition. Comment on the qualitative behaviour of the solution as t increases.

$$\frac{dy}{dt} = 3y - 60; \quad y(0) = 30$$

The results of the previous example and Practice Problem 3 can be summarized as:

- if $m < 0$ then $y(t)$ converges
- if $m > 0$ then $y(t)$ diverges.

We say that an economic model is ***stable*** whenever the variables converge as t increases. The above results indicate that an economic system represented by

$$\frac{dy}{dt} = my + c$$

is stable if the coefficient of y is negative and unstable if it is positive. Of course, it could happen that m is zero. The differential equation then becomes

$$\frac{dy}{dt} = c$$

which can be integrated directly to get

$$y(t) = \int c\,dt = ct + d$$

for some arbitrary constant d. The corresponding model is therefore unstable unless c is also zero, in which case $y(t)$ takes the constant value of d for all t.

We now investigate two applications of differential equations taken from macroeconomics and microeconomics respectively:

- national income determination
- supply and demand analysis.

We consider each of these in turn.

9.2.1 National income determination

The defining equations of the usual two-sector model are

$$Y = C + I \tag{1}$$

$$C = aY + b \tag{2}$$

$$I = I^* \tag{3}$$

The first of these is simply a statement that the economy is already in balance. The left-hand side of equation (1) is the flow of money from firms to households given as payment for the factors of production. The right-hand side is the total flow of money received by firms, either in the form of investment, or as payment for goods bought by households. In practice, the equilibrium values are not immediately attained and we need to make an alternative assumption about how national income varies with time. It seems reasonable to suppose that the rate of change of Y is proportional to the excess expenditure, $C + I - Y$: that is,

$$\frac{dY}{dt} = \alpha(C + I - Y) \tag{1$'$}$$

for some positive ***adjustment coefficient***, α. This makes sense because

- if $C + I > Y$, it gives $dY/dt > 0$ and so Y rises in order to achieve a balance between expenditure and income
- if $C + I = Y$, it gives $dY/dt = 0$ and so Y is held constant at the equilibrium level
- if $C + I < Y$, it gives $dY/dt < 0$ and so Y falls in order to achieve a balance between expenditure and income.

The usual relations (2) and (3) can be substituted into the new equation (1′) to obtain

$$\frac{dY}{dt} = \alpha(aY + b + I^* - Y)$$

$$= \alpha(a - 1)Y + \alpha(b + I^*)$$

which we recognize as a differential equation of the standard form given in this section.

Example

Consider the two-sector model

$$\frac{dY}{dt} = 0.5(C + I - Y)$$

$$C = 0.8Y + 400$$

$$I = 600$$

Find an expression for $Y(t)$ when $Y(0) = 7000$. Is this system stable or unstable?

Solution

Substituting the expressions for C and I into

$$\frac{dY}{dt} = 0.5(C + I - Y)$$

gives

$$\begin{aligned}\frac{dY}{dt} &= 0.5(0.8Y + 400 + 600 - Y)\\ &= -0.1Y + 500\end{aligned}$$

The complementary function is given by

$$\text{CF} = Ae^{-0.1t}$$

and for a particular solution we try

$$Y(t) = D$$

for some constant, D. Substituting this into the differential equation gives

$$0 = -0.1D + 500$$

which has solution $D = 5000$. The general solution is therefore

$$Y(t) = Ae^{-0.1t} + 5000$$

The initial condition

$$Y(0) = 7000$$

gives

$$A + 5000 = 7000$$

and so A is 2000. The solution is

$$Y(t) = 2000e^{-0.1t} + 5000$$

The first term is a negative exponential, so it converges to zero as t increases. Consequently, $Y(t)$ eventually settles down to an equilibrium value of 5000 and the system is stable.

Practice Problem

4 Consider the two-sector model

$$\frac{dY}{dt} = 0.1(C + I - Y)$$

$$C = 0.9Y + 100$$

$$I = 300$$

Find an expression for $Y(t)$ when $Y(0) = 2000$. Is this system stable or unstable?

In the previous example and again in Practice Problem 4 we noted that the macroeconomic system is stable. If we return to the general equation

$$\frac{dY}{dt} = \alpha(a - 1)Y + \alpha(b + I^*)$$

it is easy to see that this is always the case for the simple two-sector model, since the coefficient of Y is negative. This follows because, as previously stated, $\alpha > 0$ and because the marginal propensity to consume, a, is less than 1.

9.2.2 Supply and demand analysis

The equations defining the usual linear single-commodity market model are

$$Q_S = aP - b \tag{1}$$

$$Q_D = -cP + d \tag{2}$$

for some positive constants a, b, c and d. As in Section 9.1, we have written Q in terms of P for convenience. Previously, we have calculated the equilibrium price and quantity simply by equating supply and demand: that is, by putting

$$Q_S = Q_D$$

In writing down this relation, we are implicitly assuming that equilibrium is immediately attained and, in doing so, we fail to take into account the way in which this is achieved. A reasonable assumption to make is that the rate of change of price is proportional to excess demand, $Q_D - Q_S$: that is,

$$\frac{dP}{dt} = \alpha(Q_D - Q_S) \tag{3}$$

for some positive adjustment coefficient, α. This makes sense because

- if $Q_D > Q_S$ it gives $dP/dt > 0$ and so P increases in order to achieve a balance between supply and demand
- if $Q_S = Q_D$ it gives $dP/dt = 0$ and so P is held constant at the equilibrium level
- if $Q_D < Q_S$ it gives $dP/dt < 0$ and so P decreases in order to achieve a balance between supply and demand.

Substituting equations (1) and (2) into equation (3) gives

$$\frac{dP}{dt} = \alpha[(-cP + d) - (aP - b)] = -\alpha(a + c)P + \alpha(d + b)$$

which is a differential equation of the standard form.

Example

Consider the market model

$$Q_S = 3P - 4$$

$$Q_D = -5P + 20$$

$$\frac{dP}{dt} = 0.2(Q_D - Q_S)$$

Find expressions for $P(t)$, $Q_S(t)$ and $Q_D(t)$ when $P(0) = 2$. Is this system stable or unstable?

Solution

Substituting the expressions for Q_D and Q_S into

$$\frac{dP}{dt} = 0.2(Q_D - Q_S)$$

gives

$$\frac{dP}{dt} = 0.2[(-5P + 20) - (3P - 4)] = -1.6P + 4.8$$

The complementary function is given by

$$CF = Ae^{-1.6t}$$

and for a particular solution we try

$$P(t) = D$$

for some constant D. Substituting this into the differential equation gives

$$0 = -1.6D + 4.8$$

which has solution $D = 3$. The general solution is therefore

$$P(t) = Ae^{-1.6t} + 3$$

The initial condition

$$P(0) = 2$$

gives

$$A + 3 = 2$$

and so A is -1. The solution is

$$P(t) = -e^{-1.6t} + 3$$

Corresponding expressions for $Q_S(t)$ and $Q_D(t)$ can be found from the supply and demand equations, which give

$$Q_S(t) = 3P - 4 = 3(-e^{-1.6t} + 3) - 4 = -3e^{-1.6t} + 5$$

$$Q_D(t) = -5P + 20 = -5(-e^{-1.6t} + 3) + 20 = 5e^{-1.6t} + 5$$

Note that all three expressions involve a negative exponential that converges to zero as t increases, so the system is stable. The price $P(t)$ eventually settles down to the equilibrium price, 3, and $Q_S(t)$ and $Q_D(t)$ both approach the equilibrium quantity, 5.

Practice Problem

5 Consider the market model

$$Q_S = 2P - 2$$

$$Q_D = -P + 4$$

$$\frac{dP}{dt} = \frac{1}{3}(Q_D - Q_S)$$

Find expressions for $P(t)$, $Q_S(t)$ and $Q_D(t)$ when $P(0) = 1$. Is this system stable or unstable?

In the previous example and again in Practice Problem 5 we noted that the single-commodity market model is stable. If we return again to the general equation

$$\frac{dP}{dt} = -\alpha(a + c)P + \alpha(d + b)$$

it is easy to see that this is always the case, since the coefficient of P is negative. This follows because, as previously stated, α, a and c are all positive.

Maple offers a range of facilities for investigating the behaviour of differential equations. The basic instruction is `dsolve` which can be used either with or without an initial condition. Let us suppose that we wish to solve a differential equation involving dy/dt, that is, we want to find a solution in which y is expressed as a function of t. If the equation itself is typed into Maple and named `eq1`, say, and the initial condition is named as `init`, the solution $y(t)$ is generated by typing

```
>dsolve({eq1,init},{y(t)});
```

We illustrate this in the following example.

Example MAPLE

The value of an investment portfolio, $y(t)$, varies over an 8-year period according to

$$\frac{dy}{dt} = y(1 - y) \quad (0 \le t \le 8)$$

where y is measured in millions of dollars and t is measured in years. Find expressions for the solution of this equation in the case when the initial investment is

(a) \$100 000 **(b)** \$2 million

Plot the graphs of these solutions on the same diagram and comment on the stability of the system.

Solution

Before we solve this equation using Maple, there are two special cases worthy of note. If, at some time, t, the value of the portfolio drops to zero, so that $y = 0$, then the right-hand side of the differential equation will also be zero. The differential equation then reads

$$\frac{dy}{dt} = 0$$

It follows that y is constant for all subsequent times. In other words, once the investment falls to zero, it remains at this level.

The other special case is when $y(t) = 1$. Again the right-hand side, $y(1 - y)$, reduces to zero and again the differential equation becomes just

$$\frac{dy}{dt} = 0$$

We deduce that once the value of the investment reaches 1 million, it remains at this level from that moment onwards.

(a) To obtain the general solution using Maple we need to type in, and name, both the differential equation and the initial condition. Unfortunately you cannot just write dy/dt and hope that Maple will recognize this as a derivative. Instead we make use of the command `diff` which we used in Chapters 4 and 5. The first-order derivative of y with respect to t is written as `diff(y(t),t)` and the differential equation itself is specified by typing

```
>eq:=diff(y(t),t)=y(t)*(1-y(t));
```

Maple displays this as

$$eq := \frac{\partial}{\partial t}y(t) = y(t)(1 - y(t))$$

The initial condition, $y(0) = 0.1$, is entered, and named `init` by typing

```
>init:=y(0)=0.1;
```

We are now in a position to obtain the solution using

```
>dsolve({eq,init},{y(t)});
```

which gives

$$y(t) = \frac{1}{1 + 9e^{(-t)}}$$

(b) The initial condition, $y(0) = 2$, is easily input by editing the line defining `init` to give

```
>init:=y(0)=2;
```

The problem can then be re-solved by moving the cursor to the end of each line of instructions and pressing the Enter key. The new solution is

Figure 9.8

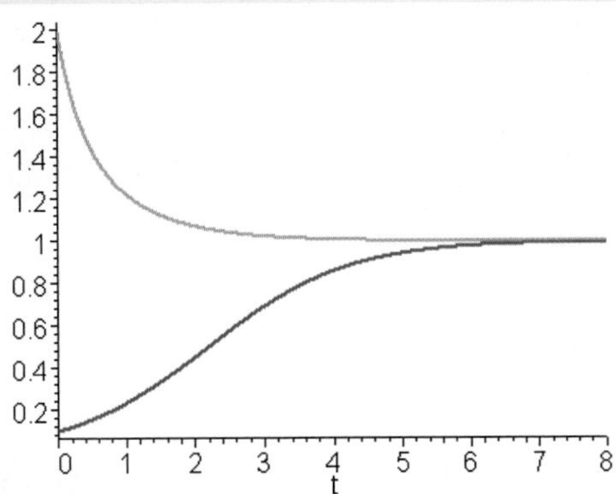

$$y(t) = \frac{1}{1 - \frac{1}{2}e^{(-t)}}$$

Figure 9.8 shows the graphs of both solutions, which is produced by typing

```
>plot({(1+9*exp(-t))^(-1),(1-0.5*exp(-t))^(-1)},t=0..8);
```

The graphs show that the system is stable with an equilibrium value of $1 million. The lower curve in Figure 9.8 shows approximate exponential growth to begin with. Between years 2 and 4, the value of the investment increases linearly before finally converging to its equilibrium value. The upper curve in Figure 9.8 shows that if the initial value exceeds the equilibrium value, then y decays exponentially to reach its equilibrium value.

Key Terms

Adjustment coefficient The constant of proportionality in the simple macroeconomic model, in which the rate of change of national income is assumed to be proportional to excess expenditure.

Arbitrary constant A letter representing an unspecified constant in the general solution of a differential equation.

Complementary function of a differential equation The solution of the differential equation, $\frac{dy}{dt} = my + c$ when the constant c is replaced by zero.

Differential equation An equation connecting derivatives of an unknown function.

Equilibrium value A solution of a differential equation that does not vary over time; it is the limiting value of $y(t)$ as t tends to infinity.

General solution of a differential equation The solution of a differential equation that contains an arbitary constant. It is the sum of the complementary function and particular solution.

Initial condition The value of $y(0)$ that needs to be specified to obtain a unique solution of a differential equation.

Particular solution of a differential equation Any one solution of a differential equation such as $\frac{dy}{dt} = my + c$.

Stable equilibrium An economic model in which the solution of the associated differential equation converges.

Practice Problems

6 Use integration to solve each of the following differential equations subject to the given initial conditions.

(a) $\frac{dy}{dt} = 2t;\ y(0) = 7$ **(b)** $\frac{dy}{dt} = e^{-3t};\ y(0) = 0$ **(c)** $\frac{dy}{dt} = t^2 + 3t - 5;\ y(0) = 1$

7 Solve the differential equation

$$\frac{dy}{dt} = -3y + 180$$

in the case when the initial condition is

(a) $y(0) = 40$ **(b)** $y(0) = 80$ **(c)** $y(0) = 60$

Comment on the qualitative behaviour of the solution in each case.

8 Consider the two-sector model

$$\frac{dY}{dt} = 0.5(C + I - Y)$$

$$C = 0.7Y + 500$$

$$I = 0.2Y + 500$$

Find an expression for $Y(t)$ when $Y(0) = 15\,000$. Is the system stable or unstable?

9 Consider the two-sector model

$$\frac{dY}{dt} = 0.3(C + I - Y)$$

$$C = 0.8Y + 300$$

$$I = 0.7Y + 600$$

Find an expression for $Y(t)$ when $Y(0) = 200$. Is this system stable or unstable?

10 Consider the market model

$$Q_S = 3P - 1$$

$$Q_D = -2P + 9$$

$$\frac{dP}{dt} = 0.5(Q_D - Q_S)$$

Find expressions for $P(t)$, $Q_S(t)$ and $Q_D(t)$ when $P(0) = 1$. Is this system stable or unstable?

11 A simple model of the growth of an economy is based on three assumptions.

(1) Savings, S, are proportional to income, Y, so that

$$S = \alpha Y \quad (\alpha > 0)$$

(2) Investment, I, is proportional to the rate of change of Y so that

$$I = \beta\frac{dY}{dt} \quad (\beta > 0)$$

(3) Investment and savings are equal so that

$$I = S$$

Use these assumptions to show that

$$\frac{dY}{dt} = \frac{\alpha}{\beta} Y$$

and hence write down a formula for $Y(t)$ in terms of $Y(0)$. Is this system stable or unstable?

12 Show, by substituting into the differential equation, that

$$y(t) = Ae^{mt} - \frac{c}{m}$$

is a solution of

$$\frac{dy}{dt} = my + c$$

13 Consider the differential equation

$$\frac{dy}{dt} = -2y + 5e^{3t}$$

(a) Find the complementary function.

(b) By substituting $y = De^{3t}$ into this equation, find a particular solution.

(c) Use your answers to parts (a) and (b) to write down the general solution and hence find the specific solution that satisfies the initial condition, $y(0) = 7$.

(d) Is the solution in part (c) stable or unstable?

14 Consider the differential equation

$$\frac{dy}{dt} = -y + 4t - 3$$

(a) Find the complementary function.

(b) By substituting $y = Dt + E$ into this equation, find a particular solution.

(c) Use your answers to parts (a) and (b) to write down the general solution and hence find the specific solution that satisfies the initial condition, $y(0) = 1$.

(d) Is the solution in part (c) stable or unstable?

15 **(Maple)** Solve the differential equation

$$\frac{dy}{dt} = \frac{ty}{t^2 + 1}$$

with initial condition $y(0) = 1$. Plot a graph of this solution on the range $0 \le t \le 5$. Hence, or otherwise, write down a simple expression for the approximate solution when t is large.

16 **(Maple)** In the absence of any withdrawals, the value of an investment fund, $y(t)$, varies according to

$$\frac{dy}{dt} = y(1 - y)$$

where y is measured in millions of dollars and t is measured in years. Money is taken out of the fund at a constant rate of \$250 000 per year so that

$$\frac{dy}{dt} = y(1 - y) - 0.25$$

(a) Find the solution of this differential equation when the initial value of the fund is \$1 million. Plot a graph of this solution over the range $0 \le t \le 20$.

➔

(b) Find the solution of this differential equation when the initial value of the fund is \$250 000. Plot a graph of this solution over the range $0 \le t \le 2$.

(c) Compare the solutions obtained in parts (a) and (b).

17 **(Maple)** The output, Q, of an Internet firm depends in the short term on capital, *K*, and time, *t*. The production and savings functions are given by

$$Q = atK \quad \text{and} \quad S = Q - bt$$

respectively, where *a* and *b* are positive constants. Assuming that capital accumulation is equal to savings, show that

$$\frac{dK}{dt} = t(aK - b)$$

If the initial capital is *c*, solve this equation to obtain an expression for *K* in terms of *t* and the constants, *a*, *b* and *c*. Write down the corresponding expression for Q. Comment on the qualitative behaviour of this solution in the case when

(a) $c = b/a$ **(b)** $c > b/a$ **(c)** $c < b/a$

Appendix 1 Differentiation from First Principles

We hinted in Section 4.1 that there was a formal way of actually proving the formulae for derivatives. This is known as 'differentiation from first principles' and we begin by illustrating the basic idea using a simple example. Figure A1.1 shows the graph of the square function $f(x) = x^2$ near $x = 3$.

The slope of the chord joining points A and B is

$$\begin{aligned}\frac{\Delta y}{\Delta x} &= \frac{(3 + \Delta x)^2 - 3^2}{\Delta x} \\ &= \frac{9 + 6\Delta x + (\Delta x)^2 - 9}{\Delta x} \\ &= \frac{6\Delta x + (\Delta x)^2}{\Delta x} \\ &= 6 + \Delta x\end{aligned}$$

Figure A1.1

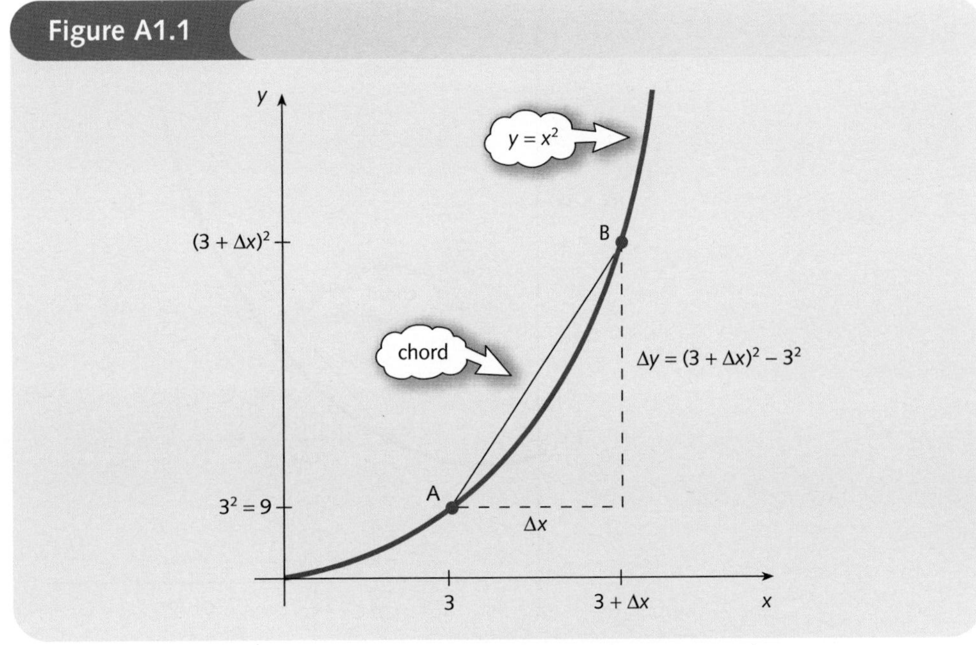

Now, as we pointed out in Section 4.1, the slope of the tangent at $x = 3$ is the limit of the slope of the chords as the width, Δx, gets smaller and smaller. In this case

$$\text{slope of tangent} = \lim_{\Delta x \to 0} (6 + \Delta x) = 6$$

In other words, the derivative of $f(x) = x^2$ at $x = 3$ is 6 (which agrees with $f'(x) = 2x$ evaluated at $x = 3$). Notice that this proof is not restricted to positive values of Δx. The chords in Figure A1.1 could equally well have been drawn to the left of $x = 3$. In both cases the slope of the chords approaches that of the tangent at $x = 3$ as the width of the interval shrinks.

Practice Problem

1 Use differentiation from first principles to find the derivative of $f(x) = x^2$ at $x = 5$.

The argument given above for the particular point, $x = 3$, can be extended quite easily to a general point, x. The details are are follows:

$$\begin{aligned}\frac{\Delta y}{\Delta x} &= \frac{(x + \Delta x)^2 - x^2}{\Delta x} \\ &= \frac{x^2 + 2x\Delta x + (\Delta x)^2 - x^2}{\Delta x} \\ &= \frac{2x\Delta x + (\Delta x)^2}{\Delta x} = 2x + \Delta x\end{aligned}$$

slope of chord

Hence

$$f'(x) = \lim_{\Delta x \to 0} (2x + \Delta x) = 2x$$

slope of tangent

In other words, we have *proved* that x^2 differentiates to $2x$.

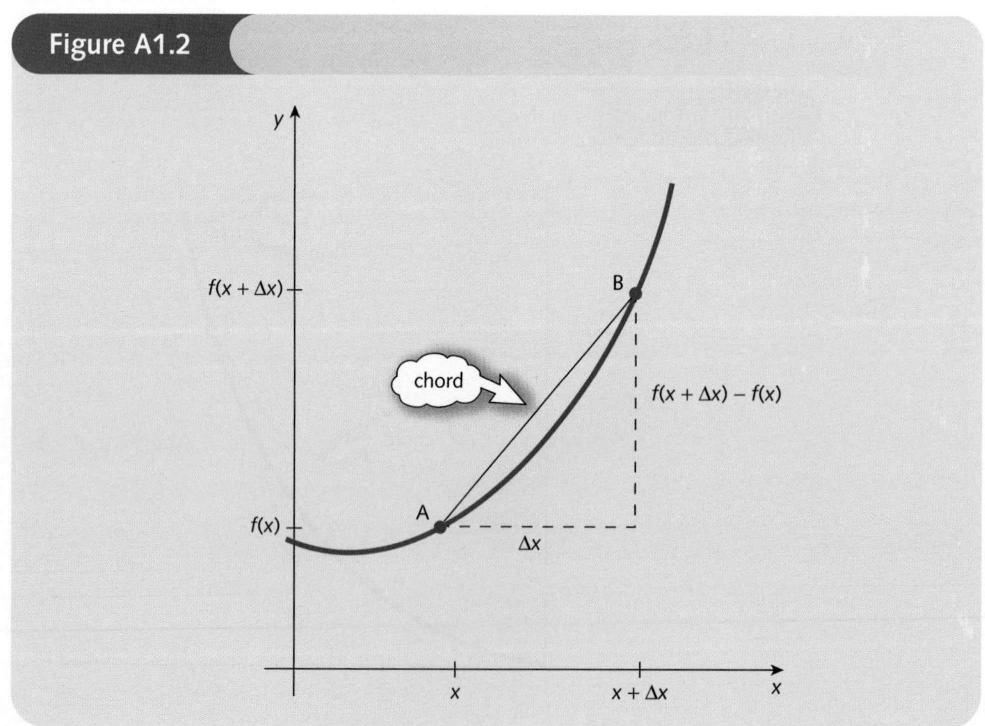

Figure A1.2

For a general function, $f(x)$, Figure A1.2 shows that we first find the slope of the chord joining A and B, i.e.

$$\frac{f(x+\Delta x)-f(x)}{\Delta x}$$

and then investigate what happens to this expression as $\Delta x \to 0$.

Example

Differentiate the following functions from first principles:

(a) $f(x) = 3x^2 + 5x - 6$ **(b)** $f(x) = \dfrac{1}{x}$

Solution

(a) The slope of the chord is

$$\begin{aligned}\frac{f(x+\Delta x)-f(x)}{\Delta x} &= \frac{[3(x+\Delta x)^2 + 5(x+\Delta x) - 6] - [3x^2 + 5x - 6]}{\Delta x} \\ &= \frac{3x^2 + 6x\Delta x + 3(\Delta x)^2 + 5x + 5\Delta x - 6 - 3x^2 - 5x + 6}{\Delta x} \\ &= \frac{6x\Delta x + 5\Delta x + 3(\Delta x)^2}{\Delta x} \\ &= 6x + 5 + 3\Delta x\end{aligned}$$

Hence the slope of the tangent is

$$f'(x) = \lim_{\Delta x \to 0}(6x + 5 + 3\Delta x) = 6x + 5$$

(b) The algebra is a little more complicated this time since we need to manipulate fractions. We begin by simplifying the numerator in the fraction

$$\frac{f(x+\Delta x)-f(x)}{\Delta x}$$

Given that $f(x) = \dfrac{1}{x}$, we see that

$$\begin{aligned}f(x+\Delta x)-f(x) &= \frac{1}{x+\Delta x} - \frac{1}{x} \\ &= \frac{x-(x+\Delta x)}{(x+\Delta x)x} \\ &= \frac{-\Delta x}{(x+\Delta x)x}\end{aligned}$$

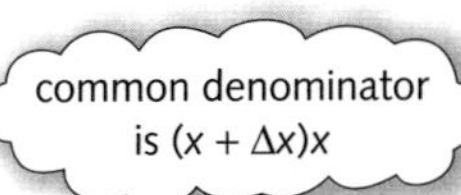

Hence

$$\frac{f(x+\Delta x)-f(x)}{\Delta x} = \frac{-1}{(x+\Delta x)x}$$

Finally, letting $\Delta x \to 0$ gives

$$f'(x) = \frac{-1}{(x+0)x} = \frac{-1}{x^2}$$

Practice Problems

2 Use first principles to differentiate each of the following functions

(a) $y = 4x^2 - 9x + 1$ **(b)** $y = \frac{1}{x^2}$

3 **(a)** By writing $(a + b)^3 = (a + b)(a + b)^2$ show that

$$(a + b)^3 = a^3 + 3a^2b + 3ab^2 + b^3$$

(b) Use the result of part (a) to prove that the cube function, x^3, differentiates to $3x^2$.

Appendix 2 Implicit Differentiation

The idea of implicit differentiation was first introduced in Section 5.1 (page 354) in the context of partial differentiation. It is possible to approach this topic via ordinary differentiation. Indeed, it can be regarded as nothing more than a simple application of the chain rule.

Example

Use implicit differentiation to find the value of $\frac{dy}{dx}$ on the curve

(a) $y^2 - 2x^3 = 25$ at the point $(-2,3)$

(b) $\ln y + 3y - x^2 = 2$ at the point $(1,1)$

Solution

(a) The first thing to do is to differentiate both sides of $y^2 - 2x^3 = 25$ with respect to x.

To differentiate the term $(y)^2$ **with respect to *x***, you first differentiate the outer 'square' function to get $2y$ and then multiply by the derivative of inner function, y, **with respect to *x***, which is dy/dx. Hence

$$\frac{d}{dx}(y^2) = 2y\frac{dy}{dx}$$

The remaining terms are more easily dealt with:

$$\frac{d}{dx}(2x^3) = 6x^2 \quad \text{and} \quad \frac{d}{dx}(25) = 0$$

Collecting these results together gives

$$2y\frac{dy}{dx} - 6x^2 = 0$$

so that

$$\frac{dy}{dx} = \frac{3x^2}{y}$$

➔

Finally, substituting $x = -2$, $y = 3$, gives

$$\frac{dy}{dx} = \frac{3(-2)^2}{3} = 4$$

(b) To differentiate the term $\ln(y)$ **with respect to** x we use the chain rule. The outer log function goes to $\frac{1}{y}$ and the inner 'y' function differentiates to $\frac{dy}{dx}$. Hence

$$\frac{d}{dx}(\ln(y)) = \frac{1}{y} \times \frac{dy}{dx}$$

Again the other terms are more straightforward:

$$\frac{d}{dx}(3y) = 3\frac{dy}{dx}, \quad \frac{d}{dx}(x^2) = 2x \quad \text{and} \quad \frac{d}{dx}(2) = 0$$

Differentiating both sides of $\ln y + 3y - x^2 = 2$ with respect to x gives:

$$\frac{1}{y}\frac{dy}{dx} + 3\frac{dy}{dx} - 2x = 0$$

To make dy/dx the subject of this equation, first multiply both sides by y to get

$$\frac{dy}{dx} + 3y\frac{dy}{dx} - 2xy = 0$$

$$(1 + 3y)\frac{dy}{dx} = 2xy \qquad \text{(add } 2xy \text{ to both sides and then factorize the left-hand side)}$$

$$\frac{dy}{dx} = \frac{2xy}{1 + 3y} \qquad \text{(divide both sides by } 1 + 3y\text{)}$$

Finally, substituting, $x = 1$, $y = 1$ gives

$$\frac{dy}{dx} = \frac{2(1)(1)}{1 + 3(1)} = \frac{1}{2}$$

Practice Problem

1 **(a)** Verify that the point (1, 2) lies on the curve $2x^2 + 3y^2 = 14$.

(b) By differentiating both sides of $2x^2 + 3y^2 = 14$ with respect to x, show that

$$\frac{dy}{dx} = -\frac{2x}{3y}$$

and hence find the gradient of the curve at (1,2).

In the previous example each of the terms involves just one of the letters x or y. It is possible to handle more complicated terms that involve both letters. For example, to differentiate the term 'xy' **with respect to** x, we use the product rule, which gives

$$\frac{d}{dx}(xy) = x\frac{dy}{dx} + 1 \times y = x\frac{dy}{dx} + y$$

This is illustrated in the following example.

Example

Find an expression for $\frac{dy}{dx}$ in terms of x and y for

$$x^2 + 3y^2 - xy = 11$$

Solution

Differentiating both sides with respect to x gives

$$2x + 6y\frac{dy}{dx} - \left(x\frac{dy}{dx} + y\right) = 0 \qquad \text{(chain and product rules)}$$

$$2x + 6y\frac{dy}{dx} - x\frac{dy}{dx} - y = 0 \qquad \text{(multiply out brackets)}$$

$$2x - y + (6y - x)\frac{dy}{dx} = 0 \qquad \text{(collect terms)}$$

$$\frac{dy}{dx} = \frac{y - 2x}{6y - x} \qquad \left(\text{make } \frac{dy}{dx} \text{ the subject}\right)$$

Practice Problem

2 By differentiating both sides of the following with respect to x, find expressions for $\frac{dy}{dx}$ in terms of x and y.

(a) $x^2 + y^2 = 16$

(b) $3y^2 + 4x^3 + 2x = 2$

(c) $e^x + 2e^y = 1$

(d) $ye^x = xy + y^2$

(e) $x^2 + 2xy^2 - 3y = 10$

(f) $\ln(x + y) = -x$

Appendix 3 Hessians

In this appendix we describe what a Hessian is, and how it can be used to classify the stationary points of an unconstrained optimization problem. In Section 5.4 (page 391) the conditions for a function $f(x, y)$ to have a minimum were stated as:

$$f_{xx} > 0, \; f_{yy} > 0 \quad \text{and} \quad f_{xx}f_{yy} - f_{xy}^2 > 0$$

where all of the partial derivatives are evaluated at a stationary point, (a, b).

It turns out that the second condition, $f_{yy} > 0$, is actually redundant. If the first and third conditions are met then the second one is automatically true. To see this notice that

$$f_{xx}f_{yy} - f_{xy}^2 > 0$$

is the same as $f_{xx}f_{yy} > f_{xy}^2$. The right-hand side is non-negative (being a square term) and so

$$f_{xx}f_{yy} > 0$$

The only way that the product of two numbers is positive is when they are either both positive or both negative. Consequently, when $f_{xx} > 0$, say, the other factor f_{yy} will also be positive.

Similarly, for a maximum point $f_{xx} < 0$, which forces the condition $f_{yy} < 0$.

The two conditions for a minimum point, $f_{xx} > 0$ and $f_{xx}f_{yy} - f_{xy}^2 > 0$ can be expressed more succinctly in matrix notation.

The 2×2 matrix, $\mathbf{H} = \begin{bmatrix} f_{xx} & f_{xy} \\ f_{yx} & f_{yy} \end{bmatrix}$ (where $f_{xy} = f_{yx}$) made from second-order partial derivatives is called a *Hessian matrix* and has determinant

$$\begin{vmatrix} f_{xx} & f_{xy} \\ f_{yx} & f_{yy} \end{vmatrix} = f_{xx}f_{yy} - f_{xy}^2$$

so the conditions for a minimum are:

(1) the number in the top left-hand corner of **H** (called the *first principal minor*) is positive

(2) the determinant of **H** (called the *second principal minor*) is positive.

For a maximum, the first principal minor is negative and the second principal minor is positive.

Example

Use Hessians to classify the stationary point of the function

$$\pi = 50Q_1 - 2Q_1^2 + 95Q_2 - 4Q_2^2 - 3Q_1Q_2$$

Solution

This profit function, considered in Practice Problem 2 in Section 5.4 (page 394), has a stationary point at $Q_1 = 5$, $Q_2 = 10$. The second-order partial derivatives are

$$\frac{\partial^2\pi}{\partial Q_1^2} = -4, \quad \frac{\partial^2\pi}{\partial Q_2^2} = -8 \quad \text{and} \quad \frac{\partial^2\pi}{\partial Q_1 \partial Q_2} = -3$$

so the Hessian matrix is

$$\mathbf{H} = \begin{bmatrix} -4 & -3 \\ -3 & -8 \end{bmatrix}$$

The first principal minor $-4 < 0$.
The second principal minor $(-4)(-8) - (-3)^2 = 23 > 0$.
Hence the stationary point is a maximum.

Practice Problems

1 The function

$$z = x^2 + y^2 - 2x - 4y + 15$$

has a stationary point at (1,2). Write down the associated Hessian matrix and hence determine the nature of this point.

[This surface was previously sketched using Maple in Practice Problem 10(a) in Section 5.4 at page 399.]

2 The profit function

$$\pi = 1000Q_1 + 800Q_2 - 2Q_1^2 - 2Q_1Q_2 - Q_2^2$$

has a stationary point at $Q_1 = 100$, $Q_2 = 300$.
Use Hessians to show that this is a maximum.

[This is the worked example on page 392 of Section 5.4.]

3 The profit function

$$\pi = 16L^{1/2} + 24K^{1/2} - 2L - K$$

has a stationary point at $L = 16$, $K = 144$.
Write down a general expression for the Hessian matrix in terms of L and K, and hence show that the stationary point is a maximum.

[This is Practice Problem 7 on page 399 of Section 5.4.]

Matrices can also be used to classify the maximum and minimum points of constrained optimization problems. In Section 5.6 the Lagrangian function was defined as

$$g(x, y, \lambda) = f(x, y) + \lambda(M - \phi(x, y))$$

Optimum points are found by applying the three first-order conditions:

$$g_x = 0, \quad g_y = 0 \quad \text{and} \quad g_\lambda = 0$$

To classify as a maximum or minimum we consider the determinant of the 3×3 matrix of second-order derivatives:

$$\bar{H} = \begin{bmatrix} g_{xx} & g_{xy} & g_{x\lambda} \\ g_{xy} & g_{yy} & g_{y\lambda} \\ g_{x\lambda} & g_{y\lambda} & g_{\lambda\lambda} \end{bmatrix}$$

If $|\bar{H}| > 0$ the optimum point is a maximum, whereas if $|\bar{H}| < 0$, the optimum point is a minimum.

Note that

$$\frac{\partial g}{\partial \lambda} = M - \phi(x, y)$$

so that

$$\frac{\partial^2 g}{\partial x \partial \lambda} = -\phi_x, \quad \frac{\partial^2 g}{\partial y \partial \lambda} = -\phi_y \quad \text{and} \quad \frac{\partial^2 g}{\partial \lambda^2} = 0$$

so $\bar{H}$ is given by

$$\begin{bmatrix} g_{xx} & g_{xy} & -\phi_x \\ g_{xy} & g_{yy} & -\phi_y \\ -\phi_x & -\phi_y & 0 \end{bmatrix}$$

This is called a *bordered Hessian* because it consists of the usual 2×2 Hessian

$$\begin{bmatrix} g_{xx} & g_{xy} \\ g_{xy} & g_{yy} \end{bmatrix}$$

'bordered' by a row and column of first-order derivatives, $-\phi_x$, $-\phi_y$ and 0.

Example

Use the bordered Hessian to classify the optimal point when the objective function

$$U = x_1^{1/2} + x_2^{1/2}$$

is subject to the budgetary constraint

$$P_1x_1 + P_2x_2 = M$$

Solution

The optimal point has already been found in Practice Problem 3 of Section 5.6 (page 418). The first-order conditions

$$\frac{\partial g}{\partial x_1} = \frac{1}{2}x_1^{-1/2} - \lambda P_1 = 0, \quad \frac{\partial g}{\partial x_2} = \frac{1}{2}x_2^{-1/2} - \lambda P_2 = 0, \quad \frac{\partial g}{\partial \lambda} = M - P_1x_1 - P_2x_2 = 0$$

were seen to have solution

$$x_1 = \frac{P_2 M}{P_1(P_1 + P_2)} \quad \text{and} \quad x_2 = \frac{P_1 M}{P_2(P_1 + P_2)}$$

The bordered Hessian is

$$\bar{H} = \begin{bmatrix} -\frac{1}{4}x_1^{-3/2} & 0 & -P_1 \\ 0 & -\frac{1}{4}x_2^{-3/2} & -P_2 \\ -P_1 & -P_2 & 0 \end{bmatrix}$$

Expanding along the third row gives

$$|\bar{H}| = -P_1 \begin{vmatrix} 0 & -P_1 \\ -\frac{1}{4}x_2^{-3/2} & -P_2 \end{vmatrix} - (-P_2) \begin{vmatrix} -\frac{1}{4}x_1^{-3/2} & -P_1 \\ 0 & -P_2 \end{vmatrix}$$

$$= \frac{1}{4}P_1^2 x_2^{-3/2} + \frac{1}{4}P_2^2 x_1^{-3/2}$$

This is positive so the point is a maximum.

Practice Problems

4 Use the bordered Hessian to show that the optimal value of the Lagrangian function

$$g(Q_1, Q_2, \lambda) = 40Q_1 - Q_1^2 + 2Q_1Q_2 + 20Q_2 - Q_2^2 + \lambda(15 - Q_1 - Q_2)$$

is a maximum.

[This is the worked example on page 415 of Section 5.6.]

5 Use the bordered Hessian to classify the optimal value of the Lagrangian function

$$g(x, y, \lambda) = 2x^2 - xy + \lambda(12 - x - y)$$

[This is Practice Problem 1 on page 413 of Section 5.6.]

Key Terms

Bordered Hessian matrix A Hessian matrix augmented by an extra row and column containing partial derivatives formed from the constraint in the method of Lagrange multipliers.

First principal minor The 1×1 determinant in the top left-hand corner of a matrix; the element a_{11} of a matrix **A**.

Hessian matrix A matrix whose elements are the second-order partial derivatives of a given function.

Second principal minor The 2×2 determinant in the top left-hand corner of a matrix.

Solutions to Problems

Getting Started

1 (a) This is shown in Figure SI.1.

Figure SI.1

Book1

	A	B	C	D
1	Economics Examination Marks			
2				
3	Candidate	Section A	Section B	
4	Fofaria	20	17	
5	Bull	38	12	
6	Eoin	34	38	
7	Arefin	40	52	
8	Cantor	29	34	
9	Devaux	30	49	
10				

(b) Type the heading `Total Mark` in cell D3. Type `=B4+C4` into cell D4. Click and drag down to D9.

(c) Type the heading Average: in cell C10. Type `=(SUM(D4:D9))/6` in cell D10 and press Enter. [Note: Excel has lots of built-in functions for performing standard calculations such as this. To find the average you could just type `=AVERAGE(D4:D9)` in cell D10.]

(d) This is shown in Figure SI.2.

Figure SI.2

Economics Examination Marks

Candidate	Section A Mark	Section B Mark	Total Mark
Arefin	40	52	92
Bull	38	12	50
Cantor	29	34	63
Devaux	30	49	79
Eoin	34	38	72
Fofaria	20	17	37
		Average:	65.5

(e) Just put the cursor over cell C5 and type in the new mark of 42. Pressing the Enter key causes cells D5 and D10 to be automatically updated. The new spreadsheet is shown in Figure SI.3.

Figure SI.3

Economics Examination Marks

Candidate	Section A Mark	Section B Mark	Total Mark
Arefin	40	52	92
Bull	38	42	80
Cantor	29	34	63
Devaux	30	49	79
Eoin	34	38	72
Fofaria	20	17	37
		Average:	70.5

2 **(a)** 14

(b) 11

(c) 5

3 **(a)** 4; is the solution of the equation $2x - 8 = 0$.

(b) Figure SI.4 shows the graph of $2x - 8$ plotted between $x = 0$ and 10.

Figure SI.4

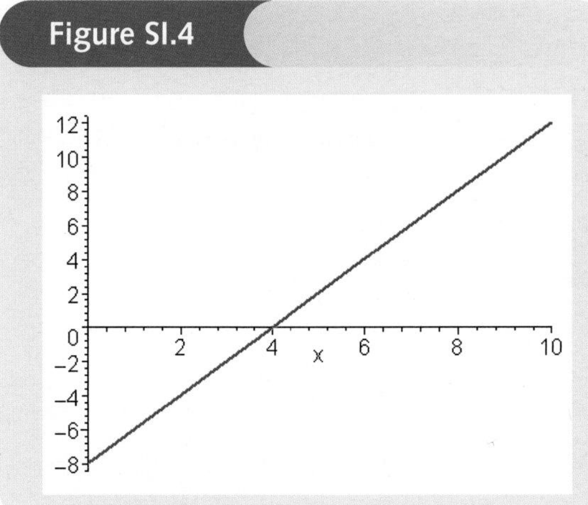

(c) $x^2 + 4x + 4$; the brackets have been 'multiplied out' in the expression $(x + 2)^2$.

(d) $7x + 4$; like terms in the expression $2x + 6 + 5x - 2$ have been collected together.

(e) Figure SI.5 shows the three-dimensional graph of the surface $x^3 - 3x + xy$ plotted between −2 and 2 in both the x and y directions.

Figure SI.5

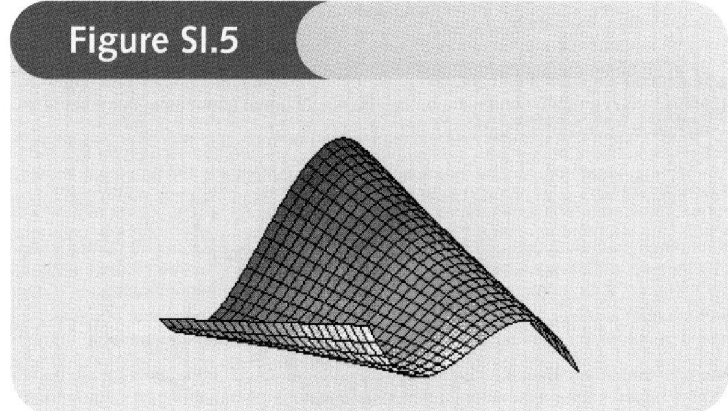

Chapter 1

Section 1.1

1 From Figure S1.1 note that all five points lie on a straight line.

Figure S1.1

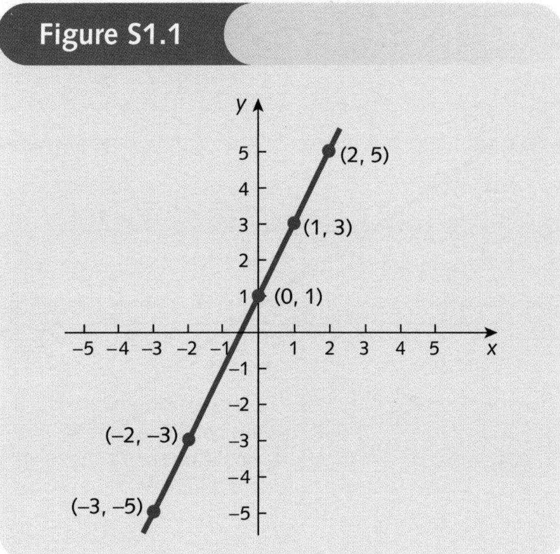

2 **(1)** **(a)** −30; **(b)** 2; **(c)** −5;
(d) 5; **(e)** 36; **(f)** −1.

(2) The key sequences (working from left to right) are

(a) [5] [×] [±] [6] [=]

(b) [±] [1] [×] [±] [2] [=]

(c) [±] [5] [0] [÷] [1] [0] [=]

(d) [±] [5] [÷] [±] [1] [=]

(e) [2] [×] [±] [1] [×] [±] [3] [×] [6] [=]

(f) same as (e) followed by [÷] [±] [2] [÷] [3] [÷] [6] [=]

3 **(1)** **(a)** −1; **(b)** −7; **(c)** 5;
(d) 0; **(e)** −91; **(f)** −5.

(2) The key sequences are

(a) [1] [−] [2] [=]

(b) [±] [3] [−] [4] [=]

(c) [1] [−] [±] [4] [=]

(d) [±] [1] [−] [±] [1] [=]

(e) [±] [7] [2] [−] [1] [9] [=]

(f) [±] [5] [3] [−] [±] [4] [8] [=]

4

Point	Check	
(−1, 2)	$2(-1) + 3(2) = -2 + 6 = 4$	✓
(−4, 4)	$2(-4) + 3(4) = -8 + 12 = 4$	✓
(5, −2)	$2(5) + 3(-2) = 10 - 6 = 4$	✓
(2, 0)	$2(2) + 3(0) = 4 + 0 = 4$	✓

The graph is sketched in Figure S1.2 (overleaf).

Figure S1.2

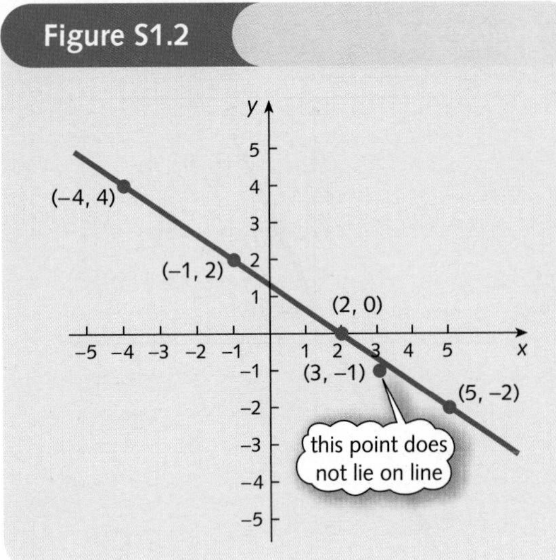

The graph shows that (3, −1) does not lie on the line. This can be verified algebraically:

$$2(3) + 3(-1) = 6 - 3 = 3 \neq 4$$

5
$$\begin{aligned} 3x - 2y &= 4 \\ 3(2) - 2y &= 4 \quad \text{(substitute } x = -2\text{)} \\ 6 - 2y &= 4 \\ -2y &= -2 \quad \text{(subtract 6 from both sides)} \\ y &= 1 \quad \text{(divide both sides by } -2\text{)} \end{aligned}$$

Hence (2, 1) lies on the line.

$$\begin{aligned} 3x - 2y &= 4 \\ 3(-2) - 2y &= 4 \\ -6 - 2y &= 4 \quad \text{(substitute } x = 2\text{)} \\ -2y &= 10 \quad \text{(add 6 to both sides)} \\ y &= -5 \quad \text{(divide both sides by } -2\text{)} \end{aligned}$$

Hence (−2, −5) lies on the line.

The line is sketched in Figure S1.3.

Figure S1.3

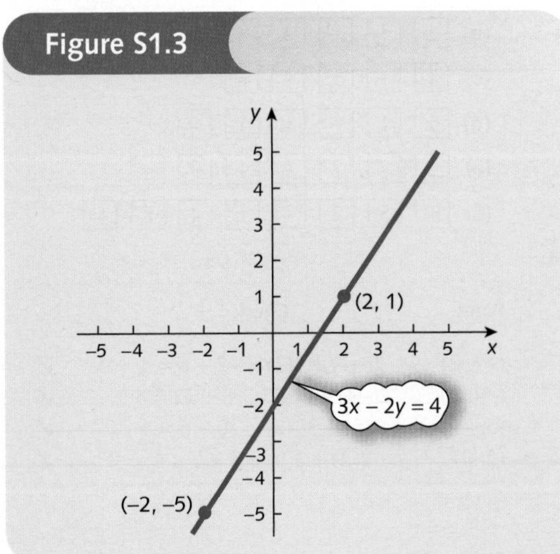

6
$$\begin{aligned} x - 2y &= 2 \\ 0 - 2y &= 2 \quad \text{(substitute } x = 0\text{)} \\ -2y &= 2 \\ y &= -1 \quad \text{(divide both sides by } -2\text{)} \end{aligned}$$

Hence (0, −1) lies on the line.

$$\begin{aligned} x - 2y &= 2 \\ x - 2(0) &= 2 \quad \text{(substitute } y = 0\text{)} \\ x - 0 &= 2 \\ x &= 2 \end{aligned}$$

Hence (2, 0) lies on the line.

The graph is sketched in Figure S1.4.

Figure S1.4

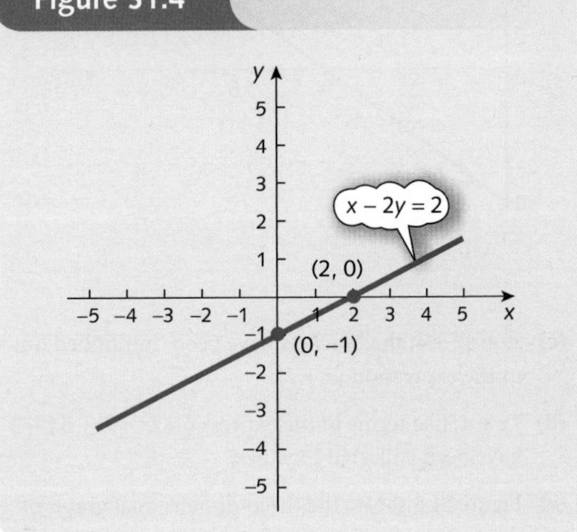

7 From Figure S1.5 the point of intersection is $(1, -\frac{1}{2})$.

Figure S1.5

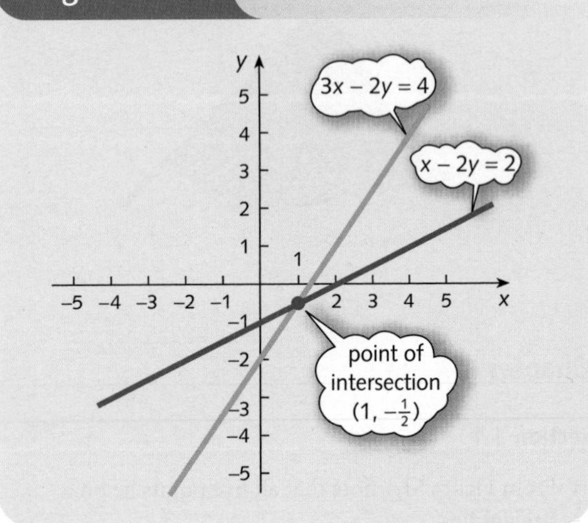

8 **(a)** $a = 1$, $b = 2$. The graph is sketched in Figure S1.6.

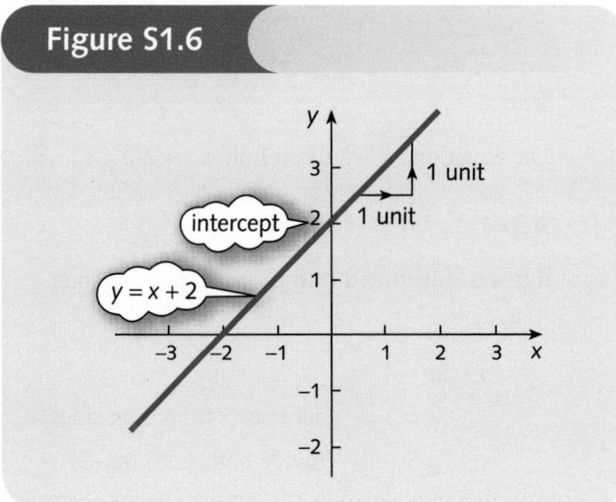

Figure S1.6

(b) $4x + 2y = 1$

$2y = 1 - 4x$ (subtract $4x$ from both sides)

$y = \frac{1}{2} - 2x$ (divide both sides by 2)

so $a = -2$, $b = \frac{1}{2}$. The graph is sketched in Figure S1.7.

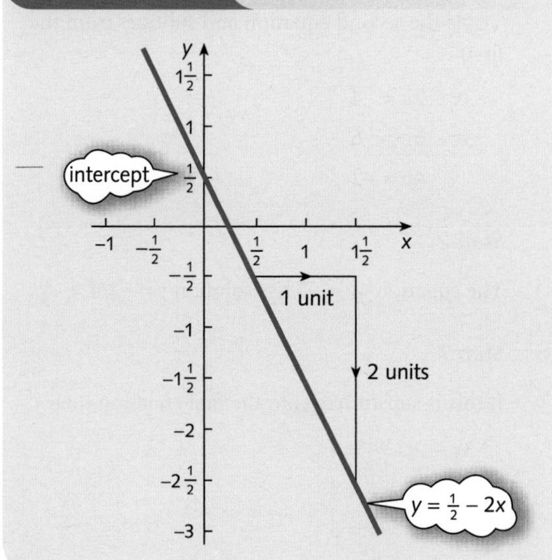

Figure S1.7

9 From Figure S1.8 the point of intersection is (2, 3).

10 **(a)** –20; **(b)** 3; **(c)** –4; **(d)** 1; **(e)** –1; **(f)** –3; **(g)** 11; **(h)** 0; **(i)** 18.

11 **(a)** 1; **(b)** 5; **(c)** –6; **(d)** –6; **(e)** –30; **(f)** 44.

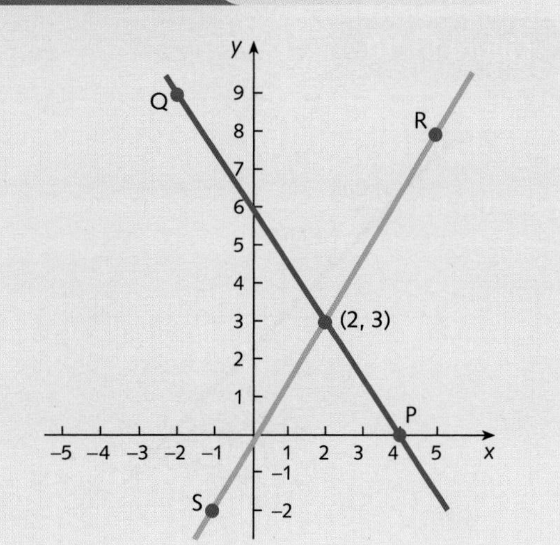

Figure S1.8

12 **(a)** 0.5; **(b)** 4; **(c)** –4; **(d)** –0.5; **(e)** 0; **(f)** 9.

13

x	y
0	8
6	0
3	4

The graph is sketched in Figure S1.9.

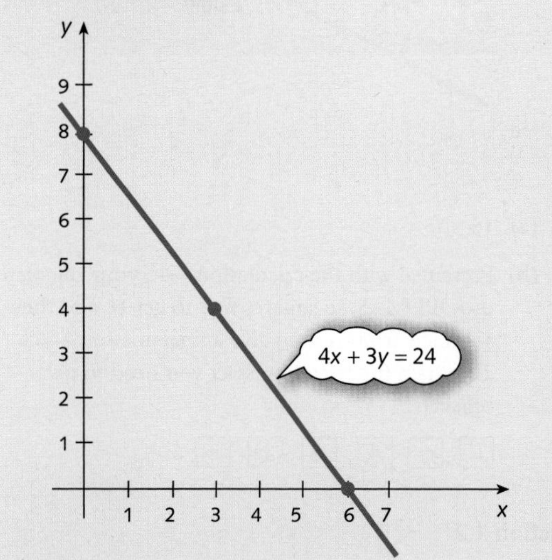

Figure S1.9

14 **(a)** (–2, –2); **(b)** $(2, 1\frac{1}{2})$; **(c)** $(1\frac{1}{2}, 1)$; **(d)** (10, –9).

15 **(a)** The graph is sketched in Figure S1.10.

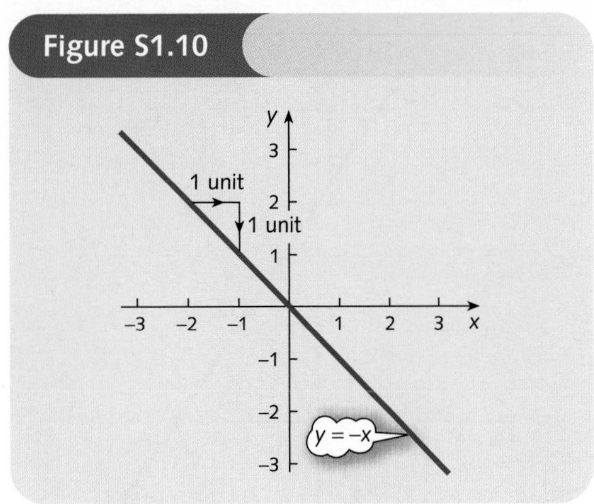

(b) The graph is sketched in Figure S1.11.

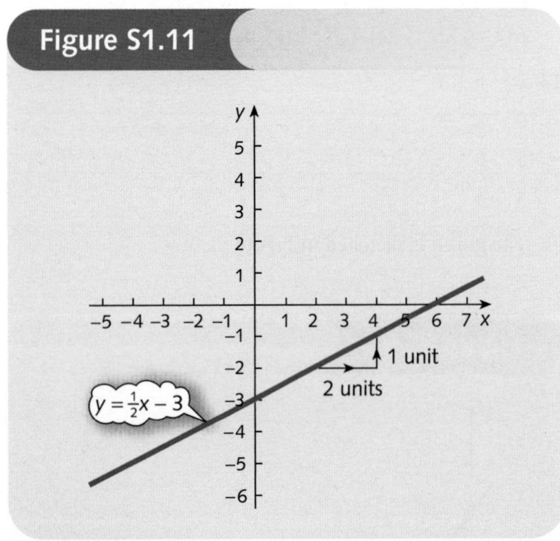

16 **(a)** 16.

(b) Presented with the calculation, -4^2, your calculator uses BIDMAS, so squares first to get 16 and then subtracts from zero to give a final answer, −16. To obtain the correct answer you need to use brackets:

Section 1.2

1 **(a)** *Step 1*

It is probably easiest to eliminate y. This can be done by subtracting the second equation from the first:

$$\begin{array}{rl} 3x - 2y = 4 & \\ x - 2y = 2 & - \\ \hline 2x \quad\; = 2 & \end{array}$$

Step 2

The equation $2x = 2$ has solution $x = 2/2 = 1$.

Step 3

If this is substituted into the first equation then

$$\begin{aligned} 3(1) - 2y &= 4 \\ 3 - 2y &= 4 \\ -2y &= 1 && \text{(subtract 3 from both sides)} \\ y &= -\tfrac{1}{2} && \text{(divide both sides by } -2\text{)} \end{aligned}$$

Step 4

As a check the second equation gives

$$\begin{aligned} x - 2y &= 1 - 2(-\tfrac{1}{2}) \\ &= 1 - (-1) = 2 \quad ✓ \end{aligned}$$

Hence the solution is $x = 1$, $y = -\tfrac{1}{2}$.

If you decide to eliminate x then the corresponding steps are as follows:

Step 1

Triple the second equation and subtract from the first:

$$\begin{array}{rl} 3x - 2y = \;\; 4 & \\ 3x - 6y = \;\; 6 & - \\ \hline 4y = -2 & \end{array}$$

Step 2

The equation $4y = -2$ has solution $y = -2/4 = -\tfrac{1}{2}$.

Step 3

If this is substituted into the first equation then

$$\begin{aligned} 3x - 2(-\tfrac{1}{2}) &= 4 \\ 3x + 1 &= 4 \\ 3x &= 3 \end{aligned}$$

(subtract 1 from both sides)

$$x = 1$$

(divide both sides by 3)

(b) *Step 1*

It is immaterial which variable is eliminated. To eliminate x multiply the first equation by 5, multiply the second by 3 and add:

$$\begin{aligned} 15x + 25y &= 95 \\ -15x + 6y &= -33 \quad + \\ \hline 31y &= 62 \end{aligned}$$

Step 2

The equation $31y = 62$ has solution $y = 62/31 = 2$.

Step 3

If this is substituted into the first equation then

$$3x + 5(2) = 19$$
$$3x + 10 = 19$$
$$3x = 9$$

(subtract 10 from both sides)

$$x = 3$$

(divide both sides by 3)

Step 4

As a check the second equation gives

$$\begin{aligned} -5x + 2y &= -5(3) + 2(2) \\ &= -15 + 4 = -11 \quad ✓ \end{aligned}$$

Hence the solution is $x = 3$, $y = 2$.

2 (a) *Step 1*

To eliminate x multiply the first equation by 4, multiply the second equation by 3 and add:

$$\begin{aligned} 12x - 24y &= -8 \\ -12x + 24y &= -3 \quad + \\ \hline 0y &= -11 \end{aligned}$$

Step 2

This is impossible, so there are no solutions.

(b) *Step 1*

To eliminate x multiply the first equation by 2 and add to the second:

$$\begin{aligned} -10x + 2y &= 8 \\ 10x - 2y &= -8 \quad + \\ \hline 0y &= 0 \end{aligned}$$

Step 2

This is true for any value of y, so there are infinitely many solutions.

3 *Step 1*

To eliminate x from the second equation multiply equation (2) by 2 and subtract from equation (1):

$$\begin{aligned} 2x + 2y - 5z &= -5 \\ 2x - 2y + 2z &= 6 \quad - \\ \hline 4y - 7z &= -11 \qquad (4) \end{aligned}$$

To eliminate x from the third equation multiply equation (1) by 3, multiply equation (3) by 2 and add:

$$\begin{aligned} 6x + 6y - 15z &= -15 \\ -6x + 2y + 4z &= -4 \quad + \\ \hline 8y - 11z &= -19 \qquad (5) \end{aligned}$$

The new system is

$$\begin{aligned} 2x + 2y - 5z &= -5 \qquad (1) \\ 4y - 7z &= -11 \qquad (4) \\ 8y - 11z &= -19 \qquad (5) \end{aligned}$$

Step 2

To eliminate y from the third equation multiply equation (4) by 2 and subtract equation (5):

$$\begin{aligned} 8y - 14z &= -22 \\ 8y - 11z &= -19 \\ -3z &= -3 \qquad (6) \end{aligned}$$

The new system is

$$\begin{aligned} 2x + 2y - 5z &= -5 \qquad (1) \\ 4y - 7z &= -11 \qquad (4) \\ -3z &= -3 \qquad (6) \end{aligned}$$

Step 3

Equation (6) gives $z = -3/-3 = 1$. If this is substituted into equation (4) then

$$\begin{aligned} 4y - 7(1) &= -11 \\ 4y - 7 &= -11 \\ 4y &= -4 \quad \text{(add 7 to both sides)} \\ y &= -1 \quad \text{(divide both sides by 4)} \end{aligned}$$

Finally, substituting $y = -1$ and $z = 1$ into equation (1) produces

$$\begin{aligned} 2x + 2(-1) - 5(1) &= -5 \\ 2x - 7 &= -5 \\ 2x &= 2 \end{aligned}$$

(add 7 to both sides)

$$x = 1$$

(divide both sides by 2)

Step 4

As a check the original equations (1), (2) and (3) give

$$\begin{aligned} 2(1) + 2(-1) - 5(1) &= -5 \quad ✓ \\ 1 - (-1) + 1 &= 3 \quad ✓ \\ -3(1) + (-1) + 2(1) &= -2 \quad ✓ \end{aligned}$$

Hence the solution is $x = 1$, $y = -1$, $z = 1$.

4 **(a)** $x = -2, y = -2$;

(b) $x = 2, y = 3/2$;

(c) $x = 3/2, y = 1$;

(d) $x = 10, y = -9$.

5 The lines are sketched in Figure S1.12.

Figure S1.12

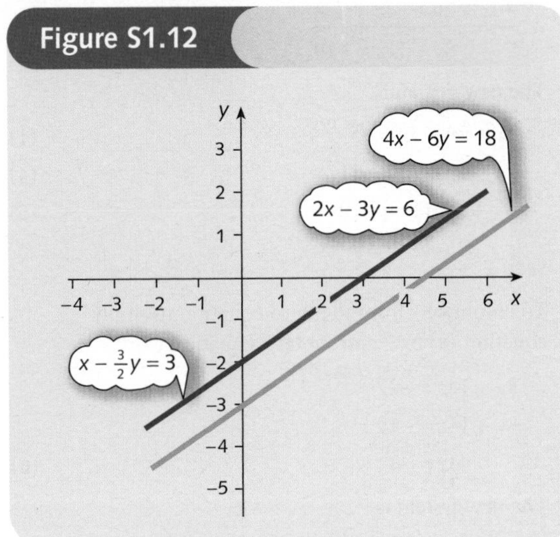

(a) Infinitely many. **(b)** No solution.

6 **(a)** Infinitely many. **(b)** No solution.

7 **(a)** $x = 3, y = -2, z = -1$;

(b) $x = -1, y = 3, z = 4$.

8 **(a)** No solution.

(b) Infinitely many solutions.

Section 1.3

1 **(a)** 0; **(b)** 48; **(c)** 16; **(d)** 25; **(e)** 1; **(f)** 17.

The function g reverses the effect of f and takes you back to where you started. For example, if 25 is put into the function f, the outgoing number is 0; and when 0 is put into g, the original number, 25, is produced. We describe this by saying that g is the inverse of f (and vice versa).

2 The demand curve that passes through (0, 75) and (25, 0) is sketched in Figure S1.13. From this diagram we see that

(a) $P = 6$ when $Q = 23$

(b) $Q = 19$ when $P = 18$

Alternatively, using algebra:

(a) Substituting $Q = 23$ gives

$$P = -3(23) + 75 = 6$$

Figure S1.13

P
75
P = −3Q + 75
18
6
19 23 25
Q

(b) Substituting $P = 18$ gives $18 = -3Q + 75$ with solution $Q = 19$

3 **(a)** In equilibrium, $Q_S = Q_D = Q$, so

$$P = -4Q + 120$$

$$P = \tfrac{1}{3}Q + 29$$

Hence

$$-4Q + 120 = \tfrac{1}{3}Q + 29$$

(since both sides equal P)

$$-4\tfrac{1}{3}Q + 120 = 29$$

(subtract $\tfrac{1}{3}Q$ from both sides)

$$-4\tfrac{1}{3}Q = -91$$

(subtract 120 from both sides)

$$Q = 21$$

(divide both sides by $-4\tfrac{1}{3}$)

Substituting this value into either the demand or supply equations gives $P = 36$.

(b) After the imposition of a \$13 tax the supply equation becomes

$$P - 13 = \tfrac{1}{3}Q_S + 29$$

$$P = \tfrac{1}{3}Q_S + 42$$

(add 13 to both sides)

The demand equation remains unchanged, so, in equilibrium,

$$P = -4Q + 120$$

$$P = \tfrac{1}{3}Q + 42$$

Hence

$$-4Q + 120 = \tfrac{1}{3}Q + 42$$

This equation can now be solved as before to get $Q = 18$ and the corresponding price is $P = 48$. The equilibrium price rises from \$36 to \$48, so the consumer pays an additional \$12. The remaining \$1 of the tax is paid by the firm.

4 For good 1, $Q_{D_1} = Q_{S_1} = Q_1$ in equilibrium, so the demand and supply equations become

$$Q_1 = 40 - 5P_1 - P_2$$
$$Q_1 = -3 + 4P_1$$

Hence

$$40 - 5P_1 - P_2 = -3 + 4P_1$$

(since both sides equal Q_1)

$$40 - 9P_1 - P_2 = -3$$

(subtract $4P_1$ from both sides)

$$-9P_1 - P_2 = -43$$

(subtract 40 from both sides)

For good 2, $Q_{D_2} = Q_{S_2} = Q_2$ in equilibrium, so the demand and supply equations become

$$Q_2 = 50 - 2P_1 - 4P_2$$
$$Q_2 = -7 + 3P_2$$

Hence

$$50 - 2P_1 - 4P_2 = -7 + 3P_2$$

(since both sides equal Q_2)

$$50 - 2P_1 - 7P_2 = -7$$

(subtract $3P_2$ from both sides)

$$-2P_1 - 7P_2 = -57$$

(subtract 50 from both sides)

The equilibrium prices therefore satisfy the simultaneous equations

$$-9P_1 - P_2 = -43 \quad (1)$$
$$-2P_1 - 7P_2 = -57 \quad (2)$$

Step 1

Multiply equation (1) by 2 and (2) by 9 and subtract to get

$$61P_2 = 427 \quad (3)$$

Step 2

Divide both sides of equation (3) by 61 to get $P_2 = 7$.

Step 3

Substitute P_2 into equation (1) to get $P_1 = 4$.

If these equilibrium prices are substituted into either the demand or the supply equations then $Q_1 = 13$ and $Q_2 = 14$.

The goods are complementary because the coefficient of P_2 in the demand equation for good 1 is negative, and likewise for the coefficient of P_1 in the demand equation for good 2.

5 **(a)** 21; **(b)** 45; **(c)** 15; **(d)** 2; **(e)** 10; **(f)** 0; inverse.

6 The supply curve is sketched in Figure S1.14.

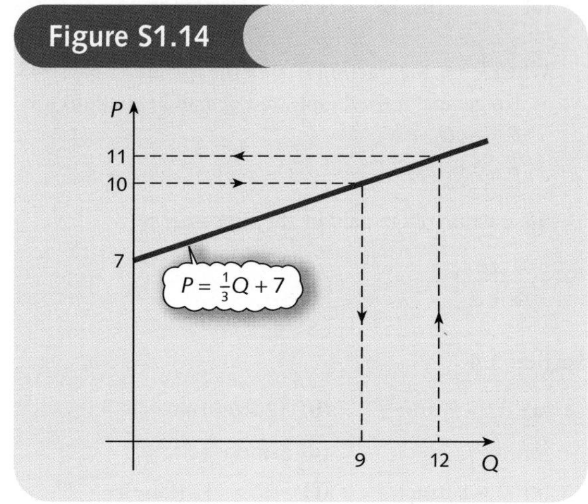

(a) 11; **(b)** 9;
(c) 0; once the price falls below 7 the firm does not plan to produce any goods.

7 **(a)** As P_S rises, consumers are likely to switch to the good under consideration, so demand for this good also rises: that is, the graph shifts to the right.

(b) As P_C rises, demand for the bundle of goods as a whole is likely to fall, so the graph shifts to the left.

(c) Assuming that advertising promotes the good and is successful, demand rises and the graph shifts to the right. For some goods, such as drugs, advertising campaigns are intended to discourage consumption, so the graph shifts to the left.

8 **(a)** Demand is 173. Additional advertising expenditure is 12.

(b) Superior.

9 **(a)** 53;

(b) Substitutable; a rise in P_A leads to an increase in Q;

(c) 6.

10 **(1)** $P = 30$, $Q = 10$.

(2) New supply equation is $0.85P = 2Q_S + 10$; $P = 33.6$, $Q = 9.28$.

11 **(a)** 17, 9; **(b)** \$324.

12 $P_1 = 40, P_2 = 10; Q_1 = 30, Q_2 = 55.$

13 $P_1 = 20, P_2 = 5, P_3 = 8; Q_1 = 13, Q_2 = 16, Q_3 = 11.$

14 0 and 30.

15 Superior; graph of (b) lies above that of (a).

Substitutable; graph of (c) lies below that of (a).

16 \$180, \$200, $^1/_3$.

(a) $^1/_2$ **(b)** $^3/_5$ **(c)** $^2/_3$; fraction is $\frac{k}{k+2}$.

When $k = 6$, the fraction is $^3/_4$ so the consumer pays \$45. [In general if the supply and demand equations are

$P = -aQ_D + b$

$P = cQ_S + d$

the fraction of tax paid by the consumer is

$\frac{a}{a+d}$.]

Section 1.4

1 **(a)** $12 > 9$ (true); **(b)** $12 > 6$ (true); **(c)** $3 > 0$ (true); **(d)** same as (c); **(e)** $2 > 1$ (true); **(f)** $-24 > -12$ (false); **(g)** $-6 > -3$ (false); **(h)** $-2 > -1$ (false); **(i)** $-4 > -7$ (true).

2 **(a)** $2x < 3x + 7$

$-x < 7$ (subtract $3x$ from both sides)

$x > -7$ (divide both sides by -1 changing sense because $-1 < 0$)

(b) $21x - 19 \geq 4x + 15$

$17x - 19 \geq 15$

(subtract $4x$ from both sides)

$17x \geq 34$

(add 19 to both sides)

$x \geq 2$

(divide both sides by 17, leaving inequality unchanged because $17 > 0$)

3 **(1)** **(a)** 8; **(b)** -12; **(c)** 14; **(d)** 4; **(e)** 56; **(f)** 1. In part (f) the innermost pair of brackets is evaluated first.

(2) All scientific calculators have a bracket facility. There are two keys typically denoted by [(] and [)] for 'opening' and 'closing' brackets. It is usually possible to have several levels (that is, brackets within brackets) so part (f) should cause no additional problem. Possible key sequences are as follows.

(a) [(] [1] [−] [3] [)] [+] [1] [0] [=].
Notice that when [)] is pressed your calculator automatically evaluates the expression inside the brackets and displays the intermediate result.

(b) [1] [−] [(] [3] [+] [1] [0] [)] [=].

(c) [2] [×] [(] [3] [+] [4] [)] [=]. Note that it is essential to press the [×] key explicitly on most calculators.

(d) [8] [−] [7] [+] [3] [=].

(e) [(] [1] [5] [−] [8] [)] [×] [(] [2] [+] [6] [)] [=].

(f) [(] [(] [2] [−] [3] [)] [+] [7] [)] [÷] [6] [)].
Notice the effect of pressing the three [)] keys. There is no need to press the [=] key at the end since the expression terminates with [)].

4 **(a)** $5z - 2z^2$

(b) $6x - 6y + 3y - 6x = -3y$

(c) $x - y + z - x^2 - x + y = z - x^2$

5 **(a)** $x^2 - 2x + 3x - 6 = x^2 + x - 6$

(b) $x^2 - xy + yx - y^2 = x^2 - y^2$

(c) $x^2 + xy + yx + y^2 = x^2 + 2xy + y^2$

(d) $5x^2 - 5xy + 5x + 2yx - 2y^2 + 2y$
$= 5x^2 - 3xy + 5x - 2y^2 + 2y$

6 **(a)** $(x + 8)(x - 8)$;

(b) $(2x + 9)(2x - 9)$.

7 **(1)** **(a)** $\frac{1}{2} \times \frac{3}{4} = \frac{1 \times 3}{2 \times 4} = \frac{3}{8}$

(b) $\not{7} \times \frac{1}{\not{14}_2} = \frac{1}{2}$

(c) $\frac{2}{3} \div \frac{8}{9} = \frac{\not{2}}{\not{3}} \times \frac{\not{9}^3}{\not{8}_4} = \frac{3}{4}$

(d) $\frac{8}{9} \div 16 = \frac{\not{8}}{9} \times \frac{1}{\not{16}_2} = \frac{1}{18}$

(2) Most scientific calculators have a fractions facility typically labelled [$a^b/_c$]. To enter a number such as $^3/_4$ you press [3] [$a^b/_c$] [4]. Try this for yourself. The display should read something like 3 ⌋ 4. To enter a number such as $3^5/_8$ you press [3] [$a^b/_c$] [5] [$a^b/_c$] [8] which then displays 3 ⌋ 5 ⌋ 8. Possible key sequences for part (1) are

(a) [1] [$a^b/_c$] [2] [×] [3] [$a^b/_c$] [4] [=]

(b) [7] [×] [1] [$a^b/_c$] [1] [4] [=]

(c) [2] [$a^b/_c$] [3] [÷] [8] [$a^b/_c$] [9] [=]

(d) [8] [$a^b/_c$] [9] [÷] [1] [6] [=]

8 **(1)** **(a)** $\frac{3}{7} - \frac{1}{7} = \frac{2}{7}$

(b) $\frac{1}{3} + \frac{2}{5} = \frac{5}{15} + \frac{6}{15} = \frac{11}{15}$

(c) $\frac{7}{18} - \frac{1}{4} = \frac{14}{36} - \frac{9}{36} = \frac{5}{36}$

(2) The key sequences are

(a) [3] [$a^b/_c$] [7] [−] [1] [$a^b/_c$] [7] [=]

(b) [1] [$a^b/_c$] [3] [+] [2] [$a^b/_c$] [5] [=]

(c) [7] [$a^b/_c$] [1] [8] [−] [1] [$a^b/_c$] [4] [=]

9 **(a)** $\frac{5}{\cancel{x-1}} \times \frac{\cancel{x-1}}{x+2} = \frac{5}{x+2}$

(b) $\frac{x^2}{x+10} \div \frac{x}{x+1} = \frac{{}^{x}\cancel{x^2}}{x+10} \times \frac{x+1}{\cancel{x}} = \frac{x(x+1)}{x+10}$

(c) $\frac{4}{x+1} + \frac{1}{x+1} = \frac{4+1}{x+1} = \frac{5}{x+1}$

(d) $\frac{2}{x+1} - \frac{1}{x+2}$

$= \frac{2(x+2)}{(x+1)(x+2)} - \frac{(1)(x+1)}{(x+1)(x+2)}$

$= \frac{(2x+4)-(x+1)}{(x+1)(x+2)} = \frac{x+3}{(x+1)(x+2)}$

10 **(a)** $4x + 5 = 5x - 7$

$5 = x - 7$ (subtract $4x$ from both sides)

$12 = x$ (add 7 to both sides)

(b) $3(3 - 2x) + 2(x - 1) = 10$

$9 - 6x + 2x - 2 = 10$

(multiply out brackets)

$7 - 4x = 10$

(collect like terms)

$-4x = 3$

(subtract 7 from both sides)

$x = -\frac{3}{4}$

(divide both sides by −4)

(c) $\frac{4}{x-1} = 5$

$4 = 5(x - 1)$ (multiply both sides by $x - 1$)

$4 = 5x - 5$ (multiply out brackets)

$9 = 5x$ (add 5 to both sides)

$\frac{9}{5} = x$ (divide both sides by 5)

(d) $\frac{3}{x} = \frac{5}{x-1}$

$3(x - 1) = 5x$ (cross-multiplication)

$3x - 3 = 5x$ (multiply out brackets)

$-3 = 2x$ (subtract $3x$ from both sides)

$-\frac{3}{2} = x$ (divide both sides by 2)

11 **(a)**, **(d)**, **(e)**, **(f)**.

12 **(a)** $x > 1$; **(b)** $x \le 3$; **(c)** $x \le -3$; **(d)** $x > 2$.

13 **(a)** 9; **(b)** 21; no.

14 **(a)** 43.96; **(b)** 1.13; **(c)** 10.34; **(d)** 0.17; **(e)** 27.38; **(f)** 3.72; **(g)** 62.70; **(h)** 2.39.

15 **(a)** $6x + 2y$; **(b)** $11x^2 - 3x - 3$; **(c)** $14xy + 2x$; **(d)** $6xyz + 2xy$; **(e)** $10a - 2b$; **(f)** $17x + 22y$; **(g)** $11 - 3p$; **(h)** $x - {}^1/_4$.

16 **(a)** $7x - 7y$; **(b)** $5xz - 2yz$; **(c)** $-5y + 4z - 2x$; **(d)** $x^2 - 7x + 10$; **(e)** $x^2 - xy + 7x$; **(f)** $x^3 + 3x^2 + 2x$; **(g)** $x^2 - xy - 1 + y$.

Note: in part (f), x^3 is an abbreviation for xxx.

17 **(1)** **(a)** $(x + 2)(x - 2)$; **(b)** $(x + y)(x - y)$; **(c)** $(3x + 10y)(3x - 10y)$; **(d)** $(ab + 5)(ab - 5)$.

(2) **(a)** 112 600 000; **(b)** 1.799 99; **(c)** 283 400; **(d)** 246 913 577.

18 **(a)** $\frac{5}{7}$; **(b)** $\frac{1}{10}$; **(c)** $\frac{3}{2}$; **(d)** $\frac{5}{48}$; **(e)** $\frac{8}{13}$; **(f)** $\frac{11}{9}$; **(g)** $\frac{141}{35}$; **(h)** $\frac{34}{5}$; **(i)** 6 **(j)** $\frac{7}{10}$ **(k)** $\frac{7}{9}$ **(l)** 4.

19 **(a)** $x + 6$

(b) $\frac{x+1}{x}$ or equivalently $1 + \frac{1}{x}$

(c) $\frac{5}{xy}$

(d) $\frac{5x+2}{6}$

(e) $\frac{7x+3}{x(x+1)}$

(f) $\frac{3x+5}{x^2}$

(g) $\frac{x^2+x-2}{x+1}$

(h) $\frac{x+3}{x(x+1)}$

20 **(a)** $-\frac{11}{7}$; **(b)** 1; **(c)** $-\frac{35}{9}$; **(d)** 8;

(e) $\frac{4}{5}$; **(f)** $\frac{1}{4}$; **(g)** $-\frac{11}{7}$; **(h)** 8;

(i) 9; **(j)** $\frac{71}{21}$; **(k)** 7; **(l)** −9;

(m) 1; **(n)** −5; **(o)** 3; **(p)** 5.

Section 1.5

1 **(a)** $\frac{1}{2}Q = 4$ (subtract 13 from both sides)

$Q = 8$ (multiply both sides by 2)

(b) $\frac{1}{2}Q = P - 13$ (subtract 13 from both sides)

$Q = 2(P - 13)$ (multiply both sides by 2)

$Q = 2P - 26$ (multiply out brackets)

(c) $Q = 2 \times 17 - 26 = 8$

2 **(a)**

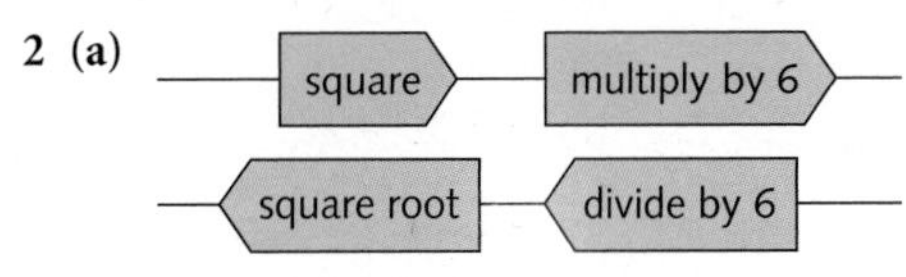

$6x^2 = y$

$x^2 = \frac{y}{6}$ (divide both sides by 6)

$x = \sqrt{\frac{y}{6}}$ (square root both sides)

(b)

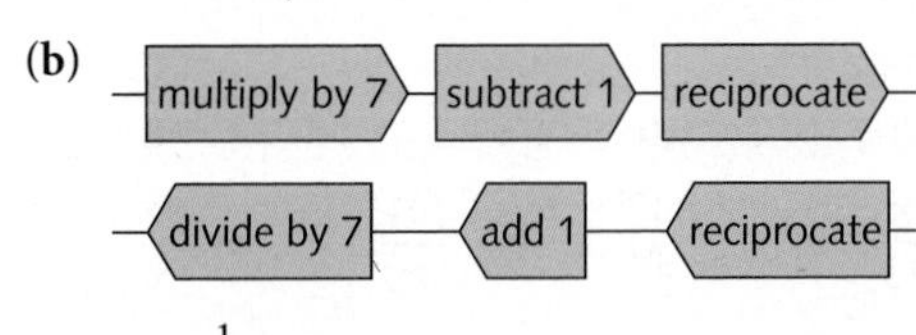

$\frac{1}{7x - 1} = y$

$7x - 1 = \frac{1}{y}$ (reciprocate both sides)

$7x = \frac{1}{y} + 1$ (add 1 to both sides)

$x = \frac{1}{7}\left(\frac{1}{y} + 1\right)$ (divide both sides by 7)

3 **(a)** $x - ay = cx + y$

$x = cx + y + ay$

(add ay to both sides)

$x - cx = y + ay$

(subtract cx from both sides)

$(1 - c)x = (1 + a)y$

(factorize both sides)

$x = \left(\frac{1 + a}{1 - c}\right)y$

(divide both sides by $1 - c$)

(b) $y = \frac{x - 2}{x + 4}$

$(x + 4)y = x - 2$

(multiply both sides by $x + 4$)

$xy + 4y = x - 2$

(multiply out the brackets)

$xy = x - 2 - 4y$

(subtract $4y$ from both sides)

$xy - x = -2 - 4y$

(subtract x from both sides)

$(y - 1)x = -2 - 4y$

(factorize left-hand side)

$x = \frac{-2 - 4y}{y - 1}$

(divide both sides by $y - 1$)

4 $Q = \frac{1}{2}P - 4$; 22.

5 **(a)** $y = 2x + 5$; **(b)** $y = 2(x + 5)$; **(c)** $y = \frac{5}{x^2}$;

(d) $y = 2(x + 4)^2 - 3$.

6 **(a)** multiply by 5 → add 3

(b) add 3 → multiply by 5

(c) square → multiply by 4 → subtract 6

(d) square → add 8 → reciprocate → multiply by 4

7 **(a)** $x = \frac{1}{9}(y + 6)$; **(b)** $x = 3y - 4$; **(c)** $x = 2y$;

(d) $x = 5(y - 8)$; **(e)** $x = \frac{1}{y} - 2$; **(f)** $x = \frac{1}{3}\left(\frac{4}{y} + 7\right)$.

8 **(a)** $P = \frac{Q}{a} - \frac{b}{a}$; **(b)** $Y = \frac{b + I}{1 - a}$;

(c) $P = \frac{1}{aQ} - \frac{b}{a}$; **(d)** $t = \frac{V + 1}{V - 5}$.

9 **(a)** $x = \frac{c - a}{b}$; **(b)** $x = \frac{a^2 - b}{a + 1}$; **(c)** $x = (g - e)^2 - f$;

(d) $x = \frac{ma^2}{b^2} + n$; **(e)** $x = \frac{n^2}{m^2} + m$; **(f)** $x = \left(\frac{a^2 + b^2}{b - a}\right)^2$.

Section 1.6

1 $S = Y - C$

$= Y - (0.8Y + 25)$ (substitute expression for C)

$= Y - 0.8Y - 25$ (multiply out brackets)

$= 0.2Y - 25$ (collect terms)

2 $Y = C + I$ (from theory)

$C = 0.8Y + 25$ (given in question)

$I = 17$ (given in question)

Substituting the given value of I into the first equation gives

$$Y = C + 17$$

and if the expression for C is substituted into this then

$$Y = 0.8Y + 42$$

$0.2Y = 42$ (subtract $0.8Y$ from both sides)

$Y = 210$ (divide both sides by 0.2)

Repeating the calculations with $I = 18$ gives $Y = 215$, so a 1 unit increase in investment leads to a 5 unit increase in income. The scale factor, 5, is called the investment multiplier. In general, the investment multiplier is given by $1/(1 - a)$, where a is the marginal propensity to consume. The foregoing is a special case of this with $a = 0.8$.

3 $Y = C + I + G$ (1)

$G = 40$ (2)

$I = 55$ (3)

$C = 0.8Y_d + 25$ (4)

$T = 0.1Y + 10$ (5)

$Y_d = Y - T$ (6)

Substituting equations (2) and (3) into equation (1) gives

$Y = C + 95$ (7)

Substituting equation (5) into (6) gives

$$Y_d = Y - (0.1Y + 10)$$
$$= 0.9Y - 10$$

so from equation (4),

$$C = 0.8(0.9Y - 10) + 25$$

$= 0.72Y + 17$ (8)

Finally, substituting equation (8) into (7) gives

$$Y = 0.72Y + 112$$

which has solution $Y = 400$.

4 The commodity market is in equilibrium when

$$Y = C + I$$

so we can substitute the given expressions for consumption ($C = 0.7Y + 85$) and investment ($I = -50r + 1200$) to deduce that

$$Y = 0.7Y - 50r + 1285$$

which rearranges to give the IS schedule,

$0.3Y + 50r = 1285$ (1)

The money market is in equilibrium when

$$M_S = M_D$$

Now we are given that $M_S = 500$ and that total demand,

$$M_D = L_1 + L_2 = 0.2Y - 40r + 230$$

so that

$$500 = 0.2Y - 40r + 230$$

which rearranges to give the LM schedule,

$0.2Y - 40r = 270$ (2)

We now solve equations (1) and (2) as a pair of simultaneous equations.

Step 1

Multiply equation (1) by 0.2 and (2) by 0.3 and subtract to get

$$22r = 176$$

Step 2

Divide through by 22 to get $r = 8$.

Step 3

Substitute $r = 8$ into equation (1) to give $Y = 2950$.

The IS and LM curves shown in Figure S1.15 confirm this, since the point of intersection has coordinates (8, 2950). A change in I does not affect the LM schedule. However, if the autonomous level of investment increases from its current level of 1200 then the right-hand side of the IS schedule (1) will rise. The IS curve moves upwards, causing both r and Y to increase.

Figure S1.15

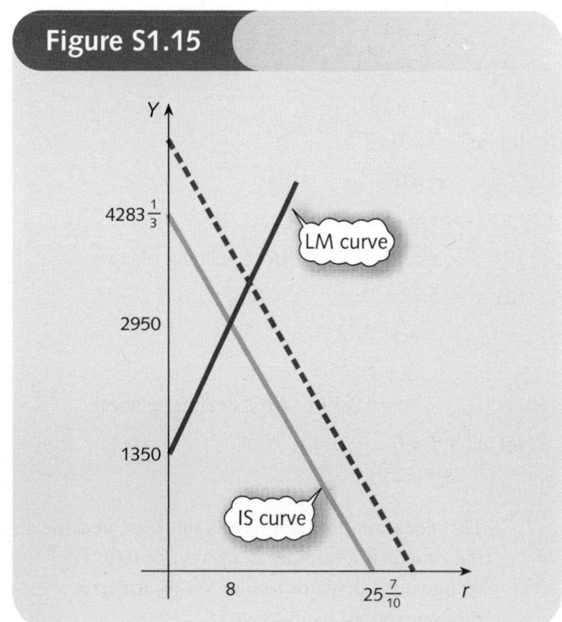

5 **(a)** 40; **(b)** 0.7; $\frac{10}{7}(C-40)$; 100.

6 **(a)** $S = 0.1Y - 72$; **(b)** $S = \frac{10Y - 500}{Y + 10}$.

7 **(a)** 325; **(b)** 225; **(c)** 100.

8 $C = \frac{aI^* + b}{1 - a}$

9 825.

10 $Y = 2500$, $r = 10$.

11 **(a)** $C = 120 + 0.8Y$;
(b) $C = 40 + 0.8Y$;
(c) $C = 120 + 0.6Y$.

With a lump sum tax, the graph has the same slope but has been shifted downwards.

With a proportional tax, the graph has the same intercept but is less steep.

(a) 600; **(b)** 200; **(c)** 300.

12 $0.9Y + 30$; 300.

(a) Slope decreases; 150. **(b)** Shifts up 5 units; 350.

Chapter 2

Section 2.1

1 **(a)** $x^2 - 100 = 0$

$$x^2 = 100$$
$$x = \pm\sqrt{100}$$
$$x = \pm 10$$

(b) $2x^2 - 8 = 0$

$$2x^2 = 8$$
$$x^2 = 4$$
$$x = \pm\sqrt{4}$$
$$x = \pm 2$$

(c) $x^2 - 3 = 0$

$$x^2 = 3$$
$$x = \pm\sqrt{3}$$

$x = \pm 1.73$ (to 2 decimal places)

(d) $x^2 - 5.72 = 0$

$$x^2 = 5.72$$
$$x = \pm\sqrt{5.72}$$

$x = \pm 2.39$ (to 2 decimal places)

(e) $x^2 + 1 = 0$

$$x^2 = -1$$

This equation does not have a solution, because the square of a number is always positive. Try using your calculator to find $\sqrt{(-1)}$. An error message should be displayed.

(f) $3x^2 + 6.21 = 0$

$$3x^2 = -6.21$$
$$x^2 = -2.07$$

This equation does not have a solution, because it is impossible to find the square root of a negative number.

(g) $x^2 = 0$

This equation has exactly one solution, $x = 0$.

2 **(a)** $a = 2$, $b = -19$, $c = -10$.

$$x = \frac{-(-19) \pm \sqrt{((-19)^2 - 4(2)(-10))}}{2(2)}$$
$$= \frac{19 \pm \sqrt{(361 + 80)}}{4}$$
$$= \frac{19 \pm \sqrt{441}}{4} = \frac{19 \pm 21}{4}$$

This equation has two solutions:

$$x = \frac{19 + 21}{4} = 10$$
$$x = \frac{19 - 21}{4} = -\frac{1}{2}$$

(b) $a = 4$, $b = 12$, $c = 9$.

$$x = \frac{-12 \pm \sqrt{((12)^2 - 4(4)(9))}}{2(4)}$$
$$= \frac{-12 \pm \sqrt{(144 - 144)}}{8}$$
$$= \frac{-12 \pm 0}{8}$$

This equation has one solution, $x = -3/2$.

(c) $a = 1$, $b = 1$, $c = 1$.

$$x = \frac{-1 \pm \sqrt{((1)^2 - 4(1)(1))}}{2(1)}$$
$$= \frac{-1 \pm \sqrt{(1 - 4)}}{2}$$
$$= \frac{-1 \pm \sqrt{(-3)}}{2}$$

This equation has no solutions, because $\sqrt{(-3)}$ does not exist.

(d) We first need to collect like terms to convert

$$x^2 - 3x + 10 = 2x + 4$$

into the standard form

$$ax^2 + bx + c = 0$$

Subtracting $2x + 4$ from both sides gives

$$x^2 - 5x + 6 = 0$$
$$a = 1, b = -5, c = 6.$$

$$x = \frac{-(-5) \pm \sqrt{((-5)^2 - 4(1)(6))}}{2(1)}$$

$$= \frac{5 \pm \sqrt{(25 - 24)}}{2}$$

$$= \frac{5 \pm \sqrt{1}}{2}$$

$$= \frac{5 \pm 1}{2}$$

This equation has two solutions:

$$x = \frac{5+1}{2} = 3$$

$$x = \frac{5-1}{2} = 2$$

3 (a) If $(x - 4)(x + 3) = 0$ then either

$x - 4 = 0$ with solution $x = 4$

or

$x + 3 = 0$ with solution $x = -3$

This equation has two solutions, $x = 4$ and $x = -3$.

(b) If $x(10 - 2x) = 0$ then either

$x = 0$

or

$10 - 2x = 0$ with solution $x = 5$

This equation has two solutions, $x = 0$ and $x = 5$.

(c) If $(2x - 6)(2x - 6) = 0$ then

$2x - 6 = 0$ with solution $x = 3$

This equation has one solution, $x = 3$.

4 (a)

x	−1	0	1	2	3	4
$f(x)$	21	5	−3	−3	5	21

The graph is sketched in Figure S2.1.

Figure S2.1

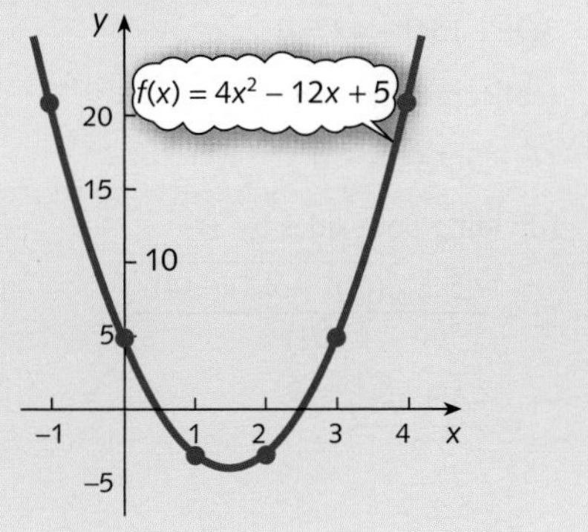

(b)

x	0	1	2	3	4	5	6
$f(x)$	−9	−4	−1	0	−1	−4	−9

The graph is sketched in Figure S2.2.

Figure S2.2

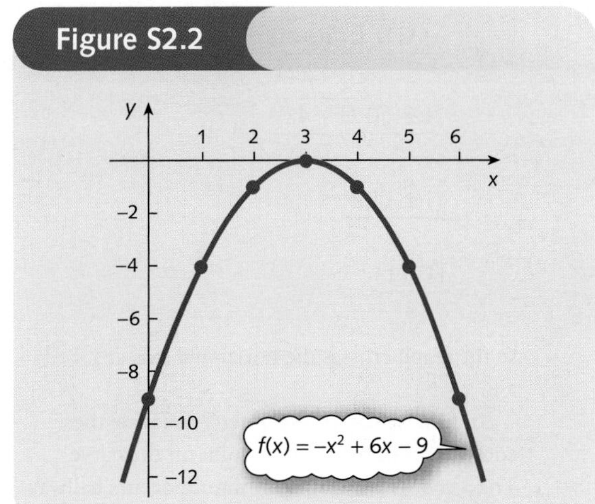

(c)

x	−2	−1	0	1	2	3	4
$f(x)$	−22	−12	−6	−4	−6	−12	−22

The graph is sketched in Figure S2.3.

Figure S2.3

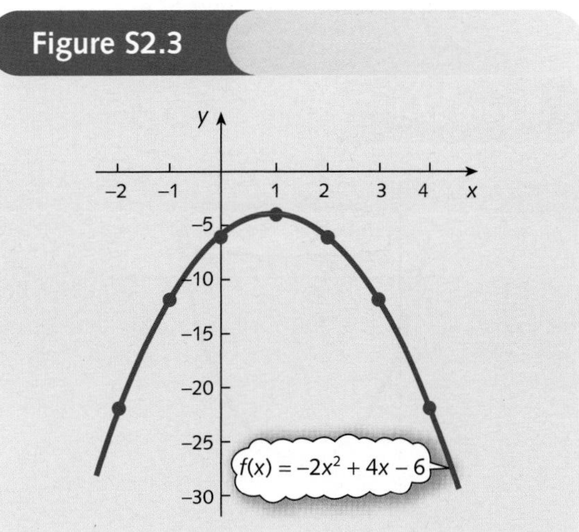

5 (a) ***Step 1***

The coefficient of x^2 is 2, which is positive, so the graph is U-shaped.

Step 2

The constant term is −6, so the graph crosses the vertical axis at $y = -6$.

Step 3

The quadratic equation

$$2x^2 - 11x - 6 = 0$$

has solution

$$x = \frac{-(-11) \pm \sqrt{((-11)^2 - 4(2)(6))}}{2(2)}$$

$$= \frac{11 \pm \sqrt{(121 + 48)}}{4}$$

$$= \frac{11 \pm \sqrt{169}}{4}$$

$$= \frac{11 \pm 13}{4}$$

so the graph crosses the horizontal axis at $x = -\frac{1}{2}$ and $x = 6$.

In fact, we can use symmetry to locate the coordinates of the turning point on the curve. The x coordinate of the minimum occurs halfway between $x = -\frac{1}{2}$ and $x = 6$ at

$$x = \frac{1}{2}\left(-\frac{1}{2} + 6\right) = \frac{11}{4}$$

The corresponding y coordinate is

$$2\left(\frac{11}{4}\right)^2 - 11\left(\frac{11}{4}\right) - 6 = -\frac{169}{8}$$

The graph is sketched in Figure S2.4.

Figure S2.4

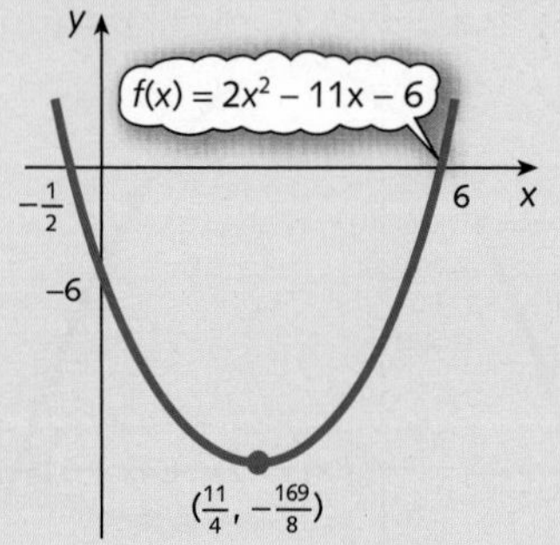

(b) *Step 1*

The coefficient of x is 1, which is positive, so the graph is U-shaped.

Step 2

The constant term is 9, so the graph crosses the vertical axis at $y = 9$.

Step 3

The quadratic equation

$$x^2 - 6x + 9 = 0$$

has solution

$$x = \frac{-(-6) \pm \sqrt{((-6)^2 - 4(1)(9))}}{2(1)}$$

$$= \frac{6 \pm \sqrt{(36 - 36)}}{2}$$

$$= \frac{6 \pm \sqrt{0}}{2} = 3$$

so the graph crosses the x axis at $x = 3$.

The graph is sketched in Figure S2.5.

Figure S2.5

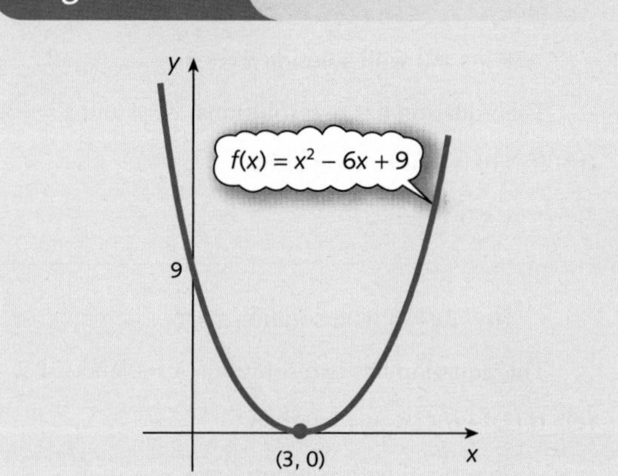

6 In equilibrium, $Q_S = Q_D = Q$, so the supply and demand equations become

$$P = 2Q^2 + 10Q + 10$$

$$P = -Q^2 - 5Q + 52$$

Hence

$$2Q^2 + 10Q + 10 = -Q^2 - 5Q + 52$$

$$3Q^2 + 15Q - 42 = 0$$

(collecting like terms)

$$Q^2 + 5Q - 14 = 0$$

(dividing both sides by 3)

$$Q = \frac{-5 \pm \sqrt{((5)^2 - 4(1)(-14))}}{2(1)}$$

$$= \frac{-5 \pm \sqrt{81}}{2}$$

$$= \frac{-5 \pm 9}{2}$$

so $Q = -7$ and $Q = 2$. Ignoring the negative solution gives $Q = 2$. From the supply equation, the corresponding equilibrium price is

$$P = 2(2)^2 + 10(2) + 10 = 38$$

As a check, the demand equation gives

$$P = -(2)^2 - 5(2) + 52 = 38$$

7 **(a)** ±9; **(b)** ±6; **(c)** ±2; **(d)** −2, 4; **(e)** −9, −1; **(f)** −2, 9.

8 **(a)** 1, −3; **(b)** $\frac{1}{2}$, −10; **(c)** 0, −5; **(d)** $-\frac{5}{3}$, $\frac{9}{4}$; **(e)** $\frac{5}{4}$, 5; **(f)** 2, −1, 4.

9 **(a)** 0.44, 4.56; **(b)** −2.28, 0.22; **(c)** −0.26, 2.59; **(d)** −0.30, 3.30; **(e)** −2; **(f)** no solutions.

10 **(a)** −4, 4; **(b)** 0, 100; **(c)** 5, 17; **(d)** 9; **(e)** no solution.

11 The graphs are sketched in Figure S2.6.

Figure S2.6

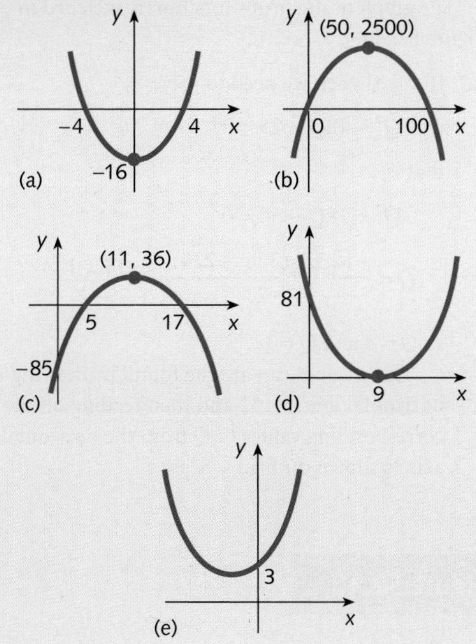

12 $c = 12$; 6.

13 $Q = 4$, $P = 36$.

14 $P = 22$, $Q = 3$.

Section 2.2

1 $\text{TR} = PQ = (1000 - Q)Q = 1000Q - Q^2$

Step 1

The coefficient of Q^2 is negative, so the graph has an inverted U shape.

Step 2

The constant term is zero, so the graph crosses the vertical axis at the origin.

Step 3

From the factorization

$$\text{TR} = (1000 - Q)Q$$

the graph crosses the horizontal axis at $Q = 0$ and $Q = 1000$.

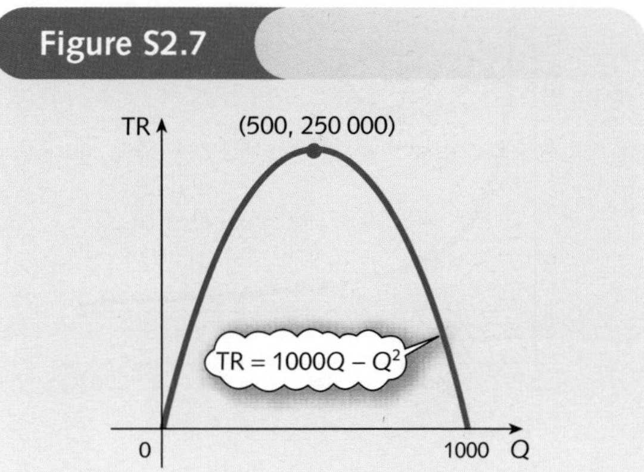

Figure S2.7

The graph is sketched in Figure S2.7. By symmetry the parabola reaches its maximum halfway between 0 and 1000 at $Q = 500$. The corresponding value of TR is

$$\text{TR} = 1000(500) - (500)^2 = 250\,000$$

From the demand equation, when $Q = 500$,

$$P = 1000 - 500 = 500$$

2 $\text{TC} = 100 + 2Q$

$$\text{AC} = \frac{100 + 2Q}{Q} = \frac{100}{Q} + 2$$

The graph of the total cost function is sketched in Figure S2.8.

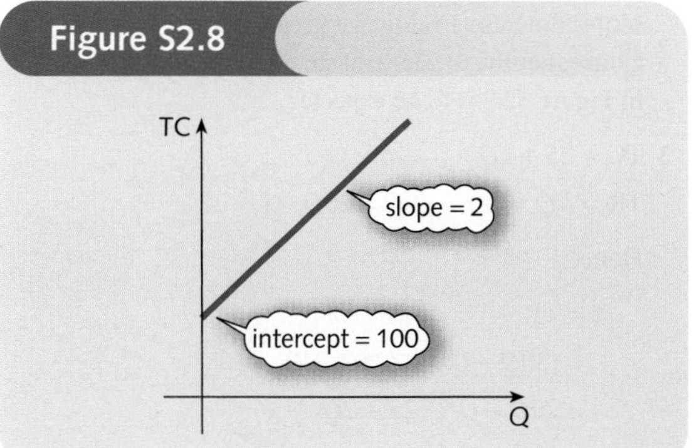

Figure S2.8

One possible table of function values for the average cost function is

Q	10	25	50	100	200
AC	12	6	4	3	2.5

The graph of the average cost function is sketched in Figure S2.9.

Figure S2.9

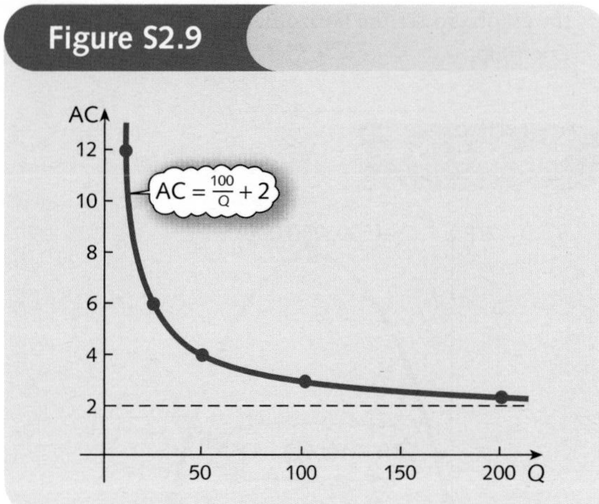

In fact, it is not necessary to plot the tabulated values if all that is required is a rough sketch. It is obvious that if a very small number is put into the AC function then a very large number is produced because of the term 100/Q. For example, when Q = 0.1

$$AC = \frac{100}{0.1} + 2 = 1002$$

It should also be apparent that if a very large number is put into the average cost function then the term $100/Q$ is insignificant, so AC is approximately 2. For example, when Q = 10 000

$$AC = \frac{100}{10\,000} + 2 = 2.01$$

The graph of AC therefore 'blows up' near $Q = 0$ but settles down to a value just greater than 2 for large Q. Consequently, the general shape of the graph shown in Figure S2.9 is to be expected.

3 $TC = 25 + 2Q$

$TR = PQ = (20 - Q)Q = 20Q - Q^2$

Hence

$$\begin{aligned}\pi &= TR - TC \\ &= (20Q - Q^2) - (25 + 2Q) \\ &= 20Q - Q^2 - 25 - 2Q \\ &= -Q^2 + 18Q - 25\end{aligned}$$

Step 1

The coefficient of Q^2 is negative, so the graph has an inverted U shape.

Step 2

The constant term is −25, so the graph crosses the vertical axis at −25.

Step 3

The quadratic equation

$$-Q^2 + 18Q - 25 = 0$$

has solutions

$$\begin{aligned}Q &= \frac{-18 \pm \sqrt{(324 - 100)}}{-2} \\ &= \frac{-18 \pm 14.97}{-2}\end{aligned}$$

so the graph crosses the horizontal axis at $Q = 1.52$ and $Q = 16.48$.

The graph of the profit function is sketched in Figure S2.10.

(a) If $\pi = 31$ then we need to solve

$$-Q^2 + 18Q - 25 = 31$$

that is,

$$-Q^2 + 18Q - 56 = 0$$

$$Q = \frac{-18 \pm \sqrt{(324 - 224)}}{-2} = -\frac{18 \pm 10}{-2}$$

so $Q = 4$ and $Q = 14$.

These values can also be found by drawing a horizontal line $\pi = 31$ and then reading off the corresponding values of Q from the horizontal axis as shown on Figure S2.10.

Figure S2.10

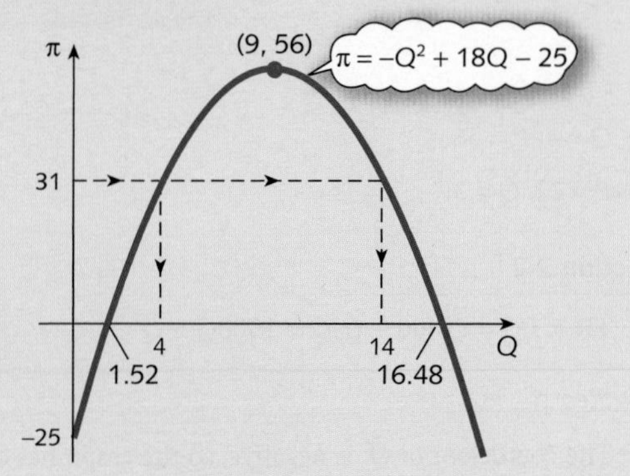

(b) By symmetry the parabola reaches its maximum halfway between 1.52 and 16.48: that is, at

$$Q = \tfrac{1}{2}(1.52 + 16.48) = 9$$

The corresponding profit is given by

$$\pi = -(9)^2 + 18(9) - 25 = 56$$

4 **(a)** $4Q$; **(b)** 7; **(c)** $10Q - 4Q^2$.

The graphs are sketched in Figures S2.11, S2.12 and S2.13.

Figure S2.11

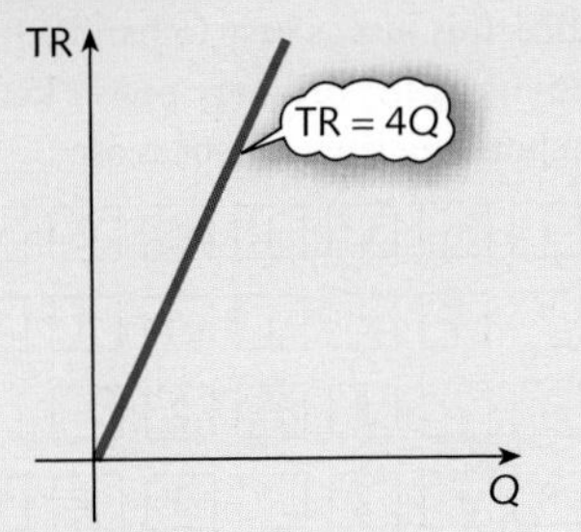

Figure S2.12

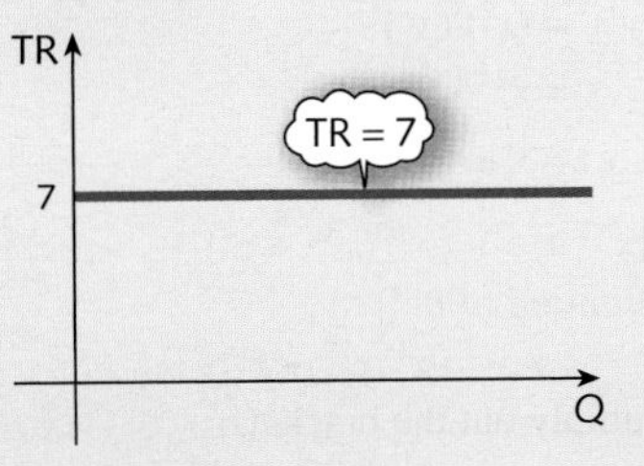

Figure S2.13

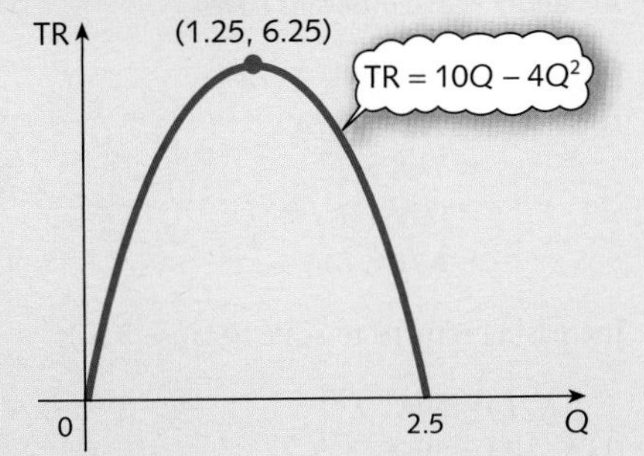

5 **(a)** $P = 50 - 4Q$;

(b) $P = \dfrac{10}{Q}$.

6 $TC = 500 + 10Q$; $AC = \dfrac{500}{Q} + 10$.

The graphs are sketched in Figures S2.14 and S2.15.

Figure S2.14

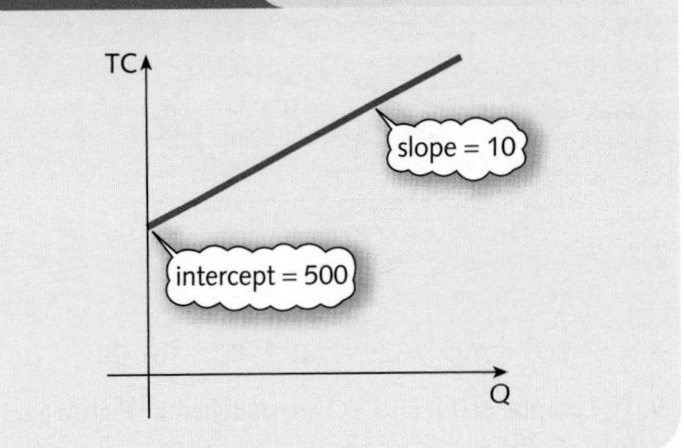

Figure S2.15

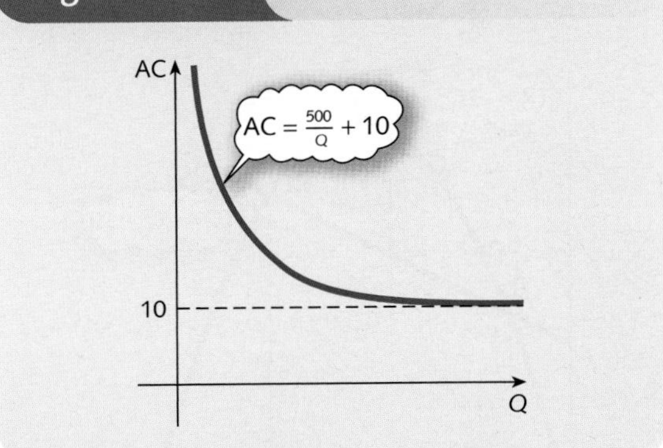

7 $TC = Q^2 + Q + 1$; $AC = Q + 1 + \dfrac{1}{Q}$.

The graphs are sketched in Figures S2.16 and S2.17.

Figure S2.16

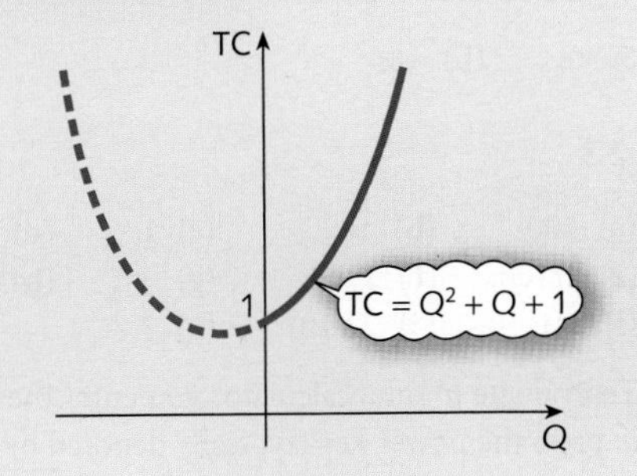

Figure S2.17

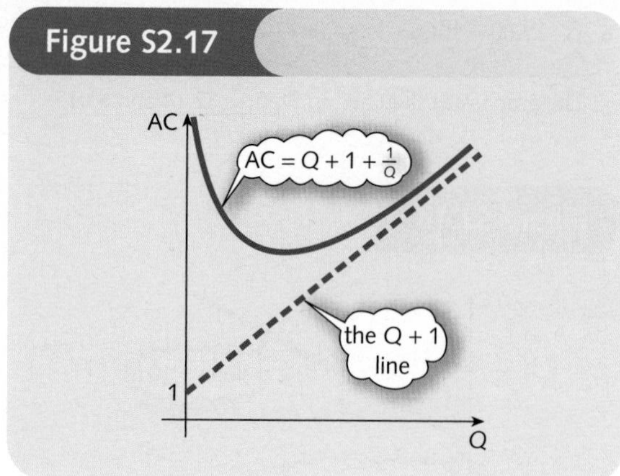

8 $\pi = -2Q^2 + 20Q - 32$; **(a)** 2, 8; **(b)** 20; **(c)** 5.

9 The graphs of TR and TC are sketched in Figure S2.18. **(a)** 1, 5; **(b)** 3.

Figure S2.18

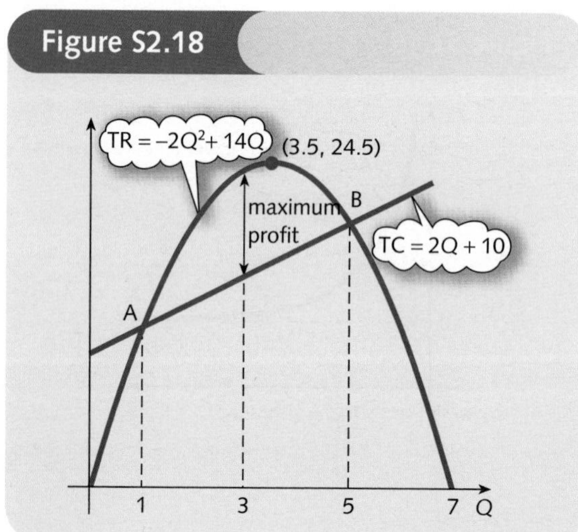

10 $a + b + c = 9$

$4a + 2b + c = 34$

$9a + 3b + c = 19$

$a = -20$, $b = 85$, $c = -56$; $\pi = -36$

11 1000, 1004.08, . . . , 1750; 22.8.

12 **(a)** 5, 30; **(b)** 18.

Section 2.3

1 **(1)** **(a)** 100; **(b)** 10; **(c)** 1; **(d)** 1/10; **(e)** 1/100; **(f)** 1; **(g)** −1; **(h)** 1/343; **(i)** 81; **(j)** 72 101; **(k)** 1.

(2) To evaluate b^n on a calculator you enter the base, b, press the power key (typically denoted by $\boxed{x^y}$), enter the power, n, and finally press the $\boxed{=}$ key to get the result. For example to calculate 10^2 the key sequence is $\boxed{1}\ \boxed{0}\ \boxed{x^y}\ \boxed{2}\ \boxed{=}$.

2 **(1)** **(a)** 4, because $4^2 = 16$.

(b) 3, because $3^3 = 27$.

(c) 32, because $4^{5/2} = (4^{1/2})^5 = 2^5$.

(d) $^1/_4$, because $8^{-2/3} = (8^{1/3})^{-2} = 2^{-2} = 1/2^2$.

(e) 1, because $1^n = 1$ for any index, n.

(2) Some calculators have a function key (typically labelled $\boxed{x^{1/y}}$) which can be used to evaluate $b^{1/n}$ directly. However, this does not result in any real reduction in the number of key presses. In practice, it is just as easy to handle fractional indices using the ordinary power key $\boxed{x^y}$. The corresponding key sequences are

(a) $\boxed{1}\ \boxed{6}\ \boxed{x^y}\ \boxed{1}\ \boxed{a^b/_c}\ \boxed{2}\ \boxed{=}$

(b) $\boxed{2}\ \boxed{7}\ \boxed{x^y}\ \boxed{1}\ \boxed{a^b/_c}\ \boxed{3}\ \boxed{=}$

(c) $\boxed{4}\ \boxed{x^y}\ \boxed{5}\ \boxed{a^b/_c}\ \boxed{2}\ \boxed{=}$

(d) $\boxed{8}\ \boxed{x^y}\ \boxed{2}\ \boxed{a^b/_c}\ \boxed{3}\ \boxed{\pm}\ \boxed{=}$

(e) $\boxed{1}\ \boxed{x^y}\ \boxed{1}\ \boxed{7}\ \boxed{a^b/_c}\ \boxed{2}\ \boxed{5}\ \boxed{\pm}\ \boxed{=}$

3 **(a)** $(x^{3/4})^8 = x^{(3/4)\times 8} = x^6$ (rule 3)

(b) $x^2 \div x^{3/2} = x^{2-(3/2)} = x^{1/2}$ (rule 2)

(c) $(x^2y^4)^3 = (x^2)^3(y^4)^3$ (rule 4)

$= x^{2\times 3}y^{4\times 3}$ (rule 3)

$= x^6y^{12}$

(d) $\sqrt{x}(x^{5/2} + y^3) = x^{1/2}(x^{5/2} + y^3)$
(definition of $b^{1/n}$)

$= x^{1/2}x^{5/2} + x^{1/2}y^3$
(multiply out the brackets)

$= x^{(1/2)+(5/2)} + x^{1/2}y^3$
(rule 1)

$= x^3 + x^{1/2}y^3$

The term $x^{1/2}y^3$ cannot be simplified, because $x^{1/2}$ and y^3 have different bases.

4 **(a)** $f(K, L) = 7KL^2$

$f(\lambda K, \lambda L) = 7(\lambda K)(\lambda L)^2$

$= 7\lambda K\, \lambda^2L^2$ (rule 4)

$= (\lambda\lambda^2)(7KL^2)$

$= \lambda^3 f(K, L)$ (rule 1)

Increasing returns to scale because 3 > 1.

(b) $f(K, L) = 50K^{1/4}L^{3/4}$

$f(\lambda K, \lambda L) = 50(\lambda K)^{1/4}(\lambda L)^{3/4}$

$= 50\lambda^{1/4}K^{1/4}\lambda^{3/4}L^{3/4}$ (rule 4)

$= (\lambda^{1/4}\lambda^{3/4})(50K^{1/4}L^{3/4})$

$= \lambda^1 f(K, L)$ (rule 1)

Constant returns to scale.

5 (1) **(a)** 3; **(b)** 2; **(c)** 1; **(d)** 0; **(e)** −1; **(f)** −2.

(2) Same as part (1), because if $M = 10^n$ then $\log_{10} M = n$.

(3) On most calculators there are two logarithm function keys, $\log_{10}$ (possibly labelled log or $\log_{10}$) and ln (possibly labelled ln or $\log_e$). The latter is known as the natural logarithm and we introduce this function in the next section. This question wants you to evaluate logarithms to base 10, so we use the key [log].

Warning: there is no standard layout for the keyboard of a calculator. It may be necessary for you first to use the shift key (sometimes called the inverse function or second function key) to activate the $\log_{10}$ function.

6 (a) $\log_b\left(\frac{x}{y}\right) + \log_b z$ (rule 2)

$= \log_b\left(\frac{xz}{y}\right)$ (rule 1)

(b) $\log_b x^4 + \log_b y^2$ (rule 3)

$= \log_b(x^4y^2)$ (rule 1)

7 (a) $3^x = 7$

$\log(3^x) = \log 7$

(take logarithms of both sides)

$x \log 3 = \log 7$

(rule 3)

$$x = \frac{\log 7}{\log 3}$$

(divide both sides by log 3)

$$x = \frac{0.845\ 098\ 040}{0.477\ 121\ 255}$$

(using base 10 on a calculator)

$x = 1.77$

(to two decimal places)

(b) $5(2)^x = 10^x$

$\log[5(2)^x] = \log(10)^x$

(take logarithms of both sides)

$\log 5 + \log(2^x) = \log(10)^x$

(rule 1)

$\log 5 + x \log 2 = x \log 10$

(rule 3)

$x(\log 10 - \log 2) = \log 5$

(collect terms and factorize)

$x \log 5 = \log 5$

(rule 2)

$x = 1$

(divide both sides by log 5)

which is, of course, the obvious solution to the original equation! Did you manage to spot this for yourself *before* you started taking logs?

8 (a) 64; **(b)** 2; **(c)** 1/3; **(d)** 1; **(e)** 1; **(f)** 6; **(g)** 4; **(h)** 1/343.

9 (a) 8; **(b)** 1/32; **(c)** 625; **(d)** 9/4; **(e)** 2/3.

10 (a) y^2; **(b)** xy^2; **(c)** x^4y^2; **(d)** 1; **(e)** 2; **(f)** $5pq^2$.

11 (a) x^{-4}; **(b)** $5x^{1/2}$; **(c)** $x^{-1/2}$; **(d)** $2x^{3/2}$; **(e)** $8x^{-4/3}$.

12 (a) 3600; **(b)** 200 000.

13 The functions in parts (a) and (b) are homogeneous of degree 7/12 and 2 respectively, so (a) displays decreasing returns to scale and (b) displays increasing returns to scale. The function in part (c) is not homogeneous.

14 $A[b(\lambda K)^\alpha + (1-b)(\lambda L)^\alpha]^{1/\alpha}$

$= A[b\lambda^\alpha K^\alpha + (1-b)\lambda^\alpha L^\alpha]^{1/\alpha}$ (rule 4)

$= A[(\lambda^\alpha)(bK^\alpha + (1-b)L^\alpha)]^{1/\alpha}$ (factorize)

$= A(\lambda^\alpha)^{1/\alpha}[bK^\alpha + (1-b)L^\alpha]^{1/\alpha}$ (rule 4)

$= \lambda A[bK^\alpha + (1-b)L^\alpha]^{1/\alpha}$ (rule 3)

so $f(\lambda K, \lambda L) = \lambda^1 f(K, L)$ as required. This is known as the constant elasticity of substitution (CES) production function.

15 (a) 2; **(b)** −1; **(c)** −3; **(d)** 6; **(e)** ½; **(f)** 4/3.

16 (a) 2/3; **(b)** 3; **(c)** ¼.

17 (a) 2; **(b)** 1; **(c)** 0; **(d)** ½; **(e)** −1.

18 (a) 0; **(b)** $\log_b\left(\frac{x^3}{y^2}\right)$; **(c)** $\log_b\left(\frac{x^5y}{z^2}\right)$.

19 (a) $2\log_b x + 3\log_b y + 4\log_b z$

(b) $4\log_b x - 2\log_b y - 5\log_b z$

(c) $\log_b x - \frac{1}{2}\log_b y - \frac{1}{2}\log_b z$

20 (a) $-q$; **(b)** $2p + q$; **(c)** $q - 4r$; **(d)** $p + q + 2r$.

21 (a) 78.31; **(b)** 1.48; **(c)** 3; **(d)** 0.23.

22 (a) $x \le 0.386$ (3 dp)

(b) $x > 14.425$ (Notice that the inequality is > here.)

23 $x = 3$ (Note that the second solution of your quadratic, $x = -5$ is not valid.)

Section 2.4

1

x	−3	−2	−1	0	1	2	3
3^x	0.04	0.11	0.33	1	3	9	27
3^{-x}	27	9	3	1	0.33	0.11	0.04

The graphs of 3^x and 3^{-x} are sketched in Figures S2.19 and S2.20 (overleaf) respectively.

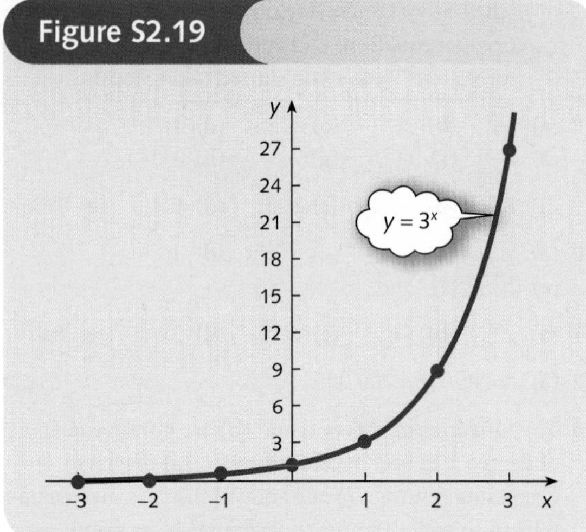

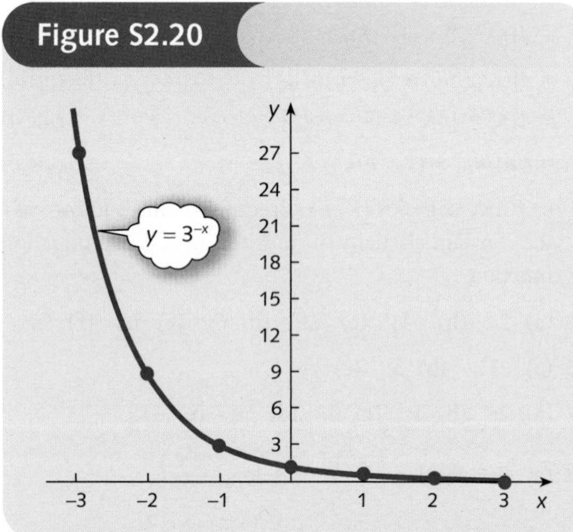

2 (a) 2.718 145 927, 2.718 268 237, 2.718 280 469.

(b) 2.718 281 828 ; values in part (a) are getting closer to that of part (b).

3 (1) Substituting $t = 0$, 10, 20 and 30 gives

(a) $y(0) = \dfrac{55}{1 + 800e^{0}} = 0.07\%$

(b) $y(10) = \dfrac{55}{1 + 800e^{-3}} = 1.35\%$

(c) $y(20) = \dfrac{55}{1 + 800e^{-6}} = 18.44\%$

(d) $y(30) = \dfrac{55}{1 + 800e^{-9}} = 50.06\%$

(2) As t increases, $e^{-0.3t}$ goes to zero, so y approaches

$$\frac{55}{1 + 800(0)} = 55\%$$

(3) A graph of y against t, based on the information obtained in parts (1) and (2), is sketched in Figure S2.21. This shows that, after a slow start, camcorder ownership grows rapidly between $t = 10$ and 30. However, the rate of growth then decreases as the market approaches its saturation level of 55%.

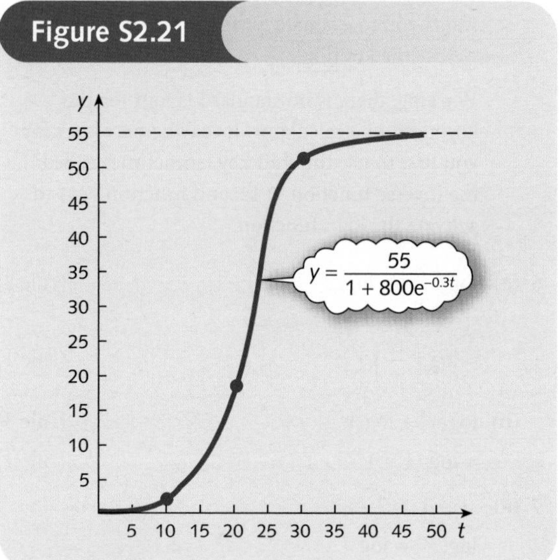

4 (a) $\ln a^2 + \ln b^3$ (rule 1)

$= 2 \ln a + 3 \ln b$ (rule 3)

(b) $\ln x^{1/2} - \ln y^3$ (rule 3)

$= \ln\left(\dfrac{x^{1/2}}{y^3}\right)$ (rule 2)

5 (a) Putting $t = 0$ and 2 into the expression for TR gives

$\text{TR} = 5e^{0} = \$5$ million

$\text{TR} = 5e^{-0.3} = \$3.7$ million

(b) To solve $5e^{-0.15t} = 2.7$ we divide by 5 to get $e^{-0.15t} = 0.54$ and then take natural logarithms, which gives

$$-0.15t = \ln(0.54) = -0.62$$

Hence $t = 4$ years.

6 (1) Missing numbers are 0.99 and 2.80.

(2) The graph is sketched in Figure S2.22.

Intercept, 0.41; slope, 0.20.

(3) $A = 0.2$, $B = e^{0.41} = 1.5$.

(4) **(a)** 9100; **(b)** 2.4×10^8; answer to part **(b)** is unreliable since $t = 60$ is well outside the range of given data.

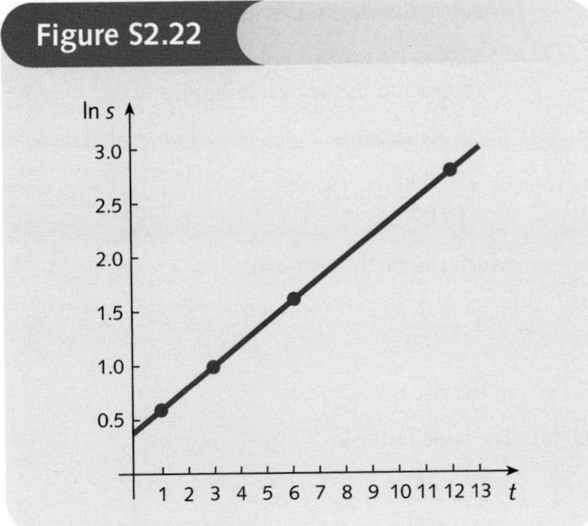

7 **(1)** **(a)** 33; **(b)** 55; **(c)** 98.

(2) 100.

(3) The graph of N against t is sketched in Figure S2.23.

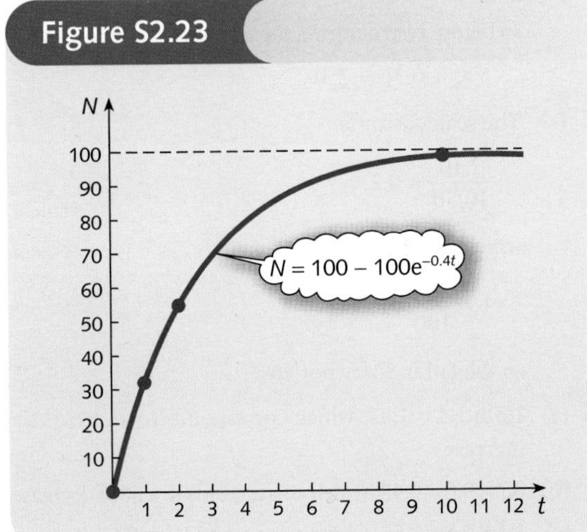

The graph sketched in Figure S2.23 is called a learning curve. It shows that immediately after training the worker can produce only a small number of items. However, with practice, output quickly increases and finally settles down at a daily rate of 100 items.

8 **(a)** $\ln x + \ln y$; **(b)** $\ln x + 4\ln y$; **(c)** $2\ln x + 2\ln y$; **(d)** $5\ln x - 7\ln y$; **(e)** $^1/_2 \ln x - {}^1/_2 \ln y$; **(f)** $^1/_2 \ln x + {}^3/_2 \ln y - {}^1/_2 \ln z$.

9 **(a)** $\ln x^3$; **(b)** $\ln\left(\frac{x^4z^5}{x^3}\right)$.

10 **(a)** 1.77; **(b)** −0.80; **(c)** no solution; **(d)** 0.87; **(e)** 0.22; **(f)** 0.35.

11 $A = 50\ 000$, $a = 0.137$; **(a)** \$25 205; **(b)** \$0.

12 The equation

$$N = c(1 - e^{-kt}) = c - ce^{-kt}$$

rearranges as

$$e^{-kt} = \frac{c - N}{c}$$

Taking logarithms gives

$$-kt = \ln\left(\frac{c-N}{c}\right)$$

$$t = -\frac{1}{k}\ln\left(\frac{c-N}{c}\right)$$

$$= \frac{1}{k}\ln\left(\frac{c-N}{c}\right)^{-1} = \frac{1}{k}\ln\left(\frac{c}{c-N}\right)$$

(a) 350 000.

(b) 60 days.

(c) Market saturation level is 700 000, which is less than the three-quarters of a million copies needed to make a profit, so the proprietor should sell.

13 Putting $y = \ln Q$, $x = \ln K$ gives

$$y = {}^1/_3 x + (\ln 3 + {}^1/_2 \ln L)$$

which is of the form '$y = ax + b$'.

Slope = $^1/_3$, intercept = $\ln 3 + {}^1/_2 \ln L$.

14 **(a)** $\ln Q = \ln(AL^n) = \ln A + \ln L^n = \ln A + n \ln L$.

(b)

$\ln L$	0	0.69	1.10	1.39	1.61
$\ln Q$	−0.69	−0.46	−0.33	−0.22	−0.16

(c) $n = 0.34$, $A = 0.50$.

15 All five graphs are increasing throughout.

Graphs of x^2, x^3 and e^x bend upwards, i.e. the slope increases with increasing x, whereas x has a constant slope and $\sqrt{x}$ has a decreasing slope.

With the exception of e^x, all graphs pass through (0, 0) and (1, 1). On $0 < x < 1$, $x^3 < x^2 < x < \sqrt{x}$ whereas on $x > 1$ the order is reversed.

16 All five graphs have the same basic shape and pass through (1, 0).

On $0.2 < x < 1$, $\ln x < \log_6 x < \log_{10} x$ whereas on $x > 1$, the order is reversed.

17 **(a)**

$$P - 100 = -{}^2/_3 Q^n$$
(subtract 100 from both sides)

$$-{}^3/_2 (P - 100) = Q^n$$
(divide both sides by $-^2/_3$)

$$150 - 1.5P = Q^n$$
(multiply out the brackets)

$$\ln(150 - 1.5P) = \ln(Q^n) = n\ln Q$$
(take logs and use rule 3)

which is of the form $y = ax + b$ with $a = n$ and $b = 0$.

(b)

ln Q	2.30	3.91	4.09
ln(150–1.5P)	2.01	3.11	3.40

ln Q	4.61	5.29	5.99
ln(150–1.5P)	3.81	4.32	4.79

$n = 0.8$.

Chapter 3

Section 3.1

1 (a) $\frac{10}{100} \times 2.90 = 0.1 \times 2.90 = \$0.29$

(b) $\frac{75}{100} \times 1250 = 0.75 \times 1250 = \$937.50$

(c) $\frac{24}{100} \times 580 = 0.24 \times 580 = \$139.20$

2 (a) The rise in annual sales is

$$55\,000 - 50\,000 = 5000$$

As a fraction of the original this is

$$\frac{5000}{50\,000} = \frac{10}{100}$$

so the percentage rise is 10%.

(b) As a fraction

15% is the same as $\frac{15}{100} = 0.15$

so the tax is

$$0.15 \times 1360 = 204$$

Hence the consumer pays

$$1360 + 204 = \$1564$$

(c) As a fraction

7% is the same as $\frac{7}{100} = 0.07$

so the fall in value is

$$0.07 \times 9500 = 665$$

Hence the final value is

$$9500 - 665 = \$8835$$

3 (a) The scale factor is

$$1 + \frac{13}{100} = 1.13$$

We are going forwards in time, so we *multiply* to get

$$6.5 \times 1.13 = \$7.345 \text{ million}$$

(b) The scale factor is

$$1 + \frac{63}{100} = 1.63$$

We are going backwards in time, so we *divide* to get

$$1.24 \div 1.63 = \$76 \text{ billion}$$

(correct to the nearest billion)

(c) The scale factor is

$$\frac{123\,050}{115\,000} = 1.07$$

which can be thought of as

$$1 + \frac{7}{100}$$

so the rise is 7%.

4 (a) The scale factor is

$$1 - \frac{65}{100} = 0.35$$

so the new level of output is

$$25\,000 \times 0.35 = 8750$$

(b) The scale factor is

$$1 - \frac{24}{100} = 0.76$$

so before restructuring the number of employees was

$$570 \div 0.76 = 750$$

(c) The scale factor is

$$\frac{2.10}{10.50} = 0.2$$

which can be thought of as

$$1 - \frac{80}{100}$$

so the fall is 80% (not 20%!).

5 (a) $1.3 \times 1.4 = 1.82$, which corresponds to an 82% increase.

(b) $0.7 \times 0.6 = 0.42$, which corresponds to a 58% decrease.

(c) $1.1 \times 0.5 = 0.55$, which corresponds to a 45% decrease.

6 100, 101.5, 105.4, 104.3, 106.9.

7 (a) 5.7% increase.

(b) $\frac{105.7}{89.3} = 1.184$ so 18.4% increase.

(c) $\frac{100}{89.3} = 1.120$ so 12.0% increase.

8 The 1990 adjusted salary is

$$17.3 \times 1.049 = 18.1$$

The 1992 adjusted salary is

$$\frac{19.8}{1.043} = 19.0$$

and so on. The complete set of 'constant 1991 prices' is listed in Table S3.1.

Table S3.1

	Year				
	90	91	92	93	94
Real salaries	18.1	18.1	19.0	21.7	23.2

During 1990/91 salaries remain unchanged in real terms. However, since 1991 salaries have outpaced inflation with steady increases in real terms.

9 (a) $\frac{7}{20}$; (b) $\frac{22}{25}$; (c) $2\frac{1}{2}$; (d) $\frac{7}{40}$; (e) $\frac{1}{500}$.

10 (a) 1.2; (b) 7.04; (c) 2190.24; (d) 62.72.

11 (a) 1.19; (b) 3.5; (c) 0.98; (d) 0.57.

12 (a) \$18.20; (b) 119 244; (c) \$101.09; (d) \$1610; (e) \$15 640.

13 (a) \$15.50; (b) \$10.54; (c) 32%.

14 \$864.

15 (a) \$26 100; (b) 31% (nearest percentage).

16 (a) 37.5% increase; (b) 8.9% increase; (c) 6.25% decrease.

17 (1) 1985.

(2) (a) 30%; (b) 52.3%; (c) 13.1%; (d) 9.4%.

(3) (a) 25%; (b) 44%; (c) 10.6%; (d) 11.1%.

(4) Public transport costs have risen at a faster rate than private transport throughout the period 1985–2000. However, for the final 5 years there are signs that the trend has stopped and has possibly reversed.

18 964, 100, 179, 750; e.g. seasonal variations.

19 (a) 100, 101.7, 113.1, 116.9.

(b) Real output: 236, 229.2, 244.7, 244.7.

Index: 100, 97.1, 103.7, 103.7.

(c) In real terms, spending on education fell by 2.9% in 1995, increased by 6.8% in 1996 and remained unchanged in 1997.

20 (a) 1 and 6 respectively.

(b) 142, 150.

(c) 94, 87, 83, 75, 79.

(d) 1.1 million and 1.6 million.

21 108.4, 119.5. These values reflect the rises given to the bulk of employees who fall into categories B and C. The generous rises given to senior management have had little effect on the index because there are only 7 (out of 390) employees in category D.

22 111.7, 173.6. These indices are higher than before. Although the total number of employees has remained almost unchanged, many of these have been promoted to the senior management team, thereby increasing the total wage bill.

Paasche index uses up-to-date information whereas Laspeyre uses only quantities relating to the base year, which become more irrelevant over time.

Laspeyre index is easier to calculate and interpret. Also we can compare two or more Laspeyre indices since they relate to the same basket of goods. It may be impossible to calculate the Paasche index since data about current performance may not be readily available at the time.

Section 3.2

1 The calculations are summarized in Table S3.2.

Table S3.2

End of year	Interest (\$)	Investment (\$)
1	80	1080
2	86.40	1166.40
3	93.31	1259.71
4	100.78	1360.49
5	108.84	1469.33
6	117.55	1586.88
7	126.95	1713.83
8	137.11	1850.94
9	148.08	1999.02
10	159.92	2158.94

2 $S = 1000(1.08)^{10} = \$2158.92$

The slight discrepancy between the two answers obtained in Problems 1 and 2 arises because the intermediate results in Problem 1 are rounded to 2 decimal places.

3
$$9000(1.03)^n = 10\,000$$
$$(1.03)^n = 1.11$$
$$\log(1.03)^n = \log(1.11)$$
$$n\log(1.03) = \log(1.11)$$
$$n = \frac{\log(1.11)}{\log(1.03)} = 3.53$$

so the firm makes a profit for the first time after 4 years.

4 (1) (a) $S = 30(1.06)^2 = \$33.71$

(b) $S = 30(1.03)^4 = \$33.77$

(c) $S = 30(1.015)^8 = \$33.79$

(d) $S = 30(1.005)^{24} = \$33.81$

(e) $S = 30(1.001\,15)^{104} = \$33.82$

(f) $S = 30(1.000\,164)^{730} = \$33.82$

(2) $S = 30e^{0.12} = \$33.82$

The results in part (1) are settling down at this value.

5 $4000 = 1000e^{0.1r}$

$$4 = e^{0.1r}$$

$$0.1r = \ln 4 = 1.386$$

$$r = 13.86\%$$

6 The quarterly scale factor is 1.03, so the overall scale factor for a year is

$$1.03^4 = 1.1255$$

which corresponds to a 12.55% increase.

7 After n years, the annual turnover of A will be $560(1.015)^n$ and the corresponding expression for B is $480(1.034)^n$. To find when supermarket B overtakes A we need to solve the equation

$$480(1.034)^n = 560(1.015)^n$$

$$\frac{1.034^n}{1.015^n} = \frac{560}{480}$$

$$\left(\frac{1.034}{1.015}\right)^n = \frac{7}{6}$$

$$n \log\left(\frac{1.034}{1.015}\right) = \log\left(\frac{7}{6}\right)$$

$$n = \frac{\log(7/6)}{\log(1.034/1.015)} = 8.31$$

so it will take 9 years.

8 \$6753.29; 50%.

9 \$23 433.19.

10 (a) Midwest; (b) BFB.

11 (a) \$619 173.64; (b) 13.

12 15 years.

13 \$50 000 $(0.95)^3$ = \$42 868.75.

14 (a) \$13 947.94; (b) \$14 156.59; (c) \$14 342.45; (d) \$14 381.03.

15 (a) \$35 000; (b) 7 years.

16 \$205.44.

17 36.6 years.

18 17.3 years.

19 We are charged interest on the interest; 26.82%.

20 7.25%.

21 (a) 6 years; (b) 5.19%.

22 7.67%.

23 (a) Interest is $(r/k)\%$ per period and there are kt periods in t years, so

$$S = P\left(1 + \frac{r}{100k}\right)^{kt}$$

(b) If $m = \frac{100k}{r}$ then $\frac{r}{100k} = \frac{1}{m}$ and

$kt = \frac{mrt}{100k}$ so

$$S = P\left(1 + \frac{1}{m}\right)^{rtm/100} = P\left[\left(1 + \frac{1}{m}\right)^m\right]^{rt/100}$$

by rule 3 of indices.

(c) Now since $m = 100k/r$ we see that if the frequency increases (i.e. if $k \to \infty$) then $m \to \infty$, causing

$$\left(1 + \frac{1}{m}\right)^m$$

to approach e. Substituting this into the result of part (b) gives

$$S = Pe^{rt/100}$$

24 (a) 112, 125.44, . . . , 964.63.

(b) 112.55, 126.68, . . . , 1064.09.

(c) 112.68, 126.97, . . . , 1089.26.

(d) 112.75, 127.12, . . . , 1102.32.

All four graphs have the same basic shape, and pass through (0, 100). As expected, as the frequency of compounding increases the values approach that of continuous compounding in (d).

25 6.17%, 7.44%, . . . , 42.58%.

Almost a straight line, but with a slight upward curvature.

Section 3.3

1 The geometric ratios of (a), (c), (d) and (e) are 2, –3, ½ and 1.07 respectively. The sequence in part (b) is an arithmetic progression, not a geometric progression, because to go from one number to the next we *add* on the fixed value of 5.

2 (a) The geometric ratio is 2, so the next term is $8 \times 2 = 16$.

$$1 + 2 + 4 + 8 + 16 = 31$$

For this series, $a = 1$, $r = 2$ and $n = 5$, so its value is

$$(1)\left(\frac{2^5 - 1}{2 - 1}\right) = 32 - 1 = 31$$

(b) For this series, $a = 100(1.07)$, $r = 1.07$ and $n = 20$, so its value is

$$100(1.07)\left(\frac{1.07^{20} - 1}{1.07 - 1}\right) = 4386.52$$

3 (a) The first \$1000 payment is invested for the full 10 years at 8% interest compounded annually, so its future value is

$$1000(1.08)^{10}$$

The second \$1000 payment is invested for 9 years, so its future value is

$$1000(1.08)^{9}$$

and so on.

The final payment of \$1000 is invested for just 1 year, so its future value is

$$1000(1.08)^{1}$$

Total savings

$$= 1000(1.08)^{10} + 1000(1.08)^{9} + \ldots + 1000(1.08)$$

$$= 1000(1.08)\left(\frac{1.08^{10} - 1}{1.08 - 1}\right)$$

$$= \$15\,645.49$$

(b) After n years,

$$\text{total savings} = 1000(1.08)\left(\frac{1.08^{n} - 1}{1.08 - 1}\right) = 13\,500(1.08^{n} - 1)$$

so we need to solve

$$13\,500(1.08^{n} - 1) = 20\,000$$

This can be done by taking logarithms following the strategy described in Section 2.3. The corresponding value of n is 11.8, so it takes 12 years.

4 If \$$x$ denotes the monthly repayment, the amount owed at the end of the first month is

$$2000(1.01) - x$$

After 2 months the amount owed is

$$[2000(1.01) - x](1.01) - x$$
$$= 2000(1.01)^2 - x(1.01) - x$$

Each month we multiply by 1.01 to add on the interest and subtract x to deduct the repayment, so after 12 months the outstanding debt is

$$2000(1.01)^{12} - x[1.01^{11} + 1.01^{10} + \ldots + 1]$$

$$= 2253.650 - x\left(\frac{1.01^{12} - 1}{1.01 - 1}\right)$$

$$= 2253.650 - 12.683x$$

If the debt is to be cleared then

$$x = \frac{2253.650}{12.683} = \$177.69$$

5 If the annual percentage rise is 2.6%, the scale factor is 1.026, so after n years the total amount of oil extracted (in billions of units) will be

$$45.5 + 45.5(1.026) + 45.5(1.026)^2 + \ldots + 45.5(1.026)^{n-1}$$

$$= 45.5\left(\frac{1.026^{n} - 1}{1.026 - 1}\right) = 1750(1.026^{n} - 1)$$

Oil reserves are exhausted after n years where n satisfies the equation

$$1750(1.026^{n} - 1) = 2625$$
$$1.026^{n} - 1 = 1.5$$
$$n \log(1.026) = \log(2.5)$$
$$n = 35.7$$

so oil will run out after 36 years.

6 11 463.88.

7 **(a)** \$9280.71; **(b)** \$9028.14.

8 140 040.

9 \$313 238.

10 \$424.19; **(a)** \$459.03; **(b)** \$456.44.

11 $rS_n = r(a + ar + ar^2 + \ldots + ar^{n-1})$

$= ar + ar^2 + ar^3 + \ldots + ar^n$

which is very similar to the given expression for S_n except that the first term, a, is missing and we have the extra term, ar^n. Consequently, when S_n is subtracted from rS_n the rest of the terms cancel, leaving

$$rS_n - S_n = ar^n - a$$

$$(r - 1)S_n = a(r^n - 1)$$

(factorize both sides)

$$S_n = a\left(\frac{r^n - 1}{r - 1}\right)$$

(divide through by $r - 1$)

The expression for S_n denotes the sum of the first n terms of a geometric series because the powers of r run from 0 to $n - 1$, making n terms in total. Notice that we are not allowed to divide by zero, so the last step is not valid for $r = 1$.

12 18 years.

13 200 million tonnes.

14 **(a)** \$480.

(b) \$3024.52.

(c) After n payments, debt is

$$((((8480 - A)R - A)R - A)R \ldots - A)R$$
$$= 8480R^n - AR(1 + R + R^2 + \ldots + R^{n-1})$$
$$= 8480R^n - AR\left(\frac{R^n - 1}{R - 1}\right)$$

Finally, setting this expression equal to zero gives the desired formula for *A*.

(d) \$637.43.

15 First row: 430.33, 291.59, . . . , 96.66.

Thirteenth row: 438.65, 299.36, . . . , 105.25.

16 \$3656.33; \$492 374.04, \$484 195.12, . . . , \$0.

17 $$4+4\left(1+\frac{r}{100}\right)+4\left(1+\frac{r}{100}\right)^2+\ldots+4\left(1+\frac{r}{100}\right)^{10}=60$$

$$4\left[\frac{1-\left(1+\frac{r}{100}\right)^{10}}{1-\left(1+\frac{r}{100}\right)}\right]=60$$

$$\frac{1-\left(1+\frac{r}{100}\right)^{10}}{-\frac{r}{100}}=15$$

$$1-\left(1+\frac{r}{100}\right)^{10}=-0.15r$$

$$\left(1+\frac{r}{100}\right)^{10}-0.15r-1=0$$

$$r=8.8$$

Section 3.4

1 **(a)** $P = 100\ 000(1.06)^{-10} = \$55\ 839.48$

(b) $P = 100\ 000e^{-0.6} = \$54\ 881.16$

2 **(a)** $\text{NPV} = \$17\ 000(1.15)^{-5} - \$8000 = \$452$

Worthwhile since this is positive.

(b) The IRR, *r*, is the solution of

$$8000\left(1+\frac{r}{100}\right)^5=17\ 000$$

$$\left(1+\frac{r}{100}\right)^5=2.125$$

(divide by 8000)

$$1+\frac{r}{100}=1.16$$

(take fifth roots)

$$r=16\%$$

Worthwhile since the IRR exceeds the market rate.

3 NPV of Project A is

$$\text{NPV}_A = \$18\ 000(1.07)^{-2} - \$13\ 500 = \$2221.90$$

NPV of Project B is

$$\text{NPV}_B = \$13\ 000(1.07)^{-2} - \$9000 = \$2354.70$$

Project B is to be preferred since

$$\text{NPV}_B > \text{NPV}_A$$

4 Rate of interest per month is ½%, so the present value, *P*, of \$*S* in *t* months' time is

$$P = S(1.005)^{-t}$$

The total present value is

$$2000(1.005)^{-1} + 2000(1.005)^{-2} + \ldots + 2000(1.005)^{-120}$$

because there are 120 months in 10 years. Using the formula for a geometric series gives

$$2000(1.005)^{-1}\left(\frac{1.005^{-120}-1}{1.005^{-1}-1}\right)=\$180\ 146.91$$

5 The formula for discounting is

$$P = S(1.15)^{-t}$$

The results are given in Table S3.3.

There is very little to choose between these two projects. Both present values are considerably less than the original expenditure of \$10 000. Consequently, neither project is to be recommended, since the net present values are negative. The firm would be better off just investing the \$10 000 at 15% interest!

Table S3.3

	Discounted revenue (\$)	
End of year	**Project 1**	**Project 2**
1	1739.13	869.57
2	1512.29	756.14
3	1972.55	1315.03
4	1715.26	3430.52
5	1491.53	1988.71
Total	8430.76	8359.97

6 The IRR satisfies the equation

$$12\ 000 = 8000\left(1+\frac{r}{100}\right)^{-1} + 2000\left(1+\frac{r}{100}\right)^{-2} + 2000\left(1+\frac{r}{100}\right)^{-3} + 2000\left(1+\frac{r}{100}\right)^{-4}$$

Values of the right-hand side of this equation corresponding to $r = 5, 6, \ldots, 10$ are listed in the table below:

r	5	6	7	8	9	10
value	12 806	12 591	12 382	12 180	11 984	11 794

This shows that r is between 8 and 9. To decide which of these to go for, we evaluate $r = 8\frac{1}{2}$, which gives 12 081, which is greater than 12 000, so $r = 9\%$ to the nearest percentage. This exceeds the market rate, so the project is worthwhile.

7 If the yield is 7% then each year the income is \$70 with the exception of the last year, when it is \$1070 because the bond is redeemed for its original value.

The present values of this income stream are listed in Table S3.4 and are calculated using the formula

$$P = S(1.08)^{-t}$$

Table S3.4

End of year	Cash flow (\$)	Present value (\$)
1	70	64.81
2	70	60.01
3	1070	849.40
Total present vlaue		974.22

8 **(a)** \$5974.43; **(b)** \$5965.01.

9 **(a)** 7%; **(b)** yes, provided there are no risks.

10 NPA_A = \$1595.94; NPV_B = \$1961.99;

NPV_C = \$1069.00, so Project B is best.

11 **(a)** \$379.08; **(b)** \$1000.

12 \$61 672.67.

13 \$38 887.69.

14 27%.

15 \$349.15.

16 **(a)** \$400 000; **(b)** \$92 550.98; **(c)** \$307 449.02.

17 **(a)** \$333 million; **(b)** $5\left(\frac{1-(1.06)^{1-n}}{1.06^2-1.06}\right)$;

(c) same expression as (b) but with 5 replaced by 50;

(d) 12.

18 For n years,

$$\frac{a}{1+\frac{r}{100}} + \frac{a}{\left(1+\frac{r}{100}\right)^2} + \ldots + \frac{a}{\left(1+\frac{r}{100}\right)^n}$$

$$= \frac{a}{1+r/100}\left(\frac{1-(1+r/100)^{-n}}{1-(1+r/100)^{-1}}\right)$$

Now, as n gets larger and larger, the number $(1 + r/100)^{-n}$ tends to zero, leaving

$$\frac{a}{1+r/100}\left(\frac{1}{1-(1+r/100)^{-1}}\right)$$

$$= \frac{a}{(1+r/100)-1} = \frac{a}{r/100} = \frac{100a}{r}$$

19 12% to the nearest whole number and 11.6% to 1 decimal place.

20 \$22 177 and \$16 659, so choose A.

\$21 702 and \$16 002, so very little difference in NPVs and no difference in choice.

Chapter 4

Section 4.1

1 **(a)** $\frac{11-3}{3-(-1)} = \frac{8}{4} = 2$

(b) $\frac{-2-3}{4-(-1)} + \frac{-5}{5} = -1$

(c) $\frac{3-3}{49-(-1)} = \frac{0}{50} = 0$

2 Using the power key $\boxed{x^y}$ on a calculator, the values of the cube function, correct to 2 decimal places, are

x	−1.50	−1.25	−1.00	−0.75
$f(x)$	−3.38	−1.95	−1.00	−0.42

x	−0.50	−0.25	0.00	0.25	0.50
$f(x)$	−0.13	−0.02	0.00	0.02	0.13

x	0.75	1.00	1.25	1.50
$f(x)$	0.42	1.00	1.95	3.38

The graph of the cube function is sketched in Figure S4.1 (overleaf).

$$f'(-1) = \frac{1.5}{0.5} = 3.0$$

$f'(0) = 0$ (because the tangent is horizontal at $x = 0$)

$$f'(1) = \frac{1.5}{0.5} = 3.0$$

[Note: $f'(-1) = f'(1)$ because of the symmetry of the graph.]

3 If $n = 3$ then the general formula gives

$$f'(x) = 3x^{3-1} = 3x^2$$

Hence

$$f'(-1) = 3(-1)^2 = 3$$
$$f'(0) = 3(0)^2 = 0$$
$$f'(1) = 3(1)^2 = 3$$

Figure S4.1

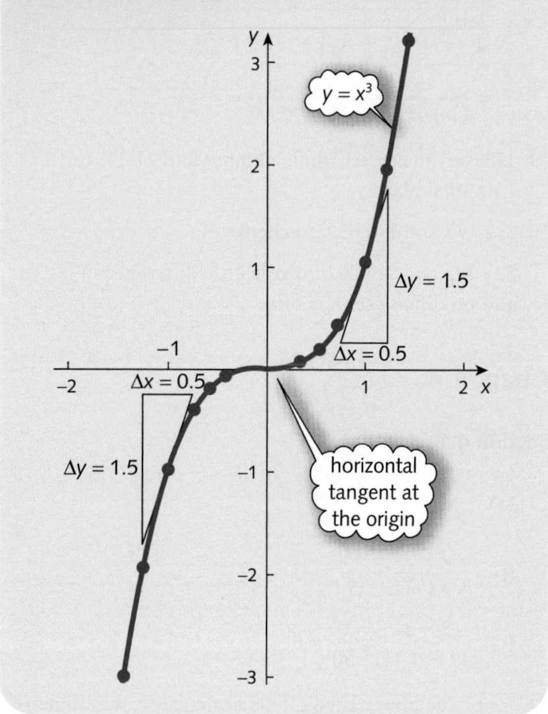

4 (a) $5x^4$; (b) $6x^5$; (c) $100x^{99}$;
(d) $-x^{-2}$ (that is, $-1/x^2$);
(e) $-2x^{-3}$ (that is, $-2/x^3$).

5 (a) 2; (b) −1; (c) 0.

6 $-^2/_3$; downhill.

7 When $x = 0$, $y = a(0) + b = b$ ✓

When $x = 1$, $y = a(1) + b = a + b$ ✓

$$\text{Slope} = \frac{(a+b)-b}{1-0} = a$$

8 The graph of $f(x) = 5$ is sketched in Figure S4.2. The graph is horizontal, so has zero slope at all values of x.

Figure S4.2

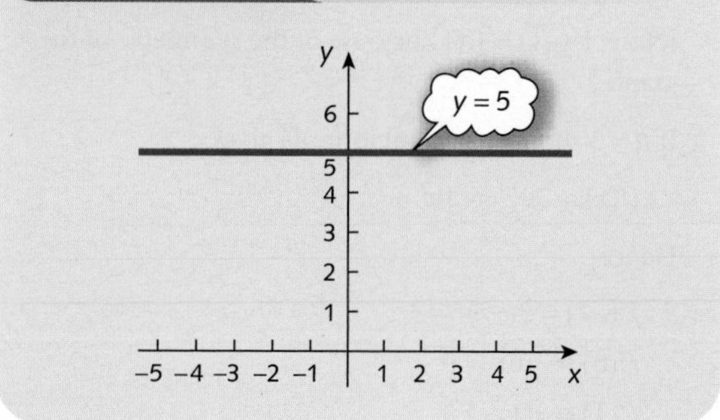

9 $7x^6$; 448.

10 (a) $8x^7$; (b) $50x^{49}$; (c) $19x^{18}$; (d) $999x^{998}$.

11 (a) $-\frac{3}{x^4}$; (b) $\frac{1}{3\sqrt[3]{x^2}}$; (c) $-\frac{1}{4\sqrt[4]{x^5}}$; (d) $\frac{3\sqrt{x}}{2}$.

12 3, 1.25, 0, −0.75, −1, −0.75, 0, 1.25;
(a) −3; (b) 0; (c) 1.

13 (a) (i) 2, 2.0248 . . . (ii) 0.24845 . . . (iii) 0.25
(b) (i) 8, 8.3018 . . . (ii) 3.01867 . . . (iii) 3
(c) (i) 0.5, 0.4986 . . . (ii) −0.06135 . . .
(iii) −0.0625

In all three cases the gradient of the chord gives a good approximation to that of the tangent.

14 (a) 8; (b) ±3; (c) $-\frac{1}{2}$; (d) 4.

Section 4.2

1 (a) $4(3x^2) = 12x^2$.
(b) $-2/x^3$ because $1/x = x^{-1}$, which differentiates to $-x^{-2}$.

2 (a) $5x^4 + 1$; (b) $2x + 0 = 2x$.

3 (a) $2x - 3x^2$; (b) $0 - (-3x^{-4}) = \frac{3}{x^4}$.

4 (a) $9(5x^4) + 2(2x) = 45x^4 + 4x$
(b) $5(8x^7) - 3(-1)x^{-2} = 40x^7 + 3/x^2$
(c) $2x + 6(1) + 0 = 2x + 6$
(d) $2(4x^3) + 12(3x^2) - 4(2x) + 7(1) - 0$
$= 8x^3 + 36x^2 - 8x + 7$

5 $f'(x) = 4(3x^2) - 5(2x) = 12x^2 - 10x$
$f''(x) = 12(2x) - 10(1) = 24x - 10$
$f''(6) = 24(6) - 10 = 134$

6 (a) $10x$; (b) $-3/x^2$; (c) 2; (d) $2x + 1$;
(e) $2x - 3$; (f) $3 + 7/x^2$; (g) $6x^2 - 12x + 49$;
(h) a; (i) $2ax + b$; (j) $2/\sqrt{x} + 3/x^2 - 14/x^3$.

7 (a) 27; (b) 4; (c) 2; (d) −36; (e) 3/8.

8 $4x^3 + 6x$;
(a) $9x^2 - 8x$; (b) $12x^3 - 6x^2 + 12x - 7$; (c) $2x - 5$;
(d) $1 + \frac{3}{x^2}$; (e) $-\frac{2}{x^3} + \frac{4}{x^2}$; (f) $\frac{3}{x^2} - \frac{10}{x^3}$.

9 (a) 14; (b) $6/x^4$; (c) 0.

10 4.

11 $f''(x) = 6ax + 2b > 0$ gives $x > -b/3a$
$f''(x) = 6ax + 2b > 0$ gives $x < -b/3a$

12 0; horizontal tangent, i.e. vertex of parabola must be at $x = 3$.

13 $\frac{1}{\sqrt{x}}$; **(a)** $\frac{5}{2\sqrt{x}}$; **(b)** $x^{-2/3}$; **(c)** $\frac{3}{2}x^{-1/4}$; **(d)** $-\frac{5}{2}x^{-3/2}$.

14 **(a)** $2P + 1$; **(b)** $50 - 6Q$; **(c)** $-30/Q^2$; **(d)** 3; **(e)** $5/\sqrt{L}$; **(f)** $-6Q^2 + 30Q - 24$.

Section 4.3

1 $TR = PQ = (60 - Q)Q = 60Q - Q^2$

(1) $MR = 60 - 2Q$

When $Q = 50$

$MR = 60 - 2(50) = -40$

(2) **(a)** $TR = 60(50) - (50)^2 = 500$

(b) $TR = 60(51) - (51)^2 = 459$

so TR changes by -41, which is approximately the same as the exact value obtained in part (1).

2 $MR = 1000 - 8Q$, so when $Q = 30$

$MR = 1000 - 8(30) = 760$

(a) $\Delta(TR) \simeq MR \times \Delta Q = 760 \times 3 = 2280$, so total revenue rises by about 2280.

(b) $\Delta(TR) \simeq MR \times \Delta Q = 760 \times (-2)$

$= -1520$

so total revenue falls by about 1520.

3 $TC = (AC)Q = \left(\frac{100}{Q} + 2\right)Q = 100 + 2Q$

This function differentiates to give MC = 2, so a 1 unit increase in Q always leads to a 2 unit increase in TC irrespective of the level of output.

4 If $K = 100$ then

$$Q = 5L^{1/2}(100)^{1/2} = 50L^{1/2}$$

because $\sqrt{100} = 10$. Differentiating gives

$$MP_L = 50(\tfrac{1}{2}L^{-1/2}) = \frac{25}{\sqrt{L}}$$

(a) $\frac{25}{\sqrt{1}} = 25$ **(b)** $\frac{25}{\sqrt{9}} = \frac{25}{3} = 8.3$;

(c) $\frac{25}{\sqrt{10\,000}} = \frac{25}{100} = 0.25$.

The fact that these values decrease as L increases suggests that the law of diminishing marginal productivity holds for this function. This can be confirmed by differentiating a second time to get

$$\frac{d^2Q}{dL^2} = 25(-\tfrac{1}{2}L^{-3/2}) = \frac{-25}{2L^{3/2}}$$

which is negative for all values of L.

5 The savings function is given, so we begin by finding MPS. Differentiating S with respect to Y gives

$$MPS = 0.04Y - 1$$

so when $Y = 40$,

$$MPS = 0.04(40) - 1 = 0.6$$

To find MPC we use the formula

$$MPC + MPS = 1$$

that is,

$$MPC = 1 - MPS = 1 - 0.6 = 0.4$$

This indicates that, at the current level of income, a 1 unit increase in national income causes a rise of about 0.6 units in savings and 0.4 units in consumption.

6 $TR = 100Q - 4Q^2$, $MR = 100 - 8Q$; 1.2.

7 $TR = 80Q - 3Q^2$, so $MR = 80 - 6Q = 80 - 6(80 - P)/3 = 2P - 80$.

8 $TR = 100Q - Q^2$; $MR = 100 - 2Q$. Graphs of TR and MR are sketched in Figures S4.3 and S4.4 respectively. MR = 0 when $Q = 50$. This is the value of Q at which TR is a maximum.

Figure S4.3

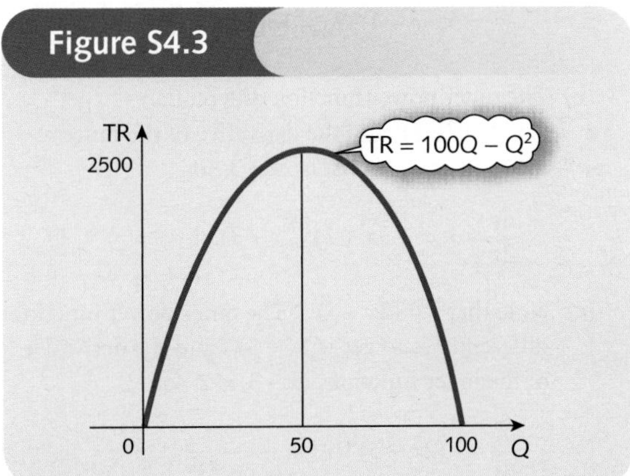

Figure S4.4

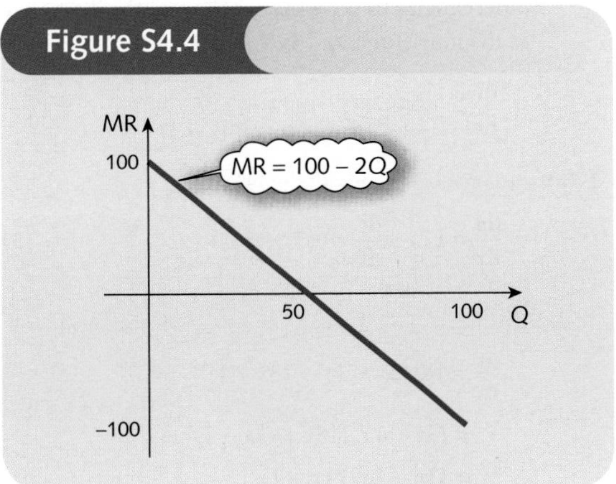

9 **(a)** TC $= 100 + 2Q + Q^2/10$;

MC $= 2 + Q/5$.

(b) MC $= 8$; $\Delta(\text{TC}) \simeq 16$.

(c) 100.

10 $\text{TC} = 15 + 2Q^2 + 9Q$; 15; $4Q + 9$.

11 **(a)** 49.98; **(b)** 49.8;

(c) 48; **(d)** 30. Yes, because $d^2Q/dL^2 = -0.02 < 0$.

12 $\dfrac{d^2Q}{dL^2} = 12 - 1.2L < 0$ for all $L > 10$.

13 MPC = 1/6 and MPS = 5/6. If national income rises by 1 unit, the approximate increase in consumption and savings is 1/6 and 5/6 respectively.

14 **(a)** MPC = 0.96, MPS = 0.04;

(b) $S = 0.2Y - 100 - 0.01Y^2$.

Section 4.4

1 **(a)** The outer power function differentiates to get $5(3x - 4)^4$ and the derivative of the inner function, $3x - 4$, is 3, so

$$\frac{dy}{dx} = 5(3x-4)^4(3) = 15(3x-4)^4$$

(b) The outer power function differentiates to get $3(x^2 + 3x + 5)^2$ and the derivative of the inner function, $x^2 + 3x + 5$, is $2x + 3$, so

$$\frac{dy}{dx} = 3(x^2+3x+5)^2(2x+3)$$

(c) Note that $y = (2x - 3)^{-1}$. The outer power function differentiates to get $-(2x - 3)^{-2}$ and the derivative of the inner function, $2x - 3$, is 2, so

$$\frac{dy}{dx} = -(2x-3)^{-2}(2) = \frac{-2}{(2x-3)^2}$$

(d) Note that $y = (4x - 3)^{1/2}$. The outer power function differentiates to get $\tfrac{1}{2}(4x - 3)^{-1/2}$ and the derivative of the inner function, $4x - 3$, is 4, so

$$\frac{dy}{dx} = \tfrac{1}{2}(4x-3)^{-1/2}(4) = \frac{2}{\sqrt{(4x-3)}}$$

2 **(a)**

$$u = x \qquad v = (3x-1)^6$$

$$\frac{du}{dx} = 1 \qquad \frac{dv}{dx} = 6(3x-1)^5 \qquad (3)$$

So

$$\frac{dy}{dx} = 18x(3x-1)^5 + (3x-1)^6$$

$$= (3x-1)^5\,[18x + (3x-1)]$$

$$= (3x-1)^5(21x-1)$$

(b)

$$u = x^3 \qquad v = (2x+3)^{1/2}$$

$$\frac{du}{dx} = 3x^2 \qquad \frac{dv}{dx} = \tfrac{1}{2}(2x+3)^{-1/2} \qquad (2)$$

$$= \frac{1}{\sqrt{(2x+3)}}$$

So

$$\frac{dy}{dx} = \frac{x^3}{\sqrt{(2x+3)}} + 3x^2\sqrt{(2x+3)}$$

(c)

$$u = x \qquad v = (x-2)^{-2}$$

$$\frac{du}{dx} = 1 \qquad \frac{dv}{dx} = -(x-2)^{-2}$$

So

$$\frac{dy}{dx} = \frac{-x}{(x-2)^2} + \frac{1}{x-2}$$

$$= \frac{-x + (x-2)}{(x-2)^2}$$

$$= \frac{-2}{(x-2)^2}$$

3 **(a)**

$$u = x \qquad v = x - 2$$

$$\frac{du}{dx} = 1 \qquad \frac{dv}{dx} = 1$$

So

$$\frac{dy}{dx} = \frac{(x-2) - x}{(x-2)^2}$$

$$= \frac{-2}{(x-2)^2}$$

(b)

$$u = x - 1 \qquad v = x + 2$$

$$\frac{du}{dx} = 1 \qquad \frac{dv}{dx} = 1$$

So

$$\frac{dy}{dx} = \frac{(x+1) - (x-1)}{(x+1)^2}$$

$$= \frac{2}{(x+1)^2}$$

4 **(a)** $20(2x + 1)^9$;

(b) $3(x^2 + 3x - 5)^2(2x + 3)$;

(c) $-7/(7x - 3)^2$;

(d) $-2x/(x^2 + 1)^2$;

(e) $4/\sqrt{(8x - 1)}$.

5 **(a)** $5x\,(x + 2)(x + 5)^2$;

(b) $x^4(4x + 5)(28x + 25)$;

(c) $\dfrac{x^3(9x+8)}{2\sqrt{(x+1)}}$.

6 **(a)** $\dfrac{x^2+8x}{(x+4)^2}$; **(b)** $\dfrac{3}{(x+1)^2}$; **(c)** $\dfrac{x^2(5x-6)}{2(x-1)^{3/2}}$.

7 **(a)** $4x(x-3)^3 + (x-3)^4 = (x-3)^3(5x-3)$;

(b) $\dfrac{x}{\sqrt{(2x-3)}} + \sqrt{(2x-3)} = \dfrac{3x-3}{\sqrt{2x-3}}$;

(c) $\dfrac{5}{(x+5)^2}$; **(d)** $\dfrac{1-x^2}{(x^2+1)^2}$.

8 $10(5x+7) = 50x + 70$

9 $5x^4(x+2)^2 + 2x^5(x+2) = 7x^6 + 24x^5 + 20x^4$

10 **(a)** $\dfrac{100-3Q}{\sqrt{(100-2Q)}}$

(b) $\dfrac{1000(2+Q)^{1/2} - 500Q(2+Q)^{-1/2}}{(2+Q)}$

$= \dfrac{2000+500Q}{(2+Q)^{3/2}}$

11 MPC = 1.8, MPS = −0.8. If national income rises by 1 unit, consumption rises by 1.8 units, whereas savings actually fall by 0.8 units.

Section 4.5

1 We are given that $P_1 = 210$ and $P_2 = 200$. Substituting $P = 210$ into the demand equation gives

$$1000 - 2Q_1 = 210$$
$$-2Q_1 = -790$$
$$Q_1 = 395$$

Similarly, putting $P = 200$ gives $Q_2 = 400$. Hence

$$\Delta P = 200 - 210 = -10$$
$$\Delta Q = 400 - 395 = 5$$

Averaging the P values gives

$$P = ½(210 + 200) = 205$$

Averaging the Q values gives

$$Q = ½(395 + 400) = 397.5$$

Hence, arc elasticity is

$$-\left(\frac{205}{397.5}\right) \times \left(\frac{5}{-10}\right) = 0.26$$

2 The quickest way of solving this problem is to find a general expression for E in terms of P and then just to replace P by 10, 50 and 90 in turn. The equation

$$P = 100 - Q$$

rearranges as

$$Q = 100 - P$$

so

$$\frac{dQ}{dP} = -1$$

Hence

$$E = -\frac{P}{Q} \times \frac{dQ}{dP} = \frac{-P}{100-P} \times (-1)$$
$$= \frac{P}{100-P}$$

(a) If $P = 10$ then $E = 1/9 < 1$ so inelastic.

(b) If $P = 50$ then $E = 1$ so unit elastic.

(c) If $P = 90$ then $E = 9$ so elastic.

At the end of Section 4.5 it is shown quite generally that the price elasticity of demand for a linear function

$$P = aQ + b$$

is given by

$$E = \frac{P}{b-P}$$

The above is a special case of this with $b = 100$.

3 Substituting $Q = 4$ into the demand equation gives

$$P = -(4)^2 - 10(4) + 150 = 94$$

Differentiating the demand equation with respect to Q gives

$$\frac{dP}{dQ} = -2Q - 10$$

so

$$\frac{dQ}{dP} = \frac{1}{-2Q-10}$$

When $Q = 4$

$$\frac{dQ}{dP} = -\frac{1}{18}$$

The price elasticity of demand is then

$$-\left(\frac{94}{4}\right) \times \left(-\frac{1}{18}\right) = \frac{47}{36}$$

From the definition

$$E = -\frac{\text{percentage change in demand}}{\text{percentage change in price}}$$

we have

$$\frac{47}{36} = -\frac{10}{\text{percentage change in price}}$$

Hence the percentage change in price is $-10 \times 36/47 = -7.7\%$: that is, the firm must reduce prices by 7.7% to achieve a 10% increase in demand.

4 **(a)** Putting $P = 9$ and 11 directly into the supply equation gives $Q = 203.1$ and 217.1 respectively, so

$$\Delta P = 11 - 9 = 2$$
$$\Delta Q = 217.1 - 203.1 = 14$$

Averaging the P values gives

$$P = ½(9 + 11) = 10$$

Averaging the Q values gives

$$Q = \tfrac{1}{2}(203.1 + 217.1) = 210.1$$

Arc elasticity is

$$\frac{10}{210.1} \times \frac{14}{2} = 0.333\ 175$$

(b) Putting $P = 10$ directly into the supply equation, we get $Q = 210$. Differentiating the supply equation immediately gives

$$\frac{dQ}{dP} = 5 + 0.2P$$

so when $P = 10$, $dQ/dP = 7$. Hence

$$E = \frac{10}{210} \times 7 = \frac{1}{3}$$

Note that, as expected, the results in parts (a) and (b) are similar. They are not identical, because in part (a) the elasticity is 'averaged' over the arc from $P = 9$ to $P = 11$, whereas in part (b) the elasticity is evaluated exactly at the midpoint, $P = 10$.

5 43/162 = 0.27.

6 22/81 = 0.27; agree to 2 decimal places.

7 **(a)** 1/4; **(b)** 1/4; **(c)** 9/8.

8 $4P/(60 - 4P)$; 7.5.

9 **(a)** $0.2P$.

(b) $0.1P^2 = Q - 4$

(subtract 4 from both sides)

$$P^2 = 10(Q - 4) = 10Q - 40$$

(multiply both sides by 10)

$$P = \sqrt{(10Q - 40)}$$

(square root both sides)

$$\frac{dP}{dQ} = \frac{5}{\sqrt{(10Q - 40)}}$$

(c) $$\frac{1}{dP/dQ} = \frac{\sqrt{(10Q - 40)}}{5}$$

$$= \frac{P}{5} = 0.2P = \frac{dQ}{dP}$$

(d) $E = 10/7$.

10 1.46; **(a)** elastic; **(b)** 7.3%.

11 If $P = AQ^{-n}$ then

$$\frac{dP}{dQ} = -nAQ^{-(n+1)}$$

so

$$\frac{dQ}{dP} = \frac{1}{-nAQ^{-(n+1)}}$$

Hence

$$E = -\frac{P}{Q} \times \frac{dQ}{dP}$$

$$= -\left(\frac{AQ^{-n}}{Q}\right) \times \left(\frac{1}{-nAQ^{-(n+1)}}\right)$$

$$= -Q^{-n} \times \left(\frac{1}{-nQ^{-n}}\right) = \frac{1}{n}$$

which is a constant.

12 $$E = \frac{P}{Q} \times \frac{dQ}{dP} = \frac{P}{Q} \times a = \frac{Pa}{Q} = \frac{aP}{aP + b}$$

(a) if $b = 0$ then $E = \frac{Pa}{Pa} = 1$

(b) if $b > 0$ then $aP + b > aP$ so $E = \frac{aP}{aP + b} < 1$

Assuming that the line is sketched with quantity on the horizontal axis and price on the vertical axis, supply is unit elastic when the graph passes through the origin, and inelastic when the vertical intercept is positive.

Section 4.6

1 **(a)** *Step 1*

$$\frac{dy}{dx} = 6x + 12 = 0$$

has solution $x = -2$.

Step 2

$$\frac{d^2y}{dx^2} = 6 > 0$$

so minimum.

Finally, note that when $x = -2$, $y = -47$, so the minimum point has coordinates (−2, −47). A graph is sketched in Figure S4.5.

Figure S4.5

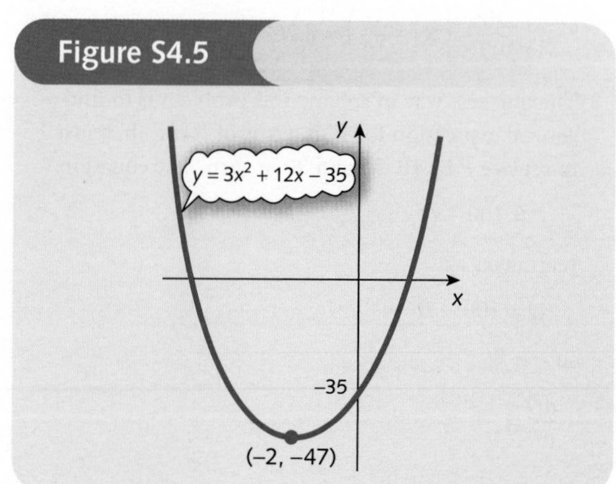

(b) ***Step 1***

$$\frac{dy}{dx} = -6x^2 + 30x - 36 = 0$$

has solutions $x = 2$ and $x = 3$.

Step 2

$$\frac{d^2y}{dx^2} = -12x + 30x$$

which takes the values 6 and −6 at $x = 2$ and $x = 3$ respectively. Hence minimum at $x = 2$ and maximum at $x = 3$.

Figure S4.6

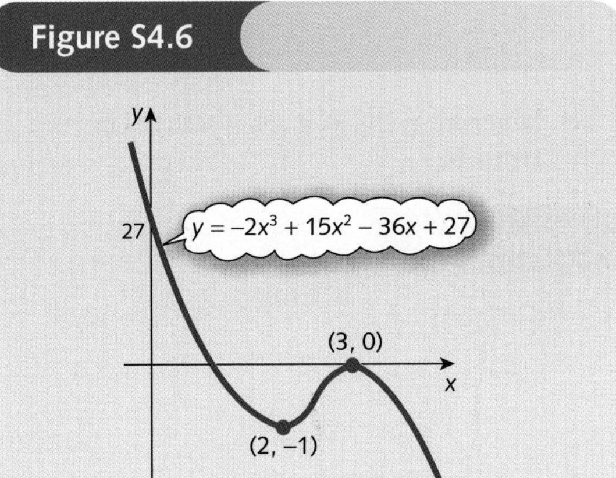

A graph is sketched in Figure S4.6 based on the following table of function values:

x	−10	0	2	3	10
$f(x)$	3887	27	−1	0	−833

2 $AP_L = \frac{Q}{L} = \frac{300L^2 - L^4}{L} = 300L - L^3$

Step 1

$$\frac{d(AP_L)}{dL} = 300 - 3L^2 = 0$$

has solution $L = \pm 10$. We can ignore −10 because it does not make sense to employ a negative number of workers.

Step 2

$$\frac{d^2(AP_L)}{dL^2} = -6L$$

which takes the value $-60 < 0$ at $L = 10$. Hence $L = 10$ is a maximum.

Now

$$MP_L = \frac{dQ}{dL} = 600L - 4L^3$$

so at $L = 10$

$$MP_L = 600(10) - 4(10)^3 = 2000$$

$$AP_L = 300(10) - (10)^3 = 2000$$

that is, $MP_L = AP_L$.

3 (a) $TR = PQ = (20 - 2Q)Q = 20Q - 2Q^2$

Step 1

$$\frac{d(TR)}{dQ} = 20 - 4Q = 0$$

has solution $Q = 5$.

Step 2

$$\frac{d^2(TR)}{dQ^2} = -2 < 0$$

so maximum.

(b) $\pi = TR - TC$

$$= (20Q - 2Q^2) - (Q^3 - 8Q^2 + 20Q + 2)$$

$$= -Q^3 + 6Q^2 - 2$$

Step 1

$$\frac{d\pi}{dQ} = -3Q^2 + 12Q = 0$$

has solutions $Q = 0$ and $Q = 4$.

Step 2

$$\frac{d^2\pi}{dQ^2} = -6Q + 12$$

which takes the values 12 and −12 when $Q = 0$ and $Q = 4$, respectively. Hence minimum at $Q = 0$ and maximum at $Q = 4$.

Finally, evaluating π at $Q = 4$ gives the maximum profit, $\pi = 30$. Now

$$MR = \frac{d(TR)}{dQ} = 20 - 4Q$$

so at $Q = 4$, $MR = 4$;

$$MC = \frac{d(TC)}{dQ} = 3Q^2 - 16Q + 20$$

so at $Q = 4$, $MC = 4$.

4 $AC = Q + 3 + \frac{36}{Q}$

Step 1

$$\frac{d(AC)}{dQ} = 1 - \frac{36}{Q^2} = 0$$

has solution $Q = \pm 6$. A negative value of Q does not make sense, so we just take $Q = 6$.

Step 2

$$\frac{d^2(AC)}{dQ^2} = \frac{72}{Q^3}$$

is positive when $Q = 6$, so it is a minimum.

Now when $Q = 6$, AC = 15. Also

$$MC = \frac{d(TC)}{dQ} = 2Q + 3$$

which takes the value 15 at $Q = 6$. We observe that the values of AC and MC are the same: that is, at the point of minimum average cost

average cost = marginal cost

There is nothing special about this example and in the next section we show that this result is true for any average cost function.

5 After tax the supply equation becomes

$$P = \tfrac{1}{2}Q_S + 25 + t$$

In equilibrium, $Q_S = Q_D = Q$, so

$$P = \tfrac{1}{2}Q + 25 + t$$
$$P = -2Q + 50$$

Hence

$$\tfrac{1}{2}Q + 25 + t = -2Q + 50$$

which rearranges to give

$$Q = 10 - \tfrac{2}{5}t$$

Hence the tax revenue, T, is

$$T = tQ = 10t - \tfrac{2}{5}t^2$$

Step 1

$$\frac{dT}{dt} = 10 - \tfrac{4}{5}t^2 = 0$$

has solution $t = 12.5$.

Step 2

$$\frac{d^2T}{dt^2} = \frac{-4}{5} < 0$$

so maximum. Government should therefore impose a tax of \$12.50 per good.

6 (a) Maximum at (1/2, 5/4); graph is sketched in Figure S4.7.

Figure S4.7

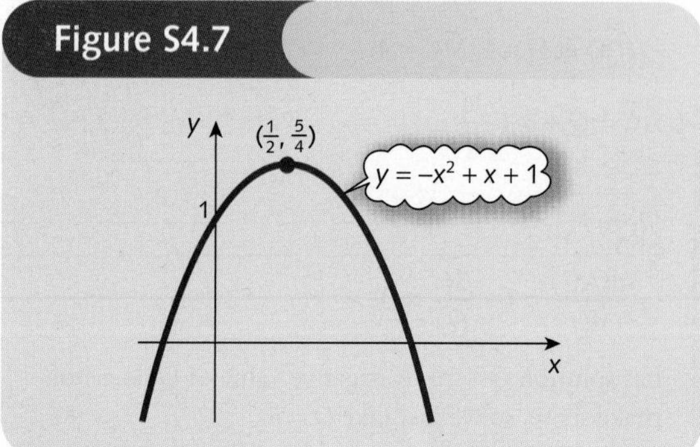

(b) Minimum at (2, 0); graph is sketched in Figure S4.8.

Figure S4.8

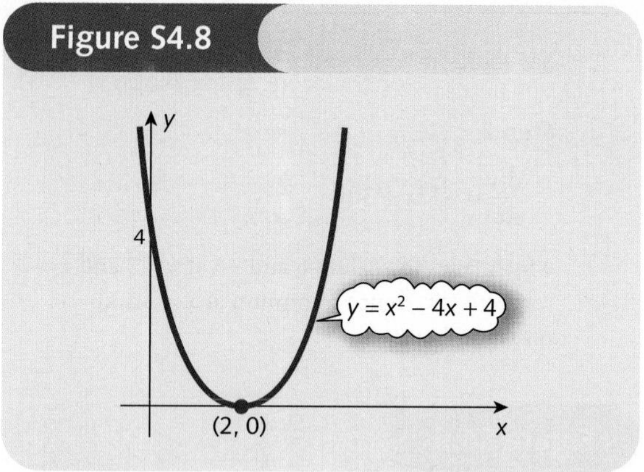

(c) Minimum at (10, 5); graph is sketched in Figure S4.9.

Figure S4.9

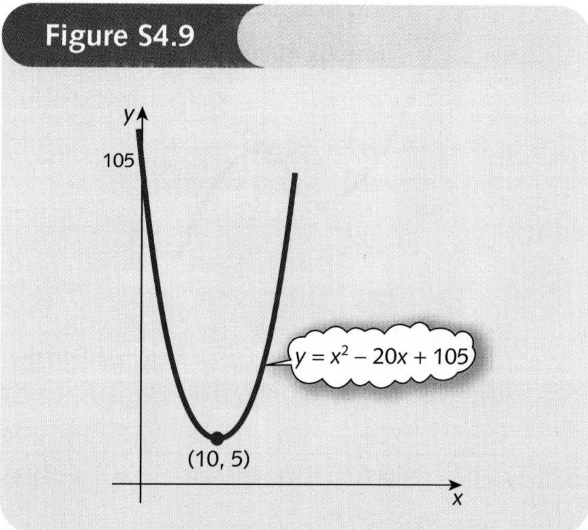

(d) Maximum at (1, 2), minimum at (−1, −2); graph is sketched in Figure S4.10.

Figure S4.10

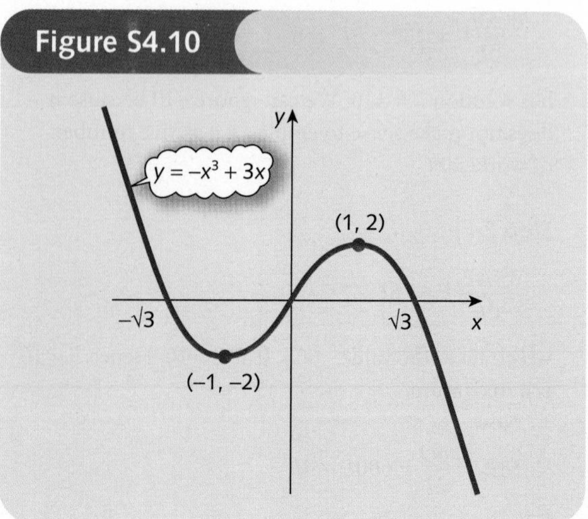

7 Graphs of the three functions are sketched in Figure S4.11, which shows that the stationary points in (a), (b) and (c) are a point of inflection, minimum and maximum, respectively.

Figure S4.11

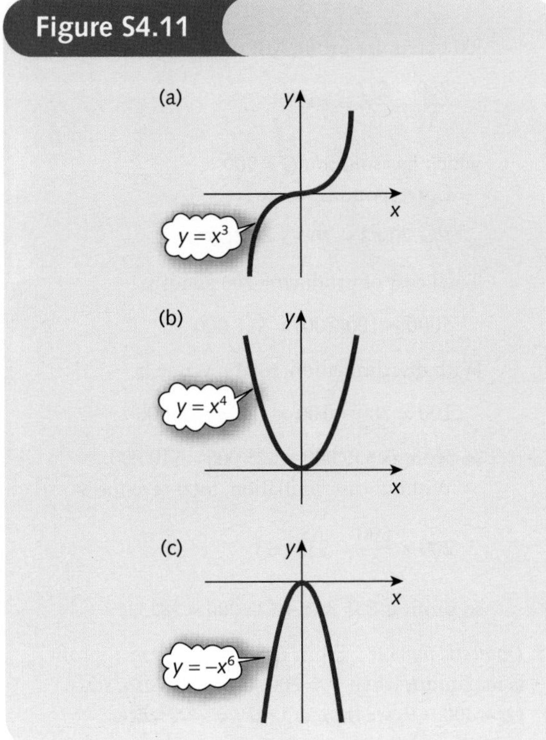

8 10.

9 TC $= 2Q^2 + 15$, AC $= 2Q + \frac{15}{Q}$,

MC $= 4Q$; $\sqrt{7.5}$; AC = 11 = MC.

10 30; $MP_L = 450 = AP_L$.

11 **(a)** TC $= 13 + (Q + 2)Q$

$= 13 + Q^2 + 2Q$

$$AC = \frac{TC}{Q} = \frac{13}{Q} + Q + 2$$

Q	1	2	3	4	5	6
AC	16	10.5	9.3	9.3	9.6	10.2

The graph of AC is sketched in Figure S4.12.

(b) From Figure S4.12 minimum average cost ≏ 9.2.

(c) Minimum at $Q = \sqrt{13}$, which gives AC = 9.21.

12 **(a)** 37 037 after 333 days.

(b) 167.

13 167.

Figure S4.12

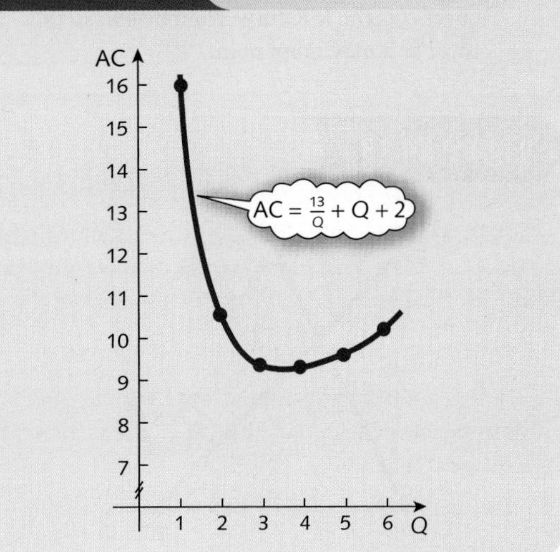

14 **(a)** TR $= 4Q - \frac{Q^2}{4}$

$$\pi = \frac{-Q^3}{20} + \frac{Q^2}{20} + 2Q - 4$$

$$MR = 4 - \frac{Q}{2}$$

$$MC = 2 - \frac{3Q}{5} + \frac{3Q^2}{20}$$

(b) 4. **(c)** MR = 2 = MC.

15 \$3.

16 **(a)** Minimum at (1, 7) and (4, −34), maximum at (2, −2).

(b) Minimum at (0, −10), inflection at (3, 17).

(c) Minimum at (−1, −½), maximum at (1, ½).

17 **(a)** Maple chooses a poor range on the y axis so that you can barely distinguish the graph from the axes. This is because the graph is undefined at $x = 0$, resulting in very large (positive and negative) values of y when x is close to 0.

(c) Minimum at (1, 1), maximum at (5, $\frac{1}{9}$).

18 $\pi = -30Q + \frac{13}{2}Q^2 - \frac{1}{3}Q^3$

$MC = 80 - 15Q + Q^2$

$MR = 50 - 2Q$

19 29.54.

Section 4.7

1 **(a)** TR $= (25 - 0.5Q)Q = 25Q - 0.5Q^2$

TC $= 7 + (Q + 1)Q = Q^2 + Q + 7$

MR $= 25 - Q$

MC $= 2Q + 1$

(b) From Figure S4.13 the point of intersection of the MR and MC curves occurs at $Q = 8$. The MC curve cuts the MR curve from below, so this must be a maximum point.

Figure S4.13

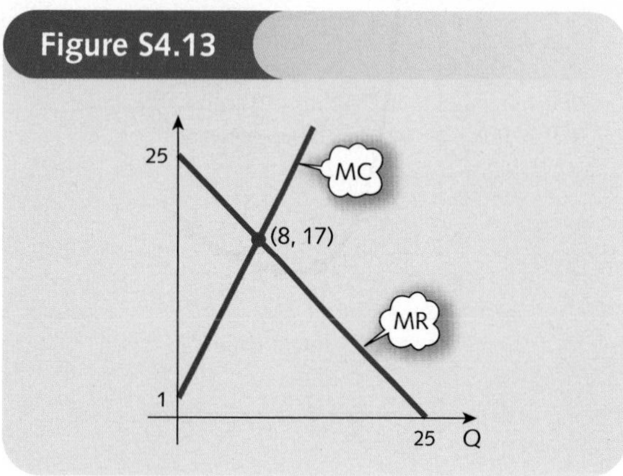

2 MC = 100.

(a) *Domestic market* $P_1 = 300 - Q_1$

$$TR_1 = 300Q_1 - Q_1^2$$

so

$$MR_1 = 200 - 2Q_1$$

To maximize profit, $MR_1 = MC$: that is,

$$300 - 2Q_1 = 100$$

which has solution $Q_1 = 100$.

Corresponding price is

$$P_1 = 300 - 100 = \$200$$

Foreign market $P_2 = 200 - \frac{1}{2}Q_2$

$$TR_2 = 200Q_2 - \tfrac{1}{2}Q_2^2$$

so

$$MR_2 = 200 - Q_2$$

To maximize profit, $MR_2 = MC$: that is,

$$200 - Q_2 = 100$$

which has solution $Q_2 = 100$.

Corresponding price is

$$P_2 = 200 - \tfrac{1}{2}(100) = \$150$$

(b) Without discrimination, $P_1 = P_2 = P$, say, so individual demand equations become

$$Q_1 = 300 - P$$

$$Q_2 = 400 - 2P$$

Adding shows that the demand equation for combined market is

$$Q = 700 - 3P$$

where $Q = Q_1 + Q_2$.

$$TR = \frac{700}{3}Q - \frac{Q^2}{3}$$

so

$$MR = \frac{700}{3} - \frac{2Q}{3}$$

To maximize profit, MR = MC: that is,

$$\frac{700}{3} - \frac{2Q}{3} = 100$$

which has solution $Q = 200$.

Corresponding price is

$$P = 700/3 - 200/3 = \$500/3$$

Total cost of producing 200 goods is

$$5000 + 100(200) = \$25\ 000$$

With discrimination, total revenue is

$$100 \times 200 + 100 \times 150 = \$35\ 000$$

so profit is \$35 000 – \$25 000 = \$10 000.

Without discrimination, total revenue is

$$200 \times \frac{500}{3} = \$33\ 333$$

so profit is \$33 333 – \$25 000 = \$8333.

3 *Domestic market* From Practice Problem 2, profit is maximum when $P_1 = 200$, $Q_1 = 100$. Also, since $Q_1 = 300 - P_1$ we have $dQ_1/dP_1 = -1$. Hence

$$E_1 = -\frac{P_1}{Q_1} \times \frac{dQ_1}{dP_1}$$

$$= -\frac{200}{100} \times (-1) = 2$$

Foreign market From Practice Problem 2, profit is maximum when $P_2 = 150$, $Q_2 = 100$. Also, since $Q_2 = 400 - 2P_2$ we have $dQ_2/dP_2 = -2$. Hence

$$E_2 = -\frac{P_2}{Q_2} \times \frac{dQ_2}{dP_2}$$

$$= -\frac{150}{100} \times (-2) = 3$$

We see that the firm charges the higher price in the domestic market, which has the lower elasticity of demand.

4 Argument is similar to that given in text but with < replaced by >.

5 **(a)** $TR = aQ^2 + bQ$, $TC = dQ + c$;

(b) $MR = 2aQ + b$, $MC = d$.

(c) The equation $2aQ + b = d$ has solution

$$Q = \frac{d - b}{2a}$$

6 **(a)** At the point of maximum total revenue

$$MR = \frac{d(TR)}{dQ} = 0$$

so $E = 1$.

(b) Maximum occurs when $Q = 10$.

7
$$TC = ACO + ACC$$
$$= \frac{(ARU)(CO)}{EOQ} + (CU)(CC)\frac{(EOQ)}{2}$$

At a stationary point

$$\frac{d(TC)}{d(EOQ)} = -\frac{(ARU)(CO)}{(EOQ)^2} + \frac{(CU)(CC)}{2} = 0$$

which has solution

$$EOQ = \sqrt{\frac{2(ARU)(CO)}{(CU)(CC)}}$$

Also

$$\frac{d^2(TC)}{d(EOQ)^2} = \frac{2(ARU)(CO)}{(EOQ)^3} > 0$$

so minimum.

8 **(a)** $P_1 = \$30, P_2 = \$20$. **(b)** $P = \$24.44$.

The profits in parts (a) and (b) are \$95 and \$83.89 respectively.

9 The argument is similar to that given in the text for AP_L.

10 The new supply equation is

$$P = aQ_S + b + t$$

In equilibrium

$$aQ + b + t = -cQ + d$$

which has solution

$$Q = \frac{d - b - t}{a + c}$$

Hence

$$tQ = \frac{(d - b)t - t^2}{a + c}$$

which differentiates to give

$$\frac{d - b - 2t}{a + c}$$

This is zero when

$$t = \frac{d - b}{2}$$

Also the second derivative is

$$\frac{-2}{a + c} < 0 \quad \text{(since } a \text{ and } c \text{ are both positive)}$$

which confirms that the stationary point is a maximum.

Section 4.8

1

x	0.50	1.00	1.50	2.00
$f(x)$	−0.69	0.00	0.41	0.69
x	2.50	3.00	3.50	4.00
$f(x)$	0.92	1.10	1.25	1.39

The graph of the natural logarithm function is sketched in Figure S4.14.

Figure S4.14

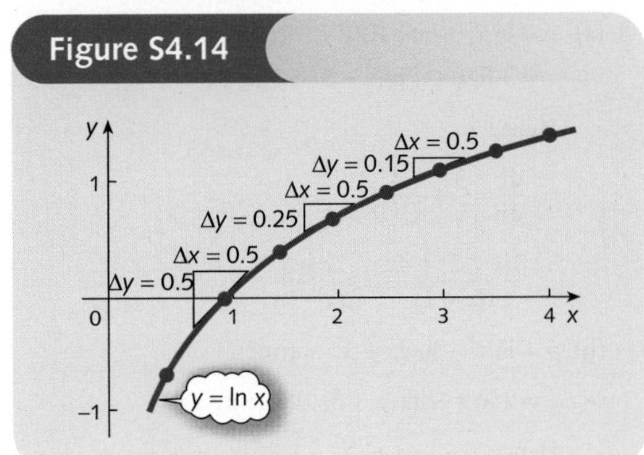

$$f'(1) = \frac{0.50}{0.50} = 1.00 \simeq 1$$

$$f'(2) = \frac{0.25}{0.50} = 0.50 \simeq \frac{1}{2}$$

$$f'(3) = \frac{0.15}{0.50} = 0.30 \simeq \frac{1}{3}$$

These results suggest that $f'(x) = 1/x$.

2 **(a)** $3e^{3x}$; **(b)** $-e^{-x}$; **(c)** $1/x$; **(d)** $1/x$.

3 **(a)** For the product rule we put

$$u = x^4 \quad \text{and} \quad v = \ln x$$

for which

$$\frac{du}{dx} = 4x^3 \quad \text{and} \quad \frac{dv}{dx} = \frac{1}{x}$$

By the product rule

$$\frac{dy}{dx} = x^4 \times \frac{1}{x} + \ln x \times 4x^3$$
$$= x^3 + 4x^3 \ln x$$
$$= x^3(1 + 4 \ln x)$$

(b) By the chain rule

$$\frac{dy}{dx} = e^{x^2} \times 2x = 2xe^{x^2}$$

(c) If

$$u = \ln x \quad \text{and} \quad v = x + 2$$

then

$$\frac{du}{dx} = \frac{1}{x} \quad \text{and} \quad \frac{dv}{dx} = 1$$

By the quotient rule

$$\frac{dy}{dx} = \frac{(x+2) \times \frac{1}{x} - (\ln x) \times 1}{(x+2)^2}$$

$$\frac{x + 2 - x \ln x}{x(x+2)^2} \quad \text{(multiply top and bottom by } x\text{)}$$

4 (a) $y = \ln x^3 + \ln(x+2)^4$ (rule 1)

$= 3 \ln x + 4 \ln(x + 2)$ (rule 3)

Hence

$$\frac{dy}{dx} = \frac{3}{x} + \frac{4}{x+2}$$

$$\frac{3(x+2) + 4x}{x(x+2)} = \frac{7x+6}{x(x+2)}$$

(b) $y = \ln x^2 - \ln(2x + 3)$ (rule 2)

$= 2 \ln x - \ln(2x + 3)$ (rule 3)

Hence

$$\frac{dy}{dx} = \frac{2}{x} - \frac{2}{2x+3} \quad \text{(chain rule)}$$

$$= \frac{2(2x+3) - 2x}{x(2x+3)}$$

$$= \frac{2x+6}{x(2x+3)}$$

5 In terms of P the total revenue function is given by

$$TR = PQ = 1000Pe^{-0.2P}$$

and the total cost function is

$$TC = 100 + 2Q = 100 + 2000e^{-0.2P}$$

Hence

$$\pi = TR - TC$$
$$= 1000Pe^{-0.2P} - 2000e^{-0.2P} - 100$$

Step 1

At a stationary point

$$\frac{d\pi}{dP} = 0$$

To differentiate the first term, $1000Pe^{-0.2P}$, we use the product rule with

$$u = 1000P \quad \text{and} \quad v = e^{-0.2P}$$

for which

$$\frac{du}{dP} = 1000 \quad \text{and} \quad \frac{dv}{dP} = -0.2e^{-0.2P}$$

Hence the derivative of $1000Pe^{-0.2P}$ is

$$u\frac{dv}{dP} + v\frac{du}{dP}$$
$$= 1000P(-0.2e^{-0.2P}) + e^{-0.2P}(1000)$$
$$= e^{-0.2P}(1000 - 200P)$$

Now

$$\pi = 1000Pe^{-0.2P} - 2000e^{-0.2P} - 100$$

so

$$\frac{d\pi}{dP} = e^{-0.2P}(1000 - 200P) - 2000(-0.2e^{-0.2P})$$
$$= e^{-0.2P}(1400 - 200P)$$

This is zero when

$$1400 - 200P = 0$$

because $e^{-0.2P} \neq 0$.

Hence $P = 7$.

Step 2

To find $\frac{d^2\pi}{dP^2}$ we differentiate

$$\frac{d\pi}{dP} = e^{-0.2P}(1400 - 200P)$$

using the product rule. Taking

$$u = e^{-0.2P} \quad \text{and} \quad v = 1400 - 200P$$

gives

$$\frac{du}{dP} = -0.2e^{-0.2P} \quad \text{and} \quad \frac{dv}{dP} = -200$$

Hence

$$\frac{d^2\pi}{dP^2} = u\frac{dv}{dP} + v\frac{du}{dP}$$
$$= e^{-0.2P}(-200) + (1400 - 200P)(-0.2e^{-0.2P})$$
$$= e^{-0.2P}(10P - 480)$$

Putting $P = 7$ gives

$$\frac{d^2\pi}{dP^2} = -200e^{-1.4}$$

This is negative, so the stationary point is a maximum.

6 To find the price elasticity of demand we need to calculate the values of P, Q and dQ/dP. We are given that $Q = 20$ and the demand equation gives

$$P = 200 - 40 \ln(20 + 1) = 78.22$$

The demand equation expresses P in terms of Q, so we first evaluate dP/dQ and then use the result

$$\frac{dQ}{dP} = \frac{1}{dP/dQ}$$

To differentiate $\ln(Q + 1)$ by the chain rule we differentiate the outer log function to get

$$\frac{1}{Q+1}$$

and then multiply by the derivative of the inner function, $Q + 1$, to get 1. Hence the derivative of $\ln(Q + 1)$ is

$$\frac{1}{Q+1}$$

and so

$$\frac{dP}{dQ} = \frac{-40}{Q+1}$$

Putting $Q = 20$ gives $dP/dQ = -40/21$, so that $dQ/dP = -21/40$. Finally, we use the formula

$$E = -\frac{P}{Q} \times \frac{dQ}{dP}$$

to calculate the price elasticity of demand as

$$E = -\frac{78.22}{20} \times \left(\frac{-21}{40}\right) = 2.05$$

7 **(a)** $6e^{6x}$; **(b)** $-342e^{-342x}$; **(c)** $-2e^{-x} + 4e^{x}$; **(d)** $40e^{4x} - 4x$.

8 **(a)** $\frac{1}{x}$; **(b)** $\frac{1}{x}$.

9 **(a)** $3x^2e^{x^3}$; **(b)** $\frac{4x^3 + 6x}{x^4 + 3x^2}$

10 **(a)** $(4x^3 + 2x^4)e^{2x}$; **(b)** $\ln x + 1$.

11 **(a)** $\frac{2e^{4x}(2x^2 - x + 4)}{(x^2 + 2)^2}$ **(b)** $\frac{e^x(x \ln x - 1)}{x(\ln x)^2}$

12 **(a)** $\frac{1}{x(1-x)}$; **(b)** $\frac{9x - 2}{2x(3x - 1)}$; **(c)** $\frac{1}{1 - x^2}$.

13 **(a)** Maximum at $(1, e^{-1})$; the graph is sketched in Figure S4.15.

Figure S4.15

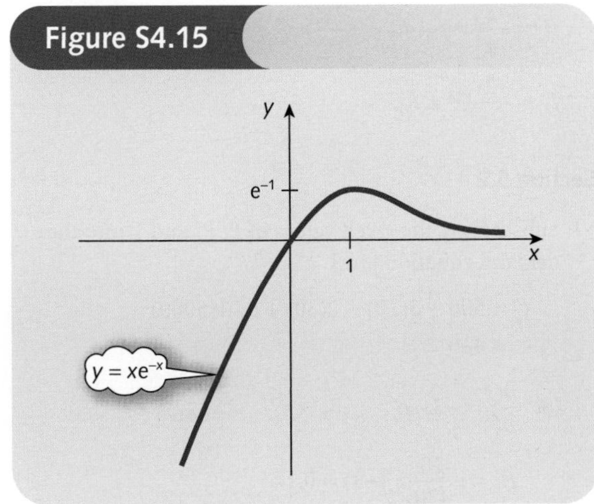

(b) Maximum at $(1, -1)$; the graph is sketched in Figure S4.16.

Figure S4.16

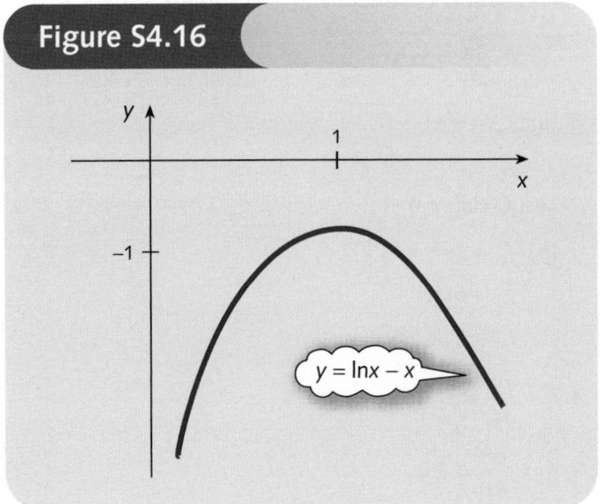

14 49.

15 50.

16 $E = \frac{10}{Q}$, which is 1 when $Q = 10$.

17 $\frac{dy}{dx} \div y = Ake^{kt} \div Ae^{kt} = k$

Chapter 5

Section 5.1

1 **(a)** -10; **(b)** -1; **(c)** 2; **(d)** 21; **(e)** 0;
(f) 21. The value of g is independent of the ordering of the variables. Such a function is said to be ***symmetric***.

2 **(a)** Differentiating $5x^4$ with respect to x gives $20x^3$ and, since y is held constant, y^2 differentiates to zero. Hence

$$\frac{\partial f}{\partial x} = 20x^3 - 0 = 20x^3$$

Differentiating $5x^4$ with respect to y gives zero because x is held fixed. Also differentiating y^2 with respect to y gives $2y$, so

$$\frac{\partial f}{\partial y} = 0 - 2y = -2y$$

(b) To differentiate the first term, x^2y^3, with respect to x we regard it as a constant multiple of x^2 (where the constant is y^3), so we get $2xy^3$. The second term obviously gives -10, so

$$\frac{\partial f}{\partial x} = 2xy^3 - 10$$

To differentiate the first term, x^2y^3, with respect to y we regard it as a constant multiple of y^3 (where the constant is x^2), so we get $3x^2y^2$. The second term is a constant and goes to zero, so

$$\frac{\partial f}{\partial y} = 3x^2y^2 - 0 = 3x^2y^2$$

3 **(a)** $f_{xx} = 60x^2$

$f_{yy} = 2$

$f_{yx} = f_{xy} = 0$

(b) $f_{xx} = 2y^3$

$f_{yy} = 6x^2y$

$f_{yx} = f_{xy} = 6xy^2$

4 $f_1 = \dfrac{\partial f}{\partial x_1} = x_2 + 5x_1^4$

$f_{11} = \dfrac{\partial^2 f}{\partial x_1^2} = 20x_1^3$

$f_{21} = \dfrac{\partial^2 f}{\partial x_2 \partial x_1} = 1$

5 $\dfrac{\partial z}{\partial x} = y - 5$, $\dfrac{\partial z}{\partial y} = x + 2$, so, at (2, 6),

$\dfrac{\partial z}{\partial x} = 1$, $\dfrac{\partial z}{\partial y} = 4$

(a) $\Delta x = -0.1$, $\Delta y = 0.1$;

$z \simeq 1(-0.1) + 4(0.1) = 0.3$, so z increases by approximately 0.3.

(b) At (2, 6), $z = 14$, and at (1.9, 6.1), $z = 14.29$, so the exact increase is 0.29.

6 **(a)** $\dfrac{dy}{dx} = \dfrac{-y}{x - 3y^2 + 1}$

(b) $\dfrac{dy}{dx} = \dfrac{y^2}{5y^4 - 2xy}$

7 324; 75; 0.

8 $85 \neq 91$; $(0, y)$ for any y.

9 **(a)** $f_x = 2x$, $f_y = 20y^4$;

(b) $f_x = 9x^2$, $f_y = -2e^y$;

(c) $f_x = y$, $f_y = x + 6$;

(d) $f_x = 6x^5y^2$, $f_y = 2x^6y + 15y^2$.

10 $f_x = 4x^3y^5 - 2x$

$f_y = 5x^4y^4 + 2y$

$f_x(1, 0) = -2$

$f_y(1, 1) = 7$

11

	f_x	f_y	f_{xx}	f_{yy}	f_{yx}	f_{xy}
(a)	y	x	0	0	1	1
(b)	e^xy	e^x	e^xy	0	e^x	e^x
(c)	$2x + 2$	1	2	0	0	0
(d)	$4x^{-3/4}y^{3/4}$	$12x^{1/4}y^{-1/4}$	$-3x^{-7/4}y^{3/4}$	$-3x^{1/4}y^{-5/4}$	$3x^{-3/4}y^{-1/4}$	$3x^{-3/4}y^{-1/4}$
(e)	$\dfrac{-2y}{x^3} + \dfrac{1}{y}$	$\dfrac{1}{x^2} - \dfrac{x}{y^2}$	$\dfrac{6y}{x^4}$	$\dfrac{2x}{y^3}$	$\dfrac{-2}{x^3} - \dfrac{1}{y^2}$	$\dfrac{-2}{x^3} - \dfrac{1}{y^2}$

12 **(a)** −0.6; **(b)** −2; **(c)** −2.6.

13 78; 94; 6.2.

14 **(a)** $f_x = -3x^2 + 2$, $f_y = 1$

$\dfrac{dy}{dx} = -\dfrac{-3x^2 + 2}{1} = 3x^2 - 2$

(b) $y = x^3 - 2x + 1$, so

$\dfrac{dy}{dx} = 3x^2 - 2$ ✓

15 1/3.

16 $f_1 = \dfrac{x_3^2}{x_2}$; $f_2 = -\dfrac{x_1x_3^2}{x_2^2} + \dfrac{1}{x_2}$; $f_3 = \dfrac{2x_1x_3}{x_2} + \dfrac{1}{x_3}$;

$f_{11} = 0$; $f_{22} = \dfrac{2x_1x_3^2}{x_2^3} - \dfrac{1}{x_2^2}$; $f_{33} = \dfrac{2x_1}{x_2} - \dfrac{1}{x_3^2}$;

$f_{12} = -\dfrac{x_3^2}{x_2^2} = f_{21}$; $f_{13} = -\dfrac{2x_3}{x_2} = f_{31}$;

$f_{23} = \dfrac{-2x_1x_3}{x_2^2} = f_{32}$.

Section 5.2

1 Substituting the given values of P, P_A and Y into the demand equation gives

$$Q = 500 - 3(20) - 2(30) + 0.01(5000) = 430$$

(a) $\dfrac{\partial Q}{\partial P} = -3$, so

$$E_P = -\frac{20}{430} \times (-3) = 0.14$$

(b) $\frac{\partial Q}{\partial P_A} = -2$, so

$$E_{P_A} = \frac{30}{430} \times (-2) = 0.14$$

(c) $\frac{\partial Q}{\partial Y} = 0.01$, so

$$E_Y = \frac{5000}{430} \times 0.01 = 0.12$$

By definition,

$$E_Y = \frac{\text{percentage change in } Q}{\text{percentage change in } Y}$$

so demand rises by $0.12 \times 5 = 0.6\%$. A rise in income causes a rise in demand, so good is superior.

2 $\frac{\partial U}{\partial x_1} = 1000 + 5x_2 - 4x_1$

$\frac{\partial U}{\partial x_2} = 450 + 5x_1 - 2x_2$, so at (138, 500)

$\frac{\partial U}{\partial x_1} = 2948$ and $\frac{\partial U}{\partial x_2} = 140$

If working time increases by 1 hour then leisure time decreases by 1 hour, so $\Delta x_1 = -1$. Also $\Delta x_2 = 15$. By the small increments formula

$$\Delta U \simeq 2948(-1) + 140(15) = -848$$

The law of diminishing marginal utility holds for both x_1 and x_2 because

$$\frac{\partial^2 U}{\partial x_1^2} = -4 < 0$$

and

$$\frac{\partial^2 U}{\partial x_2^2} = -2 < 0$$

3 Using the numerical results in Practice Problem 2,

$$\text{MRCS} = \frac{2948}{140} = 21.06$$

This represents the increase in x_2 required to maintain the current level of utility when x_1 falls by 1 unit. Hence if x_1 falls by 2 units, the increase in x_2 is approximately

$$21.06 \times 2 = \$42.12$$

4 $\text{MP}_K = 2K$ and $\text{MP}_L = 4L$

(a) $\text{MRTS} = \frac{\text{MP}_L}{\text{MP}_K} = \frac{4L}{2K} = \frac{2L}{K}$

(b) $K\frac{\partial Q}{\partial K} + L\frac{\partial Q}{\partial L} = K(2K) + L(4L)$

$$= 2(K^2 + 2L^2) = 2Q \checkmark$$

5 (a) 20/1165; (b) −15/1165;

(c) 2000/1165; −0.04%; complementary.

6 2; 1%.

7 $\frac{\partial U}{\partial x_1} = \frac{1}{5}$ and $\frac{\partial U}{\partial x_2} = \frac{5}{12}$

(a) 37/60; (b) 12/25.

8 $\text{MP}_K = 8$, $\text{MP}_L = 14\frac{1}{4}$; (a) $1\frac{25}{32}$; (b) $1\frac{25}{32}$.

9 $K(6K^2 + 3L^2) + L(6LK) = 6K^3 + 9L^2K = 3(K^3 + 3L^2K)$

10 $\frac{\partial Q}{\partial K} = \alpha A K^{\alpha-1} L^{\beta}$ and

$\frac{\partial Q}{\partial L} = \beta A K^{\alpha} L^{\beta-1}$, so

$$K\frac{\partial Q}{\partial K} + L\frac{\partial Q}{\partial L} = \alpha A K^{\alpha} L^{\beta} + \beta A K^{\alpha} L^{\beta}$$
$$= (\alpha + \beta)(A K^{\alpha} L^{\beta})$$
$$= (\alpha + \beta)Q \checkmark$$

11 The graph is sketched in Figure S5.1, which shows that $\text{MRTS} = -(-5/7) = 5/7$.

Figure S5.1

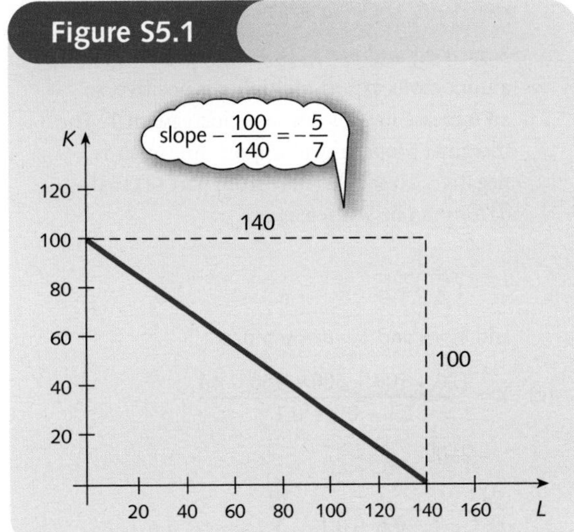

12 (b) $\frac{0.4(5\sqrt{K} + 1.5\sqrt{L})\sqrt{K}}{L}$

(c) 2.97.

13 (c) $\frac{7K^4}{3L^4}$; (d) −0.033.

Section 5.3

1 $C = a\left(\frac{b + I^*}{1 - a}\right) + b$

$$\frac{\partial C}{\partial I^*} = \frac{a}{1 - a} > 0$$

because

$$0 < a < 1$$

Hence an increase in I^* leads to an increase in C. If $a = \frac{1}{2}$ then

$$\frac{\partial C}{\partial I^*} = \frac{\frac{1}{2}}{1 - \frac{1}{2}} = 1$$

Change in C is

$$1 \times 2 = 2$$

2 **(a)** Substitute C, I, G, X and M into the Y equation to get

$$Y = aY + b + I^* + G^* + X^* - (mY + M^*)$$

Collecting like terms gives

$$(1 - a + m)Y = b + I^* + G^* + X^* - M^*$$

so

$$Y = \frac{b + I^* + G^* + X^* - M^*}{1 - a + m}$$

(b) $$\frac{\partial Y}{\partial X^*} = \frac{1}{1 - a + m}$$

$$\frac{\partial Y}{\partial m} = -\frac{b + I^* + G^* + X^* - M^*}{(1 - a + m)^2}$$

Now $a < 1$ and $m > 0$, so $1 - a + m > 0$. The autonomous export multiplier is positive, so an increase in X^* leads to an increase in Y. The marginal propensity to import multiplier is negative. To see this note from part (a) that $\partial Y/\partial m$ can be written as

$$-\frac{Y}{1 - a + m}$$

and $Y > 0$ and $1 - a + m > 0$.

(c) $$Y = \frac{120 + 100 + 300 + 150 - 40}{1 - 0.8 + 0.1}$$

$$= 2100$$

$$\frac{\partial Y}{\partial X^*} = \frac{1}{1 - 0.8 + 0.1} = \frac{10}{3}$$

and

$$\Delta X^* = 10$$

so

$$\Delta Y = \frac{10}{3} \times 10 = \frac{100}{3}$$

3 If d increases by a small amount then the intercept increases and the demand curve shifts upwards slightly. Figure S5.2 shows that the effect is to increase the equilibrium quantity from Q_1 to Q_2, confirming that $\partial Q/\partial d > 0$.

Figure S5.2

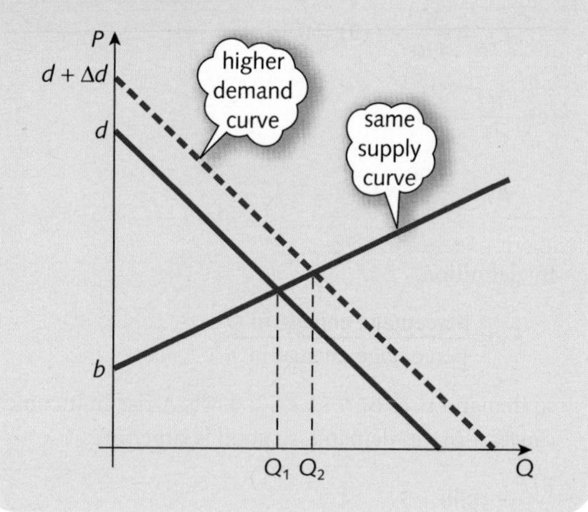

4 **(a)** Substituting equations (3) and (4) into (2) gives

$$C = a(Y - T^*) + b = aY - aT^* + b \qquad (7)$$

Substituting (5), (6) and (7) into (1) gives

$$Y = aY - aT^* + b + I^* + G^*$$

so that

$$Y = \frac{-aT^* + b + I^* + G^*}{1 - a}$$

Finally, from (7), we see that

$$C = a\left(\frac{-aT^* + b + I^* + G^*}{1 - a}\right) - aT^* + b$$

$$= \frac{a(-aT^* + b + I^* + G^*) + (1 - a)(-aT^* + b)}{1 - a}$$

$$= \frac{aI^* + aG^* - aT^* + b}{1 - a}$$

(b) $\frac{a}{1 - a} > 0$; C increases. **(c)** 1520; rise of 18.

5 **(1)** From the relations

$$C = aY_d + b$$
$$Y_d = Y - T$$
$$T = tY + T^*$$

we see that

$$C = a(Y - tY - T^*) + b$$

Similarly,

$$M = m(Y - tY - T^*) + M^*$$

Substitute these together with I, G and X into the Y equation to get the desired result.

(2) **(a)** $$\frac{\partial Y}{\partial T^*} = \frac{m - a}{1 - a + at + m - mt}$$

Numerator is negative because $m < a$. Denominator can be written as

$$(1 - a) + at + m(1 - t)$$

which represents the sum of three positive numbers, so is positive. Hence the autonomous taxation multiplier is negative.

(b) $\dfrac{\partial Y}{\partial G^*} = \dfrac{1}{1 - a + at + m - mt} > 0$

(3) (a) 1000; (b) $\Delta Y = 20$; (c) $\Delta T^* = 33\tfrac{1}{3}$.

6 From text, equilibrium quantity is

$$\frac{d - b}{a + c}$$

Substituting this into either the supply or demand equation gives the desired result.

$$\frac{\partial P}{\partial a} = \frac{c(d - b)}{(a + c)^2} > 0, \quad \frac{\partial P}{\partial b} = \frac{c}{a + c} > 0$$

$$\frac{\partial P}{\partial c} = -\frac{a(d - b)}{(a + c)^2} < 0, \quad \frac{\partial P}{\partial d} = \frac{a}{a + c} > 0$$

where the quotient rule is used to obtain $\partial P/\partial a$ and $\partial P/\partial c$. An increase in a, b or d leads to an increase in P, whereas an increase in c leads to a decrease in P.

7 (1) Substituting second and third equations into first gives

$$Y = aY + b + cr + d$$

so that

$$(1 - a)Y - cr = b + d \qquad (1)$$

(2) Substituting first and second equations into third gives

$$k_1Y + k_2r + k_3 = M_S^*$$

so that

$$k_1Y + k_2r = M_S^* - k_3 \qquad (2)$$

(3) (a) Working out $c \times (2) + k_2 \times (1)$ eliminates r to give

$$ck_1Y + k_2(1 - a)Y$$
$$= c(M_S^* - k_3) + k_2(b + d)$$

Dividing both sides by $ck_1 + k_2(1 - a)$ gives result.

(b) $\dfrac{c}{(1 - a)k_2 + ck_1}$, which is positive because the top and bottom of this fraction are both negative.

Section 5.4

1 $f_x = 2x$, $f_y = 6 - 6y$, $f_{xx} = 2$, $f_{yy} = -6$, $f_{xy} = 0$.

Step 1

At a stationary point

$$2x = 0$$
$$6 - 6y = 0$$

which shows that there is just one stationary point at (0, 1).

Step 2

$$f_{xx}f_{yy} - f_{xy}^2 = 2(-6) - 0^2 = -12 < 0$$

so it is a saddle point.

2 Total revenue from the sale of G1 is

$$TR_1 = P_1Q_1 = (50 - Q_1)Q_1 = 50Q_1 - Q_1^2$$

Total revenue from the sale of G2 is

$$TR_2 = P_2Q_2 = (95 - 3Q_2)Q_2$$
$$= 95Q_2 - 3Q_2^2$$

Total revenue from the sale of both goods is

$$TR = TR_1 + TR_2$$
$$= 50Q_1 - Q_1^2 + 95Q_2 - 3Q_2^2$$

Profit is

$$\pi = TR - TC$$
$$= (50Q_1 - Q_1^2 + 95Q_2 - 3Q_2^2) - (Q_1^2 + 3Q_1Q_2 + Q_2^2)$$
$$= 50Q_1 - 2Q_1^2 + 95Q_2 - 4Q_2^2 - 3Q_1Q_2$$

Now

$$\frac{\partial \pi}{\partial Q_1} = 50 - 4Q_1 - 3Q_2,$$

$$\frac{\partial \pi}{\partial Q_2} = 95 - 8Q_2 - 3Q_1$$

$$\frac{\partial^2 \pi}{\partial Q_1^2} = -4, \quad \frac{\partial^2 \pi}{\partial Q_1 \partial Q_2} = -3,$$

$$\frac{\partial^2 \pi}{\partial Q_2^2} = -8$$

Step 1

At a stationary point

$$50 - 4Q_1 - 3Q_2 = 0$$
$$95 - 3Q_1 - 8Q_2 = 0$$

that is,

$$4Q_1 + 3Q_2 = 50 \qquad (1)$$
$$3Q_1 + 8Q_2 = 95 \qquad (2)$$

Multiply equation (1) by 3, and equation (2) by 4 and subtract to get

$$23Q_2 = 230$$

so $Q_2 = 10$. Substituting this into either equation (1) or equation (2) gives $Q_1 = 5$.

Step 2

This is a maximum because

$$\frac{\partial^2 \pi}{\partial Q_1^2} = -4 < 0, \quad \frac{\partial^2 \pi}{\partial Q_2^2} = -8 < 0$$

and

$$\left(\frac{\partial^2\pi}{\partial Q_1^2}\right)\left(\frac{\partial^2\pi}{\partial Q_2^2}\right)-\left(\frac{\partial^2\pi}{\partial Q_1\partial Q_2}\right)^2$$
$$=(-4)(-8)-(-3)^2=23>0$$

Corresponding prices are found by substituting $Q_1 = 5$ and $Q_2 = 10$ into the original demand equations to obtain $P_1 = 45$ and $P_2 = 65$.

3 For the domestic market, $P_1 = 300 - Q_1$, so

$$TR_1 = P_1Q_1 = 300Q_1 - Q_1^2$$

For the foreign market, $P_2 = 200 - \frac{1}{2}Q_2$, so

$$TR_2 = P_2Q_2 = 200Q_2 - \tfrac{1}{2}Q_2^2$$

Hence

$$TR = TR_1 + TR_2$$
$$= 300Q_1 - Q_1^2 + 200Q_2 - \tfrac{1}{2}Q_2^2$$

We are given that

$$TC = 5000 + 100(Q_1 + Q_2)$$
$$= 5000 + 100Q_1 + 100Q_2$$

so

$$\pi = TR - TC$$
$$= (300Q_1 - Q_1^2 + 200Q_2 - \tfrac{1}{2}Q_2^2) - (5000 + 100Q_1 + 100Q_2)$$
$$= 200Q_1 - Q_1^2 + 100Q_2 - \tfrac{1}{2}Q_2^2 - 5000$$

Now

$$\frac{\partial\pi}{\partial Q_1} = 200 - 2Q_1,\ \frac{\partial\pi}{\partial Q_2} = 100 - 2Q_2$$

$$\frac{\partial^2\pi}{\partial Q_1^2} = -2,\ \frac{\partial^2\pi}{\partial Q_1^2\partial Q_2} = 0,\ \frac{\partial^2\pi}{\partial Q_2^2} = -1$$

Step 1

At a stationary point

$$200 - 2Q_1 = 0$$
$$100 - Q_2 = 0$$

which have solution $Q_1 = 100$, $Q_2 = 100$.

Step 2

This is a maximum because

$$\frac{\partial^2\pi}{\partial Q_1^2} = -2 < 0,$$

$$\frac{\partial^2\pi}{\partial Q_2^2} = -1 < 0$$

and

$$\left(\frac{\partial^2\pi}{\partial Q_1^2}\right)\left(\frac{\partial^2\pi}{\partial Q_2^2}\right)-\left(\frac{\partial^2\pi}{\partial Q_1\partial Q_2}\right)^2$$
$$=(-2)(-1)-0^2=2>0$$

Substitute $Q_1 = 100$, $Q_2 = 100$, into the demand and profit functions to get $P_1 = 200$, $P_2 = 150$ and $\pi = 10\,000$.

4 (a) Minimum at (1, 1), maximum at (−1, −1), and saddle points at (1, −1) and (−1, 1).

(b) Minimum at (2, 0), maximum at (0, 0), and saddle points at (1, 1) and (1, −1).

5 Maximum profit is \$1300 when $Q_1 = 30$ and $Q_2 = 10$.

6 Maximum profit is \$95 when $P_1 = 30$ and $P_2 = 20$.

7 Maximum profit is \$176 when $L = 16$ and $K = 144$.

8 $x_1 = 138$, $x_2 = 500$; \$16.67 per hour.

9 $Q_1 = 19$, $Q_2 = 4$.

10 (a) Minimum at (1, 2).

(b) Maximum at (0, 1).

(c) Saddle point at (2, 2).

11 (a) $P_1 = 78$, $P_2 = 68$.

(b) Rotate the box so that the Q_1 axis comes straight out of the screen. The graph increases steadily as Q_2 rises from 0 to 2.

$Q_1 = 24$. Profit in (a) and (b) is 1340 and 1300 respectively.

Section 5.5

1 *Step 1*

We are given that $y = x$, so no rearrangement is necessary.

Step 2

Substituting $y = x$ into the objective function

$$z = 2x^2 - 3xy + 2y + 10$$

gives

$$z = 2x^2 - 3x^2 + 2x + 10$$
$$= -x^2 + 2x + 10$$

Step 3

At a stationary point

$$\frac{dz}{dx} = 0$$

that is,

$$-2x + 2 = 0$$

which has solution $x = 1$. Differentiating a second time gives

$$\frac{d^2z}{dx^2} = -2$$

confirming that the stationary point is a maximum.

The value of z can be found by substituting $x = 1$ into

$$z = -2x^2 + 2x + 10$$

to get $z = 11$. Finally, putting $x = 1$ into the constraint $y = x$ gives $y = 1$. The constrained function therefore has a maximum value of 11 at the point (1, 1).

2 We want to maximize the objective function

$$U = x_1x_2$$

subject to the budgetary constraint

$$2x_1 + 10x_2 = 400$$

Step 1

$$x_1 = 200 - 5x_2$$

Step 2

$$U = 200x_2 - 5x_2^2$$

Step 3

$$\frac{dU}{dx_1} = 200 - 10x_2 = 0$$

has solution $x_2 = 20$.

$$\frac{d^2U}{dx_2^2} = -10 < 0$$

so maximum.

Putting $x_2 = 20$ into constraint gives $x_1 = 100$.

$$U_1 = \frac{\partial U}{\partial x_1} = x_2 = 20$$

and

$$U_2 = \frac{\partial U}{\partial x_2} = x_2 = 100$$

so the ratios of marginal utilities to prices are

$$\frac{U_1}{P_1} = \frac{20}{2} = 10$$

and

$$\frac{U_2}{P_2} = \frac{100}{10} = 10$$

which are the same.

3 We want to minimize the objective function

$$TC = 3x_1^2 + 2x_1x_2 + 7x_2^2$$

subject to the production constraint

$$x_1 + x_2 = 40$$

Step 1

$$x_1 = 40 - x_2$$

Step 2

$$TC = 3(40 - x_2)^2 + 2(40 - x_2)x_2 + 7x_2^2$$
$$= 4800 - 160x_2 + 8x_2^2$$

Step 3

$$\frac{d(TC)}{dx_2} = -160 + 16x_2 = 0$$

has solution $x_2 = 10$.

$$\frac{d^2(TC)}{dx_2^2} = 16 > 0$$

so minimum.

Finally, putting $x_2 = 10$ into constraint gives $x_1 = 30$.

4 Maximum value of z is 13, which occurs at (3, 11).

5 27 000.

6 $K = 10$ and $L = 4$.

7 $K = 6$ and $L = 4$.

8 Maximum profit is \$165, which is achieved when $K = 81$ and $L = 9$.

9 $x_1 = 3$, $x_2 = 4$.

Section 5.6

1 *Step 1*

$$g(x, y, \lambda) = 2x^2 - xy + \lambda(12 - x - y)$$

Step 2

$$\frac{\partial g}{\partial x} = 4x - y - \lambda = 0$$

$$\frac{\partial g}{\partial y} = -x - \lambda = 0$$

$$\frac{\partial g}{\partial \lambda} = 12 - x - y = 0$$

that is,

$$4x - y - \lambda = 0 \quad (1)$$
$$-x \quad - \lambda = 0 \quad (2)$$
$$x + y \quad = 12 \quad (3)$$

Multiply equation (2) by 4 and add equation (1), multiply equation (3) by 4 and subtract from equation (1) to get

$$-y - 5\lambda = 0 \quad (4)$$
$$-5y - \lambda = -48 \quad (5)$$

Multiply equation (4) by 5 and subtract equation (5) to get

$$-24\lambda = 48 \quad (6)$$

Equations (6), (5) and (1) can be solved in turn to get

$$\lambda = -2, y = 10, x = 2$$

so the optimal point has coordinates (2, 10). The corresponding value of the objective function is

$$2(2)^2 - 2(10) = -12$$

2 Maximize

$$U = 2x_1x_2 + 3x_1$$

subject to

$$x_1 + 2x_2 = 83$$

Step 1

$$g(x_1, x_2, \lambda) = 2x_1x_2 + 3x_1 + \lambda(83 - x_1 - 2x_2)$$

Step 2

$$\frac{\partial g}{\partial x_1} = 2x_2 + 3 - \lambda = 0$$

$$\frac{\partial g}{\partial x_2} = 2x_1 - 2\lambda = 0$$

$$\frac{\partial g}{\partial \lambda} = 83 - x_1 - 2x_2 = 0$$

that is,

$$2x_2 - \lambda = -3 \quad (1)$$

$$2x_1 - 2\lambda = 0 \quad (2)$$

$$x_1 + 2x_2 = 83 \quad (3)$$

The easiest way of solving this system is to use equations (1) and (2) to get

$$\lambda = 2x_2 + 3 \quad \text{and} \quad \lambda = x_1$$

respectively. Hence

$$x_1 = 2x_2 + 3$$

Substituting this into equation (3) gives

$$4x_2 + 3 = 83$$

which has solution $x_2 = 20$ and so $x_1 = \lambda = 43$.
The corresponding value of U is

$$2(43)(20) + 3(43) = 1849$$

The value of λ is 43, so when income rises by 1 unit, utility increases by approximately 43 to 1892.

3 *Step 1*

$$g(x_1, x_2, \lambda) = x_1^{1/2} + x_2^{1/2} + \lambda(M - P_1x_1 - P_2x_2)$$

Step 2

$$\frac{\partial g}{\partial x_1} = \frac{1}{2}x_1^{-1/2} - \lambda P_1 = 0 \quad (1)$$

$$\frac{\partial g}{\partial x_2} = \frac{1}{2}x_2^{-1/2} - \lambda P_2 = 0 \quad (2)$$

$$\frac{\partial g}{\partial \lambda} = M - P_1x_1 - P_2x_2 = 0 \quad (3)$$

From equations (1) and (2)

$$\lambda = \frac{1}{2x_1^{1/2}P_1} \quad \text{and} \quad \lambda = \frac{1}{2x_2^{1/2}P_2}$$

respectively. Hence

$$\frac{1}{2x_1^{1/2}P_1} = \frac{1}{2x_2^{1/2}P_2}$$

that is,

$$x_1P_1^2 = x_2P_2^2$$

so

$$x_1 = \frac{x_2P_2^2}{P_1^2} \quad (4)$$

Substituting this into equation (3) gives

$$M - \frac{x_2P_2^2}{P_1} - P_2x_2 = 0$$

which rearranges as

$$x_2 = \frac{P_1M}{P_2(P_1 + P_2)}$$

Substitute this into equation (4) to get

$$x_1 = \frac{P_2M}{P_1(P_1 + P_2)}$$

4 9.

5 There are two wheels per frame, so the constraint is $y = 2x$. Maximum profit is \$4800 at $x = 40$, $y = 80$.

6 Maximum profit is \$600 at $Q_1 = 10$, $Q_2 = 5$. Lagrange multiplier is 4, so profit rises to \$604 when total cost increases by 1 unit.

7 40; 2.5.

8 $x_1 = \dfrac{\alpha M}{(\alpha + \beta)P_1}$ and $x_2 = \dfrac{\beta M}{(\alpha + \beta)P_2}$

Chapter 6

Section 6.1

1 **(a)** x^2; **(b)** x^4; **(c)** x^{100}; **(d)** $\frac{1}{4}x^4$; **(e)** $\frac{1}{19}x^{19}$.

2 **(a)** $\frac{1}{5}x^5 + c$; **(b)** $-\frac{1}{2x^2} + c$; **(c)** $\frac{3}{4}x^{4/3} + c$;

(d) $\frac{1}{3}e^{3x} + c$; **(e)** $x + c$;

(f) $\frac{x^2}{2} + c$; **(g)** $\ln x + c$.

3 **(a)** $x^2 - x^4 + c$; **(b)** $2x^5 - \frac{5}{x} + c$;

(c) $\frac{7}{3}x^3 - \frac{3}{2}x^2 + 2x + c$.

4 (a) $\text{TC} = \int 2\text{d}Q = 2Q + c$

Fixed costs are 500, so $c = 500$. Hence

$\text{TC} = 2Q + 500$

Put $Q = 40$ to get TC = 580.

(b) $\text{TR} = \int (100 - 6Q)\text{d}Q$

$= 100Q - 3Q^2 + c$

Revenue is zero when $Q = 0$, so $c = 0$. Hence

$\text{TR} = 100Q - 3Q^2$

$P = \frac{\text{TR}}{Q} = \frac{100Q - 3Q^2}{Q}$

$= 100 - 3Q$

so demand equation is $P = 100 - 3Q$.

(c) $S = \int (0.4 - 0.1Y^{-1/2})\text{d}Y$

$= 0.4Y - 0.2Y^{1/2} + c$

The condition $S = 0$ when $Y = 100$ gives

$0 = 0.4(100) - 0.2(100)^{1/2} + c$

$= 38 + c$

so $c = -38$. Hence

$S = 0.4Y - 0.2Y^{1/2} - 38$

5 (a) $x^6 + c$; **(b)** $\frac{1}{5}x^5 + c$; **(c)** $\text{e}^{10x} + c$; **(d)** $\ln x + c$;

(e) $\frac{2}{5}x^{5/2} + c$; **(f)** $\frac{1}{2}x^4 - 3x^2 + c$;

(g) $\frac{1}{3}x^3 - 4x^2 + 3x + c$; **(h)** $\frac{ax^2}{2} + bx + c$;

(i) $\frac{7}{4}x^4 - 2\text{e}^{-2x} + \frac{3}{x} + c$.

6 (1) $F'(x) = 10(2x + 1)^4$, which is 10 times too big, so the integral is

$\frac{1}{10}(2x + 1)^5 + c$

(2) (a) $\frac{1}{24}(3x - 2)^8 + c$; **(b)** $-\frac{1}{40}(2 - 4x)^{10} + c$;

(c) $\frac{1}{a(x + 1)}(ax + b)^{n+1} + c$; **(d)** $\frac{1}{7}\ln(7x + 3) + c$.

7 (a) $\text{TC} = \frac{Q^2}{2} + 5Q + 20$

(b) $\text{TC} = 6\text{e}^{0.5Q} + 4$

8 (a) $\text{TR} = 20Q - Q^2$; $P = 20 - Q$

(b) $\text{TR} = 12\sqrt{Q}$; $P = \frac{12}{\sqrt{Q}}$

9 $C = 0.6Y + 7$, $S = 0.4Y - 7$

10 (a) $1000L - L^3$

(b) $12\sqrt{L} - 0.01L$

11 (a) $\frac{1}{2}x^2 + \frac{2}{7}x^{7/2} + c$

(b) $\frac{1}{11}x^{11} + \frac{1}{3}x^3$; $\frac{1}{5}\text{e}^{5x} + \text{e}^x + \frac{3}{2}\text{e}^{2x} + c$; $\frac{1}{3}x^3 - \frac{1}{2}x^2 + c$.

12 (a) $\frac{1}{4}x^4 - \frac{1}{2}x^2 + 2^{1/2}x + c$

(b) $\ln x + \frac{1}{x} + c$; $-\text{e}^{-x} + \frac{1}{3}\text{e}^{-3x} + c$; $x - x + \frac{2}{3}x^{3/2} + c$

13 $\frac{Y}{3} + \sqrt{Y} + 3$; $k = 9$.

14 (a) $500L\text{e}^{-0.02L}$; **(b)** $\ln(1 + 50Q^2)$.

Section 6.2

1 (a) $\int_0^1 x^3\text{d}x = \left[\frac{1}{4}x^4\right]_0^1$

$= \frac{1}{4}(1)^4 - \frac{1}{4}(0)^4 = \frac{1}{4}$

(b) $\int_2^5 (2x - 1)\text{d}x = [x^2 - x]_2^5 = (5^2 - 5) - (2^2 - 2) = 18$

(c) $\int_1^4 (x^2 - x + 1)\text{d}x$

$= \left[\frac{1}{3}x^3 - \frac{1}{2}x^2 + x\right]_1^4$

$= \left[\frac{1}{3}(4)^3 - \frac{1}{2}(4)^2 + 4\right] - \left[\frac{1}{3}(1)^3 - \frac{1}{2}(1)^2 + 1\right]$

$= 16.5$

(d) $\int_0^1 \text{e}^x\text{d}x = [\text{e}^x]_0^1 = \text{e}^1 - \text{e}^0 = \text{e} - 1 = 1.718\,28\ldots$

2 Substitute $Q = 8$ to get

$P = 100 - 8^2 = 36$

$\text{CS} = \int_0^8 (100 - Q^2)\text{d}Q - 8(36)$

$= \left[100Q - \frac{1}{3}Q^3\right]_0^8 - 288$

$= \left[100(8) - \frac{1}{3}(8)^3\right] - \left[100(0) - \frac{1}{3}(0)^2\right] - 288$

$= 341.33$

3 In equilibrium, $Q_S = Q_D = Q$, so

$P = 50 - 2Q$

$P = 10 + 2Q$

Hence

$$50 - 2Q = 10 + 2Q$$

which has solution $Q = 10$. The demand equation gives

$$P = 50 - 2(10) = 30$$

(a) $\text{CS} = \int_0^{10} (50 - 2Q)\text{d}Q - 10(30)$

$= [50Q - Q^2]_0^{10} - 300$

$= [50(10) - (10)^2] - [50(0) - 0^2] - 300$

$= 100$

(b) $\text{PS} = 10(30) - \int_0^{10} (10 + 2Q)\text{d}Q$

$= 300 - [10Q + Q^2]_0^{10}$

$= 300 - \{[10(10) + (10)^2] - [10(0) + 0^2]\}$

$= 100$

4 (a) $\int_1^8 800t^{1/3}\text{d}t = 800\left[\frac{3}{4}t^{4/3}\right]_1^8$

$= 800\left[\frac{3}{4}(8)^{4/3} - \frac{3}{4}(1)^{4/3}\right]$

$= 9000$

(b) $\int_1^T 800t^{1/3}\text{d}t = 800\left[\frac{3}{4}t^{4/3}\right]_0^T$

$= 800\left[\frac{3}{4}T^{4/3} - \frac{3}{4}(0)^{4/3}\right]$

$= 600T^{4/3}$

We need to solve

$$600T^{4/3} = 48\ 600$$

that is,

$$T^{4/3} = 81$$

so

$$T = 81^{3/4} = 27$$

5 $P = \int_0^{10} 5000\text{e}^{-0.06t}\text{d}t$

$= 5000\int_0^{10} \text{e}^{-0.06t}\text{d}t$

$= 5000\left[-\frac{1}{0.06}\text{e}^{-0.06t}\right]_0^{10}$

$= -\frac{5000}{0.06}(\text{e}^{-0.6} - 1)$

$= \$37\ 599.03$

6 Area is 16.

7 (a) 4; (b) 0. The graph is sketched in Figure S6.1. Integration gives a positive value when the graph is above the x axis and a negative value when it is below the x axis. In this case there are equal amounts of positive and negative area which cancel out. Actual area is twice that between 0 and 2, so is 8.

Figure S6.1

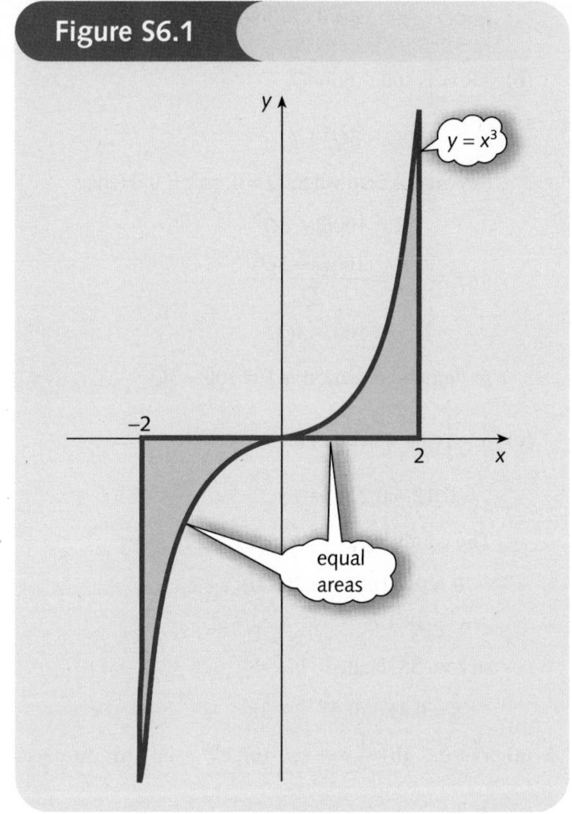

8 (a) $\frac{1}{2}, \frac{19}{20}, \frac{199}{200}$; 1.

(b) $2\sqrt{2} - 2$, $2\sqrt{20} - 2$, $2\sqrt{200} - 2$; integral does not exist because these numbers are increasing without bound.

9 (a) 1.367 544 468, 1.8, 1.936 754 441; 2.

(b) 9, 99, 999; integral does not exist because these numbers are increasing without bound.

10 (a) 100; **(b)** 20.

11 (a) 81; **(b)** 180.

12 (a) 74.67; **(b)** 58.67.

13 (a) \$427.32; **(b)** During the 47th year.

14 (a) $\frac{AT^{\alpha+1}}{\alpha + 1}$; **(b)** $\frac{A}{\alpha}(\text{e}^{\alpha T} - 1)$

15 (a) \$2785.84; **(b)** \$7869.39; **(c)** \$19 865.24; **(d)** \$20 000.

16 6.9 years.

Chapter 7

Section 7.1

1 (a) 2×2, 1×5, 3×5, 1×1.

(b) 1, 4, 6, 2, 6, ?, 6; the value of c_{43} does not exist, because **C** has only three rows.

2 $$\mathbf{A}^{\mathrm{T}} = \begin{bmatrix} 1 & 3 & 2 & 2 \\ 4 & 7 & 1 & -5 \\ 0 & 6 & 3 & 1 \\ 1 & 1 & 5 & 8 \\ 2 & 4 & -1 & 0 \end{bmatrix}$$

$$\mathbf{B}^{\mathrm{T}} = \begin{bmatrix} 1 \\ 5 \\ 7 \\ 9 \end{bmatrix}$$

$$\mathbf{C}^{\mathrm{T}} = \begin{bmatrix} 1 & 2 & 3 \\ 2 & 4 & 5 \\ 3 & 5 & 6 \end{bmatrix} = \mathbf{C}$$

Matrices with the property that $\mathbf{C}^{\mathrm{T}} = \mathbf{C}$ are called **symmetric**. Elements in the top right-hand corner are a mirror image of those in the bottom left-hand corner.

3 (a) $\begin{bmatrix} 1 & 7 \\ 3 & -8 \end{bmatrix}$; **(c)** $\begin{bmatrix} 3 \\ 2 \end{bmatrix}$; **(d)** $\begin{bmatrix} 2 \\ 2 \end{bmatrix}$; **(e)** $\begin{bmatrix} 0 & 0 \\ 0 & 0 \end{bmatrix}$.

Part (b) is impossible because **A** and **C** have different orders.

4 (1) (a) $\begin{bmatrix} 2 & -4 \\ 6 & 10 \\ 0 & 8 \end{bmatrix}$; **(b)** $\begin{bmatrix} 2 & -2 \\ 4 & 14 \\ 2 & 12 \end{bmatrix}$;

(c) $\begin{bmatrix} 1 & -3 \\ 5 & 12 \\ 1 & 10 \end{bmatrix}$; **(d)** $\begin{bmatrix} 2 & -6 \\ 10 & 24 \\ 2 & 20 \end{bmatrix}$.

From (a) and (b)

$$2\mathbf{A} + 2\mathbf{B} = \begin{bmatrix} 2 & -4 \\ 6 & 10 \\ 0 & 8 \end{bmatrix} + \begin{bmatrix} 0 & -2 \\ 4 & 14 \\ 2 & 12 \end{bmatrix} = \begin{bmatrix} 2 & -6 \\ 10 & 24 \\ 2 & 20 \end{bmatrix}$$

which is the same as (d), so

$$2(\mathbf{A} + \mathbf{B}) = 2\mathbf{A} + 2\mathbf{B}$$

(2) (a) $\begin{bmatrix} 3 & -6 \\ 9 & 15 \\ 0 & 12 \end{bmatrix}$; **(b)** $\begin{bmatrix} -6 & 12 \\ -18 & -30 \\ 0 & -24 \end{bmatrix}$.

From (a),

$$-2(3\mathbf{A}) = -2\begin{bmatrix} 3 & -6 \\ 9 & 15 \\ 0 & 12 \end{bmatrix} = \begin{bmatrix} -6 & 12 \\ -18 & -30 \\ 0 & -24 \end{bmatrix}$$

which is the same as (b), so

$$-2(3\mathbf{A}) = -6\mathbf{A}$$

5 (a) [8] because

$$1(0) + (-1)(-1) + 0(1) + 3(1) + 2(2) = 8$$

(b) [0] because $1(-2) + 2(1) + 9(0) = 0$.

(c) This is impossible, because **a** and **d** have different numbers of elements.

6

A [3 × 2] B [2 × 2]

equal so can multiply

order of **C** is 3 × 2

$$\mathbf{AB} = \begin{bmatrix} 1 & 2 \\ 0 & 1 \\ 3 & 1 \end{bmatrix}\begin{bmatrix} 1 & 2 \\ 3 & 4 \end{bmatrix} = \begin{bmatrix} c_{11} & c_{12} \\ c_{21} & c_{22} \\ c_{31} & c_{32} \end{bmatrix}$$

$$\mathbf{AB} = \begin{bmatrix} \boxed{1 \quad 2} \\ 0 \quad 1 \\ 3 \quad 1 \end{bmatrix}\begin{bmatrix} \boxed{\begin{matrix}1\\3\end{matrix}} & \begin{matrix}2\\4\end{matrix} \end{bmatrix} = \begin{bmatrix} \boxed{7} & c_{12} \\ c_{21} & c_{22} \\ c_{31} & c_{32} \end{bmatrix}$$

$$\mathbf{AB} = \begin{bmatrix} \boxed{1 \quad 2} \\ 0 \quad 1 \\ 3 \quad 1 \end{bmatrix}\begin{bmatrix} \begin{matrix}1\\3\end{matrix} & \boxed{\begin{matrix}2\\4\end{matrix}} \end{bmatrix} = \begin{bmatrix} 7 & \boxed{10} \\ c_{21} & c_{22} \\ c_{31} & c_{32} \end{bmatrix}$$

$$\mathbf{AB} = \begin{bmatrix} 1 \quad 2 \\ \boxed{0 \quad 1} \\ 3 \quad 1 \end{bmatrix}\begin{bmatrix} \boxed{\begin{matrix}1\\3\end{matrix}} & \begin{matrix}2\\4\end{matrix} \end{bmatrix} = \begin{bmatrix} 7 & 10 \\ \boxed{3} & c_{22} \\ c_{31} & c_{32} \end{bmatrix}$$

$$\mathbf{AB} = \begin{bmatrix} 1 \quad 2 \\ \boxed{0 \quad 1} \\ 3 \quad 1 \end{bmatrix}\begin{bmatrix} \begin{matrix}1\\3\end{matrix} & \boxed{\begin{matrix}2\\4\end{matrix}} \end{bmatrix} = \begin{bmatrix} 7 & 10 \\ 3 & \boxed{4} \\ c_{31} & c_{32} \end{bmatrix}$$

$$\mathbf{AB} = \begin{bmatrix} 1 \quad 2 \\ 0 \quad 1 \\ \boxed{3 \quad 1} \end{bmatrix}\begin{bmatrix} \boxed{\begin{matrix}1\\3\end{matrix}} & \begin{matrix}2\\4\end{matrix} \end{bmatrix} = \begin{bmatrix} 7 & 10 \\ 3 & 4 \\ \boxed{6} & c_{32} \end{bmatrix}$$

$$\mathbf{AB} = \begin{bmatrix} 1 \quad 2 \\ 0 \quad 1 \\ \boxed{3 \quad 1} \end{bmatrix}\begin{bmatrix} \begin{matrix}1\\3\end{matrix} & \boxed{\begin{matrix}2\\4\end{matrix}} \end{bmatrix} = \begin{bmatrix} 7 & 10 \\ 3 & 4 \\ 6 & \boxed{10} \end{bmatrix}$$

7 **(a)** $\begin{bmatrix} 5 \\ 7 \\ 5 \end{bmatrix}$; **(d)** $\begin{bmatrix} 4 & 3 \\ 2 & -1 \\ 5 & 5 \end{bmatrix}$; **(f)** $\begin{bmatrix} 9 & 6 & 13 \\ 27 & 15 & 28 \end{bmatrix}$;

(g) $\begin{bmatrix} 5 & 7 & 9 \\ 3 & 3 & 3 \\ 6 & 9 & 12 \end{bmatrix}$; **(h)** $\begin{bmatrix} 5 & 6 \\ 11 & 15 \end{bmatrix}$.

Parts (b), (c) and (e) are impossible because, in each case, the number of columns in the first matrix is not equal to the number of rows in the second.

8 **Ax** is the 3×1 matrix

$$\begin{bmatrix} x + 4y + 7z \\ 2z + 6y + 5z \\ 8x + 9y + 5z \end{bmatrix}$$

However, $x + 4y + 7z = -3$, $2x + 6y + 5z = 10$ and $8x + 9y + 5z = 1$, so this matrix is just

$$\begin{bmatrix} -3 \\ 10 \\ 1 \end{bmatrix}$$

which is **b**. Hence **Ax** = **b**.

9 **(a)** $\mathbf{J} = \begin{bmatrix} 35 & 27 & 13 \\ 42 & 39 & 24 \end{bmatrix}$; $\mathbf{F} = \begin{bmatrix} 31 & 17 & 3 \\ 25 & 29 & 16 \end{bmatrix}$.

(b) $\begin{bmatrix} 66 & 44 & 16 \\ 67 & 68 & 40 \end{bmatrix}$

(c) $\begin{bmatrix} 4 & 10 & 10 \\ 17 & 10 & 8 \end{bmatrix}$

10 **(a)** $\begin{bmatrix} 4 & 6 & 2 & 18 \\ 2 & 0 & 10 & 0 \\ 12 & 14 & 16 & 8 \end{bmatrix}$

(b) $\begin{bmatrix} 2 & 14 & 18 & 12 \\ 4 & 2 & 0 & 10 \\ 12 & 8 & 10 & 6 \end{bmatrix}$

(c) $\begin{bmatrix} 4 & 20 & 20 & 30 \\ 6 & 2 & 10 & 10 \\ 24 & 22 & 26 & 14 \end{bmatrix}$

(d) Same answer as (c).

11 **(a)** $\begin{bmatrix} 5900 \\ 1100 \end{bmatrix}$

Total cost charged to each customer.

(b) $\begin{bmatrix} 13 & 7 & 23 & 22 \\ 3 & 1 & 4 & 5 \end{bmatrix}$

Amount of raw materials used to manufacture each customer's goods.

(c) $\begin{bmatrix} 35 \\ 75 \\ 30 \end{bmatrix}$

Total raw material costs to manufacture one item of each good.

(d) $\begin{bmatrix} 1005 \\ 205 \end{bmatrix}$

Total raw material costs to manufacture requisite number of goods for each customer.

(e) [7000]

Total revenue received from customers.

(f) [1210]

Total cost of raw materials.

(g) [5790]

Profit before deduction of labour, capital and overheads.

12 **(1)** **(a)** $\begin{bmatrix} 1 & 3 & 5 \\ 2 & 4 & 6 \end{bmatrix}$

(b) $\begin{bmatrix} 1 & 2 & -3 \\ -1 & 1 & 4 \end{bmatrix}$

(c) $\begin{bmatrix} 2 & 1 \\ 5 & 5 \\ 2 & 10 \end{bmatrix}$

(d) $\begin{bmatrix} 2 & 5 & 2 \\ 1 & 5 & 10 \end{bmatrix}$

$(\mathbf{A} + \mathbf{B})^{\mathrm{T}} = \mathbf{A}^{\mathrm{T}} + \mathbf{B}^{\mathrm{T}}$: that is, 'transpose of the sum is the sum of the transposes'.

(2) **(a)** $\begin{bmatrix} 1 & 5 \\ 4 & 9 \end{bmatrix}$

(b) $\begin{bmatrix} 2 & -1 \\ 1 & 0 \\ 0 & 1 \end{bmatrix}$

(c) $\begin{bmatrix} -2 & 1 & 4 \\ 1 & 5 & 9 \end{bmatrix}$

(d) $\begin{bmatrix} -2 & 1 \\ 1 & 5 \\ 4 & 9 \end{bmatrix}$

$(\mathbf{CD})^{\mathrm{T}} = \mathbf{D}^{\mathrm{T}}\mathbf{C}^{\mathrm{T}}$: that is 'transpose of a product is the product of the transposes multiplied in reverse order'.

13 (a) $\mathbf{B}+\mathbf{C}=\begin{bmatrix}0 & 6\\5 & 2\end{bmatrix}$

so $\mathbf{A}(\mathbf{B}+\mathbf{C})=\begin{bmatrix}-15 & 24\\5 & 14\end{bmatrix}$

$\mathbf{AB}=\begin{bmatrix}-7 & 25\\6 & 10\end{bmatrix}$ and

$\mathbf{AC}=\begin{bmatrix}-8 & -1\\-1 & 4\end{bmatrix}$, so

$\mathbf{AB}+\mathbf{AC}=\begin{bmatrix}-15 & 24\\5 & 14\end{bmatrix}$

(b) $\mathbf{AB}=\begin{bmatrix}-7 & 25\\6 & 10\end{bmatrix}$, so

$(\mathbf{AB})\mathbf{C}=\begin{bmatrix}32 & 43\\4 & 26\end{bmatrix}$

$\mathbf{BC}=\begin{bmatrix}4 & 11\\-4 & 4\end{bmatrix}$, so

$\mathbf{A}(\mathbf{BC})=\begin{bmatrix}32 & 43\\4 & 26\end{bmatrix}$

14 $\mathbf{AB}=[-3]$; $\quad \mathbf{BA}=\begin{bmatrix}1 & 2 & -4 & 0\\7 & 14 & -28 & 21\\3 & 6 & -12 & 9\\-2 & -4 & 8 & -6\end{bmatrix}$

15 (a) $\mathbf{AI}=\begin{bmatrix}a & b\\c & d\end{bmatrix}\begin{bmatrix}1 & 0\\0 & 1\end{bmatrix}=\begin{bmatrix}a & b\\c & d\end{bmatrix}=\mathbf{A}$

Similarly, $\mathbf{IA}=\mathbf{A}$.

(b) $\mathbf{A}^{-1}\mathbf{A}$

$$=\frac{1}{ad-bc}\begin{bmatrix}d & -b\\-c & a\end{bmatrix}\begin{bmatrix}a & b\\c & d\end{bmatrix}$$

$$=\frac{1}{ad-bc}\begin{bmatrix}da-bc & db-bd\\-ca+ac & -cb+ad\end{bmatrix}$$

$$=\frac{1}{ad-bc}\begin{bmatrix}ad-bc & 0\\0 & ad-bc\end{bmatrix}=\begin{bmatrix}1 & 0\\0 & 1\end{bmatrix}$$

Similarly, $\mathbf{AA}^{-1}=\mathbf{I}$.

(c) $\mathbf{Ix}=\begin{bmatrix}1 & 0\\0 & 1\end{bmatrix}\begin{bmatrix}x\\y\end{bmatrix}=\begin{bmatrix}x\\y\end{bmatrix}=\mathbf{x}$

16 (a) $7x+5y$

$x+3y$

(b) $\mathbf{A}=\begin{bmatrix}2 & 3 & -2\\1 & -1 & 2\\4 & 2 & 5\end{bmatrix}$, $\mathbf{x}=\begin{bmatrix}x\\y\\z\end{bmatrix}$, $\mathbf{b}=\begin{bmatrix}6\\3\\1\end{bmatrix}$.

Section 7.2

1 $|\mathbf{A}|=6(2)-4(1)=8\neq 0$

so **A** is non-singular and its inverse is given by

$$\frac{1}{8}\begin{bmatrix}2 & -4\\-1 & -6\end{bmatrix}=\begin{bmatrix}1/4 & -1/2\\-1/8 & 3/4\end{bmatrix}$$

$|\mathbf{B}|=6(2)-4(3)=0$

so **B** is singular and its inverse does not exist.

2 We need to solve $\mathbf{Ax}=\mathbf{b}$, where

$$\mathbf{A}=\begin{bmatrix}9 & 1\\2 & 7\end{bmatrix}\quad \mathbf{x}=\begin{bmatrix}P_1\\P_2\end{bmatrix}\quad \mathbf{b}=\begin{bmatrix}43\\57\end{bmatrix}$$

Now

$$\mathbf{A}^{-1}=\frac{1}{61}\begin{bmatrix}7 & -1\\-2 & 9\end{bmatrix}$$

so

$$\begin{bmatrix}P_1\\P_2\end{bmatrix}=\frac{1}{61}\begin{bmatrix}7 & -1\\-2 & 9\end{bmatrix}\begin{bmatrix}43\\57\end{bmatrix}=\begin{bmatrix}4\\7\end{bmatrix}$$

3 In equilibrium, $Q_S=Q_D=Q$, say, so the supply equation becomes

$$P=aQ+b$$

Subtracting aQ from both sides gives

$$P-aQ=b \qquad (1)$$

Similarly, the demand equation leads to

$$P+cQ=d \qquad (2)$$

In matrix notation equations (1) and (2) become

$$\begin{bmatrix}1 & -a\\1 & c\end{bmatrix}\begin{bmatrix}P\\Q\end{bmatrix}=\begin{bmatrix}b\\d\end{bmatrix}$$

The coefficient matrix has an inverse,

$$\frac{1}{c+a}\begin{bmatrix}c & a\\-1 & 1\end{bmatrix}$$

so that

$$\begin{bmatrix}P\\Q\end{bmatrix}=\frac{1}{c+a}\begin{bmatrix}c & a\\-1 & 1\end{bmatrix}\begin{bmatrix}b\\d\end{bmatrix}$$

that is,

$$P=\frac{cb+ad}{c+a}\quad\text{and}\quad Q=\frac{-b+d}{c+a}$$

The multiplier for Q due to changes in b is given by the (2, 1) element of the inverse matrix so is

$$\frac{-1}{c+a}$$

Given that c and a are both positive it follows that the multiplier is negative. Consequently, an increase in b leads to a decrease in Q.

4 $A_{11} = +\begin{vmatrix} 4 & 3 \\ 3 & 4 \end{vmatrix} = 7$

$A_{12} = -\begin{vmatrix} 1 & 3 \\ 1 & 4 \end{vmatrix} = -1$

$A_{13} = +\begin{vmatrix} 1 & 4 \\ 1 & 3 \end{vmatrix} = -1$

$A_{21} = -\begin{vmatrix} 3 & 3 \\ 3 & 4 \end{vmatrix} = -3$

$A_{22} = +\begin{vmatrix} 1 & 3 \\ 1 & 4 \end{vmatrix} = 1$

$A_{23} = -\begin{vmatrix} 1 & 3 \\ 1 & 3 \end{vmatrix} = 0$

$A_{31} = +\begin{vmatrix} 3 & 3 \\ 4 & 3 \end{vmatrix} = -3$

$A_{32} = -\begin{vmatrix} 1 & 3 \\ 1 & 3 \end{vmatrix} = 0$

$A_{33} = +\begin{vmatrix} 1 & 3 \\ 1 & 4 \end{vmatrix} = 1$

5 Expanding along the top row of **A** gives

$$\begin{aligned} |\mathbf{A}| &= a_{11}A_{11} + a_{12}A_{12} + a_{13}A_{13} \\ &= 1(7) + 3(-1) + 3(-1) = 1 \end{aligned}$$

using the values of A_{11}, A_{12} and A_{13} from Practice Problem 4. Other rows and columns are treated similarly. Expanding down the last column of **B** gives

$$\begin{aligned} |\mathbf{B}| &= b_{13}B_{13} + b_{23}B_{23} + b_{33}B_{33} \\ &= 0(B_{13}) + 0(B_{23}) + 0(B_{33}) = 0 \end{aligned}$$

6 The cofactors of **A** have already been found in Practice Problem 4. Stacking them in their natural positions gives the adjugate matrix

$$\begin{bmatrix} 7 & -1 & -1 \\ -3 & 1 & 0 \\ -3 & 0 & 1 \end{bmatrix}$$

Transposing gives the adjoint matrix

$$\begin{bmatrix} 7 & -3 & -3 \\ -1 & 1 & 0 \\ -1 & 0 & 1 \end{bmatrix}$$

The determinant of **A** has already been found in Practice Problem 5 to be 1, so the inverse matrix is the same as the adjoint matrix.

The determinant of **B** has already been found in Practice Problem 5 to be 0, so **B** is singular and does not have an inverse.

7 Using the inverse matrix in Practice Problem 6,

$$\begin{bmatrix} P_1 \\ P_2 \\ P_3 \end{bmatrix} = \begin{bmatrix} 7 & -3 & -3 \\ -1 & 1 & 0 \\ -1 & 0 & 1 \end{bmatrix} \begin{bmatrix} 32 \\ 37 \\ 35 \end{bmatrix} = \begin{bmatrix} 8 \\ 5 \\ 3 \end{bmatrix}$$

8 **(1)** **(a)** $|\mathbf{A}| = -3$;

(b) $|\mathbf{B}| = 4$;

(c) $\mathbf{AB} = \begin{bmatrix} 4 & 4 \\ 7 & 4 \end{bmatrix}$

so $|\mathbf{AB}| = -12$. These results give $|\mathbf{AB}| = |\mathbf{A}||\mathbf{B}|$: that is, 'determinant of a product is the product of the determinants'.

(2) **(a)** $\mathbf{A}^{-1} = \begin{bmatrix} -1/3 & 1/3 \\ 5/3 & -2/3 \end{bmatrix}$

(b) $\mathbf{B}^{-1} = \begin{bmatrix} 1 & 0 \\ -1/2 & -1/4 \end{bmatrix}$

(c) $(\mathbf{AB})^{-1} = \begin{bmatrix} -1/3 & 1/3 \\ 7/12 & -1/3 \end{bmatrix}$

These results give $(\mathbf{AB})^{-1} = \mathbf{B}^{-1}\mathbf{A}^{-1}$: that is, 'inverse of a product is the product of the inverses multiplied in reverse order'.

9 $a = -3/2$, $b = -8/3$.

10 **(a)** $x = 1, y = -1$;

(b) $x = 2, y = 2$.

11 $\frac{1}{25}\begin{bmatrix} -9 & -1 \\ -2 & -3 \end{bmatrix}$; $\begin{bmatrix} P_1 \\ P_2 \end{bmatrix} = \begin{bmatrix} 40 \\ 10 \end{bmatrix}$.

12 Commodity market is in equilibrium when $Y = C + I$, so $Y = aY + b + cr + d$, which rearranges as

$$(1 - a)Y - cr = b + d \qquad (1)$$

Money market is in equilibrium when $M_S = M_D$, so $M_S^* = k_1Y + k_2r + k_3$, which rearranges as

$$k_1Y + k_2r = M_S^* - k_3 \qquad (2)$$

In matrix notation, equations (1) and (2) become

$$\begin{bmatrix} 1-a & -c \\ k_1 & k_2 \end{bmatrix} \begin{bmatrix} Y \\ r \end{bmatrix} = \begin{bmatrix} b+d \\ M_S^* - k_3 \end{bmatrix}$$

Using the inverse of the coefficient matrix,

$$\begin{bmatrix} Y \\ r \end{bmatrix} = \frac{1}{k_2(1-a) + ck_1} \times \begin{bmatrix} k_2 & c \\ -k_1 & 1-a \end{bmatrix} \begin{bmatrix} b+d \\ M_S^* - k_3 \end{bmatrix}$$

$$Y = \frac{k_2(b+d) + c(M_S^* - k_3)}{k_2(1-a) + ck_1}$$

and

$$r = \frac{k_1(b+d) + (1-a)(M_s^* - k_3)}{k_2(1-a) + ck_1}$$

The required multiplier is

$$\frac{\partial r}{\partial M_s^*} = \frac{1-a}{k_2(1-a) + ck_1}$$

Now $1 - a > 0$ since $a < 1$, so numerator is positive. Also $k_2 < 0$, $1 - a > 0$, gives $k_2(1 - a) < 0$ and $c < 0$, $k_1 > 0$ gives $ck_1 < 0$, so the denominator is negative.

13 The determinant of **A** is $-10 \neq 0$, so matrix is non-singular.

$$\mathbf{A}^{-1} = \begin{bmatrix} 1/10 & 3/10 & -1/2 \\ 3/10 & -1/10 & 1/2 \\ -1/2 & 1/2 & -1/2 \end{bmatrix}$$

It is interesting to notice that because the original matrix **A** is symmetric, so is $\mathbf{A}^{-1}$. The determinant of **B** is 0, so it is singular and does not have an inverse.

14 $a - 1$, which is non-zero provided $a \neq 1$

$$\frac{1}{a-1}\begin{bmatrix} -a & -1 & a \\ 3a-4 & -1 & 3-2a \\ 1 & 1 & -1 \end{bmatrix}$$

15 $\mathbf{A}^{-1} = \frac{1}{-41}\begin{bmatrix} 29 & 11 & 3 \\ 4 & 10 & -1 \\ 9 & 2 & 8 \end{bmatrix}; \begin{bmatrix} P_1 \\ P_2 \\ P_3 \end{bmatrix} = \begin{bmatrix} 20 \\ 5 \\ 8 \end{bmatrix}.$

16 $-95a + 110$; $a = \frac{22}{19}$.

17 **(a)** $aei - afh - dbi + dch + gbf - gce$

Substituting $d = ka$, $e = kb$, $f = kc$ into this gives

$$akbi - akch - kabi + kach + gbkc - gckb = 0$$

(b) $\frac{1}{\det(A)}\begin{bmatrix} ei - fh & -(bi - ch) & bf - ce \\ -(di - fg) & ai - cg & -(af - cd) \\ -(-dh + eg) & -(ah - bg) & ae - bd \end{bmatrix}$

18 **(a)** $\frac{1}{1-a+at}\begin{bmatrix} 1 & 1 & a \\ -a(-1+t) & 1 & a \\ t & t & -1+a \end{bmatrix}$

Autonomous consumption multiplier for Y.

(b) $Y = \frac{b + I^* + G^*}{1 - a + at}$

$$C = -\frac{-I^*a - aG^* + I^*ta + taG^* - b}{1 - a + at}$$

$$T = \frac{bt + I^*t + G^*t}{1 - a + at}$$

Section 7.3

1 **(a)** By Cramer's rule

$$x_2 = \frac{\det(\mathbf{A}_2)}{\det(\mathbf{A})}$$

where

$$\det(\mathbf{A}_2) = \begin{vmatrix} 2 & 16 \\ 3 & -9 \end{vmatrix} = -66$$

$$\det(\mathbf{A}) = \begin{vmatrix} 2 & 4 \\ 3 & -5 \end{vmatrix} = -22$$

Hence

$$x_2 = \frac{-66}{-22} = 3$$

(b) By Cramer's rule

$$x_3 = \frac{\det(\mathbf{A}_3)}{\det(\mathbf{A})}$$

where

$$\det(\mathbf{A}_3) = \begin{vmatrix} 4 & 1 & 8 \\ -2 & 5 & 4 \\ 3 & 2 & 9 \end{vmatrix}$$

$$= 4\begin{vmatrix} 5 & 4 \\ 2 & 9 \end{vmatrix} - 1\begin{vmatrix} -2 & 4 \\ 3 & 9 \end{vmatrix} + 8\begin{vmatrix} -2 & 5 \\ 3 & 2 \end{vmatrix}$$

$$= 4(37) - 1(-30) + 8(-19)$$

$$= 26$$

and

$$\det(\mathbf{A}) = \begin{vmatrix} 4 & 1 & 3 \\ -2 & 5 & 1 \\ 3 & 2 & 4 \end{vmatrix}$$

$$= 4\begin{vmatrix} 5 & 1 \\ 2 & 4 \end{vmatrix} - 1\begin{vmatrix} -2 & 1 \\ 3 & 4 \end{vmatrix} + 3\begin{vmatrix} -2 & 5 \\ 3 & 2 \end{vmatrix}$$

$$= 4(18) - 1(-11) + 3(-19)$$

$$= 26$$

Hence

$$x_3 = \frac{26}{26} = 1$$

2 The variable Y_d is the third, so Cramer's rule gives

$$Y_d = \frac{\det(\mathbf{A}_3)}{\det(\mathbf{A})}$$

where

$$\mathbf{A}_3 = \begin{bmatrix} 1 & -1 & I^* + G^* & 0 \\ 0 & 1 & b & 0 \\ -1 & 0 & 0 & 1 \\ -t & 0 & T^* & 1 \end{bmatrix}$$

Expanding along the second row gives

$$\det(\mathbf{A}_3) = 1\begin{vmatrix} 1 & I^*+G^* & 0 \\ -1 & 0 & 1 \\ -t & T^* & 1 \end{vmatrix} - b\begin{vmatrix} 1 & -1 & 0 \\ -1 & 0 & 1 \\ -t & 0 & 0 \end{vmatrix}$$

since along the second row the pattern is '− + − +'.
Now

$$\begin{vmatrix} 1 & I^*+G^* & 0 \\ -1 & 0 & 1 \\ -t & T^* & 1 \end{vmatrix} = 1\begin{vmatrix} 0 & 4 \\ T^* & 1 \end{vmatrix} - (I^*+G^*)\begin{vmatrix} -1 & 1 \\ -t & 1 \end{vmatrix}$$
$$= T^* - (I^* + G^*)(-1 + t)$$

(expanding along the first row) and

$$\begin{vmatrix} 1 & -1 & 0 \\ -1 & 0 & 1 \\ -t & 0 & 1 \end{vmatrix} = -(-1)\begin{vmatrix} -1 & 1 \\ -t & 1 \end{vmatrix} = -1 + t$$

(expanding down the second column).
Hence

$$\det(\mathbf{A}_3) = -T^* - (I^* + G^*)(-1 + t) - b(-1 + t)$$

From the worked example given in the text,

$$\det(\mathbf{A}) = 1 - a + at$$

Hence

$$Y_d = \frac{-T^* - (I^* + G^*)(-1 + t) - b(-1 + t)}{1 - a + at}$$

3 Substituting C_1, M_1 and I_1^* into the equation for Y_1 gives

$$Y_1 = 0.7Y_1 + 50 + 200 + X_1 - 0.3Y_1$$

Also, since $X_1 = M_2 = 0.1Y_2$, we get

$$Y_1 = 0.7Y_1 + 50 + 200 + 0.1Y_2 - 0.3Y_1$$

which rearranges as

$$0.6Y_1 - 0.1Y_2 = 250$$

In the same way, the second set of equations leads to

$$-0.3Y_1 + 0.3Y_2 = 400$$

Hence

$$\begin{bmatrix} 0.6 & -0.1 \\ -0.3 & 0.6 \end{bmatrix}\begin{bmatrix} Y_1 \\ Y_2 \end{bmatrix} = \begin{bmatrix} 250 \\ 400 \end{bmatrix}$$

In this question both Y_1 and Y_2 are required, so it is easier to solve using matrix inverses rather than Cramer's rule, which gives

$$Y_1 = \frac{1}{0.15}\begin{bmatrix} 0.3 & 0.1 \\ 0.1 & 0.6 \end{bmatrix}\begin{bmatrix} 250 \\ 400 \end{bmatrix}$$
$$= \frac{1}{0.15}\begin{bmatrix} 115 \\ 315 \end{bmatrix}$$

Hence $Y_1 = 766.67$ and $Y_2 = 2100$. The balance of payments for country 1 is

$$X_1 - M_1 = M_2 - M_1$$
$$= 0.1Y_2 - 0.3Y_1$$
$$= 0.1(2100) - 0.3(766.67)$$
$$= -20$$

Moreover, since only two countries are involved, it follows that country 2 will have a surplus of 20.

4 **(a)** $x_1 = \dfrac{\det(\mathbf{A}_1)}{\det(\mathbf{A})} = \dfrac{72}{18} = 4$

(b) $x_2 = \dfrac{\det(\mathbf{A}_2)}{\det(\mathbf{A})} = \dfrac{126}{42} = 3$

(c) $x_4 = \dfrac{\det(\mathbf{A}_4)}{\det(\mathbf{A})} = \dfrac{-1425}{475} = -3$

5 The equations can be rearranged as

$$Y - C + M = I^* + G^* + X^*$$
$$-aY + C + 0M = b$$
$$-mY + 0C + M = M^*$$

as required.

Autonomous investment multiplier, $\dfrac{1}{1 - a + m}$, is positive because $1 - a$ and m are both positive.

6 The multiplier is

$$\frac{-k_1}{k_2(1 - a) + ck_1}$$

which is positive since the top and bottom of this fraction are both negative. To see that the bottom is negative, note that $k_2(1 - a) < 0$ because $k_2 < 0$ and $a < 1$, and $ck_1 < 0$ because $c < 0$ and $k_1 > 0$.

7 The equations are

$$0.6Y_1 - 0.1Y_2 - I_1^* = 50$$
$$-0.2Y_1 + 0.3Y_2 = 150$$
$$0.2Y_1 - 0.1Y_2 = 0$$

The third equation follows from the fact that if the balance of payments is 0 then $M_1 = X_1$, or equivalently, $M_1 = M_2$. Cramer's rule gives

$$I_1^* = \frac{\det(\mathbf{A}_3)}{\det(\mathbf{A})} = \frac{4}{0.04} = 100$$

8 $$\begin{bmatrix} 1 - a_1 + m_1 & -m_2 \\ -m_1 & 1 - a_2 + m_2 \end{bmatrix}\begin{bmatrix} Y_1 \\ Y_2 \end{bmatrix} = \begin{bmatrix} b_1 + I_1^* \\ b_2 + I_2^* \end{bmatrix}$$

$$Y_1 = \frac{(b_1 + I_1^*)(1 - a_2 + m_2) + m_2(b_2 + I_2^*)}{(1 - a_1 + m_1)(1 - a_2 + m_2) - m_1m_2}$$

The multiplier is

$$\frac{m_2}{(1 - a_1 + m_1)(1 - a_2 + m_2) - m_1m_2}$$

which is positive since the top and bottom of the fraction are both positive. To see that the bottom is positive, note that since $a_1 < 1$, $1 - a_i + m_i > m_i$, so that $(1 - a_1 + m_1)(1 - a_2 + m_2) > m_1m_2$. Hence the national income of one country rises as the investment in the other country rises.

Section 7.4

1 $\mathbf{d} = \begin{bmatrix} 1000 \\ 300 \\ 700 \end{bmatrix} - \begin{bmatrix} 0.2 & 0.4 & 0.2 \\ 0.1 & 0.2 & 0.1 \\ 0.1 & 0.1 & 0 \end{bmatrix} \begin{bmatrix} 1000 \\ 300 \\ 700 \end{bmatrix} = \begin{bmatrix} 540 \\ 70 \\ 570 \end{bmatrix}$

2 The matrix of technical coefficients is

$$\mathbf{A} = \begin{bmatrix} 0.2 & 0.2 \\ 0.4 & 0.1 \end{bmatrix}$$

so

$$\mathbf{I} - \mathbf{A} = \begin{bmatrix} 0.8 & -0.2 \\ -0.4 & 0.9 \end{bmatrix}$$

which has inverse

$$(\mathbf{I} - \mathbf{A})^{-1} = \frac{1}{0.64}\begin{bmatrix} 0.9 & 0.2 \\ 0.4 & 0.8 \end{bmatrix}$$

Hence

$$\mathbf{x} = \frac{1}{0.64}\begin{bmatrix} 0.9 & 0.2 \\ 0.4 & 0.8 \end{bmatrix}\begin{bmatrix} 760 \\ 420 \end{bmatrix} = \begin{bmatrix} 1200 \\ 1000 \end{bmatrix}$$

so total output is 1200 units for engineering and 1000 units for transport.

3 Total outputs for I1, I2, I3 and I4 are found by summing along each row to get 1000, 500, 2000 and 1000, respectively. Matrix of technical coefficients is obtained by dividing the columns of the inter-industrial flow table for these numbers to get

$$\mathbf{A} = \begin{bmatrix} 0 & 0.6 & 0.05 & 0.1 \\ 0.1 & 0 & 0.1 & 0.1 \\ 0.2 & 0.2 & 0 & 0.4 \\ 0.3 & 0 & 0.05 & 0 \end{bmatrix}$$

4 $\mathbf{I} - \mathbf{A} = \begin{bmatrix} 0.9 & -0.2 & -0.2 \\ -0.1 & 0.9 & -0.1 \\ -0.1 & -0.3 & 0.9 \end{bmatrix}$

so

$$(\mathbf{I} - \mathbf{A})^{-1} = \frac{1}{0.658}\begin{bmatrix} 0.78 & 0.24 & 0.20 \\ 0.10 & 0.79 & 0.11 \\ 0.12 & 0.29 & 0.79 \end{bmatrix}$$

We are given that

$$\Delta\mathbf{d} = \begin{bmatrix} 1000 \\ 0 \\ -800 \end{bmatrix}$$

so

$$\Delta\mathbf{x} = \frac{1}{0.658}\begin{bmatrix} 0.78 & 0.24 & 0.20 \\ 0.10 & 0.79 & 0.11 \\ 0.12 & 0.29 & 0.79 \end{bmatrix} \times \begin{bmatrix} 1000 \\ 0 \\ -800 \end{bmatrix} = \begin{bmatrix} 942 \\ 18 \\ -778 \end{bmatrix}$$

Hence, total outputs for I1 and I2 rise by 942 and 18 respectively, and total output for I3 falls by 778 (to the nearest whole number).

5 $[450 \quad 900 \quad 400 \quad 4350 \quad 50]^{\mathrm{T}}$

6 $\mathbf{A} = \begin{bmatrix} 0 & 0.1 & 0.2 \\ 0.1 & 0 & 0.1 \\ 0.2 & 0.2 & 0 \end{bmatrix}$

$$(\mathbf{I} - \mathbf{A})^{-1} = \frac{1}{0.924}\begin{bmatrix} 0.98 & 0.14 & 0.21 \\ 0.12 & 0.96 & 0.12 \\ 0.22 & 0.22 & 0.99 \end{bmatrix}$$

(a) 1000 units of water, 500 units of steel and 1000 units of electricity.

(b) The element in the first row and third column of $(\mathbf{I} - \mathbf{A})^{-1}$ is

$$\frac{0.21}{0.924} = 0.23$$

Change in water output is $0.23 \times 100 = 23$.

7 **(a)** $[500 \quad 1000]^{\mathrm{T}}$; **(b)** $\begin{bmatrix} 0.2 & 0.1 \\ 0.4 & 0.5 \end{bmatrix}$;

(c) $\frac{1}{36}\begin{bmatrix} 0.5 & 0.1 \\ 0.4 & 0.8 \end{bmatrix}$; **(d)** $[694 \quad 1056]^{\mathrm{T}}$.

8 **(a)** $\mathbf{A} = \begin{bmatrix} 0.2 & 0.2 & 0.3 \\ 0.3 & 0.4 & 0.3 \\ 0.4 & 0.2 & 0.1 \end{bmatrix}$

$$(\mathbf{I} - \mathbf{A})^{-1} = \frac{1}{0.216}\begin{bmatrix} 0.48 & 0.24 & 0.24 \\ 0.39 & 0.60 & 0.33 \\ 0.30 & 0.24 & 0.42 \end{bmatrix}$$

(b) 100 000, 162 500, 125 000.

(c) 278 (nearest unit).

Chapter 8

Section 8.1

1 The line $-x + 3y = 6$ passes through (0, 2) and (−6, 0). Substituting $x = 1$, $y = 4$ into the equation gives

$$-1 + 3(4) = 11$$

This is greater than 6, so the test point satisfies the inequality. The corresponding region is shown in Figure S8.1 (overleaf).

Figure S8.1

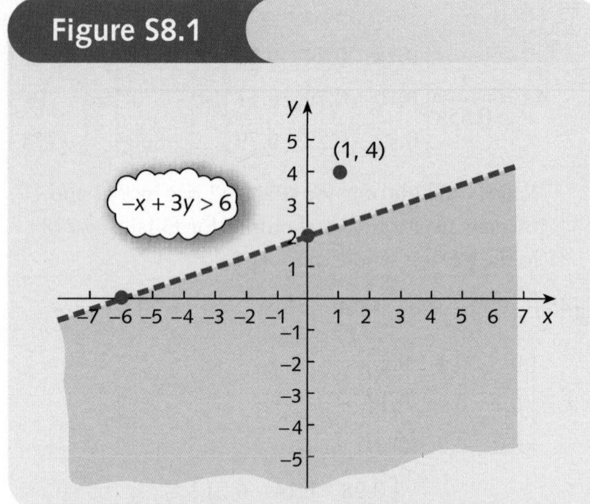

2 The non-negativity constraints indicate that we restrict our attention to the positive quadrant.

The line $x + 2y = 10$ passes through (0, 5) and (10, 0).

The line $3x + y = 10$ passes through (0, 10) and ($^{10}/_3$, 0).

Also the test point (0, 0) satisfies both of the corresponding inequalities, so we are interested in the region below both lines as shown in Figure S8.2.

Figure S8.2

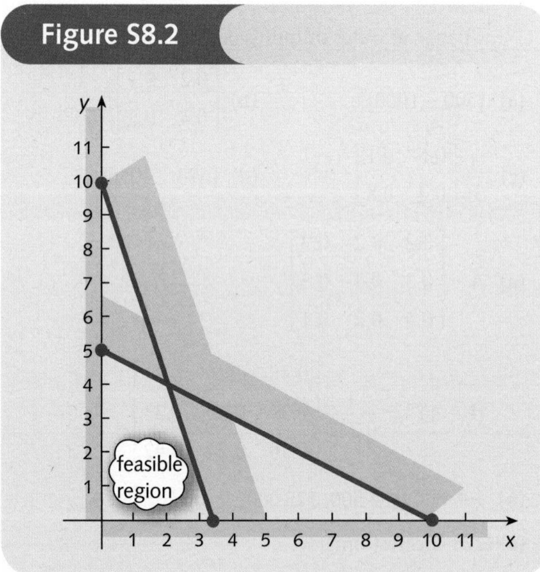

3 The answers to parts (a) and (b) are shown in Figure S8.3.

(c) Once c becomes greater than 3, the lines no longer intersect the feasible region. The maximum value of c (that is, the objective function) is therefore 3, which occurs at the corner (0, 3), when $x = 0, y = 3$.

4 *Step 1*

The feasible region is sketched in Figure S8.4.

Figure S8.3

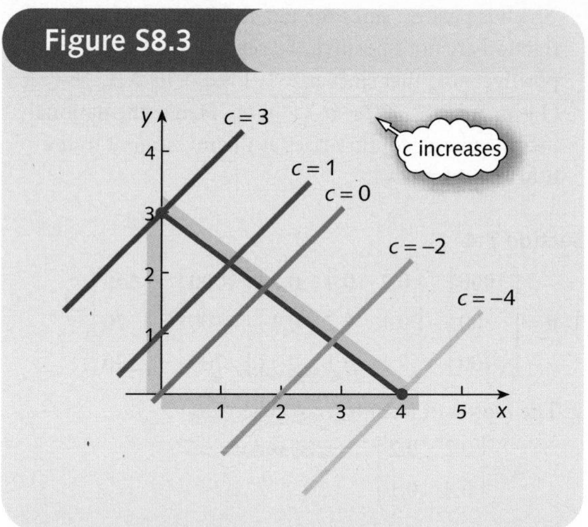

Figure S8.4

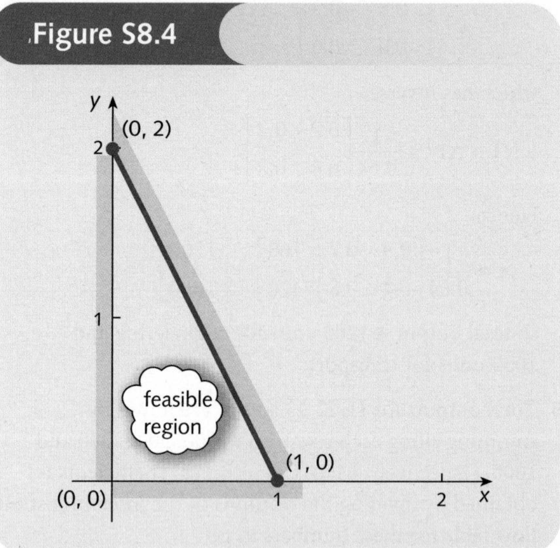

Step 2

Corners are (0, 0) (1, 0) and (0, 2).

Step 3

Corner	Objective function
(0, 0)	$0 - 0 = 0$
(1, 0)	$1 - 0 = 1$
(0, 2)	$0 - 2 = -2$

Minimum is −2, which occurs at (0, 2).

Step 1

5 The feasible region is sketched in Figure S8.2.

Step 2

Corners are (0, 0), (0, 5), (2, 4) and ($^{10}/_3$, 0).

Step 3

Corner	Objective function
(0, 0)	$3(0) + 5(0) = 0$
(0, 5)	$3(0) + 5(5) = 25$
(2, 4)	$3(2) + 5(4) = 26$
$(^{10}/_3, 0)$	$3(^{10}/_3) + 5(0) = 25$

Maximum is 26, which occurs at (2, 4).

6 The feasible regions for parts (a), (b) and (c) are sketched in Figures S8.5, S8.6 and S8.7, respectively.

Figure S8.5

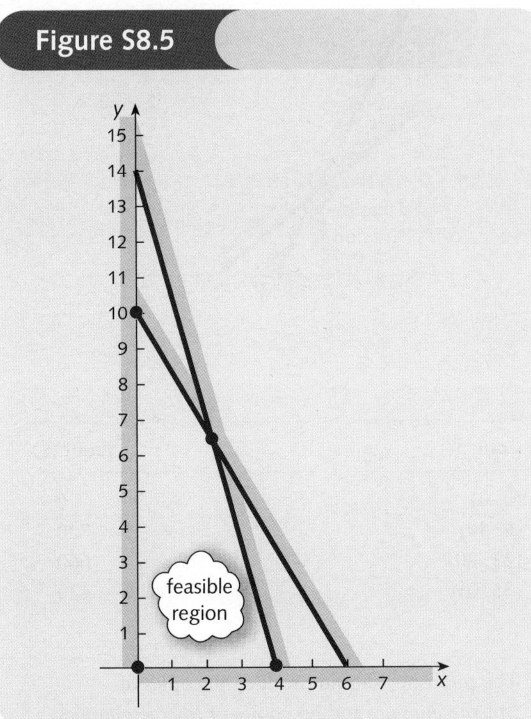

Figure S8.6

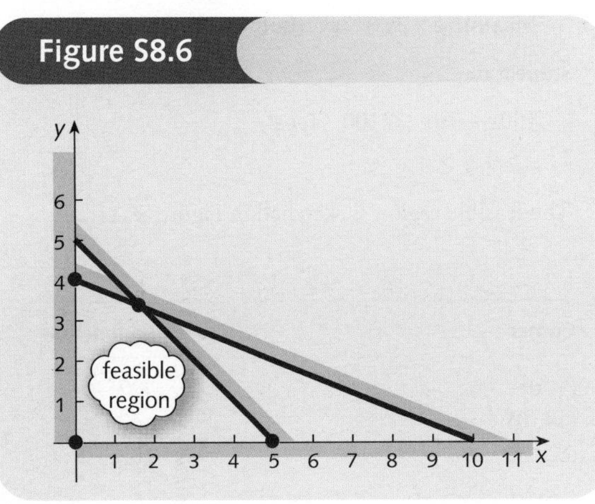

Figure S8.7

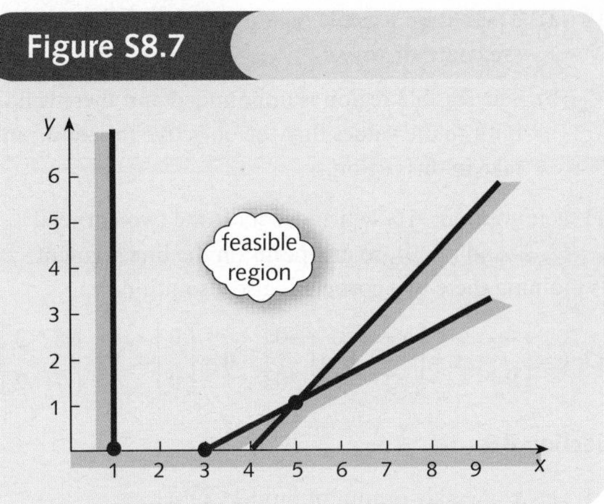

7 **(a)** Maximum is 90, which occurs at (0, 10).

(b) Maximum is 25, which occurs at $(^5/_3, ^{10}/_3)$. Note that the exact coordinates can be found by solving the simultaneous equations

$$2x + 5y = 20$$
$$x + y = 5$$

using an algebraic method.

(c) Minimum is 1, which occurs at (1, 0).

8 Figure S8.8 shows that the problem does not have a finite solution. The lines $x + y = c$ pass through $(c, 0)$ and $(0, c)$. As c increases, the lines move across the region to the right without bound.

Figure S8.8

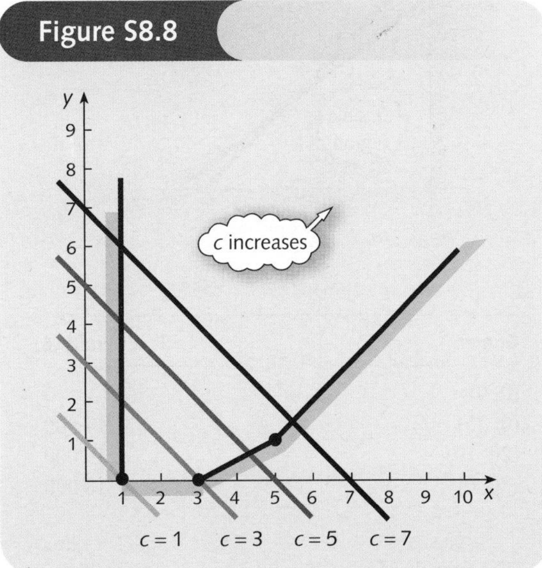

9 **(a)** Maximum is 16, which occurs at (2, 4).

(b) Maximum is 12, which occurs at any point on the line joining (0, 3) and (1, 5).

10 (**a**) There is no feasible region, since the constraints are contradictory.

(**b**) The feasible region is unbounded and there is no limit to the values that the objective function can take in this region.

11 Minimum is −16, which occurs at the two corners (2, 2) and ($^8/_3$, 0), so any point on the line segment joining these two corners is also a solution.

12 $\mathbf{c} = \begin{bmatrix} 4 \\ 9 \end{bmatrix}$ $\mathbf{x} = \begin{bmatrix} x \\ y \end{bmatrix}$ $\mathbf{b} = \begin{bmatrix} 30 \\ 28 \end{bmatrix}$ $\mathbf{0} = \begin{bmatrix} 0 \\ 0 \end{bmatrix}$ $\mathbf{A} = \begin{bmatrix} 5 & 3 \\ 7 & 2 \end{bmatrix}$

Section 8.2

1 Let x = weekly output of model COM1,
y = weekly output of model COM2.

Maximize $600x + 700y$ (profit)

subject to

$1200x + 1600y \le 40\,000$ (production costs)

$x + y \le 30$ (total output)

$x \ge 0, y \ge 0$ (non-negativity constraints)

The feasible region is sketched in Figure S8.9.

Figure S8.9

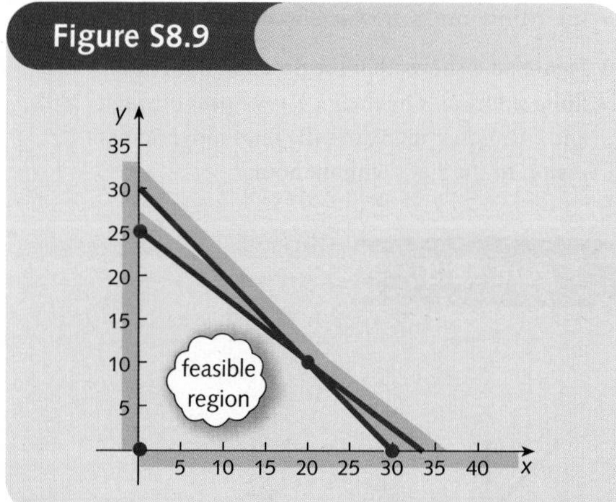

Corner	Profit ($)
(0, 0)	0
(0, 25)	17 500
(20, 10)	19 000
(30, 0)	18 000

The firm should produce 20 computers of model COM1 and 10 of model COM2 to achieve a maximum profit of $19 000.

2 Let x = number of copies of *Microeconomics*,
y = number of copies of *Macroeconomics*.

Maximize $12x + 18y$ (profit)

subject to

$12x + 15y \le 600$ (printing time)

$18x + 9y \le 630$ (binding time)

$x \ge 0, y \ge 0$ (non-negativity constraints)

The feasible region is sketched in Figure S8.10.

Figure S8.10

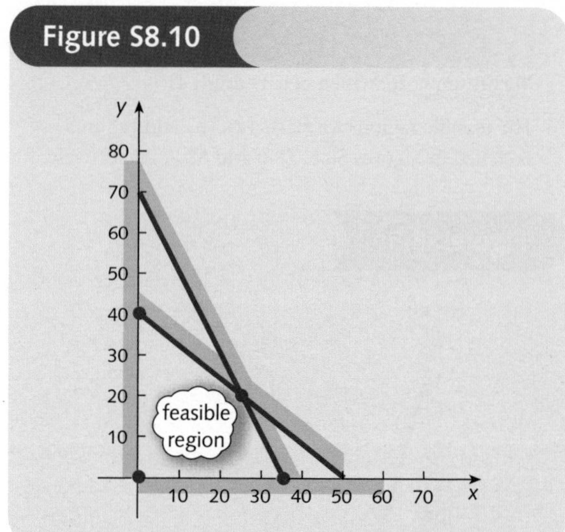

Corner	Profit ($)
(0, 0)	0
(0, 40)	720
(25, 20)	660
(35, 0)	420

The publisher should produce 40 copies of *Macroeconomics* and no copies of *Microeconomics* to achieve a maximum profit of $720.

3 Maximize $3x + 7y$ (utility)

subject to

$150x + 70y \le 2100$ (cost)

$x \ge 9, y \ge 0$

The feasible region is sketched in Figure S8.11.

Corner	Objective function
(9, 0)	27
(14, 0)	42
(9, $^{75}/_7$)	102

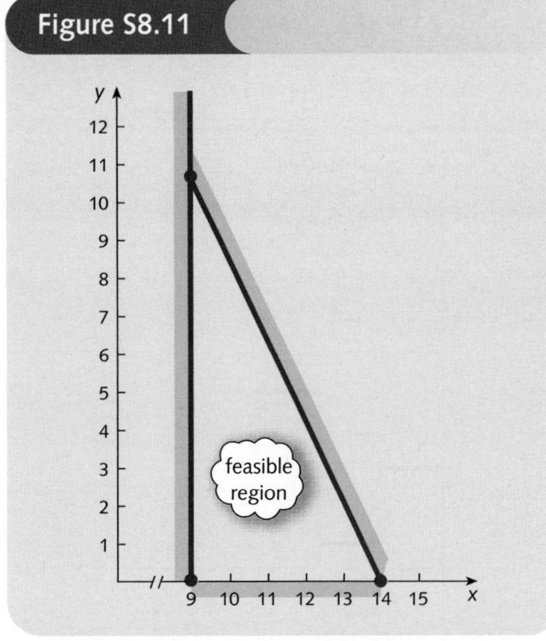

The maximum value of U occurs at $(9, {}^{75}\!/_7)$. However, it is impossible to visit the theatre ${}^{75}\!/_7$ times. The point in the feasible region with whole-number coordinates which maximizes utility is (9, 10), so we need to buy 9 items of clothing and visit the theatre 10 times per year.

4 The manufacturer should produce 10 bikes of type B and 15 of type C each month to achieve a maximum profit of \$5100.

5 The firm should produce 720 cartons of 'The Caribbean' and 630 cartons of 'Mr Fruity' each week to give a maximum profit of \$650.70.

6 The student should order a quarterpounder served with 6 oz chips to consume a minimum of 860 calories. Note that the unbounded feasible region causes no difficulty here, because the problem is one of minimization.

7 **(a)** The firm should make 30 jackets and 6 pairs of trousers each week to achieve a maximum profit of \$444.

(b) The profit margin on a pair of trousers should be between \$8 and \$14.

8 The optimal diet consists of 1.167 kg of fish meal and 1.800 kg of meat scraps, which gives a minimum cost of \$1.69 per pig per day.

9 $x = 40$, $y = 0$, $z = 100$; don't forget to type in the command `with(simplex):`

10 x_1 = number of hectares for barley in large field in year 1

x_2 = number of hectares for barley in small field in year 1

x_3 = number of hectares for cattle in large field in year 1

x_4 = number of hectares for cattle in small field in year 1

x_5, x_6, x_7, x_8 denote corresponding areas for year 2

$$\text{Maximize} \quad 400x_1 + 220x_2 + 350x_3 + 200x_4 + 420x_5 + 240x_6 + 540x_7 + 320x_8$$

subject to

$$x_1 + x_3 \le 1400$$
$$x_2 + x_4 \le 800$$
$$x_5 + x_7 \le 1400$$
$$x_6 + x_8 \le 800$$
$$2x_1 + 2x_2 - x_3 - x_4 \ge 0$$
$$2x_5 + 2x_6 - x_7 - x_8 \ge 0$$
$$6x_3 + 6x_4 - 5x_7 - 5x_8 \ge 0$$

together with the eight non-negativity constraints, $x_i \ge 0$.

$$x_1 = \frac{8800}{9},\ x_2 = 0,\ x_3 = \frac{3800}{9},\ x_4 = 800,$$

$$x_5 = 0,\ x_6 = \frac{2200}{3},\ x_7 = 1400,\ x_8 = \frac{200}{3}$$

Chapter 9

Section 9.1

1 **(1)** **(a)** $1, 3, 9, 27; 3^t$;

(b) $7, 21, 63, 189; 7(3^t)$;

(c) $A, 3A, 9A, 27A; A(3^t)$.

(2) **(a)** $1, \frac{1}{2}, \frac{1}{4}, \frac{1}{8}; \left(\frac{1}{2}\right)^t$

(b) $7, 7\left(\frac{1}{2}\right), 7\left(\frac{1}{4}\right), 7\left(\frac{1}{8}\right); 7\left(\frac{1}{2}\right)^t$

(c) $A, A\left(\frac{1}{2}\right), A\left(\frac{1}{4}\right), A\left(\frac{1}{8}\right); A\left(\frac{1}{2}\right)^t$

(3) $A, Ab, Ab^2, Ab^3; A(b^t)$.

2 **(a)** The complementary function is the solution of

$$Y_t = -\frac{1}{2}Y_{t-1}$$

so is given by

$$\text{CF} = A\left(-\frac{1}{2}\right)^t$$

For a particular solution we try

$$Y_t = D$$

Substituting this into

$$Y_t = -\frac{1}{2}Y_{t-1} + 6$$

gives

$$D = -\frac{1}{2}D + 6$$

which has solution $D = 4$, so

$$\text{PS} = 4$$

The general solution is

$$Y_t = A\left(-\frac{1}{2}\right)^t + 4$$

The initial condition, $Y_0 = 0$, gives

$$0 = A + 4$$

so A is -4. The solution is

$$Y_t = -4\left(-\frac{1}{2}\right)^t + 4$$

From the staircase diagram shown in Figure S9.1 we see that Y_t oscillates about $Y_t = 4$. Moreover, as t increases, these oscillations damp down and Y_t converges to 4. Oscillatory convergence can be expected for any solution

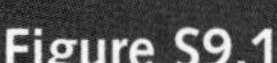

Figure S9.1

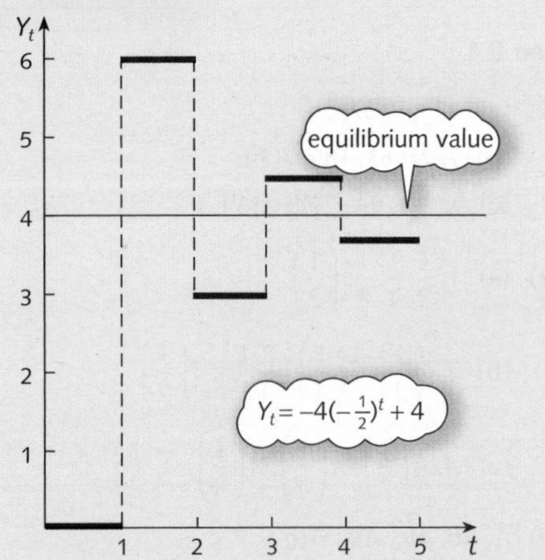

$$Y_t = A(b^t) + \text{PS}$$

when $-1 < b < 0$.

(b) CF $= A(-2)^t$ and PS $= 3$ so $Y_t = A(-2)^t + 3$. Initial condition gives $A = 1$, so $Y_t = (-2)^t + 3$. From Figure S9.2 we see that Y_t oscillates about 3 and that these oscillations explode with increasing t. Oscillatory divergence can be expected for any solution

$$Y_t = A(b^t) + \text{PS}$$

when $b < -1$.

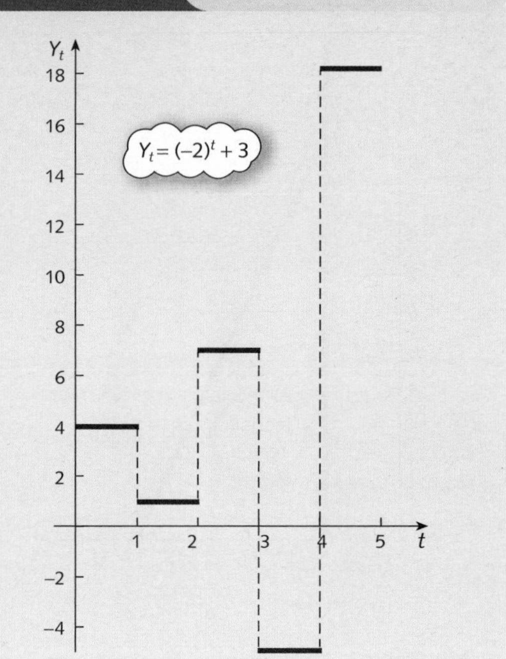

3 $Y_t = C_t + I_t$

$$= 0.9Y_{t-1} + 250 + 350$$

$$= 0.9Y_{t-1} + 600$$

This has solution

$$Y_t = A(0.9)^t + 6000$$

The initial condition, $Y_0 = 6500$, gives $A = 500$, so

$$Y_t = 500(0.9)^t + 6000$$

The system is stable because $-1 < 0.9 < 1$. In fact, Y_t converges uniformly to the equilibrium value, 6000.

4 $-2P_t + 22 = P_{t-1} - 8$

rearranges to give

$$P_t = -\frac{1}{2}P_{t-1} + 15$$

so has solution

$$P_t = A\left(-\frac{1}{2}\right)^t + 10$$

The initial condition, $P_0 = 11$, gives $A = 1$, so

$$P_t = \left(-\frac{1}{2}\right)^t + 10$$

From the demand equation,

$$Q_t = -2P_t + 22$$

we have

$$Q_t = -2\left[\left(-\frac{1}{2}\right)^t + 10\right] + 22 = -2\left(-\frac{1}{2}\right)^t + 2$$

The system is stable because $-1 < -\frac{1}{2} < 1$. In fact, P_t and Q_t display oscillatory convergence and approach the equilibrium values of 2 and 10 respectively as t increases.

5 $-2P_t + 80 = 3P_{t-1} - 20$

rearranges to give

$$P_t = -1.5P_{t-1} + 50$$

so has solution

$$P_t = A(-1.5)^t + 20$$

The initial condition, $P_0 = 8$, gives $A = -12$, so

$$P_t = -12(-1.5)^t + 40$$

From the demand equation

$$\begin{aligned} Q_t &= -2P_t + 80 \\ &= -2[-12(-1.5)^t + 20] + 80 \\ &= 24(-1.5)^t + 40 \end{aligned}$$

The system is unstable because $-1.5 < -1$. In fact, P_t and Q_t display oscillatory divergence as t increases.

6 (a) $Y_0 = 0$, $Y_1 = 2 = 2 \times 1$,

$Y_2 = 4 = 2 \times 2$, $Y_3 = 6 = 2 \times 3, \ldots$

Hence $Y_t = 2t$ and displays uniform divergence as shown in Figure S9.3.

Figure S9.3

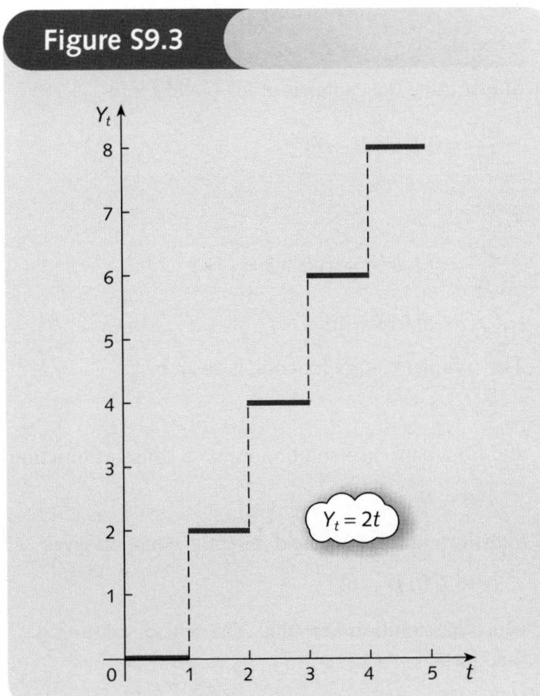

(b) $Y_0 = 4$, $Y_1 = 2$, $Y_2 = 4$, $Y_3 = 2, \ldots$

So Y_t is 4 when t is even and 2 when t is odd. Hence Y_t oscillates with equal oscillations as shown in Figure S9.4.

Figure S9.4

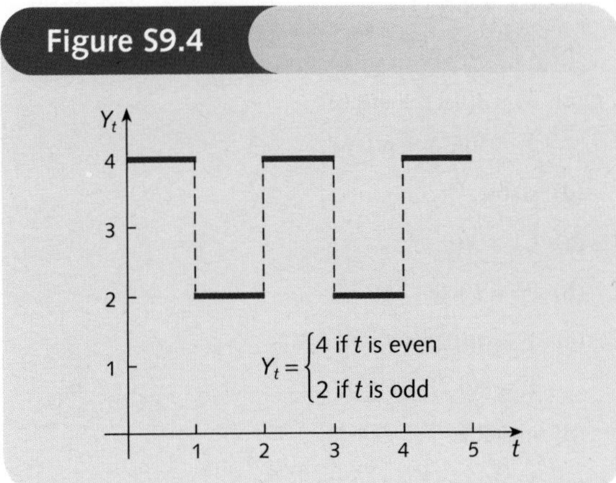

(c) $Y_0 = 3$, $Y_1 = 3$, $Y_2 = 3$, $Y_3 = 3, \ldots$

Hence $Y_t = 3$ for all t and remains fixed at this value.

7 (a) $Y_t = -7\left(\frac{1}{4}\right)^t + 8$; uniform convergence to 8.

(b) $Y_t = (-4)^t + 1$; oscillatory divergence.

8 The right-hand side, $bY_{t-1} + c$, becomes

$$\begin{aligned} &b\left(A(b^{t-1}) + \frac{c}{1-b}\right) + c \\ &= A(b^t) + \frac{bc}{1-b} + c \\ &= A(b^t) + \frac{bc + c(1-b)}{1-b} \\ &= A(b^t) + \frac{c}{1-b} \\ &= A(b^t) + D \\ &= Y_t \end{aligned}$$

which is the left-hand side.

9 $Y_t = 500(0.8)^t + 2500$; stable.

10 $P_t = 10(-0.5)^t + 60$; stable.

11 Substitute assumptions (1) and (2) into (3) to get

$$\beta(Y_t - Y_{t-1}) = \alpha Y_t$$

which rearranges as

$$Y_t = \left(\frac{\beta}{\beta - \alpha}\right)Y_{t-1}$$

with solution

$$Y_t = \left(\frac{\beta}{\beta - \alpha}\right)^t Y_0$$

If $\alpha = 0.1$ and $\beta = 1.4$ then $Y_t = (1.08)^t Y_0$.
As t increases, Y_t diverges uniformly, so unstable.

12 **(a)** $CF = A(0.1)^t$;

(b) $PS = 6(0.6)^t$;

(c) $Y_t = A(0.1)^t + 6(0.6)^t$,
$Y_t = 3(0.1)^t + 6(0.6)^t$;

(d) stable.

13 **(a)** $CF = A(0.2)^t$;

(b) $PS = t + 6$;

(c) $Y_t = A(0.2)^t + t + 6$,
$Y_t = 4(0.2)^t + t + 6$;

(d) unstable.

14 **(a)** $Y_t = 0.6Y_{t-1}^{0.8} + 160$

(b) 163.8, 195.4, 200.8, 201.7, 201.9, 201.9, 201.9, 201.9; stable.

(c) No.

15 **(a)** $k_{t+1} = 0.99k_t + 0.2k_t^{0.6}$.

(b) 1789; uniform; same behaviour.

Section 9.2

1 **(a)** $5e^{5t}$; **(b)** $-6e^{-3t}$; **(c)** mAe^{mt}.

2 **(a)** The function that differentiates to 4 times itself is $y = Ae^{4t}$. The condition

$y(0) = 6$ gives $A = 6$, so the solution is $y = 6e^{4t}$.

(b) The function that differentiates to -5 times itself is $y = Ae^{-5t}$. The condition

$y(0) = 2$ gives $A = 2$, so the solution is $y = 2e^{-5t}$.

3 The complementary function is the solution of

$$\frac{dy}{dt} = 3y$$

and is given by

$$CF = Ae^{3t}$$

For a particular solution we try a constant function

$$y(t) = D$$

Substituting this into the original equation,

$$\frac{dy}{dt} = 3y - 60$$

gives

$$0 = 3D - 60$$

which has solution $D = 20$. The general solution is therefore

$$y(t) = Ae^{3t} + 20$$

Finally, substituting $t = 0$ gives

$$y(0) = A + 20 = 30$$

and so A is 10. Hence

$$y(t) = 10e^{3t} + 20$$

A graph of y against t is sketched in Figure S9.5, which indicates that $y(t)$ rapidly diverges. We would expect divergence to occur for any solution

Figure S9.5

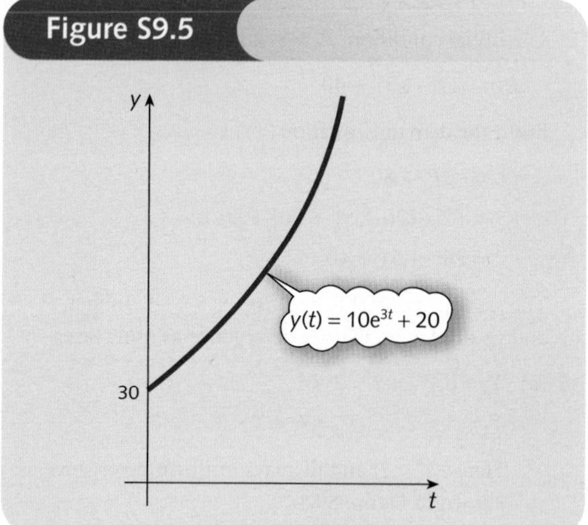

$$y(t) = Ae^{mt} + D \quad (A \neq 0)$$

when $m > 0$.

4 Substituting the expressions for C and I into

$$\frac{dY}{dt} = 0.1(C + I - Y)$$

gives

$$\frac{dY}{dt} = 0.1(0.9Y + 100 + 300 - Y)$$
$$= -0.01Y + 40$$

The complementary function is given by

$$CF = Ae^{-0.01t}$$

and for a particular solution we try a constant function

$$Y(t) = D$$

Substituting this into the differential equation gives

$$0 = -0.01D + 40$$

which has solution $D = 4000$. The general solution is therefore

$$Y(t) = Ae^{-0.01t} + 4000$$

The initial condition, $Y(0) = 2000$, gives

$$A + 4000 = 2000$$

and so A is -2000. Hence

$$Y(t) = -2000e^{-0.01t} + 4000$$

This system is stable because the complementary function is a negative exponential and so $Y(t)$ converges to its equilibrium value of 4000 as t increases.

5 Substituting the expressions for Q_S and Q_D into

$$\frac{dP}{dt} = \tfrac{1}{3}(Q_D - Q_S)$$

gives

$$\frac{dP}{dt} = \tfrac{1}{3}[(-P + 4) - (2P - 2)] = -P + 2$$

The complementary function is given by

$$CF = Ae^{-t}$$

and for a particular solution we try a constant function

$$P(t) = D$$

Substituting this into the differential equation gives

$$0 = -D + 2$$

which has solution $D = 2$. The general solution is therefore

$$P(t) = Ae^{-t} + 2$$

The initial condition, $P(0) = 1$, gives

$$A + 2 = 1$$

and so A is -1. Hence

$$P(t) = -e^{-t} + 2$$

From the supply and demand equations

$$Q_S(t) = 2P - 2 = 2(-e^{-t} + 2) - 2 = -2e^{-t} + 2$$
$$Q_D(t) = -P + 4 = -(-e^{-t} + 2) + 4 = e^{-t} + 2$$

All three functions involve a negative exponential, so the system is stable.

6 **(a)** $t^2 + 7$; **(b)** $-\tfrac{1}{3}e^{-3t} + \tfrac{1}{3}$; **(c)** $\tfrac{1}{3}t^3 + \tfrac{3}{2}t^2 - 5t + 1$.

7 **(a)** $-20e^{-3t} + 60$; starting at 40, $y(t)$ increases uniformly to 60.

(b) $20e^{-3t} + 60$; starting at 80, $y(t)$ decreases uniformly to 60.

(c) 60; $y(t)$ remains at the equilibrium level of 60 for all time.

8 $Y(t) = 5000e^{-0.05t} + 10\,000$; stable.

9 $Y(t) = 2000e^{0.15t} - 1800$; unstable.

10 $P(t) = -e^{-2.5t} + 2$; $Q_S(t) = -3e^{-2.5t} + 5$; $Q_D(t) = 2e^{-2.5t} + 5$; stable.

11 Substitute assumptions (1) and (3) into (2) to get

$$\beta\frac{dY}{dt} = \alpha Y$$

which rearranges as

$$\frac{dY}{dt} = \frac{\alpha}{\beta}Y$$

with solution

$$Y(t) = Y(0)e^{(\alpha/\beta)t}$$

This is unstable because $\alpha/\beta > 0$.

12 The right-hand side is

$$my + c = m\left(Ae^{mt} - \frac{c}{m}\right) + c$$
$$= Ame^{mt} - c + c = Ame^{mt}$$

which we recognize as the derivative of $y(t)$.

13 **(a)** Ae^{-2t}; **(b)** e^{3t}; **(c)** $Ae^{-2t} + e^{3t}$; $6e^{-2t} + e^{3t}$; **(d)** unstable.

14 **(a)** Ae^{-t}; **(b)** $4t - 7$;

(c) $y = Ae^{-t} + 4t - 7$; $y = 8e^{-t} + 4t - 7$; **(d)** unstable.

15 $y(t) = \sqrt{(t^2 + 1)}$. For large t, the solution is well approximated by a straight line with equation $y(t) = t$.

16 **(a)** $y(t) = 0.5\left(\dfrac{t+4}{t+2}\right)$

(b) $y(t) = 0.5\left(\dfrac{t-2}{t-4}\right)$

(c) In (a) the fund decreases and approaches the equilibrium value of \$500 000; in (b) the fund decreases and is completely exhausted after 2 years.

17 $K(t) = \dfrac{(ac - b)e^{a^2/2} + b}{a}$

(a) Capital remains constant, at the value c, for all time.

(b) Capital grows without bound.

(c) Capital decreases to zero.

Appendix 1

1 Slope of chord $= \dfrac{(5 + \Delta x)^2 - 5^2}{\Delta x}$

$$= \frac{25 + 10\Delta x + (\Delta x)^2 - 25}{\Delta x}$$

$$= 10 + \Delta x$$

Slope of tangent $= \lim\limits_{\Delta x \to 0}(10 + \Delta x) = 10$

2 (a) Slope of chord

$$= \frac{4(x + \Delta x)^2 - 9(x + \Delta x) + 1] - (4x^2 - 9x + 1)}{\Delta x}$$

$= 8x + 4\Delta x - 9$

Slope of tangent $= \lim_{\Delta x \to 0} (8x + 4\Delta x - 9) = 8x - 9$

(b) $f(x + \Delta x) - f(x) = \frac{1}{(x + \Delta x)^2} - \frac{1}{x^2}$

$$= \frac{x^2 - (x + \Delta x)^2}{(x + \Delta x)^2 x^2}$$

$$= \frac{-2x\Delta x - (\Delta x)^2}{(x + \Delta x)^2 x^2}$$

Slope of chord $= \frac{-2x - \Delta x}{(x + \Delta x)^2 x^2}$

Slope of tangent $= \lim_{\Delta x \to 0} \left(\frac{-2x - \Delta x}{(x + \Delta x)^2 x^2} \right) = \frac{-2}{x^3}$

3 (a) $(a + b)(a + b)^2 = (a + b)(a^2 + 2ab + b^2)$

$= a^3 + 3a^2b + 3ab^2 + b^3$

(b) Slope of chord $= \frac{(x + \Delta x)^3 + x^3}{\Delta x}$

$= 3x^2 + 3x(\Delta x) + (\Delta x)^2$

Slope of tangent

$= \lim_{\Delta x \to 0} (3x^2 + 3x\Delta x + (\Delta x)^2)$

$= 3x^2$

Appendix 2

1 (a) $2 \times 1^2 + 3 \times 2^2 = 2 + 12 = 14$

(b) $4x + 6y\frac{dy}{dx} = 0$

$6y\frac{dy}{dx} = -4x$

$\frac{dy}{dx} = \frac{-4x}{6y} = -\frac{2x}{3y}$

so when $x = 1, y = 2$,

$\frac{dy}{dx} = -\frac{1}{3}$

2 (a) $-\frac{x}{y}$; **(b)** $-\frac{1 + 6x^2}{3y}$; **(c)** $-\frac{e^{x-y}}{2}$;

(d) $\frac{(1 - e^x)y}{e^x - x - 2y}$; **(e)** $\frac{2x + 2y^2}{3 - 4xy}$; **(f)** $-(1 + x + y)$.

Appendix 3

1 $\mathbf{H} = \begin{bmatrix} 2 & 0 \\ 0 & 2 \end{bmatrix}$

The first principal minor is $2 > 0$.

The second principal minor is $2 \times 2 - 0 \times 0 = 4 > 0$.

Hence the stationary point is a minimum.

2 $\mathbf{H} = \begin{bmatrix} -4 & -2 \\ -2 & -2 \end{bmatrix}$

Principal minors are $-4 < 0$ and $4 > 0$, respectively, so maximum.

3 $\mathbf{H} = \begin{bmatrix} -4L^{-3/2} & 0 \\ 0 & -6K^{-3/2} \end{bmatrix}$

assuming that L is chosen as the first variable.

At the stationary point,

$\mathbf{H} = \begin{bmatrix} -1/16 & 0 \\ 0 & -1/288 \end{bmatrix}$

Principal minors are $-1/16 < 0$ and $1/4608 > 0$, respectively, so maximum.

4 $\bar{H} = \begin{bmatrix} -2 & 2 & -1 \\ 2 & -2 & -1 \\ -1 & -1 & 0 \end{bmatrix}$ has determinant $8 > 0$, so maximum.

5 $\bar{H} = \begin{bmatrix} 4 & -1 & -1 \\ -1 & 0 & -1 \\ -1 & -1 & 0 \end{bmatrix}$ has determinant $-6 < 0$, so minimum.

Glossary

Adjustment coefficient The constant of proportionality in the simple macroeconomic model in which the rate of change of national income is assumed to be proportional to excess expenditure.

Algebraic fraction Ratio of two expressions: $p(x)/q(x)$ where $p(x)$ and $q(x)$ are algebraic expressions such as $ax^2 + bx + c$ or $dx + e$.

Annual percentage rate (APR) This is the equivalent annual interest paid for a loan, taking into account the compounding over a variety of time periods.

Annual rate of inflation The percentage increase in the level of prices over a 12-month period.

Annuity A lump-sum investment designed to produce a sequence of equal regular payments over time.

Anti-derivative A function whose derivative is a given function.

Arbitrary constant A letter representing an unspecified constant in the general solution of a differential equation.

Arc elasticity Elasticity measured between two points on a curve.

Arithmetic progression A sequence of numbers with a constant difference between consecutive terms; the nth term takes the form $a + bn$.

Autonomous consumption and savings The levels of consumption and savings when there is no income.

Autonomous consumption multiplier The number by which you multiply the change in autonomous consumption to deduce the corresponding change in, say, national income.

Average cost Total cost per unit of output: AC = TC/Q.

Average product of labour Output per worker: $AP_L = Q/L$.

Average revenue Total revenue divided by quantity: TR/Q = $PQ/Q = P$.

Balanced budget multiplier The number by which you multiply the change in government expenditure to deduce the corresponding change in, say, national income, assuming that this change is financed entirely by a change in taxation.

Bordered Hessian matrix A Hessian matrix augmented by an extra row and column containing partial derivatives formed from the constraint in the method of Lagrange multipliers.

Capital Man-made assets used in the production of goods and services.

Chord A straight line joining two points on a curve.

Cobb–Douglas production function A production function of the form $Q = AK^{\alpha}L^{\beta}$.

Coefficient A numerical multiplier of the variables in an algebraic term, such as the numbers 4 and 7 in the expression, $4x + 7yz^2$.

Cofactor The cofactor of the element a_{ij} is the determinant of the matrix left when row i and column j are deleted, multiplied by +1 or −1, depending on whether $i + j$ is even or odd, respectively.

Column vector A matrix with one column.

Comparative statics Examination of the effect on equilibrium values due to changes in the parameters of an economic model.

Complementary function of a difference equation The solution of the difference equation $Y_t = bY_{t-1} + c$ when the constant c is replaced by zero.

Complementary function of a differential equation The solution of the differential equation $\frac{dy}{dt} = my + c$ when the constant c is replaced by zero.

Complementary goods A pair of goods consumed together. As the price of either goes up, the demand for both goods goes down.

Compound interest The interest that is added on to the initial investment, so that this will itself gain interest in subsequent time periods.

Constant of integration The arbitrary constant that appears in an expression when finding an indefinite integral.

Constant returns to scale Exhibited by a production function when a given percentage increase in input leads to the same percentage increase in output: $f(\lambda K, \lambda L) = \lambda f(K, L)$.

Constrained optimisation Maximisation (or mimimisation) of an objective function subject to at least one constraint.

Constraint A restriction on the values of an objective function.

Consumer's surplus The excess cost that a person would have been prepared to pay for goods over and above what is actually paid.

Consumption The flow of money from households to firms as payment for goods and services.

Continuous compounding The limiting value when interest is compounded with ever-increasing frequency.

Coordinates A set of numbers which determine the position of a point relative to a set of axes.

Cramer's rule A method of solving simultaneous equations, $\mathbf{Ax} = \mathbf{b}$, by the use of determinants. The ith variable x_i can be computed using $\det(\mathbf{A}_i)/\det(\mathbf{A})$ where $\mathbf{A}_i$ is the determinant of the matrix obtained from $\mathbf{A}$ by replacing the ith column by $\mathbf{b}$.

Cross-price elasticity of demand The responsiveness of demand for one good to a change in the price of another: (percentage change in quantity) ÷ (percentage change in the price of the alternative good).

Decreasing function A function, $y = f(x)$, in which y decreases as x increases.

Decreasing returns to scale Exhibited by a production function when a given percentage increase in input leads to a smaller percentage increase in output: $f(\lambda K, \lambda L) = \lambda^n f(K, L)$ where $0 < n < 1$.

Definite integration The process of finding the area under a graph by subtracting the values obtained when the limits are substituted into the anti-derivative.

Degree of homogeneity The number n in the relation $f(\lambda K, \lambda L) = \lambda^n f(K, L)$.

Demand function A relationship between the quantity demanded and various factors that affect demand, including price.

Denominator The number (or expression) on the bottom of a fraction.

Dependent variable A variable whose value is determined by that taken by the independent variables; in $y = f(x)$, the dependent variable is y.

Derivative The gradient of the tangent to a curve at a point. The derivative at $x = a$ is written $f'(a)$.

Derived function The rule, f', which gives the gradient of a function, f, at a general point.

Determinant of a matrix A determinant can be expanded as the sum of the products of the elements in any one row or column and their respective cofactors.

Difference equation An equation that relates consecutive terms of a sequence of numbers.

Differential equation An equation connecting derivatives of an unknown function.

Differentials Limiting values of incremental changes. In the limit the approximation $\Delta z \simeq \frac{\partial z}{\partial x} \times \Delta x$ becomes $\mathrm{d}z = \frac{\partial z}{\partial x} \times \mathrm{d}x$ where $\mathrm{d}z$ and $\mathrm{d}x$ are the differentials.

Differentiation The process or operation of determining the first derivative of a function.

Discount rate The interest rate that is used when going backwards in time to calculate the present value from a future value.

Discounting The process of working backwards in time to find the present values from a future value.

Discriminant The number, $b^2 - 4ac$, which is used to indicate the number of solutions of the quadratic equation $ax^2 + bx + c = 0$.

Disposable income Household income after the deduction of taxes and the addition of benefits.

Distributive law The rule which states that $a(b + c) = ab + ac$, for any numbers a, b and c.

Dynamics Analysis of how equilibrium values vary over time.

Elastic demand Where the percentage change in demand is more than the corresponding change in price: $E > 1$.

Elements The individual numbers inside a matrix. (Also called entries.)

Elimination method The method in which variables are removed from a system of simultaneous equations by adding (or subtracting) a multiple of one equation to (or from) a multiple of another.

Endogenous variable A variable whose value is determined within a model.

Equilibrium (market) This state occurs when quantity supplied and quantity demanded are equal.

Equilibrium value of a difference equation A solution of a difference equation that does not vary over time; it is the limiting value of Y_n as n tends to infinity.

Equilibrium value of a differential equation A solution of a differential equation that does not vary over time; it is the limiting value of $y(t)$ as t tends to infinity.

Euler's theorem If each input is paid the value of its marginal product, the total cost of these inputs is equal to total output, provided there are constant returns to scale.

Exogenous variable A variable whose value is determined outside a model.

Exponent A superscript attached to a variable; the number 5 is the exponent in the expression, $2x^5$.

Exponential form A representation of a number which is written using powers. For example, 2^5 is the exponential form of the number 32.

Exponential function The function $f(x) = e^x$; an exponential function in which the base is the number e = 2.718 28 . . .

External demand Output that is used by households.

Factors of production The inputs to the production of goods and services: land, capital, labour and raw materials.

Feasible region The set of points which satisfy all of the constraints in a linear programming problem.

Final demand An alternative to 'external demand'.

First-order derivative The rate of change of a function with respect to its independent variable. It is the same as the 'derivative' of a function, $y = f(x)$, and is written as $f'(x)$or dy/dx.

First principal minor The 1×1 determinant in the top left-hand corner of a matrix; the element, a_{11} of a matrix, **A**.

Fixed costs Total costs that are independent of output.

Flow chart A diagram consisting of boxes of instructions indicating the sequence of operations and their order.

Function (of one variable) A rule that assigns to each incoming number, x, a uniquely defined outgoing number, y.

Function of two variables A rule which assigns to each pair of incoming numbers, x and y, a uniquely defined outgoing number, z.

Future value The final value of an investment after one or more time periods.

General solution of a difference equation The solution of a difference equation that contains an arbitary constant. It is the sum of the complementary function and particular solution.

General solution of a differential equation The solution of a differential equation that contains an arbitary constant. It is the sum of the complementary function and particular solution.

Geometric progression A sequence of numbers with a constant ratio between consecutive terms; the nth term takes the form, ar^{n-1}.

Geometric ratio The constant multiplier in a geometric series.

Geometric series A sum of the consecutive terms of a geometric progression.

Government expenditure The total amount of money spent by government on defence, education, health, police, etc.

Gradient The gradient of a line measures steepness and is the vertical change divided by the horizontal change between any two points on the line. The gradient of a curve at a point is that of the tangent at that point.

Hessian A matrix whose elements are the second-order partial derivatives of a given function.

Homogeneous function A function with the property that when all of the inputs are multiplied by a constant, λ, the output is multiplied by λ^n where n is the degree of homogeneity.

Identity matrix An $n \times n$ matrix, **I**, in which every element on the main diagonal is 1 and the other elements are all 0. If **A** is any $n \times n$ matrix then **AI** = **I** = **IA**.

Implicit differentiation The process of obtaining dy/dx where the function is not given explicitly as an expression for y in terms of x.

Income elasticity of demand The responsiveness of demand for one good to a change in income: (percentage change in quantity) ÷ (percentage change in income).

Increasing function A function, $y = f(x)$, in which y increases as x increases.

Increasing returns to scale Exhibited by a production function when a given percentage increase in input leads to a larger percentage increase in output: $f(\lambda K, \lambda L) = \lambda^n f(K, L)$ where $n > 1$.

Indefinite integration The process of obtaining an anti-derivative.

Independent variable A variable whose value determines that of the dependent variable; in $y = f(x)$, the independent variable is x.

Index Alternative word for exponent or power.

Index number The scale factor of a variable measured from the base year multiplied by 100.

Indifference curve A curve indicating all combinations of two goods which give the same level of utility.

Indifference map A diagram showing the graphs of a set of indifference curves. The further the curve is from the origin, the greater the level of utility.

Inelastic demand Where the percentage change in demand is less than the corresponding change in price: $E < 1$.

Inferior good A good whose demand decreases as income increases.

Inflation The percentage increase in the level of prices over a 12-month period.

Initial condition The value of Y_0 (or $y(0)$) which needs to be specified to obtain a unique solution of a difference (or differential) equation.

Input–output analysis Examination of how inputs and outputs from various sectors of the economy are matched to the total resources available.

Integer programming A linear programming problem in which the search for solution is restricted to points in the feasible region with whole-number coordinates.

Integral The number $\int_a^b f(x)\mathrm{d}x$ (definite integral) or the function $\int f(x)\mathrm{d}x$ (indefinite integral).

Integration The generic name for the evaluation of definite or indefinite integrals.

Intercept Points where a graph crosses one of the coordinate axes.

Intermediate output Output from one sector which is used as input by another (or the same) sector.

Internal rate of return (IRR) The interest rate for which the net present value is zero.

Inverse function A function, written f^{-1}, which reverses the effect of a given function, f, so that $x = f^{-1}(y)$ when $y = f(x)$.

Inverse matrix A matrix $\mathbf{A}^{-1}$ with the property that $\mathbf{A}^{-1}\mathbf{A} = \mathbf{I} = \mathbf{A}\mathbf{A}^{-1}$.

Inverse (operation) The operation that reverses the effect of a given operation and takes you back to the original. For example, the inverse of halving is doubling.

Investment The creation of output not for immediate consumption.

Investment multiplier The number by which you multiply the change in investment to deduce the corresponding change in, say, national income

Isocost curve A line showing all combinations of two factors that can be bought for a fixed cost.

IS schedule The equation relating national income and interest rate based on the assumption of equilibrium in the goods market.

Isoquant A curve indicating all combinations of two factors that give the same level of output.

L-shaped curve A term used by economists to describe the graph of a function, such as $f(x) = a + \frac{b}{x}$, which bends roughly like the letter L.

Labour All forms of human input to the production process.

Labour productivity Output per worker: $AP_L = Q/L$.

Lagrange multiplier The number λ which is used in the Lagrangian function. In economics this gives the change in the value of the objective function when the value of the constraint is increased by 1 unit.

Lagrangian The function $f(x, y) + \lambda[M - \varphi(x, y)]$, where $f(x, y)$ is the objective function and $\phi(x, y) = M$ is the constraint. The stationary point of this function is the solution of the associated constrained optimization problem.

Laspeyre index An index number for groups of data that are weighted by the quantities used in the base year.

Law of diminishing marginal productivity (law of diminishing returns) Once the size of the workforce exceeds a particular value, the increase in output due to a 1 unit increase in labour will decline: $d^2Q/dL^2 < 0$ for sufficiently large L.

Leontief inverse The inverse of $\mathbf{I} - \mathbf{A}$, where $\mathbf{A}$ is the matrix of technical coefficients.

Limits of integration The numbers a and b which appear in the definite integral, $\int_a^b f(x)\mathrm{d}x$.

Linear equation An equation of the form $y = ax + b$.

Linear programming Optimization of a linear objective function subject to linear constraints.

LM schedule The equation relating national income and interest rate based on the assumption of equilibrium in the money market.

Logarithm The power to which a base must be raised to yield a particular number.

Marginal cost The cost of producing 1 more unit of output: $MC = d(TC)/dQ$.

Marginal product of capital The extra output produced by 1 more unit of capital: $MP_K = \partial Q/\partial K$.

Marginal product of labour The extra output produced by 1 more unit of labour: $MP_L = \partial Q/\partial L$.

Marginal propensity to consume The fraction of a rise in national income which goes on consumption. It is the slope of the consumption function: $MPC = dC/dY$.

Marginal propensity to consume multiplier The number by which you multiply the change in MPC to deduce the corresponding change in, say, national income.

Marginal propensity to save The fraction of a rise in national income which goes into savings. It is the slope of the savings function: $MPS = dS/dY$.

Marginal rate of commodity substitution (MRCS) The amount by which one input needs to increase to maintain a constant value of utility when the other input decreases by 1 unit: $MRTS = \partial U/\partial x_1 \div \partial U/\partial x_2$.

Marginal rate of technical substitution (MRTS) The amount by which capital needs to rise to maintain a constant level of output when labour decreases by 1 unit: $MRTS = MP_L/MP_K$.

Marginal revenue The extra revenue gained by selling 1 more unit of a good: $MR = d(TR)/dQ$.

Marginal utility The extra satisfaction gained by consuming 1 extra unit of a good: $\partial U/\partial x_i$.

Matrix A rectangular array of numbers, set out in rows and columns, surrounded by a pair of brackets. (Plural matrices.)

Matrix of technical coefficients A square matrix in which element a_{ij} is the input required from the ith sector to produce 1 unit of output for the jth sector.

Maximum (local) point A point on a curve which has the highest function value in comparison with other values in its neighbourhood; at such a point the first-order derivative is zero and the second-order derivative is either zero or negative.

Maximum point (of a function of two variables) A point on a surface which has the highest function value in comparison with other values in its neighbourhood; at such a point the surface looks like the top of a mountain.

Method of substitution (for constrained optimization problems) The method of solving constrained optimization problems whereby the constraint is used to eliminate one of the variables in the objective function.

Minimum (local) point A point on a curve which has the lowest function value in comparison with other values in its neighbourhood; at such a point the first-order derivative is zero and the second-order derivative is either zero or positive

Minimum point (of a function of two variables) A point on a surface which has the lowest function value in comparison with other values in its neighbourhood; at such a point the surface looks like the bottom of a valley or bowl.

Modelling The creation of a piece of mathematical theory which represents (a simplification of) some aspect of practical economics.

Money supply The notes and coins in circulation together with money held in bank deposits.

Monopolist The only firm in the industry.

Multiplier The number by which you multiply the change in an independent variable to find the change in the dependent variable.

National income The flow of money from firms to households.

Natural logarithm A logarithm to base e; if $M = e^n$ then n is the natural logarithm of M.

Net investment Rate of change of capital stock over time: $I = dK/dt$.

Net present value (NPV) The present value of a revenue flow minus the original cost.

Nominal data Monetary values prevailing at the time that they were measured.

Non-negativity constraints The constraints $x \geq 0$, $y \geq 0$, etc.

Non-singular matrix A square matrix with a non-zero determinant.

Number line An infinite line on which the points represent real numbers by their (signed) distance from the origin.

Numerator The number (or expression) on the top of a fraction.

Objective function A function that one seeks to optimize (usually) subject to constraints.

Optimization The determination of the optimal (usually stationary) points of a function.

Order of a matrix The dimensions of a matrix. A matrix with m rows and n columns has order $m \times n$.

Origin The point where the coordinate axes intersect.

Paasche index An index number for groups of data which are weighted by the quantities used in the current year.

Parabola The shape of the graph of a quadratic function.

Parameter A constant whose value affects the specific values but not the general form of a mathematical expression, such as the constants a, b and c in $ax^2 + bx + c$.

Partial derivative The derivative of a function of two or more variables with respect to one of these variables, the others being regarded as constant.

Particular solution of a difference equation Any one solution of a difference equation such as $Y_t = bY_{t-1} + c$.

Particular solution of a differential equation Any one solution of a difference equation such as $\frac{dy}{dt} = my + c$.

Perfect competition A situation in which there are no barriers to entry in the industry and where there are many firms selling an identical product at the market price.

Point elasticity Elasticity measured at a particular point on a curve, e.g. for a supply curve, $E = \frac{P}{Q} \times \frac{dQ}{dP}$.

Point of inflection A stationary point that is not a maximum or minimum.

Precautionary demand for money Money held in reserve by individuals or firms to fund unforeseen future expenditure.

Present value The amount that is invested initially to produce a specified future value after a given period of time.

Price elasticity of demand A measure of the responsiveness of the change in demand due to a change in price: – (percentage change in demand) ÷ (percentage change in price).

Price elasticity of supply A measure of the responsiveness of the change in supply due to a change in price: (percentage change in supply) ÷ (percentage change in price).

Primitive An alternative word for an anti-derivative.

Principal The value of the original sum invested.

Producer's surplus The excess revenue that a producer has actually received over and above the lower revenue that it was prepared to accept for the supply of its goods.

Production function The relationship between the output of a good and the inputs used to produce it.

Power Another word for exponent. If this is a positive integer then it gives the number of times a number is multiplied by itself.

Profit Total revenue minus total cost: $\pi = TR - TC$.

Quadratic function A function of the form $f(x) = ax^2 + bx + c$ where $a \neq 0$.

Real data Monetary values adjusted to take inflation into account.

Rectangular hyperbola A term used by mathematicians to describe the graph of a function, such as $f(x) = a + \frac{b}{x}$, which is a hyperbola with horizontal and vertical asymptotes.

Recurrence relation An alternative term for a difference equation. It is an expression for Y_n in terms of Y_{n-1} (and possibly Y_{n-2}, Y_{n-3}, etc).

Reduced form The final equation obtained when exogenous variables are eliminated in the course of solving a set of structural equations in a macroeconomic model.

Reverse flow chart A flow chart indicating the inverse of the original sequence of operations in reverse order.

Row vector A matrix with one row.

Saddle point A stationary point which is neither a maximum nor a minimum and at which the surface looks like the middle of a horse's saddle.

Scale factor The multiplier that gives the final value in percentage problems.

Second-order derivative The derivative of the first-order derivative. The expression obtained when the original function, $y = f(x)$, is differentiated twice in succession and is written as $f''(x)$ or d^2y/dx^2.

Second-order partial derivative The partial derivative of a first-order partial derivative. For example, f_{xy} is the second-order partial derivative when f is differentiated first with respect to y and then with respect to x.

Second principal minor The 2×2 determinant in the top left-hand corner of a matrix.

Simple interest The interest that is paid direct to the investor instead of being added to the original amount.

Simultaneous equations A set of linear equations in which there are (usually) the same number of equations and unknowns. The solution consists of values of the unknowns which satisfy all of the equations at the same time.

Singular matrix A square matrix with a zero determinant. A singular matrix fails to possess an inverse.

Sinking fund A fixed sum of money saved at regular intervals which is used to fund some future financial commitment.

Slope of a line Also known as the gradient, it is the change in the value of y when x increases by 1 unit.

Small increments formula The result $\Delta z \simeq \frac{\partial z}{\partial x}\Delta x \times \frac{\partial z}{\partial y}\Delta y$.

Speculative demand for money Money held by back by firms or individuals for the purpose of investing in alternative assets, such as government bonds, at some future date.

Square matrix A matrix with the same number of rows as columns.

Square root A number which when multiplied by itself equals a given number; the solutions of the equation $x^2 = c$, which are written $\pm\sqrt{x}$.

Stable (unstable) equilibrium An economic model in which the solution of the associated difference (or differential) equation converges (diverges).

Statics The determination of the equilibrium values of variables in an economic model which do not change over time.

Stationary points (of a function of one variable) Points on a graph at which the tangent is horizontal; at a stationary point the first-order derivative is zero.

Structural equations A collection of equations that describe the equilibrium conditions of a macroeconomic model.

Substitutable goods A pair of goods that are alternatives to each other. As the price of one of them goes up, the demand for the other rises.

Superior good A good whose demand increases as income increases.

Supply function A relationship between the quantity supplied and various factors that affect demand, including price.

Symmetric function A function of two or more variables which is unchanged by any permutation of the variables. A function of two variables is symmetric when $f(x, y) = f(y, x)$.

Tangent A line that just touches a curve at a point.

Taxation Money paid to government based on an individual's income and wealth (direct taxation) together with money paid by suppliers of goods or services based on expenditure (indirect taxation).

Technology matrix A square matrix in which element a_{ij} is the input required from the ith sector to produce 1 unit of output for the jth sector.

Time series A sequence of numbers indicating the variation of data over time.

Total cost The sum of the total variable and fixed costs: TC = TVC + FC.

Total revenue A firm's total earnings from the sales of a good: TR = PQ.

Transactions demand for money Money used for everyday transactions of goods and services.

Transpose formula The rearrangement of a formula to make one of the other letters the subject.

Transpose matrix The matrix obtained from a given matrix by interchanging rows and columns. The transpose of a matrix **A** is written $\mathbf{A}^T$.

U-shaped curve A term used by economists to describe a curve, such as a parabola, which bends upwards, like the letter U.

Unbounded region A feasible region that is not completely enclosed by a polygon. The associated linear programming problem may not have a finite solution.

Unconstrained optimisation Maximisation (or mimimisation) of an objective function without any constraints.

Uniformly convergent sequence A sequence of numbers that progressively increases (or decreases) to a finite limit.

Uniformly divergent sequence A sequence of numbers that progressively increases (or decreases) without a finite limit.

Unit elasticity of demand Where the percentage change in demand is the same as the percentage change in price: $E = 1$.

Unstable equilibrium An economic model in which the solution of the associated difference (or differential) equation diverges.

Utility The satisfaction gained from the consumption of a good.

Variable costs Total costs that change according to the amount of output produced.

***x* axis** The horizontal coordinate axis pointing from left to right.

***y* axis** The vertical coordinate axis pointing upwards.

Zero matrix A matrix in which every element is zero.

Index